THEORY OF STRUCTURAL TRANSFORMATIONS IN SOLIDS

ARMEN G. KHACHATURYAN
Department of Materials Science and Engineering
Rutgers University

DOVER PUBLICATIONS, INC.
Mineola, New York

Bibliographical Note

This Dover edition, first published in 2008, is an unabridged republication of the work originally published by John Wiley & Sons, Inc., New York, in 1983. A new Preface and an Errata list have been added to the Dover edition.

Library of Congress Cataloging-in-Publication Data

Khachaturyan, A. G. (Armen Gurgenovich), 1935–
Theory of structural transformations in solids / Armen G. Khachaturyan.
p. cm.
Originally published: New York : Wiley, c1983.
Includes bibliographical references and index.
ISBN-13: 978-0-486-46280-6 (pbk.)
ISBN-10: 0-486-46280-3 (pbk.)
1. Alloys. 2. Physical metallurgy. 3. Phase transformations (Statistical physics). I. Title.

TN690.K453 2008
669—dc22

2007034502

Manufactured in the United States of America
Dover Publications, Inc., 31 East 2nd Street, Mineola, N.Y. 11501

PREFACE TO THE DOVER EDITION

When *Theory of Structural Transformations in Solids* was published by John Wiley & Sons, Inc., in 1983, it was the first collection of results related to the theory of submicron, nano, and atomic scale structures formulated in terms of spatially continuous density functions. These density functions, today called Phase Fields, are chosen to identify the structural characteristics of different phases, domains, and atomic configurations. This approach as applied to structural transformations was completely novel twenty-five years ago, and placed the physics of structural materials on the same footing as ferroelectrics and ferromagnets where density fields, magnetization, and polarizations have been always used. The Phase Field approach as applied to structural transformations is very convenient because it characterizes microstructure formation in solids via the temporal and spatial dependence of evolving continuous functions. Since no *a priori* constraints on the possible transformation pathways are required, it has the power to predict the geometry of complex structures and their evolution involving a change of their topology. This is where the Phase Field approach provides its greatest advantage, and this is the reason why its use has dramatically increased in recent years.

The intimate relation between the transformation-induced elastic strain and microstructure is one of the focal points of the book. The strain energy plays a very special role in microstructure formation. Virtually any solid-solid transformation generates elastic strain caused either by a crystal lattice incompatibility of the coexisting phases, their volumetric mismatch, or both. The elastic strain energy generated by a phase transformation is distinct from the conventional ("chemical") free energy considered in classical thermodynamics that is caused by finite-range interatomic interactions. Instead, the elastic energy is the sum of the infinite-range (dipole-dipole like) strain-induced interaction of all finite elements of elastically incompatible phases or domains and is dependent on their spatial architecture. The elastic energy is thus the driving force for establishing the optimal microstructure geometry wherein configurational features of the microstructure are the relaxing internal thermodynamic parameters. The stress-accommodating microstructures formed in multiphase or multidomain systems are, in fact, domain structures whose origin is conceptually similar to that of domain structures in ferroelectric/ferromagnetic systems. In particular, this similarity is reflected in the fact that the Fourier space representation of the electrostatic/magnetostatic energy of arbitrary ferroelectric/magnetic domain structures as formulated in the book is similar to that for the elastic energy of multiphase/multidomain structures. These equations have recently been rediscovered and successfully used in the theory of magnetic domains.

The book also provides a full treatment of the micromechanics of structurally heterogeneous states formulated in terms of density functions. This approach is today called Phase Field Microelasticity. This approach is different from that used in the classic solid state mechanics of coherent systems where the morphology of a multi-phase system is described by the topology of interface boundaries separating homogeneous particles of different phases. In the traditional approach the elastic strain is found by solving the elasticity equation with boundary conditions defined at the interfacial surfaces that reflect the phase incompatibility. Phase Field Microelasticity treats the problem of structurally inhomogeneous systems with continuously changing structural parameters (density functions) whose values distinguish the structural heterogeneities and determine their spatial location. This method is especially convenient for predicting the orientation relations and shapes of particles of constituent phases comprising the system. The general principles of Phase Field Microelasticity are formulated and examples of how to attack selected problems are presented. The method presented is not outdated because the continuously changing density functions are naturally suited for a characterization of the evolving heterogeneous structures. This approach is a building block of the Phase Field theories and modeling methods used today for predicting microstructure evolution.

In addition the book provides probably one of the most detailed accounts of the Static Concentration Wave approach and its applications to problems of atomic ordering and decomposition. The Concentration Wave interpretation of atomic ordering reveals the profound relationships between atomic structures of ordered and disordered phases, their symmetries, thermodynamics, and diffraction. It is also a powerful method for modeling the kinetics of atomic scale rearrangements in decomposition and ordering. The use of Concentration Waves has now become a quite standard way to interpret the diffraction patterns of ordering phenomena. The Static Concentration Wave approach is essentially an application of the Phase Field method extended to the atomic scale: Concentration Waves are Fourier representations of Atomic Density Fields that are density functions of the crystal lattice site coordinates—e.g. the occupation probabilities of constituent atoms at crystal lattice sites of the underlying lattice. Unlike the conventional thermodynamic treatment of ordering phenomena involving atomic interchange between sublattices that must be specified *a priori*, the Atomic Density Field method is able to predict the geometry of these sublattices, the distribution of constituent atomic components between them, and atomic features of ordered domains and their interfaces.

Given the development of the Phase Field theories during recent years, it is important to mention the recent extension of the theory of displacive (martensitic, ferroelastic) phase transformations to the cases of dislocation dynamics and fracture. Similarities between these phase transformations and dislocation dynamics was noticed years ago. However, the formation and development of dislocations under applied stress is also a structural transformation caused by crystal lattice instability under applied stress. This instability is similar to the instability resulting in displacive phase transformations under a thermodynamic driving force. This similarity permits the application of Phase Field Microelasticity to dislocation dynamics in a way that is analogous to its application to martensitic transformations. That results is a generalized 3-dimensional Peierls-Nabarro theory of multidislocation systems; it exactly reproduces the results of this theory in the case of a single straight dislocation. In fact, the Peierls-Nabarro theory was the first example of an application of the Phase Field approach to a problem of the crystal lattice rearrangement, in which the Burgers vector is a function of coordinates and is considered as a continuous and energy minimizing density field.

Formation of cracks is another example of crystal lattice instability under the applied stress. As with dislocations, the formation and evolution of multiple cracks have been recently described by Phase Field Microelasticity methods. However in the case of cracks, the micromechanics has been further extended to take into consideration the multiple connectivity of a material with developing cracks. Despite of this complication, the theory is not more complicated than the Phase Field dislocation theory, and the modeling is not more time consuming than the modeling of dislocation dynamics. The similarity between the microstructure development in displacive phase transformations, plastic deformation, and fracture follows from the fact that all three phenomena are particular cases of the same effect—instability of the initial homogeneous solid state under the thermodynamic or mechanical driving forces with a resulting crystal lattice rearrangement. The theoretical basis presented in the book for these extensions remains relevant today.

In the preface to the first edition, I expressed optimism that the further development of our understanding of a microstructure formation might eventually reach such a level of maturity that a quantitative analysis would become an integral starting point for materials design. At least a part of this optimism turns out to be justified, and this happened mainly due to application of computers to problems of materials science. Prior to the emergence of scientific supercomputing, the traditional analytical methods were stretched almost to the limit. Even the combination of tedious calculations and ingenuity was not sufficient to resolve the most interesting real life problems. The use of computers has now dramatically changed the landscape of materials theory. In particular, Phase Field Microelasticity methods have been able to reproduce complex microstructure morphologies and their evolution in remarkable agreement with experiment. The current bottleneck for further progress appears to be mostly the number-crunching capability of supercomputers, and the good news is that these capabilities dramatically increase year after year.

The theoretical framework for Phase Field methods provided in the book therefore should be of interest to graduate students specializing in materials science and condensed matter physics, as well as researchers in these fields. Some of the sections, especially in the first part of the book, also provide a wider audience an especially simple and brief formulation of the relevant parts of crystallography of phase transformation and diffraction theory without any great sacrifice of rigor or accuracy.

Rutgers, The State University of New Jersey
February 2008

ARMEN G. KHACHATURYAN

PREFACE TO THE FIRST EDITION

The idea to write a book on phase transformations in alloys belongs to my friend, Prof. J. W. Morris. M. Kutz, Editor, at Wiley and Sons, Inc., kindly offered to publish it. That is how I found myself involved in the exhausting and exciting project of writing this book. The contents roughly follow the lecture courses I delivered in 1977 and 1979 in the Department of Material Science at the University of California-Berkeley. Many details were elucidated then in fruitful discussions with Prof. Morris and Dr. S. Wen. Some of the results the book have come from our joint research program. The book deals with phase transformations in alloys and the dramatic changes they produce in their structures.

Experimental studies of the structure of alloys pose a lot of exciting problems, the solution of which may give great satisfaction. I am referring, for instance, to the enigmatic changes in the tetragonal axial ratio of an irradiated martensite of plain carbon-steel which reversibly increases upon heating and drops with cooling, that is, behaves in exactly the opposite way to what could be expected in the case of tempering. For some reason this effect is also highly sensitive to the carbon content and is observed only about 1 percent C by weight.

Another example concerns the origin of the extraordinarily beautiful patterns formed by regularly spaced new phase precipitates in "tweed" and modulated structures in two-phase alloys, as well as domain structures that arise in ordering and martensitic transformations. These phenomena—and there are many others—are discussed in this book.

As is known, the physical properties of virtually all important materials of modern technology are strongly dependent on their microstructure. The most efficient way to obtain the desirable microstructures is by phase transformations. Today phase transformation is the basis for the thermal treatment in the processing of almost all modern materials. The structure resulting from a solid state phase transformation depends, sometimes in an intricate way, on the crystallographic relationship between the lattices of the initial and product phases, on the elastic module of the phases, and on the kinetics of transformation. If it were possible to predict the features of phase transformations precisely, beginning at a fundamental starting point, then it would be possible to predict the structure and many other important properties of materials theoretically. The dream of material scientists to replace intuition with a quantitative approach to materials design would thus come true. Extensive development of the theory of phase transformation during the last decade, accompanied by major progress in the experimental characterization of materials and phenomenological characterization of phase transformations, give real grounds for optimism that these hopes may eventually be realized.

This book is an attempt to provide a quantitative treatment of the problem of the formation of heterogeneous materials based on recent developments in the statistic and mechanistic theory of phase transformations. The quantitative answers to the central questions arising from the use of phase transformations to establish microstructure are sought: How, when, and in what sequence the structure of a heterogenous materials is formed? What are the most important quantitative characteristics of the morphology of heterogenous materials? And, how can these characteristics be predicted?

The mathematics in the book reflect very recent theoretical advances. Several of the most significant of these deal with the treatment of the elastic strain that develops when the transformation connects two solid structures that have different lattices. The work on this subject is scattered in a great many articles in a variety of different journals. Many of the relevant papers appear in the Russian literature, and are either unavailable in English or poorly translated. This book collects the results of these diverse works and reformulates them from a central standpoint to create a unified approach to phase transformations in real materials. The theoretical work has been structured so as to facilitate the contact between the theoretical treatment, employing elastic theory, matrix algebra, and statistical thermodynamics of phase transformations, and the phenomenological results and insights resulting from experimentation, including x-ray and electron diffraction and other structural studies of phase transformations in solids.

The first five chapters of the book constitute an introduction to the field. They contain a general introduction to the crystallography of phase transformations, structure analysis, statistical thermodynamics of ordering formulated in terms of concentration waves, phenomenological theory of decomposition, and crystal lattice site kinetics.

The results presented in Chapters 6 through 12 concern the structural changes in two-phase alloys controlled by elastic strain induced by phase transformations. The theoretical analysis developed in these sections enables one to predict the morphology of both single and modulated structures comprising two cubic phases and cubic and tetragonal phases ("tweed" and martensitelike structures). The theoretical results are applied to a specific set of examples of important systems. Also formulated and discussed are the consequences of structural changes for x-ray and electron diffraction.

The new and prospective development in the field of phase transformations concerns the direct computer simulation of strain-controlled processes in real solids which are too complex to be treated manually. Specific examples of the computer simulations of phase transformations include the simulations of both martensitic transformation and precipitation reactions in solids. Computer simulation, though treating idealized models, sheds new light on some aspects of the mechanisms of martensitic transformations and the formation of modulated and "tweed" structures in decomposed materials.

The final part of the book (Chapters 13 and 14) is devoted to the problems of microscopic elasticity and its application to the strain-induced ordering and decomposition in interstitial alloys. The emphasis is laid on the processes in bcc interstitial solutions based on V, Nb, Ta, and α-Fe and, especially, on the processes in the iron-carbon martensite.

The book is essentially a monograph. I have, nevertheless, also tried to make it serve as a textbook for certain advanced problems of the theory of phase transformations. The reader may judge whether I have managed to succeed at this.

I have tried to make the account logically self-contained, so that no additional sources should be required for its understanding. It is, however, assumed that the reader has some knowledge of elementary crystallography, mathematics at an undergraduate level, and statistical mechanics. Because the reader may have some trouble with the tensor and matrix algebra which is extensively used in the book, a description of the most basic definitions and matrix manipulations used in the book may be found in Appendixes 1 and 2.

As far as I know, with the exception of martensitic transformation and the concentration wave approach to the theory of order-disorder transitions, the subject matter of this book has not been covered by other books. Some books that should be mentioned are *Introduction to the Crystallography of Martensitic Transformations* by C. M. Wayman (1964), *The Theory of Transformations in Metals and Alloys* by J. W. Christian (1965, 1975), and *Transformations in Iron and Steel* by G. V. Kurdjumov, L. M. Utevsky, and R. I. Entin (Moscow 1977, in Russian)—where certain problems concerning martensitic transformation have been discussed—and the review article by A. L. Roitburd, "Martensitic Transformation as a Typical Phase Transformation in Solids" in *Solid State Physics Series*, edited by F. Seitz and D. Turnbull (1978). The concentration wave approach to ordering was discussed in my review in the *Progress in Material Science Series*, edited by J. W. Christian, P. Haasen, and T. B. Massalski (1978) and in my book in Russian, *The Theory of Phase Transformations and Structure of Solid Solutions* (1974).

The major part of the material in this book is either original or has come from the papers published in various journals. The book may be of interest to researchers and students majoring in the crystallography of diffussionless transformations, including martensitic transformations, structure transformations, and ferroelectric transitions, as well as for investigators and students studying decomposition reactions in metal alloys, high-coercive magnetic materials, ceramics, and geology.

Certain parts of the book may be of interest for those dealing with the theory of elasticity and students of the theory of mechanics concerned with allied problems, such as the state of strain within a polycrystalline aggregate under an external load.

Some of the results are intended for postgraduate students and students who are interested in the application of computers to material science. The last section of the book should be of interest to scientists concerned with the treatment of steels and development of modern solids based on interstitial solutions such as hydrides, deuterides, oxides and so on.

I am painfully aware that the book has a lot of defects that I have not managed to remove. Sometimes the material reflects my own interests rather than the relative importance of the problem. A lot of interesting advances have not been covered, and some references are lacking. I apologize for this. The only excuse is that I have at least tried to review papers containing new ideas or meaningful experimental results.

I would like to express my deep gratitude to Prof. G. V. Kurdjumov, my teacher, for many valuable discussions, moral support, and inspiring guidance that he has been giving me ever since I was making my first steps in science.

I am extremely grateful to my family who have unflinchingly borne the heavy burden of preparing the manuscript. My wife, Dr. S. V. Semenovskaya, has edited the manuscript at the cost of her own research. Owing to her, a great number of improvements have been made, many errors in the equations were corrected.

I would like to express my warm gratitude to Prof. J. W. Morris. His cooperation and many useful discussions had a great affect on the formulation of the problems. I am grateful to Dr. Sheree Wen, together with whom Prof. Morris and I worked on the computer simulation of the phase transformations entering the book. I am deeply impressed by her selfless devotion to the work, kindness, and friendly attitude.

I am very obliged to Prof. G. Thomas, C. M. Wayman, J. van Landuyt, and D. E. Laughlin for sending me their beautiful electron microscopic pictures and for valuable discussions.

I am also grateful to Dr. B. I. Pokrovski and Dr. M. S. Blanter for many fruitful discussions on ordering, and Dr. K. J. de Vos and Dr. E. G. Knizhnik who sent me electron micrographs of ALNICO alloys.

Moscow, USSR
May 1983

A. G. KHACHATURYAN

CONTENTS

1

CRYSTALLOGRAPHY OF PHASE TRANSFORMATIONS

1.1. ATOMIC STRUCTURE OF CRYSTALS

The atomic structure of an ideal crystal is described in terms of a group of atoms or a basis translated periodically along three noncoplanar directions by unit translations $\mathbf{a}_1$, $\mathbf{a}_2$, and $\mathbf{a}_3$. The geometrical image of the system of translations is a periodic lattice whose sites can be generated by translating a single point. The lattice produced by translations of a single point is called the Bravais lattice. According to this definition, any site of the Bravais lattice may be described by a reference vector

$$\mathbf{r}=n_1\mathbf{a}_1+n_2\mathbf{a}_2+n_3\mathbf{a}_3 \tag{1.1.1}$$

where n_1, n_2, and n_3 are integers.

Geometrical considerations show that only fourteen different Bravais lattices may be generated by translating a single point.

The crystal structure formed by Bravais translations of a basis may be treated as several interpenetrating Bravais lattices. Their number is equal to the number of atoms in the basis since translations of each atom generate a Bravais lattice of its own.

The Bravais lattice concept is of utmost importance in crystallography. It gives the geometrical image of an infinite number of translational symmetry operations which make up the subgroups of translations. The set of all the symmetry operations including rotations, reflections, and translations that bring the crystal lattice in coincidence with itself forms the space group of the crystal. The space group contains an infinite number of elements since it is the product of a finite number of point-group elements of the basis and infinite number of translations. The space group describes the microscopic (atomic) structure of the crystal. Sometimes, however, the macroscopic characteristics are of major importance. A number of physical properties such as elasticity, conductivity, and refraction depend on the macroscopic symmetry of a crystal rather than its atomic structure.

In treating the macroscopic symmetry properties of a crystal, translations should be regarded as identity operations since they do not change the crystallographic directions and therefore do not produce macroscopic effects. As a consequence the symmetry operations such as screw rotations and glide reflections that are combinations of translations and either rotations or reflections are just simple rotations and reflections from the macroscopic standpoint. Such a "degeneracy" of space-group elements results in the degeneracy of a space group into a point group. The point group describes the macroscopic symmetry of a crystal and is usually referred to as a crystallographic class.

Returning to crystal microscopic symmetry, we have two more definitions to introduce, those of the primitive and unit cells. The primitive cell is the smallest component of a crystal lattice that can be translated to reproduce the whole crystal. There are many ways of visualizing a primitive cell in a Bravais lattice (see Fig. 1), but there is only one requirement: a correct representation must have one lattice site per primitive cell. The usual way to avoid the ambiguity is to choose a primitive cell as a parallelepiped built on three shortest noncoplanar Bravais translations, $\mathbf{a}_1$, $\mathbf{a}_2$, and $\mathbf{a}_3$.

The unit cell, as distinguished from the primitive cell (the smallest translational component of the lattice), is defined as the smallest crystal component whose symmetry coincides with a given crystallographic class. A unit cell as a rule contains several primitive cells. For instance, the primitive cell of a face-centered-cubic (fcc) Bravais lattice is a parallelepiped built on noncoplanar vectors linking the nearest lattice sites. This primitive cell lacks certain symmetry elements of the cubic crystallographic class, but the unit cell of the lattice which is a face-centered cube composed of four primitive cells has the cubic point-group symmetry.

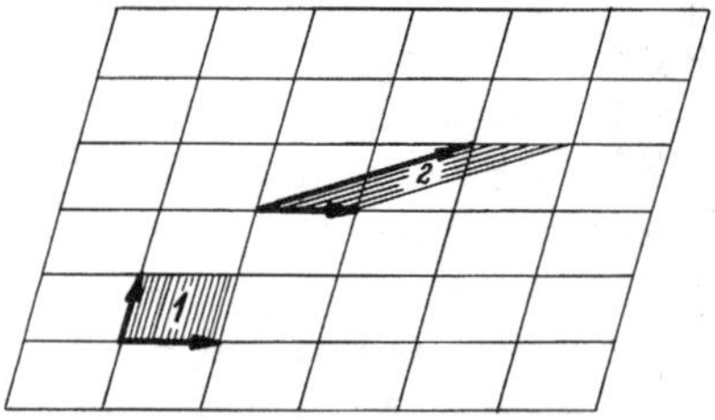

Figure 1. A primitive cell in a Bravais lattice. Cell 1 shows a primitive unit cell based on the shortest translations; cell 2, an arbitrary primitive unit cell.

1.2. CRYSTAL LATTICE AND RECIPROCAL LATTICE

The reciprocal lattice concept considerably simplifies the analysis of diffraction patterns of crystals. This concept directly follows from the periodicity condition characterizing the crystalline state, viz. the spatial periodicity of any local physical property—electronic density, internal crystal field, and so on.

For instance, the spatial periodicity of electronic density may be written in the form of three relations:

$$\rho(\boldsymbol{\xi}+\mathbf{a}_p)\equiv\rho(\boldsymbol{\xi}) \tag{1.2.1}$$

where the index, $p=1, 2, 3$, refers to three noncoplanar unit translations and $\rho(\boldsymbol{\xi})$ is the electronic density at the point $\boldsymbol{\xi}$ of the crystal.

The function $\rho(\boldsymbol{\xi})$, which is a finite function of continuous coordinates $\boldsymbol{\xi}$, can always be represented as the superposition of static plane electron density waves:

$$\rho(\boldsymbol{\xi})=\sum_{\mathbf{H}} F(\mathbf{H})e^{2\pi i\mathbf{H}\boldsymbol{\xi}} \tag{1.2.2}$$

where $F(\mathbf{H})$ are the amplitudes of the waves whose wave vectors are $2\pi\mathbf{H}$. The quantity, $F(\mathbf{H})$, is usually called the structure amplitude. Since Eq. (1.2.2) is a Fourier series expansion of electron density, the structure amplitude is a Fourier coefficient and can be written in the usual form:

$$F(\mathbf{H})=\int_{\Omega}\rho(\boldsymbol{\xi})e^{-i2\pi\mathbf{H}\boldsymbol{\xi}}d^3\xi \tag{1.2.2a}$$

where integration is over the unit cell volume.

When applied to the plane wave representation of electron density, (1.2.2), the condition (1.2.1) yields

$$\sum_{\mathbf{H}} F(\mathbf{H})e^{i2\pi\mathbf{H}(\boldsymbol{\xi}+\mathbf{a}_p)}\equiv\sum_{\mathbf{H}} F(\mathbf{H})e^{i2\pi\mathbf{H}\boldsymbol{\xi}}$$

or

$$\sum_{\mathbf{H}}(F(\mathbf{H})e^{i2\pi\mathbf{H}\mathbf{a}_p})e^{i2\pi\mathbf{H}\boldsymbol{\xi}}\equiv\sum_{\mathbf{H}} F(\mathbf{H})e^{i2\pi\mathbf{H}\boldsymbol{\xi}} \tag{1.2.3}$$

Identity (1.2.3) holds if

$$e^{2\pi i\mathbf{H}\mathbf{a}_p}\equiv 1 \quad \text{at } p=1, 2, 3 \tag{1.2.4}$$

for any $\mathbf{H}$ vector. Hence the scalar products $\mathbf{H}\mathbf{a}_p$ should be integers:

$$(\mathbf{H}\mathbf{a}_1)=h \qquad (\mathbf{H}\mathbf{a}_2)=k \qquad (\mathbf{H}\mathbf{a}_3)=l \tag{1.2.4a}$$

where h, k, and l are integers. Eqs. (1.2.4a) make up a set of linear equations in three unknown coordinates of the $\mathbf{H}$ vector. It is easy to see that the solution of Eqs. (1.2.4a) is given by

$$\mathbf{H}_{hkl}=h\mathbf{a}_1^*+k\mathbf{a}_2^*+l\mathbf{a}_3^* \tag{1.2.5}$$

where

$$\mathbf{a}_1^*=\frac{\mathbf{a}_2\times\mathbf{a}_3}{\mathbf{a}_1\cdot\mathbf{a}_2\times\mathbf{a}_3}$$

$$\mathbf{a}_2^*=\frac{\mathbf{a}_3\times\mathbf{a}_1}{\mathbf{a}_1\cdot\mathbf{a}_2\times\mathbf{a}_3}$$

$$\mathbf{a}_3^*=\frac{\mathbf{a}_1\times\mathbf{a}_2}{\mathbf{a}_1\cdot\mathbf{a}_2\times\mathbf{a}_3} \tag{1.2.6}$$

Since h, k, and l are integers and $\mathbf{a}_1^*$, $\mathbf{a}_2^*$, and $\mathbf{a}_3^*$ noncoplanar vectors, Eqs. (1.2.5) define a periodic lattice, usually referred to as a reciprocal lattice. Vectors $\mathbf{H}_{hkl}$ generating the reciprocal lattice points are called the reciprocal lattice vectors. They are defined unambiguously by the (hkl) integers which may then be employed to designate the reciprocal lattice points. It is easy to show that (hkl) are the Miller indexes of the crystal lattice plane normal to $\mathbf{H}$.

It follows from Eqs. (1.2.5). and (1.2.6) that the product

$$\mathbf{H}\mathbf{r} = (h\mathbf{a}_1^* + k\mathbf{a}_2^* + l\mathbf{a}_3^*)(n_1\mathbf{a}_1 + n_2\mathbf{a}_2 + n_3\mathbf{a}_3) = n_1 h + n_2 k + n_3 l = m \tag{1.2.7}$$

is an integer since n_1, n_2, n_3, h, k, and l are integers. Eq. (1.2.7) may be rewritten in the identical form:

$$\frac{\mathbf{H}_{hkl}}{H_{hkl}}\mathbf{r} = \frac{m}{H_{hkl}} \quad \text{for } m = 0, \pm 1, \pm 2, \pm 3, \ldots \tag{1.2.8}$$

Since $\mathbf{H}_{hkl}/H_{hkl}$ is a unit vector, Eq. (1.2.8) is a standard equation describing the mth plane normal to $\mathbf{H}_{hkl}$ and removed by m/H_{hkl} from the origin. All the crystal lattice planes normal to the same reciprocal lattice vector, $\mathbf{H}_{hkl}$, are parallel to each other. The distance between the mth and $(m+1)$th planes, that between the nearest crystal lattice planes, is equal to $1/H_{hkl}$. The latter quantity represents the interplanar distance d_{hkl}:

$$d_{hkl} = \frac{1}{H_{hkl}} \tag{1.2.9}$$

1.3. DIFFRACTION AND RECIPROCAL LATTICE

X-ray diffraction by an ideal crystal will be considered in this section. Since x-ray photon energies are incommensurable with typical elementary excitation energies in a crystal, x-ray scattering proceeds without energy losses and can be considered as a perfectly elastic process. The conservation of photon energy in scattering results in scattered and incident waves that have the same length. Hence the moduli of the corresponding wave vectors should be equal:

$$|\mathbf{k}_1| = |\mathbf{k}_2| = \frac{2\pi}{\lambda} \tag{1.3.1}$$

where $\mathbf{k}_1$ and $\mathbf{k}_2$ are the wave vectors of the incident and scattered waves, respectively, and λ is the wave length.

X-ray scattering may be treated in terms of the interaction between the incident wave and the static plane wave of electron density. Since space is isotropic with respect to infinitesimal translations, the interaction of the incident wave, scattered wave, and electron density wave should meet the momentum conservation condition

$$\mathbf{q} = \mathbf{k}_1 - \mathbf{k}_2 = 2\pi\mathbf{H} \tag{1.3.2}$$

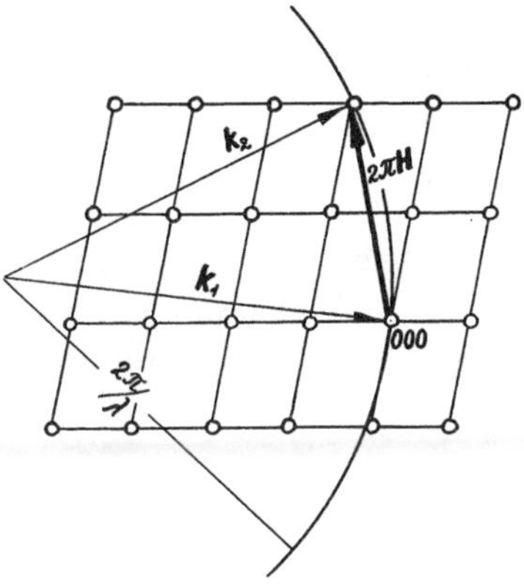

Figure 2. Ewald's construction. The incident wave propagates in the direction of the wave vector $\mathbf{k}_1$; the direction of the scattered wave is parallel to the wave vector $\mathbf{k}_2$; and λ is x-ray wavelength. The reciprocal lattice vector $2\pi\mathbf{H}$ closes the triangle formed by the vectors $\mathbf{k}_1$ and $\mathbf{k}_2$.

where, as in Section 1.2, $2\pi\mathbf{H}$ is the wave vector of the static electron density wave, and $\mathbf{q}$ the momentum transfer from the incident wave to the lattice. Scattering therefore occurs when the momentum transfer, $\mathbf{q}$, is exactly equal to one of the reciprocal lattice vectors, $2\pi\mathbf{H}$.*

In this connection note that the well-known Laue conditions for x-ray scattering by an ideal crystal coincides with the momentum conservation Eq. (1.3.2). Eq. (1.3.2) can also be given a geometrical interpretation using the so-called Ewald construction (see Fig. 2):

1. Draw a vector $2\pi/\lambda$ long in the direction of the incident wave (the wave vector $\mathbf{k}_1$), ending it at the origin of the reciprocal lattice.
2. Draw a sphere of a $2\pi/\lambda$ radius having the same origin as the $\mathbf{k}_1$ vector. This sphere is called the Ewald sphere. By construction, it goes through the reciprocal lattice origin. If some other reciprocal lattice point lies on the same sphere, the Laue condition (1.3.2) is met, and scattering associated with the reciprocal lattice vector $\mathbf{H}$ occurs.

The geometrical condition for scattering (Fig. 2) is a necessary condition, but it does not say anything of scattering intensities. The latter depend on the specific dynamics of radiation-scatterer interactions. Since the plane electron density wave scale is determined by only one parameter, that is, its amplitude, $F(\mathbf{H})$, which is the structure amplitude [see Eq. (1.2.2)], the scattered wave amplitude, Y, should be a function of $F(\mathbf{H})$. The scattered wave amplitude is nonzero only if the Laue condition (1.3.2) is met. We have therefore

$$Y(\mathbf{q}) = \begin{cases} Y(F(\mathbf{H})) & \text{if } \mathbf{q} = 2\pi\mathbf{H} \\ 0 & \text{otherwise} \end{cases} \tag{1.3.3}$$

In weak radiation-scatterer interactions such as those we are considering, Y as

*According to the de Broglie formula, the momentum vector is equal to the wave vector with accuracy to the Planck constant $\hbar$: $\mathbf{p} = \hbar\mathbf{k}$.

function of $F(\mathbf{H})$ may be represented by the first nonvanishing term in its Taylor expansion in powers of $F(\mathbf{H})$. With all other terms neglected, (1.3.3) becomes

$$Y(\mathbf{q})=\begin{cases}\alpha F(\mathbf{H}) & \text{if } \mathbf{q}=2\pi\mathbf{H}\\ 0 & \text{otherwise}\end{cases} \tag{1.3.4}$$

where α is the first-order expansion coefficient depending on the radiation-solid interaction mechanism. It is usually set equal to unity. With x-ray scattering this is equivalent to the use of so-called electron units. Bearing this in mind, Eq. (1.3.4) may be rewritten in the form

$$Y(\mathbf{q})=\begin{cases}F(\mathbf{H}) & \text{if } \mathbf{q}=2\pi\mathbf{H}\\ 0 & \text{otherwise}\end{cases} \tag{1.3.5}$$

The scattered radiation intensities are proportional to squares of scattered wave amplitude moduli. The intensity of reflection generated by the reciprocal lattice vector $\mathbf{H}$ is thus given by

$$I(\mathbf{q})=\begin{cases}|F(\mathbf{H})|^2 & \text{if } \mathbf{q}=2\pi\mathbf{H}\\ 0 & \text{otherwise}\end{cases} \tag{1.3.6}$$

With imperfect crystals, Eq. (1.3.6) should be modified. As opposed to the electron density distribution in an ideal crystal produced by a set of static electron density waves whose wave vectors form a periodic reciprocal lattice, an arbitrary electron density distribution characterizing imperfect crystals is described by a continuous superposition of static electron waves whose wave vectors $\mathbf{K}$ fill all reciprocal space with varying densities:

$$\rho(\xi)=\int\frac{d^3K}{(2\pi)^3}\,F(\mathbf{K})e^{i\mathbf{K}\xi}$$

Their amplitudes are given by the Fourier integral

$$F(\mathbf{K})=\iiint_{-\infty}^{\infty} d^3\xi\cdot\rho(\xi)e^{-i\mathbf{K}\xi} \tag{1.3.7}$$

rather than the Fourier series (1.2.2a) which describes the electron distribution in an ideal crystal.

According to the momentum conservation condition, x-ray scattering always occurs if the diffraction vector, $\mathbf{q}$, is equal to one of the wave vectors, $\mathbf{K}$, of the wave packet which approximates the arbitrary (nonperiodic) electron density distribution, $\rho(\xi)$:

$$\mathbf{q}=\mathbf{k}_2-\mathbf{k}_1=\mathbf{K} \tag{1.3.8}$$

Eq. (1.3.8) is fully analogous to (1.3.2). Following the same line of reasoning as with ideal crystal scattering, one obtains

$$Y(\mathbf{q})=F(\mathbf{K}) \quad \text{at } \mathbf{q}=\mathbf{K} \tag{1.3.9}$$

and

$$I(\mathbf{q})=|Y(\mathbf{q})|^2=|F(\mathbf{K})|^2 \quad \text{at } \mathbf{q}=\mathbf{K} \tag{1.3.10}$$

[see Eqs. (1.3.5) and (1.3.6), respectively].

Scattering intensities may thus be treated as distributed in the K-space of wave vectors or in the reciprocal lattice space. One may vary the direction and magnitude of the diffraction vector $\mathbf{q}$ (by varying the scattering geometry, i.e., the directions of the incident and scattered beams) and thus probe various regions of the reciprocal space by measuring the scattering intensities or, in other words, the distribution of squares of moduli of Fourier components of the electron density distribution.

Assuming that the total electron density distribution is the superposition of unperturbed atomic electron densities, the crystal electron density may be written

$$\rho(\boldsymbol{\xi})=\sum_{\alpha,n} \rho_\alpha(\boldsymbol{\xi}-\mathbf{r}_{n,\alpha}) \tag{1.3.11}$$

where $\rho_\alpha(\mathbf{r})$ is the unperturbed electron density of an atom of type α positioned at $\mathbf{r}=0$ and $\mathbf{r}_{n,\alpha}$ the radius vector of the nth atom of type α. Summation in Eq. (1.3.11) is over all the atoms comprising the system.

Substitution of Eq. (1.3.11) into (1.3.9) yields

$$\begin{aligned} F(\mathbf{q}) &= \iiint_{-\infty}^{\infty} d^3\xi \sum_{\alpha,n} \rho_\alpha(\boldsymbol{\xi}-\mathbf{r}_{n,\alpha})e^{-i\mathbf{q}\boldsymbol{\xi}} \\ &= \sum_{\alpha,n} e^{-i\mathbf{q}\mathbf{r}_{n,\alpha}} \iiint_{-\infty}^{\infty} \rho_\alpha(\boldsymbol{\xi}')e^{-i\mathbf{q}\boldsymbol{\xi}'}d^3\xi' \\ &= \sum_{\alpha,n} f_\alpha(\mathbf{q})e^{-i\mathbf{q}\mathbf{r}_{n,\alpha}} \end{aligned} \tag{1.3.12}$$

where $\boldsymbol{\xi}'=\boldsymbol{\xi}-\mathbf{r}_{n,\alpha}$, $f_\alpha(\mathbf{q})$ is the atomic factor of α-atoms given by the equation

$$f_\alpha(\mathbf{q})=\iiint_{-\infty}^{\infty} d^3\xi' \cdot \rho_\alpha(\boldsymbol{\xi}')e^{-i\mathbf{q}\boldsymbol{\xi}'} \tag{1.3.13}$$

Substitution of (1.3.12) into (1.3.9) and (1.3.10) gives

$$Y(\mathbf{q})=\sum_{\alpha,n} f_\alpha(\mathbf{q})e^{-i\mathbf{q}\mathbf{r}_{n,\alpha}} \tag{1.3.14}$$

and

$$I(\mathbf{q})=\left|\sum_{\alpha,n} f_\alpha(\mathbf{q})e^{-i\mathbf{q}\mathbf{r}_{n,\alpha}}\right|^2 \tag{1.3.15}$$

To exemplify the application of Eq. (1.3.15), let us consider kinematic scattering by a binary substitutional solution. In a binary substitutional solution atoms occupy only the crystal lattice sites with the vector coordinates

$$\mathbf{r}+\mathbf{u}(\mathbf{r}) \tag{1.3.16}$$

where **r** runs over the sites of the undisturbed (perfect) crystal lattice and $\mathbf{u}(\mathbf{r})$ describe atomic displacements from the perfect crystal lattice sites.

With the help of the distribution function for solute atoms,

$$c(\mathbf{r})=\begin{cases}1 & \text{when a solute atom occupies the site } \mathbf{r}\\ 0 & \text{otherwise (the site is occupied by a solvent atom)}\end{cases} \tag{1.3.17}$$

the "atomic factor" of a crystal lattice site may be written as

$$f(\mathbf{r})=f_A(1-c(\mathbf{r}))+f_B\,c(\mathbf{r})=\bar{f}+(f_B-f_A)(c(\mathbf{r})-c) \tag{1.3.18}$$

where f_A and f_B are the atomic factors of solvent and solute atoms, c is the atomic fraction of solute atoms in the alloy, and

$$\bar{f}=(1-c)f_A+cf_B \tag{1.3.19}$$

is the average atomic factor.

Substituting Eqs. (1.3.16) and (1.3.18) into Eq. (1.3.14) yields the structure amplitude for the binary alloy:

$$Y(\mathbf{q})=\sum_{\mathbf{r}}[\bar{f}+(f_B-f_A)(c(\mathbf{r})-c)]e^{-i\mathbf{q}\mathbf{u}(\mathbf{r})}e^{-i\mathbf{q}\mathbf{r}} \tag{1.3.20a}$$

or

$$Y(\mathbf{q})=\sum_{\mathbf{r}}\phi(\mathbf{r})e^{-i\mathbf{q}\mathbf{r}} \tag{1.3.20b}$$

where

$$\phi(\mathbf{r})=[\bar{f}+(f_B-f_A)(c(\mathbf{r})-c)]e^{-i\mathbf{q}\mathbf{u}(\mathbf{r})} \tag{1.3.21}$$

is the effective atomic factor of a crystal lattice site for a system involving atomic displacements.

The structure amplitude (1.3.20) may also be rewritten as

$$Y(\mathbf{q})=\langle Y(\mathbf{q})\rangle+\delta Y(\mathbf{q}) \tag{1.3.22}$$

where the symbol $\langle\cdots\rangle$ denotes averaging:

$$\langle Y(\mathbf{q})\rangle=\sum_{\mathbf{r}}\langle\phi(\mathbf{r})\rangle\exp(-i\mathbf{q}\mathbf{r}) \tag{1.3.23}$$

$$\delta Y(\mathbf{q})=\sum_{\mathbf{r}}\delta\phi(\mathbf{r})\exp(-i\mathbf{q}\mathbf{r}) \tag{1.3.24}$$

$$\delta\phi(\mathbf{r})=\phi(\mathbf{r})-\langle\phi(\mathbf{r})\rangle \tag{1.3.25}$$

Eq. (1.3.22) gives the structure amplitude (1.3.20) as the sum of the average (coherent) structure amplitude, $\langle Y(\mathbf{q})\rangle$, and the fluctuating part of the structure amplitude, $\delta Y(\mathbf{q})$.

According to Eq. (1.3.15), the total mean intensity is

$$I(\mathbf{q}) = \langle |Y(\mathbf{q})|^2 \rangle = \left\langle \left| \sum_{\mathbf{r}} [\langle \phi(\mathbf{r}) \rangle + \delta\phi(\mathbf{r})] e^{-i\mathbf{q}\mathbf{r}} \right|^2 \right\rangle$$
$$= I_{\text{coh}}(\mathbf{q}) + I_{\text{diff}}(\mathbf{q}) \tag{1.3.26}$$

where

$$I_{\text{coh}}(\mathbf{q}) = |\langle Y(\mathbf{q}) \rangle|^2 = \left| \sum_{\mathbf{r}} \langle \phi(\mathbf{r}) \rangle e^{-i\mathbf{q}\mathbf{r}} \right|^2 \tag{1.3.27}$$

is the coherent scattering intensity,

$$I_{\text{diff}}(\mathbf{q}) = \langle |\delta Y(\mathbf{q})|^2 \rangle = \sum_{\mathbf{r},\mathbf{r}'} \langle \delta\phi(\mathbf{r}) \delta\phi(\mathbf{r}') \rangle e^{-i\mathbf{q}(\mathbf{r}-\mathbf{r}')} \tag{1.3.28}$$

is the diffuse scattering intensity associated with the atomic factor fluctuations (short-range order).

The cross term in Eq. (1.3.26) vanishes because it is proportional to the average value:

$$\langle \delta\phi(\mathbf{r}) \langle \phi(\mathbf{r}) \rangle \rangle = \langle \phi(\mathbf{r}) \rangle \langle \delta\phi(\mathbf{r}) \rangle \tag{1.3.29}$$

which vanishes by the definition (1.3.25), $\langle \delta\phi(\mathbf{r}) \rangle = 0$.

If the atomic displacements are ignored,

$$\phi(\mathbf{r}) \to f(\mathbf{r})$$

[see Eqs. (1.3.18) and (1.3.21)], and we have

$$\langle \phi(\mathbf{r}) \rangle \to \langle f(\mathbf{r}) \rangle$$
$$= \bar{f} + (f_B - f_A)(\langle c(\mathbf{r}) \rangle - c)$$
$$= \bar{f} + (f_B - f_A)(n(\mathbf{r}) - c) \tag{1.3.30}$$

where $n(\mathbf{r}) = \langle c(\mathbf{r}) \rangle$ is the occupation probability of finding a solute atom at the site $\mathbf{r}$, and

$$\delta\phi(\mathbf{r}) \to f(\mathbf{r}) - \langle f(\mathbf{r}) \rangle$$
$$= (f_B - f_A)(c(\mathbf{r}) - n(\mathbf{r})) \tag{1.3.31}$$

Substituting (1.3.30) into (1.3.23) yields

$$\langle Y(\mathbf{q}) \rangle = \bar{f} \sum_{\mathbf{r}} e^{-i\mathbf{q}\mathbf{r}} + (f_B - f_A) \sum_{\mathbf{r}} (n(\mathbf{r}) - c) e^{-i\mathbf{q}\mathbf{r}} \tag{1.3.32}$$

The first term in (1.3.32) assumes nonzero values equal to $\bar{f} \cdot N$ (N for the number of crystal lattice sites) only at the fundamental reflection points

$$\mathbf{q} = 2\pi\mathbf{H} \tag{1.3.33}$$

where $\mathbf{H}$ is the reciprocal lattice vector related to the set of crystal lattice sites $\{\mathbf{r}\}$. This vector describes the fundamental Laue reflections. The second term in Eq. (1.3.32) gives the amplitude of coherent scattering in other than fundamental reflection points (it is easy to see that the second term vanishes at

$\mathbf{q}=2\pi\mathbf{H}$). This term describes coherent scattering associated with concentration heterogeneities. It can be rewritten in the form

$$\langle Y_{\text{heter}}(\mathbf{q})\rangle=(f_B-f_A)\sum_{\mathbf{r}}(n(\mathbf{r})-c)e^{-i\mathbf{q}\mathbf{r}}=(f_B-f_A)\Delta\tilde{n}(\mathbf{q}) \qquad (1.3.34)$$

where

$$\Delta\tilde{n}(\mathbf{q})=\sum_{\mathbf{r}}(n(\mathbf{r})-c)e^{-i\mathbf{q}\mathbf{r}} \qquad (1.3.35)$$

Since $\langle Y_{\text{heter}}(\mathbf{q})\rangle$ and $\bar{f}\sum_{\mathbf{r}} e^{-i\mathbf{q}\mathbf{r}}$ take on nonzero values at different points of the reciprocal space, the coherent intensity $I_{\text{coh}}(\mathbf{q})$ can be presented as

$$I_{\text{coh}}(\mathbf{q})=\left|\bar{f}\sum_{\mathbf{r}} e^{-i\mathbf{q}\mathbf{r}}\right|^2+(f_B-f_A)^2\cdot|\Delta\tilde{n}(\mathbf{q})|^2 \qquad (1.3.36)$$

Finally, the third contribution to the scattering comes from diffuse scattering. Its intensity is described by Eq. (1.3.28). Diffuse scattering is associated with the short-range order effects. Substitution of Eq. (1.3.31) into (1.3.28) yields

$$I_{\text{diff}}(\mathbf{q})=(f_B-f_A)^2\sum_{\mathbf{r},\mathbf{r}'}\langle(c(\mathbf{r})-n(\mathbf{r}))(c(\mathbf{r}')-n(\mathbf{r}'))\rangle e^{-i\mathbf{q}(\mathbf{r}-\mathbf{r}')}$$

or

$$I_{\text{diff}}(\mathbf{q})=(f_B-f_A)^2\langle|\delta\tilde{c}(\mathbf{q})|^2\rangle \qquad (1.3.37)$$

where

$$\delta\tilde{c}(\mathbf{q})=\sum_{\mathbf{r}}(c(\mathbf{r})-n(\mathbf{r}))e^{-i\mathbf{q}\mathbf{r}}$$

For a homogeneous solid solution Eq. (1.3.37) becomes simply

$$I_{\text{diff}}(\mathbf{q})=(f_B-f_A)^2\sum_{\mathbf{r},\mathbf{r}'}\langle(c(\mathbf{r})-c)(c(\mathbf{r}')-c)\rangle e^{-i\mathbf{q}(\mathbf{r}-\mathbf{r}')} \qquad (1.3.38)$$

The two-particle correlator

$$K(\mathbf{r},\mathbf{r}')=\langle(c(\mathbf{r})-c)(c(\mathbf{r}')-c)\rangle \qquad (1.3.39)$$

of a homogeneous solution depends on the coordinate difference $\mathbf{r}-\mathbf{r}'$ rather than on the coordinates themselves:

$$K(\mathbf{r},\mathbf{r}')=K(\mathbf{r}-\mathbf{r}') \qquad (1.3.40)$$

The reason is that any coordinate-dependent function describing a homogeneous solution would be invariant under arbitrary crystal lattice translations

$$\mathbf{r}\to\mathbf{r}+\mathbf{T} \qquad \mathbf{r}'\to\mathbf{r}'+\mathbf{T}$$

where $\mathbf{T}$ is an arbitrary reference vector. Substituting Eq. (1.3.40) into (1.3.38) gives

$$I_{\text{diff}}(\vec{q})=N(f_B-f_A)^2\sum_{\rho}K(\rho)e^{-i\mathbf{q}\rho} \qquad (1.3.41)$$

where $\rho=\mathbf{r}-\mathbf{r}'$. The form (1.3.41) is obtained after the change of variables $(\mathbf{r},\mathbf{r}')\to(\mathbf{r},\rho)$.

The so-called Warren short-range-order parameters, $\alpha(\boldsymbol{\rho})$, are often used in the theory of scattering. They are defined by the equation

$$K(\boldsymbol{\rho})=c(1-c)\alpha(\boldsymbol{\rho}) \tag{1.3.42}$$

Substituting Eq. (1.3.42) into (1.3.41) yields

$$I_{\text{diff}}(\mathbf{q})=N(f_B-f_A)^2 c(1-c)\sum_{\mathbf{r}}\alpha(\mathbf{r})e^{-i\mathbf{qr}} \tag{1.3.43}$$

It follows from Eq. (1.3.39) that

$$K(\mathbf{r})=\langle(c(\mathbf{r})-c)(c(0)-c)\rangle=\langle c(\mathbf{r})c(0)\rangle-c^2 \tag{1.3.44}$$

If $\mathbf{r}=0$,

$$\begin{aligned}K(0)&=\langle c(0)c(0)\rangle-c^2=\langle c(0)\rangle-c^2=c-c^2\\&=c(1-c)\end{aligned} \tag{1.3.45}$$

because by the definition of the function $c(\mathbf{r})$, $c(0)^2\equiv c(0)$ and $\langle c(\mathbf{r})\rangle=c$ (for a disordered alloy). If $\mathbf{r}\neq 0$, the average $\langle c(\mathbf{r})c(0)\rangle$ is a probability of finding simultaneously one B-atom at the site 0 and another B-atom at the site $\mathbf{r}$. This probability is a product of the a priori probability, c, of finding a B-atom at the site 0 and the a posteriori probability, $n(B\mathbf{r}|B0)$, of finding a B-atom at the site $\mathbf{r}$ under the condition that a B-atom is definitely occupying the site 0:

$$\langle c(\mathbf{r})c(0)\rangle=c\cdot n(B\mathbf{r}|B0) \tag{1.3.45}$$

Taking into account the latter and Eq. (1.3.45), one has

$$K(\mathbf{r})=c(1-c)\alpha(\mathbf{r})=\begin{cases}c\cdot n(B\mathbf{r}|B0)-c^2 & \text{if } \mathbf{r}\neq 0\\ c(1-c) & \text{if } \mathbf{r}=0\end{cases} \tag{1.3.47}$$

We conclude this section by stressing once more that the theory described here is only valid for weak, single-scattering interactions between radiation and solids. This requirement is equivalent to the Born approximation, or the approximation of kinematic scattering.

1.4. PHASE TRANSFORMATIONS AND CRYSTAL LATTICE REARRANGEMENTS

As a rule phase transformations in solids are accompanied by crystal lattice rearrangements. The atomic arrangement of the new crystal lattice is different from that of the initial phase. The effects involved in a crystal lattice rearrangement are caused by both homogeneous strains affecting the macroscopic shape of a crystal and heterogeneous displacements of atoms which do not produce visible macroscopic effects. Heterogeneous displacements in phase transformations can always be described as superpositions of displacement static waves.

Below we shall consider the homogeneous component of crystal lattice rearrangement since homogeneous strain determines the macroscopic shape of transformed "islands" of a new phase formed within the parent phase and thus affects the morphology of two-phase mixtures.

Under homogeneous strain, each initial crystal lattice site vector, **r**, undergoes transformation into **r**′:

$$\mathbf{r} \rightarrow \mathbf{r}'$$

The initial and transformed lattice sites are related by the equation

$$r_i' = A_{ij} r_j \tag{1.4.1}$$

where A_{ij} is the homogeneous distortion matrix, and i and j are the Cartesian indexes (here and below, sums taken over all twice repeated indexes are implied). The symbolic form of Eq. (1.4.1) is

$$\mathbf{r}' = \hat{\mathbf{A}}\mathbf{r} \tag{1.4.2}$$

(The caret ^above a letter indicates that it is an operator.)

According to Eq. (1.4.2), crystal lattice site displacements, **u**, caused by homogeneous strain are given by

$$\mathbf{u} = \mathbf{r}' - \mathbf{r} = \hat{\mathbf{A}}\mathbf{r} - \mathbf{r} = (\hat{\mathbf{A}} - \hat{\mathbf{I}})\mathbf{r} \quad \text{or}$$
$$\mathbf{u} = \hat{\mathbf{U}}\mathbf{r} \tag{1.4.3}$$

where $\hat{\mathbf{I}}$ is the identity operator and

$$\hat{\mathbf{U}} = \hat{\mathbf{A}} - \hat{\mathbf{I}} \tag{1.4.4}$$

is the matrix of distortions. Eq. (1.4.3) may be rewritten in the suffix form

$$u_i = U_{ij} r_j = (A_{ij} - \delta_{ij}) r_j \tag{1.4.5}$$

where δ_{ij} is the Kronecker symbol (the matrix representation of the identity operator, $\hat{\mathbf{I}}$).

The $\hat{\mathbf{A}}$ and $\hat{\mathbf{U}}$ matrices can be determined if the crystal lattice parameters of the initial and transformed phases and the orientational relationships between the lattices are known; in other words, one has to learn how the unit Bravais translations of the initial phase are transformed to obtain the Bravais translations of the new phase. The corresponding procedure is exemplified for the *bcc* → *hcp* rearrangement in Problem 1 in Section 1.9.

Any homogeneous crystal lattice distortion can be reduced to two operations: lattice deformation involving extension and contraction along some mutually orthogonal axes (so-called principal axes) and subsequent rigid body rotation. Mathematically, the latter statement may be written as

$$\hat{\mathbf{A}} = \hat{\mathbf{R}}\hat{\mathbf{F}} \tag{1.4.6}$$

which means that homogeneous distortion $\hat{\mathbf{A}}$ is the product of a Hermitian matrix $\hat{\mathbf{F}}$ ($\hat{\mathbf{F}}^{+} = \hat{\mathbf{F}}$, where $\hat{\mathbf{F}}^{+}$ is the Hermitian conjugate of $\hat{\mathbf{F}}$) describing the deformation and a unitary matrix of rigid body rotation $\hat{\mathbf{R}}$*.

*Unitary matrices correspond to operators that do not change vector lengths. Further details and the principal definitions of the matrix algebra are given in Appendix 1.

The homogeneous deformation matrix $\hat{\mathbf{F}}$ should be Hermitian because only a Hermitian matrix can always be diagonalized in an orthogonal basis and has real eigenvalues. Its effect on the crystal lattice translations will thus be reduced to extension and contraction along the orthogonal basis axes (see Appendix 1). Let the corresponding matrix $\hat{\mathbf{F}}$ eigenvalues and eigenvectors be λ_1, λ_2, and λ_3 and $\mathbf{e}_1$, $\mathbf{e}_2$, and $\mathbf{e}_3$, respectively ($\mathbf{e}_\sigma \mathbf{e}_{\sigma'} = \delta_{\sigma,\sigma'}$ where $\sigma, \sigma' = 1, 2, 3$). The matrix $\hat{\mathbf{F}}$ may then be written as a bilinear form (see Appendix 2):

$$F_{ij} = \lambda_1 e_1^i e_1^j + \lambda_2 e_2^i e_2^j + \lambda_3 e_3^i e_3^j \tag{1.4.7}$$

or in a symbolic form

$$\mathbf{F} = \lambda_1 \mathbf{e}_1 * \mathbf{e}_1 + \lambda_2 \mathbf{e}_2 * \mathbf{e}_2 + \lambda_3 \mathbf{e}_3 * \mathbf{e}_3 \tag{1.4.8}$$

where $*$ denotes the dyadic (tensor) products.

Since the matrix $\hat{\mathbf{F}}$ describes extensions and contractions along the principal directions $\mathbf{e}_1$, $\mathbf{e}_2$, and $\mathbf{e}_3$, its eigenvalues should be positive. A negative eigenvalue makes no physical sense, as it requires that the crystal lattice vectors directed along the corresponding principal direction change direction upon the deformation.

The invariant tensor form for the rigid body rotation matrix $\hat{\mathbf{R}}$ is as follows:

$$(\hat{\mathbf{R}})_{ij} = R_{ij} = \delta_{ij} \cos\phi + p_i p_j (1 - \cos\phi) + \delta_{ijk} p_k |\sin\phi| \tag{1.4.9}$$

Here ϕ is the rotation angle about the unit vector, $\mathbf{p}$, in the direction defined by the right-hand screw rule (see Fig. 3) and δ_{ijk} is the completely antisymmetric third-rank unit tensor.

If the crystal lattice rearrangement matrix, $\hat{\mathbf{A}}$, is known, the deformation matrix, $\hat{\mathbf{F}}$, may be found from the equation

$$\hat{\mathbf{A}}^+ \hat{\mathbf{A}} = (\hat{\mathbf{R}}\hat{\mathbf{F}})^+ (\hat{\mathbf{R}}\hat{\mathbf{F}}) = \hat{\mathbf{F}}^+ \hat{\mathbf{R}}^+ \hat{\mathbf{R}} \hat{\mathbf{F}} = \hat{\mathbf{F}}^+ \hat{\mathbf{F}} = \hat{\mathbf{F}}^2 \tag{1.4.10}$$

This follows from the equality $(\hat{\mathbf{R}}\hat{\mathbf{F}})^+ = \hat{\mathbf{F}}^+ \hat{\mathbf{R}}^+$ (see Appendix 1), the Hermitian character of $\hat{\mathbf{F}}(\hat{\mathbf{F}}^+ = \hat{\mathbf{F}})$, and the identity

$$\hat{\mathbf{R}}^+ = \hat{\mathbf{R}}^{-1} \tag{1.4.11}$$

valid for any unitary matrix (see Appendix 1). The $\hat{\mathbf{R}}^{-1}$ matrix is an inverse matrix to $\hat{\mathbf{R}}$ and is defined by

$$\hat{\mathbf{R}}^{-1} \hat{\mathbf{R}} = \hat{\mathbf{I}} \tag{1.4.12}$$

Since the matrices $\hat{\mathbf{F}}$ and

$$\hat{\mathbf{F}}^2 = \hat{\mathbf{A}}^+ \hat{\mathbf{A}}$$

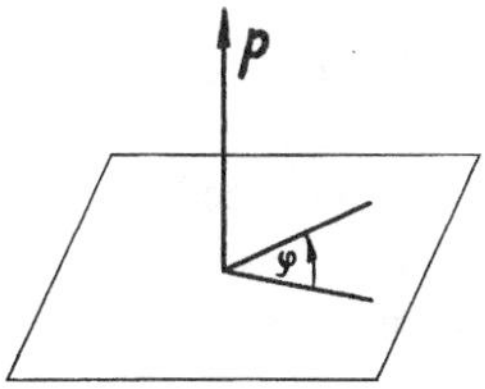

Figure 3. Relation between the rotation axis direction, $\mathbf{p}$, and rotational angle, ϕ.

are commutative, they have the same eigenvectors and hence are diagonalized in the same basis (Appendix 1). From the diagonal representation of a Hermitian matrix, one can easily ascertain that squaring a Hermitian matrix results in merely squaring its eigenvalues. With this in mind, we may write $\hat{\mathbf{F}}^2$ as a bilinear form:

$$\hat{\mathbf{F}}^2 = \lambda_1^2 \mathbf{e}_1 * \mathbf{e}_1 + \lambda_2^2 \mathbf{e}_2 * \mathbf{e}_2 + \lambda_3^2 \mathbf{e}_3 * \mathbf{e}_3 \tag{1.4.13}$$

[see Eq. (1.4.8)].

If the general matrix of homogeneous distortion is known, the application of Eqs. (1.4.8), (1.4.10), and (1.4.13) yields the corresponding Hermitian deformation matrix $\hat{\mathbf{F}}$. The procedure involves the following steps:

1. The matrix multiplication of $\hat{\mathbf{A}}^+$ by $\hat{\mathbf{A}}$ to find the Hermitian matrix, $\hat{\mathbf{F}}^2 = \hat{\mathbf{A}}^+\hat{\mathbf{A}}$.
2. The diagonalization of the $\hat{\mathbf{F}}^2$ matrix to find the eigenvalues λ_1^2, λ_2^2, and λ_3^2 and the corresponding eigenvectors, $\mathbf{e}_1$, $\mathbf{e}_2$, and $\mathbf{e}_3$.
3. Using $\hat{\mathbf{F}}^2$ matrix eigenvalues to determine the $\hat{\mathbf{F}}$ matrix eigenvalues, λ_1, λ_2, and λ_3. The later matrix can then be written according to Eq. (1.4.8).

The solution of Eq. (1.4.6) with respect to $\hat{\mathbf{R}}$ yields the rigid body rotation matrix if the matrices $\hat{\mathbf{A}}$ and $\hat{\mathbf{F}}$ are known:

$$\hat{\mathbf{R}} = \hat{\mathbf{A}}\hat{\mathbf{F}}^{-1} \tag{1.4.14}$$

where $\hat{\mathbf{F}}^{-1}$ is the inverse matrix to $\hat{\mathbf{F}}$. The $\hat{\mathbf{F}}^{-1}$ matrix may also be found from the bilinear expansion:

$$\hat{\mathbf{F}}^{-1} = \frac{1}{\lambda_1}\mathbf{e}_1 * \mathbf{e}_1 + \frac{1}{\lambda_2}\mathbf{e}_2 * \mathbf{e}_2 + \frac{1}{\lambda_3}\mathbf{e}_3 * \mathbf{e}_3 \tag{1.4.15}$$

(see Appendix 2).

1.5. EFFECT OF CRYSTAL LATTICE REARRANGEMENT ON GEOMETRY OF CRYSTAL LATTICE PLANES

Section 1.4, described the macroscopic crystal shape changes caused by phase transformations using the matrix $\hat{\mathbf{A}}$ which is the product of homogeneous deformation and rigid body rotation. Matrix $\hat{\mathbf{A}}$ relates the initial and transformed lattice site positions unambiguously, that is, it determines atomic displacements involved in the rearrangement.

In this section we show that changes in interplanar distances and plane orientations may also be found from the homogeneous distortion matrix $\hat{\mathbf{A}}$. All the information about crystal lattice planes in fact is contained in the reciprocal lattice geometry and can be easily determined if the reciprocal lattice transformation associated with the crystal lattice rearrangement is known. In other words, the problem is reduced to the determination of the reciprocal lattice rearrangement caused by the crystal lattice rearrangement.

We shall proceed from crystal lattice rearrangement Eq. (1.4.2)

$$\mathbf{r}' = \hat{\mathbf{A}}\mathbf{r} \tag{1.5.1}$$

and Eq. (1.2.7)

$$\mathbf{H}\mathbf{r} = m \tag{1.5.2}$$

where m is an integer and $\mathbf{H}$ the reciprocal lattice vector. Eq. (1.5.2) can be rewritten in the form

$$\mathbf{H}(\hat{\mathbf{A}}^{-1}\hat{\mathbf{A}})\mathbf{r} = m \tag{1.5.3}$$

for, by definition, $\hat{\mathbf{A}}^{-1}\hat{\mathbf{A}} = \hat{\mathbf{I}}$. The equation

$$(\mathbf{H}\hat{\mathbf{A}}^{-1})(\hat{\mathbf{A}}\mathbf{r}) = m \tag{1.5.4}$$

is identical to (1.5.3). From (1.5.1) and (1.5.4) we have

$$\mathbf{H}'\mathbf{r}' = m$$

where

$$\mathbf{H}' = \mathbf{H}\hat{\mathbf{A}}^{-1} = (\hat{\mathbf{A}}^{-1})^{+}\mathbf{H} \tag{1.5.5a}$$

The corresponding suffix form is

$$H_i' = H_j(\hat{\mathbf{A}}^{-1})_{ji} = (\hat{\mathbf{A}}^{-1})_{ij}^{+} H_j \tag{1.5.5b}$$

This leads us to conclude that the reciprocal lattice transformation corresponding to the crystal lattice rearrangement represented by matrix $\hat{\mathbf{A}}$ is given by the matrix that is the transposed inverse of $\hat{\mathbf{A}}$, $(\hat{\mathbf{A}}^{-1})^{+}$.

Eqs. (1.5.5) solve the problem of determining interplanar distances and plane orientations in the transformed crystal lattice: the planes should be normal to $\mathbf{H}'$, defined by (1.5.5), and the interplanar distances should be equal to $1/H'$.

The rotation angle between the initial and transformed lattice planes is the angle between the corresponding reciprocal lattice vectors $\mathbf{H}$ and $\mathbf{H}'$:

$$\cos\phi = \frac{\mathbf{H}'}{|\mathbf{H}'|}\frac{\mathbf{H}}{|\mathbf{H}|} = \frac{(\mathbf{H}\hat{\mathbf{A}}^{-1})\mathbf{H}}{|\mathbf{H}'|\,|\mathbf{H}|} \tag{1.5.6}$$

where ϕ is the rotation angle.

1.6. VARIOUS ORIENTATIONS PRODUCED BY PHASE TRANSFORMATIONS

As a rule in phase transformations parent phase symmetry results in new phase crystals having different orientations from the parent phase. The reason for this is different, though crystallographically equivalent, crystal lattice rearrangements.

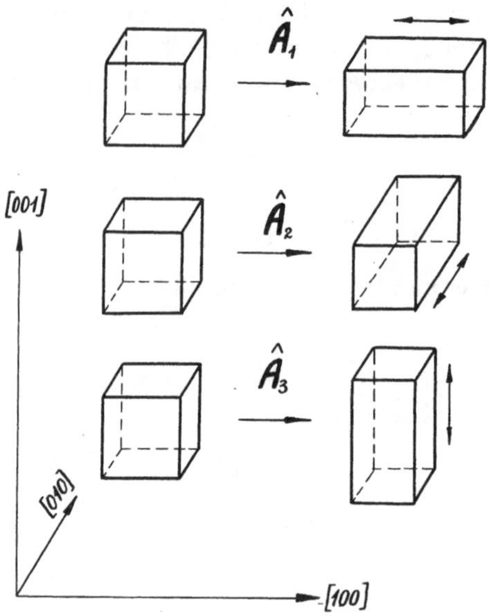

Figure 4. Three types of the crystal lattice rearrangement corresponding to the cubic→tetragonal phase transformation.

Consider, for example, a phase transformation corresponding to the cubic-to-tetragonal lattice rearrangement. The resulting tetragonal crystals may have three different orientations relative to the parent phase, with tetragonal axes along the [100], [010], and [001] cubic phase directions (see Fig. 4). The corresponding rearrangements are described by the matrices

$$\hat{\mathbf{A}}_1=\begin{pmatrix}\eta_3 & 0 & 0\\ 0 & \eta_1 & 0\\ 0 & 0 & \eta_1\end{pmatrix}\qquad \hat{\mathbf{A}}_2=\begin{pmatrix}\eta_1 & 0 & 0\\ 0 & \eta_3 & 0\\ 0 & 0 & \eta_1\end{pmatrix}\qquad \hat{\mathbf{A}}_3=\begin{pmatrix}\eta_1 & 0 & 0\\ 0 & \eta_1 & 0\\ 0 & 0 & \eta_3\end{pmatrix} \tag{1.6.1}$$

given in the Cartesian coordinates with x-, y-, and z-axes coinciding with the [100], [010], and [001] cubic phase directions. The values η_1 and η_3 are the extension (contraction) factors along the principal matrix directions, [100], [010], and [001].

One more example is the cubic → rhombohedral crystal lattice rearrangement. This may be treated for instance as a contraction of each cubic phase elementary cell along its [111] direction. As the contraction proceeds along the $[\bar{1}11]$, $[1\bar{1}1]$, and $[11\bar{1}]$ directions, identical rhombohedral lattices are produced but with different orientations relative to the parent phase lattice (see Fig. 5).

In general, all the possible crystal lattice rearrangements that result in new phase orientations can be determined from parent crystal symmetry. Let, for example, a new phase be generated by the matrix $\hat{\mathbf{A}}_1$. The initial and resulting lattice sites are then related by the equation

$$\mathbf{r}'_1=\hat{\mathbf{A}}_1\mathbf{r}_1 \tag{1.6.2}$$

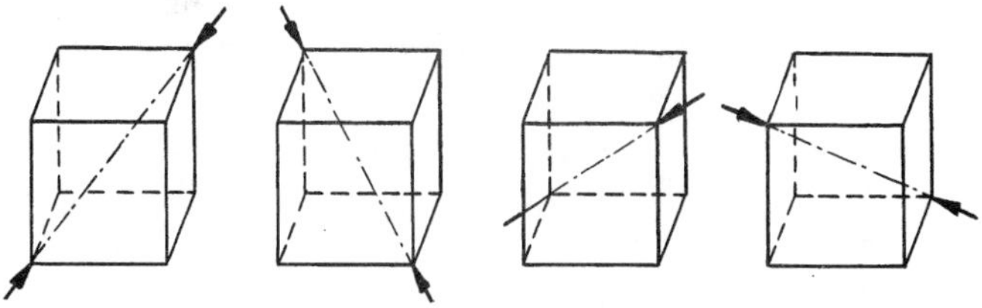

Figure 5. Examples of the crystal lattice rearrangements of a cubic phase into a rhombohedral one. Arrows show the contraction directions.

where the sets of points $\{\mathbf{r}_1\}$ and $\{\mathbf{r}_1'\}$ belong to the initial and transformed lattices, respectively. The application of one of the symmetry operations (reflections or rotations) of the parent phase crystallographic class, $\hat{\mathbf{G}}_n$, to Eq. (1.6.2) yields

$$\mathbf{r}_n' = \hat{\mathbf{G}}_n \hat{\mathbf{A}}_1 \mathbf{r}_1 \tag{1.6.3}$$

where

$$\mathbf{r}_n' = \hat{\mathbf{G}}_n \mathbf{r}_1' \tag{1.6.4}$$

Since symmetry operations are unitary,

$$\hat{\mathbf{G}}_n^+ = \hat{\mathbf{G}}_n^{-1} \quad \text{and} \quad \hat{\mathbf{G}}_n^+ \hat{\mathbf{G}}_n = \hat{\mathbf{I}} \tag{1.6.5}$$

The substitution of Eq. (1.6.5) into (1.6.3) gives

$$\mathbf{r}_n' = \hat{\mathbf{G}}_n \hat{\mathbf{A}}_1 \hat{\mathbf{I}} \mathbf{r}_1 = \hat{\mathbf{G}}_n \hat{\mathbf{A}}_1 (\hat{\mathbf{G}}_n^+ \hat{\mathbf{G}}_n) \mathbf{r}_1 = \hat{\mathbf{G}}_n \hat{\mathbf{A}}_1 \hat{\mathbf{G}}_n^+ (\hat{\mathbf{G}}_n \mathbf{r}_1) \tag{1.6.6}$$

or

$$\mathbf{r}_n' = \hat{\mathbf{G}}_n \hat{\mathbf{A}}_1 \hat{\mathbf{G}}_n^+ \mathbf{r}_n = \hat{\mathbf{A}}_n \mathbf{r}_n$$

where

$$\mathbf{r}_n = \hat{\mathbf{G}}_n \mathbf{r}_1 \tag{1.6.7}$$

and

$$\hat{\mathbf{A}}_n = \hat{\mathbf{G}}_n \hat{\mathbf{A}}_1 \hat{\mathbf{G}}_n^+ \tag{1.6.8}$$

The set of the parent phase crystal lattice sites, $\{\mathbf{r}_n\}$, coincides with $\{\mathbf{r}_1\}$ since $\hat{\mathbf{G}}_n$ is a symmetry operation of the parent phase.

The set of sites of the new phase lattice, $\{\mathbf{r}_n'\}$, generated by the matrix, $\hat{\mathbf{A}}_n$, can be also obtained from crystal lattice sites of the new phase, $\{\mathbf{r}_1'\}$, by the unitary matrix, $\hat{\mathbf{G}}_n$ [see Eq. (1.6.4)]. Since unitary symmetry operations do not affect relative lattice site positions, both sets, $\{\mathbf{r}_1'\}$ and $\{\mathbf{r}_n'\}$, correspond to the same new phase lattice but only differ in their orientations with respect to the initial phase set of sites. In other words, it follows from Eq. (1.6.6) that the matrix

$$\hat{\mathbf{A}}_n = \hat{\mathbf{G}}_n \hat{\mathbf{A}}_1 \hat{\mathbf{G}}_n^+$$

as well as $\hat{\mathbf{A}}_1$ describe a crystal lattice rearrangement leading to the new phase.

The crystal orientation generated by $\hat{\mathbf{A}}_n$ can be obtained from the crystal orientation generated by $\hat{\mathbf{A}}_1$ with the help of the symmetry operation, $\hat{\mathbf{G}}_n$ [see Eq. (1.6.4)].

If the crystallographic class of the parent phase includes the symmetry operations $\hat{\mathbf{I}}, \hat{\mathbf{G}}_2, \hat{\mathbf{G}}_3, \ldots, \hat{\mathbf{G}}_n, \ldots, \hat{\mathbf{G}}_\nu$, ν crystal lattice site rearrangements can be derived from matrix $\hat{\mathbf{A}}_1$ and Eq. (1.6.8):

$$\hat{\mathbf{A}}_1, \hat{\mathbf{A}}_2 = \hat{\mathbf{G}}_2\hat{\mathbf{A}}_1\hat{\mathbf{G}}_2^+, \hat{\mathbf{A}}_3 = \hat{\mathbf{G}}_3\hat{\mathbf{A}}_1\hat{\mathbf{G}}_3^+, \ldots, \hat{\mathbf{A}}_n = \hat{\mathbf{G}}_n\hat{\mathbf{A}}_1\hat{\mathbf{G}}_n^+, \ldots \tag{1.6.9}$$

Not all of the matrices (1.6.9) are necessarily different. Some may coincide with each other. The set of all different matrices (1.6.9) describes the total number of crystal lattice rearrangements with different orientations of the new phase crystals with respect to the parent phase. For instance, the $\hat{\mathbf{A}}_2$ and $\hat{\mathbf{A}}_3$ operations in (1.6.1) may be obtained from $\hat{\mathbf{A}}_1$ by rotating the new phase crystals by 90° which is a cubic phase symmetry operation involved in series (1.6.9).

1.7. INVARIANT PLANE STRAIN

The so-called invariant plane strain plays an important part in the crystallography of phase transformations. The term applies to distortions that displace a crystal lattice plane as a rigid body but not its orientation. In other words, neither relative atomic positions in the invariant plane nor its orientation change with invariant plane strain.

One further characteristic of invariant plane strain is due to the fact that the fitting together of the initial and transformed phases along an invariant plane does not produce elastic strain. For that reason the invariant plane strain is the basic concept behind the phenomenological theory of martensitic crystal morphology proposed by Wechsler, Lieberman, and Read (1) and Bowles, and Mackenzie (2). In Section 8.5. we will show that the invariant plane strain also determines the orientational relations of coherent platelike precipitates in the early stage of decomposition which is a less-known fact.

To find the general form of the invariant plane strain which reflects all its properties, we start from the equation of a plane

$$\mathbf{rn} = d = \text{constant} \tag{1.7.1}$$

where $\mathbf{n}$ is the unit vector normal to the plane and d the distance of the plane from the origin. Using Eq. (1.7.1), a rigid body displacement of a plane that remains parallel to itself may be written in the form

$$\mathbf{u} = \boldsymbol{l}\varepsilon d = \varepsilon \boldsymbol{l}(\mathbf{nr}) \tag{1.7.2}$$

where $\mathbf{u}$ is the displacement of all the points of the plane along the direction determined by the unit vector $\boldsymbol{l}$ and ε is a dimensionless constant describing the amount of strain.

It follows from Eqs. (1.7.1) and (1.7.2) that all the points of the plane, $\mathbf{r}$ (the points satisfying the equation $\mathbf{nr} = \text{constant}$), undergo the same displacement.

This means that Eq. (1.7.2) represents a rigid body translation of a plane that does not involve changes in interatomic distances within it.

Eq. (1.7.2) may be written in the suffix form

$$u_i = \varepsilon l_i n_j r_j$$

or

$$u_i = U_{ij}^{\text{inv}} r_j$$

where

$$U_{ij}^{\text{inv}} = \varepsilon l_i n_j \tag{1.7.3}$$

is the invariant plane distortion given as the dyadic (tensor) product of the vectors $\boldsymbol{l}$ and $\mathbf{n}$. The crystal lattice sites before and after the distortion are related by the equation

$$r_i' = r_i + U_{ij}^{\text{inv}} r_j = (\delta_{ij} + U_{ij}^{\text{inv}}) r_j = (\delta_{ij} + \varepsilon l_i n_j) r_j$$

or in the symbolic form

$$\mathbf{r}' = (\hat{\mathbf{I}} + \varepsilon \boldsymbol{l} * \mathbf{n})\mathbf{r}$$

The matrix of the invariant plane strain is

$$\hat{\mathbf{A}}_{\text{inv}} = \hat{\mathbf{I}} + \varepsilon \boldsymbol{l} * \mathbf{n} \tag{1.7.4}$$

[see Eq. (1.4.2)].

In applications we often encounter a problem when the invariant plane orientation (the vector $\mathbf{n}$) and the displacement direction $\boldsymbol{l}$ should be expressed in terms of the components of the matrix

$$\hat{\mathbf{A}} = \hat{\mathbf{R}}_1 \hat{\mathbf{A}}_{\text{inv}} \tag{1.7.5}$$

which differs from $\hat{\mathbf{A}}_{\text{inv}}$ by an unknown rotation $\hat{\mathbf{R}}_1$. Then

$$\hat{\mathbf{A}}^+ \hat{\mathbf{A}} = (\hat{\mathbf{R}}_1 \hat{\mathbf{A}}_{\text{inv}})^+ (\hat{\mathbf{R}}_1 \hat{\mathbf{A}}_{\text{inv}}) = \hat{\mathbf{A}}_{\text{inv}}^+ \hat{\mathbf{R}}_1^+ \hat{\mathbf{R}}_1 \hat{\mathbf{A}}_{\text{inv}} = \hat{\mathbf{A}}_{\text{inv}}^+ \hat{\mathbf{A}}_{\text{inv}} \tag{1.7.6}$$

Substituting Eq. (1.7.4) into (1.7.6) gives

$$\begin{aligned}\hat{\mathbf{A}}^+ \hat{\mathbf{A}} &= (\hat{\mathbf{I}} + \varepsilon \boldsymbol{l} * \mathbf{n})^+ (\hat{\mathbf{I}} + \varepsilon \boldsymbol{l} * \mathbf{n}) = (\hat{\mathbf{I}} + \varepsilon \mathbf{n} * \boldsymbol{l})(\hat{\mathbf{I}} + \varepsilon \boldsymbol{l} * \mathbf{n}) \\ &= \hat{\mathbf{I}} + [(\varepsilon \boldsymbol{l} + \tfrac{1}{2}\varepsilon^2 \mathbf{n}) * \mathbf{n} + \mathbf{n} * (\varepsilon \boldsymbol{l} + \tfrac{1}{2}\varepsilon^2 \mathbf{n})]\end{aligned} \tag{1.7.7}$$

The $\hat{\mathbf{A}}^+ \hat{\mathbf{A}}$ matrix is a Hermitian one and may therefore be written as a bilinear form [see Eq. (1.4.13)]:

$$\hat{\mathbf{A}}^+ \hat{\mathbf{A}} = \lambda_1^2 \mathbf{e}_1 * \mathbf{e}_1 + \lambda_2^2 \mathbf{e}_2 * \mathbf{e}_2 + \lambda_3^2 \mathbf{e}_3 * \mathbf{e}_3 \tag{1.7.8}$$

where λ_1^2, λ_2^2, and λ_3^2 are the eigenvalues and $\mathbf{e}_1$, $\mathbf{e}_2$, and $\mathbf{e}_3$ the eigenvectors of the $\hat{\mathbf{A}}^+ \hat{\mathbf{A}}$ matrix.

Using the identity (A.2.5)

$$\mathbf{e}_1 * \mathbf{e}_1 + \mathbf{e}_2 * \mathbf{e}_2 + \mathbf{e}_3 * \mathbf{e}_3 = \hat{\mathbf{I}} \tag{1.7.9}$$

which always holds for a set of mutually orthogonal unit vectors, Eq. (1.7.8) may be rewritten in the form

$$\hat{\mathbf{A}}^+\hat{\mathbf{A}}=\hat{\mathbf{I}}+(\lambda_1^2-1)\mathbf{e}_1*\mathbf{e}_1+(\lambda_2^2-1)\mathbf{e}_2*\mathbf{e}_2+(\lambda_3^2-1)\mathbf{e}_3*\mathbf{e}_3 \qquad (1.7.10)$$

The matrix (1.7.10) can be presented in the form (1.7.7) if one of its eigenvalues is less than 1, the other equal to 1, and the third greater than 1. Let

$$\lambda_1^2<1, \quad \lambda_2^2=1, \quad \lambda_3^2>1 \qquad (1.7.11)$$

Eq. (1.7.10) can then be written as

$$\hat{\mathbf{A}}^+\hat{\mathbf{A}}=\hat{\mathbf{I}}+(\lambda_1^2-1)\mathbf{e}_1*\mathbf{e}_1+(\lambda_3^2-1)\mathbf{e}_3*\mathbf{e}_3=\hat{\mathbf{I}}+(\mathbf{B}*\mathbf{n}+\mathbf{n}*\mathbf{B}) \qquad (1.7.12)$$

where

$$\mathbf{n}=\sqrt{\frac{\lambda_3^2-1}{\lambda_3^2-\lambda_1^2}}\mathbf{e}_3+\sqrt{\frac{1-\lambda_1^2}{\lambda_3^2-\lambda_1^2}}\mathbf{e}_1 \qquad (1.7.13)$$

$$\mathbf{B}=\tfrac{1}{2}(\lambda_3^2-\lambda_1^2)\left(\sqrt{\frac{\lambda_3^2-1}{\lambda_3^2-\lambda_1^2}}\mathbf{e}_3-\sqrt{\frac{1-\lambda_1^2}{\lambda_3^2-\lambda_1^2}}\mathbf{e}_1\right) \qquad (1.7.14)$$

Comparison of Eqs. (1.7.12) and (1.7.14) with (1.7.7) yields

$$\varepsilon\boldsymbol{l}+\frac{\varepsilon^2}{2}\mathbf{n}=\tfrac{1}{2}(\lambda_3^2-\lambda_1^2)\left(\sqrt{\frac{\lambda_3^2-1}{\lambda_3^2-\lambda_1^2}}\mathbf{e}_3-\sqrt{\frac{1-\lambda_1^2}{\lambda_3^2-\lambda_1^2}}\mathbf{e}_1\right) \qquad (1.7.15)$$

The solution of (1.7.15) and (1.7.13) with respect to $\boldsymbol{l}$ gives

$$\varepsilon\boldsymbol{l}=(\lambda_3-\lambda_1)\left[\lambda_1\sqrt{\frac{\lambda_3^2-1}{\lambda_3^2-\lambda_1^2}}\mathbf{e}_3-\lambda_3\sqrt{\frac{1-\lambda_1^2}{\lambda_3^2-\lambda_1^2}}\mathbf{e}_1\right] \qquad (1.7.16)$$

and

$$\varepsilon=\lambda_3-\lambda_1 \qquad (1.7.17)$$

1.8. INVARIANT PLANE STRAIN AND CRYSTAL LATTICE PLANE ORIENTATIONS

In Eq. (1.7.4) the invariant plane strain is given by the dyadic product of two vectors. One of the vectors is perpendicular to the plane and the other one is collinear with the macroscopic shear direction. Eqs. (1.5.5) are used to determine the invariant plane strain effects on crystal planes. It requires an inversion of matrix (1.7.4):

$$(\hat{\mathbf{A}}_{\text{inv}})^{-1}=(\hat{\mathbf{I}}+\varepsilon\boldsymbol{l}*\mathbf{n})^{-1}=\hat{\mathbf{I}}-\frac{\varepsilon\boldsymbol{l}*\mathbf{n}}{1+\varepsilon(\boldsymbol{l}\mathbf{n})} \qquad (1.8.1)$$

It follows from (1.8.1) that

$$(\hat{\mathbf{A}}_{\text{inv}}^{-1})^+=\hat{\mathbf{I}}-\frac{\varepsilon\mathbf{n}*\boldsymbol{l}}{1+\varepsilon(\boldsymbol{l}\mathbf{n})} \qquad (1.8.2)$$

Substituting Eq. (1.8.2) into (1.5.5a) yields the equation describing the reciprocal lattice rearrangement associated with invariant plane strain (1.7.4):

$$\mathbf{H}' = (\hat{\mathbf{A}}_{\text{inv}}^{-1})^{+}\mathbf{H} = \left(\hat{\mathbf{I}} - \frac{\varepsilon \mathbf{n} * \boldsymbol{l}}{1+\varepsilon(\mathbf{n}\boldsymbol{l})}\right)\mathbf{H} = \mathbf{H} - \varepsilon \frac{\mathbf{n}(\mathbf{H}\boldsymbol{l})}{1+\varepsilon(\mathbf{n}\boldsymbol{l})} \tag{1.8.3}$$

or

$$\Delta\mathbf{H} = -\varepsilon\mathbf{n}\frac{\mathbf{H}\boldsymbol{l}}{1+\varepsilon(\mathbf{n}\boldsymbol{l})} \tag{1.8.4}$$

where $\Delta\mathbf{H}$ is the displacement of reciprocal lattice points due to invariant plane strain (1.7.4). The displacement $\Delta\mathbf{H}$ is always collinear with the vector $\mathbf{n}$ normal to the invariant plane.

It is quite obvious that the operator $(\hat{\mathbf{A}}_{\text{inv}}^{-1})^{+}$ in Eq. (1.8.2) responsible for the reciprocal lattice distortion describes invariant plane strain in the reciprocal lattice as well. The order of vectors $\boldsymbol{l}$ and $\mathbf{n}$, however, is inverted in this equation, with $\boldsymbol{l}$ being the vector perpendicular to the reciprocal lattice invariant plane and $\mathbf{n}$ the shear direction in the reciprocal lattice. A schematic of the reciprocal lattice point displacements is given in Fig. 6.

Let us introduce in the reciprocal lattice the interplanar distance $d^*(\boldsymbol{l})$ between parallel planes normal to $\boldsymbol{l}$. The equation for the reciprocal lattice plane perpendicular to $\boldsymbol{l}$ and separated from the origin by the distance $d^*(\boldsymbol{l})m$ where m is an integer is

$$(\mathbf{H}\boldsymbol{l}) = md^*(\boldsymbol{l}) \quad \text{for } m = 0, \pm 1, \pm 2, \ldots \tag{1.8.5}$$

It follows from Eqs. (1.8.4) and (1.8.5) that all the reciprocal lattice points lying in the mth plane undergo the same displacement

$$\Delta\mathbf{H}_m = -\mathbf{n}\varepsilon\frac{d^*(\boldsymbol{l})}{1+\varepsilon\mathbf{n}\boldsymbol{l}}m \quad \text{for } m = 0, \pm 1, \pm 2, \ldots \tag{1.8.6}$$

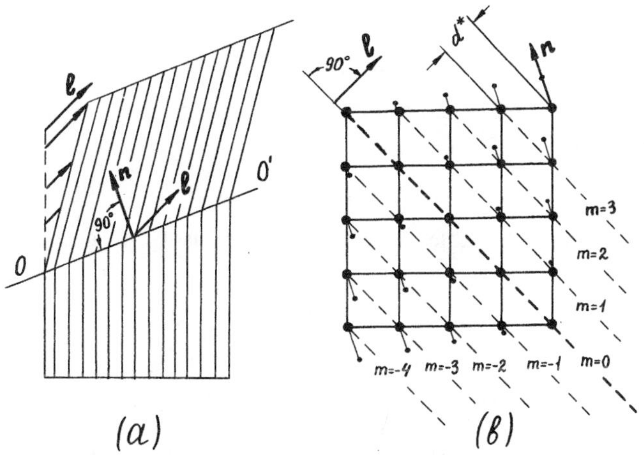

Figure 6. (*a*) The invariant plane strain; (*b*) the corresponding diffraction pattern. Diffraction pattern (*b*) superposes the diffraction patterns from the matrix and the transformed phase: ● a reciprocal lattice point of the matrix; · a reciprocal lattice point of the transformed part of the crystal.

The displacement vanishes at $m=0$, that is, for the reciprocal lattice planes containing the origin. All the crystal lattice planes corresponding to the reciprocal lattice points that lie in the reciprocal lattice plane containing the origin and are perpendicular to $\boldsymbol{l}$ retain their orientations and interplanar distances and thus are not affected by the invariant plane strain (see Fig. 6).

Eq. (1.8.4) is in perfect agreement with a well-known experimental fact that in diffraction patterns the splitting of reciprocal lattice points caused by twinning and shear (both are described as invariant plane strain) occurs in the direction perpendicular to the twin or shear plane.

1.9 WORKED EXAMPLES

1. *Find the matrix* $\hat{\mathbf{A}}$ *of the fcc→hcp crystal lattice rearrangement for Co in the representation related to the Cartesian system of the fcc lattice axes* $[100]_{fcc}$, $[010]_{fcc}$, and $[001]_{fcc}$.

The $fcc \rightarrow hcp$ transformation is a combination of the $\langle 11\bar{2}\rangle$ $(111)_{fcc}$ shuffling along $(111)_{fcc}$ planes, homogeneous isotropic strain in the $(111)_{fcc}$ planes and change of the interplanar distance $d_{(111)}$ between the $(111)_{fcc}$ planes. The macroscopic effect of the shuffling on the shape of a transformed crystal is eliminated by alternating of the shuffling directions $\langle 11\bar{2}\rangle_{fcc}$ and therefore can be neglected. Thus the macroscopic shape change is determined by the isotropic strain in the $(111)_{fcc}$ planes and the change of the interplanar distance of these planes.

Taking into account the crystal lattice correspondence

$$[\bar{1}/2 \quad 1/2 \quad 0]_{fcc} \rightarrow [100]_{hcp}$$

we have

$$a_{fcc}/\sqrt{2} \rightarrow a_{hcp} \tag{1.9.1}$$

Since the interplanar distance $d_{(111)}$ of the fcc lattice after the $fcc \rightarrow hcp$ transformation becomes $1/2 \cdot c_{hcp}$, we have

$$d_{(111)} = a_{fcc}/\sqrt{3} \rightarrow 1/2 \cdot c_{hcp} \tag{1.9.2}$$

The relation (1.9.2) enables us to calculate the isotropic strain in the $(111)_{fcc}$ planes:

$$\varepsilon_1 = \frac{a_{hcp} - a_{fcc}/\sqrt{2}}{a_{fcc}/\sqrt{2}} = \frac{a_{hcp}\sqrt{2}}{a_{fcc}} - 1$$

The invariant form of an isotropic strain in the $(111)_{fcc}$ planes is

$$\varepsilon_{ij}^{(1)} = \varepsilon_1(\delta_{ij} - e_i^{[111]} \cdot e_j^{[111]}) = \left(\frac{a_{hcp} \cdot \sqrt{2}}{a_{fcc}} - 1\right) \cdot (\delta_{ij} - e_i^{[111]} \cdot e_j^{[111]}) \tag{1.9.3}$$

where $\mathbf{e}^{[111]}$ is the unit vector normal to a $(111)_{fcc}$ plane. The relation (1.9.2) allows to obtain the strain value along the $[111]_{fcc}$ direction normal to the

$(111)_{fcc}$ planes:

$$\varepsilon_2=\frac{\frac{1}{2}c_{hcp}-a_{fcc}/\sqrt{3}}{a_{fcc}/\sqrt{3}}=\frac{\sqrt{3}}{2}\frac{c_{hcp}}{a_{fcc}}-1 \tag{1.9.4}$$

The invariant form of this strain is

$$\varepsilon_{ij}^{(2)}=\varepsilon_2 e_i^{[111]}\cdot e_j^{[111]}=\left(\frac{\sqrt{3}}{2}\frac{c_{hcp}}{a_{fcc}}-1\right)\cdot e_i^{[111]}\cdot e_j^{[111]} \tag{1.9.5}$$

According to Eqs. (1.9.3) and (1.9.5) the total strain is

$$\begin{aligned} u_{ij}&=\varepsilon_{ij}^{(1)}+\varepsilon_{ij}^{(2)}=\varepsilon_1(\delta_{ij}-e_i^{[111]}\cdot e_j^{[111]})+\varepsilon_2 e_i^{[111]}\cdot e_j^{[111]}\\ &=\varepsilon_1\delta_{ij}+(\varepsilon_2-\varepsilon_1)e_i^{[111]}\cdot e_j^{[111]}\\ &=\left(\frac{a_{hcp}\sqrt{2}}{a_{fcc}}-1\right)\delta_{ij}+\left(\frac{\sqrt{3}}{2}\frac{c_{hcp}}{a_{fcc}}-\frac{a_{hcp}\sqrt{2}}{a_{fcc}}\right)\cdot e_i^{[111]}\cdot e_j^{[111]} \end{aligned} \tag{1.9.6}$$

Using the crystal lattice parameters for pure Co:

$$\begin{aligned} a_{fcc}&=3.552\ \text{Å}\\ a_{hcp}&=2.507\ \text{Å}\\ c_{hcp}&=4.069\ \text{Å} \end{aligned} \tag{1.9.7}$$

in Eq. (1.9.6) we have

$$u_{ij}=-0.00185\delta_{ij}-0.00607 e_i^{[111]}\cdot e_j^{[111]} \tag{1.9.8}$$

Since $\mathbf{e}^{[111]}=(1/\sqrt{3}, 1/\sqrt{3}, 1/\sqrt{3})_{fcc}$, Eq. (1.9.8) may be rewritten as

$$\begin{aligned} u_{ij}&=-0.00185\begin{pmatrix}1&0&0\\0&1&0\\0&0&1\end{pmatrix}-0.00607\begin{pmatrix}1/3&1/3&1/3\\1/3&1/3&1/3\\1/3&1/3&1/3\end{pmatrix}\\ &=-\begin{pmatrix}0.0039&0.002&0.002\\0.002&0.0039&0.002\\0.002&0.002&0.0039\end{pmatrix} \end{aligned} \tag{1.9.9}$$

According to Eq. (1.4.4) a crystal lattice rearrangement is described by the matrix

$$A_{ij}=\delta_{ij}+u_{ij} \tag{1.9.10}$$

Substitution of the matrix (1.9.9) to (1.9.10) results in

$$A_{ij}=\begin{pmatrix}0.9961&-0.0020&-0.0020\\-0.0020&0.9961&-0.0020\\-0.0020&-0.0020&0.9961\end{pmatrix}$$

2. *Let the crystal lattice rearrangement be described by the matrix*

$$\hat{\mathbf{A}}=\begin{pmatrix}1.2165&0&0.3968\\0&0.9353&0\\-0.2806&0&0.9259\end{pmatrix} \tag{1.9.11}$$

Find the deformation matrix $\hat{\mathbf{F}}$, *the rigid body rotation* $\hat{\mathbf{R}}$ *and the matrix which describes reciprocal lattice site displacements generated by the crystal lattice rearrangement* $\hat{\mathbf{A}}$.

Employing the matrix (1.9.11) in Eq. (1.4.10) we have

$$\hat{\mathbf{F}}^2=\hat{\mathbf{A}}^+\hat{\mathbf{A}}=\begin{pmatrix}1.2165 & 0 & -0.2806\\ 0 & 0.9353 & 0\\ 0.3968 & 0 & 0.9259\end{pmatrix}\begin{pmatrix}1.2165 & 0 & 0.3968\\ 0 & 0.9353 & 0\\ -0.2806 & 0 & 0.9259\end{pmatrix}$$

$$=\begin{pmatrix}1.5586 & 0 & 0.2229\\ 0 & 0.8748 & 0\\ 0.2229 & 0 & 1.0147\end{pmatrix} \quad (1.9.12)$$

The eigenvalues λ^2 of the matrix (1.9.12) is determined from Eq. (A.1.20):

$$\begin{vmatrix}1.5586-\lambda^2 & 0 & 0.2229\\ 0 & 0.8748-\lambda^2 & 0\\ 0.2229 & 0 & 1.0147-\lambda^2\end{vmatrix}=0 \quad (1.9.13)$$

The determinant (1.9.13) is

$$(1.5586-\lambda^2)\cdot(0.8748-\lambda^2)\cdot(1.0147-\lambda^2)-0.2229\cdot 0.2229\cdot(0.8748-\lambda^2)=0$$

It is reduced to the cubic equation

$$(\lambda^2)^3-3.4481\cdot(\lambda^2)^2+3.7830\cdot\lambda^2-1.3401=0$$

whose solution results in three roots:

$$\lambda_1^2=1.6383 \quad \lambda_2^2=0.8751 \quad \lambda_3^2=0.9348 \quad (1.9.14)$$

The normalized eigenvectors $\mathbf{e}_s$ can be found from the set of equations

$$\hat{\mathbf{F}}^2\mathbf{e}=\hat{\mathbf{A}}^+\hat{\mathbf{A}}\mathbf{e}=\lambda_s\mathbf{e}_s \quad (1.9.15)$$

if we use the eigenvalues (1.9.14) in Eq. (1.9.15). In this case we have

$$\begin{aligned}\mathbf{e}_1&=(\ 0.9418,\ 0,\ 0.3363)\\ \mathbf{e}_2&=(\ 0\ ,\ 1,\ 0\ )\\ \mathbf{e}_3&=(-0.3363,\ 0,\ 0.9418)\end{aligned} \quad (1.9.16)$$

According to Eqs. (1.4.7), (1.9.14), and (1.9.16), the matrix $\hat{\mathbf{F}}$ related to the same Cartesian axes as the initial matrix $\hat{\mathbf{A}}$ is

$$F_{ij}=\lambda_1 e_1^i e_1^j+\lambda_2 e_2^i e_2^j+\lambda_3 e_3^i e_3^j$$

$$=\begin{pmatrix}1.2447 & 0 & 0.0992\\ 0 & 0.9353 & 0\\ 0.0992 & 0 & 1.0025\end{pmatrix} \quad (1.9.17)$$

Substitution of the numerical values (1.9.14) and (1.9.16) to Eq. (1.4.15) yields

$$F_{ij}^{-1}=\begin{pmatrix} 0.8098 & 0 & -0.0801 \\ 0 & 1.0690 & 0 \\ -0.0801 & 0 & 1.0050 \end{pmatrix} \tag{1.9.18}$$

Knowing the matrix $\hat{\mathbf{F}}^{-1}$ and $\hat{\mathbf{A}}$ we can find the rigid body rotation matrix $\hat{\mathbf{R}}$ utilizing Eq. (1.4.14):

$$\hat{\mathbf{R}}=\hat{\mathbf{A}}\hat{\mathbf{F}}^{-1}=\begin{pmatrix} 1.2165 & 0 & 0.3968 \\ 0 & 0.9353 & 0 \\ -0.2806 & 0 & 0.9259 \end{pmatrix}\begin{pmatrix} 0.8098 & 0 & -0.0801 \\ 0 & 1.0690 & 0 \\ -0.0801 & 0 & 1.0050 \end{pmatrix}$$

$$=\begin{pmatrix} 0.9534 & 0 & 0.3014 \\ 0 & 1 & 0 \\ -0.3014 & 0 & 0.9534 \end{pmatrix} \tag{1.9.19}$$

According to Eq. (1.5.5), the problem of finding the matrix which describes the reciprocal lattice site rearrangement caused by the crystal lattice rearrangement $\hat{\mathbf{A}}$ is reduced to the finding of the matrix $\hat{\mathbf{A}}^{-1}$. Using Eq. (A.1.10) and (1.9.11), we have

$$(\hat{\mathbf{A}}^{-1})^{+}=\frac{1}{\det\|\hat{\mathbf{A}}\|}\cdot\text{cofactor }\hat{\mathbf{A}}=\frac{1}{1.1576}\begin{pmatrix} 0.8659 & 0 & -0.3711 \\ 0 & 1.2377 & 0 \\ 0.2624 & 0 & 1.1378 \end{pmatrix}$$

$$=\begin{pmatrix} 0.7480 & 0 & -0.3206 \\ 0 & 1.0692 & 0 \\ 0.2267 & 0 & 0.9829 \end{pmatrix} \tag{1.9.20}$$

The matrix (1.9.20) being applied to a reciprocal lattice vector $\mathbf{H}$ of the crystal lattice before the transformation $\hat{\mathbf{A}}$ gives the reciprocal lattice vector $\mathbf{H}'$ after the phase transformation.

2

STABILITY OF HOMOGENEOUS SOLID SOLUTIONS

2.1. INFINITESIMAL FLUCTUATIONS AND THE CONCEPT OF METASTABILITY

There are two types of crystal solid solutions. One is the substitutional solid solution. The lattice sites of the substitutional solid solution are fully occupied by several different kinds of atoms. Any atomic redistribution and variation in the composition are accomplished by atoms of one sort replacing another and by permutations of dissimilar atoms.

The other, the interstitial solid solution, can accept interstitial solute atoms into cavities between the host solvent atoms (interstices). The interstitial solution can be considered as a kind of "lattice gas," a gas whose atoms are permitted to occupy some definite lattice sites rather than arbitrary points. The analogy is based on the fact that interstitial atoms can only occupy interstices within the framework of the host lattice.

Despite the different geometries of substitutional and interstitial solutions, both may be described by the same mathematical model. This in fact may be justified physically by regarding the interstitial solution as a kind of substitutional solution, with the atoms and their vacancies as two sorts of constituent particles. In this connection the host atoms forming an immobile frame are ignored. For instance, consider a binary interstitial solution with atoms occupying octahedral interstices in the fcc host atom lattice. These interstices form one fcc lattice with some sites containing interstitials and others not. This interstitial solution may be treated as a substitutional solution composed of interstitials and vacancies (the host lattice atoms being neglected).

Following this line of reasoning, a thermodynamic analysis of stability and phase transformations in substitutional solutions may be applied to interstitial solutions. All features that characterize the behavior and properties of substitutional solutions such as order-disorder reactions, decomposition reactions, short- and long-range order should also be expected to manifest themselves in interstitial solutions.

For simplicity, we will consider the case of a binary substitutional solution. Of course all our conclusions are also valid for interstitial binary solutions.

At high temperatures, when typical interchange energies W are much smaller than the thermal energy, κT ($W/\kappa T \ll 1$ where κ is the Boltzmann constant, T the absolute temperature), the interaction energy may be neglected and alloys may be treated as ideal solutions. In this case atoms are randomly distributed over all the crystal lattice sites. The high-temperature phase is thus fully disordered, and the probability of finding a particular atom at a particular crystal lattice site (occupation probability) is equal to the corresponding atomic fraction.

At low temperatures, when the typical value of the interaction energy W is much larger than the thermal energy, κT ($W/\kappa T \gg 1$), the atomic arrangement is determined from the minimum internal energy condition.

The phase transformation between a disordered and an ordered phase take place at intermediate temperature when $W/\kappa T \sim 1$.

Depending on the nature of interatomic interactions, the low-temperature state may correspond to either an ordered phase or a mixture of ordered phases, disordered phases, or ordered and disordered phases. In certain cases, in ordering or decomposition involving a redistribution of atoms over the crystal lattice sites accompanied by crystal lattice rearrangements, the situation is even more complex.

Returning to the binary substitutional solution, its state may be described by the occupation probability $n(\mathbf{r})$, that is, probability of finding a solute atom at a crystal lattice site, $\mathbf{r}$. In a disordered solution occupation probabilities, $n(\mathbf{r})$, are the same for all sites that can be occupied. They are equal to the atomic fraction, c, of a solute:

$$n(\mathbf{r}) = c$$

In an ordered solution the function $n(\mathbf{r})$ becomes dependent on the site coordinates, $\mathbf{r}$, and varies over a range commensurate with a few interatomic distances. This function can be expanded in a Fourier series; that is, it can be represented as the sum of static concentration waves whose amplitudes are the Fourier coefficients and whose wave vectors determine the superstructure periods:

$$n(\mathbf{r}) = c + \sum_j Q(\mathbf{k}_j)e^{i\mathbf{k}_j\mathbf{r}} \tag{2.1.1}$$

where $\mathbf{k}_j$ are the wave vectors and $Q(\mathbf{k}_j)$ the Fourier coefficients. The amplitudes, $Q(\mathbf{k}_j)$, are treated as long-range order parameters since the disordered state, $n(\mathbf{r}) = c$, can be obtained if the amplitudes, $Q(\mathbf{k}_j)$, vanish. It will be shown below that the amplitudes, $Q(\mathbf{k}_j)$, are equal to the traditional long-range order parameters with a constant factor.

If a disordered solid solution decomposes to give a mixture of two disordered phases, $n(\mathbf{r})$ also become a function of crystal lattice site coordinates. However unlike ordering, the decomposition results in the coordinate dependence of $n(\mathbf{r})$ such that a typical distance, L, of the distribution function, $n(\mathbf{r})$, is macroscopically large; that is, it is commensurable with macroscopical dimensions of new phase domains. The conclusion follows that the static wave representation of $n(\mathbf{r})$, Eq. (2.1.1), describing the distribution of the decomposition products is mainly determined by long waves. The scale of their wave vectors, k^0, is related to the scale of heterogeneity, L, by the uncertainty relationship

$$k^0 L \sim 2\pi \tag{2.1.2}$$

From (2.1.2) the typical wave vector scale is estimated as

$$k^0 \sim \frac{2\pi}{L}$$

Since L has a macroscopic length ($L \to \infty$), k^0 is a macroscopically small value

$$k^0 \to 0$$

It thus follows that ordering and decomposition may be interpreted as the loss of stability of a disordered solution with respect to static concentration waves. The loss of stability with respect to concentration waves characterized by asymptotically small wave vectors generates the so-called spinodal instability and results in a homogeneous decomposition of the disordered solid solution. The loss of stability with respect to concentration waves with finite wave vectors leads to the order-disorder transformation.

The ordering is a first-order phase transformation if the equilibrium concentration wave amplitudes have finite values at the transformation temperature. By contrast, for infinitesimal amplitudes the ordering is a second-order transformation. It will be shown in Section 3.3 that decomposition is always a first-order phase transformation.

According to the foregoing definitions, a disordered state is unstable with respect to finite amplitude concentration waves at the temperature of a first-order phase transformation and is stable with respect to infinitesimal fluctuations. It follows that small supercooling of a disordered alloy does not affect its stability with respect to infinitesimal fluctuations but rather makes it unstable with respect to finite fluctuations, that is, results in a metastable state.

In general, any phase occurring within single-phase regions of equilibrium T-c diagrams is stable with respect to both finite and infinitesimal fluctuations. If the phase transformation is a first-order one, there are supercooled and overheated metastable states that are unstable with respect to finite fluctuations, which serve as critical nucleation centers, and stable with respect to infinitesimal fluctuations. In this temperature range phase transformations follow the conventional nucleation-and-growth mechanism which requires fluctuations of finite magnitude.

By increasing supercooling or overheating of a metastable phase, absolute instability (instability with respect to infinitesimal fluctuations) may be attained. At the absolute instability temperature a qualitative change of the physical properties of a system occurs. In particular, the thermodynamic functions of the system make no physical sense at that point because they cannot be expressed analytically by means of such external intensive parameters as temperature, pressure, and composition.

First-order transformations are characterized by two absolute instability temperatures, T_c^+ and T_c^-, which refer to overheated and supercooled metastable phases, respectively. The equilibrium of the first-order transformation temperature, T_o, always falls in between T_c^- and T_c^+:

$$T_c^- < T_o < T_c^+$$

These are the supercooling and overheating limits of the temperature hysteresis range.

Second-order transformations result from instability with respect to specific infinitesimal fluctuations whose scales determine the magnitudes of the equilibrium long-range order parameters. The specific fluctuation may be a static concentration wave (an order-disorder transformation), magnetization (a ferromagnetic transformation), the average phase of the wave function of helium atoms or electrons (superfluidity and superconductivity transformations, respectively), and so on. Since second-order transformations occur because of absolute instability with respect to infinitesimal fluctuations, the corresponding metastable phases cannot exist at all, and the second-order transformation temperature coincides with T_c^- and T_c^+, so that $T_c^- = T_o = T_c^+$. Second-order transformations therefore do not involve hysteresis phenomena. For that reason the two phases involved in a second-order transformation (an ordered and disordered phase) cannot occur on the same side of a second-order transformation temperature, and neither supercooling nor overheating are possible.

2.2. STABILITY OF DISORDERED ALLOYS WITH RESPECT TO INFINITESIMAL FLUCTUATIONS

In analyzing the problem of stability of disordered alloys with respect to infinitesimal fluctuations, we will introduce arbitrary heterogeneity and evaluate the corresponding change in the Halmholtz free energy of the alloy. Let the occupation probability be written

$$n(\mathbf{r}) = c + \Delta(\mathbf{r}) \qquad (2.2.1)$$

where $\Delta(\mathbf{r})$ is the heterogeneous part of the probability of finding a solute atom at the site $\mathbf{r}$ and c the atomic fraction of the solute. Since the free energy is a functional of the atomic distribution, it may be written in the form

$$F = F(\{\Delta(\mathbf{r})\}) \qquad (2.2.2)$$

In the case of infinitesimal heterogeneity we are discussing, the free energy may be expanded in powers of $\Delta(\mathbf{r})$:

$$F=F_0+\sum_{\mathbf{r}} A(\mathbf{r})\Delta(\mathbf{r})+\frac{1}{2}\sum_{\mathbf{r},\mathbf{r}'} B(\mathbf{r},\mathbf{r}')\Delta(\mathbf{r})\Delta(\mathbf{r}')$$

$$+\frac{1}{3!}\sum_{\mathbf{r},\mathbf{r}',\mathbf{r}''} C(\mathbf{r},\mathbf{r}',\mathbf{r}'')\Delta(\mathbf{r})\Delta(\mathbf{r}')\Delta(\mathbf{r}'')+\cdots \qquad (2.2.3)$$

where

$$A(\mathbf{r})=\left(\frac{\delta F}{\delta\Delta(\mathbf{r})}\right)_{n=c}$$

$$B(\mathbf{r},\mathbf{r}')=\left(\frac{\delta^2 F}{\delta\Delta(\mathbf{r})\delta\Delta(\mathbf{r}')}\right)_{n=c}$$

$$C(\mathbf{r},\mathbf{r}',\mathbf{r}'')=\left(\frac{\delta^3 F}{\delta\Delta(\mathbf{r})\delta\Delta(\mathbf{r}')\delta\Delta(\mathbf{r}'')}\right)_{n=c}$$

are the expansion coefficients corresponding to the disordered state at temperature T (by definition, they are determined at $\Delta(\mathbf{r})\equiv 0$). The summations in (2.2.3) are over all the crystal lattice sites.

All disordered crystal lattice sites are crystallographically equivalent, hence

$$A(\mathbf{r})=\text{constant}$$

The first-order term in series (2.2.3) therefore vanishes

$$\sum_{\mathbf{r}} A(\mathbf{r})\Delta(\mathbf{r})=\text{constant}\sum_{\mathbf{r}}\Delta(\mathbf{r})=0 \qquad (2.2.4)$$

since

$$\sum_{\mathbf{r}}\Delta(\mathbf{r})=0 \qquad (2.2.5)$$

according to definition (2.2.1).

Bearing this in mind, one can rewrite Eq. (2.2.3) in the form

$$\Delta F=\frac{1}{2}\sum_{\mathbf{r},\mathbf{r}'} B(\mathbf{r},\mathbf{r}')\Delta(\mathbf{r})\Delta(\mathbf{r}')$$

$$+\frac{1}{3!}\sum_{\mathbf{r},\mathbf{r}',\mathbf{r}''} C(\mathbf{r},\mathbf{r}',\mathbf{r}'')\Delta(\mathbf{r})\Delta(\mathbf{r}')\Delta(\mathbf{r}'')+\cdots \qquad (2.2.6)$$

where $\Delta F=F-F_0$ is the free energy change due to heterogeneity $\Delta(\mathbf{r})$.

In analyzing disordered state stability with respect to infinitesimal fluctuations, $\Delta(\mathbf{r})$, the first nonvanishing term in (2.2.6)

$$\Delta F=\tfrac{1}{2}\sum_{\mathbf{r},\mathbf{r}'} B(\mathbf{r},\mathbf{r}')\Delta(\mathbf{r})\Delta(\mathbf{r}') \qquad (2.2.7)$$

should only be considered. Since the $B(\mathbf{r},\mathbf{r}')$ coefficients refer to the disordered state invariant under Bravais translations $\mathbf{T}$ ($\mathbf{r}\to\mathbf{r}+\mathbf{T}$, $\mathbf{r}'\to\mathbf{r}'+\mathbf{T}$), we have

$$B(\mathbf{r},\mathbf{r}')\equiv B(\mathbf{r}+\mathbf{T},\mathbf{r}'+\mathbf{T}) \qquad (2.2.8)$$

The identity (2.2.8) reflecting the translational symmetry of the disordered phase holds if

$$B(\mathbf{r}, \mathbf{r}') = B(\mathbf{r} - \mathbf{r}') \tag{2.2.9}$$

Eq. (2.2.7) may then be rewritten as

$$\Delta F = \frac{1}{2} \sum_{\mathbf{r},\mathbf{r}'} B(\mathbf{r} - \mathbf{r}')\Delta(\mathbf{r})\Delta(\mathbf{r}') \tag{2.2.10}$$

Using the concentration-wave representation

$$\Delta(\mathbf{r}) = \sum_{\mathbf{k}}{}' Q(\mathbf{k})e^{i\mathbf{kr}} \tag{2.2.11}$$

we obtain Eq. (2.2.10) in the quadratic (diagonal) form

$$\Delta F = \frac{N}{2} \sum_{\mathbf{k}}{}' b(\mathbf{k})|Q(\mathbf{k})|^2 \tag{2.2.12}$$

where

$$b(\mathbf{k}) = \sum_{\mathbf{r}} B(\mathbf{r})e^{-i\mathbf{kr}} \tag{2.2.13}$$

is the characteristic function of the disordered state, $Q(\mathbf{k})$ are the concentration wave amplitudes, and N is the total number of crystal lattice sites. The summation in Eq. (2.2.12) is over all the N wave vectors within the first Brillouin zone allowed by the cyclic boundary conditions. The prime in Eq. (2.2.12) means exclusion of the term corresponding to $\mathbf{k}=0$. The coefficients $b(\mathbf{k})$ characterize the properties of the disordered state.

If all amplitudes, $Q(\mathbf{k})$, in Eq. (2.2.11) vanish, the state is fully disordered (perfectly homogeneous), $\Delta(r) \equiv 0$. The appearance of a heterogeneity results in that certain amplitudes, $Q(\mathbf{k})$, become nonzero.

If all coefficients, $b(\mathbf{k})$, are positive, the arising of any nonzero amplitudes, $Q(\mathbf{k})$, results in an increase of the free energy (2.2.12). The latter conclusion may be reformulated as follows: if all coefficients, $b(\mathbf{k})$, are positive, any infinitesimal fluctuations result in an increase of the free energy. The latter means that the disordered state is stable since it corresponds to the free energy minimum.

In fact, if at least one of the coefficients, $b(\mathbf{k})$, either vanishes or is less than zero, that is, if $b(\mathbf{k}_0) \leqslant 0$, the uniform solution becomes unstable. One can always choose the heterogeneity

$$\Delta_0(\mathbf{r}) = \tfrac{1}{2}[Q(\mathbf{k}_0)e^{i\mathbf{k}_0\mathbf{r}} + Q^*(\mathbf{k}_0)e^{-i\mathbf{k}_0\mathbf{r}}] \tag{2.2.14}$$

which, according to Eq. (2.2.12), results in the free energy change

$$\Delta F = \tfrac{1}{2}Nb(\mathbf{k}_0)|Q(\mathbf{k}_0)|^2 \tag{2.2.15}$$

(All the other terms in the sum (2.2.12) vanish since, according to (2.2.14), all amplitudes besides $Q(\mathbf{k}_0)$ and $Q^*(\mathbf{k}_0)$ are equal to zero). When $b(\mathbf{k}_0) \leqslant 0$, the free energy ΔF is equal to zero or less than zero, i.e. when at least one of the coefficients, $b(\mathbf{k})$, has a nonpositive value, the homogeneous state of the solid solution is unstable with respect to heterogeneity (2.2.14).

We have thus proved that a disordered solution is stable with respect to infinitesimal fluctuations when all the $b(\mathbf{k})$ values are positive, and, conversely, it is unstable when the characteristic function $b(\mathbf{k})$ takes on negative values. According to definitions (2.2.3) and (2.2.13), the characteristic function $b(\mathbf{k})$ is a thermodynamic function of the disordered state and therefore depends on temperature, T, composition, c, and pressure, p:

$$b(\mathbf{k})=b(\mathbf{k};\ T, c, p) \tag{2.2.16}$$

At high temperatures, when the disordered state is stable, all $b(\mathbf{k})$ are positive. Upon temperature lowering to the absolute instability temperature, T_c^-, or T_0 (for first- and second-order transformations, respectively), the minimum value of $b(\mathbf{k};\ T, c, p)$ corresponding to a certain wave vector $\mathbf{k}=\mathbf{k}_0$ vanishes (see Fig. 7). The condition of vanishing of the minimum value of the characteristic function is reduced to two equations:

$$b(\mathbf{k}_0;\ T, c, p)=0 \tag{2.2.17a}$$

and

$$\left(\frac{\partial b(\mathbf{k};\ T, c, p)}{\partial \mathbf{k}}\right)_{\mathbf{k}=\mathbf{k}_0}=0 \tag{2.2.17b}$$

The second equation gives the necessary condition for the function $b(\mathbf{k})$ to reach its minimum at $\mathbf{k}=\mathbf{k}_0$.

These two equations determine the instability wave vector, $\mathbf{k}_0$, and absolute instability temperature, T_c^-. As mentioned in Section 2.1, the situation with $\mathbf{k}_0=0$ corresponds to a spinodal instability in alloys with a miscibility gap.

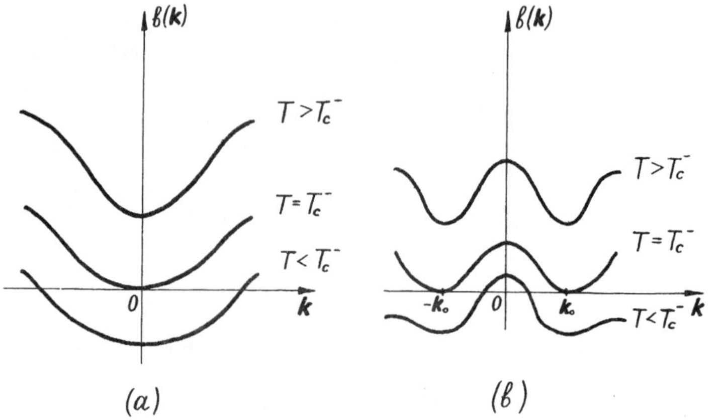

Figure 7. The typical plots $b(\mathbf{k})$ with respect to $\mathbf{k}$ in different temperature ranges. (*a*) The case of decomposition; (*b*) the case of ordering. The decomposition case (*a*) arises when the minimum of $b(\mathbf{k})$ falls at $\mathbf{k}=0$. The case of ordering corresponds to the situation when $b(\mathbf{k})$ assumes its minimum value at $\mathbf{k}\neq 0$.

Nonzero $\mathbf{k}_0$'s correspond to a homogeneous ordering in the whole volume of the disordered phase which is different in principle from the conventional nucleation-and-growth mechanism.

The characteristic function $b(\mathbf{k}; T, c, p)$ thus contains information sufficient to predict whether the phase transformation is an ordering or decomposition reaction: the transformation is a decomposition when $b(\mathbf{k})$ takes on a minimum value at $\mathbf{k}=0$, and an ordering otherwise.

It has already been mentioned that with second-order transformations, the absolute instability temperature, T_c^-, is at the same time the phase transformation temperature, T_o. Conversely, a first-order transformation of a disordered phase occurs at the equilibrium temperature, T_o, which exceeds the corresponding absolute instability temperature, T_c^-:

$$T_o > T_c^-$$

As shown in Section 2.1, metastable homogeneous solid solutions are stable with respect to infinitesimal heterogeneities within the range

$$T_c^- < T < T_c^+ \tag{2.2.18}$$

where the appearance of infinitesimal heterogeneities is associated with a free energy increase, $\Delta F > 0$. To overcome the barrier whose height depends on the trajectory of the system from the initial metastable homogeneous state to the final stable state corresponding to the global free energy minimum, some expenditure of energy is required. The minimum barrier height, ΔF_0, gives the work of the formation of a critical nucleus of the new phase. The critical nucleus is an atomic configuration corresponding to the minimum height barrier. As the critical nucleus formation increases the free energy of the system, the realization of the process fully depends on thermal fluctuations.

The probability of fluctuation overcoming a barrier, ΔF_0, is given by the conventional equation for the thermodynamics of fluctuations:

$$W \sim e^{-\Delta F_0/\kappa T} \tag{2.2.19}$$

We thus arrive at the conclusion that a supercooled disordered solution in the range (2.2.18) is stable for infinitesimal heterogeneities and unstable for finite heterogeneities which are critical nuclei.

To form a clearer idea of the metastable state, one may turn to the following geometrical interpretation. An arbitrary state of a binary solid solution may be described by N occupation probabilities, $n(\mathbf{r}_1)$, $n(\mathbf{r}_2), \ldots, n(\mathbf{r}_N)$, that is, the probabilities of finding solute atoms at the respective lattice sites (N is the total number of crystal lattice sites). Any solid solution state (atomic distribution) may also be given by a point in the N-dimensional phase space whose N coordinates are the occupation probabilities:

$$\{n(\mathbf{r}_1), n(\mathbf{r}_2), \ldots, n(\mathbf{r}_N)\}$$

Since the free energy is a functional of the atomic distribution

$$F = F(\{n(\mathbf{r}_1), \ldots, n(\mathbf{r}_N)\}) = F(\{n(\mathbf{r})\}) \tag{2.2.20}$$

we may introduce the free energy hypersurface in the N-dimensional phase space of occupation probabilities. This hypersurface is described by Eq. (2.2.20).

Using this geometrical image, one can form a visual idea of what occurs when a supercooled solution undergoes transition to the stable equilibrium state. The condition of stability of the disordered solution requires that the solution be stable with respect to any infinitesimal fluctuation in the atomic distribution. In other words, infinitesimal deviations from the disordered state result in a free energy increase. Using the geometrical interpretation given above, the metastable as well as stable states may be put in correspondence to the points in the phase space of occupation probabilities where the free energy hypersurface has its minimum.

At $T > T_o$, when the disordered state is stable, this is the absolute minimum. Supercooling in the range $T_c^- < T < T_o$ makes the disordered solution a metastable phase; that is, the free energy minimum corresponding to the disordered states becomes only a relative minimum while the absolute hypersurface minimum shifts to the heterogeneous state (see Fig. 8a).

Geometrically, it is clear that the representative point should overcome a barrier separating the relative and absolute free energy minimums in passing from the former to the latter one. Since a barrier transition always requires energy expenditures, the process may be due only to fluctuations. The probability of the corresponding fluctuation taking place is given by Eq. (2.2.19).

Usually, ΔF_0 far exceeds κT, $\Delta F_0 \gg \kappa T$. The transformation pathway should therefore be across the lowest barrier separating the states involved since the probability of overcoming the lower barrier is larger than the probability of realizing other pathways. Topologically, the lowest barrier should be a saddle point of the free energy hypersurface. The conclusion follows that a transformation from the metastable to stable state should occur through the lowest saddle point which is a "watershed" between the relative and absolute minimums of the free energy hypersurface.

The set of N occupation probabilities, $\{n^*(\mathbf{r}_1), n^*(\mathbf{r}_2), \ldots, n^*(\mathbf{r}_N)\}$, that gives the coordinates of the lowest saddle point describes the atomic distribution in the transformation state that corresponds to the critical nucleus of the stable phase. This state is called a critical nucleus because, as soon as it is attained, there is no need in further fluctuations to reach the equilibrium state and the transformation to the equilibrium is accompanied by a monotonous decrease of free energy.

Further cooling of a metastable alloy to the absolute instability temperature, T_c^-, radically changes the topology of the free energy hypersurface. The lowest barrier separating the two minimums is now removed, and the relative minimum corresponding to the metastable state becomes a saddle point (Fig. 8b). System passage to the absolute free energy minimum may be accomplished with a monotonous decrease of free energy and without fluctuation overcoming

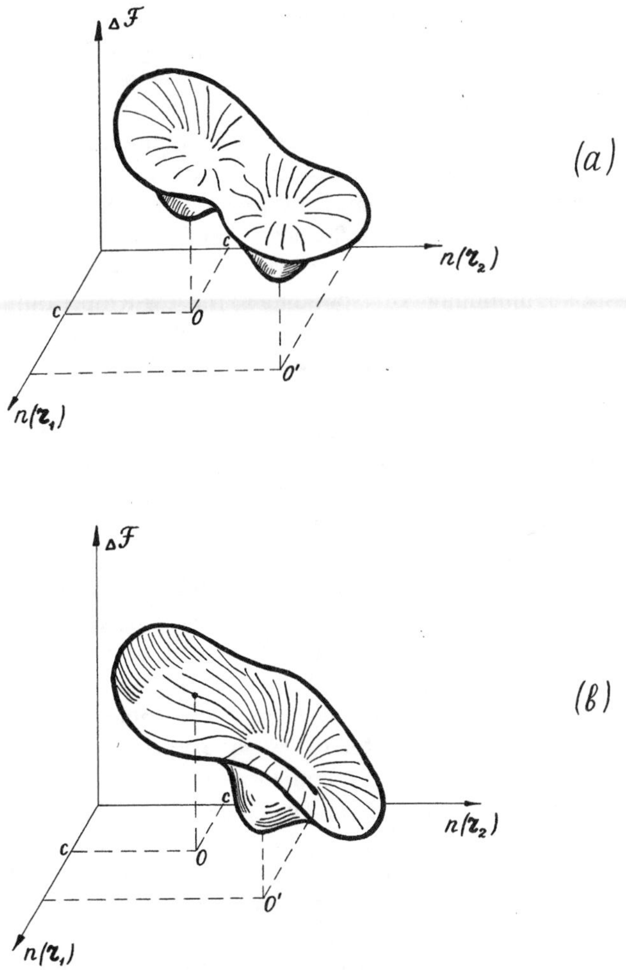

Figure 8. A schematic of the hypersurface $\Delta F = \Delta F(\{n(\mathbf{r})\})$ in the particular case of $N=3$. Since $n(\mathbf{r}_1)+n(\mathbf{r}_2)+n(\mathbf{r}_3)=3c$, $\Delta F=\Delta F(\{n(\mathbf{r})\})$ is the function of only two independent variables. (*a*) The case $T_c^- < T < T_0$. The metastable homogeneous state is described by the point 0 in the phase space. Its coordinates are $n(\mathbf{r}_1)=n(\mathbf{r}_2)=c$. The point 0′ corresponds to the stable heterogeneous state when $n(\mathbf{r}_1)\neq n(\mathbf{r}_2)$, which is the absolute minimum of the free energy. (*b*) The case $T<T_c^-$. The homogeneous state $n(\mathbf{r}_1)=n(\mathbf{r}_2)=c$ corresponds to a saddle point 0 on the surface $\Delta F=\Delta F(\{n(\mathbf{r})\})$. This point corresponds to the absolute instability situation. The equilibrium state 0′ corresponds to the absolute minimum on the free energy hypersurface.

barriers to transformation, and the phase transformation occurs without nucleation at $T \leqslant T_c^-$. Spinodal decomposition provides a well-known example of a phase transformation of this type that does not involve nucleation.

Topologically, the treatment of second-order transformations and decomposition reactions occurring at the critical point (at the *T-c* diagram point corresponding to the top of the miscibility gap) is simpler than that of first-order transformations. The temperature of both a second-order transformation and a decomposition occurring at the critical point is the absolute instability temperature; that is, the corresponding disordered phase becomes unstable with respect to certain infinitesimal fluctuations at $T \leqslant T_o$. At $T > T_o$ the disordered state is stable and corresponds to the absolute minimum of the free energy hypersurface. When temperature, T, decreases to the phase transformation point, T_o, the disordered state becomes unstable, and the corresponding absolute free energy minimum becomes an inflection point. Further cooling converts the inflection to a saddle point as a new extremum appears near it, the absolute free energy surface minimum corresponding to the low-temperature phase. The distance separating the disordered state saddle point and ordered state absolute energy minimum in the phase space vanishes at $T = T_o$ and increases with supercooling.

Second-order phase transformations and decomposition reactions occurring at the critical point therefore always proceed with monotonous decrease of free energy and do not require fluctuation formation of critical nuclei. This is one of the most significant features distinguishing these reactions from first-order phase transformations described above. It is, however, often given an insufficient attention in textbooks dealing with phase transformations.

After this short consideration of the phase space geometry of second-order transformations, we shall return to the first-order ones. According to the foregoing discussion, the atomic distribution leading to the formation of the critical nucleus may be attained by varying a macroscopic number of variables (occupation probabilities) to find the lowest saddle point separating the stable and metastable phases. In this general formulation the problem, however, proves to be a very complicated one and can hardly be solved without considerable simplifications.

The most natural approach is to reduce the number of degrees of freedom, the number of variables determining the free energy of an alloy. This may be done by introducing more or less realistic relationships between the occupation probabilities. Each relation removes one degree of freedom. The classical theory treats new phase nuclei in terms of the simplest model possible involving only one degree of freedom (independent variable), the nucleus radius.

Following this line of reasoning and using occupation probabilities as phase identity characteristics, we must impose severe restrictions on the variable values. The occupation probabilities within the new phase spherical nucleus should be set equal to the equilibrium occupation probabilities of the new phase. Conversely, the occupation probabilities outside the nucleus should be the same as those characterizing the parent phase. These requirements reduce the number of independent variables to one, the nucleus radius.

In the classical approach each nucleus possesses all properties of a macroscopic inclusion of the new phase. In particular, the total free energy change associated with the nucleus formation includes two terms. One of these is proportional to the difference between the specific free energies of the new and parent phases and the nucleus volume. The other term is proportional to the interphase surface tension coefficient and the surface area of the nucleus.

In other words, a small nucleus is treated as macroscopic new phase inclusion since it is described using parameters such as interphase surface energy and specific free energy which are of entirely macroscopic nature.

The appearance of a nucleus thus results in a free energy change equal to

$$\Delta F(r) = (f_1 - f_0)\frac{4\pi}{3}r^3 + 4\pi r^2 \gamma_s \tag{2.2.21}$$

where f_1 and f_0 are the new and parent phase free energies per unit volume, respectively, γ_s is the surface tension energy coefficient (free energy per unit area), r the nucleus radius playing the role of the only system variable, and $(4\pi/3)r^3$ and $4\pi r^2$ are the spherical nucleus volume and surface area, respectively.

As already mentioned, Eq. (2.2.21) is based on an approximation that reduces the N-dimensional phase space to the one-dimensional one. Figure 9 shows a one-dimensional free energy "hypersurface" in the phase space in the temperature range, $T_c^- < T < T_o$, corresponding to the supercooled state of the parent phase. At $T < T_o$ the plot of $\Delta F(r)$ with relation to r has a maximum since $\gamma_s > 0$ and $f_1 - f_o < 0$ [see Eq. (2.2.21)]. At the one-dimensional phase space the saddle point of the free energy surface degenerates to a mere maximum. According to the classical theory, the new phase nucleus whose radius corresponds to the maximum of the $\Delta F(r)$ curve is the critical nucleus.

Naturally, the validity of the method used to derive Eq. (2.2.21) and of all the approximations involved in the classical treatment of the nucleation phenomenon should be questioned. In fact use of Eq. (2.2.21) may be justified only if the dimensions of the nucleus are large compared with the interphase layer thickness. The latter condition is the necessary condition for treating the total energy in macroscopic terms, as a sum of the volume and surface energies.

However, neither the critical nucleus's radius nor its surface layer thickness are typical parameters of length. Indeed, the radius is determined by the saddle point's position on the free energy surface. The surface layer thickness is also determined from rather a complicated consideration of the atomic distribution within the intermediate interphase region. The size and interphase layer thickness of the critical nucleus should therefore be expressed in terms of some independent parameter of length. This parameter represents the correlation length. The correlation length is a typical distance from a local heterogeneity at which heterogeneity effects on its environment diminish to the extent that

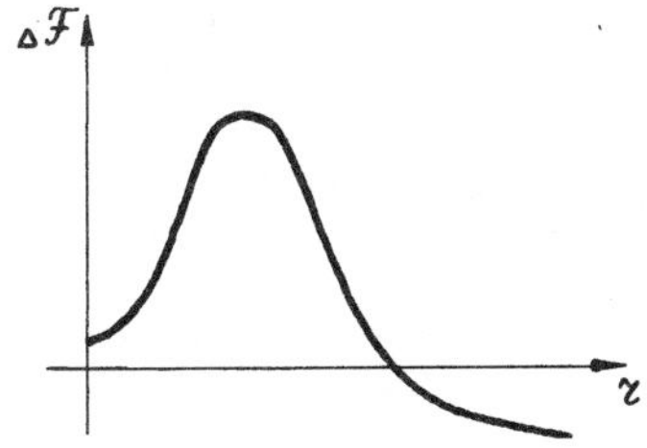

Figure 9. Dependence of the free energy of a supercooled alloy on the radius of a nucleus. The maximum of the plot ΔF with respect to r corresponds to the critical nucleus radius.

they may be neglected. The surface layer thickness and the radius of the critical nucleus should be of the same order of magnitude as the correlation length. They should therefore be commensurate with each other.* This conclusion is in disagreement with the macroscopical approximation that requires the critical nucleus's radius be far larger than its surface layer thickness. For this reason the classical approach to the nucleation problem may only claim to give a qualitative rather than quantitative description of the phenomenon.

*This problem will be discussed in more detail in Section 4.4.

3

ORDERING IN ALLOYS

Size limitations make it impossible to cover all aspects of the order-disorder phenomena in alloys in this book. We shall only discuss those problems of ordering that are necessary for understanding its applications described as follows. First, in this connection belongs the method of static concentration waves which gives a very efficient way of dealing with various problems of the theory. The method was proposed by Khachaturyan in 1962 to 1963 (3–5) to extend the microscopic theory of ordering to long-range interatomic interactions that cannot in principle be described by the conventional theories. It was further developed by Badalyan and Khachaturyan in 1969 (6) who considered ordering in multicomponent alloys. The same authors applied the concentration wave approach to the correlation effects within the arbitrary-range interaction model of ordering in 1970 and 1973 (7, 8).*

The concentration wave approach has made it possible to overcome some serious difficulties insuperable within the conventional microscopic theories of ordering by Gorsky, Bragg and Williams (10–12), Guggenheim and Fouler (13), Bethe (14), Peierls (15), Kirkwood and Kikuchi (16, 17). The traditional theories have been extended in the following directions:

1. To take into account interatomic interactions at arbitrary distances.
2. To reject usual a priori assumptions on the atomic structure of ordered phases involved in order-disorder transformations and to provide the possibility of predicting the ordered phase structures from only the known pairwise interatomic interaction energies.
3. To establish connections between the Landau-Lifshitz thermodynamical theory of second-order transformations in ordering of alloys and the statistical theory.
4. To take into account correlation effects in the arbitrary-range interaction model.

*The concentration wave formalism is described in detail in the review (9).

There are problems such as strain-induced ordering in interstitial solutions and ordering in ionic ceramics that cannot be handled without concentration wave formalism.

To sum up, any theory that pretends to use more or less realistic interatomic potentials should be based on the static concentration wave method since all realistic potentials known, such as Coulomb interactions, strain-induced interactions, and interatomic interactions in metals associated with conduction electrons, are essentially of a long-range nature and cannot therefore be included in the traditional scheme of the conventional ordering theories.

3.1. STATIC CONCENTRATION WAVE REPRESENTATION OF ORDERED PHASE STRUCTURES

We shall consider (with a few exceptions) the cases of ordering in binary alloys. Atomic distribution in a binary alloy can be described by means of one function $n(\mathbf{r})$. In the case of substitutional solution the occupation probability, $n(\mathbf{r})$, is the probability of finding an atom of some definite kind (of the kind A) at the site $\mathbf{r}$ of the crystal lattice. We can describe the atomic distribution in a substitutional binary solution by one function $n(\mathbf{r})$ since the occupation probabilities $n_A(\mathbf{r})$ and $n_B(\mathbf{r})$ for A and B atoms, respectively, are not independent. They must satisfy the identity

$$n_A(\mathbf{r})+n_B(\mathbf{r})=1$$

which reflects the fact that a crystal lattice site is occupied either by atom A or by atom B.

In the case of an interstitial solution the function $n(\mathbf{r})$ is the occupation probability of finding an interstitial atom in the interstitial site, $\mathbf{r}$. We will consider only the cases where all interstitial sites are crystallographically equivalent. In a disordered phase the probabilities, $n(\mathbf{r})$ are the same for all sites that can be occupied. They are equal to the atomic fraction c of the relevant component in a binary substitutional solution, and in an interstitial solution they equal the fraction of the interstitial sites that are occupied. A more complex situation occurs in an ordered phase. The function $n(\mathbf{r})$ becomes dependent on the site coordinate, $\mathbf{r}$. This dependence describes the sublattices that arise during the ordering process. If, for instance, the function $n(\mathbf{r})$ assumes t values $n_1, n_2, \ldots, n_t$ on a set of crystal lattice sites, $\{\mathbf{r}\}$, the function $n(\mathbf{r})$ describes t sublattices into which the crystal lattice of the disordered alloy is subdivided as a result of the ordering reaction. The locus of the sites making up each of the t sublattices is determined by the following t equations:

$$\begin{aligned} n(\mathbf{r}) &= n_1 \\ n(\mathbf{r}) &= n_2 \\ &\cdots\cdots \\ n(\mathbf{r}) &= n_t \end{aligned} \tag{3.1.1}$$

where the values $n_1, n_2, \ldots, n_t$ are the occupation probabilities of sites of the 1st, 2nd, ..., tth sublattice, respectively.*

Let us consider the case where all positions of crystal lattice sites (or interstices in the interstitial solution) $\{\mathbf{r}\}$ are described by one Bravais lattice. In this case the function $n(\mathbf{r})$ that determines the distribution of the solute atoms in an ordered phase can be expanded in a Fourier series; that is, it can be represented as a superposition of static concentration waves:

$$n(\mathbf{r}) = c + \frac{1}{2}\sum_j [Q(\mathbf{k}_j)e^{i\mathbf{k}_j\mathbf{r}} + Q^*(\mathbf{k}_j)e^{-i\mathbf{k}_j\mathbf{r}}] \tag{3.1.2}$$

where $\exp(i\mathbf{k}_j\mathbf{r})$ is a static concentration wave, $\mathbf{k}_j$ is a nonzero wave vector defined in the first Brillouin zone of the disordered alloy, $\mathbf{r}$ is a site vector of the lattice $\{\mathbf{r}\}$ describing the positions that can be occupied by an atom of the alloying element, index j denotes the wave vectors in the Brillouin zone, $Q(\mathbf{k}_j)$ is a static concentration wave amplitude, and c is the atomic fraction of the alloying element.

The representation (3.1.2) gives the transformation from the relevant atomic distribution in an ordered phase in terms of N probabilities, $n(\mathbf{r})$ (N being the number of crystal lattice sites) to N amplitudes $Q(\mathbf{k})$. (It follows from cyclic boundary conditions for the function $n(\mathbf{r})$ that the first Brillouin zone contains N vectors $\mathbf{k}$.) The sum (3.1.2) can be rewritten in another form by combining those terms whose wave vectors, $\mathbf{k}_{j_s}$, enter into the same star, s (the index j_s refers to the vectors of the star, s):†

$$n(\mathbf{r}) = c + \sum_s \eta_s E_s(\mathbf{r}) \tag{3.1.3}$$

where

$$E_s(\mathbf{r}) = \frac{1}{2}\sum_{j_s} (\gamma_s(j_s)e^{i\mathbf{k}_{j_s}\mathbf{r}} + \gamma_s^*(j_s)e^{-i\mathbf{k}_{j_s}\mathbf{r}}) \tag{3.1.4}$$

and

$$Q(\mathbf{k}_{j_s}) = \eta_s \gamma_s(j_s) \tag{3.1.5}$$

The summation in Eqs. (3.1.3) and (3.1.4) is carried out over all vectors, $\{j_s\}$, of the star, s. Here η_s are the long-range-order parameters, and $\gamma_s(j_s)$ are coefficients that determine the symmetry of the occupation probabilities $n(\mathbf{r})$ (the symmetry of the superstructure) with respect to rotation and reflection symmetry operations.

*The term "sublattice" refers to crystal lattice sites forming several interpenetrating Bravais lattices that can be brought in coincidence with each other by the superlattice symmetry operations (usually, this term refers only to sites forming the Bravais lattice).

†The star is a set of wave vectors $\mathbf{k}_{j_s}$ that may be obtained from one wave vector by applying to it all operations of the symmetry group of the disordered solution. It is assumed that the vectors $\mathbf{k}_{j_s}$ differing from each other by a fundamental reciprocal lattice vector $2\pi\mathbf{H}$ are identical.

Actually, one can easily prove that the rotation and reflection operations result in the mutual permutations of the exponents in Eqs. (3.1.3) and (3.1.4) entering each of the functions $E_s(\mathbf{r})$ or, similarly, that they result in the mutual permutation of the corresponding coefficients, $\gamma_s(j_s)$. If the relationship between the coefficients, $\gamma_s(j_s)$, provides the invariance of all the functions $E_s(\mathbf{r})$ under the permutations of $\gamma_s(j_s)$, the rotation and reflection operations yielding these permutations are symmetry operations of the ordered distribution (3.1.3) and therefore the operations form a subgroup of the relevant superlattice.

The vectors, $\mathbf{k}_{j_s}$, in Eq. (3.1.4) are the superlattice vectors of the reciprocal lattice situated in the first Brilluoin zone of the disordered phase. They determine new unit translations in the reciprocal lattice arising from the reduction of the translation symmetry caused by ordering. All the remaining superlattice vectors in the reciprocal lattice of the ordered phase can be found by adding to the vectors $\mathbf{k}_{j_s}$ the fundamental reciprocal lattice vectors $2\pi\mathbf{H}$.

It follows from Eq. (3.1.5) that the long-range order parameters η_s are proportional to the amplitudes of the static concentration waves but are defined ambiguously. To avoid this ambiguity, it is necessary to introduce an additional condition. It may be, for instance, the normalization condition

$$\sum_{j_s} |\gamma_s(j_s)|^2 = 1 \tag{3.1.6}$$

or the more frequently used condition that in a completely ordered state, where occupation probabilities, $n(\mathbf{r})$, are either unity or zero on all the lattice sites, $\{\mathbf{r}\}$, all the parameters, η_s, should be equal to unity. This requirement completely defines the values of the constants, $\gamma_s(j_s)$. The latter definition of the long-range order parameters coincides with the conventional definition of the long-range order parameters in terms of the occupation probabilities of sites in the different sublattices.

We now give a few examples of the application of Eq. (3.1.3) to structures of various ordered phases (superstructures). First, consider layer superstructures. The atomic arrangement in a layer superstructure is described by alternating parallel planes filled preferentially by A or B atoms in a binary substitutional solution (interstitial atoms or their vacancies in an interstitial solution). The wave vectors of the concentration waves describing a layer superstructure should be perpendicular to the layer planes. Since we deal with periodic layer structures, the wave vectors should be multiples of the shortest one. Eq. (3.1.3) may therefore be rewritten in the form

$$n(\mathbf{r}) = c + \frac{1}{2} \sum_{s=1}^{s_{max}-1} \eta_s(\gamma_s e^{i\mathbf{k}_0 s\mathbf{r}} + \gamma_s^* e^{-i\mathbf{k}_0 s\mathbf{r}}) \tag{3.1.7}$$

where $\mathbf{k}_0$ is the shortest vector, s is an integer, and s_{max} is the minimal integer that makes the vector $s_{max}\mathbf{k}_0/2\pi$ equal to the disordered phase reciprocal vector. A few examples of the concentration wave representation of layer superstructures follows.

3.1.1. Ni_4Mo (D1a)

The D1a structure occurs in fcc solutions. It is a layer structure generated by the wave vector $\mathbf{k}_0$

$$\mathbf{k}_0 = 2\pi(\tfrac{4}{5}\mathbf{a}_1^* + \tfrac{2}{5}\mathbf{a}_3^*) \tag{3.1.8}$$

which corresponds to the point $(\frac{4}{5}\,0\,\frac{2}{5})$ in the reciprocal lattice. Here $\mathbf{a}_1^*$, $\mathbf{a}_2^*$, and $\mathbf{a}_3^*$ are the reciprocal lattice vectors in the [100], [010], and [001] directions: the reciprocal lattice translations $\mathbf{a}_1^*$, $\mathbf{a}_2^*$, and $\mathbf{a}_3^*$ are defined by Eqs. (1.2.6) in terms of the smallest crystal lattice translations $\mathbf{a}_1$, $\mathbf{a}_2$, and $\mathbf{a}_3$ along the [100], [010], and [001] directions, respectively. Eq. (3.1.7) gives

$$n(\mathbf{r}) = c + \tfrac{1}{2}\eta_1(\gamma_1 e^{i\mathbf{k}_0\mathbf{r}} + \gamma_1^* e^{-i\mathbf{k}_0\mathbf{r}}) + \tfrac{1}{2}\eta_2(\gamma_2 e^{i2\mathbf{k}_0\mathbf{r}} + \gamma_2^* e^{-i2\mathbf{k}_0\mathbf{r}}) \tag{3.1.9}$$

where $\eta = \eta_1 = \eta_2$ is the long-range-order parameter. We put $\eta_1 = \eta_2 = \eta$ since the vector $2\mathbf{k}_0 = 2\pi(\frac{8}{5}\mathbf{a}_1^* + \frac{4}{5}\mathbf{a}_3^*)$ corresponding to the point $(\frac{8}{5}\,0\,\frac{4}{5})$ of the reciprocal lattice belongs to the same star as the initial vector $\mathbf{k}_0 = 2\pi(\frac{4}{5}\mathbf{a}_1^* + \frac{2}{5}\mathbf{a}_3^*)$. This may be proved by subtracting the fundamental reciprocal lattice vector (200) from $(\frac{8}{5}0\frac{4}{5})$:

$$(\tfrac{8}{5}\,0\,\tfrac{4}{5}) - (200) = (\tfrac{\bar{2}}{5}\,0\,\tfrac{4}{5})$$

One may see that the resulting vector $(\frac{\bar{2}}{5}\,0\,\frac{4}{5})$ belongs to the same star as the initial vector $(\frac{4}{5}\,0\,\frac{2}{5})$. With $\gamma_1 = \gamma_2 = \frac{2}{5}$, Eq. (3.1.9) describes the D1a superstructure:

$$n(\mathbf{r}) = c + \tfrac{2}{5}\eta(\cos \mathbf{k}_0\mathbf{r} + \cos 2\mathbf{k}_0\mathbf{r}) \tag{3.1.10}$$

The fcc crystal lattice site positions may be expressed in terms of the elementary crystal lattice translations $\mathbf{a}_1$, $\mathbf{a}_2$, and $\mathbf{a}_3$

$$\mathbf{r} = x\mathbf{a}_1 + y\mathbf{a}_2 + z\mathbf{a}_3 \tag{3.1.11}$$

where (x, y, z) are the site coordinates [with an fcc lattice (x, y, z) are integers or half-integers whose sum $x+y+z$ should be an integer].

According to the definition (1.2.6)

$$\mathbf{a}_p\mathbf{a}_q^* = \delta_{pq} \qquad p, q = 1, 2, 3 \tag{3.1.12}$$

where δ_{pq} is the Kronecker symbol. Using Eqs. (3.1.8), (3.1.11), and (3.1.12), one obtains the scalar product, $\mathbf{k}_0\mathbf{r}$, in the form

$$\begin{aligned}\mathbf{k}_0\mathbf{r} &= 2\pi(\tfrac{4}{5}\mathbf{a}_1^* + \tfrac{2}{5}\mathbf{a}_3^*)(x\mathbf{a}_1 + y\mathbf{a}_2 + z\mathbf{a}_3)\\ &= 2\pi(\tfrac{4}{5}x + \tfrac{2}{5}z)\end{aligned} \tag{3.1.13}$$

Substitution of (3.1.13) into (3.1.10) yields

$$n(x, y, z) = c + \tfrac{2}{5}\eta\left(\cos\frac{4\pi}{5}(2x+z) + \cos\frac{8\pi}{5}(2x+z)\right) \tag{3.1.14}$$

In going over alternating (201) fcc lattice planes, function (3.1.14) assumes two values

$$n_1 = c + \frac{4}{5}\eta$$

$$n_2 = c - \frac{1}{5}\eta \tag{3.1.15}$$

Ordering thus results in subdivision of the fcc lattice sites into two sublattices characterized by different Mo-atom occupation probabilities. At the stoichiometric composition of $c = c_{st} = \frac{1}{5}$ and $\eta = 1$, a fully ordered phase is formed with $n(\mathbf{r})$ equal to either 1 or 0 ($\cos 2\pi/5 - \cos \pi/5 = -\frac{1}{2}$). Assuming that $n(\mathbf{r})$ are the Mo-atom occupation probabilities, one comes to the conclusion that the (201) planes occupied by Mo and Ni alternate as follows: MoNiNiNiNiMoNiNiNiNi.... The corresponding superlattice is depicted in Fig. 10.

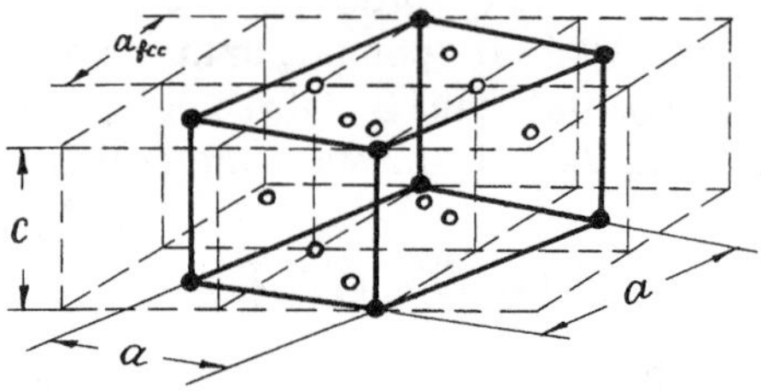

Figure 10. The structure of the fcc-based substitutional ordered phase Ni_4Mo (D1a).

3.1.2. Ni_2V, Pt_2Mo (D_{2h}^{25}-Immm)

The Ni_2V layer superstructure occurs in fcc solutions and is generated by the wave vector

$$\mathbf{k}_0 = 2\pi(\tfrac{4}{3}\mathbf{a}_1^* + \tfrac{2}{3}\mathbf{a}_3^*) \tag{3.1.16}$$

This vector describes the $(\frac{4}{3}\,0\,\frac{2}{3})$ point of the reciprocal lattice of a disordered solution. Eq. (3.1.7) gives

$$n(\mathbf{r}) = c + \tfrac{1}{2}\eta(\gamma_1 e^{i\mathbf{k}_0\mathbf{r}} + \gamma_1^* e^{-i\mathbf{k}_0\mathbf{r}}) \tag{3.1.17}$$

since $2\mathbf{k}_0 \equiv -\mathbf{k}_0$ to within the shift by the reciprocal lattice vector $2\pi(4\mathbf{a}_1^* + 2\mathbf{a}_3^*)$ (vector (402)). The distribution (3.1.17) describes the Ni_2V superstructure if $\gamma_1 = \frac{1}{3}$. Then

$$n(\mathbf{r}) = c + \tfrac{2}{3}\eta \cos \mathbf{k}_0\mathbf{r} \tag{3.1.18}$$

As

$$\mathbf{k}_0\mathbf{r} = 2\pi\left(\frac{4}{3}\mathbf{a}_1^* + \frac{2}{3}\mathbf{a}_3^*\right)(x\mathbf{a}_1 + y\mathbf{a}_2 + z\mathbf{a}_3) = 4\frac{\pi}{3}(2x + z)$$

Eq. (3.1.18) may be rewritten in the form

$$n(x, y, z) = c + \frac{2}{3}\eta \cos \frac{4\pi}{3}(2x + z) \tag{3.1.19}$$

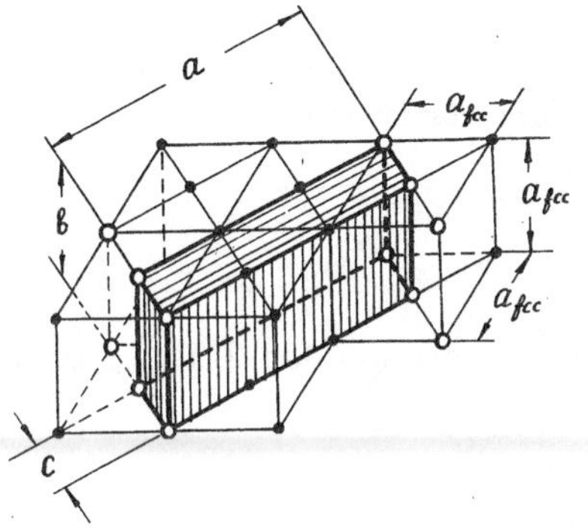

Figure 11. The structure of the fcc-based substitutional ordered phase Ni_2V (D_{2h}^{25}—I mmm).

At $c=\frac{1}{3}$ and $\eta=1$, Eq. (3.1.19) describes the fully ordered state with vanadium occupation probabilities equal to either unity (on the sites occupied by V-atoms) or zero (on all the other sites occupied by Ni). The corresponding structure is shown in Fig. 11.

The other layer superstructures in the fcc lattice generated by the vectors $2\pi\mathbf{a}_3^*$, $2\pi(\frac{1}{2}\mathbf{a}_1^*+\frac{1}{2}\mathbf{a}_2^*+\frac{1}{2}\mathbf{a}_3^*)$ and $2\pi(\mathbf{a}_1^*+\frac{1}{2}\mathbf{a}_3^*)$ are CuAuI ($L1_0$), CuPt (L1) and Al_3Ti (DO_{22}), respectively.*

Ordering in bcc structures may also lead to layer superstructures, as is shown next.

3.1.3. βCuZn (B2)

The first example is the formation of the B2 superstructure generated by the vector

$$\mathbf{k}_0=2\pi(\mathbf{a}_1^*+\mathbf{a}_2^*+\mathbf{a}_3^*) \tag{3.1.20}$$

which corresponds to the (111) point of the reciprocal lattice of the bcc crystal.

Eq. (3.1.7) for the B2 superstructure reads

$$n(\mathbf{r})=c+\tfrac{1}{2}\eta e^{i\mathbf{k}_0\mathbf{r}} \tag{3.1.21}$$

because $e^{i\mathbf{k}_0\mathbf{r}}=e^{-i\mathbf{k}_0\mathbf{r}}$. Since

$$\mathbf{k}_0\mathbf{r}=2\pi(\mathbf{a}_1^*+\mathbf{a}_2^*+\mathbf{a}_3^*)(x\mathbf{a}_1+y\mathbf{a}_2+z\mathbf{a}_3)=2\pi(x+y+z) \tag{3.1.22}$$

Eq. (3.1.21) can be rewritten as

$$n(\mathbf{r})=n(x,\ y,\ z)=c+\tfrac{1}{2}\eta e^{i2\pi(x+y+z)} \tag{3.1.23}$$

where x, y, z are the coordinates of the bcc lattice sites. At $c=\frac{1}{2}$ and $\eta=1$ Eq. (3.1.23) describes the completely ordered phase shown in Fig. 12.

*See reference 9 for details.

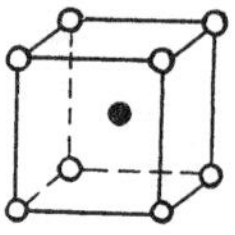

Figure 12. The structure of the bcc-based substitutional ordered phase β-CuZn (B2).

3.1.4. Fe_3Al (DO_3)

The DO_3 layer superstructure derives from the bcc solution. It is generated by the wave vector $(\frac{1}{2}\frac{1}{2}\frac{1}{2})$:

$$\mathbf{k}_1=\tfrac{1}{2}\mathbf{k}_0=2\pi(\tfrac{1}{2}\mathbf{a}_1^*+\tfrac{1}{2}\mathbf{a}_2^*+\tfrac{1}{2}\mathbf{a}_3^*) \tag{3.1.24}$$

Eq. (3.1.7) for the DO_3 superstructure has the form

$$n(\mathbf{r})=c-\frac{1}{4}\eta_1 e^{i\mathbf{k}_0\mathbf{r}}+\eta_2\left(\frac{1}{4i}e^{i(\mathbf{k}_0\mathbf{r}/2)}-\frac{1}{4i}e^{-i(\mathbf{k}_0\mathbf{r}/2)}\right) \tag{3.1.25a}$$

or

$$n(\mathbf{r})=c-\tfrac{1}{4}\eta_1 e^{i\mathbf{k}_0\mathbf{r}}+\tfrac{1}{2}\eta_2\sin(\tfrac{1}{2}\mathbf{k}_0\mathbf{r}) \tag{3.1.25b}$$

Substitution of (3.1.24) into (3.1.25) gives

$$n(\mathbf{r})=n(x,y,z)=c-\tfrac{1}{4}\eta_1 e^{i2\pi(x+y+z)}+\tfrac{1}{2}\eta_2\sin\pi(x+y+z) \tag{3.1.26}$$

At $c=\frac{1}{4}$ and $\eta_1=\eta_2=1$ Eq. (3.1.26) describes the fully ordered DO_3 superstructure depicted in Fig. 13.

The Cu_3Au ($L1_2$), $CuPt_7$, and $CuPt_3$ structures may serve as examples of nonlayer superstructures formed in the fcc lattice.

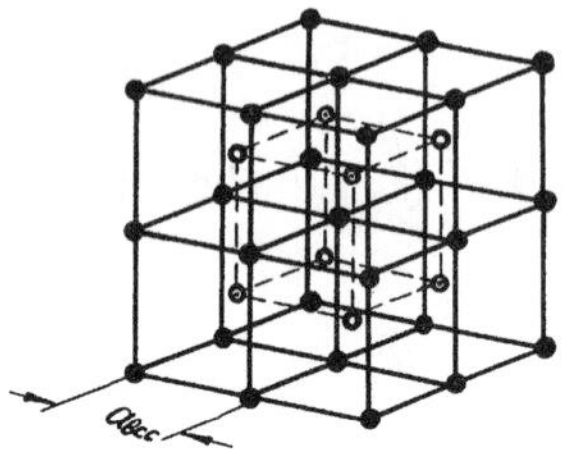

Figure 13. The structure of the bcc-based substitutional ordered phase Fe_3Al (DO_3); ● = Fe, ○ = Al, ⊙ = Fe. Symbols ● and ⊙ designate crystallographically nonequivalent positions filled by Fe-atoms.

3.1.5. Cu_3Au ($L1_2$)

The probability to find an Au-atom in the site **r** in the Cu_3Au superstructure is

$$n(\mathbf{r})=c+\tfrac{1}{4}\eta_1(e^{i2\pi\mathbf{a}_1^*\mathbf{r}}+e^{i2\pi\mathbf{a}_2^*\mathbf{r}}+e^{i2\pi\mathbf{a}_3^*\mathbf{r}}) \tag{3.1.27}$$

or

$$n(x,y,z)=c+\tfrac{1}{4}\eta_1(e^{i2\pi x}+e^{i2\pi y}+e^{i2\pi z}) \tag{3.1.28}$$

The function $n(\mathbf{r})$ takes two values on fcc lattice sites, $n_1=c+\frac{3}{4}\eta_1$ and $n_2=c-\frac{1}{4}\eta_1$, and describes the ordered phase presented in Fig. 14.

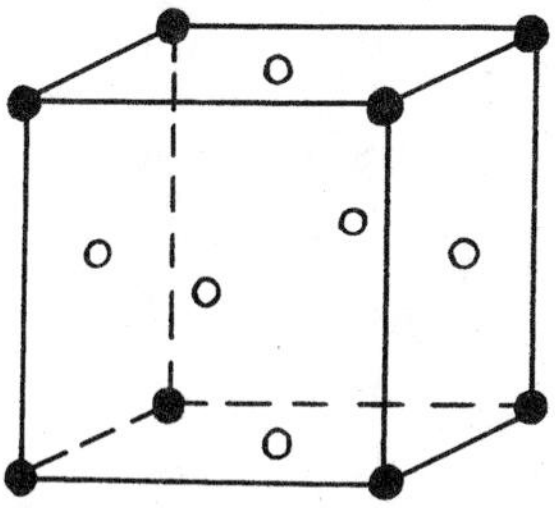

Figure 14. The structure of the fcc-based substitutional ordered phase Cu_3Au ($L1_2$).

3.1.6. $CuPt_7$

The concentration wave representation of the occupation probabilities in $CuPt_7$ is

$$
\begin{aligned}
n(\mathbf{r}) = c + \tfrac{1}{8}\eta_1\left(e^{i2\pi \mathbf{a}_1^*\mathbf{r}} + e^{i2\pi \mathbf{a}_2^*\mathbf{r}} + e^{i2\pi \mathbf{a}_3^*\mathbf{r}}\right) \\
+ \tfrac{1}{8}\eta_2\left(e^{i\pi(\mathbf{a}_1^*+\mathbf{a}_2^*+\mathbf{a}_3^*)\mathbf{r}} + e^{i\pi(-\mathbf{a}_1^*+\mathbf{a}_2^*+\mathbf{a}_3^*)\mathbf{r}}\right. \\
\left. + e^{i\pi(\mathbf{a}_1^*-\mathbf{a}_2^*+\mathbf{a}_3^*)\mathbf{r}} + e^{i\pi(\mathbf{a}_1^*+\mathbf{a}_2^*-\mathbf{a}_3^*)\mathbf{r}}\right)
\end{aligned} \quad (3.1.29)
$$

or

$$
\begin{aligned}
n(x, y, z) = c + \tfrac{1}{8}\eta_1\left(e^{i2\pi x} + e^{i2\pi y} + e^{i2\pi z}\right) \\
+ \tfrac{1}{8}\eta_2\left(e^{i\pi(x+y+z)} + e^{i\pi(-x+y+z)}\right. \\
\left. + e^{i\pi(x-y+z)} + e^{i\pi(x+y-z)}\right)
\end{aligned} \quad (3.1.29a)
$$

Function (3.1.29) describing the distribution of Cu-atoms assumes three values on the set of all fcc lattice sites:

$$
\begin{aligned}
n_1 &= c + \tfrac{3}{8}\eta_1 + \tfrac{1}{2}\eta_2 \\
n_2 &= c + \tfrac{3}{8}\eta_1 - \tfrac{1}{2}\eta_2 \\
n_3 &= c - \tfrac{1}{8}\eta_1
\end{aligned}
$$

The corresponding structure is shown in Fig. 15.

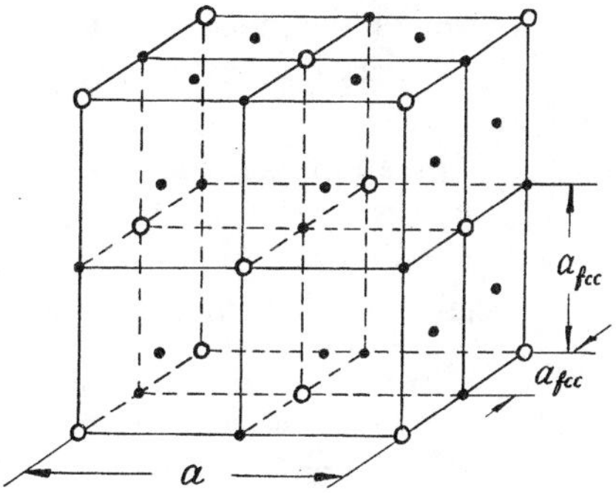

Figure 15. The structure of the fcc-based substitutional ordered phase $CuPt_7$: O-Cu, •-Pt. Pt-atoms are shown only on the faces visible to the reader.

3.1.7. $CuPt_3$

The $CuPt_3$ superstructure is the superposition of the concentration waves

$$n(\mathbf{r})=c+\tfrac{1}{4}\eta_1 e^{i2\pi \mathbf{a}_3^*\mathbf{r}}+\frac{1}{4}\eta_2\left(e^{i\pi(\mathbf{a}_1^*+\mathbf{a}_2^*+\mathbf{a}_3^*)\mathbf{r}}+e^{i\pi(\mathbf{a}_1^*+\mathbf{a}_2^*-\mathbf{a}_3^*)\mathbf{r}}\right) \tag{3.1.30}$$

or

$$n(x, y, z)=c+\frac{1}{4}\eta_1 e^{i2\pi z}+\tfrac{1}{4}\eta_2\left(e^{i\pi(x+y+z)}+e^{i\pi(x+y-z)}\right) \tag{3.1.30a}$$

This superstructure is depicted in Fig. 16.

More examples of concentration wave treatment of ordered phases arising from fcc and bcc lattices may be found in review (9).

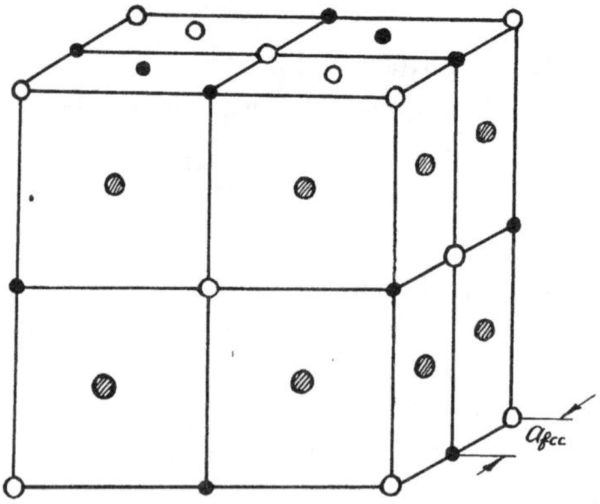

Figure 16. The structure of the fcc-based substitutional ordered phase $CuPt_3$: ○ = Cu, ● = Pt, ◍ = Pt. Positions ● and ◍ are crystallographically nonequivalent.

3.2. SECOND-ORDER TRANSFORMATIONS IN THE PHENOMENOLOGICAL THEORY OF ORDERING

Prior to further discussion of the concentration wave formalism as applied to the microscopic theory of ordering, we shall dwell on the phenomenological theory of second-order transformations in alloys which in effect involves the idea of the concentration waves in an implicit form.*

This theory was developed by Landau in 1937 (18) and Lifshitz in 1942 (19). A different theory formulation from that suggested by these authors will be given. We will not use the theory of the space-group representations because it is difficult to follow unless the reader has a special mathematical background.

*Considering the original formulation of the phenomenological theory of second-order transformations in terms of the concentration wave approach, one may easily see that the role of concentration waves in that theory is played by the functions determining the irreducible representation of the space group that generates the second-order transformation.

For simplicity, we will again consider one of the fourteen Bravais lattices.

As shown in Section 2.2, the necessary condition for the second-order transformation is for the minimum of the coefficient $b(\mathbf{k};\ T,\ c,\ p)$ to vanish. Mathematically, this condition is represented by Eqs. (2.2.17). Since the function $b(\mathbf{k};\ T, c, p)$ of $\mathbf{k}$ has the symmetry of the disordered phase, it takes the same minimum value $b(\mathbf{k}_0;\ T,\ c,\ p)$ not just at one wave vector, $\mathbf{k}_0$, but at several wave vectors, $\mathbf{k}_{01}, \mathbf{k}_{02}, \mathbf{k}_{03}, \ldots, \mathbf{k}_{0j}$. All these vectors can be obtained from one vector (for example, $\mathbf{k}_{01}$) by means of applying to it all the operations of the point-symmetry group of the disordered solution lattice. This conclusion is illustrated by Fig. 17, where the example of a typical plot of the function $b=b(\mathbf{k})$ for the cubic disordered solution is shown. According to the definition of the characteristic function (2.2.13) and the identity $\exp(i2\pi\mathbf{Hr})\equiv 1$, the function $b(\mathbf{k})$ has the periodicity coinciding with the periodicity of the reciprocal lattice of the disordered phase:

$$b(\mathbf{k}+2\pi\mathbf{H})=b(\mathbf{k})$$

The vectors $\mathbf{k}_{01}$, $\mathbf{k}_{02}$, $\mathbf{k}_{03}$, and $\mathbf{k}_{04}$ as well as the vectors $\mathbf{k}'_{01}$, $\mathbf{k}'_{02}$, $\mathbf{k}'_{03}$, and $\mathbf{k}'_{04}$ shown in Fig. 17 mark the points in the reciprocal space where the function $b(\mathbf{k})$ takes the minimum values $b(\mathbf{k}_0)$ and $b(\mathbf{k}'_0)$, respectively: it follows from the symmetry of the disordered crystal that

$$b(\mathbf{k}_{01})=b(\mathbf{k}_{02})=b(\mathbf{k}_{03})=b(\mathbf{k}_{04})=b(\mathbf{k}_0)$$

Figure 17. A typical relief of the characteristic function $b(\mathbf{k})$ in the reciprocal space of a disordered cubic solution. Contours are described by $b(\mathbf{k})=$constant. ● reciprocal lattice point.

and

$$b(\mathbf{k}'_{01})=b(\mathbf{k}'_{02})=b(\mathbf{k}'_{03})=b(\mathbf{k}'_{04})=b(\mathbf{k}'_{0})$$

In the general case not all vectors from the set $\mathbf{k}_{01} \ldots \mathbf{k}_{0j}$ are crystallographically different. Some of them differ from each other by a fundamental reciprocal lattice vector $2\pi\mathbf{H}$ and consequently can be considered as the same wave vector. Such a set of the wave vectors, $\{\mathbf{k}_0\}$, is termed a star.*

Therefore the vanishing of $b(\mathbf{k})$ at the point $\mathbf{k}=\mathbf{k}_{01}$ results in a simultaneous vanishing of $b(\mathbf{k})$ at the points $\mathbf{k}_{02}$, $\mathbf{k}_{03}$, and $\mathbf{k}_{0j}$ and consequently in the loss of stability of the disordered phase against a few infinitesimal concentration waves with all wave vectors, $\mathbf{k}_{01}, \mathbf{k}_{02}, \ldots, \mathbf{k}_{0j}$. This means that Eq. (2.2.14) can be represented as

$$n(\mathbf{r})=c+\tfrac{1}{2}\sum_j \left[Q(\mathbf{k}_{0j})e^{i\mathbf{k}_{0j}\mathbf{r}}+Q(\mathbf{k}_{0j})^* e^{-i\mathbf{k}_{0j}\mathbf{r}}\right] \tag{3.2.1}$$

where the summation is carried out over all wave vectors of the star, $\{\mathbf{k}_0\}$. Wave vectors $\mathbf{k}_{0j}$ in (3.2.1) determine the positions of the superstructure reflections situated in the first Brillouin zone and consequently are superlattice vectors (ordering vectors). The amplitudes $Q(\mathbf{k}_{0j})$ are proportional to long-range order parameters. Since the amplitudes $Q(\mathbf{k}_{0j})$ are small values in the vicinity of the second-order transition point, it is possible to expand the free energy F into series of amplitudes $Q(\mathbf{k}_{0j})$. This expansion is

$$\begin{aligned} F=F_{\text{disord}}&+\frac{N}{2}\sum_j b(\mathbf{k}_{0j})|Q(\mathbf{k}_{0j})|^2+\frac{1}{3!}\sum_{j_1j_2j_3} C(\mathbf{k}_{0j_1},\mathbf{k}_{0j_2},\mathbf{k}_{0j_3}) \\ &\times Q(\mathbf{k}_{0j_1})Q(\mathbf{k}_{0j_2})Q(\mathbf{k}_{0j_3})+\frac{1}{4!}\sum_{j_1j_2j_3j_4} D(\mathbf{k}_{0j_1},\mathbf{k}_{0j_2},\mathbf{k}_{0j_3},\mathbf{k}_{0j_4}) \\ &\times Q(\mathbf{k}_{0j_1})Q(\mathbf{k}_{0j_2})Q(\mathbf{k}_{0j_3})Q(\mathbf{k}_{0j_4})+\cdots \end{aligned} \tag{3.2.2}$$

where $C(\mathbf{k}_{0j_1}, \mathbf{k}_{0j_2}, \mathbf{k}_{0j_3})$ and $D(\mathbf{k}_{0j_1}, \mathbf{k}_{0j_2}, \mathbf{k}_{0j_3}, \mathbf{k}_{0j_4})$ are the third-order and the fourth-order terms of the Taylor expansion. The indexes j_1, j_2, j_3, and j_4 refer to the vectors of the star, $\{\mathbf{k}_0\}$, labeling all vectors of the star. For the second-order term of Eq. (3.2.2), Eq. (2.2.12) has been used.

It should be born in mind that the Taylor expansion (3.2.2) about the second-order transformation point might not be mathematically correct because of the singularity of the free energy at this point. One can use, however, the Taylor expansion (3.2.2) outside the small interval (usually a fraction of a degree) in the vicinity of the order-disorder transformation temperature T_0 where all following considerations are valid.

The free energy (3.2.2) is evidently invariant under a displacement of the ordered distribution $n(\mathbf{r})$ as a rigid body by any translation $\mathbf{T}$ of the disordered phase, that is, at

$$n(\mathbf{r}) \rightarrow n(\mathbf{r}+\mathbf{T}) \tag{3.2.3}$$

*In particular, the set of four vectors, $\{\mathbf{k}'_0\}$, in Fig. 17 contains only one independent wave vector. One can easily ascertain that other three vectors differ from the original one by the fundamental reciprocal lattice vector. It means that the star, $\{\mathbf{k}'_0\}$ consists of a single wave vector.

This invariance leads to the limitations for the coefficients of the free energy expansion (3.2.2). To find these limitations, we should first observe how the amplitudes of concentration wave transform at the transformation (3.2.3). It follows from Eq. (3.2.1) that

$$n(\mathbf{r}+\mathbf{T})=c+\tfrac{1}{2}\sum_j [(Q(\mathbf{k}_{0j})e^{i\mathbf{k}_{0j}\mathbf{T}})e^{i\mathbf{k}_{0j}\mathbf{r}}+(Q^*(\mathbf{k}_{0j})e^{-i\mathbf{k}_{0j}\mathbf{T}})e^{-i\mathbf{k}_{0j}\mathbf{r}}] \tag{3.2.4}$$

Eqs. (3.2.1) and (3.2.4) differ from each other by only their amplitude values: the amplitudes in Eq. (3.2.4) differ from those in Eq. (3.2.1) by factors $\exp(i\mathbf{k}_{0j}\mathbf{T})$. Therefore the transformation (3.2.3) corresponds to the transformation

$$Q(\mathbf{k}_{0j}) \rightarrow Q(\mathbf{k}_{0j}) \exp(i\mathbf{k}_{0j}\mathbf{T}) \tag{3.2.5}$$

for the amplitudes $Q(\mathbf{k}_{0j})$.

Using the transformation (3.2.5) in Eq. (3.2.2), one has

$$\begin{aligned}\Delta F=&\frac{1}{2}\sum_j b(\mathbf{k}_{0j})|Q(\mathbf{k}_{0j})|^2+\frac{1}{3!}\sum_{j_1 j_2 j_3}[C(\mathbf{k}_{0j_1},\mathbf{k}_{0j_2},\mathbf{k}_{0j_3})\\ &\times e^{i(\mathbf{k}_{0j_1}+\mathbf{k}_{0j_2}+\mathbf{k}_{0j_3})\mathbf{T}}]\times Q(\mathbf{k}_{0j_1})Q(\mathbf{k}_{0j_2})Q(\mathbf{k}_{0j_3})\\ &+\frac{1}{4!}\sum_{j_1 j_2 j_3 j_4}[D(\mathbf{k}_{0j_1},\mathbf{k}_{0j_2},\mathbf{k}_{0j_3},\mathbf{k}_{0j_4})e^{i(\mathbf{k}_{0j_1}+\mathbf{k}_{0j_2}+\mathbf{k}_{0j_3}+\mathbf{k}_{0j_4})\mathbf{T}}]\\ &\times Q(\mathbf{k}_{0j_1})Q(\mathbf{k}_{0j_2})Q(\mathbf{k}_{0j_3})Q(\mathbf{k}_{0j_4})+\cdots\end{aligned} \tag{3.2.6}$$

Comparing Eqs. (3.2.2) and (3.2.6), one can easily see that the invariance of free energy is possible only if the exponents

$$\exp(i(\mathbf{k}_{0j_1}+\mathbf{k}_{0j_2}+\mathbf{k}_{0j_3})\mathbf{T}),\ \exp(i(\mathbf{k}_{0j_1}+\mathbf{k}_{0j_2}+\mathbf{k}_{0j_3}+\mathbf{k}_{0j_4})\mathbf{T})\ldots \tag{3.2.7}$$

and so on, are identically equal to unity for any Bravais translation $\mathbf{T}$. The coefficients $C(\mathbf{k}_{0j_1}, \mathbf{k}_{0j_2}, \mathbf{k}_{0j_3})$ and $D(\mathbf{k}_{0j_1}, \mathbf{k}_{0j_2}, \mathbf{k}_{0j_3}, \mathbf{k}_{0j_4})$ can then take nonzero values. Otherwise, if the exponentials (3.2.7) are not equal to unity identically for any reference vector $\mathbf{T}$ of the disordered phase, the coefficients $C, D, \ldots$ must vanish to preserve the free energy invariance.

The condition that the exponentials (3.2.7) be identically equal to unity at any reference vector $\mathbf{T}$ is equivalent to the conservation of "quasi-momentum" conditions

$$\begin{gathered}\mathbf{k}_{0j_1}+\mathbf{k}_{0j_2}+\mathbf{k}_{0j_3}=2\pi\mathbf{H}_1\\ \mathbf{k}_{0j_1}+\mathbf{k}_{0j_2}+\mathbf{k}_{0j_3}+\mathbf{k}_{0j_4}=2\pi\mathbf{H}_2\ldots\end{gathered}$$

where $\mathbf{H}_1, \mathbf{H}_2, \ldots$ are the fundamental reciprocal lattice vectors since, according to the reciprocal lattice definition (1.2.4),

$$e^{i2\pi\mathbf{H}\mathbf{T}}\equiv 1$$

The free energy expansion (3.2.6) is invariant under any translation $\mathbf{T}$, provided that the expansion coefficients satisfy the following conditions:

$$C(\mathbf{k}_{0j_1}, \mathbf{k}_{0j_2}, \mathbf{k}_{0j_3}) = \begin{cases} \neq 0 & \text{if } \mathbf{k}_{0j_1} + \mathbf{k}_{0j_2} + \mathbf{k}_{0j_3} = 2\pi\mathbf{H} \\ = 0 & \text{otherwise} \end{cases}$$

$$D(\mathbf{k}_{0j_1}, \mathbf{k}_{0j_2}, \mathbf{k}_{0j_3}, \mathbf{k}_{0j_4}) = \begin{cases} \neq 0 & \text{if } \mathbf{k}_{0j_1} + \mathbf{k}_{0j_2} + \mathbf{k}_{0j_3} + \mathbf{k}_{0j_4} = 2\pi\mathbf{H} \\ = 0 & \text{otherwise} \end{cases} \tag{3.2.8}$$

and so on, where **H** is any reciprocal lattice vector.

Let us introduce the long-range order parameter η defined by the relation

$$Q(\mathbf{k}_{0j}) = \eta\gamma_j \tag{3.2.9}$$

where $Q(\mathbf{k}_{0j})$ is the concentration wave amplitude from Eq. (3.2.1) and γ_j are the coefficients that determine the symmetry of the occupation probabilities $n(\mathbf{r})$ and meet the normalization condition

$$\sum_j |\gamma_j|^2 = 1 \tag{3.2.10}$$

where the summation is over all the vectors of the star.

Making use of the definition (3.2.9) of the long-range-order parameter, the normalization condition (3.2.10), and the restrictions (3.2.8) in Eq. (3.2.2), we have

$$F = F_{\text{disord}} + \frac{1}{2} Nb(\mathbf{k}_0; T, c)\eta^2 + \frac{1}{3!} C(T, c)\eta^3 + \frac{1}{4!} D(T, c)\eta^4 + \cdots \tag{3.2.11}$$

where

$$C(T, c) = \sum_{\substack{j_1 j_2 j_3 \\ \mathbf{k}_{0j_1} + \mathbf{k}_{0j_2} + \mathbf{k}_{0j_3} = 2\pi\mathbf{H}}} C(\mathbf{k}_{0j_1}, \mathbf{k}_{0j_2}, \mathbf{k}_{0j_3})\gamma(\mathbf{k}_{0j_1})\gamma(\mathbf{k}_{0j_2})\gamma(\mathbf{k}_{0j_3}) \tag{3.2.12}$$

$$D(T, c) = \sum_{\substack{j_1 j_2 j_3 j_4 \\ \mathbf{k}_{0j_1} + \mathbf{k}_{0j_2} + \mathbf{k}_{0j_3} + \mathbf{k}_{0j_4} = 2\pi\mathbf{H}}} D(\mathbf{k}_{0j_1}, \mathbf{k}_{0j_2}, \mathbf{k}_{0j_3}, \mathbf{k}_{0j_4})\gamma(\mathbf{k}_{0j_1})\gamma(\mathbf{k}_{0j_2})\gamma(\mathbf{k}_{0j_3})\gamma(\mathbf{k}_{0j_4}) \tag{3.2.13}$$

are the coefficients of the third- and fourth-order terms of the expansion (3.2.11). Since these coefficients are state functions of the disordered solution, they depend on the temperature, T, and composition, c.

At a second-order transformation point ($T = T_0$) the equilibrium value of the long-range order parameter η vanishes. Therefore at $T = T_0$ the values of coefficients $b(\mathbf{k}_0; T, c)$, $C(T, c)$, and $D(T, c)$ must provide a minimum of the free energy (3.2.11) at $\eta = 0$.

It follows from Eq. (2.2.17) that at $T = T_0$, $b(\mathbf{k}_0; T_0, c) = 0$, and hence

$$\Delta F = F - F_{\text{disord}} = \frac{1}{3!} C(T_0, c)\eta^3 + \frac{1}{4!} D(T_0, c)\eta^4 + \cdots \tag{3.2.14}$$

The value ΔF in Eq. (3.2.14) takes on its minimum value at $\eta = 0$ if

$$C(T_0, c) = 0 \tag{3.2.15}$$

and

$$D(T_0, c) > 0 \tag{3.2.16}$$

Otherwise, if $C(T_0, c) \neq 0$, one can always choose such a small long-range order parameter (positive or negative) that the free energy (3.2.11) would be smaller as compared with the value corresponding to $\eta = 0$. If the coefficient $C(T_0, c)$ is not equal to zero identically due to the symmetry, Eq. (3.2.15) is an additional equation with respect to the variables T_0 and c. Together with Eq. (2.2.17) the latter forms a set of two simultaneous equations. The solution yields an isolated point at the equilibrium of the T-c diagram, but the usual situation under consideration corresponds to the case where the second-order transformation temperatures form a line, $T_0 = T_0(c)$, in the T-c diagram. Therefore one can conclude that due to the symmetry of the disordered phase the coefficient $C(T, c)$ should be identically equal to zero. This takes place when all coefficients $C(\mathbf{k}_{0j_1}, \mathbf{k}_{0j_2}, \mathbf{k}_{0j_3})$ entering Eq. (3.2.12) for $C(T, c)$ are also identically equal to zero. According to the selection rule (3.2.8), this is realized if

$$\mathbf{k}_{0j_1} + \mathbf{k}_{0j_2} + \mathbf{k}_{0j_3} \neq 2\pi\mathbf{H} \tag{3.2.17}$$

where $\mathbf{k}_{0j_1}$, $\mathbf{k}_{0j_2}$, and $\mathbf{k}_{0j_3}$ are any (possibly equal) vectors of the star, $\{\mathbf{k}_0\}$. Even if only one coefficient, $C(\mathbf{k}_{0j_1}, \mathbf{k}_{0j_2}, \mathbf{k}_{0j_3})$, entering the expression for $C(T, c)$ does not vanish identically due to the symmetry of the disordered phase, one can always choose coefficients, $\gamma(\mathbf{k}_{0j})$, so that they provide a nonzero value of $C(T, c)$.

The necessary condition for the second-order phase transformation is defined by Eq. (3.2.17): *a second-order transformation in an ordering alloy is possible only if a fundamental reciprocal lattice vector* $2\pi\mathbf{H}$ *cannot be obtained by combining any three wave vectors of the star* $\{\mathbf{k}_0\}$, *which generates the ordering.* This definition is another formulation of the necessary condition of second-order transformations, as obtained in the original papers (18, 19). According to these papers, this is a condition of impossibility for constructing a third-order invariant of the coefficients $\gamma(\mathbf{k}_{0j})$. In the theory of group representations it means that a cube of the space-group representation generating the ordered structure does not contain the identity representation.

When the ordered phase is connected with a nonzero star, $\{\mathbf{k}_0\}$, the criterion (3.2.17) and the Landau criterion are equivalent. The criterion (3.2.17), however, is more convenient for practical applications. To analyze this type of order-disorder phase transformation, it is necessary to know the star, $\{\mathbf{k}_0\}$, whose wave vectors generate the given ordered structure. Later it will be shown that the star, $\{\mathbf{k}_0\}$, can be found by means of x-ray, neutron, or electron diffraction in a selected area, since the vectors of the star, $\{\mathbf{k}_0\}$, link the superlattice reflection points to the nearest fundamental reflection.

3.3. EXAMPLES FOR DETERMINING PARTICULAR PHASE TRANSFORMATION

Eq. (3.2.17) which is the necessary condition for a second-order transformation can be used to arrive at certain conclusions about the kind of phase transformations. The examples below illustrate this possibility.

3.3.1. Decomposition

Any Decomposition Reaction Is a First-Order Transformation.
To prove this, recall that decomposition occurs when concentration wave instability is generated by the star $\mathbf{k}_0=0$ (see Section 2.1). Hence

$$\mathbf{k}_0+\mathbf{k}_0+\mathbf{k}_0=3\mathbf{k}_0=0$$

Since the condition (3.2.17) is not fulfilled, decomposition cannot be a second-order transformation; it is always a first-order one.

3.3.2. Crystallization

Any Crystallization Reaction Is Always a First-Order Transformation.
The necessary condition (3.2.17) for a second-order transformation leads to this very important finding by Landau (20). The high-temperature disordered state is an isotropic liquid. All rotation operations of the point group of an isotropic liquid form a three dimensional rotation group. This is the reason why any star is a set of infinite wave vectors whose origin is at the center of the sphere of radius $|\mathbf{k}_0|$ and end at the surface of this sphere (see Fig. 18). Figure 18 demonstrates that such a star always contains three vectors whose sum is equal to zero; that is, these three vectors form an equilateral triangle. Therefore the necessary condition for a second-order transformation (3.2.17) does not hold, and consequently crystallization is always a first-order transformation.

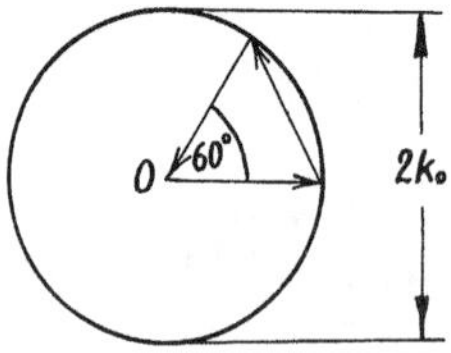

Figure 18. Relations between vectors of a star generating the liquid to solid phase transition. The vectors are shown by arrows.

3.3.3. CuAuI ($L1_0$) and Cu_3Au ($L1_2$) Superlattices

The CuAuI and Cu_3Au ordered phases arise in fcc solid solution. The star, $\{\mathbf{k}_0\}$, associated with these superstructures includes three vectors:

$$(100), (010), (001) \tag{3.3.1}$$

(from here on the wave vector coordinates are given in the fcc and bcc reciprocal lattice basis $2\pi\mathbf{a}_1^*$, $2\pi\mathbf{a}_2^*$, $2\pi\mathbf{a}_3^*$).

One can clearly see that the sum of three vectors (3.3.1) gives the fundamental reciprocal lattice vector (111) of the fcc crystal lattice:

$$(100)+(010)+(001)=(111) \tag{3.3.2}$$

Therefore, in accordance with Eq. (3.2.17), $L1_0$ and $L1_2$ superlattices must always arise due to the first-order transformation.

It should be noted that the constant $C(T, c)$ (3.2.12) for the CuAuI type superstructure vanishes. This follows from the fact that the product

$$C(T, c) = C(2\pi\mathbf{a}_1^*, 2\pi\mathbf{a}_2^*, 2\pi\mathbf{a}_3^*)\gamma(2\pi\mathbf{a}_1^*)\gamma(2\pi\mathbf{a}_2^*)\gamma(2\pi\mathbf{a}_3^*) \tag{3.3.3}$$

is zero since $\gamma(2\pi\mathbf{a}_1^*) = \gamma(2\pi\mathbf{a}_2^*) = 0$ for the CuAuI superlattice. This does not contradict the conclusion that this order-disorder reaction should be a first-order transformation. If the CuAuI phase were due to a second-order transformation, its structure would become unstable with the formation of the additional concentration waves

$$\eta[\delta\gamma(2\pi\mathbf{a}_1^*)e^{i2\pi\mathbf{a}_1^*\mathbf{r}} + \delta\gamma(2\pi\mathbf{a}_2^*)e^{i2\pi\mathbf{a}_2^*\mathbf{r}}]$$

since the third-order term constant (3.3.3) would assume the nonzero value

$$C(T, c) = C(2\pi\mathbf{a}_1^*, 2\pi\mathbf{a}_2^*, 2\pi\mathbf{a}_3^*)\delta\gamma(2\pi\mathbf{a}_1^*)\delta\gamma(2\pi\mathbf{a}_2^*)\delta\gamma(2\pi\mathbf{a}_3^*)$$

3.3.4. CuPt Superlattice

The CuPt fcc-based superlattice is generated by the star $\{\mathbf{k}_0\}$ that includes four wave vectors (see Section 3.1):

$$(\tfrac{1}{2}\,\tfrac{1}{2}\,\tfrac{1}{2}),\ (\tfrac{\bar{1}}{2}\,\tfrac{1}{2}\,\tfrac{1}{2}),\ (\tfrac{1}{2}\,\tfrac{\bar{1}}{2}\,\tfrac{1}{2}),\ (\tfrac{1}{2}\,\tfrac{1}{2}\,\tfrac{\bar{1}}{2}) \tag{3.3.4}$$

Clearly, it is not possible to construct a fundamental reciprocal lattice vector from any three vectors of (3.3.4). Therefore in this case there is no symmetry exclusion for the second-order transformation.

3.4. THE EQUILIBRIUM EQUATION IN THE LONG-RANGE INTERACTION MODEL

The phenomenological approach enables one to find general properties of ordered phases without the use of specific models. But this approach is only valid for a narrow interval of temperatures in the vicinity of the second-order transformation point. If one is interested in studying the properties of a solid solution within a wide-range interval of temperatures, or in finding the magnitudes of thermodynamic parameters, one has to utilize more or less simple models that permit statistical calculations. Such calculations would be based on the generally known statistical theories of Gorsky, Bragg, and Williams (10–12), Guggenheim and Fowler (13), Bethe (14), Peierls (15), Kirkwood (16), and Kikuchi (17). As mentioned before, however, there are some shortcomings in these applications which are difficult to obviate.

The first shortcoming arises if one has to take into account more long-range interactions than those of the nearest and next-nearest neighboring atoms. In such an attempt there are enormous technical difficulties caused by the necessity of solving a large set of transcendental equations.

The second shortcoming is that the theories (10–17) do not deal with the most important question of how to determine the structure of an ordered phase if the atomic interaction energies are known; the structure of the ordered phase is always assumed to be known beforehand.

These shortcomings may be overcome by the method of static concentration waves. In formulating the statistical theory of order-disorder transformations, we shall accept the conventional model. According to this model, (1) the configuration-dependent fraction of the energy of an alloy is adequately approximated by pairwise interatomic interactions, and (2) atoms in both substitutional and interstitial solutions can occupy only the sites of a certain rigid lattice.

In the model of binary substitutional and interstitial solutions the configurational part of the Hamiltonian is

$$H=\frac{1}{2}\sum_{\mathbf{r}\mathbf{r}'} \tilde{V}(\mathbf{r}, \mathbf{r}')c(\mathbf{r})c(\mathbf{r}') \tag{3.4.1}$$

where the summation is over all Ising lattice sites $\mathbf{r}$ and $\mathbf{r}'$:

$$c(\mathbf{r})=\begin{cases}1 & \text{if there is a solute atom in the site } \mathbf{r}\\ 0 & \text{otherwise}\end{cases} \tag{3.4.2}$$

In a binary substitutional solution $\tilde{V}(\mathbf{r}, \mathbf{r}')$ is the interchange energy

$$\tilde{V}(\mathbf{r}, \mathbf{r}')=\tilde{V}_{AA}(\mathbf{r}, \mathbf{r}')+\tilde{V}_{BB}(\mathbf{r}, \mathbf{r}')-2\tilde{V}_{AB}(\mathbf{r}, \mathbf{r}') \tag{3.4.3}$$

where $\tilde{V}_{AA}(\mathbf{r}, \mathbf{r}')$, $\tilde{V}_{BB}(\mathbf{r}, \mathbf{r}')$, $\tilde{V}_{AB}(\mathbf{r}, \mathbf{r}')$ are pairwise interaction energies of A-A, B-B, A-B pairs of atoms placed at the Ising lattice sites $\mathbf{r}$ and $\mathbf{r}'$, respectively.

In an interstitial solution $\tilde{V}(\mathbf{r}, \mathbf{r}')$ is the interaction energy of two interstitials placed at the lattice sites (interstices) $\mathbf{r}$ and $\mathbf{r}'$.

As shown above, the atomic distribution in a binary substitutional solution can be described by means of the single-site density function, $n(\mathbf{r})=\langle c(\mathbf{r})\rangle$, that is, by the occupation probabilities of finding a solute atom in the site $\mathbf{r}$ of the Ising lattice. We consider the case where all N sites of the lattice are crystallographically equivalent, that is, they can be obtained from each other by the symmetry operations of the lattice.

It should be recalled that in a disordered crystal $n(\mathbf{r})=c$ where c is the atomic fraction of the solute atoms with respect to the total number of the lattice sites. In an ordered crystal the function $n(\mathbf{r})$ become dependent on the site position, $\mathbf{r}$. It describes the mutual arrangement of sublattices into which the original lattice is subdivided during the ordering process. The locus of the sites making up each of the sublattices is defined by Eq. (3.1.1).

Thus the problem of finding the ordered phase structure reduces to the determination of the function $n(\mathbf{r})$. To solve it, one should first find an equation for the function $n(\mathbf{r})$.

It is seen that there is a kind of Pauli exclusion principle for a binary solution: *each lattice site can be occupied by either one or zero of the atoms of some definite type*. This is the reason that the occupation probability $n(\mathbf{r})$ looks like the

Fermi-Dirac function:

$$n(\mathbf{r}) = \frac{1}{\exp((-\mu + \Phi(\mathbf{r}))/\kappa T) + 1} \tag{3.4.4}$$

where μ is the chemical potential determined by the condition of conservation of the number of the solute atoms N_1:

$$\sum_{\mathbf{r}} n(\mathbf{r}) = \sum_{\mathbf{r}} \frac{1}{\exp((-\mu + \Phi(\mathbf{r}))/\kappa T) + 1} = N_1 \tag{3.4.5}$$

where $\Phi(\mathbf{r})$ is the resulting potential formed by all atoms on the site $\mathbf{r}$. The potential $\Phi(\mathbf{r})$ is a functional with respect to the function $n(\mathbf{r})$ and thus depends on the temperature and the composition. In the self-consistent field approximation (the mean-field approximation) $\Phi(\mathbf{r})$ is

$$\Phi(\mathbf{r}) = \sum_{\mathbf{r}} \tilde{V}(\mathbf{r}, \mathbf{r}')n(\mathbf{r}') \tag{3.4.6}$$

The approximation (3.4.6) does not take into account the change in the atomic distribution around the site $\mathbf{r}$ (the latter forms the potential $\Phi(\mathbf{r})$) due to a solute atom at the site $\mathbf{r}$. That is, in Eq. (3.4.6) the correlation effects are neglected.

Substituting Eq. (3.4.6) into Eq. (3.4.4), one obtains the equation for the function $n(\mathbf{r})$.

$$n(\mathbf{r}) = \frac{1}{\exp\left[\left(-\mu + \sum_{\mathbf{r}'} \tilde{V}(\mathbf{r}, \mathbf{r}')n(\mathbf{r}')\right) \Big/ (\kappa T)\right] + 1} \tag{3.4.7}$$

which is valid in the self-consistent field (mean-field) approximation and has been proposed by Khachaturyan (3–5). This approximation corresponds to the thermodynamic potential

$$F = U - TS - \mu \sum_{\mathbf{r}} n(\mathbf{r}) \tag{3.4.8}$$

where

$$U = \frac{1}{2} \sum_{\mathbf{r},\mathbf{r}'} \tilde{V}(\mathbf{r}, \mathbf{r}')n(\mathbf{r})n(\mathbf{r}') \tag{3.4.9}$$

is the internal energy and

$$S = -\kappa \sum_{\mathbf{r}} [n(\mathbf{r}) \ln n(\mathbf{r}) + (1 - n(\mathbf{r})) \ln (1 - n(\mathbf{r}))] \tag{3.4.10}$$

is the entropy; in other words, Eq. (3.4.7) can be obtained by setting the first variation of the free energy (3.4.8) with respect to $n(\mathbf{r})$ equal to zero.

The Helmholtz free energy is

$$F = U - TS \tag{3.4.11}$$

where U and S are given by Eqs. (3.4.9) and (3.4.10), respectively. It has been shown in (21) that the self-consistent field Eq. (3.4.7) is asymptotically correct in the limiting cases of high and low temperatures but not correct in the range in the vicinity of the second-order transformation temperature. The width of this range becomes smaller, the larger the radius of interatomic interactions (22).

The nonlinear difference equation (3.4.7) has several solutions. Each provides an extreme of free energy (3.4.8). The solution $n(\mathbf{r})$ which provides the absolute minimum of the free energy (3.4.8) at given T and c describes the atomic structure of the stable superlattice. Changes of T and c lead to the transformation from one of these solutions to another. This transformation describes an order-disorder or a phase transformation between two superlattices. Therefore we conclude that Eq. (3.4.7) contains all the information concerning the order-disorder phase transformations; a complete analysis of the order-disorder transitions requires the solution of the nonlinear Eq. (3.4.7).

Eq. (3.4.7) takes on a slightly more complicated form if solute atoms can occupy several crystallographically nonequivalent sites in the host lattice, such as octahedral and tetrahedral interstices in an fcc interstitial solution. Additional phase transformation paths associated with the transitions of interstitial atoms between different types of interstices then arise. The same is the case with Frenkel defects (interstitial atom-vacancy in the host lattice) produced by a passage of an atom from a regular lattice site to an interstice. The interatomic interaction potential chosen may be such that Frenkel defects are formed spontaneously at a certain temperature. This situation is of considerable interest, for instance, for the analysis of superionic transformations in some ionic crystals where defects are charge carriers.

When the occupation sites are nonequivalent crystallographically and cannot be brought into coincidence with each other by the host lattice symmetry operations, some local energy term that takes different values at nonequivalent positions should be included into the local potential $\Phi(\mathbf{r})$. In other words, Eq. (3.4.6) should be replaced with

$$\Phi(\mathbf{r}) = e_0(\mathbf{r}) + \sum_{\mathbf{r}'} \tilde{V}(\mathbf{r}, \mathbf{r}')n(\mathbf{r}') \qquad (3.4.12)$$

where $e_0(\mathbf{r})$ is the local energy associated with occupation sites, $\mathbf{r}$, in a dilute solution ($n(\mathbf{r}) \to 0$), that is, the quantity describing the energy gap between the nonequivalent sites.

Substituting (3.4.12) into (3.4.4) gives

$$n(\mathbf{r}) = \left\{ \exp\left[\frac{-\mu + e_0(\mathbf{r}) + \sum_{\mathbf{r}'} \tilde{V}(\mathbf{r}, \mathbf{r}')n(\mathbf{r}')}{\kappa T} \right] + 1 \right\}^{-1} \qquad (3.4.13)$$

If all occupation sites are equivalent crystallographically, i.e. can be brought into coincidence with each other by symmetry operations, e_0 is a constant and may therefore be included into the chemical potential μ in Eq. (3.4.13). Its effect is then reduced to renormalization of the chemical potential:

$$\mu \to \mu + e_0.$$

3.5. STATIC CONCENTRATION WAVES AND DIFFRACTION

One of the advantages of the static concentration wave description of ordered phase structures is that it reveals very simple and direct relations between the concentration waves and diffraction patterns. Every static concentration wave generates a corresponding superlattice Laue reflection. This will be demonstrated for a binary substitutional solution, for example.

The mean scattering amplitude (1.3.32) over all the atomic configurations yields the coherent scattering amplitude

$$\langle Y(\mathbf{q})\rangle = \bar{f}\sum_{\mathbf{r}} e^{-i\mathbf{q}\mathbf{r}} + (f_B - f_A)\sum_{\mathbf{r}} (n(\mathbf{r}) - c)e^{-i\mathbf{q}\mathbf{r}} \tag{3.5.1}$$

where the symbol $\langle\cdots\rangle$ implies thermodynamic averaging and

$$\langle c(\mathbf{r})\rangle = n(\mathbf{r}) \tag{3.5.2}$$

The latter equality follows from the definition of stochastic variables $c(\mathbf{r})$ (1.3.17) and occupation probabilities $n(\mathbf{r})$.

Consider a simple case of crystal lattice sites forming a Bravais lattice. The first term in Eq. (3.5.1)

$$Y_0(\mathbf{q}) = \bar{f}\sum_{\mathbf{r}} e^{-i\mathbf{q}\mathbf{r}} \tag{3.5.3}$$

then describes the coherent amplitude responsible for the disordered crystal reflections (fundamental reflections) since $Y_0(\mathbf{q})$ is independent of atomic configurations and

$$\sum_{\mathbf{r}} e^{-i\mathbf{q}\mathbf{r}} = \begin{cases} N & \text{if } \mathbf{q} = 2\pi\mathbf{H} \\ 0 & \text{otherwise} \end{cases} \tag{3.5.4}$$

that is, the sum (3.5.4) takes nonzero values at fundamental reciprocal lattice points only.

The second term in (3.5.1)

$$\langle Y_{\text{heter}}(\mathbf{q})\rangle = (f_B - f_A)\sum_{\mathbf{r}} (n(\mathbf{r}) - c)e^{-i\mathbf{q}\mathbf{r}} \tag{3.5.5}$$

describes the coherent scattering amplitude corresponding to the superlattice reflections that vanish in the disordered state, when $n(\mathbf{r}) = c$.

By definition, the average scattering amplitude does not include fluctuations of the occupation numbers and therefore cannot describe short-range order effects.

There is a very simple relationship between static concentration waves generating the atomic distribution in the ordered phase and its x-ray pattern. This may be found by substituting (2.1.1) into (3.5.5). The result is

$$\langle Y_{\text{heter}}(\mathbf{q})\rangle = (f_B - f_A)\sum_j Q(\mathbf{k}_j)\sum_{\mathbf{r}} e^{-i(\mathbf{q}-\mathbf{k}_j)\mathbf{r}} \tag{3.5.6}$$

One can see that, since

$$\sum_{\mathbf{r}} e^{-i(\mathbf{q}-\mathbf{k}_j)\mathbf{r}} = \begin{cases} N & \text{if } \mathbf{q}-\mathbf{k}_j = 2\pi\mathbf{H} \\ 0 & \text{otherwise} \end{cases}$$

the scattering amplitude is nonzero only at reciprocal lattice points

$$\mathbf{q} = 2\pi\mathbf{H} + \mathbf{k}_j$$

The latter expression describes the superlattice reflections (the corresponding expression for the fundamental lattice reflections is $\mathbf{q} = 2\pi\mathbf{H}$).

According to Eq. (3.5.6), the structure amplitude of a superlattice reflection (the scattering amplitude per one scattering crystal lattice site at $2\pi\mathbf{H} + \mathbf{k}_j$ reciprocal lattice point) is

$$F(2\pi\mathbf{H} + \mathbf{k}_j) = (f_B - f_A)Q(\mathbf{k}_j). \tag{3.5.7}$$

Eq. (3.5.7) reveals a very important fact: the amplitude of a superlattice reflection is always proportional to the amplitude of the corresponding static concentration wave $Q(\mathbf{k}_j)$ or, taking into account the definition (3.1.5), to the corresponding long-range order parameter η_s.

From this one can conclude that the *static concentration waves generating the structure of an ordered phase are completely determined by the x-ray diffraction pattern: the wave vector of every concentration wave is equal to the reciprocal lattice reflection which is within the first Brillouin zone. The amplitude of the concentration wave is proportional to the structure amplitude of this reflection.*

After this brief digression into the problem of diffraction patterns generated by solid solutions and their relations to concentration waves, we will return to the statistical thermodynamics of ordered alloys.

3.6. APPLICATION OF THE CONCENTRATION WAVE METHOD TO THE SOLUTION OF THE MEAN-FIELD EQUATION (SIMPLE LATTICE)

We will first consider a simple lattice. We define a "simple lattice" as one whose sites make up one Bravais lattice. The pairwise interatomic interaction energies depend in a special way on the coordinates of interacting atoms in this case:

$$\tilde{V}(\mathbf{r}, \mathbf{r}') = W(\mathbf{r} - \mathbf{r}') \tag{3.6.1}$$

This is due to invariance of pairwise potentials under translational operations $\mathbf{r} \to \mathbf{r} + \mathbf{T}$. Indeed

$$\tilde{V}(\mathbf{r} + \mathbf{T}, \mathbf{r}' + \mathbf{T}) \equiv \tilde{V}(\mathbf{r}, \mathbf{r}')$$

if $V(\mathbf{r}, \mathbf{r}') = W(\mathbf{r} - \mathbf{r}')$.

Substituting (3.6.1) into (3.4.7) and (3.4.9) yields

$$n(\mathbf{r}) = \frac{1}{\exp\left(\left(-\mu + \sum_{\mathbf{r}'} W(\mathbf{r} - \mathbf{r}')n(\mathbf{r}')\right)\Big/\kappa T\right) + 1} \tag{3.6.2}$$

and

$$U = \frac{1}{2}\sum_{\mathbf{r},\mathbf{r}'} W(\mathbf{r} - \mathbf{r}')n(\mathbf{r})n(\mathbf{r}') \tag{3.6.3a}$$

$$F=\frac{1}{2}\sum_{\mathbf{r},\mathbf{r}'} W(\mathbf{r}-\mathbf{r}')n(\mathbf{r})n(\mathbf{r}')$$
$$+T\kappa\sum_{\mathbf{r}}[n(\mathbf{r})\ln n(\mathbf{r})+(1-n(\mathbf{r}))\ln(1-n(\mathbf{r}))] \tag{3.6.3b}$$

In the arbitrary-range interaction model, Eq. (3.6.2) can only be solved by the method of static concentration waves. Following (4), we may in fact substitute the concentration wave representation of $n(\mathbf{r})$ (3.1.3) into Eq. (3.6.2) to obtain the set of transcendental equations:

$$c+\sum_{s=1}^{t-1}\eta_s E_s(\mathbf{r})=\left[\exp\left(\frac{-\mu+V(0)c+\sum_{s=1}^{t-1}V(\mathbf{k}_s)\eta_s E_s(\mathbf{r})}{\kappa T}\right)+1\right]^{-1} \tag{3.6.4}$$

where

$$V(\mathbf{k})=\sum_{\mathbf{r}}W(\mathbf{r})e^{i\mathbf{kr}} \tag{3.6.5}$$

is the Fourier transform of the interatomic potential, $t-1$ the number of the nonzero stars $\{\mathbf{k}_s\}$ generating the structure of the superlattice, and η_s the long-range order parameters. We have also used the identities

$$V(\mathbf{k}_1)=V(\mathbf{k}_2)=\cdots=V(\mathbf{k}_j)=\cdots=V(\mathbf{k}_s)$$

following from symmetry of the disordered solution lattice and the definition of the star $\{\mathbf{k}_s\}$.

Substituting the numerical values for all the crystal lattice site coordinates, $\mathbf{r}=(x, y, z)$, into (3.6.4) gives a set of transcendental equations involving t unknowns: the atomic fraction of the solute, c, and long-range order parameters

$$\eta_1, \eta_2, \ldots, \eta_{t-1}.$$

The number of the equations is equal to the number of different values assumed by the function $n(\mathbf{r})$ (left-hand side of Eq. (3.6.4)) on all crystal lattice sites at arbitrary c and η_s values. The set may be solved if that number is equal to the number of the unknowns, t. Hence the number of different occupation probability values over the lattice should also be equal to t. This drives us to the important conclusion, later referred to as

Criterion 1. *The function $n(\mathbf{r})$ in the form* (3.1.3) *may be a solution of Eq. (3.6.2) if the total number of long-range order parameters, $\eta_1, \eta_2, \ldots, \eta_{t-1}$, is less by one than the total number of different values taken by that function on all crystal lattice sites.*

Since, according to (3.1.1), the number of sublattices arising from ordering is equal to the number of different $n(\mathbf{r})$ values, Criterion 1 may be reformulated as follows: the number of sublattices formed in ordering is always larger by one than the total number of long-range order parameters describing the relevant superstructure.

Criterion 1 also enables one to find the coefficients $\gamma_s(j_s)$ in Eq. (3.1.3) since fulfilment of the criterion depends on the choice of their values.

Criterion 1 may be used to determine the ordered phase structure, provided the stars generating the superstructure are known. Two examples given below show how this may be done.

Consider the case of an fcc binary substitutional solution where ordering is generated by the star $\{100\}$. This star comprises three vectors

$$\{\mathbf{k}_0\}=\{100\}=(2\pi\mathbf{a}_1^*,\ 2\pi\mathbf{a}_2^*,\ 2\pi\mathbf{a}_3^*) \tag{3.6.6}$$

corresponding to the (100), (010), and (001) points of the reciprocal lattice of the fcc crystal. According to Eq. (3.1.3), the general expression for the superstructure generated by the star (3.6.6) is

$$n(\mathbf{r})=c+\eta(\gamma_1 e^{i2\pi\mathbf{a}_1^*\mathbf{r}}+\gamma_2 e^{i2\pi\mathbf{a}_2^*\mathbf{r}}+\gamma_3 e^{i2\pi\mathbf{a}_3^*\mathbf{r}}) \tag{3.6.7}$$

where

$$\mathbf{r}=x\mathbf{a}_1+y\mathbf{a}_2+z\mathbf{a}_3 \tag{3.6.8}$$

is the crystal lattice site vector, $\mathbf{a}_1$, $\mathbf{a}_2$, $\mathbf{a}_3$ are the elementary translations of the fcc lattice along the [100], [010], [001] directions, respectively, x, y, z are the fcc lattice site coordinates (integers or half-integers whose sum is an integer), and $\mathbf{a}_1^*$, $\mathbf{a}_2^*$, $\mathbf{a}_3^*$ are the reciprocal lattice elementary translations related to $\mathbf{a}_1$, $\mathbf{a}_2$, $\mathbf{a}_3$ by (1.2.6).

It follows from definition (1.2.6) that

$$\mathbf{a}_p\mathbf{a}_q^*=\delta_{pq} \qquad p,\ q=1,\ 2,\ 3 \tag{3.6.9}$$

Eqs. (3.6.8) and (3.6.9) yield

$$\begin{aligned}\mathbf{r}\mathbf{a}_1^*&=x\\ \mathbf{r}\mathbf{a}_2^*&=y\\ \mathbf{r}\mathbf{a}_3^*&=z\end{aligned} \tag{3.6.10}$$

Substituting (3.6.10) into (3.6.7) gives

$$n(\mathbf{r})=c+\eta(\gamma_1 e^{i2\pi x}+\gamma_2 e^{i2\pi y}+\gamma_3 e^{i2\pi z}) \tag{3.6.11}$$

With the permitted x, y, z values the function $n(x, y, z)$ only assumes four different values:

$$\begin{aligned}n_1&=c+\eta(\gamma_1+\gamma_2+\gamma_3)\\ n_2&=c+\eta(-\gamma_1+\gamma_2+\gamma_3)\\ n_3&=c+\eta(\gamma_1-\gamma_2+\gamma_3)\\ n_4&=c+\eta(\gamma_1+\gamma_2-\gamma_3)\end{aligned} \tag{3.6.12}$$

Criterion 1 may then be used to find the coefficients $\gamma_1, \gamma_2, \gamma_3$. Since the function $n(\mathbf{r})$ depends on the single long-range order parameter η, Criterion 1 requires that the function $n(\mathbf{r})$ assume only two different values on all the fcc lattice sites.

This is possible if

$$n_2 = n_3 = n_4 \neq n_1 \tag{3.6.13a}$$

$$n_1 = n_2, \quad n_2 = n_3 \tag{3.6.13b}$$

if certain values from the set n_1, n_2, n_3, n_4 defined by (3.6.12) are equal to each other. All the other variants of Eqs. (3.6.13) do not describe new superstructures but give only various orientations of the superstructures defined by Eqs. (3.6.13). We shall show below how the coefficients γ_1, γ_2, and γ_3 may be found from Eqs. (3.6.13).

Substituting (3.6.12) into (3.6.13a) yields immediately

$$\gamma_1 = \gamma_2 = \gamma_3 = \gamma \tag{3.6.14}$$

for one of the two superstructures. Substituting (3.6.14) into (3.6.11) gives

$$n(x, y, z) = c + \eta\gamma(e^{i2\pi x} + e^{i2\pi y} + e^{i2\pi z}) \tag{3.6.15}$$

The distribution (3.6.15) gives two different occupation probabilities,

$$n_1 = c + 3\eta\gamma$$

on all vertices of fcc cubes and

$$n_2 = c - \eta\gamma$$

on all centers of a cube face. That is, the distribution (3.6.15) describes the $L1_2$ (Cu_3Au) superstructure characterized by the probability n_1 of finding Au at a cube vertex and the probability n_2 of finding it at a center of a cube face (see Fig. 14). Eq. (3.6.15) is also applicable to the completely ordered state when the probabilities are either zero or unity:

$$\begin{aligned} c + 3\eta\gamma &= 1 \\ c - \eta\gamma &= 0 \end{aligned} \tag{3.6.16}$$

As is known, long-range order parameters may be normalized in a number of ways to remove ambiguity inherent in their definition (see Section 3.1). Eq. (3.1.6) provides one of the possible normalization conditions. One other condition used more frequently is the requirement that the long-range order parameter value in a fully ordered system be equal to unity. To meet this requirement, we must substitute $\eta = 1$ into the set (3.6.16):

$$\begin{aligned} c + 3\gamma &= 1 \\ c - \gamma &= 0 \end{aligned} \tag{3.6.17}$$

The solution of (3.6.17) is

$$c = c_{st} = \tfrac{1}{4} \qquad \gamma = \tfrac{1}{4} \tag{3.6.18}$$

where $c_{st} = \frac{1}{4}$ is the stoichiometric composition corresponding to the chemical formula A_3B. When substituted into (3.6.15), (3.6.18) gives

$$n(x, y, z) = \tfrac{1}{4} + \tfrac{1}{4}\eta(e^{i2\pi x} + e^{i2\pi y} + e^{i2\pi z}) \tag{3.6.19}$$

The second superstructure generated by the star $\{100\}$ may be derived from equality (3.6.13b). Substituting (3.6.12) into it gives

$$\gamma_1 = \gamma_2 = 0$$

thus reducing (3.6.11) to

$$n(x, y, z) = c + \eta\gamma_3 e^{i2\pi z} \tag{3.6.20}$$

The distribution (3.6.20) gives two different occupation probabilities

$$n_1 = c + \eta\gamma_3 \quad \text{and} \quad n_2 = c - \eta\gamma_3$$

on alternating (001) fcc crystal lattice planes. It describes the superstructure $L1_0$ (CuAuI) (see Fig. 19). The analysis analogous to that performed for Cu_3Au shows that the fully ordered state is characterized by

$$c = c_{st} = \tfrac{1}{2} \quad \text{and} \quad \gamma_3 = \tfrac{1}{2}$$

The conclusion thus follows that star $\{100\}$ can only generate two superstructures (Cu_3Au and CuAuI).

The two examples given illustrate the applicability of Criterion 1. We shall now proceed with a general discussion of the solutions to Eq. (3.6.2).

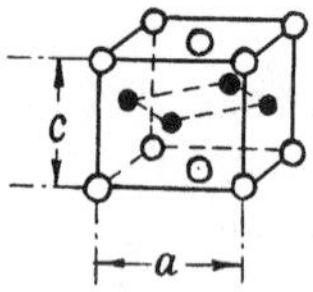

Figure 19. The structure of the fcc-based substitutional ordered structure CuAuI ($L1_0$).

It is easy to see that the symmetry of $n(\mathbf{r})$ in Eq. (3.1.3) depends solely on the functions $E_s(\mathbf{r})$. Changes of the long-range order parameters that do not enter the expressions for $E_s(\mathbf{r})$ have no affect on $n(\mathbf{r})$ symmetry and thus on ordered phase symmetry. Conversely, variations in the coefficients $\gamma_s(j_s)$ entering into $E_s(\mathbf{r})$ result in changes in the symmetry of $n(\mathbf{r})$ and thus in changes in the ordered phase symmetry, since the relationships between the coefficients $\gamma_s(j_s)$ determine the symmetry of the corresponding function $E_s(\mathbf{r})$ with respect to reflections and rotations. This is a fundamental reason why it is not necessary to solve the self-consistent field equation to determine the coefficients $\gamma_s(j_s)$.

Since the symmetry of the function $n(\mathbf{r})$ given by (3.1.3) is determined by the constants $\gamma_s(j_3)$ and does not depend on the long-range order parameters, η_s, any function $\bar{\phi}(\mathbf{r})$ invariant under the superstructure symmetry operations can be obtained from $n(\mathbf{r})$ by mere replacement of the quantities c and η_s in (3.1.3) with some other quantities $\tilde{c}$ and $\tilde{\eta}_s$:

$$\bar{\phi}(\mathbf{r}) = \tilde{c} + \sum_{s=1}^{t-1} \tilde{\eta}_s E_s(\mathbf{r}) \tag{3.6.21}$$

In particular, the function $n^2(\mathbf{r})$ should look like the function $n(\mathbf{r})$. The latter can be directly verified, for example, for the function (3.6.15) describing the superlattice Cu_3Au:

$$n^2(\mathbf{r})=[c^2+\tfrac{3}{16}\eta^2]+\frac{c\eta}{8}(e^{i2\pi\mathbf{a}_1^*\mathbf{r}}+e^{i2\pi\mathbf{a}_2^*\mathbf{r}}+e^{i2\pi\mathbf{a}_3^*\mathbf{r}})$$

$$+\tfrac{1}{8}\eta^2(e^{i2\pi(\mathbf{a}_1^*+\mathbf{a}_2^*)\mathbf{r}}+e^{i2\pi(\mathbf{a}_1^*+\mathbf{a}_3^*)\mathbf{r}}+e^{i2\pi(\mathbf{a}_2^*+\mathbf{a}_3^*)\mathbf{r}}) \qquad (3.6.22)$$

Since $e^{i2\pi(\mathbf{a}_1^*+\mathbf{a}_2^*)\mathbf{r}}\equiv e^{i2\pi\mathbf{a}_3^*\mathbf{r}}$, $e^{i2\pi(\mathbf{a}_1^*+\mathbf{a}_3^*)\mathbf{r}}\equiv e^{i2\pi\mathbf{a}_2^*\mathbf{r}}$, $e^{i2\pi(\mathbf{a}_2^*+\mathbf{a}_3^*)\mathbf{r}}\equiv e^{i2\pi\mathbf{a}_1^*\mathbf{r}}$ (it follows from the identity $e^{i(\mathbf{k}+2\pi\mathbf{H})\mathbf{r}}\equiv e^{i\mathbf{k}\mathbf{r}}$, where $\mathbf{H}$ is any fundamental reciprocal lattice vector), one can combine the second and the third terms of Eq. (3.6.22) to obtain

$$n^2(\mathbf{r})=\tilde{c}+\tilde{\eta}(e^{i2\pi\mathbf{a}_1^*\mathbf{r}}+e^{i2\pi\mathbf{a}_2^*\mathbf{r}}+e^{i2\pi\mathbf{a}_3^*\mathbf{r}}) \qquad (3.6.23)$$

where

$$\tilde{c}=c^2+\tfrac{3}{16}\eta^2, \qquad \tilde{\eta}=\tfrac{1}{8}(c\eta+\eta^2) \qquad (3.6.24)$$

Returning to the problem of constructing the structure of an ordered phase, it is necessary to keep in mind that in addition to Criterion 1 there is one more criterion that holds as a rule. We shall refer to it as

Criterion 2. *The sum of any two ordering wave vectors* $\mathbf{k}_{j_s}$ *(which may be equal to each other) entering into* $n(\mathbf{r})$ *must be equal to either a fundamental reciprocal lattice vector* $2\pi\mathbf{H}$ *or the third ordering vector also entering into* $n(\mathbf{r})$.

The meaning of Criterion 2 becomes obvious if one notes that the vectors $\mathbf{k}_{j_s}$ are Bravais translations of the reciprocal lattice of the ordered phase. Criterion 2 reflects the trivial fact that a sum of any two Bravais translations is a Bravais translation as well.

Criterion 2 may prove to be very useful in constructing the function $n(\mathbf{r})$ that satisfies Criterion 1. Contrary to Criterion 1, which is a necessary condition for a distribution $n(\mathbf{r})$, Criterion 2 does not need to be satisfied by any superlattice. There are a few examples of superlattices that do not satisfy Criterion 2; all these exceptions are connected with ordering wave vectors that are a quarternary of a fundamental reciprocal lattice vector $2\pi\mathbf{H}$. They give rise to special extinction rules for superstructure amplitudes.

It should be remembered that there are situations where the construction of $n(\mathbf{r})$, generated by wave vectors that provide the absolute minimum of $V(\mathbf{k})$, proves to be impossible. That means that employment of these wave vectors in $n(\mathbf{r})$ does not satisfy Criteria 1 and 2. In such cases additional stars are involved for these criteria to be satisfied.

3.7. HOW TO FIND THE ATOMIC ARRANGEMENT OF THE MOST STABLE SUPERSTRUCTURE

The first step toward solving the problem was made in Section 3.6. It was shown that, if stars generating the most stable superstructure in an alloy are known, the atomic arrangement of the superstructure may be determined using Criterion 1. The stars may be found from x-ray, neutron, and electron diffraction patterns. In Section 3.5 star vectors were said to be just segments linking the superlattice and fundamental reflections nearest to each other. Criterion 1 makes it possible to obviate the traditional time-consuming procedure based on treatment of Laue reflection intensities. In other words, an ordered phase structure can be found solely from the information contained in the diffraction pattern geometry. The efficiency of this approach has been demonstrated by Usikov and Khachaturyan (23) for Ta-O alloys. The structures of the interstitial ordered phases in Ta-O alloys were determined from the geometries of selected areas in the diffraction patterns. The same approach was utilized in the analysis of ordering in ceramics by Pokrovskii who derived the structures of the compounds from x-ray Debye-Sherrer patterns (24, 25).

For the sake of consistency it would be desirable to have a method for determining stars of stable superstructures without recourse to diffraction experiments. The stars are derived from interatomic interaction energies which, as we know, determine all structural and thermodynamic characteristics of phase transformations.

We shall begin with a rather trivial point that the highest-temperature superstructure is at the same time the stablest one. The problem of finding the stablest high-temperature superstructure is then reduced to the problem of finding the highest order-disorder transformation temperature. An unambiguous solution of mean-field Eqs. (3.6.2) and (3.6.4) at high temperatures is

$$n(\mathbf{r})=c \tag{3.7.1}$$

This distribution describes a disordered state with all crystal lattice sites characterized by the same occupation probabilities. Solution (3.7.1) implies that the disordered state is stable with respect to any heterogeneous perturbation of distribution (3.7.1). Let us consider a perturbed distribution

$$n(\mathbf{r})=c+\delta n(\mathbf{r}) \tag{3.7.2}$$

where $\delta n(\mathbf{r})$ is an infinitesimal heterogeneous perturbation. At elevated temperatures, when the disordered state is stable, the perturbation $\delta n(\mathbf{r})$ is zero. Temperature decrease results in the loss of stability of the disordered state at some critical point that manifests itself by the appearance of stable nonzero perturbation $\delta n(\mathbf{r})$. It thus follows that, to find the critical point that is the lower-temperature limit of existence of the disordered phase, we must investigate mean-field Eq. (3.6.2) and determine the temperature at which the nonzero $\delta n(\mathbf{r})$ term appears. The critical temperatures are the bifurcation points of the nonlinear integral-type Eq. (3.6.2).

Substituting (3.7.2) into (3.6.2) followed by expansion of (3.6.2) in powers of $\delta n(\mathbf{r})$ and truncation of higher than first-order terms gives the linear equation

$$\delta n(\mathbf{r})= -\frac{c(1-c)}{\kappa T}\sum_{\mathbf{r}'} W(\mathbf{r}-\mathbf{r}')\delta n(\mathbf{r}') \tag{3.7.3}$$

with respect to $\delta n(\mathbf{r})$.

The Fourier transform of Eq. (3.7.3) is

$$\delta\tilde{n}(\mathbf{k}) = -\frac{c(1-c)}{\kappa T} V(\mathbf{k})\delta\tilde{n}(\mathbf{k}) \tag{3.7.4}$$

where

$$\delta\tilde{n}(\mathbf{k}) = \sum_{\mathbf{r}} \delta n(\mathbf{r}) e^{-i\mathbf{k}\mathbf{r}} \tag{3.7.5}$$

$$V(\mathbf{k}) = \sum_{\mathbf{r}} W(\mathbf{r}) e^{-i\mathbf{k}\mathbf{r}} \tag{3.7.6}$$

The linear equation (3.7.4) has a nontrivial solution if

$$-\frac{c(1-c)}{\kappa T} V(\mathbf{k}) = 1 \tag{3.7.7}$$

that is, if

$$T = T(\mathbf{k}) = -\frac{c(1-c)}{\kappa} V(\mathbf{k}) \tag{3.7.8}$$

At $V(\mathbf{k}) < 0$ the function $T = T(\mathbf{k})$ (3.7.8) defines bifurcation points of Eq. (3.6.2). Each temperature $T(\mathbf{k})$ corresponds to a point of loss of stability of the random distribution $n(\mathbf{r}) = c$ with respect to a lattice generated by the star $\{\mathbf{k}\}$ of the ordering wave vector $\mathbf{k}$. The function $V(\mathbf{k})$ takes within the first Brillouin zone of the lattice of the disordered phase both negative and positive values since

$$\sum_{\mathbf{k}} V(\mathbf{k}) \equiv 0 \tag{3.7.9}$$

where the summation is carried out over all N quasi-continuum points $\mathbf{k}$ within the first Brillouin zone allowed by the periodic boundary conditions. The latter result is due to the natural definition $W(0) = 0$ which excludes from Hamiltonian (3.4.1) the physically meaningless situation where two interacting solute atoms occupy the same site. In fact, the back Fourier transformation

$$\frac{1}{N} \sum_{\mathbf{k}} V(\mathbf{k}) e^{i\mathbf{k}\mathbf{r}} = W(\mathbf{r})$$

gives

$$\frac{1}{N} \sum_{\mathbf{k}} V(\mathbf{k}) = W(0) = 0 \tag{3.7.10}$$

Therefore $T(\mathbf{k})$ of (3.7.8) lies within the range

$$0 < T(\mathbf{k}) < T_0$$

where T_0 is maximum of the function $T(\mathbf{k})$:

$$T_0 = \max\left[-\frac{c(1-c)}{\kappa} V(\mathbf{k})\right] = -\frac{c(1-c)}{\kappa} \min V(\mathbf{k}) \tag{3.7.11}$$

Here min $V(\mathbf{k})$ is the absolute minimum of $V(\mathbf{k})$. According to (3.7.11), the quantity T_0 is the maximum temperature of the bifurcation of the nonlinear equation (3.6.2), that is, at $T > T_0$ Eq. (3.6.2) has the only solution $n(\mathbf{r}) = c$ corresponding to the disordered state. At $T = T_0$ a new solution arises that describes the ordered phase. The structure of the ordered phase is generated by the star $\{\mathbf{k}_s\}$ whose wave vectors $\mathbf{k}_{j_s}$ fall at the absolute minimum of $V(\mathbf{k})$.*

Thus the temperature T_0, which is determined by Eq. (3.7.11), is the temperature of the order-disorder phase transformation or decomposition.†

Therefore the structure of the highest-temperature (and consequently the most stable) superstructure is generated by the star $\{\mathbf{k}_s\}$ *whose ordering wave vectors* $\mathbf{k}_{j_s}$ *provide the absolute minimum of* $V(\mathbf{k})$.

This statement gives the simplest way of determining the star generating the structure of the most stable superstructure at a given choice of all long-range interaction energies and hence solves the problem stated in the title of this section. With the knowledge of the star generating the stablest ordered phase, one may determine its structure using Criterion 1 derived in Section 3.6.

There are two kinds of $V(\mathbf{k})$ minima. The minima of the first kind may correspond to arbitrary points of the reciprocal space. Their positions depend on the type of interatomic interaction potential $W(\mathbf{r} - \mathbf{r}')$, and they shift when the latter changes. The minima of the second kind are realized in the "singular" points of the reciprocal space of the disordered phase and do not shift under small variations of the interatomic potential since their positions depend solely on the symmetry of the disordered phase.

A simple criterion formulated below may be applied to determine the "singular" special points in the reciprocal space. To do so, it is necessary to find the points $\mathbf{k}$ in which the identity

$$\frac{\partial V(\mathbf{k})}{\partial \mathbf{k}} \equiv 0 \tag{3.7.12}$$

holds due to symmetry only.

Let a $\mathbf{k}$-vector group $\hat{\mathbf{G}}(\mathbf{k})$ of the disordered phase contain the symmetry operations $\hat{\mathbf{g}}_1(\mathbf{k}), \hat{\mathbf{g}}_2(\mathbf{k}), \ldots, \hat{\mathbf{g}}_n(\mathbf{k})$ (these operations bring the crystal lattice of the disordered phase into coincidence with itself and do not change the wave vector).‡ Then the group $\hat{\mathbf{G}}(\mathbf{k})$ is a subgroup of the space group of the disordered phase.

On one hand, since both the crystal lattice and the vector $\mathbf{k}$ are invariant under the $\hat{\mathbf{g}}_1(\mathbf{k}), \hat{\mathbf{g}}_2(\mathbf{k}), \ldots, \hat{\mathbf{g}}_n(\mathbf{k})$ operations, the gradient $\partial b(\mathbf{k})/\partial \mathbf{k}$ defined on the crystal lattice must also be invariant. On the other hand, if two or more symmetry operations of the point group $\hat{\mathbf{G}}(\mathbf{k})$ intersect at one point, these symmetry

*If the absolute minimum of $V(\mathbf{k})$ falls at the point $\mathbf{k} = 0$, a decomposition reaction occurs.

†Strictly speaking, the bifurcation point T_0 of Eq. (3.6.2) is the point of the absolute loss of stability of a disordered phase. The temperature T_0 coincides with the second-order transformation temperature but is slightly less than the first-order transformation temperature.

‡It should be borne in mind that changes of the vector $\mathbf{k}$ by any fundamental reciprocal vector $2\pi\mathbf{H}$ are regarded as identity transformations.

operations must necessarily change the direction of any finite-magnitude vector including the vector $\partial V(\mathbf{k})/\partial \mathbf{k}$. This contradiction can be solved if $\partial V(\mathbf{k})/\partial \mathbf{k} \equiv 0$.

Therefore the special point $\mathbf{k}$ providing the identity (3.7.12) must satisfy the following requirement: *the point group of the special point* $\mathbf{k}$ *in the reciprocal space of the disordered phase contains two or more symmetry operations intersecting in the same point.*

The later condition defines the positions of all special points in the reciprocal lattice of the disordered phase. They have been found by Lifshitz (19) by means of the space-group representation analysis. The more simple analysis considered here differs from that given by Lifshitz but leads to the same results.

The requirement that the ends of all ordering vectors $\mathbf{k}_{0j}$ generating the structure of the ordered phase fall on the special points was also first formulated by Lifshitz (19), and it is referred to as Lifshitz's criterion.

As has been shown in (26) the Lifshitz criterion has a very general meaning: it is the necessary condition of ordered phase stability with respect to antiphase domain formation. The criterion is valid for first-order transformations as well as for second-order ones, and its validity does not depend on the temperature, which can be well below that of the phase transformation temperature.

3.8. EXAMPLES OF SOLUTION OF MEAN-FIELD EQUATIONS FOR OCCUPATION PROBABILITIES

As shown in Section 3.7, the occupation probabilities describing a superstructure may be written in terms of concentration waves (3.1.3) if the star generating the superstructure is known. The concentration wave representation (3.1.3) makes it possible to reduce nonlinear mean-field Eq. (3.6.2) to a set of transcendental equations with respect to the composition and long-range order parameters.

This approach enables one to solve the mean-field equation even with arbitrary long-range order interactions, and moreover the mathematical treatment of ordering in the long-range interaction case proves to be not more complex than with the usual case of nearest-neighbor interactions. Actually, substitution of the concentration-wave representation (3.1.3) into (3.6.2) leads to the set of transcendental equations:

$$c + \sum_{s=1}^{t-1} \eta_s E_s(\mathbf{r}) = \left[\exp\left(\frac{-\mu + V(0)c + \sum_{s=1}^{t-1} V(\mathbf{k}_s)\eta_s E_s(\mathbf{r})}{\kappa T} \right) + 1 \right]^{-1} \qquad (3.8.1)$$

which is much simpler than the original nonlinear difference Eq. (3.6.2). Here

$$V(\mathbf{k}) = \sum_{\mathbf{r}} W(\mathbf{r}) e^{i\mathbf{k}\mathbf{r}} \qquad (3.8.2)$$

is the Fourier transform of interatomic pairwise energies $W(\mathbf{r})$. The identities

$$V(\mathbf{k}_{1_s})=V(\mathbf{k}_{2_s})=\cdots=V(\mathbf{k}_{j_s})=\cdots=V(\mathbf{k}_s)$$

have also been used in the derivation of (3.8.1). They hold for vectors $\mathbf{k}_{j_s}$ making up one star and follow from the symmetry of the disordered phase.

For example, according to (3.8.2), the function $V(\mathbf{k})$ for a bcc lattice is

$$\begin{aligned}V(\mathbf{k})=8W_1 \cos \pi h\cdot\cos \pi k\cdot\cos \pi l+2W_2(\cos 2\pi h+\cos 2\pi k+\cos 2\pi l)\\ +4W_3(\cos 2\pi h\cdot\cos 2\pi k+\cos 2\pi h\cdot\cos 2\pi l+\cos 2\pi k\cdot\cos 2\pi l)\\ +8W_4[\cos 3\pi h\cdot\cos \pi k\cdot\cos \pi l+\cos 3\pi k\cdot\cos \pi h\cdot\cos \pi l\\ +\cos 3\pi l\cdot\cos \pi h\cdot\cos \pi k]+8W_5 \cos 2\pi h\cdot\cos 2\pi k\cdot\cos 2\pi l \quad (3.8.3)\end{aligned}$$

where

$$\mathbf{k}=2\pi(h\mathbf{a}_1^*+k\mathbf{a}_2^*+l\mathbf{a}_3^*)$$

$W_1, W_2, W_3, \ldots$ are the interchange energies for the 1st, 2nd, 3rd, ... coordination shells, respectively.

A set of transcendental equations with respect to the unknowns c; $\eta_1, \eta_2, \ldots, \eta_{t-1}$ arises if one substitutes the allowed numerical values of coordinates $\mathbf{r}(x, y, z)$ of the crystal lattice sites into Eq. (3.8.1). The number of these equations naturally is equal to the number of values that are taken by the function $n(\mathbf{r})$ on all the different crystal lattice sites. If the structure of an ordered phase is defined correctly, the number of values taken by $n(\mathbf{r})$ is just equal to the number of unknowns, c; $\eta_1, \eta_2, \ldots, \eta_{t-1}$, in other words, is equal to t.

It follows from Eq. (3.8.1) that even in the arbitrary long-range interaction model, the thermodynamics of a solid solution is determined completely by $t-1$ energy parameters $V(\mathbf{k}_s)$ and one parameter $V(0)$, where t is the number of sublattices formed in ordering.

Two examples will be given to show how the nonlinear mean-field Eq. (3.6.2) may be reduced to a set of transcendental equations.

3.8.1. Layer-Ordered Phases B2(βCuZn), $L1_0$(CuAuI), $L1_1$(CuPt)

The $L1_0$(CuAuI) and $L1_1$(CuPt) fcc-based layer superstructures and the B2(β-CuZn) bcc-based superstructure are described by the same simplest equation:

$$n(\mathbf{r})=c+\tfrac{1}{2}\eta e^{i\mathbf{k}_0\mathbf{r}} \qquad (3.8.4)$$

where $\mathbf{k}_0$ is the wave vector equal to half the fundamental reciprocal lattice vector. All the three superstructures may therefore be analyzed together. The corresponding transcendental equation describing the long-range order parameter η should be equally applicable to them.

The wave vector $\mathbf{k}_0$ in (3.8.4) is

$$\mathbf{k}_0=\begin{cases}2\pi(\mathbf{a}_1^*+\mathbf{a}_2^*+\mathbf{a}_3^*) & \text{for } \beta\text{-CuZn superstructure, point (111)}\\ \pi(\mathbf{a}_1^*+\mathbf{a}_2^*+\mathbf{a}_3^*) & \text{for CuPt superstructure, point } (\tfrac{1}{2}\tfrac{1}{2}\tfrac{1}{2})\\ 2\pi\mathbf{a}_3^* & \text{for CuAuI superstructure, point (001)}\end{cases} \qquad (3.8.5)$$

Substituting Eq. (3.8.4) into (3.8.1) gives

$$c+\tfrac{1}{2}\eta e^{i\mathbf{k}_0\mathbf{r}}=\frac{1}{\exp\left[\left(-\mu+V(0)c+V(\mathbf{k}_0)\tfrac{1}{2}\eta e^{i\mathbf{k}_0\mathbf{r}}\right)/\kappa T\right]+1} \tag{3.8.6}$$

[see Eq. (3.8.1)].

The function $e^{i\mathbf{k}_0\mathbf{r}}$ with the ordering vector $\mathbf{k}_0$ defined by Eq. (3.8.5) takes either $+1$ or -1 on the variety of all fcc (CuAuI and CuPt) and bcc (β-CuZn) lattice sites. Taking into account this, one may reduce Eq. (3.8.6) to the set of two equations

$$\begin{aligned} c+\tfrac{1}{2}\eta&=\left[\exp\left(\frac{-\mu+V(0)c+V(\mathbf{k}_0)\frac{1}{2}\eta}{\kappa T}\right)+1\right]^{-1}\\ c-\tfrac{1}{2}\eta&=\left[\exp\left(\frac{-\mu+V(0)c-V(\mathbf{k}_0)\frac{1}{2}\eta}{\kappa T}\right)+1\right]^{-1}\end{aligned} \tag{3.8.7}$$

By excluding the term $-\mu+V(0)c$ from the set (3.8.7), one can reduce this set to one transcendental equation

$$\ln\frac{(c-\frac{1}{2}\eta)(1-c-\frac{1}{2}\eta)}{(c+\frac{1}{2}\eta)(1-c+\frac{1}{2}\eta)}=\frac{V(\mathbf{k}_0)}{\kappa T}\eta \tag{3.8.8}$$

The same Eq. (3.8.8) describes the dependence of the long-range order parameter η on the temperature T and the composition c for the different superstructures, β-CuZn, CuAuI, and CuPt. The only difference is in the dependence of the energy parameter $V(\mathbf{k}_0)$ on the interchange energies. Thus according to Eq. (3.8.2) and (3.8.3)

$$V(\mathbf{k}_0)=V(2\pi(\mathbf{a}_1^*+\mathbf{a}_2^*+\mathbf{a}_3^*))=-8W_1+6W_2+12W_3-24W_4+8W_5+\cdots \tag{3.8.9}$$

for the bcc-based superstructure B2(β-CuZn). For the fcc-based superstructure CuPt

$$V(\mathbf{k}_0)=V(\pi(\mathbf{a}_1^*+\mathbf{a}_2^*+\mathbf{a}_3^*))=-6W_2+12W_4+\cdots \tag{3.8.10}$$

and

$$V(\mathbf{k}_0)=V(2\pi\mathbf{a}_3^*)=-4W_1+6W_2-8W_3+12W_4-16W_5+\cdots \tag{3.8.11}$$

for the bcc-based superstructure CuAuI.

It is worthwhile mentioning that Eq. (3.8.8) is a generalization of the Gorsky-Bragg-Williams equations for the superstructures in question. The latter may be obtained as the limiting case if the nearest-neighbor interaction approximation to $V(\mathbf{k})$ is adopted ($W_1\neq 0$, $W_2=W_3=\cdots=0$).

3.8.2. Fe_3Al (DO_3) BCC-Based Superlattice

It follows from Eq. (3.1.25) that

$$n(\mathbf{r})=c+\eta_1E_1(\mathbf{r})+\eta_2E_2(\mathbf{r}) \tag{3.8.12}$$

where according to Eq. (3.1.26)

$$E_1(\mathbf{r}) = -\tfrac{1}{4}e^{i2\pi(\mathbf{a}_1^*+\mathbf{a}_2^*+\mathbf{a}_3^*)\mathbf{r}} = -\tfrac{1}{4}e^{i2\pi(x+y+z)}$$
$$E_2(\mathbf{r}) = \tfrac{1}{2}\sin\pi(\mathbf{a}_1^*+\mathbf{a}_2^*+\mathbf{a}_3^*)\mathbf{r} = \tfrac{1}{2}\sin\pi(x+y+z) \tag{3.8.13}$$

and x, y, z are the coordinates of the bcc lattice sites. The bcc lattice sites are subdivided into three sublattices at the ordering. One can readily see that

$E_1(\mathbf{r}) = -\frac{1}{4}$, $E_2(\mathbf{r}) = 0$ if $\mathbf{r}$ ranges over the sites of the first sublattice

$E_1(\mathbf{r}) = \frac{1}{4}$, $E_2(\mathbf{r}) = \frac{1}{2}$ if $\mathbf{r}$ ranges over the sites of the second sublattice

$E_1(\mathbf{r}) = \frac{1}{4}$, $E_2(\mathbf{r}) = -\frac{1}{2}$ if $\mathbf{r}$ ranges over the sites of the third sublattice (3.8.14)

Making use of Eq. (3.8.14) in Eq. (3.8.1), one has the set of three transcendental equations:

$$c - \frac{1}{4}\eta_1 = \left[\exp\left(\frac{-\mu + cV(0) - V(\mathbf{k}_0)\frac{1}{4}\eta_1}{\kappa T}\right) + 1\right]^{-1}$$
$$c + \frac{1}{4}\eta_1 + \frac{1}{2}\eta_2 = \left[\exp\left(\frac{-\mu + cV(0) + V(\mathbf{k}_0)\frac{1}{4}\eta_1 + V(\mathbf{k}_1)\frac{1}{2}\eta_2}{\kappa T}\right) + 1\right]^{-1}$$
$$c + \frac{1}{4}\eta_1 - \frac{1}{2}\eta_2 = \left[\exp\left(\frac{-\mu + cV(0) + V(\mathbf{k}_0)\frac{1}{4}\eta_1 + V(\mathbf{k}_1)(-\frac{1}{2})\eta_2}{\kappa T}\right) + 1\right]^{-1} \tag{3.8.15}$$

where

$$\mathbf{k}_0 = 2\pi(\mathbf{a}_1^* + \mathbf{a}_2^* + \mathbf{a}_3^*)\ [\text{point } (111)]$$
$$\mathbf{k}_1 = \pi(\mathbf{a}_1^* + \mathbf{a}_2^* + \mathbf{a}_3^*)\ [\text{point } (\tfrac{1}{2}\tfrac{1}{2}\tfrac{1}{2})] \tag{3.8.16}$$

and where according to (3.8.3)

$$V(\mathbf{k}_0) = -8W_1 + 6W_2 + 12W_3 - 24W_4 + 8W_5 + \cdots$$
$$V(\mathbf{k}_1) = -6W_2 + 12W_3 - 8W_5 + \cdots$$

Combining the first and the second equations as well as the first and the third equations in (3.8.15), one obtains, respectively,

$$\ln\frac{(1-c-\frac{1}{4}\eta_1-\frac{1}{2}\eta_2)(c-\frac{1}{4}\eta_1)}{(c+\frac{1}{4}\eta_1+\frac{1}{2}\eta_2)(1-c+\frac{1}{4}\eta_1)} = \frac{1}{2\kappa T}V(\mathbf{k}_0)\eta_1 + \frac{1}{2\kappa T}V(\mathbf{k}_1)\eta_2$$
$$\ln\frac{(1-c-\frac{1}{4}\eta_1+\frac{1}{2}\eta_2)(c-\frac{1}{4}\eta_1)}{(c+\frac{1}{4}\eta_1-\frac{1}{2}\eta_2)(1-c+\frac{1}{4}\eta_1)} = \frac{1}{2\kappa T}V(\mathbf{k}_0)\eta_1 - \frac{1}{2\kappa T}V(\mathbf{k}_1)\eta_2 \tag{3.8.17}$$

The summation of Eqs. (3.8.17) gives

$$\ln\frac{[(1-c-\frac{1}{4}\eta_1)^2 - \frac{1}{4}\eta_2^2](c-\frac{1}{4}\eta_1)^2}{[(c+\frac{1}{4}\eta_1)^2 - \frac{1}{4}\eta_2^2](1-c+\frac{1}{4}\eta_1)^2} = \frac{V(\mathbf{k}_0)}{\kappa T}\eta_1 \tag{3.8.18}$$

The solution of Eq. (3.8.18) with respect to η_2 is

$$\frac{\eta_2}{2}=-\left[\frac{(c-\frac{1}{4}\eta_1)^2(1-c-\frac{1}{4}\eta_1)^2\exp(-(\eta_1/\kappa T)V(\mathbf{k}_0))-(c+\frac{1}{4}\eta_1)^2(1-c+\frac{1}{4}\eta_1)^2}{(c-\frac{1}{4}\eta_1)^2\exp(-(\eta_1/\kappa T)V(\mathbf{k}_0))-(1-c+\frac{1}{4}\eta_1)^2}\right]^{\frac{1}{2}} \tag{3.8.19}$$

The substitution of Eq. (3.8.19) into, for example, the first Eq. in (3.8.17) reduces the set of Eqs. (3.8.17) to the single transcendental equation with respect to η_1.

3.9. SYMMETRY OF SUPERLATTICE POINTS IN THE RECIPROCAL LATTICE AND STABILITY OF ORDERED PHASES: STABLE STRUCTURES IN FCC AND BCC SOLUTIONS

It was shown in Section 3.7 that high-temperature ordered phases are generated by static concentration waves whose wave vectors fall at the absolute minima of the Fourier transform of interchange energies, $V(\mathbf{k})$. Variations in interchange energies affect the function $V(\mathbf{k})$ and may shift its absolute minima, thus causing changes in the wave vectors.

Not all of the absolute minima, however, are affected by variations in interchange energies. The function $V(\mathbf{k})$ always has extrema at high-symmetry (special) points of the reciprocal space meeting the Lifshitz criterion, that is the points characterized by point-group symmetries including intersecting symmetry elements.

The necessary condition for an extremum of $V(\mathbf{k})$ is automatically fulfilled in the Lifshitz points owing to the disordered crystal symmetry rather than to particular interchange energy values. For that reason variations in the interchange energy values do not affect these extrema positions and hence cannot shift the absolute minimum if that minimum coincides with one of the Lifshitz points.

If, on the contrary, the absolute minimum of $V(\mathbf{k})$ falls at a point other than the Lifshitz point, even small variations in the interchange energies will affect the minimum position and the superlattice vectors generating the ordered phase structure. Variations in thermodynamic parameters (composition, temperature, or pressure) should result in variations in interchange energies and thus shift the absolute minima of $V(\mathbf{k})$ that do not coincide with Lifshitz points.

We arrive at the conclusion that a *high-temperature ordered phase whose superlattice points are not Lifshitz points of the disordered phase is unstable and can only exist in narrow temperature and composition ranges.*

Conversely, *a high-temperature ordered phase whose superlattice points are all Lifshitz points of the disordered phase reciprocal lattice can be stable within a comparatively wide field of the T-c diagram.*

These important conclusions have in fact wider applicability and are valid for any ordered phase besides the high-temperature ordered phases.

It has been shown by Khachaturyan that the structure of any ordered phase is stable with respect to the formation of antiphase domains if all ordered phase superlattice points coincide with the minima of $V(\mathbf{k})$ (26).

There is, however, a situation where stability of an ordered phase cannot be achieved though all its superlattice points correspond to $V(\mathbf{k})$ minima. In this situation at least one superlattice point is not a Lifshitz special point. Any variation in the external thermodynamic parameters displaces the minimum of $V(\mathbf{k})$ from the superlattice point, thus violating the above-mentioned condition for stability of an ordered phase with respect to formation of antiphase domains.

We therefore can make a general statement that a superstructure is always unstable with respect to the formation of antiphase domains if at least one superlattice point does not coincide with a Lifshitz point of the disordered phase reciprocal lattice.

The rigorous proof of the latter statement may be found in the works (26, 9). It is based on the well-known fact that an ordered phase superlattice vector can shift upon the introduction of antiphase domains whose out-of-step vector changes the phase of the concentration wave associated with the superlattice vector under consideration.* This shift stabilizes the ordered phase, bringing the superlattice point into coincidence with the new minimum of $V(\mathbf{k})$.

Since the number of nonequivalent special points (special stars) in the reciprocal lattice of a disordered phase is limited, it is actually possible to find "special-point" superstructures stable with respect to antiphase domain formation for each disordered phase. To do this, the techniques described in Sections 3.6 and 3.7 may be applied. Determining the special-point ordered phases, which can be made without additional assumptions about the alloy's dynamics, is important because these phases may exist within comparatively wide temperature, composition, and pressure ranges; ordered phases whose superlattice points fall at general points of the reciprocal lattice should be highly unstable under composition variations, as is the case with the Ni_4Mo superstructure and the Ti_nO_{2n-1} structure series which are not special-point superstructures in the fcc and bcc lattices, respectively. These structures have narrow concentration stability ranges compared with the special-point superstructures Ni_3Mo and TiO_2.

We will consider in this section derivation examples of all fcc- and bcc-based substitutional and interstitial special-point superstructures.

It should be mentioned that substitutional and interstitial solutions are isomorphous to each other because an interstitial phase can be regarded as a substitutional solution with interstitial atoms and their vacancies as constituents. The vacancies play the role of the second component in a substitutional solution. Together with their vacancies interstitial atoms occupy all interstices of the host atom lattice. For instance, a solution based on an fcc host lattice with solute atoms in octahedral interstices may be regarded as a binary substitutional solution in an fcc lattice comprising octahedral interstices. An isomor-

*In terms of the diffraction theory this corresponds to the statement that antiphase domains as well as stacking faults shift positions of reflections.

phous interstitial superstructure may thus be put in correspondence with any special-point substitutional superstructure.

3.9.1. "Special-Point" Fcc-Based Superstructures

There are three "special" stars meeting the Lifshitz criterion in an fcc lattice:

$$(100),\ (010),\ (001) \tag{3.9.1a}$$

$$(\tfrac{1}{2}\tfrac{1}{2}\tfrac{1}{2}),\ (\tfrac{\bar{1}}{2}\tfrac{1}{2}\tfrac{1}{2}),\ (\tfrac{1}{2}\tfrac{\bar{1}}{2}\tfrac{1}{2}),\ (\tfrac{1}{2}\tfrac{1}{2}\tfrac{\bar{1}}{2}) \tag{3.9.1b}$$

$$(\tfrac{1}{2}10),\ (10\tfrac{1}{2}),\ (0\tfrac{1}{2}1),\ (\tfrac{1}{2}\bar{1}0),\ (\bar{1}0\tfrac{1}{2}),\ (0\tfrac{1}{2}\bar{1}) \tag{3.9.1c}$$

Eqs. (3.9.1) give the coordinates of the special points in the usual basis $2\pi\mathbf{a}_1^*$, $2\pi\mathbf{a}_2^*$, and $2\pi\mathbf{a}_3^*$ where $2\mathbf{a}_1^*$, $2\mathbf{a}_2^*$, and $2\mathbf{a}_3^*$ are the reciprocal lattice translations in the [100], [010], and [001] directions, respectively.

According to (3.1.3) and (3.9.1), the general form of the probability $n(\mathbf{r})$ of the solute atom distribution in a "special" fcc-based substitutional or interstitial superlattice is determined by the equation

$$\begin{aligned} n(\mathbf{r}) = c &+ \eta_1[\gamma_1(1)e^{i2\pi\mathbf{a}_1^*\mathbf{r}} + \gamma_1(2)e^{i2\pi\mathbf{a}_2^*\mathbf{r}} + \gamma_1(3)e^{i2\pi\mathbf{a}_3^*\mathbf{r}}] \\ &+ \eta_2[\gamma_2(1)e^{i\pi(\mathbf{a}_1^*+\mathbf{a}_2^*+\mathbf{a}_3^*)\mathbf{r}} + \gamma_2(2)e^{i\pi(-\mathbf{a}_1^*+\mathbf{a}_2^*+\mathbf{a}_3^*)\mathbf{r}} \\ &+ \gamma_2(3)e^{i\pi(\mathbf{a}_1^*-\mathbf{a}_2^*+\mathbf{a}_3^*)\mathbf{r}} + \gamma_2(4)e^{i\pi(\mathbf{a}_1^*+\mathbf{a}_2^*-\mathbf{a}_3^*)\mathbf{r}}] \\ &+ \tfrac{1}{2}\eta_3[\gamma_3(1)e^{i2\pi(\frac{1}{2}\mathbf{a}_1^*+\mathbf{a}_2^*)\mathbf{r}} \\ &+ \gamma_3(2)e^{i2\pi(\mathbf{a}_1^*+\frac{1}{2}\mathbf{a}_3^*)\mathbf{r}} + \gamma_3(3)e^{i2\pi(\frac{1}{2}\mathbf{a}_2^*+\mathbf{a}_3^*)\mathbf{r}} + \text{compl. conj.}] \end{aligned} \tag{3.9.2}$$

where $\mathbf{r}$ is given by Eq. (3.1.11).

Substituting (3.1.11) into (3.9.2) gives

$$\begin{aligned} n(\mathbf{r}) = n(x, y, z) = c &+ \eta_1[\gamma_1(1)e^{i2\pi x} + \gamma_1(2)e^{i2\pi y} + \gamma_1(3)e^{i2\pi z}] \\ &+ \eta_2[\gamma_2(1)e^{i\pi(x+y+z)} + \gamma_2(2)e^{i\pi(-x+y+z)} + \gamma_2(3)e^{i\pi(x-y+z)} + \gamma_2(4)e^{i\pi(x+y-z)}] \\ &+ \eta_3\tfrac{1}{2}[\gamma_3(1)e^{i\pi(x+2y)} + \gamma_3(2)e^{i\pi(2x+z)} + \gamma_3(3)e^{i\pi(y+2z)} + \text{compl. conj.}] \end{aligned} \tag{3.9.3}$$

Using Criterion 1 (see Section 3.6), one can easily obtain all possible functions $n(\mathbf{r})$ describing "special" fcc-based substitutional and isomorphous interstitial superlattices that are stable in the formation of antiphase domains.

If the superlattice is described by only one long-range order parameter η_s, one can find the corresponding function $n(\mathbf{r})$ and hence the structure of the ordered phase by the choice of the coefficients $\gamma_s(j_s)$ giving rise to two values of $n(\mathbf{r})$ on the variety of all fcc lattice sites. The example of this procedure has been presented in Section 3.6.

1. Substitutional Cu_3Au-type (Fig. 14) and isomorphous interstitial Fe_4N-type (Fig. 20) superstructures can be obtained from Eq. (3.9.3) by setting

$$\eta_2 = \eta_3 = 0,\ \gamma_1(1) = \gamma_1(2) = \gamma_1(3) = \gamma_1$$

$$n(\mathbf{r}) = c + \eta_1\gamma_1(e^{i2\pi\mathbf{a}_1^*\mathbf{r}} + e^{i2\pi\mathbf{a}_2^*\mathbf{r}} + e^{i2\pi\mathbf{a}_3^*\mathbf{r}}) \tag{3.9.4}$$

2. Substitutional CuAuI-type and isomorphous interstitial $(Fe, Ni)_2N$-type

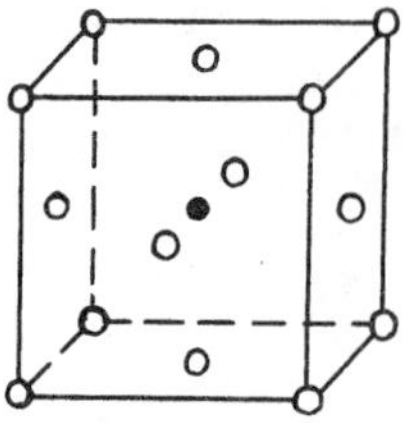

Figure 20. The structure of the fcc-based interstitial ordered phase Fe_4N. It is isomorphous to the substitutional ordered phase Cu_3Au depicted in Fig. 14: ○=Fe, ●=N.

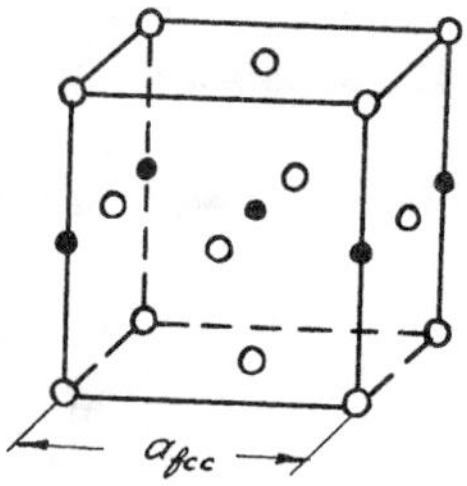

Figure 21. The structure of the fcc-based interstitial ordered phase $(Fe, Ni)_2N$. It is isomorphic to the substitutional ordered phase CuAuI depicted in Fig. 19: ○=(Fe, Ni), ●=an interstitial atom N.

(27) superstructures (Figs. 19 and 21, respectively) are obtained from (3.9.2) with

$$\gamma_1(1)=\gamma_1(2)=0,\ \eta_2=\eta_3=0$$

The result is

$$n(\mathbf{r})=c+\eta_1\gamma_1(3)e^{i2\pi\mathbf{a}_3^*\mathbf{r}} \tag{3.9.5}$$

3. CuPt-type superlattice (Fig. 22):

$$n(\mathbf{r})=c+\eta_2\gamma_2 e^{i\pi(\mathbf{a}_1^*+\mathbf{a}_2^*+\mathbf{a}_3^*)\mathbf{r}} \tag{3.9.6}$$

$(\eta_1=\eta_3=0,\ \gamma_2(2)=\gamma_2(3)=\gamma_2(4)=0)$.

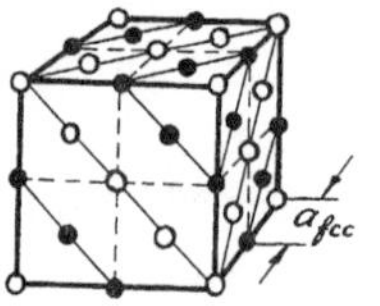

Figure 22. The structure of the fcc-based substitutional ordered phase CuPt ($L1_1$).

4. Substitutional superlattice *AB* (Fig. 23) and isomorphous interstitial superlattice (Fig. 24):

$$n(\mathbf{r})=c+\eta_2\gamma_2[-e^{i\pi(\mathbf{a}_1^*+\mathbf{a}_2^*+\mathbf{a}_3^*)\mathbf{r}}+e^{i\pi(-\mathbf{a}_1^*+\mathbf{a}_2^*+\mathbf{a}_3^*)\mathbf{r}} + e^{i\pi(\mathbf{a}_1^*-\mathbf{a}_2^*+\mathbf{a}_3^*)\mathbf{r}}+e^{i\pi(\mathbf{a}_1^*+\mathbf{a}_2^*-\mathbf{a}_3^*)\mathbf{r}}] \tag{3.9.7}$$

$(\eta_1=\eta_3=0,\ -\gamma_2(1)=\gamma_2(2)=\gamma_2(3)=\gamma_2(4)=\gamma_2)$.

5. Substitutional superlattice *AB* isomorphous interstitial superlattice Me_2X:

$$n(\mathbf{r})=c+\eta_3\gamma_3[\cos 2\pi(\mathbf{a}_1^*+\tfrac{1}{2}\mathbf{a}_3^*)\mathbf{r}+\sin 2\pi(\mathbf{a}_1^*+\tfrac{1}{2}\mathbf{a}_3^*)\mathbf{r}] \tag{3.9.8}$$

$(\eta_1=\eta_2=0,\ \gamma_3(1)=\gamma_3(3)=0,\ \gamma_3(2)=\gamma_3-i\gamma_3)$.

If the superlattice is described by two long-range order parameters, one can find the function $n(\mathbf{r})$ according to Criterion 1 by the choice of the coefficients $\gamma_s(j_s)$ providing three values of $n(\mathbf{r})$ on all fcc lattice sites.

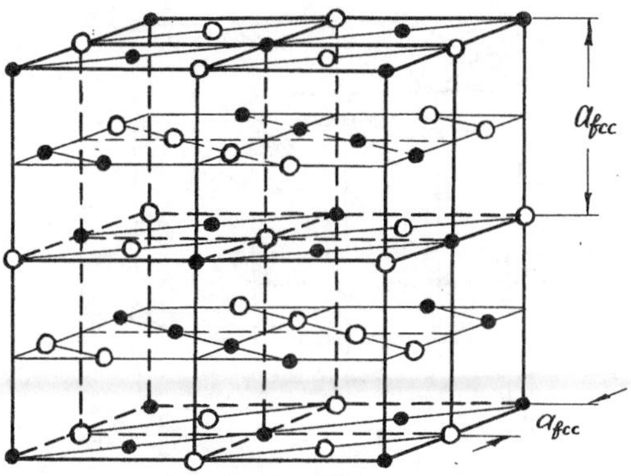

Figure 23. The structure of the fcc-based substitutional ordered phase AB generated by the star $\{\frac{1}{2}\frac{1}{2}\frac{1}{2}\}$: ○ = A, ● = B.

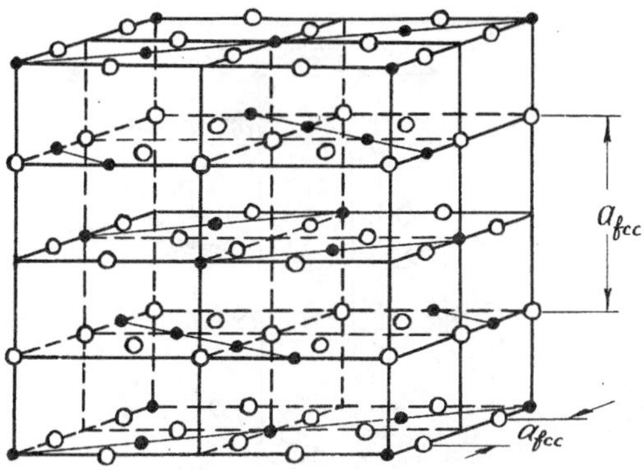

Figure 24. The structure of the fcc-based interstitial ordered phase A_2X generated by the star $\{\frac{1}{2}\frac{1}{2}\frac{1}{2}\}$. It is isomorphous to the substitutional ordered phase depicted in Fig. 23: ○-host atom A, •-interstitial atom X.

6. $CuPt_7$-type substitutional (Fig. 15) (the isomorphous interstitial superlattice Fe_8N is shown in Fig. 25):

$$n(\mathbf{r}) = c + \eta_1\gamma_1(e^{i2\pi\mathbf{a}_1^*\mathbf{r}} + e^{i2\pi\mathbf{a}_2^*\mathbf{r}} + e^{i2\pi\mathbf{a}_3^*\mathbf{r}}) + \eta_2\gamma_2[e^{i\pi(\mathbf{a}_1^*+\mathbf{a}_2^*+\mathbf{a}_3^*)\mathbf{r}} + e^{i\pi(-\mathbf{a}_1^*+\mathbf{a}_2^*+\mathbf{a}_3^*)\mathbf{r}} + e^{i\pi(\mathbf{a}_1^*-\mathbf{a}_2^*+\mathbf{a}_3^*)\mathbf{r}} + e^{i\pi(\mathbf{a}_1^*+\mathbf{a}_2^*-\mathbf{a}_3^*)\mathbf{r}}] \tag{3.9.9}$$

($\eta_3 = 0$, $\gamma_1(1) = \gamma_1(2) = \gamma_1(3) = \gamma_1$, $\gamma_2(1) = \gamma_2(2) = \gamma_2(3) = \gamma_2(4) = \gamma_2$).

7. $CuPt_3$-type substitutional (Fig. 16) and Me_4X interstitial (Fig. 26) superlattices:

$$n(\mathbf{r}) = c + \eta_1\gamma_1 e^{i2\pi\mathbf{a}_3^*\mathbf{r}} + \eta_2\gamma_2[e^{i\pi(\mathbf{a}_1^*+\mathbf{a}_2^*+\mathbf{a}_3^*)\mathbf{r}} + e^{i\pi(\mathbf{a}_1^*+\mathbf{a}_2^*-\mathbf{a}_3^*)\mathbf{r}}] \tag{3.9.10}$$

($\gamma_1(1) = \gamma_1(2) = 0$, $\gamma_1(3) = \gamma_1$, $\gamma_2(2) = \gamma_2(3) = 0$, $\gamma_2(1) = \gamma_2(4) = \gamma_2$, $\eta_3 = 0$).

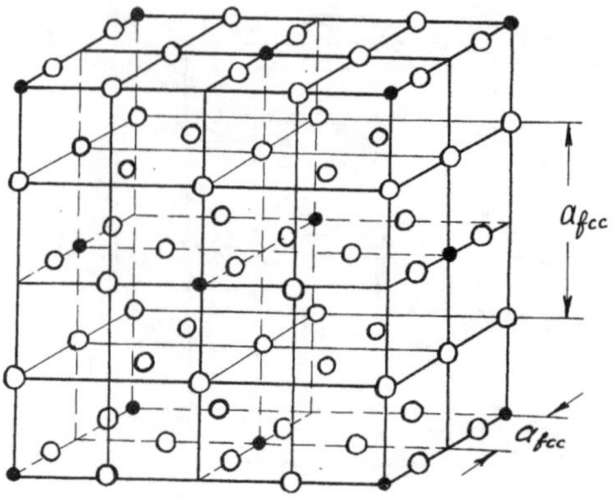

Figure 25. The structure of the fcc-based interstitial ordered phase A_8X. It is isomorphous to the substitutional ordered phase $CuPt_7$ depicted in Fig. 15: ○ = host atom A, ● = interstitial atom X.

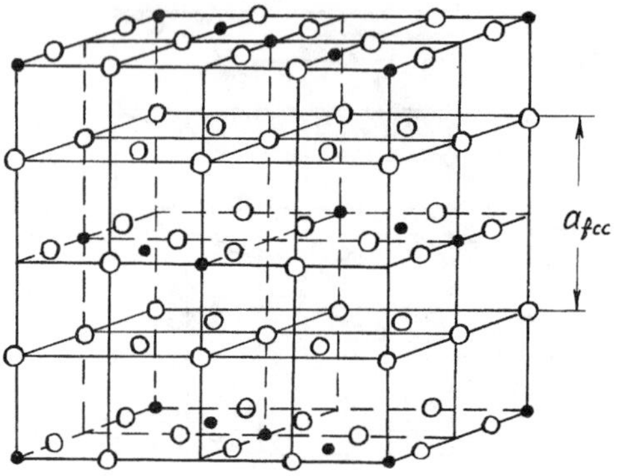

Figure 26. The structure of the fcc-based interstitial ordered phase A_4X. It is isomorphous to the substitutional structure $CuPt_3$ depicted in Fig. 16: ○ = host atom A, ● = interstitial atom X.

8. Al_3Ti-type substitutional and Ni_4NII-type interstitial superlattices (Figs. 27 and 28):

$$n(\mathbf{r})=c+\eta_1\gamma_1 e^{i2\pi\mathbf{a}_3^*\mathbf{r}}+\tfrac{1}{2}\eta_3\gamma_3\cos 2\pi(\mathbf{a}_1^*+\tfrac{1}{2}\mathbf{a}_3^*)\mathbf{r} \tag{3.9.11}$$

$(\gamma_1(1)=\gamma_1(2)=0,\ \gamma_1(3)=\gamma_1,\ \gamma_3(1)=\gamma_3(3)=0,\ \gamma_3(2)=\gamma_3,\ \eta_2=0).$

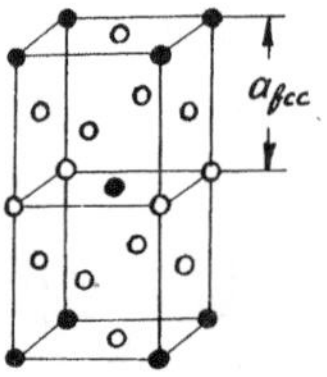

Figure 27. The structure of the fcc-base ordered phase Al_3Ti (DO_{22}). ● Ti ○ Al.

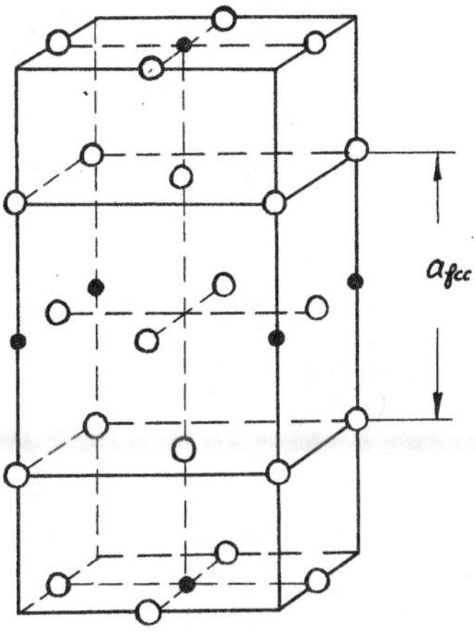

Figure 28. The structure of the fcc-based interstitial ordered phase Ni_4N11 (Nb_4N_3). It is isomorphous to the substitutional ordered phase Al_3Ti depicted in Fig. 27: ○ = Ni, ● = interstitial atom N.

3.9.2. "Special point" Bcc-Based Superlattices

Since the bcc lattice is a Bravais lattice, the distribution of the atoms in an ordered bcc substitutional solution is described by Eq. (3.1.3). The situation, however, is more complicated if one is interested in an interstitial bcc solution. In such a solution there are two species of crystallographically nonequivalent interstices—octahedral and tetrahedral interstices. They form three and six, respectively, interpenetrating bcc sublattices. Thus a bcc host lattice is a complex lattice for an interstitial solution with several interstices per unit cell. The above results, however, prove to be valid for each sublattice of interstices. For the sake of brevity, we consider only the case where the interstitial atoms occupy the bcc sublattice of O_z octahedral interstices displaced by $\frac{1}{2}\mathbf{a}_3$ with respect to the host sublattice.

There are "special" stars satisfying the Lifshitz criterion in a bcc lattice:

$$(111) \tag{3.9.12a}$$

$$(\tfrac{1}{2}\tfrac{1}{2}0),\ (0\tfrac{1}{2}\tfrac{1}{2}),\ (\tfrac{1}{2}0\tfrac{1}{2}),\ (\tfrac{1}{2}\tfrac{\bar{1}}{2}0),\ (0\tfrac{1}{2}\tfrac{\bar{1}}{2}),\ (\tfrac{1}{2}0\tfrac{\bar{1}}{2}) \tag{3.9.12b}$$

$$(\tfrac{1}{2}\tfrac{1}{2}\tfrac{1}{2}),\ (\tfrac{\bar{1}}{2}\tfrac{\bar{1}}{2}\tfrac{\bar{1}}{2}) \tag{3.9.12c}$$

Just as in the case of ordering in the fcc lattices, the coordinates of the star vectors are given in the usual basis of the reciprocal lattice, $2\pi\mathbf{a}_1^*$, $2\pi\mathbf{a}_2^*$, $2\pi\mathbf{a}_3^*$. Making use of Eq. (3.1.3), the ordering vectors (3.9.12), and Criterion 1 (see Section 3.6), one can construct only six superstructures, in the same way as was done for fcc-based ordered phases. These are

$$n(\mathbf{r}) = c + \eta_1\gamma_1 e^{i2\pi(\mathbf{a}_1^* + \mathbf{a}_2^* + \mathbf{a}_3^*)\mathbf{r}} \tag{3.9.13}$$

(β-CuZn-type, Fig. 12),

$$n(\mathbf{r}) = c + \eta_2\gamma_2 e^{i2\pi(\frac{1}{2}\mathbf{a}_1^* + \frac{1}{2}\mathbf{a}_2^*)\mathbf{r}} \tag{3.9.14}$$

(Ta_2O-layer interstitial superstructure, Fig. 29),

$$n(\mathbf{r}) = c + \eta_3\gamma_3[\cos \pi(\mathbf{a}_1^* + \mathbf{a}_2^* + \mathbf{a}_3^*)\mathbf{r} + \sin \pi(\mathbf{a}_1^* + \mathbf{a}_2^* + \mathbf{a}_3^*)\mathbf{r}] \tag{3.9.15}$$

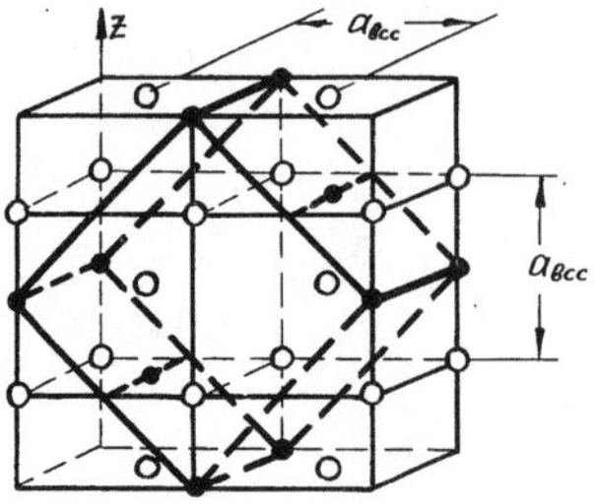

Figure 29. The structure of the bcc-based interstitial ordered phase Ta_2O: ○ = Ta, ● = an interstitial atom O.

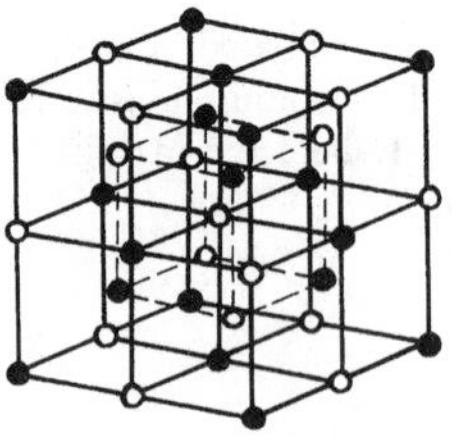

Figure 30. The structure of the bcc-based ordered phase NaTl (B32).

(NaTl-type superstructure, Fig. 30),

$$n(\mathbf{r}) = c + \eta_1 \gamma_1 e^{i2\pi(\mathbf{a}_1^* + \mathbf{a}_2^* + \mathbf{a}_3^*)\mathbf{r}} + \eta_2 \gamma_2 \left[e^{i\pi(\mathbf{a}_1^* + \mathbf{a}_2^*)\mathbf{r}} + e^{i\pi(\mathbf{a}_1^* - \mathbf{a}_2^*)\mathbf{r}} \right] \tag{3.9.16}$$

Ta_4O-type superstructure, Fig. 31),

$$n(\mathbf{r}) = c + \eta_1 \gamma_1 e^{i2\pi(\mathbf{a}_1^* + \mathbf{a}_2^* + \mathbf{a}_3^*)\mathbf{r}} + \eta_2 \gamma_2 \left[e^{i\pi(\mathbf{a}_1^* + \mathbf{a}_2^*)\mathbf{r}} + e^{i\pi(\mathbf{a}_1^* - \mathbf{a}_2^*)\mathbf{r}} + e^{i\pi(\mathbf{a}_1^* + \mathbf{a}_3^*)\mathbf{r}} + e^{i\pi(\mathbf{a}_1^* - \mathbf{a}_3^*)\mathbf{r}} + e^{i\pi(\mathbf{a}_2^* + \mathbf{a}_3^*)\mathbf{r}} + e^{i\pi(\mathbf{a}_2^* - \mathbf{a}_3^*)\mathbf{r}} \right] \tag{3.9.17}$$

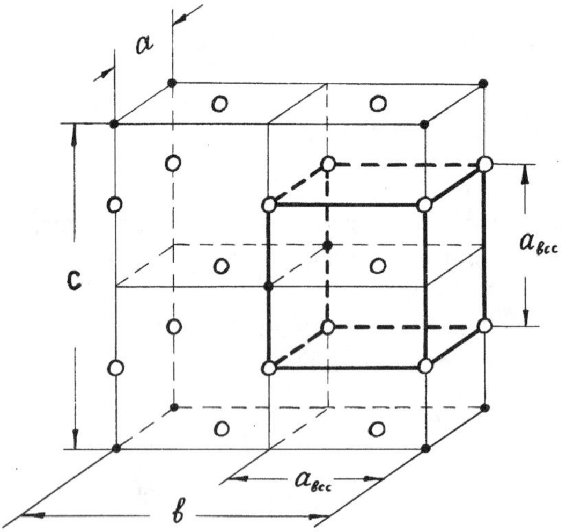

Figure 31. The structure of the bcc-based interstitial ordered phase Ta_4O: ○ = Ta host atom. ● = an interstitial atom O.

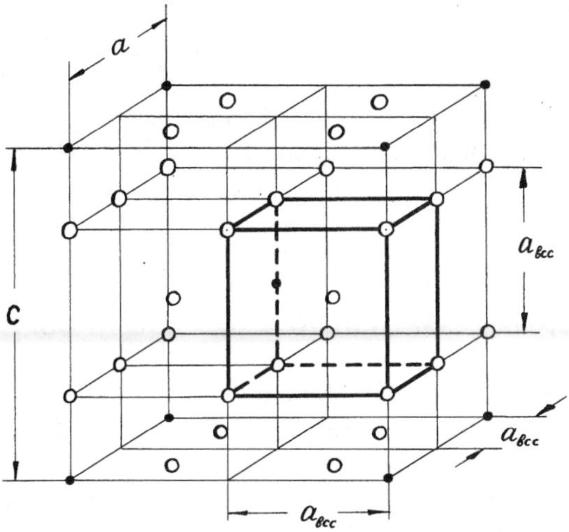

Figure 32. The structure of the bcc-based interstitial ordered phase Fe_8N: ○=Fe, •=N.

(Fe_8N bcc-based superstructure, Fig. 32),

$$n(\mathbf{r}) = c + \eta_1\gamma_1 e^{i2\pi(\mathbf{a}_1^* + \mathbf{a}_2^* + \mathbf{a}_3^*)\mathbf{r}} + \eta_3\gamma_3 \sin \pi(\mathbf{a}_1^* + \mathbf{a}_2^* + \mathbf{a}_3^*)\mathbf{r} \qquad (3.9.18)$$

(Fe_3Al-type substitutional superstructure, Fig. 13).

It should be noted that the distributions (3.9.14), (3.9.16), and (3.9.17) generated by the star (3.9.12b) cannot be attributed to any known substitutional superstructure. The reason of this was found by de Fontaine (28). De Fontaine has shown that the minimum of $V(\mathbf{k})$ at the special point $(\frac{1}{2}\frac{1}{2}0)$ in a substitutional bcc solution cannot be obtained in a realistic model of the pairwise interchange energies $W(\mathbf{r})$. This means that ordered structures based on bcc special points $\{\frac{1}{2}\frac{1}{2}0\}$ are not expected to occur in real systems. De Fontaine's arguments, however, do not forbid the existence of interstitial superlattices Ta_2O, Ta_4O, Fe_8N, and so on, generated by the $\{\frac{1}{2}\frac{1}{2}0\}$ star.

To conclude this section we emphasize once more that the special-point superstructures in fcc and bcc solutions are not merely a group of possible superstructures including some of the infinite number of superstructures that may in principle be constructed. The special-point superstructures, as opposed to all the other ordered phases, possess higher stabilities, may exist in wider fields of equilibrium diagrams, and therefore are more frequently observed in real alloys. The superstructures constructed above confirm this statement. The list of the special-point superstructures derived in this section includes all the conventional ordered phases in fcc and bcc solutions cited in the reviews on the problem of ordering.

On the other hand, whenever "nonspecial-point" superstructures occur, their instability should be expected to manifest itself by the formation of a large number of antiphase domains. This may lead to formation of unstable intermediate structures related to the parent one by a sequence of antiphase domains, or incommensurable superstructures and long-period superstructures, and so forth. For instance, the nonspecial-point superstructure Ni_2Cr (star $\{\frac{1}{3}\frac{1}{3}0\}$) decomposes to a set of antiphase domains (29). The instability of Ni_4Mo alloys with respect to antiphase domains along (210) planes was shown in (30, 31) to result in the formation of an intermediate metastable structure.

3.10. STABILITY OF NONSTOICHIOMETRIC ORDERED PHASES: SECONDARY ORDERING AND DECOMPOSITION

Consider a nonstoichiometric high-temperature ordered phase cooled to the absolute zero temperature, 0° K. Since the composition of the ordered phase differs from the stoichiometric one, the alloy cannot attain the fully ordered state when the occupation probabilities are either zero or unity, i.e. the state characterized by the zero entropy (3.4.10) value.

Such a situation, however, contradicts the Nernst theorem according to which the entropy of an equilibrium phase at zero temperature must be zero. This leads to the conclusion that the nonstoichiometric ordered phase under consideration cannot be stable at low temperatures. Cooling of this phase should be accompanied by secondary phase transformations, either secondary ordering or decomposition. Both result in equilibrium phases that conform to the Nernst theorem. The nonstoichiometric parent phase undergoes a transformation to give either an ordered phase whose stoichiometry coincides with the parent phase composition or a mixture of two stoichiometric phases. Whichever, the equilibrium state entropy vanishes at 0° K. An example of a secondary-ordering reaction is the formation of the Fe_3Al (DO_3) phase having the composition $c=\frac{1}{4}$ (75 atomic percent Fe, 25 atomic percent Al) from the high-temperature ordered phase FeAl(B2). The initial alloy composition is nonstoichiometric for the B2 and stoichiometric for the DO_3 phase. Cooling of the Fe-25%Al alloy with the B2 structure results in the secondary ordering reaction which yields the DO_3 structure at 550° C (32, 33). This reaction is a special case of the more general phenomenon of stepwise ordering. Ordering of nonstoichiometric phases may involve a sequence of ordering reactions until the completely ordered state is attained (34).

Kubo and Wayman were the first to observe the secondary decomposition for a nonstoichiometric ordered cubic Cu-Zn superstructure (β-brass) having a B2 structure that decomposes to a mixture of two superstructures with the same structure (35). They have also suggested a theory of secondary decomposition based on the formalism of static concentration waves (36).

We shall now consider the problem of phase transformations and of secondary ordered phase structures formed in secondary ordering reactions. We shall proceed from the same equation, (3.6.2), and carry out the same treatment as in Section 3.7. As before, the secondary ordering temperature will be assumed to be the maximum bifurcation point of Eq. (3.6.2).

Bifurcation points may be found from consideration of the function

$$n(\mathbf{r}) = n_0(\mathbf{r}) + \delta n(\mathbf{r}) \tag{3.10.1}$$

where $n_0(\mathbf{r})$ describes the high-temperature ordered phase, $\delta n(\mathbf{r})$ is an infinitesimal heterogeneous perturbation having a lower symmetry than the function $n_0(\mathbf{r})$ has. Substitution of (3.10.1) into (3.6.2) followed by expansion of (3.6.2) into powers of $\delta n(\mathbf{r})$ and truncation of all the higher than first-order terms gives the variation, $\delta n(\mathbf{r})$, expressed in the form of the linear equation

$$\delta n(\mathbf{r}) = -\frac{1}{\kappa T} n_0(\mathbf{r})(1 - n_0(\mathbf{r})) \sum_{\mathbf{r}'} W(\mathbf{r} - \mathbf{r}')\delta n(\mathbf{r}') \tag{3.10.2}$$

Eq. (3.10.2) is reduced to (3.7.3) if $n_0(\mathbf{r}) = c$, if the high-temperature phase is a disordered solution.

Let us write $n_0(\mathbf{r})$ as a superposition of concentration waves (3.1.3)

$$n_0(\mathbf{r}) = c + \sum_s \eta_s E_s(\mathbf{r}) = c + \sum_{s, j_s} \eta_s \gamma_s(j_s) e^{i\mathbf{k}_{j_s}\mathbf{r}} \tag{3.10.3}$$

where the summation is over all the wave vectors generating the high-temperature superstructure.

It was mentioned at the end of Section 3.6 that the introduction of any function $\bar{\phi}(\mathbf{r})$ having the same symmetry as $n_0(\mathbf{r})$ may be done by merely replacing c and η_s in (3.10.3) with certain other values depending on the choice of the function, $\bar{\phi}(\mathbf{r})$. When applied to the function

$$\bar{\phi}(\mathbf{r}) = \frac{1}{n_0(\mathbf{r})(1 - n_0(\mathbf{r}))}$$

which, by definition, has the same symmetry as $n_0(\mathbf{r})$, this yields

$$\frac{1}{n_0(\mathbf{r})(1 - n_0(\mathbf{r}))} = \tilde{c} + \sum_{s, j_s} \tilde{\eta}_s \gamma_s(j_s) e^{i\mathbf{k}_{j_s}\mathbf{r}} \tag{3.10.4}$$

The $\tilde{c}$ and $\tilde{\eta}_s$ parameters in Eq. (3.10.4) can be expressed by actual long-range order parameters η_s and composition c by substituting (3.10.3) into the left-hand side of Eq. (3.10.4) and comparing the result with its right-hand side.

Eq. (3.10.2) can be rewritten in the form

$$\sum_{\mathbf{r}'} W(\mathbf{r} - \mathbf{r}')\delta n(\mathbf{r}') + \frac{\kappa T}{n_0(\mathbf{r})(1 - n_0(\mathbf{r}))} \delta n(\mathbf{r}) = 0 \tag{3.10.5}$$

Substituting (3.10.4) into (3.10.5) yields

$$\sum_{\mathbf{r}'} W(\mathbf{r} - \mathbf{r}')\delta n(\mathbf{r}') + \kappa T \left(\tilde{c} + \sum_{s, j_s} \tilde{\eta}_s \gamma_s(j_s) e^{i\mathbf{k}_{j_s}\mathbf{r}} \right) \delta n(\mathbf{r}) = 0 \tag{3.10.6}$$

The Fourier transform of (3.10.6), that is, multiplication by $e^{-i\mathbf{k}\mathbf{r}}$ and summation over all crystal lattice sites, gives

$$(V(\mathbf{k}) + \kappa T \tilde{c})\delta \tilde{n}(\mathbf{k}) + \kappa T \sum_{s, j_s} \tilde{\eta}_s \gamma_s(j_s) \delta \tilde{n}(\mathbf{k} - \mathbf{k}_{j_s}) = 0 \tag{3.10.7}$$

where

$$\delta\tilde{n}(\mathbf{k})=\sum_{\mathbf{r}}\delta n(\mathbf{r})e^{-i\mathbf{kr}} \qquad (3.10.7a)$$

$$V(\mathbf{k})=\sum_{\mathbf{r}}W(\mathbf{r})e^{-i\mathbf{kr}} \qquad (3.10.7b)$$

Shifts of the argument in (3.10.7)

$$\mathbf{k}\rightarrow\mathbf{k}-\mathbf{k}_{l_{s'}} \qquad (3.10.8)$$

where $\mathbf{k}_{l_{s'}}$ is one of the wave vectors entering (3.10.3) gives

$$(V(\mathbf{k}-\mathbf{k}_{l_{s'}})+\kappa T\tilde{c})\delta\tilde{n}(\mathbf{k}-\mathbf{k}_{l_{s'}})+\sum_{s,j_s}\kappa T\tilde{\eta}_s\gamma_s(j_s)\delta\tilde{n}(\mathbf{k}-\mathbf{k}_{j_s}-\mathbf{k}_{l_{s'}})=0 \qquad (3.10.9)$$

As is shown at the end of Section 3.6, the sum of any two vectors entering (3.10.3) is a vector also entering this equation. Therefore the vectors $\mathbf{k}_{j_s}+\mathbf{k}_{l_{s'}}$, which appear as arguments in the second term of (3.10.9) also enter Eq. (3.10.3):

$$\mathbf{k}_{j_s}+\mathbf{k}_{l_{s'}}=\mathbf{k}_{m_{s''}} \qquad (3.10.10)$$

Shifting each wave vector in (3.10.7) by (3.10.8), we obtain a set of homogeneous linear equations in $\delta n(\mathbf{k}-\mathbf{k}_{j_s})$. The number of those equations is equal to the number of wave vectors (including $\mathbf{k}=0$) generating the high-temperature superstructure. A set of homogeneous equations has a nonzero solution when its determinant is equal to zero. The latter condition is fulfilled at certain temperatures

$$T=T(\mathbf{k})$$

which are the bifurcation points of Eq. (3.6.2). The maximum bifurcation point is the critical temperature of the secondary phase transition:

$$T_0=\max T(\mathbf{k}) \qquad (3.10.11)$$

If $T(\mathbf{k})$ has its maximum at $\mathbf{k}=\mathbf{k}_0\neq 0$, a secondary ordering takes place. Otherwise (if the maximum occurs at $\mathbf{k}=\mathbf{k}_0=0$), the phase transition is a secondary decomposition.

We now turn to the theory by Kubo and Wayman (36) who discussed the secondary decomposition of a nonstoichiometric B2 superstructure (such as that of β-brass).

The B2 superstructure is formed in bcc solid solutions. It is described by Eq. (3.8.4)

$$n_0(\mathbf{r})=c+\tfrac{1}{2}\eta e^{i\mathbf{k}_0\mathbf{r}} \qquad (3.10.12)$$

where the superstructure vector

$$\mathbf{k}_0=2\pi(\mathbf{a}_1^*+\mathbf{a}_2^*+\mathbf{a}_3^*) \qquad (3.10.12a)$$

corresponds to the point (111) in the reciprocal lattice of the bcc crystal, $2\mathbf{a}_1^*$, $2\mathbf{a}_2^*$, $2\mathbf{a}_3^*$ are the elementary reciprocal lattice translations along the [100], [010], and [001] directions, respectively. Substituting (3.10.12) into (3.10.4) yields

$$\frac{1}{n_0(\mathbf{r})(1-n_0(\mathbf{r}))}=\tilde{c}+\tfrac{1}{2}\tilde{\eta}e^{i\mathbf{k}_0\mathbf{r}} \tag{3.10.13}$$

where

$$\tilde{c}=\frac{c(1-c)-(\frac{1}{2}\eta)^2}{[c^2-(\frac{1}{2}\eta)^2][(1-c)^2-(\frac{1}{2}\eta)^2]} \tag{3.10.14a}$$

$$\tfrac{1}{2}\tilde{\eta}=2\frac{(c-\frac{1}{2})\frac{1}{2}\eta}{[c^2-(\frac{1}{2}\eta)^2][(1-c)^2-(\frac{1}{2}\eta)^2]} \tag{3.10.14b}$$

The latter equations are obtained by substituting (3.10.12) into the left-hand side of Eq. (3.10.13), and comparing both sides of that equation at $e^{i\mathbf{k}_0\mathbf{r}}=1$ and $e^{i\mathbf{k}_0\mathbf{r}}=-1$ [clearly, $e^{i\mathbf{k}_0\mathbf{r}}$ is either zero or unity if $\mathbf{k}_0$ is given by (3.10.12a)]. With Eq. (3.10.12) for the function $n_0(\mathbf{r})$, Eq. (3.10.9) becomes

$$(V(\mathbf{k})+\kappa T\tilde{c})\delta\tilde{n}(\mathbf{k})+\kappa T\tfrac{1}{2}\tilde{\eta}\delta\tilde{n}(\mathbf{k}-\mathbf{k}_0)=0 \tag{3.10.15}$$

Shifting $\mathbf{k}$ in (3.10.15), $\mathbf{k}\to\mathbf{k}-\mathbf{k}_0$, one obtains the second equation

$$(V(\mathbf{k}-\mathbf{k}_0)+\kappa T\tilde{c})\delta\tilde{n}(\mathbf{k}-\mathbf{k}_0)+\kappa T\tfrac{1}{2}\eta\delta\tilde{n}(\mathbf{k})=0 \tag{3.10.16}$$

since

$$\begin{aligned}\delta n(\mathbf{k}-2\mathbf{k}_0)&\equiv n(\mathbf{k})\\ V(\mathbf{k}-2\mathbf{k}_0)&\equiv V(\mathbf{k}),\end{aligned} \tag{3.10.17}$$

Eq. (3.10.17) follows from the definitions (3.10.7) and identity

$$e^{i2\mathbf{k}_0\mathbf{r}}\equiv 1 \tag{3.10.18}$$

which is valid because $2\mathbf{k}_0=2\pi(2\mathbf{a}_1^*+2\mathbf{a}_2^*+2\mathbf{a}_3^*)$ corresponds to the fundamental reciprocal lattice vector (222).* The set of two homogeneous equations, (3.10.15) and (3.10.16), has a nontrivial solution with respect to $\delta\tilde{n}(\mathbf{k})$ and $\delta\tilde{n}(\mathbf{k}-\mathbf{k}_0)$ if its determinant vanishes:

$$\mathrm{Det}=\left\|\begin{matrix}V(\mathbf{k})+\kappa T\tilde{c} & \kappa T\frac{1}{2}\tilde{\eta}\\ \kappa T\frac{1}{2}\tilde{\eta} & V(\mathbf{k}-\mathbf{k}_0)+\kappa T\tilde{c}\end{matrix}\right\|=0 \tag{3.10.19}$$

The latter condition may be written as the quadratic equation in $\tilde{c}\kappa T$:

$$(\tilde{c}\kappa T)^2+(\tilde{c}\kappa T)(V(\mathbf{k})+V(\mathbf{k}-\mathbf{k}_0))+V(\mathbf{k})V(\mathbf{k}-\mathbf{k}_0)-(\tfrac{1}{2}\tilde{\eta}\kappa T)^2=0 \tag{3.10.20}$$

or

$$[\tilde{c}^2-(\tfrac{1}{2}\tilde{\eta})^2](\kappa T)^2+\kappa T\tilde{c}[V(\mathbf{k})+V(\mathbf{k}-\mathbf{k}_0)]+V(\mathbf{k})V(\mathbf{k}-\mathbf{k}_0)=0 \tag{3.10.20}$$

*$\delta\tilde{n}(\mathbf{k}-2\mathbf{k}_0)=\sum_{\mathbf{r}}\delta n(\mathbf{r})e^{-i(\mathbf{k}-2\mathbf{k}_0)\mathbf{r}}=\sum_{\mathbf{r}}\delta n(\mathbf{r})e^{-i\mathbf{k}\mathbf{r}}e^{i2\mathbf{k}_0\mathbf{r}}=\sum_{\mathbf{r}}\delta n(\mathbf{r})e^{-i\mathbf{k}\mathbf{r}}=\delta\tilde{n}(\mathbf{k})$

because of the identity (3.10.18).

According to (3.10.14)

$$\tilde{c}^2-(\tfrac{1}{2}\tilde{\eta})^2=\frac{1}{[c^2-(\frac{1}{2}\eta)^2][(1-c)^2-(\frac{1}{2}\eta)^2]} \tag{3.10.21}$$

Substituting (3.10.21) and (3.10.14a) into (3.10.20) yields

$$(\kappa T)^2+\left[c(1-c)-\left(\frac{\eta}{2}\right)^2\right](V(\mathbf{k})+V(\mathbf{k}-\mathbf{k}_0))\kappa T+\left[c^2-\left(\frac{\eta}{2}\right)^2\right]\left[(1-c)^2-\left(\frac{\eta}{2}\right)^2\right] \times V(\mathbf{k})V(\mathbf{k}-\mathbf{k}_0)=0 \tag{3.10.22}$$

The solution of (3.10.22) gives two branches of bifurcation points:

$$\kappa T^{\pm}=-\tfrac{1}{2}[c(1-c)-(\tfrac{1}{2}\eta)^2][V(\mathbf{k})+V(\mathbf{k}-\mathbf{k}_0)] \pm\tfrac{1}{2}\sqrt{\begin{array}{l}[c(1-c)-(\frac{1}{2}\eta)^2]^2[V(\mathbf{k})+V(\mathbf{k}-\mathbf{k}_0)]^2\\ \qquad -4V(\mathbf{k})V(\mathbf{k}-\mathbf{k}_0)[c^2-(\frac{1}{2}\eta)^2][(1-c)^2-(\frac{1}{2}\eta)^2]\end{array}} \tag{3.10.23}$$

The temperatures corresponding to the branch T^+ are always higher than those of the branch T^-. Since the critical instability point of a secondary phase transition is the highest bifurcation point, the branch T^+ should only be considered:

$$\kappa T=\kappa T^{+}(\mathbf{k})=-\tfrac{1}{2}[c(1-c)-(\tfrac{1}{2}\eta)^2][V(\mathbf{k})+V(\mathbf{k}-\mathbf{k}_0)] +\tfrac{1}{2}\sqrt{\begin{array}{l}[c(1-c)-(\frac{1}{2}\eta)^2]^2[V(\mathbf{k})+V(\mathbf{k}-\mathbf{k}_0)]^2\\ \qquad -4V(\mathbf{k})V(\mathbf{k}-\mathbf{k}_0)[c^2-(\frac{1}{2}\eta)^2][(1-c)^2-(\frac{1}{2}\eta)^2]\end{array}} \tag{3.10.24}$$

The next problem is to find the vector $\mathbf{k}=\mathbf{k}_1$ that corresponds to the absolute maximum of the $T=T^+(\mathbf{k})$ function.

It follows from identity (3.10.17) and (3.10.24) that

$$T^+(\mathbf{k}+\mathbf{k}_0)\equiv T^+(\mathbf{k}) \tag{3.10.25}$$

Unlike the $V(\mathbf{k})$ and $\delta\tilde{n}(\mathbf{k})$ functions, $T^+(\mathbf{k})$ is a periodic function whose minimum period is equal to the superlattice vector $\mathbf{k}_0$ rather than the fundamental lattice vector $2\mathbf{k}_0$.

The fulfillment of condition (3.10.25) results in the appearance of a new Lifshitz "special point," $\mathbf{k}=\mathbf{k}_1=\frac{1}{2}\mathbf{k}_0$. The function $T^+(\mathbf{k})$ must necessarily have an extremum at $\mathbf{k}_1=\frac{1}{2}\mathbf{k}_0$. The position of this extremum is determined by the symmetry of the system and is independent of the energy parameters entering $T^+(\mathbf{k})$.

We will now assume that the absolute maximum of $T^+(\mathbf{k})$ falls at that very Lifshitz point, $\mathbf{k}=\mathbf{k}_1=\frac{1}{2}\mathbf{k}_0$. Since $\mathbf{k}_1=\frac{1}{2}\mathbf{k}_0\neq 0$, secondary ordering occurs. According to the definition (3.10.7b), and bearing in mind the bcc lattice symmetry, we may write

$$V(\mathbf{k}_1)=V(\tfrac{1}{2}\mathbf{k}_0)=V(-\tfrac{1}{2}\mathbf{k}_0) \tag{3.10.26}$$

Using this result and Eq. (3.10.24), we obtain the secondary ordering temperature in the form

$$T_0=T(\mathbf{k}_1)=-\frac{1}{2\kappa}[c(1-c)-(\tfrac{1}{2}\eta)^2]V(\mathbf{k}_1)$$
$$+\frac{1}{2\kappa}|V(\mathbf{k}_1)|\sqrt{[c(1-c)-(\tfrac{1}{2}\eta)^2]^2-[c^2-(\tfrac{1}{2}\eta)^2][(1-c)^2-(\tfrac{1}{2}\eta)^2]}$$
$$=-\frac{1}{2\kappa}[c(1-c)-(\tfrac{1}{2}\eta)^2]V(\mathbf{k}_1)+\frac{1}{2\kappa}|V(\mathbf{k}_1)|\,|1-2c|\frac{\eta}{2} \quad (3.10.27)$$

It is easy to verify that the secondary ordering temperature is positive (that is, the reaction may go) if $V(\mathbf{k}_1)<0$. That given, Eq. (3.10.27) may be transformed into

$$T_0=(c-\tfrac{1}{2}\eta)(1-c+\tfrac{1}{2}\eta)\frac{|V(\mathbf{k}_1)|}{\kappa}=-(c-\tfrac{1}{2}\eta)(1-c+\tfrac{1}{2}\eta)\frac{V(\mathbf{k}_1)}{\kappa} \quad (3.10.28)$$

Therefore the secondary ordering in the B2 high-temperature phase occurs as a result of the loss of stability with respect to the concentration waves

$$\pm\mathbf{k}_1=\pm\tfrac{1}{2}\mathbf{k}_0=\pm\pi(\mathbf{a}_1^*+\mathbf{a}_2^*+\mathbf{a}_3^*)$$

As shown in Section 3.1, these two vectors and the vector $\mathbf{k}_0=2\pi(\mathbf{a}_1^*+\mathbf{a}_2^*+\mathbf{a}_3^*)$ generate the DO_3 superstructure (Fe_3Al-type) depicted in Fig. 13.

Let us now consider a secondary decomposition that may occur in a B2 high-temperature nonstoichiometric superstructure.

Kubo and Wayman were the first to observe a secondary decomposition in a nonstoichiometric β-CuZn (35). They found that nonstoichiometric β-CuZn decomposed to a mixture of two superstructures of the same B2 symmetry but of different compositions from the parent phase. It is noteworthy that the electron micrographs and selected area diffraction patterns studied by the authors exhibited morphology and side bands that are usually associated with spinodal decomposition in cubic crystals.

The authors (35) also reconsidered the earlier results which led them to conclude that isomorphous secondary decomposition reactions were seemingly observed in the Ni-Al (37), Cu-Sn (38), Cu-Al-Ni (39) and Ag-Zn (40) alloys. They also showed that all these observations might be interpreted in terms of secondary decomposition reactions in nonstoichiometric B2 superstructures and proposed the theory (36) described below in a slightly modified form.

It has already been mentioned that secondary decomposition occurs if the Fourier transform, $V(\mathbf{k})$, of the interchange energies leads to the $T^+(\mathbf{k})$ function, assuming its absolute maximum value at $\mathbf{k}=0$. At $\mathbf{k}=0$ Eq. (3.10.24) for the absolute instability temperature (spinodal curve) is then

$$T_0=-\frac{1}{2\kappa}[c(1-c)-(\tfrac{1}{2}\eta)^2][V(0)+V(\mathbf{k}_0)]$$
$$+\frac{1}{2\kappa}\sqrt{\begin{aligned}&[c(1-c)-(\tfrac{1}{2}\eta)^2]^2[V(0)+V(\mathbf{k}_0)]^2\\&-4V(\mathbf{k}_0)V(0)[c^2-(\tfrac{1}{2}\eta)^2][(1-c)^2-(\tfrac{1}{2}\eta)^2]\end{aligned}} \quad (3.10.29)$$

The equilibrium long-range-order parameter η for the B2 superstructure is determined by Eq. (3.8.8)

$$\ln\frac{(c+\frac{1}{2}\eta)(1-c+\frac{1}{2}\eta)}{(c-\frac{1}{2}\eta)(1-c-\frac{1}{2}\eta)}=-\frac{1}{\kappa T_0}V(\mathbf{k}_0)\eta \tag{3.10.30}$$

The solution of the set of the transcendental Eqs. (3.10.29) and (3.10.30) with respect to η and T_0 results in the spinodal curve, $T=T_0(c)$. The second-order transformation curve for the high-temperature superstructure B2, $T=T_c(c)$, is determined as bifurcation point of Eq. (3.10.30)

$$T_c(c)=-\frac{c(1-c)V(\mathbf{k}_0)}{\kappa} \tag{3.10.31}$$

[compare with Eq. (3.7.11)].

The parameter, η, is equal to zero along the $T=T_c(c)$ curve given by Eq. (3.10.31). Substituting $\eta=0$ into (3.10.29) results in

$$T_0=-\frac{1}{2\kappa}c(1-c)[V(0)+V(\mathbf{k}_0)]+\frac{1}{2\kappa}|V(0)-V(\mathbf{k}_0)|c(1-c)$$

Since $V(\mathbf{k}_0)<V(0)$ (otherwise, the high-temperature phase transformation generated by the superlattice vector $\mathbf{k}_0$ would be decomposition not ordering),

$$T_0=-\frac{1}{2\kappa}c(1-c)[V(0)+V(\mathbf{k}_0)]+\frac{V(0)-V(\mathbf{k}_0)}{2\kappa}c(1-c)=-c(1-c)\frac{V(\mathbf{k}_0)}{\kappa} \tag{3.10.32}$$

Comparison of (3.10.32) with (3.10.31) yields

$$T_0(c)=T_c(c) \tag{3.10.33}$$

Equality (3.10.33) has an important implication: the *low-temperature part of the order-disorder transformation curve on the T-c equilibrium diagram is the secondary decomposition spinodal.*

This conclusion was first drawn by Allen and Cahn who studied the limiting case of spinodal decomposition at η being asymptotically small (41).

The boundary point that separates the high-temperature branch of the curve (3.10.31) describing the high-temperature ordering from the low-temperature branch describing spinodal decomposition is a tricritical point (see Fig. 33).

Since the bifurcation point lies on the curve (3.10.31) where $\eta=0$, Eq. (3.10.29) must be analyzed at a small η value. Expansion of (3.10.29) in a power series of η gives

$$T_0=-\frac{c(1-c)V(\mathbf{k}_0)}{\kappa}+\left[\frac{V(0)+V(\mathbf{k}_0)}{2\kappa}+\frac{2V(0)V(\mathbf{k}_0)[c^2+(1-c)^2]-c(1-c)[V(0)+V(\mathbf{k}_0)]^2}{2\kappa c(1-c)[V(0)-V(\mathbf{k}_0)]}\right](\tfrac{1}{2}\eta)^2+\cdots$$

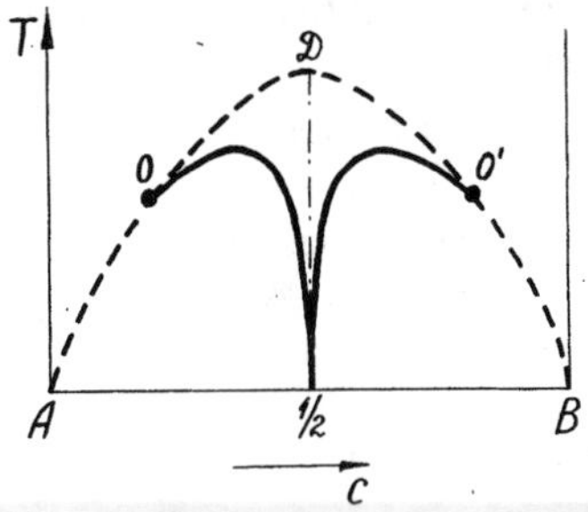

Figure 33. Critical temperature for ordering (dashed line) and for spinodal (solid line). 0 and 0′ are tricritical points, $A0$ and $0'B$ are also the spinodal lines for secondary decomposition, $0D0'$ is the order-disorder line.

or

$$T_0 - T_c = (\tfrac{1}{2}\eta)^2 \frac{1}{\kappa}\left[-\frac{V(\mathbf{k}_0)(V(\mathbf{k}_0)+3V(0))}{V(0)-V(\mathbf{k}_0)} + \frac{V(0)V(\mathbf{k}_0)}{V(0)-V(\mathbf{k}_0)}\frac{1}{c(1-c)}\right] + \cdots \tag{3.10.34}$$

The value η entering (3.10.34) for an asymptotically small supercooling $T_c - T_0$ with respect to the order-disorder transformation temperature T_0 may be determined from Eq. (3.10.30). It is given by

$$(\tfrac{1}{2}\eta)^2 \approx \frac{T_c - T_0}{T_c}\frac{3c^2(1-c)^2}{1-3c(1-c)} = -\kappa\frac{T_c - T_0}{V(\mathbf{k}_0)}\frac{3c(1-c)}{1-3c(1-c)} \tag{3.10.35}$$

Substituting (3.10.35) into (3.10.34) yields

$$T_c - T_0 = \frac{T_c - T_0}{V(\mathbf{k}_0)}\frac{3c(1-c)}{1-3c(1-c)}\left[-\frac{V(\mathbf{k}_0)(V(\mathbf{k}_0)+3V(0))}{V(0)-V(\mathbf{k}_0)} + \frac{V(0)V(\mathbf{k}_0)}{V(0)-V(\mathbf{k}_0)}\frac{1}{c(1-c)}\right] \tag{3.10.36}$$

The nontrivial solution of Eq. (3.10.36) (i.e., $T_c - T_0 \neq 0$) which characterizes the bifurcation point appears when

$$1 = \frac{3c(1-c)}{V(\mathbf{k}_0)[1-3c(1-c)]}\left[-\frac{V(\mathbf{k}_0)(V(\mathbf{k}_0)+3V(0))}{V(0)-V(\mathbf{k}_0)} + \frac{V(0)V(\mathbf{k}_0)}{V(0)-V(\mathbf{k}_0)}\frac{1}{c(1-c)}\right] \tag{3.10.37}$$

The solution of (3.10.37) yields the tricritical composition c_b corresponding to the bifurcation point of the set of the combined Eqs. (3.10.29) and (3.10.30):

$$c_b = \frac{1}{2} \pm \sqrt{\frac{1}{12}\cdot\frac{V(\mathbf{k}_0)-V(0)}{V(\mathbf{k}_0)+V(0)}} \tag{3.10.38}$$

The temperature T_b corresponding to the tricritical temperature may be obtained by substituting (3.10.38) into (3.10.31).

The calculations made above are rather tedious. There is, however, one limiting case that allows of a considerable simplification of the analysis. This is the case of a nonstoichiometric superstructure characterized by the maximum long-range order parameter possible at a given composition. The corresponding secondary ordering theory has been developed by Khachaturyan (9). When applied to secondary decomposition in nonstoichiometric B2 superstructures, the theory leads to the conclusions described below.

The occupation probabilities of solute atoms (3.10.2) in a B2 superlattice assume two values at all crystal lattice sites, either $c-\frac{1}{2}\eta$ or $c+\frac{1}{2}\eta$. Since the occupation probabilities can not be negative, $c-\frac{1}{2}\eta \geqslant 0$, we come to the inequality

$$\tfrac{1}{2}\eta \leqslant c$$

which gives for the maximum long-range order parameter, η, at a given concentration:

$$(\tfrac{1}{2}\eta)_{\max} = c \tag{3.10.39}$$

Substituting that value into Eq. (3.10.29) yields

$$T_0^+ = -[c(1-c)-c^2]\frac{V(0)+V(\mathbf{k}_0)}{2\kappa} + [c(1-c)-c^2]\frac{|V(0)+V(\mathbf{k}_0)|}{2\kappa} \tag{3.10.40}$$

The function T_0^+ (3.10.40) is only nonzero if $V(0)+V(\mathbf{k}_0)$ is negative. In that case Eq. (3.10.40) becomes

$$T_0^+ = -2c(1-2c)\frac{V(0)+V(\mathbf{k}_0)}{2\kappa} \tag{3.10.41}$$

3.11. ORDERING IN CRYSTALS COMPOSED OF SEVERAL INTERPENETRATING BRAVAIS LATTICES

Ordering in comparatively simple phases whose crystal lattices comprise but one Bravais lattice has been analyzed in the preceeding sections. However, in many cases, as in ordering of hcp binary substitutional solutions, the crystal lattice consists of several interpenetrating Bravais sublattices displaced with respect to each other by the vectors $\mathbf{h}_1, \mathbf{h}_2, \ldots, \mathbf{h}_p, \ldots, \mathbf{h}_\nu$, where ν is the number of Bravais lattices. In other words, the basis of the crystal contains ν sites having $\mathbf{h}_1, \mathbf{h}_2, \ldots, \mathbf{h}_p, \ldots, \mathbf{h}_\nu$ coordinates at the origin. If crystal lattice sites of all the Bravais sublattices are crystallographically equivalent, in other words, can be brought into coincidence with each other by one of the symmetry operations of the disordered phase, the theoretical treatment of ordering may be simplified considerably and reduced to the procedure described above—to the determination of the ordered phase structure based on only one Bravais lattice. The reason for that is twofold. First, each of the sublattices comprising the crystal is a Bravais lattice and ordering in it is described by Eq. (3.1.3).

Second, it seems obvious from physical considerations that the symmetry of the atomic arrangements in all the Bravais sublattices making up the crystal should be the same and therefore would be described by the same function of the form (3.1.3).* Indeed, let us suppose the contrary: let an atomic distribution

*In a general case, the long-range order parameters η_s and composition, c, in Eq. (3.1.3) may take different values $\eta_s(p)$ and $c(p)$ for each of the sublattices $(p=1, 2, \ldots, \nu)$.

in one of the sublattices have a lower symmetry than that in the other sublattice. Then the potential field produced by the low-symmetry atomic distribution in the former sublattice should have the same symmetry as the atomic distribution generating this field, a lower symmetry than that of the atomic distribution in the higher-symmetry sublattice. The low-symmetry field should cause a redistribution of atoms in the high-symmetry sublattice with the result that its symmetry will decrease until it fits the potential field symmetry. This effect should always be present irrespective of the potential field strength.

To determine the translational symmetry of the crystal with a given basis therefore, one need not determine the $c(p)$ and $\eta_s(p)$ parameters that do not affect the symmetry of the distribution function described by these parameters. It is sufficient to find the coefficients $\gamma_s(j_s)$ of the ordered distribution in one of the Bravais lattices.

The problem is complicated somewhat if one is interested in the numerical values of the $c(p)$ and $\eta_s(p)$ parameters. Then it is necessary to apply the theory analogous to that developed for the case of one Bravais lattice in the preceeding sections. This theory will be described below.

A crystal lattice site of a crystal formed by translations of a basis is described by two vectors $(\mathbf{h}_p, \mathbf{r})$, or, for brevity, $(p, \mathbf{r})$, where the vector $\mathbf{r}$ refers to the unit cell's origin position and $\mathbf{h}_p$ is the distance of the p-type site within the cell with respect to the origin.

Since interatomic interaction energy for a pair of atoms at $(p, \mathbf{r})$ and $(q, \mathbf{r}')$ is invariant under Bravais translations, we have

$$\tilde{V}(\mathbf{r}+\mathbf{h}_p, \mathbf{r}'+\mathbf{h}_q)=W_{pq}(\mathbf{r}-\mathbf{r}') \tag{3.11.1}$$

Substituting (3.11.1) into (3.4.7) yields

$$n(p, \mathbf{r})=\left[\exp\left(\frac{-\mu+\sum\limits_{q,\mathbf{r}'} W_{pq}(\mathbf{r}-\mathbf{r}')n(q, \mathbf{r}')}{\kappa T}\right)+1\right]^{-1} \tag{3.11.2}$$

where $n(p, \mathbf{r})$ is the probability of finding a solute atom at $(p, \mathbf{r})$. Distribution (3.11.2) is associated with the Helmholz free energy

$$F=U-TS \tag{3.11.3}$$

where

$$U=\tfrac{1}{2}\sum_{p,q}\sum_{\mathbf{r},\mathbf{r}'} W_{pq}(\mathbf{r}-\mathbf{r}')n(p, \mathbf{r})n(q, \mathbf{r}') \tag{3.11.3a}$$

is the internal energy and

$$S=-\kappa\sum_{p,\mathbf{r}}\left[n(p, \mathbf{r})\ln n(p, \mathbf{r})+(1-n(p, \mathbf{r}))\ln(1-n(p, \mathbf{r}))\right] \tag{3.11.3b}$$

is the entropy in the mean-field approximation.

Eq. (3.11.2) can be solved by the method analogous to that applied in Section 3.7 to the case of one Bravais lattice. Again, the solute atom occupation probabilities, $n(p, \mathbf{r})$, can be written as a superposition of static concentration waves, but of a more complicated nature. The static concentration waves resemble now Bloch functions:

$$\phi_{\sigma\mathbf{k}}(p, \mathbf{r}) = v_\sigma(p, \mathbf{k})e^{i\mathbf{kr}} \tag{3.11.4}$$

where $v_\sigma(p, \mathbf{k})$ is a unit "polarization vector" of the wave, $\mathbf{k}$ is a wave vector, σ is a "polarization number." Just as simple waves, exp ($i\mathbf{kr}$), are eigenfunctions of the matrix formed by interatomic pairwise energies, $W(\mathbf{r}-\mathbf{r}')$, in a simple lattice, the waves, $\phi_{\sigma\mathbf{k}}(p, \mathbf{r})$, are eigenfunctions of the matrix $W_{pq}(\mathbf{r}-\mathbf{r}')$ in a complex lattice.

$$\sum_{q=1}^{\nu} \sum_{\mathbf{r}'} W_{pq}(\mathbf{r}-\mathbf{r}')\phi_{\sigma\mathbf{k}}(q, \mathbf{r}') = \lambda_\sigma(\mathbf{k})\phi_{\sigma\mathbf{k}}(p, \mathbf{r}) \tag{3.11.5}$$

where $\lambda_\sigma(\mathbf{k})$ is an eigenvalue of the matrix $W_{pq}(\mathbf{r}-\mathbf{r}')$. The "polarization number" σ plays the role of a branch number of the spectrum $\lambda_\sigma(\mathbf{k})$.

Substituting Eq. (3.11.4) into (3.11.5) gives the secular equation

$$\sum_{q=1}^{\nu} V_{pq}(\mathbf{k})v_\sigma(q, \mathbf{k}) = \lambda_\sigma(\mathbf{k})v_\sigma(p, \mathbf{k}) \tag{3.11.6}$$

where

$$V_{pq}(\mathbf{k}) = \sum_{\mathbf{r}} W_{pq}(\mathbf{r})e^{-i\mathbf{kr}} \tag{3.11.7}$$

The set of ν homogeneous equations (3.11.6) in ν unknowns $v_\sigma(p, \mathbf{k})$ possesses nontrivial solutions if its determinant is equal to zero. The corresponding equation

$$\det \| V_{pq}(\mathbf{k}) - \lambda\delta_{pq} \| = 0 \tag{3.11.8}$$

is a characteristic equation of degree ν in λ which has solutions: $\lambda_1(\mathbf{k}), \lambda_2(\mathbf{k}), \ldots, \lambda_\nu(\mathbf{k})$. Therefore ν branches in $\lambda_\sigma(\mathbf{k})$ correspond to each wave vector $\mathbf{k}$. Since the matrix $V_{pq}(\mathbf{k})$ is Hermitian all its eigenvalues $\lambda_\sigma(\mathbf{k})$ are real values, and all eigenfunctions $v_\sigma(p, \mathbf{k})$ are orthogonal to each other

$$\sum_{p=1}^{\nu} v_\sigma(p, \mathbf{k})v_{\sigma'}(p, \mathbf{k}) = \delta_{\sigma\sigma'} \tag{3.11.9}$$

Being eigenfunctions of the Hermitian matrix $W_{pq}(\mathbf{r}-\mathbf{r}')$ concentration waves $\phi_{\sigma\mathbf{k}}(p, \mathbf{r})$ form a complete set of the orthogonal functions. Therefore the occupation probabilities, $n(p, \mathbf{r})$, can be represented as a linear superposition of these concentration waves

$$n(p, \mathbf{r}) = c + \tfrac{1}{2}\sum_{\sigma=1}^{\nu}\sum_{\mathbf{k}} [Q_\sigma(\mathbf{k})v_\sigma(p, \mathbf{k})e^{i\mathbf{kr}} + Q_\sigma^*(\mathbf{k})v_\sigma^*(p, \mathbf{k})e^{-i\mathbf{kr}}] \tag{3.11.10}$$

where $Q_\sigma(\mathbf{k})$ are amplitudes of concentration waves (the analogue of the normal coordinates).

Eigenvalues $\lambda_\sigma(\mathbf{k})$ are degenerated with respect to all eigenfunctions $v_\sigma(p, \mathbf{k})$ referring to the same branch σ, but to the different vectors of the same star $\mathbf{k}$. Therefore the sum (3.11.10) can be rewritten in another form by grouping the terms with the same branch index σ whose wave vectors $\mathbf{k}_{j_s}$ enter into the same star s:

$$n(p, \mathbf{r}) = c + \sum_s \sum_{\sigma=1}^{\nu} \eta_{s,\sigma} E_{s,\sigma}(p, \mathbf{r}) \tag{3.11.11}$$

where

$$E_{s,\sigma}(p, \mathbf{r}) = \frac{1}{2} \sum_{j_s} [\gamma_{s,\sigma}(j_s) v_{\sigma \mathbf{k}_{j_s}}(p, \mathbf{r}) + \gamma^*_{s,\sigma}(j_s) v^*_{\sigma \mathbf{k}_{j_s}}(p, \mathbf{r})] \tag{3.11.12}$$

$$Q_\sigma(\mathbf{k}_s) = \eta_{s,\sigma} \gamma_{s,\sigma}(j_s) \tag{3.11.13}$$

and $\eta_{\sigma,s}$ is a long-range-order parameter.

The index $_{j_s}$ in (3.11.12), and (3.11.13) refers to the wave vectors of the star, s. It is convenient to choose constants $\gamma_{s,\sigma}(j_s)$ so that they satisfy the normalization condition

$$\sum_{j_s} |\gamma_{s,\sigma}(j_s)|^2 = 1 \tag{3.11.14}$$

According to the definition (3.11.12) for $E_{s,\sigma}(p, \mathbf{r})$ and (3.11.5),

$$\sum_{q, \mathbf{r}'} W_{pq}(\mathbf{r} - \mathbf{r}') E_{s,\sigma}(q, \mathbf{r}') = \lambda_\sigma(\mathbf{k}_s) E_{s,\sigma}(p, \mathbf{r}) \tag{3.11.15}$$

The substituting Eq. (3.11.11) into the self-consistent field Eq. (3.11.2) gives the transcendental equation

$$c + \sum_{s,\sigma} \eta_{s,\sigma} E_{s,\sigma}(p, \mathbf{r}) = \left[\exp\left(-\frac{\mu}{\kappa T} + \frac{1}{\kappa T} \sum_{\sigma,s} \lambda_\sigma(\mathbf{k}_s) \eta_{s,\sigma} E_{s,\sigma}(p, \mathbf{r}) \right) + 1 \right]^{-1} \tag{3.11.16}$$

which is an analogue of the transcendental Eq. (3.8.1) for a crystal with a single Bravais lattice.

Substitution of the permitted numerical values of the crystal lattice site coordinates, $(p, \mathbf{r})$, into (3.11.16) leads to a set of transcendental equations with respect to unknowns, c, $\eta_{\sigma,s}$. The number of these equations is equal to the number of different values the function $n(p, \mathbf{r})$ takes on the variety of crystal lattice sites. Eqs. (3.11.16) can be solved if that number equals the total number of the variables, c, $\eta_{\sigma,s}$. As with a single Bravais lattice, we again come to the conclusion that the distribution function $n(p, \mathbf{r})$ should meet the following criterion:

Criterion 1. *The number of different values it assumes on all crystal lattice sites* $(p, \mathbf{r})$ *in a crystal consisting of several Bravais lattices should be larger by one than the total number of the long-range order parameters.*

Criterion 1 may be applied to determine the numerical values of the coefficients $\gamma_{s,\sigma}(j_s)$ and thus to determine the atomic structure of the ordered phase.

The order-disorder transition temperature can be found as the highest-temperature bifurcation point of the nonlinear Eq. (3.11.2). This bifurcation point should be found from the linearized Eq. (3.11.2)

$$\delta n(p, \mathbf{r}) = -\frac{c(1-c)}{\kappa T} \sum_{\mathbf{r}'} \sum_{q} W_{pq}(\mathbf{r}-\mathbf{r}')\delta n(q, \mathbf{r}') \tag{3.11.17}$$

which is an analogue of Eq. (3.7.3).

The Fourier transform of Eq. (3.11.17) is

$$\delta \tilde{n}(p, \mathbf{k}) = -\frac{c(1-c)}{\kappa T} \sum_{q} V_{pq}(\mathbf{k})\delta \tilde{n}(q, \mathbf{k}) \tag{3.11.18}$$

where

$$\delta \tilde{n}(p, \mathbf{k}) = \sum_{\mathbf{r}} \delta n(p, \mathbf{r}) e^{-i\mathbf{kr}}$$

Eq. (3.11.18) has nontrivial solutions if its determinant is zero:

$$\det \left\| \delta_{pq} + \frac{c(1-c)}{\kappa T} V_{pq}(\mathbf{k}) \right\| = 0 \tag{3.11.19}$$

In the terms of the diagonal representation of the matrix $V_{pq}(\mathbf{k})$ the determinant (3.11.19) is the product of all eigenvalues of the matrix

$$\delta_{pq} + \frac{c(1-c)}{\kappa T} V_{pq}(\mathbf{k})$$

which are

$$1 + \frac{c(1-c)}{\kappa T} \lambda_\sigma(\mathbf{k}) \tag{3.11.20}$$

At high temperatures all eigenvalues (3.11.20) are positive, and the determinant (3.11.19) is positive as well. The determinant (3.11.19) vanishes when the minimum eigenvalue from (3.11.20) becomes equal to zero, when

$$1 + \frac{c(1-c)}{\kappa T} \min(\lambda_\sigma(\mathbf{k})) = 0 \tag{3.11.21}$$

[min $\lambda_\sigma(\mathbf{k})$ is the absolute minimum of $\lambda_\sigma(\mathbf{k})$]. The temperature which is consistent with Eq. (3.11.21) is the temperature of the order-disorder transition. It is equal to

$$T_0 = -\frac{c(1-c)}{\kappa} \lambda_{\sigma_0}(\mathbf{k}_{j_{s_0}}) \tag{3.11.22}$$

where

$$\lambda_{\sigma_0}(\mathbf{k}_{j_{s_0}}) = \min \lambda_\sigma(\mathbf{k}) \tag{3.11.23}$$

Therefore the temperature of an order-disorder transition in the complex lattice is determined by Eq. (3.11.22). *The structure of the high-temperature ordered phase is determined by the star* $\{\mathbf{k}s_0\}$ *whose wave vectors* $\mathbf{k}_{j_{s_0}}$ *and "polarization vectors"* $v_{\sigma_0}(p, k_{j_{s_0}})$ *provide the absolute minimum of* $\lambda_{\sigma_0}(\mathbf{k})$.

Ordering is caused by instability of the disordered state with respect to the infinitesimal disturbance

$$\delta n(p, \mathbf{r})=\eta_{s_0,\sigma_0}E_{s_0,\sigma_0}(p, \mathbf{r}) \tag{3.11.24}$$

in the atomic distribution $n(p, \mathbf{r})=c$ where

$$E_{s_0,\sigma_0}(p, \mathbf{r})=\frac{1}{2}\sum_{j_{s_0}}[\gamma_{s_0,\sigma_0}(j_{s_0})v_{\sigma_0}(p, \mathbf{k}_{j_{s_0}})e^{i\mathbf{k}_{j_{s_0}}\mathbf{r}}+\text{compl. conj.}] \tag{3.11.25}$$

is a function (3.11.12) corresponding to the "polarization" σ_0 and the star s_0.

If the function

$$n(p, \mathbf{r})=c+\eta_{s_0,\sigma_0}E_{s_0,\sigma_0}(p, \mathbf{r}) \tag{3.11.26}$$

does not satisfy condition 1, as it takes more than two values of all the interstices $(p, \mathbf{r})$, one should add to the right-hand side of Eq. (3.11.26) a corresponding number of functions of the type $\eta_{s,\sigma}E_{s,\sigma}(p, \mathbf{r})$ whose symmetry is higher than that of the dominant term $\eta_{s_0\sigma_0}E_{s_0,\sigma_0}(p, \mathbf{r})$. The addition of such functions does not change the symmetry of the distribution (3.11.26) and therefore does not change the number of the values that the function (3.11.26) takes on all the interstices. However, this procedure increases the total number of the long-range order parameters $\eta_{s,\sigma}$ involved in $n(p, \mathbf{r})$. Therefore one should reach the situation where the number of values the resulting function $n(p, \mathbf{r})$ assumes on the variety of all interstices is equal to the number of the unknown parameters c and $\eta_{s,\sigma}$, where condition 1 holds.

One must always remember that the coefficients $\gamma_{s_0,\sigma_0}(j_{s_0})$ in the dominant function should be chosen so that the number of added functions $E_{\sigma,s}(p, \mathbf{r})$ be minimum (the latter provides the minimum of the internal and free energy). This means that as a rule in high-temperature superlattices the single coefficient $\gamma_{s,\sigma}(j_s)=\gamma_{s,\sigma}$ within each of the function $E_{s,\sigma}(p, \mathbf{r})$ entering into $n(p, \mathbf{r})$ has a nonzero value:

$$n(p, \mathbf{r})=c+\frac{1}{2}\sum_{s,\sigma}\eta_{s,\sigma}[\gamma_{s,\sigma}v_\sigma(p, \mathbf{k}_s)e^{i\mathbf{k}_s\mathbf{r}}+\gamma_{s,\sigma}^*v_\sigma^*(p, \mathbf{k}_s)e^{-i\mathbf{k}_s\mathbf{r}}] \tag{3.11.27}$$

4

DECOMPOSITION IN ALLOYS

In thermodynamics decomposition reactions should occur in any non-stoichiometric solid solution. This fact is based on the Nernst theorem which showed that, for entropies of fully ordered substances to vanish, they must have stoichiometric compositions at 0° K. Hence a stable state of a solid solution at 0° K is a mixture of fully ordered phases with stoichiometric compositions. This is realized by decomposition reactions resolving the solution into stable stoichiometric phases.

These arguments are applicable to any alloy if the diffusion rate is sufficient for the decomposition to occur. In certain cases, however, the diffusion rates at reduced temperatures are too low for the decomposition to be perceptible over a reasonable period of time.

One type of decomposition is decomposition into a mixture of ordered and disordered phases or several ordered phases. In that case decomposition also proceeds by interchange of atoms between crystal lattice sites retaining their relative positions. The process, however, involves both long-range diffusion to change phase compositions and short-range diffusion (over distances comparable with interatomic separations) to achieve ordering. The typical equilibrium diagram is shown in Fig. 34b.

The other type of decomposition is the most complicated one. It involves, in addition to long-range diffusion (concentration separation) and short-range diffusion (ordering), rearrangement of crystal lattice sites. Long-range diffusion, however, remains to be the slowest process controlling the decomposition kinetics irrespective of the decomposition type.

The simplest case of decomposition is the isostructural decomposition resulting in separation of a homogeneous high-temperature solution into a mixture of phases having the same structure as the initial solution but differing from it by their compositions. The isostructural decomposition only involves interchanges of atoms between various crystal lattice sites. The typical phase

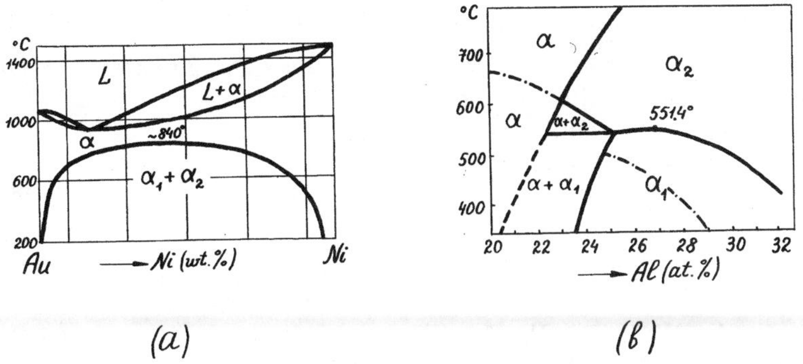

Figure 34. Typical equilibrium *T–c* diagrams involving decomposition. (*a*) The Au-Ni diagram with the miscibility gap; (*b*) Fe-Al diagram describing decomposition and ordering. The dot-dash line shows the ferromagnetic transition.

diagram of a system where isostructural decomposition occurs is given in Fig. 34a.

4.1. THERMODYNAMICS OF DECOMPOSITION

For simplicity, only the decomposition of a binary solution will be considered. According to the second law of thermodynamics the equilibrium state of a system should meet the condition of the minimal Helmholtz free energy. An analysis of equilibrium alloys should therefore be based on consideration of free energy.

Following this line of reasoning, we shall begin with the determination of the free energy of a two-phase mixture, $\alpha+\beta$. Let the atomic fraction of the solute be c_α in the phase α and c_β in the phase β. The total solute atomic fraction in the alloy will be denoted c.

The conservation condition for the solute then reads

$$c_\alpha N_\alpha + c_\beta N_\beta = cN \tag{4.1.1}$$

where N is the total number of crystal lattice sites in the alloy, N_α and N_β are the numbers of lattice sites in the phases α and β, respectively. The total number of solute atoms in the alloy and the numbers of solute atoms in the phases α and β will then be cN, $c_\alpha N_\alpha$, and $c_\beta N_\beta$.

The decomposition reaction does not change the number of crystal lattice sites, and after the decomposition, each site occurs either in the phase α or in the phase β:

$$N_\alpha + N_\beta = N \tag{4.1.2}$$

If the interphase surface energy of the two-phase mixture $\alpha+\beta$ is neglected, the free energy may be written as the sum of the bulk free energies of the constituent phases:

$$F_{\alpha+\beta}=f(c_\alpha)N_\alpha+f(c_\beta)N_\beta \tag{4.1.3}$$

where $f(c_\alpha)$ and $f(c_\beta)$ are the free energies of the α and β phases, respectively, per lattice site (specific free energies).

The division of Eqs. (4.1.1), (4.1.2), (4.1.3) by N yields

$$c_\alpha\omega_\alpha+c_\beta\omega_\beta=c \tag{4.1.4}$$

$$\omega_\alpha+\omega_\beta=1 \tag{4.1.5}$$

$$f_{\text{mix}}=f(c_\alpha)\omega_\alpha+f(c_\beta)\omega_\beta \tag{4.1.6}$$

where $\omega_\alpha=N_\alpha/N$ and $\omega_\beta=N_\beta/N$ are the volume fractions of the α and β phases, respectively, $f_{\text{mix}}=F_{\alpha+\beta}/N$.

These three equations enable one to express the free energy of the mixture $\alpha+\beta$ through a sole variable c and exclude the internal parameters of the system ω_α and ω_β.

The solution of the homogeneous set of Eqs. (4.1.4) and (4.1.5) yields

$$\omega_\alpha=\frac{c-c_\beta}{c_\alpha-c_\beta},\quad \omega_\beta=\frac{c_\alpha-c}{c_\alpha-c_\beta} \tag{4.1.7}$$

The relationships (4.1.7) give the so-called lever rule which makes it possible to find the volume fractions of the α and β phases at each alloy composition if the equilibrium diagram is known. Substituting Eq. (4.1.7) into (4.1.6) yields

$$f(c_\alpha)\,\frac{c-c_\beta}{c_\alpha-c_\beta}+f(c_\beta)\,\frac{c_\alpha-c}{c_\alpha-c_\beta}=f_{\text{mix}}$$

or

$$f_{\text{mix}}=\frac{c_\alpha f(c_\alpha)-c_\beta f(c_\beta)}{c_\alpha-c_\beta}+\frac{f(c_\alpha)-f(c_\beta)}{c_\alpha-c_\beta}\,c \tag{4.1.8}$$

Since c_α, c_β, $f(c_\alpha)$, and $f(c_\beta)$ are assumed to be constant values, Eq. (4.1.8) gives the linear dependence of the Helmholtz free energy on the alloy composition. This dependence is shown as a straight line linking the points $(c_\alpha, f(c_\alpha))$ and $(c_\beta, f(c_\beta))$ in Fig. 35.

The latter result leads right to an important conclusion: a single-phase solution cannot be stable within the composition range $c_1<c<c_2$ if the curve

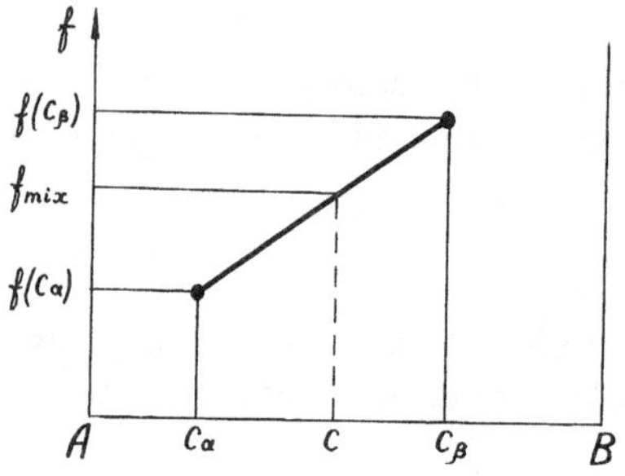

Figure 35. Concentration dependence of the free energy of a two-phase mixture ($\alpha+\beta$). c_α and c_β are the concentrations of phases α and β composing the mixture.

$f = f(c)$ describing the free energy of the single-phase alloy lies above the straight line describing the free energy of the two-phase alloy and linking the points $(c_1, f(c_1))$ and $(c_2, f(c_2))$ (see Fig. 36a). Indeed, in this case the free energy of the two-phase alloy, $f_{mix}(c)$, proves to be smaller than that of the single-phase alloy, $f(c)$, at all intermediate compositions, $c_1 < c < c_2$.

If the curve $f = f(c)$ is concave everywhere within the range $c_1 < c < c_2$, it should lie below the straight line linking the points $(c_1, f(c_1))$ and $(c_2, f(c_2))$ and therefore the single-phase state is stable within that range (Fig. 36b).

In this respect the following conclusion can be drawn: an alloy of a composition corresponding to a point on the convex part of the curve $f = f(c)$ (i.e., $d^2f/dc^2 < 0$) cannot be stable against decomposition into a two-phase mixture. We would then always choose a small range containing the point $(c, f(c))$ where $d^2f/dc^2 < 0$, that is, where the curve $f = f(c)$ is convex. However, as has just been shown above, the single-phase state is unstable in that case.

Since the range where the curve $f = f(c)$ is convex may be chosen infinitesimally small, a homogeneous solid solution is unstable at $d^2f/dc^2 < 0$ with respect to an infinitesimal concentration heterogeneity within the infinitesimal range $c_1 < c < c_2$. This instability is usually called spinodal instability, and the corresponding decomposition reaction is called spinodal decomposition. The line that describes the boundary of the spinodal instability region in the T-c diagram

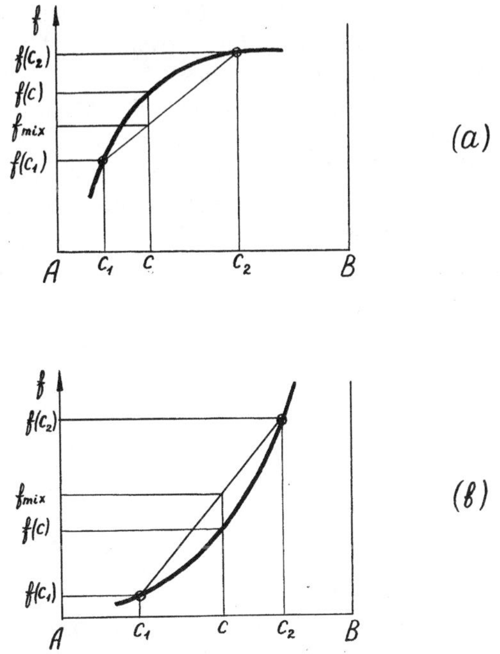

Figure 36. Concentration dependence of the free energy. (*a*) The case of spinodal decomposition; (*b*) the case of metastable and stable alloy.

where

$$\frac{d^2f(c, T)}{dc^2}=0$$

is called the spinodal curve.

On the other hand, if $d^2f/dc^2>0$, that is, the curve $f=f(c)$ is concave at point c, the homogeneous solid solution is stable with respect to infinitesimal heterogeneity. Indeed, if $d^2f/dc^2>0$, we can always choose an infinitesimal region $c_1<c<c_2$ about the point c where $d^2f/dc^2>0$, that is, where the curve $f=f(c)$ is concave (see Fig. 36b). The curve $f=f(c)$ lies below the straight line connecting the points $(c_1.\ f(c_1))$ and $(c_2,\ f(c_2))$ and therefore the homogeneous single-phase state is more stable than a mixture of two phases having an infinitesimally different composition. Therefore, if a homogeneous alloy characterized by $d^2f/dc^2>0$ at the point c is unstable with respect to the formation of a two-phase mixture with phase compositions, c_α and c_β, substantially different from the alloy composition, the alloy is stable with respect to an infinitesimally small composition heterogeneity. In other words, the decomposition reaction in that case should involve the formation of a finite composition heterogeneity and follow the nucleation-and-growth mechanism. Such an alloy is called a metastable alloy.

Let us now consider an arbitrary $f=f(c)$ curve and try to find the composition ranges where the alloy exists as a two-phase mixture. To do this, we should construct all the possible straight lines tangent to the curve at two different points simultaneously and lying below it. The plot in Fig. 37 exemplifies the situation when there are two such lines.

The line *ABDEGH* describes the free energies of the most stable alloys at various compositions. The straight line segments *BD* and *EG* describe two-phase mixtures, the curves *AB*, *DE*, and *GH* correspond to homogeneous single-phase alloys. Single-phase alloys are unstable along the *BCD* and *EFG* lines since these curve sections lie above the *BD* and *EG* straight-line sections, and therefore the corresponding single-phase states have higher free energies than the two-phase mixtures.

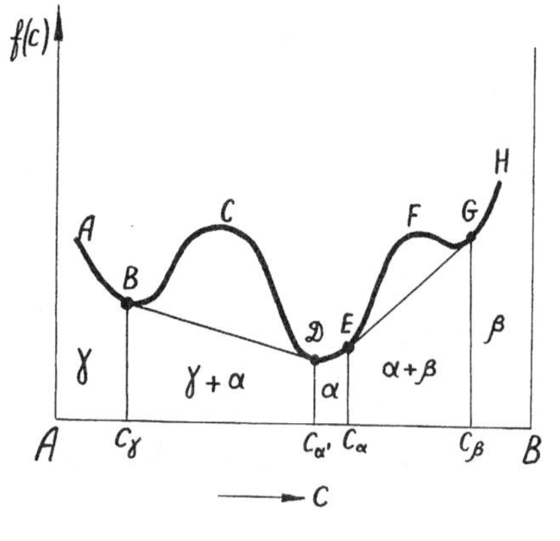

Figure 37. Concentration dependence of the free energy for an equilibrium diagram with two two-phase fields, $\gamma+\alpha$ and $\alpha+\beta$.

The tangent points c_γ, $c_{\alpha'}$, c_α, and c_β are therefore the solubility limits separating the two-phase mixtures in the T-c equilibrium diagram.

Mathematically, the condition that the straight lines be tangent to the $f=f(c)$ curve at two curve points simultaneously may be written as

$$\frac{f(c_\alpha)-f(c_\beta)}{c_\alpha-c_\beta}=\left(\frac{df}{dc}\right)_{c=c_\beta}=\left(\frac{df}{dc}\right)_{c=c_\beta}=\mu \tag{4.1.9}$$

where μ is the chemical potential.

The principles of the construction of an equilibrium diagram will be illustrated below for the particular case of the mean-field free energy (3.6.3b).

Substitution of $n(\mathbf{r})=c=\text{constant}$ into Eq. (3.6.3b) gives

$$\begin{aligned}F&=\frac{c^2}{2}\sum_{\mathbf{r},\mathbf{r}'}W(\mathbf{r}-\mathbf{r}')+\kappa TN[c\ln c+(1-c)\ln(1-c)]\\&=N\{\tfrac{1}{2}V(0)c^2+\kappa T[c\ln c+(1-c)\ln(1-c)]\}\end{aligned} \tag{4.1.10}$$

where

$$V(0)=\sum_{\mathbf{r}}W(\mathbf{r}) \tag{4.1.11}$$

Eq. (4.1.10) may be rewritten as

$$f(c)=\frac{F}{N}=\tfrac{1}{2}V(0)c^2+\kappa T[c\ln c+(1-c)\ln(1-c)] \tag{4.1.12}$$

According to (4.1.9), an equilibrium diagram remains unaffected if the free energy, $f(c)$, in (4.1.9) is replaced by the free energy, $f(c)-\mu c$, where μ is a constant. The expression for the free energy

$$f(c)=-\tfrac{1}{2}V(0)c(1-c)+\kappa T[c\ln c+(1-c)\ln(1-c)] \tag{4.1.13}$$

may therefore be employed in place of (4.1.12). The specific free energy (4.1.13) differs from (4.1.12) by an additional term $\mu c=\frac{1}{2}V(0)c$.

The expression for the free energy (4.1.13) is more convenient because of its symmetry with respect to $c=\frac{1}{2}$:

$$f(c)=f(1-c) \tag{4.1.14}$$

The decomposition reaction takes place if

$$V(0)<0$$

Eq. (4.1.13) may then be rewritten in the form

$$\frac{f(c)}{|V(0)|}=\frac{1}{2}c(1-c)+\tau[c\ln c+(1-c)\ln(1-c)] \tag{4.1.15}$$

where

$$\tau=\frac{\kappa T}{|V(0)|} \tag{4.1.16}$$

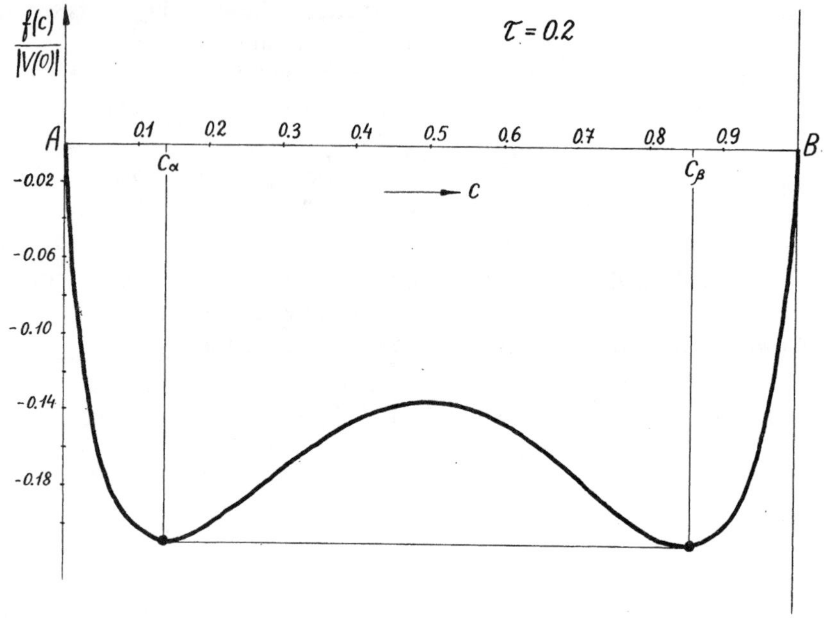

Figure 38. Concentration dependence of the free energy in the mean-field approximation. The reduced temperature is $\tau = \kappa T/|V(0)| = 0.2$.

is reduced temperature. The plot of free energy (4.1.15) with respect to c at $\tau = 0.2$ is depicted in Fig. 38. If $f(c) = f(1-c)$, the straight line tangent to and lying below $f(c)$ is a horizontal line (see Fig. 38). The solubility limits are, according to Fig. 38, the minimum points of the curve (4.1.15). The solubility curve is therefore determined by the necessary minimum condition:

$$\frac{df(c)/|V(0)|}{dc} = \frac{1}{2} - c + \tau \ln \frac{c}{1-c} = 0 \tag{4.1.17}$$

It follows from Eq. (4.1.17) that the T-c equilibrium diagram corresponding to the free energy (4.1.12) is determined by the equation

$$\tau = \frac{c - \frac{1}{2}}{\ln (c/(1-c))} \tag{4.1.18}$$

The equilibrium diagram is shown in Fig. 39. The spinodal curve $\tau = c(1-c)$ given by the equation

$$\frac{d^2 f(c)/|V(0)|}{dc^2} = -1 + \frac{\tau}{c(1-c)} = 0$$

is also presented in that figure.

It is noteworthy that the T-c diagram in Fig. 39 is correct for any long-range interaction potential, $W(\mathbf{r})$, providing a negative $V(0)$.

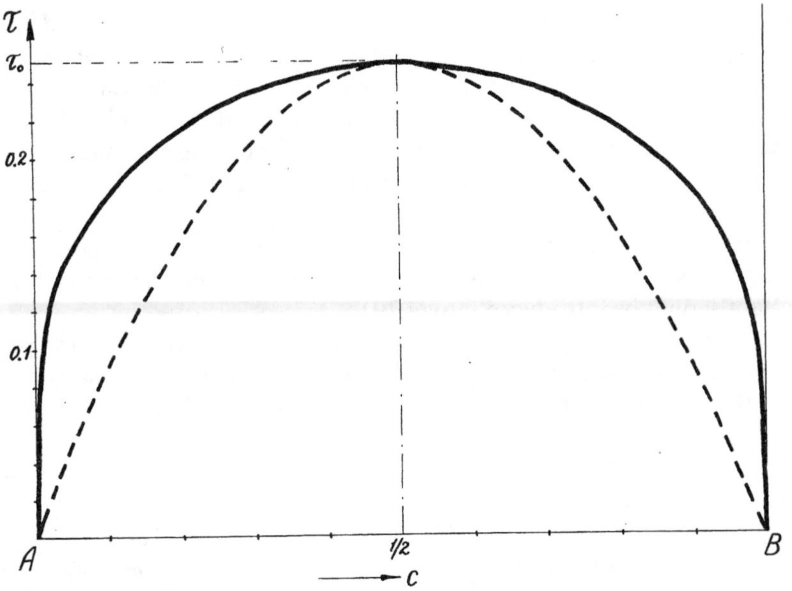

Figure 39. The equilibrium T–c diagram calculated in the mean-field approximation. The reduced temperature is $\tau = \kappa T/|V(0)|$; the solvus is designated by the solid line, and the dashed line describes the spinodal: $\tau_0 = 0.25$ is the reduced temperature corresponding to the top of the miscibility gap.

4.2. FREE ENERGY OF HETEROGENEOUS ALLOYS

The conclusions in Section 4.1 are based on the assumption that a multiphase alloy is a mixture of homogeneous phases whose free energy is equal to the sum of free energies of the constituents [see Eq. (4.1.3)].

This assumption is valid if all the concentration changes occur within interphase boundary layers whose volume is negligibly small compared with the volume of the homogeneous phases.

In certain cases, however, heterogeneous alloys cannot be treated as mixtures of homogeneous phases since composition changes occur continuously over the crystal bulk. The free energy of such an alloy does not fit Eq. (4.1.3) and further development of the theory is needed.

The assumption of the additivity of free energies (4.1.3) may be generalized in the form

$$F = \int_{(V)} f(c(\mathbf{r})) \frac{dV}{v} \tag{4.2.1}$$

where V is the alloy volume, v is the volume of the crystal unit cell, $c(\mathbf{r})$ is the atomic fraction of the solute (composition) at the point $\mathbf{r}$. Integration in (4.2.1) is carried out over the total alloy volume.

The free energy (4.2.1) cannot ensure the homogeneity of the equilibrium phases composing the two-phase mixture. It follows from that fact that all concentration distributions, $c(\mathbf{r})$, that assume only two values, c_α and c_β, and meet the conservation condition (4.1.1) provide the same value of the free energy (4.2.1). Therefore the free energy (4.2.1) cannot prevent the appearance of nonphysical concentration distributions discontinuous at each point **r**. In other words, a bicrystal of two homogeneous phases with compositions c_α and c_β and an arbitrary stepwise concentration profile assuming the values c_α and c_β will have the same free energy (4.2.1).

To remove this ambiguity, an additional nonlocal term depending on the compositions at neighboring points and vanishing when the composition is constant should be introduced into (4.2.1).

That term should depress all discontinuities along the concentration profile, creating free energy increase. It will be shown below that for a smooth concentration profile in an isotropic solution, this term is given by

$$\Delta F_{\text{het}} = \int_{(V)} \frac{1}{2} m(c)(\nabla c)^2 \frac{dV}{v} \tag{4.2.2}$$

where $m(c)$ is a positive value. This function in fact depresses all the discontinuities on the concentration profile since the function $m(\nabla c)^2$ increases infinitely at $c(\mathbf{r})$ function jump points.

The total free energy is given by the sum of terms (4.2.1) and (4.2.2):

$$F = \int_{(V)} \frac{dV}{v} \left[\frac{1}{2} m(\nabla c)^2 + f(c) \right] \tag{4.2.3}$$

The equation for the free energy in the form (4.2.3) has been used by Cahn in his theory of spinodal decomposition (42, 43) and earlier by Ornstein and Zernike in the theory of the critical opalescence (44, 45).

To derive Eq. (4.2.3), we must proceed from the general assumption that the specific free energy at point **r** is a function of all composition derivatives with respect to coordinates. The total free energy may then be written

$$F = \int_{(V)} f\left(c; \frac{\partial c}{\partial r_i}; \frac{\partial^2 c}{\partial r_i \partial r_j}, \ldots \right) \frac{dV}{v} \tag{4.2.4}$$

where i and j are the Cartesian indexes.

The introduction of the derivatives into the integrand provides the description of nonlocal characteristics of the specific free energy f. Expansion of the free energy $f(c, \partial c/\partial r_i, \partial^2 c/\partial r_i \partial r_j)$ into a Taylor series of all derivatives gives the sum of invariants

$$F = \int_{(V)} \left\{ f(c) + \mathbf{A}(c)\nabla c + \frac{1}{2} m_{ij}(c) \frac{\partial c}{\partial r_i} \frac{\partial c}{\partial r_j} + \frac{1}{2} \beta_{ij}(c) \frac{\partial^2 c}{\partial r_i \partial r_j} + \cdots \right\} \frac{dV}{v} \tag{4.2.5}$$

where $\mathbf{A}(c)$, $\beta_{ij}(c)$, and $m_{ij}(c)$ are the Taylor series coefficients. The term $\mathbf{A}\nabla c$ vanishes since the vector, $\mathbf{A}$, cannot exist in homogeneous media.

With smooth concentration profiles all the derivatives of the type $\partial^n c/\partial r_i \ldots \partial r_j$ have small values of the order of $(1/L)^n$ where L is a typical distance at which the composition, $c(\mathbf{r})$, changes considerably (in the case of a smooth concentration heterogeneity, L has a large value). The first nonvanishing terms of the expansion have values of the order $(1/L)^2$. There are two such terms, $m_{ij}\frac{\partial c}{\partial r_i}\frac{\partial c}{\partial r_j}$ and $\beta_{ij}\frac{\partial^2 c}{\partial r_i \partial r_j}$. The latter one may, however, be reduced to the form of $m_{ij}\frac{\partial c}{\partial r_i}\frac{\partial c}{\partial r_j}$ if the Gauss theorem is applied.* Therefore the expansion with the higher than first nonvanishing terms truncated is

$$F=\int_{(V)}\frac{dV}{v}\left(f(c)+\frac{1}{2}m_{ij}\frac{\partial c}{\partial r_i}\frac{\partial c}{\partial r_j}\right) \tag{4.2.6}$$

The only second-rank tensor referring to an isotropic medium is the Kronecker tensor δ_{ij}. Therefore

$$m_{ij}(c)=m(c)\delta_{ij} \tag{4.2.7}$$

Substituting Eq. (4.2.7) into (4.2.6) yields

$$F=\int_{(V)}\left[f(c)+\frac{1}{2}m(c)\frac{\partial c}{\partial r_i}\frac{\partial c}{\partial r_i}\right]\frac{dV}{v}=\int_{(V)}\left[f(c)+\frac{1}{2}m(\nabla c)^2\right]\frac{dV}{v} \tag{4.2.8}$$

The coefficient m should be positive to provide stability of the homogeneous state, $c(\mathbf{r})=$constant. Otherwise the appearance of a heterogeneity would be energetically favourable as it would make a negative contribution to the free energy.

To elucidate the microscopical meaning of the function $f(c)$ and coefficient m, Eq. (4.2.8) should be compared with the continuous limit of the free energy (3.6.3b) derived from microscopic considerations.

To carry out the limit transition to the continuous equation for the free energy, one should assume concentration heterogeneity to be smooth. Eq. (3.6.3b) may then be transformed by replacing the occupation probabilities, $n(\mathbf{r})$, with atomic fractions (compositions), $c(\mathbf{r})$, and summations with integrations over infinite space. The result of these replacements is

$$F=\tfrac{1}{2}\iint W(\mathbf{r}-\mathbf{r}')c(\mathbf{r})c(\mathbf{r}')\frac{dV}{v}\frac{dV'}{v}+\kappa T\int\frac{dV}{v}\left[c(\mathbf{r})\ln c(\mathbf{r})+(1-c(\mathbf{r}))\ln(1-c(\mathbf{r}))\right] \tag{4.2.9}$$

The integration limits are here extended to infinity since the integrands vanish outside the crystal volume.

$$\int_{(V)}\beta_{ij}\frac{\partial^2 c}{\partial r_i\partial r_j}dV=\int_{(V)}\frac{\partial}{\partial r_i}\left(\beta_{ij}(c)\frac{\partial c}{\partial r_j}\right)dV-\int_{(V)}\left(\frac{\partial}{\partial c}\beta_{ij}(c)\right)\frac{\partial c}{\partial r_i}\frac{\partial c}{\partial r_j}dV$$

$$=\oint_S\beta_{ij}(c)\frac{\partial c}{\partial r_j}dS_i-\int_{(V)}\frac{d\beta_{ij}(c)}{dc}\frac{\partial c}{\partial r_i}\frac{\partial c}{\partial r_j}dV=-\int_{(V)}\bar{\beta}_{ij}(c)\frac{\partial c}{\partial r_i}\frac{\partial c}{\partial r_j}dV$$

where $\bar{\beta}_{ij}(c)=d\beta_{ij}/dc$. The symbol $\oint_S dS_i(\cdots)$ implies integration over the surface S enveloping the integration volume V. The corresponding term has the value of the order of surface energy and may be neglected compared with the volume terms.

The first term in (4.2.9) is the internal energy [compare with (3.6.3a)]:

$$U=\frac{1}{2}\iint W(\mathbf{r}-\mathbf{r}')c(\mathbf{r})c(\mathbf{r}')\frac{dV}{v}\frac{dV'}{v} \tag{4.2.10}$$

Change of variables,

$$(\mathbf{r},\mathbf{r}')\rightarrow(\mathbf{r},\boldsymbol{\rho})$$

where $\boldsymbol{\rho}=\mathbf{r}-\mathbf{r}'$, transforms Eq. (4.2.10) into

$$U=\frac{1}{2}\int\frac{dV}{v}c(\mathbf{r})\int\frac{dV'}{v}W(\boldsymbol{\rho})c(\mathbf{r}-\boldsymbol{\rho}) \tag{4.2.11}$$

Taking advantage of the smooth character of the concentration profile $c(\mathbf{r})$, we may expand the function $c(\mathbf{r}-\boldsymbol{\rho})$ in powers of $\boldsymbol{\rho}$:

$$c(\mathbf{r}-\boldsymbol{\rho})\approx c(\mathbf{r})+\frac{\partial c(\mathbf{r})}{\partial\mathbf{r}}\boldsymbol{\rho}+\frac{1}{2}\frac{\partial^2 c(\mathbf{r})}{\partial r_i\partial r_j}\rho_i\rho_j+\cdots \tag{4.2.12}$$

The dimensionless parameter in the expansion (4.2.12) is in fact the r_c/L ratio where r_c is the range parameter of pairwise interaction energies $W(\mathbf{r}-\mathbf{r}')$, L is a typical distance at which the concentration $c(\mathbf{r})$ changes appreciably. In verifying this, one should bear in mind that the integration over $\boldsymbol{\rho}$ in (4.2.11) is in fact carried out over a small region, with dimensions comparable to the interaction range parameter, r_c. If $r_c/L\ll 1$, one may use the approximation (4.2.12) up to the second-order terms with respect to r_c/L, truncate all but three terms in the series (4.2.12). Substituting Eq. (4.2.12) into (4.2.10) yields

$$\begin{aligned}U&=\frac{1}{2}\int\frac{dV}{v}c(\mathbf{r})\int\frac{d^3\rho}{v}W(\boldsymbol{\rho})\left[c(\mathbf{r})+\frac{\partial c}{\partial\mathbf{r}}\boldsymbol{\rho}+\frac{1}{2}\frac{\partial^2 c}{\partial r_i\partial r_j}\rho_i\rho_j+\cdots\right]\\&=\frac{1}{2}\int\frac{dV}{v}c^2(\mathbf{r})V(0)+\frac{1}{2}\int\frac{dV}{v}c(\mathbf{r})\frac{\partial c}{\partial\mathbf{r}}\int\frac{d^3\rho}{v}\boldsymbol{\rho}W(\boldsymbol{\rho})\\&\quad-\frac{1}{2}\int\frac{dV}{v}m_{ij}c(\mathbf{r})\frac{\partial^2 c}{\partial r_i\partial r_j}\end{aligned} \tag{4.2.13}$$

where

$$V(0)=\int\frac{d^3\rho}{v}W(\boldsymbol{\rho}) \tag{4.2.14}$$

$$m_{ij}=-\int\frac{d^3\rho}{v}\rho_i\rho_j W(\boldsymbol{\rho}) \tag{4.2.15}$$

Since

$$W(\boldsymbol{\rho})=W(-\boldsymbol{\rho})$$

$$\int\boldsymbol{\rho}W(\boldsymbol{\rho})\frac{d^3\rho}{v}=0 \tag{4.2.16}$$

In an isotropic medium

$$m_{ij} = m\delta_{ij} \tag{4.2.17}$$

Therefore

$$m_{ij} = -\int \frac{d^3\rho}{v} \rho_i \rho_j W(\boldsymbol{\rho}) = -\frac{\delta_{ij}}{3} \int \frac{d^3\rho}{v} \rho^2 W(\rho) \tag{4.2.18}$$

that is,

$$m = -\frac{1}{3} \int \frac{d^3\rho}{v} \rho^2 W(\boldsymbol{\rho}) \tag{4.2.19}$$

Substituting Eqs. (4.2.16) and (4.2.17) into (4.2.13) results in the continuum limit

$$U = \frac{1}{2} \int \frac{dV}{v} \left[V(0)c^2 - mc\nabla^2 c\right] \tag{4.2.20}$$

where $\nabla^2 = \Delta = \partial^2/\partial r_1^2 + \partial^2/\partial r_2^2 + \partial^2/\partial r_3^2$ is the Laplacian.

The application of the Gauss theorem to Eq. (4.2.20) (see footnote on p. 105) yields

$$U = \frac{1}{2} \int \frac{dV}{v} \left[V(0)c^2 + m(\nabla c)^2\right] \tag{4.2.21}$$

Substituting (4.2.21) into the first term of (4.2.9) gives eventually

$$F = \int \frac{dV}{v} \left[f(c) + \frac{1}{2} m(\nabla c)^2 \right] \tag{4.2.22}$$

where

$$f(c) = \tfrac{1}{2} V(0)c^2 + \kappa T(c \ln c + (1-c) \ln (1-c)) \tag{4.2.23}$$

$$m = -\frac{1}{3} \int \frac{d^3\rho}{v} \rho^2 W(\boldsymbol{\rho}) \tag{4.2.24}$$

Eq. (4.2.22) coincides completely with the phenomenological Eq. (4.2.8). The former, however, has an obvious advantage. The definitions (4.2.23) and (4.2.24) in (4.2.22) reveal the microscopical meaning of both specific free energy (in the mean-field approximation) and phenomenological coefficient m which enter the phenomenological Eq. (4.2.8) as symbols.

In accordance with the previous conclusions, the coefficient m is responsible for nonlocal contributions to the free energy associated with interaction between neighboring concentration heterogeneity. The latter may be realized since according to Eq. (4.2.24) the coefficient m is proportional to the second moment of the pairwise interaction energy.

Finally, we shall prove that for a solid solution undergoing decomposition, the coefficient m has a positive value. In fact, according to the microscopic theory formulated in Chapter 3, the decomposition reaction occurs if the absolute minimum of the Fourier transform of the interchange energy $V(\mathbf{k})$ falls at the point $\mathbf{k}=0$ [see footnote to Eq. (3.7.11)]. The Taylor expansion of $V(\mathbf{k})$ in powers of $\mathbf{k}$ at $\mathbf{k}=0$ is given by

$$V(\mathbf{k})=V(0)+\frac{1}{2}\left(\frac{\partial^2 V(\mathbf{k})}{\partial k_i \partial k_j}\right)_{\mathbf{k}=0} k_i k_j+\cdots \tag{4.2.25}$$

where according to Eqs. (4.2.15), (4.2.24) and (3.7.6)

$$\left(\frac{\partial^2 V(\mathbf{k})}{\partial k_i \partial k_j}\right)_{\mathbf{k}=0}=\left(\frac{\partial^2}{\partial k_i \partial k_j}\sum_{\rho} W(\rho)e^{-i\mathbf{k}\rho}\right)_{\mathbf{k}=0}=-\sum_{\rho}\rho_i\rho_j W(\rho)$$

$$=-\frac{\delta_{ij}}{3}\sum_{\rho}\rho^2 W(\rho)=\delta_{ij}m \tag{4.2.26}$$

where

$$m=-\tfrac{1}{3}\sum_{\mathbf{r}} r^2 W(\mathbf{r}) \tag{4.2.26a}$$

and $\mathbf{r}$ is the crystal lattice site vector.

The linear term in the series (4.2.25) vanishes since

$$V(\mathbf{k})=V(-\mathbf{k})$$

Substituting (4.2.26) into (4.2.25) gives

$$V(\mathbf{k})=V(0)+\frac{m}{2}k^2+\cdots \tag{4.2.27}$$

Since $V(0)$ is the absolute minimum of $V(\mathbf{k})$ the second-order expansion coefficient m must be positive.

If, on the other hand, $V(0)$ is the maximum of $V(\mathbf{k})$, as in the case of ordering when the absolute minimum of $V(\mathbf{k})$ falls at $\mathbf{k}\neq 0$, the coefficient m should be negative.

4.3. EXTREME STATES OF SOLID SOLUTIONS

Eq. (4.2.8) makes it possible to find all the extremes of the free energy function, minima, maxima, and saddle points. This is important since the absolute free energy minimum describes the equilibrium state, the relative minima—the metastable states, and some of the saddle points—the critical nucleus states of the alloy.

Any extreme state may be determined by solving the variational problem. The necessary condition for an extremum is

$$\delta F=0 \tag{4.3.1}$$

where the additional condition of variation is the conservation of the number of solute atoms

$$\int_{(V)} \frac{dV}{v} c(\mathbf{r}) = \text{constant} \tag{4.3.2}$$

This variational problem may be solved by the Lagrange method of undetermined coefficients, by the determination of the variation of the functional

$$\Omega = F - \mu \int_{(V)} \frac{dV}{v} c(\mathbf{r}) \tag{4.3.3}$$

where μ is the Lagrange multiplier, and by setting the variation equal to zero. In other words, an extreme state may be found from the variational equation

$$\delta\Omega = \delta\left(F - \mu \int_{(V)} \frac{dV}{v} c(\mathbf{r})\right) = 0 \tag{4.3.4}$$

where the functional Ω plays the part of the thermodynamic potential. Substituting the free energy (4.2.8) into (4.3.3) gives

$$\Omega = \int_{(V)} \frac{dV}{v} \left[\frac{1}{2} m(\nabla c)^2 + f(c) - \mu c\right] \tag{4.3.5a}$$

The evaluation of the first variation of (4.3.5a) leads to

$$\delta\Omega = \int_{(V)} \frac{dV}{v} \left[\frac{df}{dc}\delta c - \mu\delta c + m \cdot \nabla c \cdot \nabla\delta c\right] = 0 \tag{4.3.5b}$$

The transformation of the third term in (4.3.5b) by application of the Gauss theorem yield

$$\begin{aligned} \int_{(V)} m\nabla c\nabla\delta c \frac{dV}{v} &= \int_{(V)} m[\nabla(\delta c\nabla c) - \delta c\nabla^2 c]\frac{dV}{v} = m\int_{(V)} \nabla(\delta c\nabla c)\frac{dV}{v} \\ &- m\int_{(V)} \delta c\nabla^2 c\frac{dV}{v} = m\oint_S d\mathbf{S}\delta c\nabla c - m\int_{(V)} \frac{dV}{v}\delta c\nabla^2 c \end{aligned} \tag{4.3.6}$$

where $d\mathbf{S}$ is an element of surface S enveloping the integration volume V. Since the boundary condition is $\delta c(\mathbf{r}) \equiv 0$ on the surface S, the surface integral in (4.3.6) vanishes. So we have

$$\int_{(V)} \frac{dV}{v} \nabla c \cdot \nabla\delta c = -m\int_{(V)} \frac{dV}{v} \delta c(\mathbf{r})\nabla^2 c \tag{4.3.7}$$

Substituting Eq. (4.3.7) into (4.3.5b) gives

$$\delta\Omega = \int_{(V)} \frac{dV}{v}\left[\frac{df}{dc} - \mu - m\nabla^2 c\right]\delta c = 0 \tag{4.3.8}$$

Since the variation, $\delta c(\mathbf{r})$, is an arbitrary infinitesimal function, Eq. (4.3.8) can only be satisfied if

$$\frac{df}{dc} - \mu - m\nabla^2 c = 0 \tag{4.3.9}$$

Mathematically, Eq. (4.3.9) is the Euler equation of the variational problem.

All solutions to (4.3.9) describe free energy extrema. This equation, however, fails to give information on the type of the extreme found, whether it is a minimum, a maximum, or a saddle point.

According to Khachaturyan and Suris (46) the latter problem may be solved by analyzing the second variation of the free energy. Since the first variation of the functional Ω is

$$\delta\Omega = \int_{(V)} \frac{dV}{v}\left[\frac{df}{dc} - \mu - m\nabla^2 c\right]\delta c(\mathbf{r})$$

the second variation is given by

$$\delta^2\Omega = \delta^2 F = \int_{(V)} \frac{dV}{v}\,\delta c(\mathbf{r})\left[\frac{d^2f}{dc^2} - m\nabla^2\right]\delta c(\mathbf{r}) \tag{4.3.10}$$

If $c = c_0(\mathbf{r})$ is a solution of Eq. (4.3.9) to be analyzed, the second variation

$$\delta^2 F = \int_{(V)} \frac{dV}{v}\,\delta c(\mathbf{r})\left[\frac{d^2f}{dc^2} - m\nabla^2\right]_{c=c_0(\mathbf{r})}\delta c(\mathbf{r}) \tag{4.3.11}$$

gives the deviation of the free energy from the extreme value caused by the deviation of the concentration profile, $\delta c(\mathbf{r})$, from the concentration profile $c = c_0(\mathbf{r})$.

If $c = c_0(\mathbf{r})$ corresponds to a free energy minimum, the free energy deviation must be positive. Conversely, if $c = c_0(\mathbf{r})$ corresponds to the energy maximum, the free energy deviation must be negative. Finally, if $c = c_0(\mathbf{r})$ corresponds to a saddle point, the deviation may be of either sign. It thus follows that, to determine the type of a given extreme, one must determine the sign of the second variation of free energy (4.3.11).

One can easily see that the second variation (4.3.11) is a quadratic form:

$$\delta^2 F = \int_{(V)} \frac{dV}{v}\,\delta c(\mathbf{r})\hat{\mathbf{H}}\delta c(\mathbf{r}) \tag{4.3.12a}$$

of the differential operator

$$\hat{\mathbf{H}} = \left[-m\nabla^2 + \frac{d^2f}{dc^2}\right]_{c=c_0(\mathbf{r})} \tag{4.3.12b}$$

The quadratic form (4.3.12a) takes positive values under any infinitesimal variation $\delta c(\mathbf{r})$ if it is a positive definite quadratic form. The latter should be the case if all the eigenvalues of the operator $\hat{\mathbf{H}}$ are positive. If all the operator $\hat{\mathbf{H}}$ eigenvalues are negative, the quadratic form (4.3.12) is negative definite. Finally, if some of the eigenvalues are positive and the others negative, the quadratic form (4.3.12) may take both positive and negative values. The sign of the second derivative depends on the choice of the variation $\delta c(\mathbf{r})$ in the latter case. In other words, the sign of the free energy increment (4.3.12) depends on the direction along which the representative point is displaced in the phase space from the extremum.

The first case corresponds to the situation where the free energy is minimum, at $c=c_0(\mathbf{r})$, the second one corresponds to a free energy maximum, and the third one to a saddle point.

The analysis is thus reduced to the determination of the operator $\hat{\mathbf{H}}$ spectrum, i.e. to the eigenvalue problem:

$$\hat{\mathbf{H}}\psi=\left[-m\nabla^2+\left(\frac{d^2f}{dc^2}\right)_{c=c_0(\mathbf{r})}\right]\psi=\varepsilon\psi \qquad (4.3.13)$$

where ε is the eigenvalue and ψ the eigenfunction. It should be noted that Eq. (4.3.13) is analogous to the Schrödinger wave equation for a particle in a potential well $(d^2f/dc^2)_{c=c_0(\mathbf{r})}$.

4.4. CRITICAL NUCLEUS IN A SOLID SOLUTION

The analysis of stability of a solid solution carried out in Section 2.2 shows that a transformation from a metastable to a stable state occurs by overcoming the smallest barrier. The top of this barrier is a saddle point of the free-energy hypersurface. It corresponds to a concentration heterogeneity usually called a critical nucleus. As the saddle point associated with a critical nucleus is an extremum of the free energy hypersurface; the concentration profile describing the critical nucleus state is given by a solution of the nonlinear Eq. (4.3.9). A critical nucleus should be a local heterogeneity. Otherwise (if heterogeneity $c=c_0(\mathbf{r})$ is extended over the whole crystal), free energy increase due to the heterogeneity would be proportional to the crystal volume. To overcome such a barrier would require a macroscopical free energy increase forbidden by the second principle of thermodynamics.

We may assume that a heterogeneity corresponding to a critical nucleus in an isotropic medium is spherically isotropic and may therefore be described by the function

$$c=c_0(|\mathbf{r}|)$$

It will be shown below that any spherical local heterogeneity $c=c_0(|\mathbf{r}|)$ obtained as a solution of Eq. (4.3.9) corresponds to a saddle-point state on the free energy hypersurface.

In the case of a spherical local heterogeneity $c=c_0(|\mathbf{r}|)$ Eq. (4.3.13) becomes identical to the well-known Schrödinger equation describing a particle in a centrosymmetrical field. Eq. (4.3.13) may be separated into the radial and angular parts, exactly like the Schrödinger equation [see, e.g. (47)]. The radial part of Eq. (4.3.13) is then obtained in the form

$$\left[-m\left(\frac{1}{r^2}\frac{d}{dr}r^2\frac{d}{dr}-\frac{l(l+1)}{r^2}\right)+U(r)-\varepsilon_n(l)\right]R_{n_r}(l,r)=0 \qquad (4.4.1)$$

where n_r is the radial "quantum number" ($n_r=0, 1, 2, \ldots, \infty$), l is the "azimuthal quantum number", $U(r)=(d^2f/dc^2)_{c=c_0(\mathbf{r})}$.

Let the azimuthal number l be equal to unity. Eq. (4.4.1) then becomes

$$\left[-m\left(\frac{1}{r^2}\frac{d}{dr}r^2\frac{d}{dr}-\frac{2}{r^2}\right)+U(r)-\varepsilon_{n_r}(1)\right]R_{n_r}(1, r)=0 \tag{4.4.2}$$

We may now compare Eqs. (4.4.2) and (4.3.9) for a spherically symmetrical function $c_0(r)$. The latter equation assumes the form

$$-m\left(\frac{1}{r^2}\frac{d}{dr}r^2\frac{d}{dr}\right)c_0(r)+\left(\frac{df(c)}{dc}\right)_{c=c_0(\mathbf{r})}-\mu=0 \tag{4.4.3}$$

where

$$\left(\frac{1}{r^2}\frac{d}{dr}r^2\frac{d}{dr}\right)=\frac{d^2}{dr^2}+\frac{2}{r}\frac{d}{dr}$$

is the radial part of the Laplacian ∇^2.

The first derivative of (4.4.3) with respect to r is

$$-m\left(\frac{1}{r^2}\frac{d}{dr}r^2\frac{d}{dr}-\frac{2}{r^2}\right)\frac{dc_0}{dr}+\left(\frac{d^2f}{dc^2}\right)_{c=c_0(\mathbf{r})}\frac{dc_0}{dr}=0 \tag{4.4.4}$$

Since $U(r)=(d^2f/dc^2)_{c=c_0(\mathbf{r})}$, comparison of Eqs. (4.4.2) and (4.4.4) enables us to make an important conclusion that the function $dc_0(r)/dr$ is an eigenfunction corresponding to the zero eigenvalue $\varepsilon_{n_r^0}(1)$:

$$\varepsilon_{n_r^0}(1)=0 \qquad R_{n_r^0}(1, r)=\frac{dc_0(r)}{dr} \tag{4.4.5}$$

The result obtained shows that it is not necessary to solve the eigenvalue problem to determine the eigenfunction of Eq. (4.4.2) corresponding to the zero eigenvalue. If the solution of Eq. (4.4.3), $c_0(r)$, is known, this eigenfunction may be found as the first derivative of $c_0(r)$ with respect to the radius, r.

We should next determine the radial number n_r^0 for which $\varepsilon_{n_r^0}(1)=0$. This may be done using the Sturm theorem which relates the radial number n_r to the number of nodal points of the eigenfunction $R_{n_r}(l, r)$.

According to the Sturm theorem eigenvalues $\varepsilon_{n_r}(l)$ increase with $n_r(n_r \geqslant 0)$, and the radial eigenfunction $R_n(l, r)$ corresponding to the n_r+1 eigenvalue $\varepsilon_{n_r}(l)$ has n_r nodal points away from the point $r=0$. If the function $c=c(r)$ describes a spherical local concentration heterogeneity that decreases monotonically as r increases, the function $dc_0(r)/dr$ vanishes at $r=0$ only, that is, does not have any nodal points at $r\neq 0$. According to the Sturm theorem that means that the radial number n_r^0 corresponding to the zero eigenvalue $\varepsilon_{n_r^0}(1)=0$ is also equal to zero: the zero eigenvalue corresponds to the state with n_r^0 equal to 0 and l equal to 1.

If the "principal quantum number"

$$n=n_r+l+1$$

is introduced, the conventional designation of the state $n_r=0$ and $l=1$ is $2p$ (we employ here the usual symbols, s, p, d, f, to describe states with $l=1, 2, 3, 4$, respectively). According to (47) the state $1s(n_r=0,\ l=0)$, however, has a lower energy than the $2p$ state ($n_r=0$, $l=1$) since the ground state of a particle in a spherically symmetrical field is always an s state ($l=0$). Hence

$$\varepsilon_0(0)<\varepsilon_0(1)=0$$

and the spectrum of the operator $\hat{\mathbf{H}}$ contains both positive and negative eigenvalues. Consequently the corresponding concentration profile, $c_0(r)$, refers to a saddle point of the free energy hypersurface.

An example of the concentration profile of a critical nucleus will be given. We shall proceed from the mean-field approximation for the specific free energy (4.1.15). Let us consider a representative point, $c=0.2$ and $\tau=\kappa T/|V(0)|=0.2$, of the diagram shown in Fig. 39. That point lies between the solubility and spinodal curves and therefore corresponds to the metastable state of the homogeneous solution.

Since the critical nucleus is an extreme state, it is described by Eq. (4.3.9). The corresponding concentration profile must meet the boundary conditions

$$\lim_{r\to\infty} c(r)=\bar{c} \qquad \lim_{r\to\infty} (\nabla^2 c)=0 \tag{4.4.6}$$

where $\bar{c}$ is the composition of the homogeneous alloy. The boundary conditions reflect the fact that the appearance of a critical nucleus should not affect the composition profile outside its vicinity.

At r approaching infinity, we therefore have, according to (4.3.9) and (4.4.6),

$$\frac{df(\bar{c})}{dc}-\mu=0 \quad \text{and} \quad \mu=\frac{df(\bar{c})}{d\bar{c}} \tag{4.4.7}$$

Substituting Eq. (4.4.7) into (4.3.9) yields

$$m\nabla^2 c-\frac{df(c)}{dc}+\frac{df(\bar{c})}{d\bar{c}}=0 \tag{4.4.8}$$

Since $c=c(r)$ depends solely on r

$$\nabla^2 c(r)=\frac{1}{r}\frac{d^2}{dr^2}(rc)$$

and Eq. (4.4.8) may be rewritten in the form

$$\frac{1}{\rho}\frac{d^2}{d\rho^2}(\rho c)-\frac{df(c)}{|V(0)|dc}+\frac{df(\bar{c})}{|V(0)|d\bar{c}}=0 \tag{4.4.9}$$

where

$$\rho=r\sqrt{|V(0)|/m} \tag{4.4.9a}$$

Figure 40 demonstrates the plot of $df(c)/dc-df(\bar{c})/d\bar{c}$ as a function of c calculated within the framework of the mean-field approximation (4.1.15). Eq. (4.4.9) is a nonlinear equation and cannot be solved analytically. If, however, the function $df(c)/dc-df(\bar{c})/d\bar{c}$ with respect to c is approximated by two straight line segments, as shown in Fig. 40, Eq. (4.4.9) may be solved. The approxima-

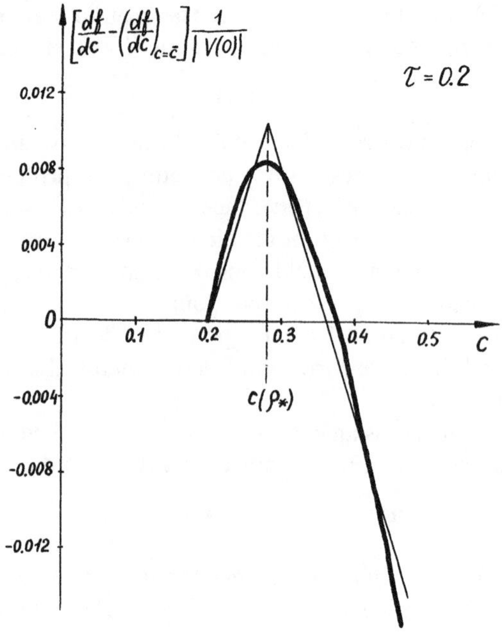

Figure 40. Concentration dependence of the function $(df/dc)-(df/dc)_{c=\bar{c}}$ calculated in the mean-field approximation [Eq. (4.1.15)] at the reduced temperature $\tau=0.2$ and $\bar{c}=0.2$. Straight lines show the linear approximation

tion used is as follows:

$$\frac{1}{|V(0)|}\left[\frac{df(c)}{dc}-\frac{df(\bar{c})}{d\bar{c}}\right]=\begin{cases}0.1333(c-0.2) & \text{at } 0.2<c\leqslant 0.28\\ -0.125(c-0.36) & \text{at } c>0.28\end{cases} \tag{4.4.10}$$

Then Eq. (4.4.9) becomes within the range $0.2<c<0.28$

$$\frac{d^2\rho c}{d\rho^2}-0.1333\rho(c-0.2)=0 \tag{4.4.11}$$

The range $0.2<c<0.28$ corresponds to $\rho>\rho_*$ where ρ_* is the nucleus radius defined by the equation

$$c(\rho_*)=0.28 \tag{4.4.12}$$

The solution of (4.4.11) is

$$c(\rho)=0.2+\frac{A}{\rho}e^{-0.3647\rho}+\frac{B}{\rho}e^{0.3647\rho} \tag{4.4.13}$$

at $\rho>\rho_*$, where A and B are integration constants.

The boundary condition (4.4.6)

$$\lim_{\rho\to\infty} c(\rho)=\bar{c}=0.2$$

is met if $B=0$. Hence

$$c(\rho)=0.2+\frac{A}{\rho}\exp(-0.3647\rho) \tag{4.4.14}$$

Using Eq. (4.4.9) and approximation (4.4.10), we obtain for the range $c>0.28$

$$\frac{d^2 c\rho}{d\rho^2}+0.125\rho(c-0.36)=0 \tag{4.4.15}$$

Eq. (4.4.15) holds in the range $0<\rho<\rho_*$. The real solution to Eq. (4.4.15) is

$$c(\rho)=0.36+\frac{Ce^{i0.3535\rho}+C^*e^{-i0.3535\rho}}{\rho} \tag{4.4.16}$$

where C is a complex integration constant. To provide a finite value for $c(r)$ at $\rho=0$, C should be set equal to $D/2i$ in Eq. (4.4.16).

Then

$$c(\rho)=0.36+\frac{D\sin 0.3535\rho}{\rho} \tag{4.4.17}$$

where D is a new real integration constant. We thus have

$$c(\rho)=\begin{cases} 0.2+\dfrac{A}{\rho}e^{-0.3647\rho} & \text{at } \rho>\rho_* \\ 0.36+\dfrac{D}{\rho}\sin 0.3535\rho & \text{at } \rho<\rho_* \end{cases} \tag{4.4.18}$$

The constants A and D may be found from the condition of continuity of the functions $c(\rho)$ and $dc/d\rho$ at the point ρ_* where $c(\rho_*)=0.28$. The continuity conditions for $c(\rho)$ at $\rho=\rho_*$ are

$$0.2+\frac{A}{\rho_*}e^{-0.3647\rho_*}=0.28$$

$$0.36+\frac{D}{\rho_*}\sin 0.3535\rho_*=0.28 \tag{4.4.19}$$

The continuity condition for $dc(\rho)/d\rho$ at $\rho=\rho_*$ is

$$A\left(\frac{d}{d\rho}\frac{e^{-0.3647\rho}}{\rho}\right)_{\rho=\rho_*}=D\left(\frac{d}{d\rho}\frac{\sin 0.3535\rho}{\rho}\right)_{\rho=\rho_*} \tag{4.4.20}$$

Eqs. (4.4.19) and (4.4.20) form a set of three transcendental equations with three unknowns, A, D, and ρ_*. The solution to this set gives

$$A=38.5122 \qquad D=1.5628 \qquad \rho_*=10.4908 \tag{4.4.21}$$

Substituting the quantities (4.4.21) into Eq. (4.4.18) and the use of the definition of ρ given by (4.4.9a),

$$\rho=\frac{r}{r_0} \quad \text{where } r_0=\sqrt{\frac{m}{|V(0)|}} \tag{4.4.22}$$

yield the concentration profile for the critical nucleus

$$c(r)=\begin{cases} 0.2+\dfrac{38.5122\exp(-0.3647\,r/r_0)}{r/r_0} & \text{at } r\geqslant 10.4908r_0 \\ 0.36+\dfrac{1.5628\sin 0.3535r/r_0}{r/r_0} & \text{at } r<10.4908r_0 \end{cases} \tag{4.4.23}$$

The concentration profile of the critical nucleus is plotted in Fig. 41.

The nearest-neighbor interaction in a fcc lattice is characterized by the following relations:

$$m=-\tfrac{1}{3}\sum_{\mathbf{r}} W(\mathbf{r})r^2=-\tfrac{1}{3}\cdot 12W_1\frac{a^2}{2}=2|W_1|a^2 \quad \text{[see Eq. (4.2.26a)]}$$

$$V(0)=\sum_{\mathbf{r}} W(\mathbf{r})=12W_1=-12|W_1| \quad \text{[see Eq. (4.1.11)]}$$

where a is the crystal lattice parameter of the fcc lattice, W_1 is the nearest-neighbor interaction energy.

Then according to (4.4.21) and (4.4.22)

$$r_0=\sqrt{\frac{m}{|V(0)|}}=\sqrt{\frac{2|W_1|a^2}{12|W_1|}}=\frac{a}{\sqrt{6}} \tag{4.4.24}$$

and the critical radius is

$$r_*=10.4908r_0=10.4908\frac{a}{\sqrt{6}}=4.28a$$

Now a few more points concerning the critical nuclei. The discussion in Section 2.2 leads one to conclude that the critical nucleus can not be described

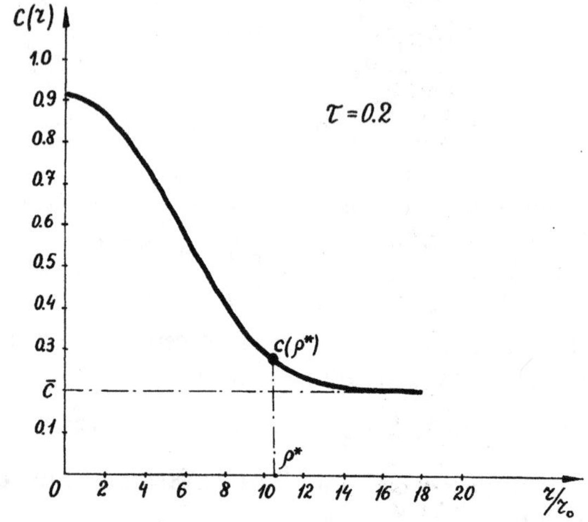

Figure 41. Concentration profile of a critical nucleus calculated in the mean-field approximation at the reduced temperature $\tau=0.2$ and composition $\bar{c}=0.2$. The radius of a critical nucleus is calculated $\rho_*=10.4908r_0$ at $c(\rho_*)=0.28$.

accurately within the framework of the conventional nucleation theory (48, 49). In fact, according to the conventional theory, the critical nucleus is a small new phase particle whose boundary layer thickness is far smaller than the particle radius. As mentioned in Section 2.2, this assumption does not hold with critical nuclei whose boundary layer thickness is comparable to its radius, and both values are commensurable with the interaction energy range. The calculated concentration profile plotted in Fig. 41 confirms this conclusion.

4.5. EXTREME STATES OF ONE-DIMENSIONAL HETEROGENEITIES IN METASTABLE ALLOYS

Khachaturyan and Suris (46) have shown that the exact analytical solution of the problem of alloy extreme states may be obtained for the one-dimensional case.* At the first thought the suggestion that the extreme states may be described by one-dimensional concentration distribution looks highly unrealistic. It is difficult to conceive of how this model can be applied to actual systems. In many cases, however, one-dimensional distributions prove to be energetically favorable. It will be shown that the formation of one-dimensional distributions along the "softest" elastic modulus direction produces the smallest increase in the elastic energy caused by heterogeneity. In cases of small interphase energies and large differences between the atomic radii of solute and solvent atoms, such heterogeneities as critical nuclei and new phase particles may be regarded as pseudo-one-dimensional ones (a rigorous criterion for the applicability of the one-dimensional approximation will be derived in Section 8.1).

For the one-dimensional case Eq. (4.3.9) describing the concentration profiles of the extreme states may be reduced to

$$m\frac{d^2c}{dx^2} - \frac{df}{dc} + \mu = 0 \tag{4.5.1}$$

where x is the coordinate in the direction of the concentration gradient.

It is interesting that Eq. (4.5.1) has the same form as the equation of motion describing the classical nonlinear oscilator. The x, c, and m quantities stand for time, coordinate, and oscillator mass, respectively, and the term $-\dfrac{df(c)}{dc}+\mu$ stands for the restoring force.

Eq. (4.5.1) may be solved after the introduction of new variables, the "momentum"

$$p = m\frac{dc}{dx}$$

and "coordinate" c. Then

$$m\frac{d^2c}{dx^2} = \frac{d}{dx}p = \frac{dc}{dx}\frac{dp}{dc} = \frac{1}{m}p\frac{dp}{dc} = \frac{d}{dc}\frac{p^2}{2m}$$

*Later their derivations were repeated by Langer (50).

and Eq. (4.5.1) may be rewritten in the form

$$\frac{d}{dc}\left[\frac{p^2}{2m} - f(c) + \mu c\right] = 0 \tag{4.5.2}$$

Integrating (4.5.2) yields

$$\frac{p^2}{2m} - f(c) + \mu c = -E \tag{4.5.3}$$

where E is an integration constant representing the integral of motion. This quantity is an analogue of the total energy of an oscillator whose potential energy is given by $-f(c)+\mu c$. The value $f(c)-\mu c$ is the specific thermodynamic potential (thermodynamic potential per crystal lattice site).

In a two-phase field of an equilibrium diagram, Eq. (4.1.9) is reduced to the combined equations

$$\left[\frac{d}{dc}(f(c)-\mu c)\right]_{c=c_\alpha} = \left[\frac{d}{dc}(f(c)-\mu c)\right]_{c=c_\beta} = 0 \tag{4.5.4a}$$

$$f(c_\alpha) - \mu c_\alpha = f(c_\beta) - \mu c_\beta \tag{4.5.4b}$$

where c_α and c_β are the equilibrium compositions of the coexisting phases α and β. Eqs. (4.5.4) show that the thermodynamic potential has two absolute minima at the points c_α and c_β [Eq. (5.5.4a)] and that the magnitudes of the function $f(c)-\mu c$ at those minima are equal to each other [Eq. (4.5.4b). The dependence of thermodynamic potential with respect to c for an equilibrium state is plotted in Fig. 42a.

The two minima of $f(c)-\mu c$ are not equal to each other in a nonequilibrium (supercooled) state (Fig. 43a).

The characteristics of the thermodynamic potential in a two-phase region considered above provide the possibility of analyzing all the solutions of the nonlinear equation (4.5.1).

It follows from (4.5.3) that

$$p = m\frac{dc}{dx} = \sqrt{2m(f(c) - \mu c - E)} \tag{4.5.5}$$

Integrating (4.5.5) yields the concentration profile for an extreme state

$$m\int \frac{dc}{\sqrt{2m(f(c)-\mu c - E)}} = x - x_0 \tag{4.5.6}$$

where x_0 is the integration constant representing the origin of the coordinate system.

Eq. (4.5.6) includes two parameters, the chemical potential μ and "energy" E. The chemical potential, μ, which is a Lagrange multiplier may be found by the usual way from the additional condition of conservation of the number of solute atoms. In the one-dimensional case the conservation condition (4.3.2) may be

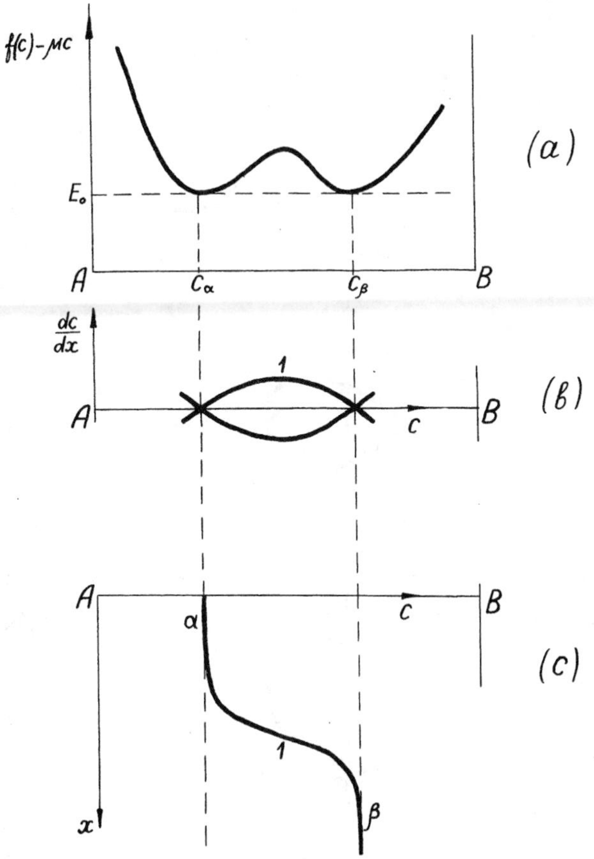

Figure 42. Determination of the equilibrium concentration profile of a two-phase alloy. (*a*) The typical plot of the thermodynamic potential $f(c)-\mu c$ with respect to c in the equilibrium state (case 1); (*b*) the phase trajectory corresponding to the equilibrium two-phase state (c_α and c_β are equilibrium compositions of equilibrium phases α and β); (*c*) the concentration profile describing the interphase boundary.

written in the form

$$S\int_{-L/2}^{L/2} c(x,\mu,E)\frac{dx}{v}=N_1 \tag{4.5.7}$$

where L is the length of the sample in the x direction, S the area of the cross section normal to x, v the volume per crystal lattice site, and N_1 the number of solute atoms. The left-hand side of Eq. (4.5.7) may be reduced to

$$S\int_{-L/2}^{L/2}\frac{dx}{v}c(x,\mu,E)=\frac{SL}{v}\frac{1}{L}\int_{-L/2}^{L/2}dx\,c(x,\mu,E)=N\frac{1}{L}\int_{-L/2}^{L/2}dx\,c(x,\mu,E) \tag{4.5.8}$$

where SL is the total sample volume and $N=SL/v$ the total number of crystal

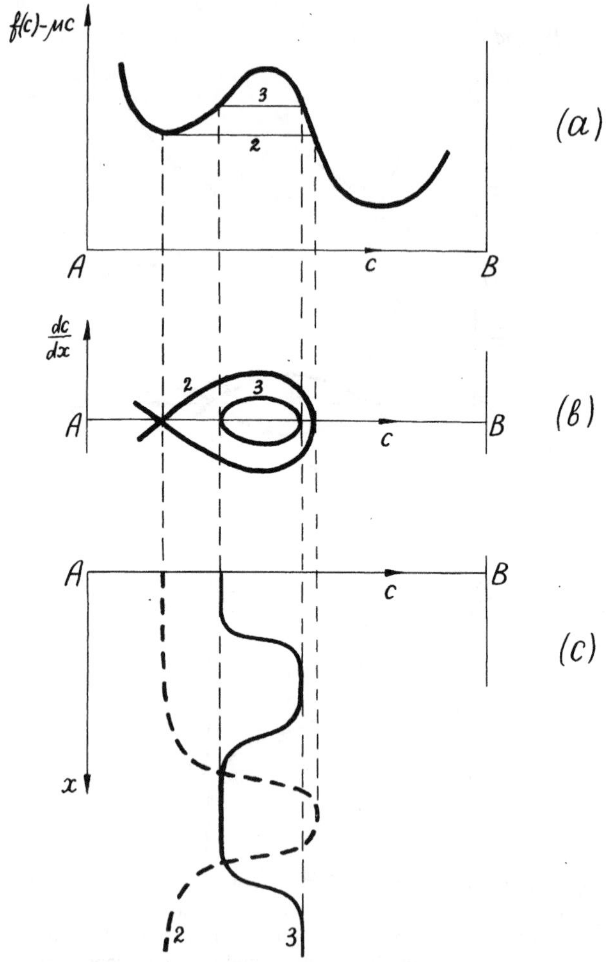

Figure 43. Determination of the concentration profiles for a one-dimensional critical nucleus and the one-dimensional periodic distribution corresponding to the free energy extremum. Curves 2 and 3 designate the critical nucleus and the periodic extreme state, respectively. (*a*) The typical plot of the thermodynamic potential $f(c)-\mu c$ with respect to c in a supercooled state; (*b*) the phase trajectories corresponding to the critical nucleus case and the case of the periodic extreme state; (*c*) the typical concentration profiles describing a critical nucleus and periodic extreme state.

lattice sites according to the definition of v. Substituting Eq. (4.5.8) into (4.5.7) yields

$$N\frac{1}{L}\int_{-L/2}^{L/2} c(x, \mu, E)dx = N_1$$

or (4.5.9)

$$\frac{1}{L}\int_{-L/2}^{L/2} c(x, \mu, E)dx = \bar{c}$$

where $\bar{c}=N_1/N$ is the atomic fraction of solute atoms. Eq. (4.5.9) thus determines the chemical potential μ. The other free parameter, E, determines the whole variety of extreme one-dimensional states.

The analysis of Eq. (4.5.5) by the method of phase trajectories in Figs. 42 and 43 shows that solutions of three types are possible for Eq. (4.5.1).*

The first case displayed in Fig. 42c describes two adjacent equilibrium phases, $\alpha+\beta$, with the equilibrium compositions c_α and c_β, separated by an intermediate interphase layer. This is the case when two absolute minima of the specific thermodynamic potential are equal to each other and $E=E_0=\min[f(c)-\mu c]$ where $\min[f(c)-\mu c]$ is the value of the absolute potential minimum.

The second case is shown in Fig. 43(b). This is the case of a one-dimensional critical nucleus. One-dimensional critical nuclei are formed when E is equal to the thermodynamic potential value at the alloy composition, $\bar{c}$, when

$$E=f(\bar{c})-\mu\bar{c} \tag{4.5.10}$$

where

$$\mu=\left(\frac{df(c)}{dc}\right)_{c=\bar{c}} \tag{4.5.11}$$

The third case presented in Fig. 43(c) corresponds to the general situation of a periodic concentration distribution.

The first case describes the equilibrium two-phase state because the thermodynamic potential then reaches its absolute minimum. For the one-dimensional model the thermodynamic potential (4.3.5a) is given by

$$\Omega=S\int_{-L/2}^{L/2}\frac{dx}{v}\left[\frac{m}{2}\left(\frac{dc}{dx}\right)^2+f(c)-\mu c\right] \tag{4.5.12}$$

To prove that potential (4.5.12) has the minimum value in the case presented in Fig. 42, we will write the function $f(c)-\mu c$ in the form

$$f(c)-\mu c=E_0+\frac{p^2}{2m}=E_0+\frac{m}{2}\left(\frac{dc}{dx}\right)^2 \tag{4.5.13}$$

using the "motion integral" (4.5.3). Substituting Eq. (4.5.13) into (4.5.12) yields

$$\Omega=\frac{SL}{v}\left[E_0+\frac{1}{L}\int_{-L/2}^{L/2}dx\, m\left(\frac{dc}{dx}\right)^2\right]=N\left[E_0+\frac{1}{L}\int_{-L/2}^{L/2}dx\frac{p^2}{m}\right] \tag{4.5.14}$$

Since the integral

$$\int_{-L/2}^{L/2}dx\cdot m\left(\frac{dc}{dx}\right)^2$$

in Eq. (4.5.14) has a finite value at L approaching infinity (the function $(dc/dx)^2$

*The details of the method of phase trajectories may be found in any book dealing with nonlinear vibrations.

does not vanish only within the finite range of the interphase layer),

$$\Omega \to NE_0 \quad \text{at} \quad L \to \infty \tag{4.5.15a}$$

Since $E_0 = \min (f(c) - \mu c)$,

$$\Omega_{\min} = NE_0 \tag{4.5.15b}$$

Therefore we have

$$\Omega \to \Omega_{\min}$$

at the microscopical limit $L \to \infty$.

This proves that the case under discussion corresponds to an equilibrium two-phase distribution.

Substituting Eq. (4.5.15b) into (4.5.14) gives at $L \to \infty$

$$\Omega = \Omega_{\min} + S \int_{-\infty}^{\infty} \frac{dx}{v} \frac{p^2}{m} \tag{4.5.16}$$

where the second term

$$\Omega_s = S \int_{-\infty}^{\infty} \frac{dx}{v} \frac{p^2}{m} = \gamma_s S$$

is proportional to the interphase area S and consequently represents the surface energy. The surface energy coefficient is

$$\gamma_s = \int_{-\infty}^{\infty} \frac{dx}{v} \frac{p^2}{m} \tag{4.5.17}$$

Eq. (4.5.17) may be reduced to the integral of composition

$$\gamma_s = \int_{-\infty}^{\infty} dx \frac{p^2}{mv} = \int_{c_\alpha}^{c_\beta} \frac{dx}{dc} \frac{p^2}{mv} dc = \int_{c_\alpha}^{c_\beta} dc \frac{p^2}{vm(dc/dx)} = \int_{c_\alpha}^{c_\beta} \frac{p}{v} dc \tag{4.5.18}$$

Substituting Eq. (4.5.5) into (4.5.18) yields

$$\gamma_s = \int_{c_\alpha}^{c_\beta} \sqrt{2m(f(c) - \mu c - E_0)} \frac{dc}{v} \tag{4.5.19}$$

where, according to (4.5.4a), $\mu = (df/dc)_{c=c_\alpha} = (df/dc)_{c=c_\beta}$ and $E_0 = \min [f(c) - \mu c]$. Eq. (4.5.19) may therefore finally be rewritten as

$$\gamma_s = \int_{c_\alpha}^{c_\beta} \sqrt{2m\{f(c) - \mu c - \min [f(c) - \mu c]\}} \frac{dc}{v} \tag{4.5.20}$$

Eq. (4.5.20) for the interphase energy was first derived by Cahn and J. E. Hilliard (51).

Returning to the critical nucleus case presented in Fig. 43 (part b), we may rewrite Eq. (4.5.6) in the form

$$m \oint \frac{dc}{\sqrt{2m\left[f(c) - c\dfrac{df(\bar{c})}{d\bar{c}} - f(\bar{c}) + \bar{c}\dfrac{df(\bar{c})}{d\bar{c}}\right]}} = x \tag{4.5.21}$$

using the definitions (4.5.10) and (4.5.11). The mathematical transformations identical to those leading from (4.5.12) to (4.5.16) give the thermodynamic potential of the critical nucleus in the form

$$\Omega = NE + S\int_{-\infty}^{\infty} \frac{p^2}{m}\,dx \tag{4.5.22}$$

where E is determined by Eq. (4.5.10). Hence

$$\Omega = \Omega_0 + S\int_{-\infty}^{\infty} \frac{p^2}{m}\,dx \tag{4.5.23}$$

where $\Omega_0 = NE = N(f(\bar{c}) - (df(\bar{c})/d\bar{c})\bar{c})$ is the thermodynamic potential of a supercooled homogeneous solution. The transformations of (4.5.23) similar to those leading from (4.5.17) to (4.5.20) yield the energy of the formation of a critical nucleus

$$\Delta\Omega = \Omega - \Omega_0 = S\oint \sqrt{2m\left[f(c) - f(\bar{c}) - \frac{df(\bar{c})}{d\bar{c}}(c-\bar{c})\right]}\,\frac{dc}{v}$$

$$= \frac{2S}{v}\int_{\bar{c}}^{c_0} \sqrt{2m\left[f(c) - f(\bar{c}) - \frac{df(\bar{c})}{d\bar{c}}(c-\bar{c})\right]}\,dc \tag{4.5.24}$$

where c_0 is the maximum concentration in the nucleus that may be found from the equation

$$f(c_0) - f(\bar{c}) - \frac{df(\bar{c})}{d\bar{c}}(c_0 - \bar{c}) = 0 \tag{4.5.25}$$

[the integration limits in (4.5.24) are the concentrations at which the integrand vanishes].

Finally, we will prove that the solutions of (4.5.1) describing the critical nucleus and periodic distribution cases are associated with saddle points of the free energy hypersurface.

In fact, according to what has been said in Section 4.3, an extremum is a saddle point if the eigenvalue problem (4.3.13) has both positive and negative solutions. In the one-dimensional case Eq. (4.3.13) is reduced to

$$-m\frac{d^2}{dx^2}\psi_n(x) + \left(\frac{d^2f}{dc^2}\right)_{c=c_0(x)}\psi_n(x) = \varepsilon_n\psi_n(x) \tag{4.5.27}$$

where $c_0(x)$ is a solution of (4.5.1) under the question, $\varepsilon_0 < \varepsilon_1 < \varepsilon_2 < \cdots < \varepsilon_n < \cdots$ are the eigenvalues, and $n = 0, 1, 2, \ldots$ is the eigenvalue number. According to the Sturm theorem, n is the number of nodal points of the eigenfunction $\psi_n(x)$.

The first derivative of (4.5.1) with respect to x is

$$-m\frac{d^2}{dx^2}\left(\frac{dc_0(x)}{dx}\right) + \left(\frac{d^2f}{dc^2}\right)_{c=c_0(x)}\left(\frac{dc_0(x)}{dx}\right) = 0 \tag{4.5.28}$$

Comparison of Eqs. (4.5.28) and (4.5.27) shows that $dc_0(x)/dx$ is always the eigenfunction corresponding to the zero eigenvalue. If the function $dc_0(x)/dx$

has several nodal points, $n_0 > 1$. Then the eigenfunction $dc_0(x)/dx = \psi_{n_0}(x)$ should correspond to the n_0th eigenvalue $\varepsilon(n_0) = 0$, and all the eigenvalues with $n < n_0$ should be smaller than $\varepsilon(n_0) = 0$:

$$\varepsilon(0) < \varepsilon(1) < \cdots < \varepsilon(n_0) = 0 \tag{4.5.29}$$

The spectrum of Eq. (4.5.27) will then include both positive and negative eigenvalues, and the distribution $c_0(x)$ will correspond to a saddle point of the free energy hypersurface.

Therefore, to determine the type of an extremum, one should determine the number of points at which the function $c_0(x)$ becomes zero. For instance, the function $dc_0(x)/dx$ corresponding to the critical nucleus becomes zero only at $x=0$ (see Fig. 43c). Hence $n_0 = 1$, and $\varepsilon(0) < \varepsilon(1) = 0$. Since $\varepsilon(0)$ is a negative value, the critical nucleus corresponds to a saddle point of the free energy hypersurface. The function $c_0(x)$ associated with the periodic extreme state (see Fig. 43c) always has many nodal points, and therefore any periodic distribution also corresponds to a saddle point. We thus arrive at the conclusion that both the critical nucleus and periodic distribution states are unstable states.

4.6. WORKED EXAMPLES

1. *Calculate the temperature dependence of the interphase energy in the mean-field approximation.*

The specific free energy is determined in the mean-field approximation by Eq. (4.1.15):

$$f(c) = |V(0)|[\tfrac{1}{2}c(1-c) + \tau(c \ln c + (1-c)\ln(1-c))] \tag{4.6.1}$$

where $\tau = \kappa T/|V(0)|$ is the reduced temperature.

Since $f(c)$ is a symmetrical function with respect to the point $c = \frac{1}{2}$, we have $c_\beta = (1-c_\alpha)$ and

$$\frac{df(c_\alpha)}{dc_\alpha} = \frac{df(c_\beta)}{dc_\beta} = \mu = 0 \tag{4.6.2}$$

(see, e.g., Fig. 38). According to (4.5.20) and (4.6.2) the surface energy coefficient is

$$\gamma_s = \int_{c_\alpha}^{c_\beta} \sqrt{2m(f(c) - \min f(c))}\frac{dc}{v} = \int_{c_\alpha}^{1-c_\alpha} \sqrt{2m(f(c) - \min f(c))}\frac{dc}{v} \tag{4.6.3}$$

$$\tag{4.6.4}$$

where $\min f(c) = f(c_\alpha) = f(c_\beta)$ and c_α is a solution to Eq. (4.1.18)

$$\tau = \frac{-0.5 + c_\alpha}{\ln[c_\alpha/(1-c_\alpha)]} \tag{4.6.5}$$

The τ with respect to c_α dependence is plotted in Fig. 39.

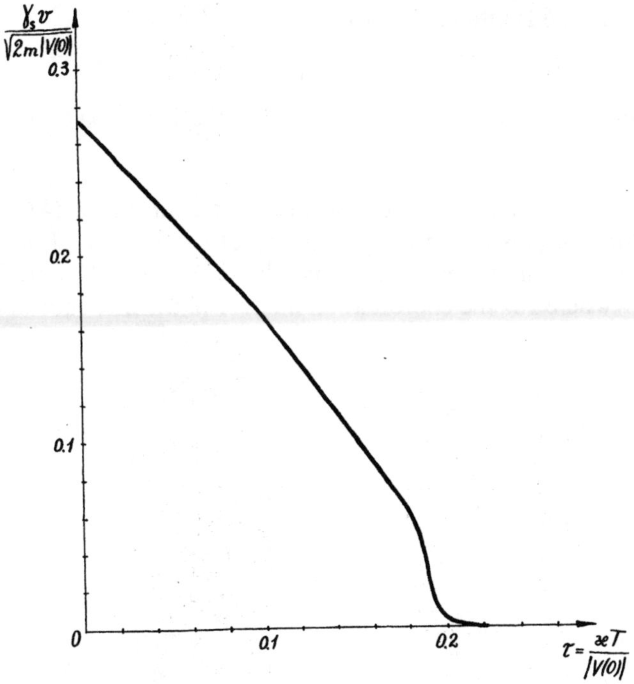

Figure 44. Calculated temperature dependence of the interphase specific free energy γ_s with respect to reduced temperature $\tau = \kappa T/|V(0)|$ (the mean-field approximation).

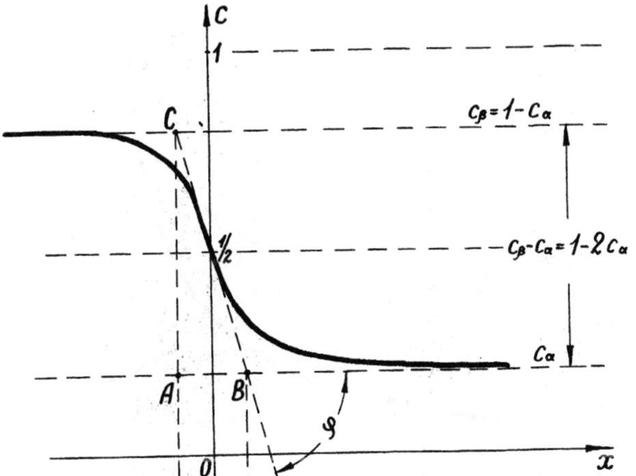

Figure 45. The typical concentration profile of the equilibrium interphase (the mean-field approximation): $D = AC \text{ ctn } \mathcal{G} = (1-2c_\alpha)/(dc/dx)_{c=\frac{1}{2}}$.

Substituting (4.6.1) into (4.6.3) yields

$$\frac{\gamma_s v}{\sqrt{2m|V(0)|}} = \int_{c_\alpha}^{1-c_\alpha} dc \sqrt{\begin{aligned}&\tfrac{1}{2}c(1-c)+\tau[c\ln c+(1-c)\ln(1-c)]\\ &\quad -\tfrac{1}{2}c_\alpha(1-c_\alpha)-\tau[c_\alpha \ln c_\alpha+(1-c_\alpha)\ln(1-c_\alpha)]\end{aligned}} \qquad (4.6.6)$$

The integration in (4.6.6) over the range determined by (4.6.5) gives the $\gamma_s v/\sqrt{2m|V(0)|}$ with respect to τ dependence plotted in Fig. 44. For the nearest-neighbor interaction model Eq. (4.2.26a) gives $m = -\frac{1}{3}d_0^2 z W_1$ and Eq. (4.1.11), $|V(0)| = zW_1$ where z is the coordination number of the first coordination shell,

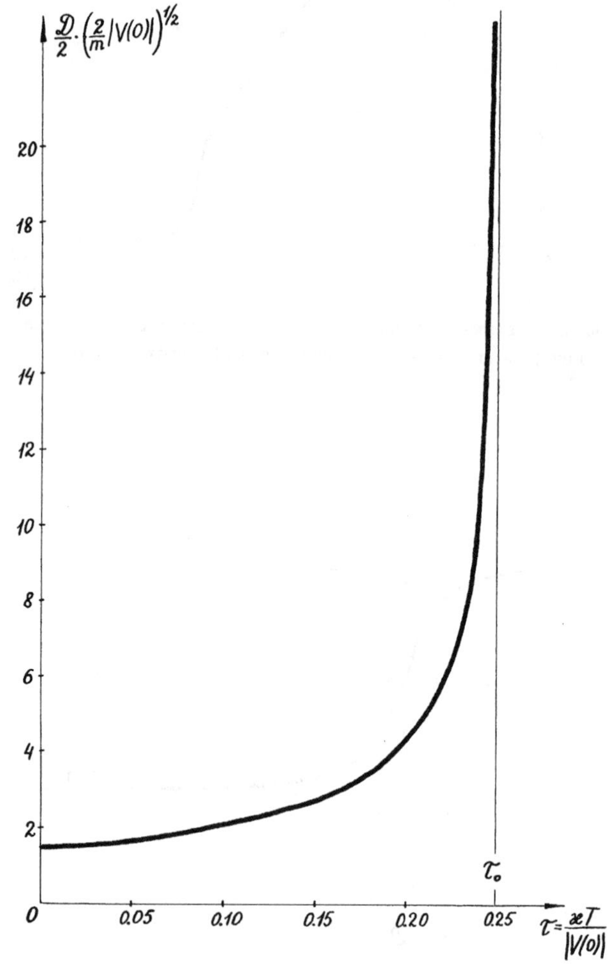

Figure 46. The temperature dependence of the interphase boundary thickness D in the mean-field approximation. $\tau_0 = \frac{1}{4}$ corresponds to the top of the miscibility gap.

d_0 is the interatomic distance. Hence $\sqrt{2m|V(0)|}=z|W_1|d_0$. For the fcc lattice

$$\frac{\gamma v}{\sqrt{2m|V(0)|}}=\frac{\gamma a^3/4}{12|W_1|a/\sqrt{2}}=\frac{\gamma a^2\sqrt{2}}{48|W_1|}$$

For the bcc lattice
$$\frac{\gamma v}{\sqrt{2m|V(0)|}}=\frac{\gamma a^3/2}{8|W_1|2a/\sqrt{3}}=\frac{\gamma a^2\sqrt{3}}{32|W_1|}$$

Fig. 44 shows that the specific interphase energy γ depends strongly on temperature and vanishes at the critical (spinodal) reduced temperature $\tau=\frac{1}{4}$.

2. *Calculate the temperature dependence of the interphase boundary layer thickness in the mean-field approximation.*

The typical concentration profile for the equilibrium interphase boundary is given in Fig. 45. The segment, AB, may be taken for the thickness of the interphase boundary. This segment is the projection of the segment BC of the straight line tangent to the curve $c=c_0(x)$ at the point $c=\frac{1}{2}$ and limited by two points of intersection with the lines $c=c_\alpha$ and $c=c_\beta=1-c_\alpha$, respectively (see Fig. 45):

$$D=AB=AC/\tan\mathscr{S}$$

Since $AC=1-2c_\alpha$ and $\tan\mathscr{S}=(dc(x)/dx)_{c=\frac{1}{2}}$, we have

$$D=AB=(1-2c_\alpha)\left[\left(\frac{dc(x)}{dx}\right)_{c=\frac{1}{2}}\right]^{-1} \tag{4.6.7}$$

It follows from Eqs. (4.5.5), (4.6.2), and (4.5.4a) that

$$\left(\frac{dc(x)}{dx}\right)_{c=\frac{1}{2}}=\frac{1}{m}\sqrt{2m(f(\tfrac{1}{2})-\min f(c))} \tag{4.6.8}$$

Substituting (4.6.1) into (4.6.8) with $\min f(c)=f(c_\alpha)$ yields

$$\left(\frac{dc(x)}{dx}\right)_{c=\frac{1}{2}}=\sqrt{\frac{2}{m}|V(0)|\left[\frac{1}{8}-\tau\ln 2-\frac{1}{2}c_\alpha(1-c_\alpha)-\tau(c_\alpha\ln c_\alpha+(1-c_\alpha)\ln(1-c_\alpha))\right]} \tag{4.6.9}$$

Substituting (4.6.1) into (4.6.8) with $\min f(c)=f(c_\alpha)$ yields

$$D=2\left(\frac{2}{m}|V(0)|\right)^{-\frac{1}{2}}\left(\frac{1}{2}-c_\alpha\right)\left(\frac{1}{8}-\tau\ln 2-\frac{1}{2}c_\alpha(1-c_\alpha)-\tau[c_\alpha\ln c_\alpha+(1-c_\alpha)\ln(1-c_\alpha)]\right)^{-\frac{1}{2}}$$

where the c_α with respect to τ dependence determined by (4.6.5) is shown in Fig. 39. The temperature dependence of the interphase boundary thickness is plotted in Fig. 46 in reduced variable coordinates, $\sqrt{(2/m)}|V(0)|D/2$ and τ.

5

DIFFUSION KINETICS IN SOLID SOLUTIONS

The statistical thermodynamics described in Chapter 3 provides the basis for considering the transformation of alloy microstructures. Theory, however, proves insufficient for predicting the sequence of intermediate states the system passes through during its evolution to the equilibrium state. Questions concerning intermediate structures and their succession in the phase transformations should be addressed to kinetic theories.

We shall describe the kinetic theory of diffusion in solid solutions. The diffusion processes that will be analyzed control kinetics of such widespread phase transformation reactions as decomposition and ordering and therefore are of utmost importance for the theory of phase transformations.

It is well known that the existing experimental methods for studying diffusion processes are based on the measurements of the dissolution of concentration inhomogeneities whose space dimensions greatly exceed interatomic ranges. For this reason the usual diffusion experiment only gives the macroscopic coefficients of continuous diffusion equations.

There are, however, some diffusion processes that occur within the atomic ranges and can be studied by the x-ray diffraction technique. Such processes are spinodal decomposition, long-range order kinetics below the stability limits of the disordered phase, homogenization of the sandwichlike deposit structures, short-range order kinetics, and so on. The first three processes should be described by the discrete theory of single-site probability kinetics (the theory of discrete diffusion). The latter process should be described by the discrete theory of two-site probability kinetics.

We shall formulate the most simple variant of the theory of discrete diffusion in a linear approximation that enables one to treat easily the results of the corresponding x-ray measurements.

5.1. CRYSTAL LATTICE SITE DIFFUSION IN SOLID SOLUTIONS

The problem of discrete diffusion (the random walk problem) in a crystal lattice of a real solid solution is rather complex since it is reduced to solving a set of nonlinear equations. In many cases, however, the more simple linear kinetic theory proves to be sufficient to analyze experimental data. In this section we shall describe the linear kinetic theory proposed by Khachaturyan in 1967 (52).*

Following (52), let us consider a multicomponent alloy and introduce the single-site occupation probability of finding an atom of type α at the crystal lattice site $\mathbf{r}$ at the moment of time t:

$$n(\alpha\mathbf{r}, t) \quad \text{for } \alpha=1, 2, \ldots, \nu$$

We shall consider the Önsager equation for diffusion relaxation of nonequilibrium single-site probability $n(\alpha\mathbf{r}, t)$:

$$\frac{dn(\alpha\mathbf{r}, t)}{dt}=\sum_{\beta=1}^{\nu}\sum_{\mathbf{r}'}\frac{1}{\kappa T}\tilde{L}_{\alpha\beta}(\mathbf{r}-\mathbf{r}')c_\alpha c_\beta\frac{\delta F}{\delta n(\beta\mathbf{r}', t)} \tag{5.1.1}$$

where $\tilde{L}_{\alpha\beta}(\mathbf{r}-\mathbf{r}')$ is a Hermitian matrix of kinetic coefficients, F is the Helmholtz free energy, c_α is the atomic fraction of atoms of the kind α. The variational derivative $\delta F/\delta n(\beta\mathbf{r}', t)$ is calculated under the condition that

$$\sum_{\beta=1}^{\nu} n(\beta\mathbf{r}, t)=1$$

at any t and $\mathbf{r}$. This identity reflects the fact that each crystal lattice site $\mathbf{r}$ is definitely occupied by one of the atoms composing the multicomponent solution.

The summation of Eq. (5.1.1) over $\mathbf{r}$ gives

$$\frac{dN_\alpha}{dt}=\frac{1}{\kappa T}L_{\alpha\beta}(0)c_\alpha c_\beta\sum_{\mathbf{r}'}\frac{\delta F}{\delta n(\beta\mathbf{r}', t)} \tag{5.1.2}$$

where N_α is the total number of α atoms,

$$L_{\alpha\beta}(0)=\sum_{\mathbf{r}}\tilde{L}_{\alpha\beta}(\mathbf{r}) \tag{5.1.3}$$

In (5.1.2), and below, the Einstein summation rule is used (if an index is repeated in a term, summation with respect to that index is to be understood). Since the number of atoms of the kind of α is constant, $dN_\alpha/dt=0$, and therefore the equation

$$L_{\alpha\beta}(0)=\sum_{\mathbf{r}}\tilde{L}_{\alpha\beta}(\mathbf{r})=0 \tag{5.1.4}$$

*For the English speaking audience the microscopic diffusion theory is known from the papers by Cook, de Fontaine, and Hilliard (53) and de Fontaine and Cook (54). The same microscopic diffusion equations were, however, obtained a few years earlier in the paper (52).

holds. The latter follows from (5.1.2) since

$$\sum_{\mathbf{r}'} \frac{\delta F}{\delta n(\beta \mathbf{r}', t)} \neq 0$$

for the nonequilibrium distribution $n(\beta\mathbf{r}', t)$. Expanding $\delta F/\delta n(\beta\mathbf{r}', t)$ in a series with respect to the deviations $\Delta n(\beta\mathbf{r}', t)$ from their equilibrium quantities, $n(\beta\mathbf{r}', t) - c_\beta$, and retaining the first nonvanishing terms of this expansion, one obtains

$$\frac{\delta F}{\delta n(\beta\mathbf{r}', t)} \simeq \sum_{\mathbf{r}''} \psi(\beta\mathbf{r}', \gamma\mathbf{r}'')\Delta n(\gamma\mathbf{r}'', t) \tag{5.1.5}$$

$$\Delta n(\gamma\mathbf{r}'', t) = n(\gamma\mathbf{r}'', t) - c_\gamma$$
$$\psi(\beta\mathbf{r}', \gamma\mathbf{r}'') = \left(\frac{\delta F}{\delta n(\beta\mathbf{r}')\delta n(\gamma\mathbf{r}'')}\right)_{\substack{n(\beta\mathbf{r}')=c_\beta \\ n(\gamma\mathbf{r}'')=c_\gamma}} \tag{5.1.6}$$

Since the equilibrium state is homogeneous,

$$\psi(\beta\mathbf{r}', \gamma\mathbf{r}'') = \psi_{\beta\gamma}(\mathbf{r}' - \mathbf{r}'') \tag{5.1.7}$$

Substituting (5.1.7) into (5.1.5) and (5.1.5) into (5.1.1), one obtains

$$\frac{d\Delta n(\alpha\mathbf{r}, t)}{dt} = -\sum_{\mathbf{r}'} \tilde{W}_{\alpha\gamma}(\mathbf{r} - \mathbf{r}')\Delta n(\gamma\mathbf{r}', t) \tag{5.1.8}$$

where

$$\tilde{W}_{\alpha\gamma}(\mathbf{r} - \mathbf{r}') = -\frac{1}{\kappa T}\sum_{\mathbf{r}'', \beta} \tilde{L}_{\alpha\beta}(\mathbf{r} - \mathbf{r}')c_\alpha c_\beta \psi_{\beta\gamma}(\mathbf{r}'' - \mathbf{r}') \tag{5.1.9}$$

The Fourier transform of Eq. (5.1.8) is

$$\frac{d\Delta\tilde{n}(\alpha\mathbf{k}, t)}{dt} = -W_{\alpha\gamma}(\mathbf{k})\Delta\tilde{n}(\gamma\mathbf{k}, t) \tag{5.1.10}$$

where

$$W_{\alpha\gamma}(\mathbf{k}) = -\frac{1}{\kappa T}\sum_{\beta} L_{\alpha\beta}(\mathbf{k})c_\alpha c_\beta \tilde{\psi}_{\beta\gamma}(\mathbf{k})$$
$$L_{\alpha\beta}(\mathbf{k}) = \sum_{\mathbf{r}} \tilde{L}_{\alpha\beta}(\mathbf{r})e^{-i\mathbf{k}\mathbf{r}}$$
$$\tilde{\psi}_{\beta\gamma}(\mathbf{k}) = \sum_{\mathbf{r}} \psi_{\beta\gamma}(\mathbf{r})e^{-i\mathbf{k}\mathbf{r}}$$
$$\Delta\tilde{n}(\alpha\mathbf{k}, t) = \sum_{\mathbf{r}} \Delta n(\alpha\mathbf{r}, t)e^{-i\mathbf{k}\mathbf{r}} \tag{5.1.11}$$

The limit transition $\mathbf{k} \to 0$ in Eq. (5.1.10) gives

$$\frac{d\Delta\tilde{n}(\alpha\mathbf{k}, t)}{dt} = -k_i k_j D_{ij}^{\alpha\gamma}\Delta\tilde{n}(\gamma\mathbf{k}, t)$$

where $W_{\alpha\gamma}(\mathbf{k}) \approx D_{ij}^{\alpha\gamma} k_i k_j$, i, j are the Cartesian indexes and $D_{ij}^{\alpha\gamma}$ are the diffusivities which enter Fick's first law for a multicomponent alloy:

$$J_\alpha^i = -D_{ij}^{\alpha\gamma} \nabla_j c_\gamma$$

(where $\mathbf{J}$ is a diffusion flux).

Linear Eqs. (5.1.10) and (5.1.11) can be represented in a matrix form:

$$\frac{d\Delta\tilde{n}(\alpha\mathbf{k}, t)}{dt} = -(\hat{W}(\mathbf{k}))_{\alpha\beta} \Delta\tilde{n}(\beta\mathbf{k}, t) \tag{5.1.12}$$

where

$$\hat{W}(\mathbf{k}) = -\hat{L}(\mathbf{k})\hat{\psi}(\mathbf{k}) \tag{5.1.13a}$$

$$(\hat{L}(\mathbf{k}))_{\alpha\beta} = \frac{1}{\kappa T} L_{\alpha\beta}(\mathbf{k}) c_\alpha c_\beta, \quad (\hat{\psi}(\mathbf{k}))_{\alpha\beta} = \tilde{\psi}_{\alpha\beta}(\mathbf{k}) \tag{5.1.13b}$$

The solution of Eq. (5.1.12) is

$$\Delta\tilde{n}(\alpha\mathbf{k}, t) = (\exp(-\hat{W}(\mathbf{k})t))_{\alpha\beta} \Delta\tilde{n}(\beta\mathbf{k}, 0) \tag{5.1.14}$$

Here $\Delta\tilde{n}(\beta\mathbf{k}, 0)$ being the Fourier transform of the function $\Delta n(\beta\mathbf{r}, 0)$ determines an initial condition.

It is convenient to introduce the eigenvectors $v_s(\alpha, \mathbf{k})$ and $v_s^+(\alpha, \mathbf{k})$, corresponding to the eigenvalues $\lambda_s(\mathbf{k})$ and $\lambda_s^+(\mathbf{k})$ of the non-Hermitian matrix $\hat{W}(\mathbf{k})$ and transposed matrix $\hat{W}^+(\mathbf{k})$, respectively:

$$\begin{aligned} W_{\alpha\beta}(\mathbf{k}) v_s(\beta, \mathbf{k}) &= \lambda_s(\mathbf{k}) v_s(\alpha, \mathbf{k}) \\ W_{\alpha\beta}^+(\mathbf{k}) v_s^+(\beta, \mathbf{k}) &= \lambda_s^+(\mathbf{k}) v_s^+(\alpha, \mathbf{k}) \end{aligned} \tag{5.1.15}$$

where $W_{\beta\alpha}^+(\mathbf{k}) = W_{\alpha\beta}(\mathbf{k})$ and s is the index of eigenvalue ($s = 1, 2, \ldots, \nu$).

For the eigenvectors of non-Hermitian operators there are known to be orthogonality relationships, which in the present case have the form:

$$\sum_\alpha v_s(\alpha, \mathbf{k}) v_{s'}^+(\alpha, \mathbf{k}) = \delta_{ss'} \tag{5.1.16}$$

One can represent the matrices $(\exp(-\hat{\mathbf{W}}(\mathbf{k})t))_{\alpha\beta}$ as a bilinear expansion:

$$(\exp(-\hat{\mathbf{W}}(\mathbf{k})t))_{\alpha\beta} = \sum_{s=1}^{\nu} \exp(-\lambda_s(\mathbf{k})t) v_s(\alpha, \mathbf{k}) v_s^+(\beta, \mathbf{k}) \tag{5.1.17}$$

It is easy to prove that the eigenvalues $\lambda_s(\mathbf{k})$ are the real values (52).

Making use of Eq. (5.1.17) in Eq. (5.1.14), one can obtain the resulting form of the solution to Eq. (5.1.10):*

$$\Delta n(\alpha\mathbf{k}, t) = \sum_{s=1}^{\nu} e^{-\lambda_s(\mathbf{k})t} v_s(\alpha, \mathbf{k}) \sum_{\beta=1}^{\nu} v_s^+(\beta, \mathbf{k}) \Delta n(\beta\mathbf{k}, 0) \tag{5.1.18}$$

Now we shall consider the important special case of a binary substitutional alloy.

*Of course the solution of the set (5.1.10) can also be obtained directly in a usual way.

For the exchange diffusion mechanism in a binary solution A-B, it is enough to consider diffusion, for example, of only the B-atoms, since an occupation probability for an A-atom, $n(A\mathbf{r})$, satisfies the identity

$$n(A\mathbf{r})+n(B\mathbf{r})\equiv 1$$

The kinetic single-site probability Eq. (5.1.1) for this case is

$$\frac{d\Delta n(B\mathbf{r},t)}{dt}=\sum_{\mathbf{r}'}\frac{\tilde{L}_0(\mathbf{r}-\mathbf{r}')}{\kappa T}c_A c_B\frac{\delta F}{\delta n(B\mathbf{r}',t)} \tag{5.1.19}$$

where $\Delta n(B\mathbf{r},t)=n(B\mathbf{r},t)-c_B$, $\tilde{L}_0(\mathbf{r}-\mathbf{r}')$ is a probability of an elementary jump from a site $\mathbf{r}$ to the site $\mathbf{r}'$ at the time interval.

The linearization of Eq. (5.1.19) yields

$$\frac{\delta F}{\delta n(B\mathbf{r}',t)}\simeq\sum_{\mathbf{r}''}B(\mathbf{r}'-\mathbf{r}'')\Delta(\mathbf{r}'',t) \tag{5.1.20}$$

where

$$B(\mathbf{r}'-\mathbf{r}'')=\left(\frac{\delta^2 F}{\delta n(B\mathbf{r}')\delta n(B\mathbf{r}'')}\right)_{\{n(B\mathbf{r})\}=c_B} \tag{5.1.21}$$

is calculated at $n(B\mathbf{r})\equiv c_B$.

The solution of the Eq. (5.1.19) with (5.1.20) gives the expression analogous to (5.1.14)

$$\Delta\tilde{n}(\mathbf{k},t)=\Delta n(\mathbf{k},0)\exp(-\lambda(\mathbf{k})t) \tag{5.1.22}$$

where

$$\lambda(\mathbf{k})=-\frac{L_0(\mathbf{k})}{\kappa T}c_B(1-c_B)b(\mathbf{k},T,c_B) \tag{5.1.23}$$

is the amplification factor and

$$b(\mathbf{k},T,c_B)=\sum_{\mathbf{r}}B(\mathbf{r})e^{-i\mathbf{k}\mathbf{r}} \tag{5.1.24}$$

is the characteristic function determined by Eq. (2.2.13).

The mean-field approximation of the characteristic function can be determined in an explicit form. Using the free energy (3.6.3b) in Eq. (5.1.21), we obtain

$$B(\mathbf{r}'-\mathbf{r}'')=W(\mathbf{r}'-\mathbf{r}'')+\frac{\kappa T}{c_B(1-c_B)}\delta_{\mathbf{r}'\mathbf{r}''} \tag{5.1.25}$$

where $\delta_{\mathbf{r}'\mathbf{r}''}$ is the Kronecker symbol, $W(\mathbf{r}'-\mathbf{r}'')$ the interchange energy.

Substituting Eq. (5.1.25) into Eq. (5.1.24) yields

$$b(\mathbf{k},T,c_B)=V(\mathbf{k})+\frac{\kappa T}{c_B(1-c_B)} \tag{5.1.26}$$

With Eq. (5.1.26), Eq. (5.1.22) obtains its final form in the mean-field approximation:

$$\Delta\tilde{n}(\mathbf{k}, t) = \Delta n(\mathbf{k}, 0) \exp\left[L_0(\mathbf{k})\left(1 + \frac{c_B(1-c_B)}{\kappa T} V(\mathbf{k})\right) t\right] \tag{5.1.27}$$

The condition (5.1.4) yields

$$\tilde{L}_0(0) + \sum_{\mathbf{r}}{}' \tilde{L}_0(\mathbf{r}) = 0 \tag{5.1.28}$$

(the prime symbol means that the term $\mathbf{r}=0$ is omitted).

Substitution of Eq. (5.1.28) into

$$L_0(\mathbf{k}) = \sum_{\mathbf{r}} \tilde{L}_0(\mathbf{r}) e^{-i\mathbf{kr}} \tag{5.1.29}$$

gives

$$L_0(\mathbf{k}) = -\sum_{\mathbf{r}}{}' \tilde{L}_0(\mathbf{r})(1 - e^{-i\mathbf{kr}}) = -2\sum_{\mathbf{r}}{}' \tilde{L}_0(\mathbf{r}) \sin^2 \tfrac{1}{2}\mathbf{kr} \tag{5.1.30}$$

With Eq. (5.1.30), Eq. (5.1.27) may be rewritten as

$$\Delta\tilde{n}(\mathbf{k}, t) = \Delta\tilde{n}(\mathbf{k}, 0) \exp\left\{-2\left(\sum_{\mathbf{r}}{}' \tilde{L}_0(\mathbf{r}) \sin^2 \tfrac{1}{2}\mathbf{kr}\right)\left(1 + \frac{c_B(1-c_B)}{\kappa T} V(\mathbf{k})\right) t\right\} \tag{5.1.31}$$

where the summation is over all the crystal lattice sites.

The solution (5.1.22) of the kinetics equation (5.1.19) demonstrates that in the linear approximation the amplitude of each concentration wave $\Delta\tilde{n}(\mathbf{k}, t)$ relaxes independently of the others. The relaxation occurs in time in accordance with the exponential decay law:

$$t_0(\mathbf{k}) = -\frac{\kappa T}{c_B(1-c_B)L_0(\mathbf{k})b(\mathbf{k}, T, c_B)}$$

Eqs. (5.1.20) to (5.1.22) are also valid for the more real cases of substitutional diffusion if the coefficients of diffusion of A and B atoms are considerably different. In such a case one can suppose that "fast" atoms of type A form a quasi-equilibrium atmosphere around "slow" B-atoms. Therefore the time evolution of the B-atoms should be only considered. Its kinetics is described by the coefficients $\tilde{L}_0(\mathbf{r}-\mathbf{r}') = \tilde{L}_B(\mathbf{r}-\mathbf{r}')$ which have the meaning of transition probabilities for "slow" atoms of type B.

The thermodynamic characteristic function $b(\mathbf{k}, T, c_B)$ and the kinetic characteristic function $L_0(\mathbf{k})$ can be found from Eqs. (5.1.22) and (5.1.23), using data from the disordering kinetics in multilayer periodically deposited films.

X-ray scattering from periodic multilayer films as well as from any specimen with a periodic concentration profile gives rise to satellites in the x-ray diffraction patterns. The satellites fall on the "reciprocal lattice points": $\pm\mathbf{k}_0, \pm 2\mathbf{k}_0, \ldots, \pm m\mathbf{k}_0$, where $\mathbf{k}_0 = (2\pi/\lambda_0)/\mathbf{n}$, λ_0 is the modulation period, m is an integer, $\mathbf{n}$ the unit vector along the modulation direction. According to Eq. (1.3.36) the intensities of the satellites are proportional to the squared modulus of $\Delta\tilde{n}(\mathbf{k}, t)$:

$$I(\mathbf{k}, t) \sim |\Delta \tilde{n}(\mathbf{k}, t)|^2 \tag{5.1.32}$$

It follows from Eqs. (5.1.32), (5.1.22), and (5.1.23) that

$$\ln \frac{I(\mathbf{k}, t)}{I(\mathbf{k}, 0)} = 2L_0(\mathbf{k}) \frac{c_B(1-c_B)}{\kappa T} b(\mathbf{k})t \tag{5.1.33}$$

where $\mathbf{k} = \mathbf{k}_0 m$.

Eq. (5.1.33) makes it possible to find the diffusivities that enter the expression for $L_0(\mathbf{k}) = -2\sum_{\mathbf{r}}' \tilde{L}_0(\mathbf{r}) \sin^2 \frac{1}{2}\mathbf{kr}$ and the value of $b(\mathbf{k})$ from time dependences of x-ray satellite intensities at various temperatures. The "disordering kinetics" measurements of this type were undertaken by Cook and Hilliard for Ag-Au (55), E. M. Philofski and Hilliard for Cu-Pd (56), Paulson for Cu-Au (57), and Tsakalakos for Cu-Ni (58).

The time dependence of ln $(I(\mathbf{k}_0, t)/I(\mathbf{k}_0, 0))$ in Cu-Ni measured at 400°C for modulations of various wavelengths is plotted in Fig. 47.

Because the characteristic kinetic function $L_0(\mathbf{k})$ involves the probabilities of diffusion jumps, the method described above provides the unique means for the direct determination of elementary diffusion events. This can be done if the modulation period λ_0 is commensurate with the interatomic distance.

The nonlinear Eq. (5, 1.19) can also be used in the analysis of long-range order

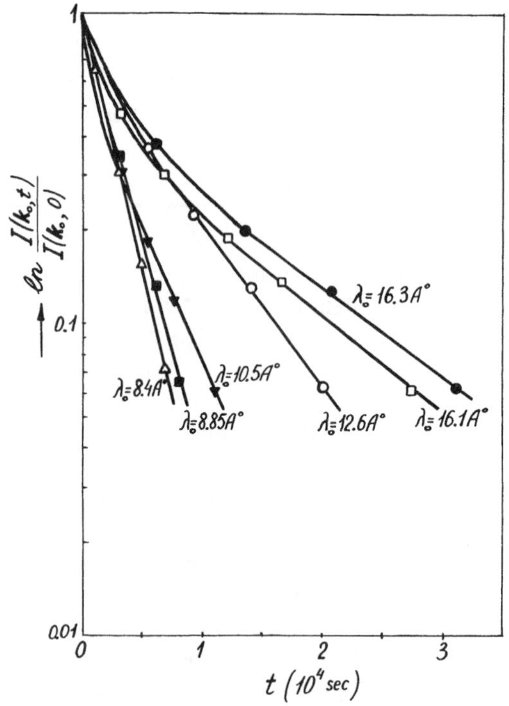

Figure 47. ln $[I(\mathbf{k}_0, t)/I(\mathbf{k}_0, 0)]$ with respect to time, t, for composition modulated Cu-Ni foils of various wavelengths λ_0 at 400°C [after T. Tsakalakos (58)].

kinetics if one takes into account the equation

$$\Delta n(\mathbf{k}_s, t) = \gamma_s \eta_s(t) \tag{5.1.34}$$

which follows from Eq. (3.1.3) where $\mathbf{k}_s$ is a superlattice vector of the ordered phase entering the star s, γ_s is a constant, and η_s a long-range-order parameter that is an amplitude of the concentration wave. In the case of the long-range order the value $\delta F/\delta n(B\mathbf{r}', t)$ as well as the value $\Delta n(B\mathbf{r})$ can be represented as a superposition of the concentration waves:

$$\frac{\delta F}{\delta n(B\mathbf{r})} = \tilde{c}(\eta_1, \ldots, \eta_s, \ldots) + \sum_s \tilde{\eta}_s(\eta_1, \eta_2, \ldots, \eta_s, \ldots) E_s(\mathbf{r})$$

$$n(B\mathbf{r}) = c + \Delta n(B\mathbf{r}) = c + \sum_s \eta_s E_s(\mathbf{r}) \tag{5.1.35}$$

where $E_s(\mathbf{r})$ is given by Eq. (3.1.4) since the value $\delta F/\delta n(B\mathbf{r})$ possesses the same symmetry as the function $n(B\mathbf{r})$ [see Eq. (3.6.21)].

Substitution of Eq. (5.1.35) into (5.1.19), followed by the Fourier transformation, gives the set of equations in the long-range order parameters:

$$\frac{d\eta_s}{dt} = \frac{L_0(\mathbf{k}_s)}{\kappa T} c_B(1-c_B)\tilde{\eta}_s(\eta_1, \ldots, \eta_s, \ldots), \quad s = 1, 2, \ldots \tag{5.1.36}$$

Eq. (5.1.36) describes long-range order kinetics if the temperature is below the absolute instability limit T_c^-, that is, if the nucleation-and-growth mechanism does not take place

Let us consider an example of a Cu_3Au alloy in the self-consistent field approximation. In the general case, using Eq. (3.6.3b), we have

$$\mathscr{S}(\mathbf{r}) = \frac{\delta F}{\delta n(B\mathbf{r})} = \sum_{\mathbf{r}} \tilde{V}(\mathbf{r}-\mathbf{r}') n(B\mathbf{r}') + \kappa T \ln \frac{n(B\mathbf{r})}{1 - n(B\mathbf{r})} \tag{5.1.37}$$

Substituting (3.6.19) into (5.1.37) results in

$$\mathscr{S}(\mathbf{r}) = \frac{\delta F}{\delta n(B\mathbf{r})} = \frac{1}{4} V(0) + \frac{\kappa T}{4} \ln \frac{(1+3\eta)(1-\eta)^2}{3(3+\eta)^3}$$
$$+ \left[V(2\pi \mathbf{a}_3^*) \frac{1}{4} \eta + \frac{\kappa T}{4} \ln \frac{(1+3\eta)(3+\eta)}{3(1-\eta)^2} \right] E(\mathbf{r}) \tag{5.1.38}$$

if $c = \frac{1}{4}$. Therefore

$$\tilde{\eta}(\eta) = \frac{1}{4} V(2\pi \mathbf{a}_3^*)\eta + \frac{\kappa T}{4} \ln \frac{(1+3\eta)(3+\eta)}{3(1-\eta)^2} \tag{5.1.39}$$

Substituting (5.1.39) into (5.1.36) gives the differential equation

$$\frac{d\eta_s}{dt} = \frac{3}{16} \frac{L_0(2\pi \mathbf{a}_3^*)}{\kappa T} \left[\frac{1}{4} V(2\pi \mathbf{a}_3^*)\eta + \frac{\kappa T}{4} \ln \frac{(1+3\eta)(3+\eta)}{3(1-\eta)^2} \right] \tag{5.1.40}$$

The solution of Eq. (5.1.40) is

$$\int_{\eta_0}^{\eta} d\eta \left[\frac{V(2\pi\mathbf{a}_3^*)}{\kappa T}\eta + \ln\frac{(1+3\eta)(3+\eta)}{3(1-\eta)^2} \right]^{-1} = \frac{3}{64} L_0(2\pi\mathbf{a}_3^*)t \qquad (5.1.41)$$

where η_0 is the magnitude of the initial long-range order parameter.

5.2. PERCOLATION MECHANISM OF "FAST" ATOM SUBSTITUTIONAL DIFFUSION IN BINARY ALLOYS

The substitutional diffusion in solid solutions has been studied in much detail. Nevertheless, there are problems that have not received due attention. One of these is that of the change of the mechanism of diffusion of "fast" atoms at low temperatures.

As has just been shown, the consideration of substitutional diffusion of "slow" atoms in a binary alloy reduces to the random walk problem in crystal lattice. The situation proves to be more complicated for the "fast" atoms at low temperatures when "slow" ones can be considered as immobile frame. In this case the mechanism of substitutional diffusion involves the direct interchange of a vacancy and a "fast" (migrating) atom on the nearest-neighboring sites. As a matter of fact it means that vacancies can migrate through "conducting" circuits of channels formed by chains of "fast" atoms, which are the nearest neighbors. The substitution in such a chain of at least one "fast" atom by immobile "slow" one would result in the disconnection of the "conducting" channel and its subdivision into two separate ones since a transition of a vacancy through a "slow" atom is impossible.

Conducting channels at low atomic fraction of migrating atoms form finite isolated close circuits (clusters). In this situation the diffusion coefficient of migrating atoms vanishes since the migration between nearest close clusters is impossible. One can easily see that the diffusion coefficient assumes nonzero values only when an infinite cluster arises for the first time. It occurs if the atomic fraction of migrating atoms becomes more than the concentration threshold $c_0(s)$, that is, if $c > c_0(s)$.

In this formulation the problem of diffusion of "fast" atoms in a binary substitutional alloy reduces to the site problem of the percolation theory [for example, see the review (59).*

The numerical calculations result in the following quantities for the concentration thresholds:

$$c_0(s) = 0.195 \quad \text{for a fcc lattice}$$
$$c_0(s) = 0.243 \quad \text{for a bcc lattice}$$

These numerical values have been obtained for the random atomic distribution.

*To date the most important application of the percolation theory is the electron theory of amorphous semiconductors.

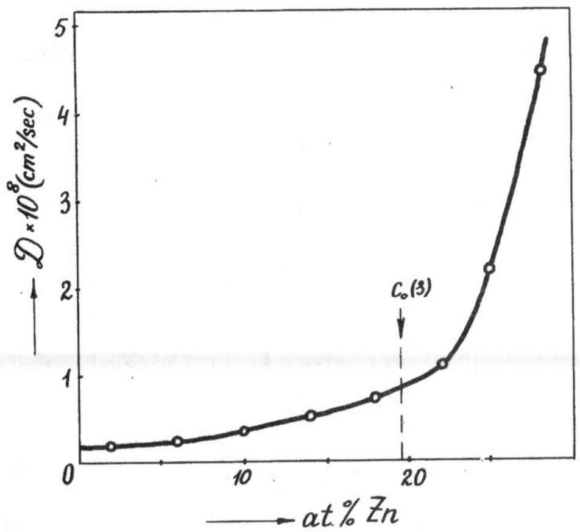

Figure 48. Chemical diffusivity with relation to concentration for alpha brass at 890°C [after R. W. Balluffi and L. L. Seigle (61)].

One should keep in mind that taking into account the short-range order might change them. The conclusion to be drawn from this is that the diffusion of "fast" atoms in a substitutional solution is a threshold phenomenon. The diffusion coefficient vanishes if the content of migrating atoms is less than the concentration threshold $c_0(s)$ and is a nonzero value if the content becomes greater than $c_0(s)$.*

In any case the measured activation energy of the "fast" atom migration should substantially depend on the concentration of the alloy and have the singular behavior near the concentration threshold $c_0(s)$.

It is likely that da Silva and Mehl (60) and Baluffi and Seigle (61) observed a transition to the percolation diffusion mechanism in alpha brass at 890° C (see Fig. 48). According to (61) the intrinsic diffusivities of Zn and Cu at 28 percent Zn are

$$D_{Zn} = 6.3\ 10^{-8}\ \text{cm}^2/\text{sec}$$
$$D_{Cu} = 1.2\ 10^{-8}\ \text{cm}^2/\text{sec}$$

In this case "slow" Cu atoms may only approximately be regarded as immobile. The contribution of "slow" atoms to diffusion should obscure the percolation effect described above. Figure 48 demonstrates that, in accordance with the percolation theory predictions, the chemical diffusivity in Cu-Zn increases rapidly above the percolation threshold $c_{Zn}(s) = 0.195$ at 890° C. It should be noted that the lower the temperature the sharper the increase of diffusivity.

Actually, if $D_{Cu}/D_{Zn} \to 0$, diffusion in an alloy becomes a threshold phenomenon. It ceases at $c < c_{Zn}(s) = 0.195$ and arises only at $c > c_{Zn}(s)$. However, in a Zn-Cu alloy where migration of "slow" Cu atoms gives a noticeable con-

*Obviously, it is true if we do not take into account the migration of "slow" atoms with respect to "fast" ones.

tribution to diffusion, the chemical diffusivity will not be zero at $c \lessapprox c_{Zn}(s)$. The chemical diffusivity raised at the threshold concentration will not in this case be as abrupt as the percolation mechanism requires.

5.3. SPINODAL DECOMPOSITION

It was shown in Section 2.2 that a homogeneous solid solution is stable with respect to infinitesimal concentration fluctuations if the characteristic function $b(\mathbf{k}, T, c)$ is positive for all wave vectors $\mathbf{k}$. The loss of stability occurs if

$$b(\mathbf{k}, T, c) < 0 \tag{5.3.1}$$

The wave vectors for which (5.3.1) holds form the instability range in the $\mathbf{k}$-space. Because growth of the concentration waves whose wave vectors belong to the instability range [satisfy the inequality (5.3.1)] results in the monotonic decrease of the free energy, the amplitudes of these waves increase spontaneously. On the other hand, concentration wave amplitudes decrease if the corresponding wave vectors fall outside the instability range ($b(\mathbf{k}, T, c) > 0$).

The absolute instability around $\mathbf{k}_0 = 0$ has been shown in Section 2.2 to result in the spinodal decomposition, whereas the absolute instability around $\mathbf{k}_0 \neq 0$ must lead to continuous ordering. It is easy to see that the same follows from the solution (5.1.22) to the linearized microscopic diffusion equation. According to Eqs. (5.1.22) and (5.1.23), the amplification factor

$$\lambda(\mathbf{k}) = -\frac{L_0(\mathbf{k})}{\kappa T} c_B(1 - c_B) b(\mathbf{k}, T, c_B) \tag{5.3.2}$$

where

$$L_0(\mathbf{k}) = -2 \sum_{\mathbf{r}}' \tilde{L}_0(\mathbf{r}) \sin^2 \tfrac{1}{2}\mathbf{kr} \tag{5.3.3}$$

determines the rate of growth ($\lambda(\mathbf{k}) < 0$) or decay ($\lambda(\mathbf{k}) > 0$) of concentration waves. The amplification and attenuation are described by the equation

$$\Delta\tilde{n}(\mathbf{k}, t) = \Delta\tilde{n}(\mathbf{k}, 0) \exp\left\{-2t\left(\sum_{\mathbf{r}}' \tilde{L}_0(\mathbf{r}) \sin^2 \frac{1}{2}\mathbf{kr}\right) c_B(1 - c_B) \frac{b(\mathbf{k}, T, c_B)}{\kappa T}\right\} \tag{5.3.4}$$

By definition, the probabilities of elementary diffusion jumps $\tilde{L}_0(\mathbf{r})$ are positive values. Hence the function $\sum_{\mathbf{r}}' \tilde{L}_0(\mathbf{r}) \sin^2 \frac{1}{2}\mathbf{kr}$ entering Eq. (5.3.4) is also positive. It thus follows that the sign of the amplification factor (5.3.2)

$$\lambda(\mathbf{k}) = 2 \sum_{\mathbf{r}}' \tilde{L}_0(\mathbf{r}) \sin^2 \frac{1}{2}\mathbf{kr} \cdot \frac{b(\mathbf{k}, T, c_B)}{\kappa T} c_B(1 - c_B) \tag{5.3.5}$$

should be the same as the sign of the characteristic thermodynamic function $b(\mathbf{k}, T, c)$. In other words, $\lambda(\mathbf{k})$ is negative within the instability range where $b(\mathbf{k}, T, c)$ is also negative. This provides the exponential growth of amplitudes of the corresponding concentration waves. If, on the other hand, $b(\mathbf{k}, T, c)$ is positive, $\lambda(\mathbf{k})$ is also positive, and the corresponding concentration waves decay.

Using Eq. (4.2.3)

$$F = \int \left[\frac{1}{2} m(\nabla n(\mathbf{r}))^2 + f(n(\mathbf{r})) \right] \frac{dV}{v} \tag{5.3.6}$$

[(5.3.6) differs from (4.2.3) in that $c(\mathbf{r})$ is replaced with $n(\mathbf{r})$ and m is assumed to be a constant] we now can calculate the characteristic function $b(\mathbf{k}, T, c)$.

Let us substitute

$$n(\mathbf{r}) = c_B + \Delta n(\mathbf{r}) \tag{5.3.7}$$

into Eq. (5.3.6). Bearing in mind that $\Delta n(\mathbf{r})$ is a small concentration fluctuation, we may expand (5.3.6) in powers of $\Delta n(\mathbf{r})$. After truncating the second non-vanishing term, we obtain

$$F \cong F_0 + \int \left[\frac{1}{2} m(\nabla \Delta n(\mathbf{r}))^2 + \frac{1}{2} \frac{d^2 f(c_B)}{dc_B^2} (\Delta n(\mathbf{r}))^2 \right] \frac{dV}{v} \tag{5.3.8}$$

where $F_0 = f(c_B)V/v$ is the free energy of a homogeneous alloy. It should be noted that the transition from Eq. (5.3.6) to Eq. (5.3.8) is equivalent to the linearization of (5.3.6).

An arbitrary fluctuation $\Delta n(\mathbf{r})$ can always be written as a Fourier series in the cyclic crystal volume V or, which is the same, as superposition of plane concentration waves

$$\Delta n(\mathbf{r}) = \frac{1}{N} \sum_{\mathbf{k}} \Delta \tilde{n}(\mathbf{k}) e^{i\mathbf{k}\mathbf{r}} \tag{5.3.9}$$

where $\mathbf{k}$ runs over all quasi-continuum points allowed by the cyclic boundary conditions (the separations between these points are macroscopically small, of the order of $V^{-\frac{1}{3}}$), N is the total number of crystal lattice sites.

Substituting Eq. (5.3.9) into (5.3.8) and integrating over the cyclic crystal volume, we have

$$\Delta F = F - F_0 = \frac{1}{2N} \sum_{\mathbf{k}} b(\mathbf{k}, T, c_B) |\Delta \tilde{n}(\mathbf{k})|^2 \tag{5.3.10}$$

where

$$b(\mathbf{k}, T, c_B) = mk^2 + \frac{d^2 f(c_B)}{dc_B^2} \tag{5.3.11}$$

Eq. (5.3.11) yields the long-wave phenomenological approximation to the characteristic function. The long-wave mean-field approximation to the characteristic function (5.1.26) yields

$$b(\mathbf{k}, T, c_B) = V(\mathbf{k}) + \frac{\kappa T}{c_B(1-c_B)} \approx V(0) + \frac{\kappa T}{c_B(1-c_B)} + m_{ij} k_i k_j \tag{5.3.12}$$

where

$$m_{ij}=\left(\frac{\partial^2 V(\mathbf{k})}{\partial k_i \partial k_j}\right)_{\mathbf{k}=0}$$

is the second-order expansion coefficient. For a cubic solution

$$m_{ij}=\delta_{ij}m \tag{5.3.13}$$

Substituting Eq. (5.3.13) into (5.3.12) yields

$$b(\mathbf{k}, T, c_B)\approx V(0)+\frac{\kappa T}{c_B(1-c_B)}+mk^2 \tag{5.3.14}$$

It is easy to see that

$$V(0)+\frac{\kappa T}{c_B(1-c_B)}=\frac{d^2 f_{mf}(c_B)}{dc_B^2} \tag{5.3.15}$$

where

$$f_{mf}(c_B)=\tfrac{1}{2}V(0)c_B^2+\kappa T[c_B \ln c_B+(1-c_B)\ln(1-c_B)] \tag{5.3.16}$$

is the specific mean-field free energy.

Comparison of Eqs. (5.3.15), (5.3.14), and (5.3.11) affords the microscopic interpretation of the phenomenological coefficients.

The long-wave approximation of the kinetic function

$$-L_0(\mathbf{k})=2\sum_{\mathbf{r}}{}' \tilde{L}_0(\mathbf{r})\sin^2\tfrac{1}{2}\mathbf{kr} \tag{5.3.17}$$

yields for a cubic solution

$$-L_0(\mathbf{k})=2\sum_{\mathbf{r}}{}' \tilde{L}_0(\mathbf{r})\sin^2\tfrac{1}{2}\mathbf{kr}\approx\tfrac{1}{6}\left(\sum_{\mathbf{r}}{}' \tilde{L}_0(\mathbf{r})r^2\right)k^2+\cdots$$

$$\simeq D_0 k^2 \tag{5.3.18a}$$

where

$$D_0=\tfrac{1}{6}\sum_{\mathbf{r}}{}' \tilde{L}_0(\mathbf{r})r^2 \tag{5.3.18b}$$

is the diffusivity.

Substituting Eq. (5.3.18a) and (5.3.11) into (5.3.4), we obtain

$$\Delta\tilde{n}(\mathbf{k}, t)=\Delta n(\mathbf{k}, 0)\exp\left[-\frac{D_0 k^2}{\kappa T}c_B(1-c_B)\left(mk^2+\frac{d^2 f(c_B)}{dc_B^2}\right)t\right] \tag{5.3.19}$$

The phenomenological continuous Eq. (5.3.19) derived by Cahn to explain the behavior of alloys in the spinodal decomposition has played a very important part in the working out of the modern approach to the problem (42). It has been the starting point of many subsequent studies, especially in the field of linear microscopic theories. As a matter of fact, these theories are extension of Cahn's concept to the case of crystal lattice site diffusion.

The coefficient D_0 in Eqs. (5.3.18a) to (5.3.19) is the diffusivity of an ideal solution. This is easy to see by setting interchange energies equal to zero. Thus, by setting $V(0)=0$ in Eq. (5.3.15), we obtain

$$\frac{d^2f(c_B)}{dc_B^2} \to \frac{\kappa T}{c_B(1-c_B)} \tag{5.3.20}$$

Substituting (5.3.20) and $m=0$ into Eq. (5.3.19) yields

$$\Delta\tilde{n}(\mathbf{k}, t)=\Delta\tilde{n}(\mathbf{k}, 0)\exp(-D_0k^2t) \tag{5.3.21}$$

Eq. (5.3.21) is in fact the Fourier transform of the solution of the diffusion equation for an infinite solid solution

$$\frac{\partial \Delta n(\mathbf{r}, t)}{\partial t}=D_0\nabla^2\Delta n(\mathbf{r}, t) \tag{5.3.22}$$

The microscopic problem of atomic diffusion considered above is related directly to the random walk problem in a crystal lattice. Actually, assuming $\tilde{n}(\mathbf{k}, 0)=1$ in Eq. (5.3.21) (this corresponds to the assumption that the traveling atom definitely occupies the crystal lattice site $\mathbf{r}=0$ at the start) and $c_B=1/N \to 0$ (the assumption of a single traveling atom) we obtain

$$\Delta\tilde{n}(\mathbf{k}, t)=\exp(L_0(\mathbf{k})t) \tag{5.3.23}$$

The back Fourier transform of Eq. (5.3.23) gives

$$\begin{aligned} n(\mathbf{r}, t)&=\frac{1}{N}\sum_{\mathbf{k}}\exp(L_0(\mathbf{k})t)e^{i\mathbf{k}\mathbf{r}} \\ &=v\int_{\text{B.Z.}}\exp(L_0(\mathbf{k})t)e^{i\mathbf{k}\mathbf{r}}\frac{d^3k}{(2\pi)^3} \end{aligned} \tag{5.3.24}$$

where the integration is over the first Brillouin zone.*

Eq. (5.3.24) solves the random walk problem. It gives the probability to find the atom on the crystal lattice site $\mathbf{r}$ at the moment t if at the start moment, $t=0$, the atom occupied the site $\mathbf{r}=0$. The elementary jump probabilities between the sites $\mathbf{r}$ and $\mathbf{r}'$ are given by $\tilde{L}_0(\mathbf{r}-\mathbf{r}')$. At large t (long-time approximation) the major contributions to the integral value come from small $\mathbf{k}$ because the function $L_0(\mathbf{k})$ in Eq. (5.3.17) has its minimum value at $\mathbf{k}=0$. Using the long-wave approximation (5.3.18a), we may rewrite Eq. (5.3.24) in the form

$$n(\mathbf{r}, t)\approx v\iiint_{-\infty}^{\infty} e^{-D_0k^2t}e^{i\mathbf{k}\mathbf{r}}\frac{d^3k}{(2\pi)^3}=\frac{v}{8(\pi D_0t)^{3/2}}e^{-r^2/4D_0t)} \tag{5.3.25}$$

The method employed to obtain the long-time approximation (5.3.25) is in fact the steepest descent method. It is easy to see that the asymptotic approxi-

*Eq. (5.3.24) is obtained by substituting integration over the first Brillouin zone for summing over the quasicontinuum of points $\mathbf{k}$ within that zone.

mation (5.3.25) is the Green function of the diffusion equation: it provides the solution of the diffusion problem for an infinite isotropic medium with a point source. This is in complete agreement with the well-known fact that the long-time asymptotic approximation to the random walk problem solution is given by the solution to the diffusion equation with a point source (62, 63).

After this brief discussion of the relation between the random walk problem and the linearized Önsager equation for single lattice site probabilities, we will return to the spinodal decomposition.

It follows from Eq. (5.3.19) that the amplification factor

$$\lambda(\mathbf{k})=\frac{D_0 k^2}{\kappa T}\, c_B(1-c_B)\left(mk^2+\frac{d^2 f(c_B)}{dc_B^2}\right) \tag{5.3.26}$$

is positive at any $\mathbf{k}$ above the spinodal curve. This follows from the fact that $m>0$, and, by definition, $d^2f/dc_B^2>0$. According to Eq. (5.1.22), we have concentration wave decay in this case. The situation changes radically below the spinodal curve. By definition $d^2f/dc_B^2<0$ in this case, and the amplification factor (5.3.26) becomes negative within the sphere of the radius k_1 with the centrum at $\mathbf{k}=0$ (see Fig. 49). By definition, this sphere is the instability range in the $\mathbf{k}$-space. The boundary of the instability range is described by the equation

$$\frac{d^2 f}{dc_B^2}+mk_1^2=0 \tag{5.3.27}$$

At $|\mathbf{k}|>k_1$, the amplification factor (5.3.26) changes its sign and turns positive. The concentration waves with the wave vectors outside the instability range ($|\mathbf{k}|>k_1$) must therefore decay.

The development of the spinodal decomposition over time deserves a detailed consideration. The highest amplification rate will correspond to the wave vectors $k=k_{\max}$ (see Fig. 49). According to Cahn (42) these waves dominate the concentration wave spectrum and determine the structure characteristics of a spinodal alloy in the early stage of the decomposition. Moreover Cahn has shown that the spinodal instability in a cubic solid solution is the instability with respect to the concentration waves whose wave vectors are not arbitrary vectors within the instability range but the vectors collinear with certain sym-

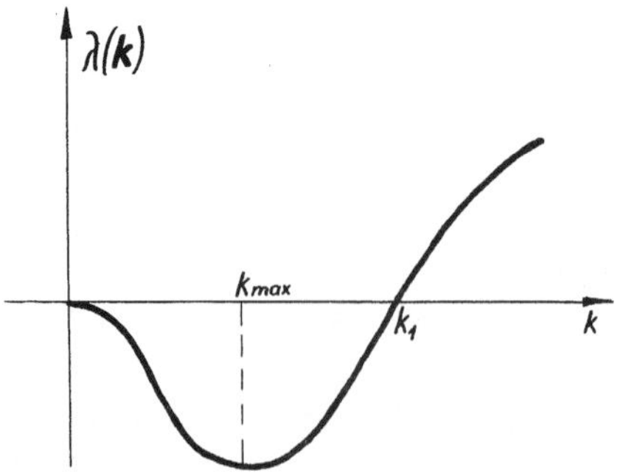

Figure 49. Typical dependence of the amplification factor $\lambda(\mathbf{k})$ with respect to wave vector $\mathbf{k}$ on the spinodal instability of a homogeneous solution. The value $k_{\max}$ corresponds to the highest amplification rate. The instability region is $0<k<k_1$.

metry directions (43). The latter conclusion follows from consideration of strain energy contributions to the decomposition thermodynamics.

One of the most elegant results in Cahn's theory of the spinodal decomposition is the periodic modulation of the concentration profile formed in the decomposition reaction. According to Cahn the concentration waves whose six wave vectors, $\pm(k_{max}, 0, 0)$, $\pm(0, k_{max}, 0)$, $\pm(0, 0, k_{max})$, lie on the $\langle 100\rangle$ directions show the highest growth rates if $c_{11}-c_{12}-2c_{44}<0$, where c_{11}, c_{12}, c_{44} are the elastic moduli. The result is a periodic concentration modulation (on the condition that other concentration waves make negligibly small contributions to concentration heterogeneity).

If $c_{11}-c_{12}-2c_{44}>0$, the modulated structure is generated by the dominant concentration wave along the $\langle 111\rangle$ directions.

Cahn's theory appears very attractive because of its utter simplicity and efficiency in the treatment of spinodal decomposition. There are, however, certain shortcomings. First, the theory is only applicable to the early stage of the decomposition when the linear approximation is valid. Second, the assumption that the contributions from concentration waves other than the dominant ones may be ignored needs verification.

To solve these problems, Cahn's theory should be extended to include the nonlinear effects in the kinetic equation. This would provide a means to handle late stages of the decomposition when the amplitudes of concentration heterogeneities are commensurate with the difference between the compositions of the equilibrium phases. An analysis of the spinodal decomposition based on the numerical solution of the nonlinear Eq. (5.1.19) (64) will be given below.

5.4. COMPUTER SIMULATION OF SPINODAL DECOMPOSITION: FORMATION OF GP ZONES

Cahn has shown (43) that strain-induced interactions in a cubic solid solution result in the domination of one-dimensional concentration heterogeneities. Starting from this point, Morris and Khachaturyan have analyzed spinodal decomposition in the one-dimensional case (64); their results are reproduced in this section. The analysis carried out in the work (64) is based on the numerical solution of the nonlinear kinetic equation (5.1.19). The authors (64) have used the mean-field approximation (3.6.3b) and the nearest-neighbor interaction model. Eq. (3.6.3b) for a fcc solid solution may then be written in the form

$$F=\frac{1}{2}\sum_{m=-N/2}^{N/2}\left[4W_1(n_m n_{m+1}+n_m n_{m-1}+n_m^2)\right.$$
$$\left.+\kappa T(n_m\ln n_m+(1-n_m)\ln(1-n_m))\right] \qquad (5.4.1)$$

where $n(\mathbf{r})=n_m$ is the occupation probability to find a solute atom at any crystal lattice site of the mth (001) plane of the fcc crystal, N is the total number of (001) planes in the crystal, and W_1 is the nearest-neighbor interaction energy.

In the one-dimensional case Eq. (5.1.19) may be rewritten

$$\frac{d\Delta n_m(t)}{dt}=\sum_{m=-N/2}^{N/2}\frac{\tilde{L}_0^{\text{interpl.}}(m-m')}{\kappa T}c_B(1-c_B)\frac{\delta F}{\delta n_{m'}(t)} \tag{5.4.2}$$

where $\Delta n_m(t)=n_m(t)-c_B$, $\tilde{L}_0^{\text{interpl.}}(m-m')$ are the probabilities of elementary diffusion jumps between the m'th and mth planes.

Let us write the transformation probabilities $\tilde{L}_0^{\text{interpl.}}(m-m')$ in terms of the diffusivity D_0. In so doing, we make the assumption that elementary diffusion jumps are only permitted between the nearest-neighbor sites of the fcc lattice. Eq. (5.3.18b) then becomes

$$D_0=\tilde{L}_0(1)a_{\text{fcc}}^2 \tag{5.4.3}$$

where a_{fcc} is the fcc crystal lattice parameter. Using Eq. (5.4.3) in (5.3.17), we obtain

$$L_0(\mathbf{k})=-2\frac{D_0}{a_{\text{fcc}}^2}\sum_{\mathbf{r}_j}\sin^2\frac{1}{2}\mathbf{k}\mathbf{r}_j \tag{5.4.4}$$

where the summation is over twelve sites of the first coordination shell of the fcc lattice. Since we consider a one-dimensional modulation along the [001] direction,

$$\mathbf{k}=(0,0,k) \tag{5.4.5}$$

Substitution of Eq. (5.4.5) into (5.4.4) and summation over $\mathbf{r}_j$ yield

$$L_0(\mathbf{k})=-8\frac{D_0}{a_{\text{fcc}}^2}(1-\cos kd) \tag{5.4.6}$$

where d is the (001) interplanar distance. Clearly, the function $L_0(\mathbf{k})$ in Eq. (5.4.6) is a one-dimensional Fourier transform of the interplanar transformation probabilities:

$$\tilde{L}_0^{\text{interpl.}}(m)=\begin{cases}-8\dfrac{D_0}{a_{\text{fcc}}^2} & \text{if } m=0\\[2ex] 4\dfrac{D_0}{a_{\text{fcc}}^2} & \text{if } m=\pm1\\[2ex] 0 & \text{otherwise}\end{cases} \tag{5.4.7}$$

With the relation (5.4.7), Eq. (5.4.2) reads

$$\frac{d\Delta n_m(t)}{dt}=-\frac{8D_0}{a_{\text{fcc}}^2}\frac{c_B(1-c_B)}{\kappa T}\left\{\frac{\delta F}{\delta n_m(t)}-\frac{1}{2}\frac{\delta F}{\delta n_{m+1}(t)}-\frac{1}{2}\frac{\delta F}{\delta n_{m-1}(t)}\right\} \tag{5.4.8}$$

Substituting Eq. (5.4.1) into (5.4.8), we have

$$\frac{d\Delta n_m(t)}{dt}=-\frac{8D_0}{a_{\text{fcc}}^2}\frac{c_B(1-c_B)}{\kappa T}\left\{4W_1\left[n_m(t)-\frac{1}{2}n_{m+1}(t)-\frac{1}{2}n_{m-1}(t)\right]\right.$$
$$\left.+\kappa T\left[\ln\frac{n_m(t)}{1-n_m(t)}-\frac{1}{2}\ln\frac{n_{m+1}(t)}{1-n_{m+1}(t)}-\frac{1}{2}\ln\frac{n_{m-1}(t)}{1-n_{m-1}(t)}\right]\right\} \tag{5.4.9}$$

Eq. (5.4.9) may be simplified by the introduction of the reduced temperature

$$\tau = \frac{\kappa T}{|V(0)|} = \frac{\kappa T}{12|W_1|}$$

and time

$$\bar{t} = t\,\frac{8D_0}{a_{\text{fcc}}^2} \tag{5.4.10}$$

It may then be written as

$$\frac{dn_m(\bar{t})}{d\bar{t}} = -c_B(1-c_B)\left\{\frac{1}{3\tau}\left[n_m(\bar{t}) - \frac{1}{2}n_{m+1}(\bar{t}) - \frac{1}{2}n_{m-1}(\bar{t})\right]\right.$$
$$\left. + \left[\ln\frac{n_m(\bar{t})}{1-n_m(\bar{t})} - \frac{1}{2}\ln\frac{n_{m+1}(\bar{t})}{1-n_{m+1}(\bar{t})} - \frac{1}{2}\ln\frac{n_{m-1}(\bar{t})}{1-n_{m-1}(\bar{t})}\right]\right\} \tag{5.4.11}$$

Below we shall describe computer simulation results for a representative point of the τ-c diagram ($\tau = 0.125$, $\bar{c} = 0.16$) which is close to the spinodal and corresponds to the asymmetrical case (see Fig. 39). The initial condition for Eq. (5.4.11), $n_m(0)$ (the atomic distribution at the start of the spinodal decomposition) has been generated by the Monte Carlo technique to insure the high-temperature short-range order that satisfies the Krivoglaz equation (66):

$$\langle|\Delta\tilde{n}(\mathbf{k}, t)|^2\rangle = \frac{c_B(1-c_B)}{1 + c_B(1-c_B)V(\mathbf{k})/\kappa T_{\text{init.}}} \tag{5.4.12}$$

where

$$V(\mathbf{k}) = 4W_1 + 8W_1 \cos kd \tag{5.4.13}$$

$T_{\text{init}} > T_{sp}$, T_{sp} is the temperature of spinodal decomposition. The Fourier transform of the solution of the linearized Eq. (5.4.11) is

$$\Delta\tilde{n}(\mathbf{k}, t) = \Delta\tilde{n}(\mathbf{k}, 0)\exp(-\lambda(\mathbf{k})\bar{t}) \tag{5.4.14}$$

where the amplification factor $\lambda(\mathbf{k})$ is

$$\lambda(\mathbf{k}) = \sin^2\frac{kd}{2}\left[\frac{c_B(1-c_B)}{3\tau}(1 + 2\cos kd) + 1\right] \tag{5.4.15}$$

[compare with Eq. (5.1.31)]. Since the wave numbers k allowed by the cyclic boundary conditions are

$$k_j = \frac{2\pi}{dN}j \quad \text{for } j = 0, 1, \ldots, N-1$$

we have the solution of the linearized Eq. (5.4.11) written in terms of the coordinates of the real crystal lattice site space as the back Fourier transform of Eq. (5.4.14)

$$n_m(\bar{t}) = c_B + \frac{1}{N}\sum_{j=0}^{N-1}\Delta n\left(\frac{2\pi}{Nd}j, 0\right)\exp\left(-\lambda\left(\frac{2\pi}{Nd}j\right)\bar{t}\right)e^{i(2\pi j/N)m} \tag{5.4.16}$$

Figures 50 and 51 illustrate the development of the concentration profile $n_m(t)$ and variations of intensities of x-ray diffuse reflections

$$I(k_j, \bar{t}) = |\Delta\tilde{n}(k_j, \bar{t})|^2$$

To visualize small concentration and intensity variations, the ordinates are given on a variable scale. The horizontal lines on the concentration profile plots are the equilibrium compositions of the two coexisting phases approached eventually by concentration heterogeneities during their growth.

Comparison of the exact numerical solution and the solution of the linearized Eq. (5.4.16) shows the two solutions almost to coincide at reduced times of about 0.1.* Since reduced times are measured in units of the elementary diffusion jump time,

$$t_0 = \frac{a_{\text{fcc}}^2}{8D_o} \tag{5.4.17}$$

[see Eq. (5.4.10)], we conclude that the first "linear" stage of spinodal decomposition only goes for $0.1t_0$ (see Figs. 50a and 51a). It follows from Figs. 50 and 51 that we must discard the linear approximation in favor of the numerical solution of the nonlinear problem if the decomposition duration is of the order of t_0. In about $500t_0$ concentration heterogeneities grow to the equilibrium concentrations (see Fig. 50d). This corresponds to the completion of the second stage of the decomposition when the solid solution is transformed into a mixture of fine precipitates of two phases of the equilibrium compositions.

The third stage is coarsening. Its duration far exceeds those of the second and, especially, the first stages. A computer simulation shows that coarsening in the symmetrical case takes less time than it does in the asymmetrical case.

It is important that the linear stage of spinodal decomposition produces a pseudo periodical distribution that gives rise to broadened satellites in diffraction patterns (Fig. 51a). Periodicity is perfected during nonlinear growth and subsequent coarsening (compare Figs. 50a and c).

The computer simulation results agree with the experimental data on 51.5Cu-33.5Ni-15Fe reported by Butler and Thomas (65). These authors observed a rapid change of the Curie temperature during aging of the alloy at 625° C: within 1 min, the Curie temperature increased from 237° C to about 454° C, whereas measurements taken in 15 min and 5 hrs showed it to vary only insignificantly in further aging.

According to (58), the activation energy of diffusion in Cu-50Ni is 64.5 kcal/mol. With the preexponential factor of the order of 1 cm^2/sec we have

$$D_0 \approx \exp\left(-\frac{64{,}500}{2T}\right)\left[\frac{\text{cm}^2}{\text{sec}}\right] \tag{5.4.18}$$

Eq. (5.4.17) makes it possible to express the characteristic interplanar transition time in terms of the diffusivity D_0 and crystal lattice parameter a_{fcc} and thus to

*Its long-wave asymptotic coincides with Cahn's equation.

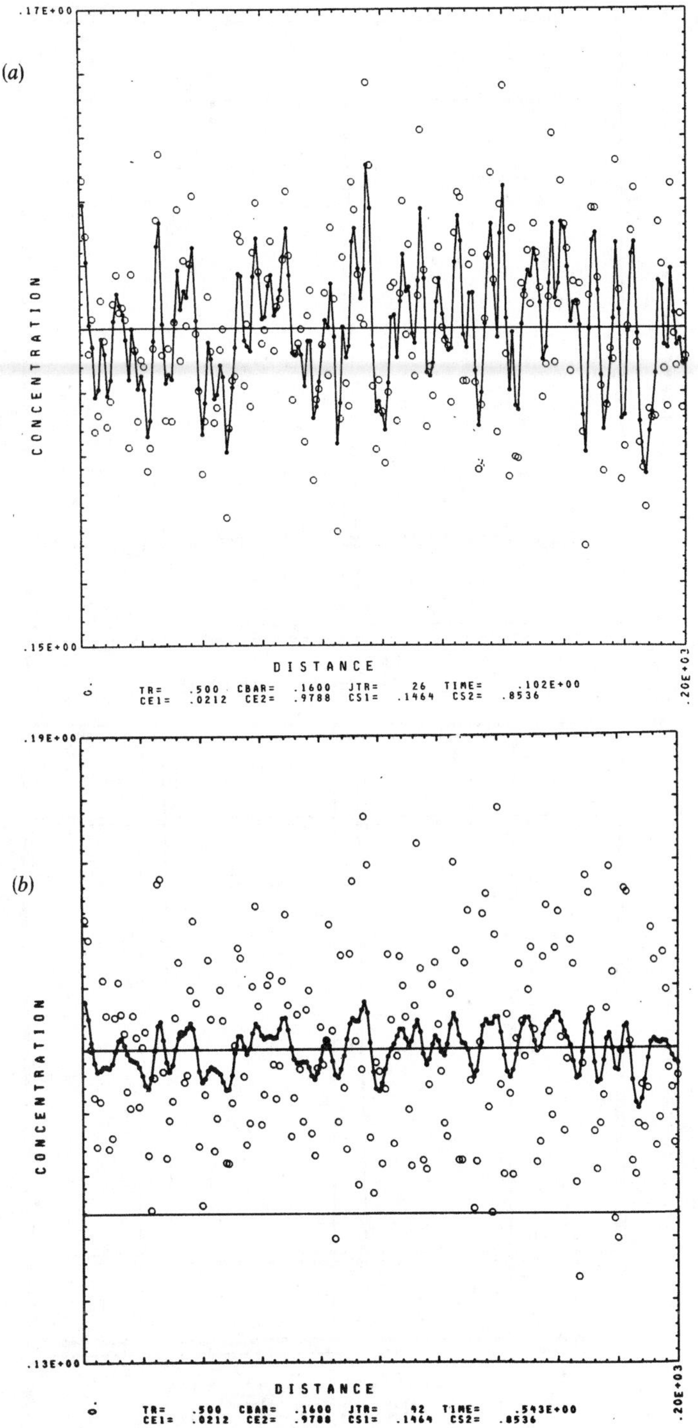

Figure 50. Concentration profiles in computer simulation for the asymmetrical case of spinodal decomposition ($\tau = 0.125$, $\bar{c} = 0.16$) (64). Concentration profiles at (*a*) $\bar{t} = 0.102$, (*b*) $\bar{t} = 0.543$, (*c*) $\bar{t} = 200$,

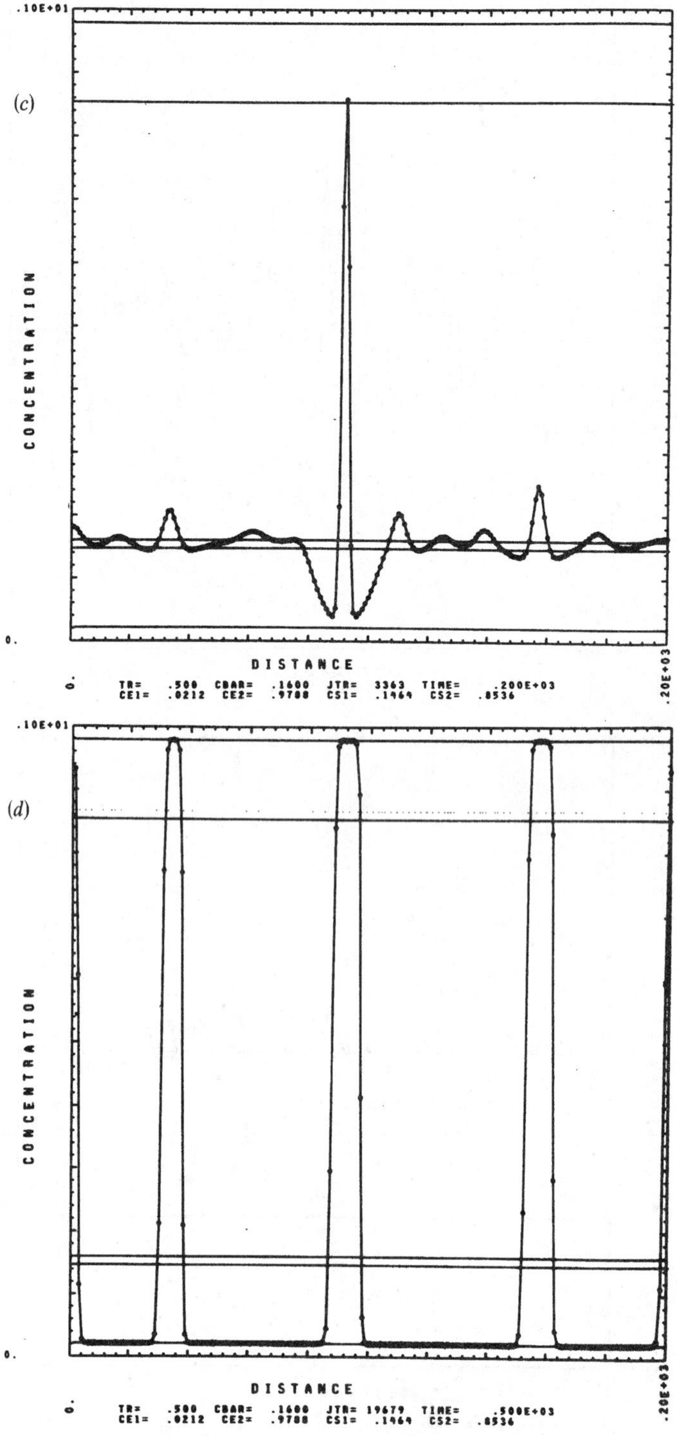

(*d*) $\bar{t} = 500$ where reduced time $\bar{t}$ is measured in units t_0. Open circles in diagrams correspond to the solution of linearized equation (5.4.11).

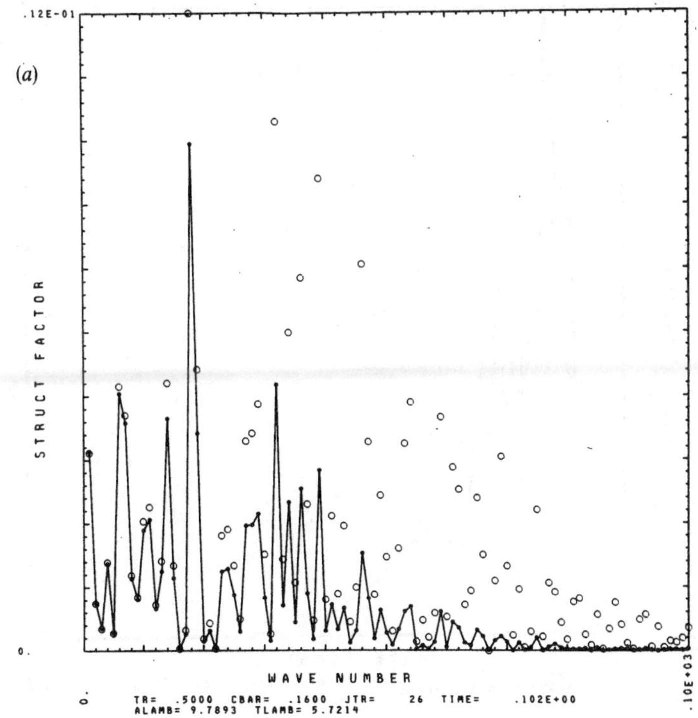

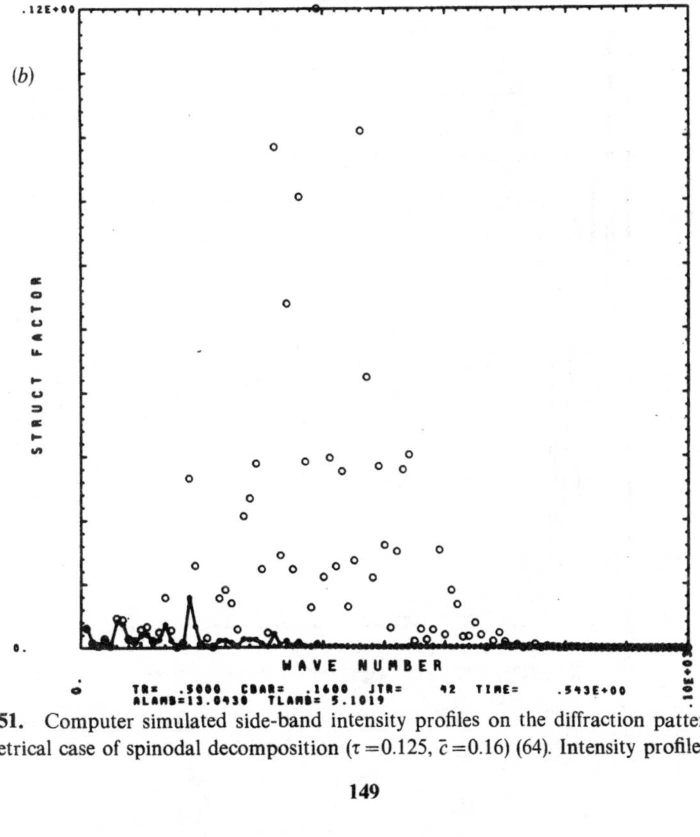

Figure 51. Computer simulated side-band intensity profiles on the diffraction patterns for the asymmetrical case of spinodal decomposition ($\tau = 0.125$, $\bar{c} = 0.16$) (64). Intensity profiles at (*a*) $t =$

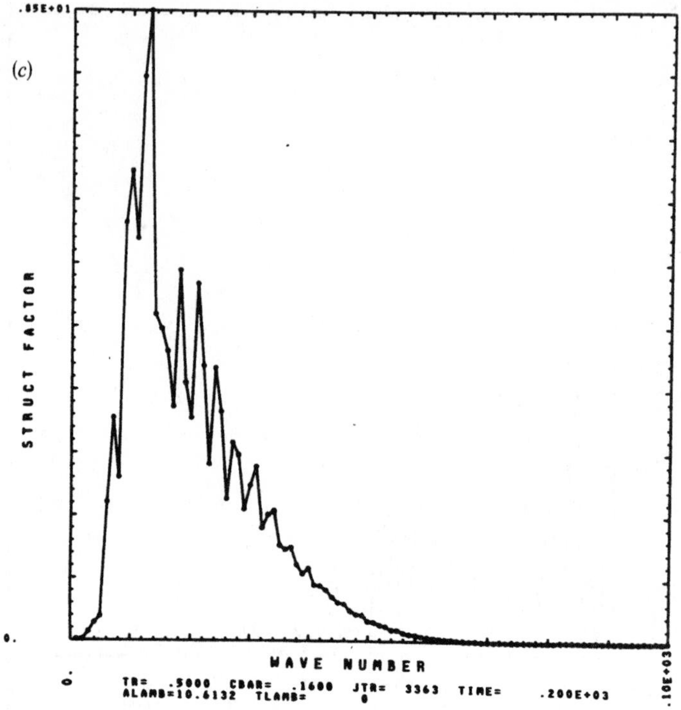

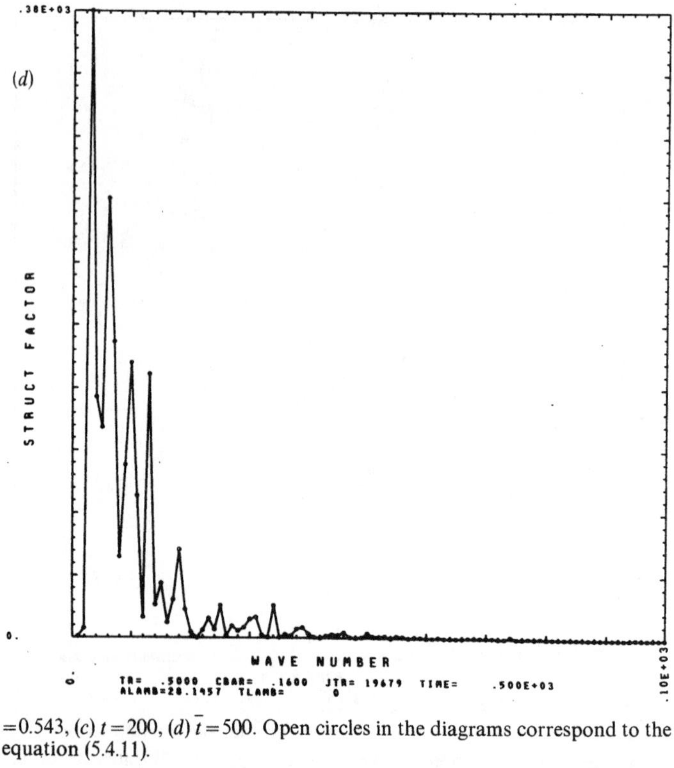

0.102, (*b*) $t = 0.543$, (*c*) $t = 200$, (*d*) $\bar{t} = 500$. Open circles in the diagrams correspond to the solution of linearized equation (5.4.11).

estimate durations of various stages of the decomposition. Using the numerical values $a_{fcc}=3.56\times10^{-8}$ cm, $T=625°+273°=898°$ K, we obtain from (5.4.17)

$$t_0=\frac{(3.56\times10^{-8})^2}{8\exp\left(-\dfrac{64{,}500}{2\times898}\right)}=0.6 \text{ sec} \tag{5.4.19}$$

This is the unit of the time scale used in the computer simulation of spinodal decomposition. It thus follows that the "linear" stage of the decomposition terminates in $0.1t_0\simeq0.06$ sec.

The second (nonlinear) stage of the decomposition takes about $500t_0$ according to computer simulation results. This is equal to 300 sec = 5 min on the usual time scale.

Taking into consideration the low accuracy of the diffusion coefficient measurements, the latter value may be regarded as being in a very good agreement with the experimental value of 1 min reported by Butler and Thomas (65).

An important result of the computer simulation is the appearance of extremely thin new phase precipitates in the asymmetrical alloy. The thickness of those precipitates is microscopic, of about three (001) planes. Their maximum occupation probabilities approach the equilibrium compositions (see Fig. 50c). Figure 51c demonstrates a broadening of the diffraction spots caused by the small precipitate thickness. The precipitates are formed after a pseudoperiodic modulated structure corresponding to the nonlinear decomposition stage appears. This phenomenon is associated with the formation of GP zones. Recall that one of the reasons a GP zone is regarded as different from a usual new phase precipitate is its extreme thinness, of the order of 1 to 3 crystal lattice planes. This was considered incongruous with conventional thermodynamics. The computer simulation of the process, which used conventional thermodynamics, however, shows, that such microscopically thin precipitates can be formed and that clearly GP zones are nothing other than precipitates (see Fig. 50c).

Comparing Figs. 50b and c, one may see that according to the computer simulation, the "satellite" stage precedes the GP zone stage. The transition from the "satellite" stage to the stage of formation of GP zones predicted by the computer simulation was in fact observed by Rioja and Laughlin in the classic GP zone alloy, Al-4.0Cu (67). Laughlin reported the appearance of a modulated structure that manifested itself by the side-band effect in the diffraction patterns and was later replaced by the usual GP zone structure.

In sum, we wish to stress the following points:

1. The "linear" stage of spinodal decomposition is far less (by a factor of 10^4) prolonged than all the other stages.
2. The nonlinear stage of decomposition and, in the majority of cases, coarsening result in a refinement of the pseudoperiodic distribution.

It seems reasonable to assume that the formation of most modulated structures observed occurs during the coarsening stage. This assumption should, however, be verified by comparing the duration of the "linear" stage of decomposition, $t \sim 0.1t_0$ with t_0 given by Eq. (5.4.17), with the typical period of time required to detect structural changes in alloys.

3. There are important reasons for regarding GP zones as extremely thin new phase precipitates formed in the intermediate stage of the decomposition (at the end of the second stage).

5.5. SHORT-RANGE ORDER RELAXATION KINETICS

Diffusion measurements are usually performed on systems where concentration heterogeneities have macroscopic dimensions far exceeding interatomic distances. This is the reason why the conventional diffusion measurements only give diffusivities entering continuous diffusion equations. The other shortcoming of diffusion experiments comes from low rates of diffusion relaxation of macroscopic concentration heterogeneities. To reduce measurement time relatively high temperatures should be used. The information available from these experiments is therefore limited to high-temperature diffusion characteristics.

Kinetic measurements of relaxation of short-range order are free from the limitations mentioned. Short-range order is in fact a naturally occurring concentration heterogeneity whose dimensions are commensurate with interatomic distances. The kinetics of relaxation of short-range order is thus determined by microscopic diffusion over distances of the order of interatomic separations. For that reason studies of short-range order kinetics provide far more detailed information on the diffusion mechanisms compared with the conventional diffusion measurements. In particular, they offer the possibility of determining the microscopic characteristics of atomic migrations, including probabilities and types of atomic jumps in an elementary diffusion event.

In addition to that, short-range order diffusion measurements can be taken at comparatively low temperatures because short-range order relaxation times are fairly small. The results thus obtained may be utilized to determine low-temperature diffusivities and activation energies.

Relaxation of short-range order to its equilibrium value can occur, for instance, after quenching of a disordered alloy. In this case the short-range order inherited from a high-temperature state changes to its new equilibrium value. Since short-range order is an unique natural inhomogeneity whose dimension is commensurable with atomic ranges, the kinetic process of attainment of equilibrium short-range order is described by the discrete mechanism of diffusion. Short-range order kinetics can be observed through any short-range order dependent phenomena such as residual resistance, the Zener effect in internal friction, and x-ray diffuse scattering. X-ray diffuse scattering is an object of especial interest since it enables one to obtain the most detailed information of both equilibrium and nonequilibrium short-range order and therefore is the most convenient instrument for investigating short-range order kinetics.

Since short-range order parameters are proportional to the pair-site probabilities (to the two-particle correlation function (1.3.42)) the problem of short-range order kinetics is reduced to finding and solving kinetic equations for a pair-site probability. The theory of short-range order kinetics was proposed by Khachaturyan in 1967 (52); the same results were again obtained by Cook in 1970 (68). Following (52), we shall derive these kinetic equations for a ν-component crystalline solid solution.

Let us consider a two-particle correlation function $p(\alpha\mathbf{r}, \beta\mathbf{r}'; t)$ which is the occupation probability of finding simultaneously an atom of the type α in the crystal lattice site $\mathbf{r}$ and an atom of the type β in the crystal lattice site $\mathbf{r}'$. We will also introduce the a posteriori probabilities $p(\alpha\mathbf{r}|\beta\mathbf{r}'; t)$ that an atom of the type α occupies the site $\mathbf{r}$ under the condition that an atom of the type β definitely occupies the site $\mathbf{r}'$.

The change in the correlation function $p(\alpha\mathbf{r}, \beta\mathbf{r}'; t)$ caused by movement of the α-atom at the site $\mathbf{r}$ under the condition that the β-atom at the site $\mathbf{r}'$ is fixed can be represented as the change in the a posteriori probability, $dp(\alpha\mathbf{r}|\beta\mathbf{r}'; t)$, multiplied by the a priori probability c_β (an atomic fraction of β-atoms) of finding a β-atom at the site $\mathbf{r}'$:

$$c_\beta dp(\alpha\mathbf{r}|\beta\mathbf{r}'; t)$$

Similarly, the change in the probability $p(\alpha\mathbf{r}, \beta\mathbf{r}'; t)$ resulting from movement of a β-atom under the condition that an α-atom at the site $\mathbf{r}$ is fixed is

$$c_\alpha dp(\beta\mathbf{r}'|\alpha\mathbf{r}; t)$$

Thus the total change in the correlation function $p(\alpha\mathbf{r}, \beta\mathbf{r}'; t)$ is

$$dp(\alpha\mathbf{r}, \beta\mathbf{r}'; t) = c_\beta dp(\alpha\mathbf{r}|\beta\mathbf{r}'; t) + c_\alpha dp(\beta\mathbf{r}'|\alpha\mathbf{r}; t) \tag{5.5.1}$$

or in time derivatives

$$\frac{dp(\alpha\mathbf{r}, \beta\mathbf{r}'; t)}{dt} = c_\beta \frac{\partial p(\alpha\mathbf{r}|\beta\mathbf{r}'; t)}{\partial t} + c_\alpha \frac{\partial p(\beta\mathbf{r}'|\alpha\mathbf{r}; t)}{\partial t} \tag{5.5.2}$$

Eq. (5.5.2) reduces the problem of the short-range order kinetics to the more simple problem of usual discrete diffusion (single-site kinetics) of an atom in the external field formed by a fixed atom. Actually, the a posteriori probability $p(\alpha\mathbf{r}|\beta\mathbf{r}'; t)$ describes discrete diffusion of an α-atom in the external field formed by a fixed β-atom. The latter enables one to employ the Önsager equation technique, which was used in Section 5.1 for the analysis of discrete diffusion. In this case the analogue of Eq. (5.1.1) can be represented in the form

$$\begin{aligned}
\frac{\partial \Delta p(\alpha\mathbf{r}|\beta\mathbf{r}'; t)}{\partial t} &= \sum_{\gamma=1}^{\nu} \sum_{\mathbf{r}''} \frac{\tilde{L}_{\alpha\gamma}(\mathbf{r}-\mathbf{r}'')}{\kappa T} c_\alpha c_\gamma \frac{\delta F(\beta\mathbf{r}')}{\delta p(\gamma\mathbf{r}''|\beta\mathbf{r}'; t)} \\
\frac{\partial \Delta p(\beta\mathbf{r}'|\alpha\mathbf{r}; t)}{\partial t} &= \sum_{\gamma=1}^{\nu} \sum_{\mathbf{r}''} \frac{\tilde{L}_{\beta\gamma}(\mathbf{r}'-\mathbf{r}'')}{\kappa T} c_\beta c_\gamma \frac{\delta F(\alpha\mathbf{r})}{\delta p(\gamma\mathbf{r}''|\alpha\mathbf{r}; t)}
\end{aligned} \tag{5.5.3}$$

where $\Delta p(\alpha\mathbf{r}|\beta\mathbf{r}'; t) = p(\alpha\mathbf{r}|\beta\mathbf{r}'; t) - c_\alpha$, $F(\alpha\mathbf{r})$ is the free energy calculated under the additional condition that an α-atom is fixed in the site $\mathbf{r}$.

Linearization of the first variational derivatives of the free energy (5.5.3) with respect to $\Delta p(\gamma\mathbf{r}''|\beta\mathbf{r}'; t)$ and $\Delta p(\gamma\mathbf{r}''|\alpha\mathbf{r}; t)$, respectively, gives the resulting kinetic equation for the correlation function:

$$\frac{d\Delta p(\alpha\mathbf{r}, \beta\mathbf{r}'; t)}{dt} = -\sum_{\mathbf{r}''} \tilde{W}_{\alpha\gamma}(\mathbf{r}-\mathbf{r}'')\Delta p(\gamma\mathbf{r}'', \beta\mathbf{r}'; t) - \sum_{\mathbf{r}''} \tilde{W}_{\beta\gamma}(\mathbf{r}-\mathbf{r}'')\Delta p(\gamma\mathbf{r}'', \alpha\mathbf{r}'; t) \tag{5.5.4}$$

where $\Delta p(\alpha\mathbf{r}, \beta\mathbf{r}'; t) = p(\alpha\mathbf{r}, \beta\mathbf{r}'; t) - p(\alpha\mathbf{r}, \beta\mathbf{r}'; \infty)$. (Trivial identities $p(\alpha\mathbf{r}, \beta\mathbf{r}'\ t) \equiv p(\alpha\mathbf{r}|\beta\mathbf{r}'; t)c_\beta \equiv p(\beta\mathbf{r}'|\alpha\mathbf{r}; t)c_\alpha$ are employed.) The quantities $\tilde{W}_{\alpha\gamma}(\mathbf{r}-\mathbf{r}'')$ coincide with the quantities (5.1.9), with accuracy up to the approximation corresponding to the decoupling of the three-particle correlation function (three-site probabilities). Since $p(\alpha\mathbf{r}, \beta\mathbf{r}'; t) = p_{\alpha\beta}(\mathbf{r}-\mathbf{r}'; t)$ (the latter is a consequence of the uniformity of the relevant disordered solution), one can carry out the Fourier transformation of Eq. (5.5.4). The result is

$$\frac{d\Delta\tilde{p}_{\alpha\beta}(\mathbf{k}, t)}{dt} = -W_{\alpha\gamma}(\mathbf{k})\Delta\tilde{p}_{\gamma\beta}(\mathbf{k}, t) - W_{\beta\gamma}(\mathbf{k})\Delta\tilde{p}_{\gamma\alpha}(\mathbf{k}, t) \tag{5.5.5}$$

Eq. (5.5.5) can be rewritten in the compact operator form

$$\frac{d\Delta\hat{\mathbf{p}}(\mathbf{k})}{dt} = -\hat{\mathbf{W}}(\mathbf{k})\Delta\hat{\mathbf{p}}(\mathbf{k}) - \Delta\hat{\mathbf{p}}^+(\mathbf{k})\hat{\mathbf{W}}^+(\mathbf{k}) \tag{5.5.6}$$

where $\hat{\mathbf{W}}^+(\mathbf{k})$ and $\Delta\hat{\mathbf{p}}^+(\mathbf{k})$ are the transposed operators with respect to $\hat{\mathbf{W}}$ and $\hat{\mathbf{p}}$. The operator solution of Eq. (5.5.6) is

$$\Delta\hat{\mathbf{p}}(\mathbf{k}, t) = \exp(-\hat{\mathbf{W}}(\mathbf{k})t)\Delta\hat{\mathbf{p}}(\mathbf{k}, 0)\exp(-\hat{\mathbf{W}}^+(\mathbf{k})t) \tag{5.5.7}$$

or in the usual matrix form

$$\Delta\tilde{p}_{\alpha\beta}(\mathbf{k}, t) = [\exp(-\hat{\mathbf{W}}(\mathbf{k})t)]_{\alpha\gamma}\Delta\tilde{p}_{\gamma\varepsilon}(\mathbf{k}, 0)[\exp(-\hat{\mathbf{W}}^+(\mathbf{k})t)]_{\varepsilon\beta} \tag{5.5.8}$$

where $\Delta\hat{p}_{\gamma\varepsilon}(\mathbf{k}, 0) = \tilde{p}_{\gamma\varepsilon}(\mathbf{k}, 0) - \tilde{p}_{\gamma\varepsilon}(\mathbf{k}, \infty)$ characterizes an initial deviation of the correlation function from its equilibrium value $\tilde{p}_{\gamma\varepsilon}(\mathbf{k}, \infty)$.

Substituting Eq. (5.1.17) into (5.5.8) gives

$$\tilde{p}_{\alpha\beta}(\mathbf{k}, t) = \tilde{p}_{\alpha\beta}(\mathbf{k}, \infty) + \sum_{s_1, s_2} e^{-\lambda_{s_1}(\mathbf{k})t}e^{-\lambda_{s_2}(\mathbf{k})t} \times v_{s_1}(\alpha, \mathbf{k})[v_{s_1}^+(\gamma, \mathbf{k})\Delta\tilde{p}_{\gamma\varepsilon}(\mathbf{k}, 0)v_{s_2}^+(\varepsilon, \mathbf{k})]v_{s_2}(\beta, \mathbf{k}) \tag{5.5.9}$$

Consider the simplest case of the short-range order relaxation in a binary substitutional solution. If the velocity of a "fast" β-atom is much greater than that of a "slow" α-atom, Eq. (5.5.9) can be represented in the simple form (69):

$$\tilde{p}_{\alpha\alpha}(\mathbf{k}, t) = \tilde{p}_{\alpha\alpha}(\mathbf{k}, \infty) + e^{-2\lambda_\alpha(\mathbf{k})t}(\tilde{p}_{\alpha\alpha}(\mathbf{k}, 0) - \tilde{p}_{\alpha\alpha}(\mathbf{k}, \infty)) \tag{5.5.10}$$

Since the correlation function $p_{\alpha\beta}(\mathbf{r}-\mathbf{r}')$ can be represented in the form

$$p_{\alpha\alpha}(\mathbf{r}-\mathbf{r}') = \langle c_\alpha(\mathbf{r})c_\alpha(\mathbf{r}')\rangle \tag{5.5.11}$$

[see the first term of Eq. (1.3.44)], it follows from Eq. (1.3.44) that

$$\tilde{p}_{\alpha\alpha}(\mathbf{k}, t) = \tilde{K}(\mathbf{k}, t) \quad \text{for } \mathbf{k} \neq 0^*$$

where $\tilde{K}(\mathbf{k}, t) = \sum_{\mathbf{r}} K(\mathbf{r}, t)e^{-i\mathbf{kr}}$, and therefore, in accordance with Eq. (1.3.41), short-range order induced x-ray diffuse scattering is

$$I_{\text{diff}}(\mathbf{k}, t) = (f_\alpha - f_\beta)^2 \tilde{p}_{\alpha\alpha}(\mathbf{k}, t) \tag{5.5.12}$$

Eqs. (5.5.10) and (5.5.12) enable us to present the x-ray diffuse scattering intensities as

$$I_{\text{diff}}(\mathbf{k}, t) = I_{\text{diff}}(\mathbf{k}, \infty) + [I_{\text{diff}}(\mathbf{k}, 0) - I_{\text{diff}}(\mathbf{k}, \infty)]e^{-2\lambda_\alpha(\mathbf{k})t} \tag{5.5.13}$$

Eq. (5.5.13) demonstrates that x-ray diffuse intensity at each point of the recip-

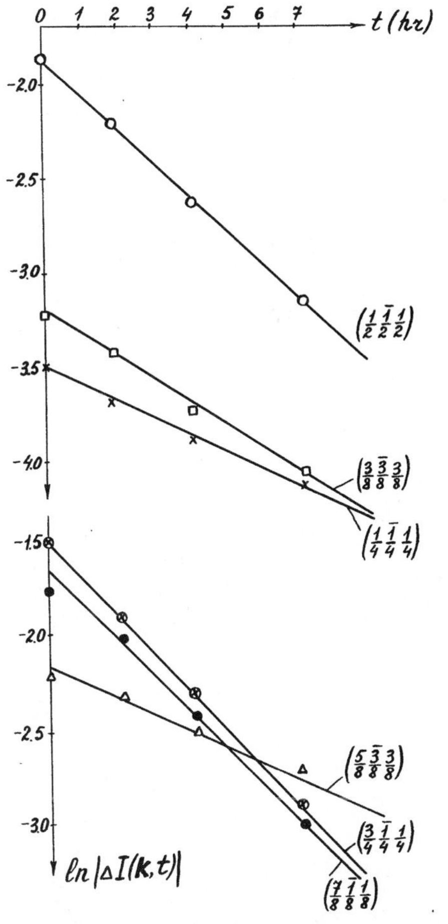

Figure 52. The plot of x-ray diffuse intensity, $\ln[I_{\text{diff}}(\mathbf{k}, t) - I_{\text{diff}}(\mathbf{k}, \infty)]$, with respect to time, t, for the Fe–16 at% Al disordered alloy at 320°C [after S. V. Semenovskaya (70)].

*The Fourier transform of the second term of the sum (1.3.44) gives the contribution only at $\mathbf{k} = 0$.

rocal space **k** tends to its equilibrium value in accordance with the exponential decay law with its own time of decay $t_0(k)=\frac{1}{2}(\lambda_\alpha(\mathbf{k}))^{-1}$.

Until now there has been but a single study (70, 71) of the relaxation of x-ray diffuse scattering intensities. It refers to the case of Fe-Al disordered alloys. Figure 52 shows that experimentally observed relaxation of x-ray intensities due to diffusion of "slow" Fe-atoms as described by Eq. (5.5.13). The dependence $\lambda_\alpha=\lambda_\alpha(\mathbf{k})$ enables one to obtain the important information on the atomic mechanism of diffusion (70). With the measured dependence of $\lambda_\alpha(\mathbf{k})$ on the temperature found, one can calculate the activation energy of diffusion of Fe-atoms ($Q_{\mathrm{Fe}}=54$ k-cal/mole) (71).

6

DIFFUSIONLESS (MARTENSITIC) TRANSFORMATIONS IN ALLOYS

6.1. WHAT IS THE MARTENSITIC TRANSFORMATION?

Apart from atomic redistributions over crystal lattice sites in an alloy, there are degrees of freedom associated with crystal lattice site rearrangements. These rearrangements may be reduced to a homogeneous distortion, a shuffling of crystal lattice planes, and static displacement waves. All these displacive modes involve cooperative movements of thousands of atoms.

Displacement of crystal lattice sites is quite a natural consequence of alloy tendency to decrease its free energy by all possible means. A phase transformation that can be treated in terms of displacement only is called a martensitic transformation. According to this definition, a martensitic transformation is diffusionless, which is to say it does not involve any diffusion migration of atoms.

In certain intermediate cases both the diffusion and crystal lattice site rearrangement mechanisms contribute to the process. Solid state reactions of this type are usually called bainite transformations.

Martensitic transformations proceed by diffusionless formation of martensitic phase "islands" within a parent phase crystal lattice. Since crystal lattices of the two phases involved differ from each other, for them to fit together requires significant atomic displacements producing elastic strain. The elastic strain may be reduced by a diffusion migration of atoms as in the usual recrystallization processes or by further transformation until location, shape, size, and orientation of the newly formed portions of the martensitic phase are such that a partial relaxation of the elastic strain produced by previously formed martensitic crystals is provided.

In typical cases of martensitic transformations, the elastic strain energy controlling the magnitude of kinetic barriers is comparatively high and may block the transformation at any stage. This is the reason why martensitic transformations occur only under high driving force conditions, that is, require significant supercooling. In fact the driving force of the process must be large to overcome all the kinetic barriers. Moreover even such supercooling sometimes fails to bring the transformation to the completion leaving a part of the parent phase unchanged, its percentage depending on the magnitude of the transformation driving force. To develop further the process requires additional supercooling. This is typical for martensitic transformations characterized by a significant crystal lattice mismatch between the parent and martensitic phases. The behavior of the system is then described by a thermoelastic equilibrium with an unambiguous dependence of the volume fraction of the martensitic phase on the temperature.

Elastic strain and coherent conjugation of the crystal lattices of parent and martensitic phases are the key features of martensitic kinetics and morphology. Martensitic transformations have been studied for a long time, and a number of morphological and kinetic characteristics common to a wide variety of martensitic transformations have been documented. These can be summarized as follows.

Morphologically, the martensitic phase is usually formed as a thin plate, needle, or lath lying along a reasonably well-defined habit plane of the parent crystal. The crystallographic axes of the martensitic phase show reproducible orientation relationships to those of the parent phase. The interior of the martensitic platelet is dense with crystal defects, either in the form of crystallographic twins or dislocations. If a martensitic platelet impinges on a free crystal surface, the latter will be deformed to exhibit a characteristic surface relief.

Kinetically, the martensitic transformation is extremely rapid, with each platelet growing at a speed that approaches the speed of sound. In general, the transformation begins with the rapid growth of a single platelet that triggers growth of further platelets by an autocatalytic process. The result is a "burst" of the martensitic transformation often producing an audible "click" in an experiment.

The martensitic transformation will not usually go to completion at its initiation temperature. To maintain the transformation and bring it to completion, it is necessary to decrease the temperature gradually. The temperature at which a martensitic transformation begins is denoted the martensitic start temperature, M_s, and that at which it is essentially completed is called the martensitic finish temperature, M_f.

The martensitic transformation may be reversed by heating, but transformation hysteresis effects are usually observed. Sometimes complete reversibility is observed when the same atomic structure is formed after the completion of a heating-cooling thermocycle. The reverse martensite transformation is initiated at a temperature, A_s, and completed at a higher temperature, A_f, which may be several hundreds degrees higher than the M_s temperature at which the martensitic transformation starts.

In keeping with the displacing character of transformation, both the temperature at which transformation occurs and morphology of the platelets formed are strongly influenced by applied stress.

It is hardly possible to give here credit to all the scientists from various countries who contributed to our modern concept of the problem. Kurdjumov made the most important discoveries in the field and should be singled out. To him we owe our modern concepts of the structure of martensitic crystals and orientation relations (72). He found that martensitic transformations occur not only in carbon-steel alloys but also in many nonferrous alloys (73). Using a Cu-Zn and Cu-Al ordered alloy as model system, he proved without doubt that the martensitic transformation is indeed diffusionless (74–75). He described the phenomenon of thermoelastic equilibrium and reversible movement of interphase boundaries (76–78). Lastly, he was the first to observe kinetics of isothermal martensitic transformation (79).

The development of martensitic transformations sometimes deviates from the scheme given above. For instance, iron-carbon alloys do not undergo reverse martensitic transformations. As a rule a transformation is irreversible if it involves additional processes such as plastic deformation, irreversible decomposition, and so forth. The absence of reverse martensitic transformations in iron-carbon alloys for example, is caused by the carbide precipitation reaction.

6.2. NUCLEATION AND GROWTH OF MARTENSITES

As shown in Section 2.2 first-order transformations at temperatures above the absolute instability points follow the nucleation-and-growth mechanism. The martensitic transformation is not an exception. Its development at temperatures exceeding the absolute instability temperature also involves the formation and growth of the nucleus. There is, however, one feature that makes martensitic transformations significantly different from transformations involving diffusion such as decomposition or ordering. In the case of decomposition the new phase embryos should be formed by the thermonucleation mechanism. As a rule, in martensitic transformations the formation of a martensite embryo, which is a displacive heterogeneity, is associated with an increase in elastic strain energy that is too high for such embryo to be formed by the thermonucleation mechanism. There is, however, no need for displacive heterogeneities (embryos) to be formed during the transformation because they are always present in the parent phase as crystal lattice defects. The latter are "frozen in" natural nuclei of the martensitic phase. Moreover, if one considers hypothetically an "ideal" crystal devoid of any defects, its surface would still play the part of a martensitic phase nucleus.*

*This situation is quite similar to solid-to-liquid phase transformations. According to (80) the rate of crystalline phase melting is not controlled by the nucleation kinetics since its surface plays the part of a natural liquid phase nucleus. Therefore fluctuations enabling the formation of liquid phase nuclei are not necessary.

The effect of defects in the parent phase crystal lattice is twofold. They may act as nuclei of the martensitic phase or as sources of elastic strain field promoting the nucleation process (the limiting situation holds even when the formation of a martensitic nucleus in elastic strain field generated by a defect does not require fluctuations). Both mechanisms appear to occur in real systems. The first one implies the development of crystal lattice defects into a martensitic crystal. It has been discussed by Olson and Cohen (81). The elastic strain assisted nucleation has been discussed by Magee (82) and Christian (83).

Since martensitic nuclei are local displacive heterogeneities, the highest rate of their growth should be of the same order of magnitude as the rate of strain disturbance propagation, i.e. the sound propagation velocity. In many respects the growth mechanism greatly resembles the mechanism of dislocation motion under applied stress. In both cases an elementary event is thermofluctuation formation of a growth embryo. In the case of dislocation, the role of an embryo is played by a double kink (see Fig. 53). The size of a growth embryo as well as of a critical nucleus of the martensitic phase is of the order of the correlation length (a typical distance at which the effect by a displacive heterogeneity on the parent phase regions in its vicinity is noticeable). With metal alloys the correlation length should be associated with conduction electrons participating in interatomic bonding. The only length characteristics of conduction electrons is the wave length of Fermi energy electrons. The latter has in turn the same order of magnitude as the crystal lattice parameters of typical metals. The typical size of a growth embryo of a martensitic phase in metal alloys should therefore be commensurate with crystal lattice spacing.

For simplicity an embryo of a martensitic phase will be called an elementary martensitic particle. Using this terminology, the martensite reaction (the growth stage) may be treated as successive creation of elementary martensite particles located in such a way that the elastic strain energy is reduced to a minimum. Some details of this mechanism were elucidated by a computer simulation of a martensitic transformation carried out by Wen, Morris, and Khachaturyan (84, 85). These details will be discussed in Sections 12.3 and 12.4.

According to (84, 85) the creation of an elementary martensite particle is controlled by the so-called thermoelastic potential which is the sum of changes in the elastic strain energy, bulk chemical free energy and interphase boundary energy caused by the formation of the elementary particle at a given parent lattice point. Since the contribution from elastic strain predominates, the formation of an elementary martensite particle is controlled by long-range strain-

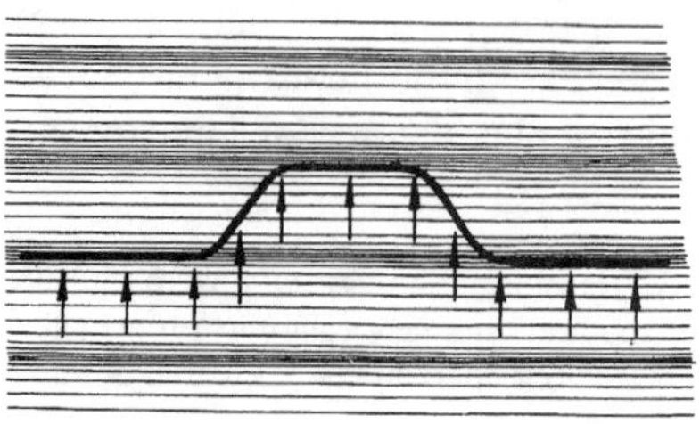

Figure 53. Schematic drawing of a double kink in the shear plane. The heavy line shows a dislocation line; the arrows indicate the shear direction. Shadow is used to show the Peierls potential relief. The area above and below the dislocation line may be formally treated as the matrix and new phase, respectively.

induced interactions of the newly born particle with the whole martensite crystal already formed. Each elementary transformation event will therefore depend on particular features of the substructure of the two-phase crystal as a whole.

One may easily see that the martensitic transformation occurs if the thermoelastic potential assumes a negative value at least at one point of the crystal. Otherwise, if the thermoelastic potential is positive over the whole crystal, the martensitic transformation will be arrested, since the creation of any elementary martensitic particle will result in a free energy increase. In that case the process may develop further because the thermofluctuation overcomes the minimum height barriers to the formation of elementary martensite particles (the process will then follow the isothermal kinetics) or because of an increase in the chemical driving force caused by supercooling which would thus bring the thermoelastic potential minima down to negative values.

It should be emphasized again that the situation, when a martensitic reaction terminates because the thermoelastic potential becomes positive at all parent phase points, may only occur due to long-range strain-induced interactions between a newly formed martensite particle and the martensitic crystals formed previously. Such a situation, with strain-induced interactions predominating, is typical for displacive transformations and has no analogue in usual diffusion transformations.

As mentioned above, evolution of a dislocation structure under applied stress in a glide plane is in many respects a two-dimensional analogue of the development of martensitic transformations. The only minor difference between the two phenomena is that the driving force of formation of a dislocation structure is applied stress whereas with martensitic transformations, it is the difference between chemical potentials of the initial and martensitic phases. The nucleation problem is solved in the same way in both cases: any crystal always contains dislocations playing the part of "frozen in" embryos. As in the case of martensite crystal growth, a dislocation glide develops by formation of double kinks which are analogues of elementary martensite particles. A dislocation glide may be either thermoactivated (small driving force stress) or activated athermally (large driving force). The same is true of martensitic transformations: thermoactivated motion of the interphase under the conditions of the isothermal martensite kinetics occurs when the chemical driving force is small whereas athermal ("burst") growth takes place when the driving force is large.

Finally, the creation of a double kink is controlled by the elastic strain potential produced by long-range strain-induced interactions between the newly formed double kink and all the previous dislocations, just like formation of an elementary martensite particle is controlled by the thermoelastic potential depending on strain-induced interactions between this particle and the martensite crystal as a whole.

All that has been said is only valid if the parent phase is stable under infinitesimal displacements and strain but unstable with respect to finite displacement heterogeneities resulting in the triggering of a martensitic reaction. Any infinitesimal homogeneous strain or atomic displacement with respect to the crystal lattice sites then gives rise to the increase of free energy by

$$\Delta F = \frac{V}{2}\lambda_{ijkl}\delta\varepsilon_{ij}\delta\varepsilon_{kl} + \frac{1}{2}\sum_{\mathbf{r},\mathbf{r}',\alpha,\beta} A_{ij}^{\alpha\beta}(\mathbf{r},\mathbf{r}')u_i(\alpha,\mathbf{r})u_j(\beta,\mathbf{r}') \tag{6.2.1}$$

where $\delta\varepsilon_{ij}$ is the infinitesimal homogeneous strain, $\mathbf{u}(\alpha, \mathbf{r})$ the displacement of an atom of the αth kind from the crystal lattice site $\mathbf{r}$, λ_{ijkl} the fourth rank tensor of the elastic constants, i, j, k, l are the indexes of the Cartesian coordinates, $A_{ij}^{\alpha\beta}(\mathbf{r}, \mathbf{r}')$ are the material constants (Born-Karman constants) characterizing rigidity of the parent phase lattice, and V is the total crystal volume. Eq. (6.2.1) gives the first nonvanishing term in the Taylor expansion of free energy in strain and displacements. It does not involve the linear strain and displacement terms since the undistorted state of the parent phase corresponds to a relative minimum of free energy. Stability of the homogeneous parent phase with respect to infinitesimal distortions requires that the first and second terms in (6.2.1) be positive definite quadratic forms (any strain and distortion produce an increase in free energy).

We shall now discuss the conditions when these terms are positive. As is known, a Hermitian (symmetric) quadratic form is positive definite if all the eigenvalues of its matrix representation are positive. According to the crystal lattice vibration theory [see, e.g., Born and Kun Huang (86)] the matrix $A_{ij}^{\alpha\beta}(\mathbf{r}, \mathbf{r}')$ eigenvalues are proportional to squares of crystal lattice vibration frequencies, $\omega_\sigma^2(\mathbf{k})$, where σ is the number of the branch of the normal vibration mode with the wave vector $\mathbf{k}$. If at least one eigenvalue, $\omega_{\sigma 0}^2(\mathbf{k}_0)$, is zero, the quadratic form is not positive definite and the parent phase is unstable with respect to the displacive mode corresponding to the zero eigenvalue. Usually this mode is referred to as "soft" mode, and the corresponding instability is called "soft" mode instability.

The first term of Eq. (6.2.1) is positive definite if all shear moduli are positive. Instability with respect to homogeneous shear arises if the corresponding shear modulus vanishes. Thus in a cubic crystal there are two shear moduli that determine stability of the crystal with respect to any shear. They are

$$\frac{c_{11}-c_{12}}{2} \quad \text{for the } \langle 1\bar{1}0\rangle(110) \text{ shear}$$

$$c_{44} \quad \text{for the } \langle 001\rangle(100) \text{ shear}$$

where c_{11}, c_{12}, and c_{44} are the elastic constants.

Soft mode instability with respect to the $\langle 1\bar{1}0\rangle(110)$ shear (the case of vanishing of the $(c_{11}-c_{12})/2$ modulus) was the first explanation of martensitic transformations (87). This type of instability is, however, not observed in steel and nonferrous alloys. It has only been found in few alloys such as In-Tl (88), V_3Si (89), Nb_3Sn (90) and certain other superconductive alloys that have the A15 structure. In all these cases the shear modulus $(c_{11}-c_{12})/2$ vanishes at the martensitic transformation temperature. In all cases of soft mode martensitic transformations the parent phase is cubic, and the martensitic phase is slightly tetragonal.

In the case of soft mode transformations the parent and martensitic phase crystal lattices differ only slightly, whereas with usual martensitic transformations the difference is significant.

6.3. SHAPE DEFORMATION PRODUCED BY MARTENSITIC TRANSFORMATION

As mentioned in Section 6.2, morphologically, martensitic phases are usually formed as thin platelets, needles, or laths with rather well-defined habit planes. X-ray observations show that martensitic transformations give rise to special orientation relationships between the crystal lattices of the parent and martensite phases (72).

They also produce shape deformations that manifest themselves as well-defined surface distortions, the so-called surface relief. Analysis of surface relief characteristics shows that a straight line in a parent phase remains a straight line in the martensite phase, and also a crystal lattice plane of the former is transformed into the crystal lattice plane of the latter. A typical substructure of a martensite crystal contains a set of transformation twins forming a more or less regular array.

A consistent crystallographic theory of martensitic transformations should give quantitative predictions of habit plane orientations, orientational relationships, shape deformations, and internal structures of martensite crystals based solely on crystal lattice parameters and crystal structures of the parent and martensite phases. Theories satisfying these requirements were proposed by Wechsler, Lieberman, Read (11) and Bowles and Mackenzie (2) independently. Both formulations are based on the matrix description of crystal lattice rearrangements. Later, Christian showed them to be mathematically identical (92). The theories (1, 2) proceed from evidence that the fitting together of the parent and martensitic phases along an invariant plane produces no elastic strain and therefore does not require energy expenditures. Complete absence of elastic strain is just the reason why the phenomenological theory of martensitic transformations can be formulated in purely geometrical terms without recourse to the theory of elasticity.

An algebraic formulation of the theory is given below. It differs somewhat from those given in the original works (1, 2) but of course produces the same final results.

The specific role played by invariant plane strain in the formation of two-phase superstructures has already been stressed in Section 1.7. At the first glance, the applicability of the invariant plane strain concept may seem problematic: it is known that a crystal lattice rearrangement associated with a phase transition may be described in terms of invariant plane strain in exceptional cases only, such as the fcc → hcp transformation in Co. The most frequent transformations cannot be described by invariant plane strain at all. The seeming controversy is, easy to solve, however, if one bears in mind that a crystal lattice rearrangement involved in a phase transition can be realized in a number of ways, each resulting in the corresponding homogeneous martensite crystal characterized by the crystallographic orientation with respect to the parent phase lattice of its own. Such homogeneous martensite crystal will be called a structure domain of the martensite phase.

For that reason there are a number of crystal lattice rearrangements (displacive modes) resulting in various structure domains (orientational variants) of the martensite phase that may be obtained from each other by applying the symmetry operations of the parent phase.

If a martensite transformation is carried out heterogeneously, with various crystal lattice rearrangements applied to various regions of the martensite crystal, a description of the corresponding macroscopic shape deformation in terms of the invariant plane strain may be obtained.

The following simple example may serve as illustration. Consider a "sandwich" consisting of alternating plates of two different structure domains of the martensite phase. These domains are produced by two crystal lattice rearrangements $\hat{\mathbf{A}}_1$ and $\hat{\mathbf{A}}_2$ (see Fig. 54). The platelike domains are characterized by different crystal lattice orientations with respect to the parent phase. To provide stress-free fitting together of adjacent domains, the boundary plane should be geometrically identical in both domains. Usually this boundary is the twinning plane of the martensite lattice, and adjacent structure domains are twin-related.

Let us draw a vector inside the "sandwich" equal to $\mathbf{R}$ before the crystal lattice rearrangement (segment OA in Fig. 54). The vector $\mathbf{R}$ may always be decomposed into the sum of collinear vectors $\mathbf{r}_j$ which are segments of $\mathbf{R}$ lying between the parallel planes [they become internal boundaries of the "sandwich"

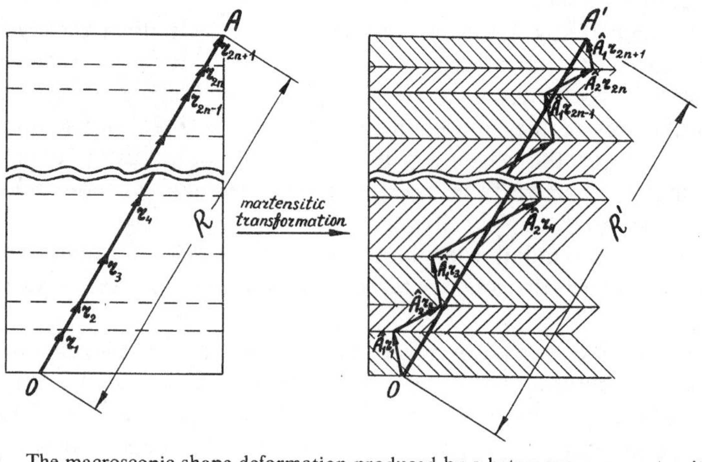

Figure 54. The macroscopic shape deformation produced by a heterogeneous martensitic transformation. Two types of domains (two orientational variants) of the martensitic phase are indicated by different hatching. $0A$ is a macroscopic vector drawn within the parent phase. $0A$ transforms into $0A'$ after the crystal lattice rearrangement.

after the phase transformation]:

$$\mathbf{R}=\mathbf{r}_1+\mathbf{r}_2+\mathbf{r}_3+\cdots+\mathbf{r}_j+\cdots+\mathbf{r}_{2n+1}$$

where $j=1, 2, \ldots, 2n+1$ are the indexes of layers in the sandwich. Without loss of generality, all $\mathbf{r}_j$ with even indexes j may be assumed to belong to the plates undergoing the $\hat{\mathbf{A}}_1$ crystal lattice rearrangement, and those with odd indexes may be taken to lie within the plates undergoing the $\hat{\mathbf{A}}_2$ rearrangement. Since the rearrangements $\hat{\mathbf{A}}_1$ and $\hat{\mathbf{A}}_2$ that are chosen do not produce any discontinuity at the internal boundaries of the sandwich (this may always be done by introducing small rigid body rotations of structure domains), the transition of $\mathbf{R}$ into $\mathbf{R}'$ caused by this heterogeneous transformation may be written as

$$\mathbf{R}'=\hat{\mathbf{A}}_1\mathbf{r}_1+\hat{\mathbf{A}}_2\mathbf{r}_2+\hat{\mathbf{A}}_1\mathbf{r}_3+\hat{\mathbf{A}}_2\mathbf{r}_4+\cdots+\hat{\mathbf{A}}_2\mathbf{r}_{2n}+\hat{\mathbf{A}}_1\mathbf{r}_{2n+1}$$

Grouping together of odd and even terms in this equation yields

$$\mathbf{R}'=\hat{\mathbf{A}}_1(\mathbf{r}_1+\mathbf{r}_3+\mathbf{r}_5+\mathbf{r}_7+\cdots+\mathbf{r}_{2n+1})+\hat{\mathbf{A}}_2(\mathbf{r}_2+\mathbf{r}_4+\mathbf{r}_6+\mathbf{r}_8+\cdots\mathbf{r}_{2n}) \quad (6.3.1)$$

Since all the vectors $\mathbf{r}_j$ are parallel to each other, their sums may be written as

$$\begin{aligned}\mathbf{r}_1+\mathbf{r}_3+\mathbf{r}_5+\mathbf{r}_7+\cdots+\mathbf{r}_{2n+1}&=x\mathbf{R}\\ \mathbf{r}_2+\mathbf{r}_4+\mathbf{r}_6+\mathbf{r}_8+\cdots+\mathbf{r}_{2n}&=(1-x)\mathbf{R}\end{aligned} \quad (6.3.2)$$

where x is the fraction of the martensite crystal occupied by domains of the first type.

Substituting (6.3.2) into (6.3.1) gives

$$[x\hat{\mathbf{A}}_1+(1-x)\hat{\mathbf{A}}_2]\mathbf{R}=\mathbf{R}'$$

The macroscopic shape deformation associated with the phase transition is thus described by the matrix

$$\langle\hat{\mathbf{A}}\rangle=x\hat{\mathbf{A}}_1+(1-x)\hat{\mathbf{A}}_2 \quad (6.3.3)$$

The point of the above consideration is that the combination of the matrices $\hat{\mathbf{A}}_1$ and $\hat{\mathbf{A}}_2$ may often describe the invariant plane strain though none of the matrices, $\hat{\mathbf{A}}_1$ and $\hat{\mathbf{A}}_2$, taken separately may. To apply the invariant plane strain theory, one must make a proper choice of the parameter x determining the volume fraction of the new phase crystal formed by the transformation $\hat{\mathbf{A}}_1$ and then use the mathematical procedure described in Section 1.7.

If the shape deformation $\langle\hat{\mathbf{A}}\rangle$ is an invariant plane strain, and the interphase habit plane coincides with the invariant plane, the total length of any macroscopical straight line segment lying in the habit plane will be invariant under the transformation. Fitting together the parent and martensitic phases along this plane then causes no elastic strain and consequently does not increase free energy. The latter is the reason why the habit plane of a martensitic crystal must be the invariant plane of the macroscopical deformation $\langle\hat{\mathbf{A}}\rangle$.

Considering the problem in more detail, one, however, finds that the heterogeneous structure of the martensite crystal makes it impossible to eliminate elastic strain entirely. The proper choice of the macroscopic deformation $\langle\hat{\mathbf{A}}\rangle$

may only eliminate the long-range strain field because structure heterogeniety may be neglected at large distances from the interphase boundary. These distances must be far greater than a typical length parameter of a martensite crystal heterogeniety which is a typical thickness of the structure domains. Therefore a more rigorous conclusion is that the proper choice of the matrix $\langle\hat{\mathbf{A}}\rangle$ and the invariant habit provides vanishing of long-range strain.

On the other hand, short-range strain does not vanish at all. In fact the habit plane is an invariant plane with respect to the macroscopic shape deformation $\langle\hat{\mathbf{A}}\rangle$, but it is not such for the $\hat{\mathbf{A}}_1$ and $\hat{\mathbf{A}}_2$ rearrangements that are responsible for formation of individual domains. For that reason local coherent conjugation of individual platelike martensite domains with the parent phase along the habit plane should give rise to short-range strain which accumulates at the interphase boundary within a thin layer whose thickness is commensurate with a typical length of the martensite crystal heterogeneity, i.e. with the thickness of martensite domains.

It will be shown in Sections 11.2 to 11.4 that long-range strain energy is proportional to the total heterogeneous martensite crystal volume whereas short-range strain energy is proportional to its interphase area. As long-range strain vanishes, it in turn eliminates the dominant term, the bulk elastic strain energy. Since long-range strain provides the major contribution to the elastic energy balance, heterogeneous domain structure of martensitic crystals responsible for short-range strain has only a weak effect on the habit plane orientation. Short-range strain determines the thickness of structure domains.

The problem of determining the invariant plane, and hence the habit plane, from an arbitrary representation of the invariant plane strain matrix has been solved in Section 1.7. The procedure for the determination of the habit plane involves the following steps:

1. The determination of the eigenvalues $\lambda_1^2(x)$, $\lambda_2^2(x)$, and $\lambda_3^2(x)$ and eigenvectors, $\mathbf{e}_1$, $\mathbf{e}_2$, $\mathbf{e}_3$, of the Hermitian matrix

$$\langle\hat{\mathbf{A}}(x)\rangle^+\langle\hat{\mathbf{A}}(x)\rangle \tag{6.3.4}$$

According to Eq. (1.7.11) the condition for invariant plane strain is that one of the eigenvalues, for example, $\lambda_2^2(x)$, be equal to unity. This condition predetermines the next step.

2. The determination of the volume fraction of domains of the first type, $x=x_0$, from the equation

$$\lambda_2^2(x)=1 \tag{6.3.5}$$

If the eigenvalues, $\lambda_1^2(x)$ and $\lambda_3^2(x)$, and the corresponding eigenvectors, $\mathbf{e}_1$ and $\mathbf{e}_3$, are known, the final step may be done.

3. The determination of the habit plane normal $\mathbf{n}$ and the macroscopical shear vector $\varepsilon_0\boldsymbol{l}$ from Eqs. (1.7.13) and (1.7.16).

Thus crystallographically, the martensite transformation theory is reduced to the completely geometrical problem of packing martensite domains in a martensite crystal without violating crystal lattice continuity and without elastic distortions. To satisfy these requirements, one must find how the parent and martensite phase lattices and the lattices of various martensite domains are to be fitted together without generating elastic strain. The first problem has already been solved in this section by reducing macroscopical shape deformations to invariant plane strain. The second one will be discussed in detail in the next section.

6.4. STRUCTURE DOMAINS OF A MARTENSITE PHASE

It was pointed out in Section 6.3 that there are several crystal lattice rearrangements leading to the various crystallographic orientations (structure domains) of the same martensite phase. Let the operators corresponding to these rearrangements be

$$\hat{\mathbf{B}}_1, \hat{\mathbf{B}}_2, \hat{\mathbf{B}}_3, \ldots \tag{6.4.1}$$

All the operators of the series are related to each other by rotations and reflections that make up the point group (crystallographic symmetry class) of the parent phase.

For instance, if we choose $\hat{\mathbf{B}}_1$ as generating operator, we have, according to (1.6.9),

$$\hat{\mathbf{B}}_2 = \hat{\mathbf{g}}_2^+ \hat{\mathbf{B}}_1 \hat{\mathbf{g}}_2, \quad \hat{\mathbf{B}}_3 = \hat{\mathbf{g}}_3^+ \hat{\mathbf{B}}_1 \hat{\mathbf{g}}_3, \quad \hat{\mathbf{B}}_4 = \hat{\mathbf{g}}_4^+ \hat{\mathbf{B}}_1 \hat{\mathbf{g}}_4, \ldots \tag{6.4.2a}$$

where $\hat{\mathbf{g}}_2, \hat{\mathbf{g}}_3, \hat{\mathbf{g}}_4, \ldots$ are the symmetry operations (rotations and reflections) of the parent lattice point group producing the operators

$$\hat{\mathbf{B}}_2, \hat{\mathbf{B}}_3, \hat{\mathbf{B}}_4, \ldots \tag{6.4.2b}$$

Each operator from (6.4.2b) generates a martensite phase structure domain of its own.

It should be noted that the application of different operators, for example, $\hat{\mathbf{B}}_1$ and $\hat{\mathbf{B}}_2$, to adjacent regions of the parent phase leads to a gap between the crystal lattices of the corresponding structure domains. The gap occurs at the boundary plane separating the domains (see Fig. 55). To remove the discontinuity, the domains should be rotated as rigid bodies. For that reason the crystal lattice rearrangement operators (6.4.1) should be combined with rigid body rotations to provide crystal lattice continuity after the martensitic transformation. This gives a series of new rearrangement operators

$$\hat{\mathbf{A}}_1 = \hat{\mathbf{R}}_1 \hat{\mathbf{B}}_1, \quad \hat{\mathbf{A}}_2 = \hat{\mathbf{R}}_2 \hat{\mathbf{B}}_2, \quad \hat{\mathbf{A}}_3 = \hat{\mathbf{R}}_3 \hat{\mathbf{B}}_3, \ldots \tag{6.4.3}$$

which include rotations $\hat{\mathbf{R}}_1, \hat{\mathbf{R}}_2, \hat{\mathbf{R}}_3, \hat{\mathbf{R}}_4, \ldots$ of domains of the 1st, 2nd, 3rd, 4th, etc. types responsible for maintaining crystal lattice continuity during the transformation.

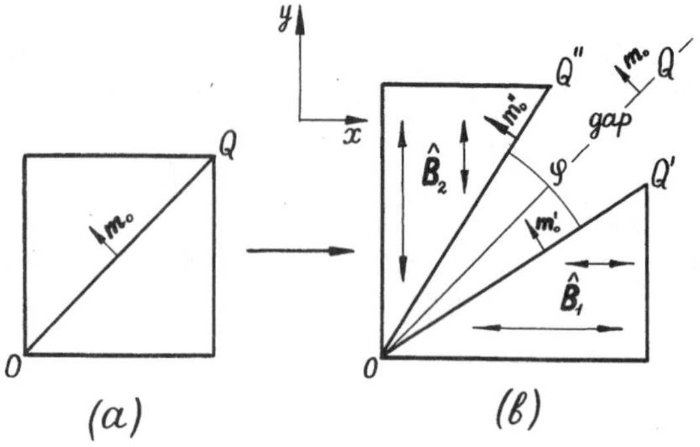

Figure 55. Formation of a gap between the adjacent structure domains produced by a cubic→ tetragonal martensitic transformation. (*a*) The parent phase before the transformation occurred; (*b*) formation of two domains of the martensitic phase after the transformation.

The cubic parent phase situated above the trace $0Q$ of the (110) plane in the drawing (*a*) undergoes the tetragonal deformation $\hat{\mathbf{B}}_2$ along the [010] tetragonal axis direction; the cubic phase situated below $0Q$ is transformed by means of the deformation $\hat{\mathbf{B}}_1$ along the [100] tetragonal axis direction. Long arrows in (*b*) show the directions of the tetragonal axes in two domains; $0Q''$ and $0Q'$ are boundary plane traces in the second and first domains after the transformation; $0Q$ is the trace of the boundary plane before the transformation occurred; $\mathbf{m}_0$, $\mathbf{m}_0''$, and $\mathbf{m}_0'$ are unit vectors normal to the boundary planes before and after the transformation.

Matrices of the type (6.4.3) rather than (6.4.1) were applied in Section 6.3 to analyze macroscopical crystal shape deformations since matrices (6.4.3) do not violate martensite crystal lattice continuity.

As already mentioned, the crystallographic theory of martensite transformations requires stress-free packing of martensite domains in the crystal without violating its continuity. Prior to formulating such a theory, we must consider stress-free coherent conjugation of various martensite domains. Obviously stress-free conjugation may only occur along the crystal lattice planes that are geometrically identical in the domains involved. The term "identical planes" as applied to different domains means that these planes, Q' and Q'', formed from some crystal lattice plane Q of the parent phase by the rearrangements $\hat{\mathbf{B}}_1$ and $\hat{\mathbf{B}}_2$, respectively, can be brought in coincidence with each other by a rigid body rotation $\hat{\mathbf{R}}$.

Mathematically, this may be written as follows:

$$\hat{\mathbf{R}}\mathbf{r}_s'' \equiv \mathbf{r}_s' \tag{6.4.4}$$

where the vectors $\mathbf{r}_s''$ and $\mathbf{r}_s'$ lie in the Q'' and Q' planes of the two adjacent martensite domains, respectively. By definition, the vectors $\mathbf{r}_s''$ and $\mathbf{r}_s'$ are derived from the parent phase vector $\mathbf{r}_s$ lying in the parent phase plane Q generating the planes Q'' and Q' of the domains:

$$\mathbf{r}_s' = \hat{\mathbf{B}}_1\mathbf{r}_s, \quad \mathbf{r}_s'' = \hat{\mathbf{B}}_2\mathbf{r}_s \tag{6.4.5}$$

Eq. (6.4.4) may therefore be rewritten in the form

$$\hat{\mathbf{R}}\hat{\mathbf{B}}_2\mathbf{r}_s \equiv \hat{\mathbf{B}}_1\mathbf{r}_s \tag{6.4.6}$$

Scalar multiplication of Eq. (6.4.6) by itself yields

$$(\hat{\mathbf{R}}\hat{\mathbf{B}}_2\mathbf{r}_s)(\hat{\mathbf{R}}\hat{\mathbf{B}}_2\mathbf{r}_s)=(\hat{\mathbf{B}}_1\mathbf{r}_s)(\hat{\mathbf{B}}_1\mathbf{r}_s)$$

or since

$$\hat{\mathbf{R}}\hat{\mathbf{B}}_2\mathbf{r}_s=\mathbf{r}_s(\hat{\mathbf{R}}\hat{\mathbf{B}}_2)^+$$

$$\hat{\mathbf{B}}_1\mathbf{r}_s=\mathbf{r}_s\hat{\mathbf{B}}_1^+, \text{ it yields}$$

$$\mathbf{r}_s(\hat{\mathbf{R}}\hat{\mathbf{B}}_2)^+(\hat{\mathbf{R}}\hat{\mathbf{B}}_2)\mathbf{r}_s\equiv\mathbf{r}_s\hat{\mathbf{B}}_1^+\hat{\mathbf{B}}_1\mathbf{r}_s \tag{6.4.7}$$

Since $(\hat{\mathbf{R}}\hat{\mathbf{B}}_2)^+=\hat{\mathbf{B}}_2^+\hat{\mathbf{R}}^+$ and $\hat{\mathbf{R}}^+=\hat{\mathbf{R}}^{-1}$ where $\hat{\mathbf{R}}^{-1}$ is the inverse matrix to $\hat{\mathbf{R}}$ (the latter relation is valid for any unitary matrix) Eq. (6.4.7) may be simplified and transformed into

$$\mathbf{r}_s(\hat{\mathbf{B}}_2^+\hat{\mathbf{B}}_2-\hat{\mathbf{B}}_1^+\hat{\mathbf{B}}_1)\mathbf{r}_s\equiv 0 \tag{6.4.8}$$

Eq. (6.4.8) which holds for any vector $\mathbf{r}_s$ lying in the plane Q is the necessary condition for the planes Q' and Q'' derivative from Q to be geometrically identical to each other; that is, this condition makes it possible to bring them into coincidence with each other by a rigid body rotation $\hat{\mathbf{R}}$.

The rotation angle ϕ bringing the planes Q' and Q'' in coincidence with each other (see Fig. 55) is easy to find as the angle made by the vectors normal to the planes Q' and Q''.

Let the unit vector normal to the parent phase plane Q be $\mathbf{m}_0$. According to Eq. (1.5.5) the rearrangements $\hat{\mathbf{B}}_1$ and $\hat{\mathbf{B}}_2$ transform the vector $\mathbf{m}_0$ into

$$\mathbf{m}_0'=\mathbf{m}_0\mathbf{B}_1^{-1}=(\hat{\mathbf{B}}_1^{-1})^+\mathbf{m}_0$$

$$\mathbf{m}_0''=\mathbf{m}_0\hat{\mathbf{B}}_2^{-1}=(\hat{\mathbf{B}}_2^{-1})^+\mathbf{m}_0 \tag{6.4.9}$$

where $\mathbf{m}_0'$ and $\mathbf{m}_0''$ are the vectors (no longer unit vectors) normal to the planes Q' and Q'', respectively.

The cosine of the angle ϕ between the unit vectors $\mathbf{m}_0'/|\mathbf{m}_0'|$ and $\mathbf{m}_0''/|\mathbf{m}_0''|$ given by the scalar product of those vectors is according to (6.4.9)

$$\cos\phi=\frac{\mathbf{m}_0'\mathbf{m}_0''}{|\mathbf{m}_0'||\mathbf{m}_0''|}=\frac{\mathbf{m}_0\hat{\mathbf{B}}_1^{-1}(\hat{\mathbf{B}}_2^{-1})^+\mathbf{m}_0}{\sqrt{(\mathbf{m}_0\hat{\mathbf{B}}_1^{-1}(\hat{\mathbf{B}}_1^{-1})^+\mathbf{m}_0)(\mathbf{m}_0\hat{\mathbf{B}}_2^{-1}(\hat{\mathbf{B}}_2^{-1})^+\mathbf{m}_0)}} \tag{6.4.10}$$

where $|\mathbf{m}_0'|=\sqrt{\mathbf{m}_0'\mathbf{m}_0'}=\sqrt{\mathbf{m}_0\hat{\mathbf{B}}_1^{-1}(\hat{\mathbf{B}}_1^{-1})^+\mathbf{m}_0}$, $|\mathbf{m}_0''|=\sqrt{\mathbf{m}_0''\mathbf{m}_0''}=\sqrt{\mathbf{m}_0\hat{\mathbf{B}}_2^{-1}(\hat{\mathbf{B}}_2^{-1})^+\mathbf{m}_0}$. The axis of rotation by ϕ bringing the vectors $\mathbf{m}_0''$ and $\mathbf{m}_0'$ in coincidence with each other is by definition perpendicular to both these vectors and therefore collinear with the vector product $\mathbf{m}_0'\times\mathbf{m}_0''$. The corresponding unit vector $\mathbf{p}$ is given by*

$$\mathbf{p}=\frac{\mathbf{m}_0''\times\mathbf{m}_0'}{|\mathbf{m}_0''\times\mathbf{m}_0'|}=\frac{\mathbf{m}_0''\times\mathbf{m}_0'}{\sqrt{[\mathbf{m}_0''\times\mathbf{m}_0'][\mathbf{m}_0''\times\mathbf{m}_0']}}=\left(\frac{\mathbf{m}_0''}{\sqrt{\mathbf{m}_0''\mathbf{m}_0''}}\times\frac{\mathbf{m}_0'}{\sqrt{\mathbf{m}_0'\mathbf{m}_0'}}\right)$$
$$\times\frac{1}{\sqrt{1-\dfrac{(\mathbf{m}_0'\mathbf{m}_0'')^2}{(\mathbf{m}_0'\mathbf{m}_0')(\mathbf{m}_0''\mathbf{m}_0'')}}} \tag{6.4.11}$$

*The rotation direction chosen brings the domain of the second type associated with the vector $\mathbf{m}_0''$ in coincidence with the first type domain which retains its orientation.

According to Eqs. (6.4.10) and (6.4.11)

$$\mathbf{p}=\left(\frac{\mathbf{m}_0''}{|\mathbf{m}_0''|}\times\frac{\mathbf{m}_0'}{|\mathbf{m}_0'|}\right)\frac{1}{\sqrt{1-\cos^2\phi}}=\frac{1}{|\sin\phi|}\left(\frac{\mathbf{m}_0''}{|\mathbf{m}_0''|}\times\frac{\mathbf{m}_0'}{|\mathbf{m}_0'|}\right) \quad (6.4.12)$$

With the knowledge of the rotation angle ϕ [from Eq. (6.4.10)] and rotation direction [from Eq. (6.4.12)] one may apply Eq. (1.4.9) to construct the rigid body rotation matrix $\hat{\mathbf{R}}$.

It thus appears that the only unknown needed to solve the problem is the vector $\mathbf{m}_0$ normal to the plane Q. In this case, because the operators $\hat{\mathbf{B}}_1$ and $\hat{\mathbf{B}}_2$ describe the parent phase crystal lattice rearrangements into the martensite lattices which have different orientations with respect to the parent phase, the problem of the determination of $\mathbf{m}_0$ (or the plane Q) is greatly simplified. It will be shown below that the plane Q is the reflection plane that transforms the operator $\hat{\mathbf{B}}_1$ into $\hat{\mathbf{B}}_2$ and that $\mathbf{m}_0$ is the vector normal to that plane.

As shown above, all the operators $\hat{\mathbf{B}}_2, \hat{\mathbf{B}}_3, \hat{\mathbf{B}}_4, \ldots$ leading to structure domains of the 2nd, 3rd, 4th, ... types are related to each other and to the generating operator $\hat{\mathbf{B}}_1$ by reflection and rotation operations making up the point group (class) of the parent phase lattice. In other words,

$$\hat{\mathbf{B}}_2=\hat{\mathbf{g}}_2^+\hat{\mathbf{B}}_1\hat{\mathbf{g}}_2,\quad \hat{\mathbf{B}}_3=\hat{\mathbf{g}}_3^+\hat{\mathbf{B}}_1\hat{\mathbf{g}}_3,\quad \hat{\mathbf{B}}_4=\hat{\mathbf{g}}_4^+\hat{\mathbf{B}}_1\hat{\mathbf{g}}_4,\ldots$$

where $\hat{\mathbf{g}}_1, \hat{\mathbf{g}}_2, \hat{\mathbf{g}}_3, \hat{\mathbf{g}}_4, \ldots$ are the symmetry operations of the parent phase lattice generating different matrices $\hat{\mathbf{B}}_2, \hat{\mathbf{B}}_3, \hat{\mathbf{B}}_4, \ldots$.

Let the symmetry operation transforming $\hat{\mathbf{B}}_1$ into $\hat{\mathbf{B}}_2$ be reflection $\hat{\mathbf{\Phi}}$:

$$\hat{\mathbf{B}}_2=\hat{\mathbf{\Phi}}^+\hat{\mathbf{B}}_1\hat{\mathbf{\Phi}} \quad (6.4.13)$$

Any vector $\mathbf{r}_s$ lying in the mirror plane corresponding to the operator $\hat{\mathbf{\Phi}}$ then remains unaffected by the reflection

$$\hat{\mathbf{\Phi}}\mathbf{r}_s=\mathbf{r}_s \quad \text{or} \quad \mathbf{r}_s\hat{\mathbf{\Phi}}^+=\mathbf{r}_s \quad (6.4.14)$$

We shall now show that Eq. (6.4.8) holds automatically for the operator $\hat{\mathbf{A}}_2$ in the form (6.4.13). In fact

$$\begin{aligned}\mathbf{r}_s(\hat{\mathbf{B}}_2^+\hat{\mathbf{B}}_2-\hat{\mathbf{B}}_1^+\hat{\mathbf{B}}_1)\mathbf{r}_s&=\mathbf{r}_s(\hat{\mathbf{\Phi}}^+\hat{\mathbf{B}}_1^+\hat{\mathbf{B}}_1\hat{\mathbf{\Phi}}-\hat{\mathbf{B}}_1^+\hat{\mathbf{B}}_1)\mathbf{r}_s\\&=\mathbf{r}_s\hat{\mathbf{\Phi}}^+(\hat{\mathbf{B}}_1^+\hat{\mathbf{B}}_1)\hat{\mathbf{\Phi}}\mathbf{r}_s-\mathbf{r}_s\hat{\mathbf{B}}_1^+\hat{\mathbf{B}}_1\mathbf{r}_s=0\end{aligned} \quad (6.4.15)$$

since according to Eq. (6.4.14)

$$\mathbf{r}_s\hat{\mathbf{\Phi}}^+(\hat{\mathbf{B}}_1^+\hat{\mathbf{B}}_1)\hat{\mathbf{\Phi}}\mathbf{r}_s=\mathbf{r}_s\hat{\mathbf{B}}_1^+\hat{\mathbf{B}}_1\mathbf{r}_s \quad (6.4.16)$$

Since any vector $\mathbf{r}_s$ that lies in the reflection plane transforming the matrix $\hat{\mathbf{B}}_1$ into $\hat{\mathbf{B}}_2$ (transforming domains of the first type into domains of the second type) provides the fulfillment of the necessary condition (6.4.8), this reflection plane is, according to the foregoing, the plane Q generating the planes Q' and Q'' along which the ideal fitting together of the martensite structure domains occurs. The vector $\mathbf{m}_0$ is the unit vector normal to Q.

It should be noted that two adjacent martensite domains related to each other by a mirror or glide-mirror plane may be regarded as a twin.* The mirror or glide-mirror plane separating them is a twinning plane.

This is the reason why a phase transition leading to alternating domains of the first and second types yields a structure that looks like a set of twins. It should, however, be clearly understood that a crystal lattice rearrangement resulting in a twin-like martensite structure in fact bears no relation to actual twinning. Rather the reverse martensite-parent phase transition restores the ideal parent phase structure containing no defects associated with plastic deformation involved in real twinning.

Let us consider, for example, a cubic to-tetragonal phase transition. This transition may involve crystal lattice rearrangements of three types and hence produce three types of structure domains. These are tetragonal distortions along the [100], [010], and [001] cubic axes, respectively (see Fig. 4). It is easy to see that these distortions are related to each other by the set of {011} mirror planes. According to what has been said above, two tetragonal distortions related to each other by reflection in the (011) parent phase plane transform this plane into the (011) plane of tetragonal martensite domains providing their stress-free fitting together. A martensite plate should therefore consist of (011) transformation twins.

Note that the fcc → bcc and bcc → fcc transformations are described as tetragonal Bain distortions (see Fig. 56), and all that has been said of cubic to tetragonal phase transitions remains valid for the fcc → bcc and bcc → fcc phase transitions. The only difference is that the (011) mirror planes of the parent fcc

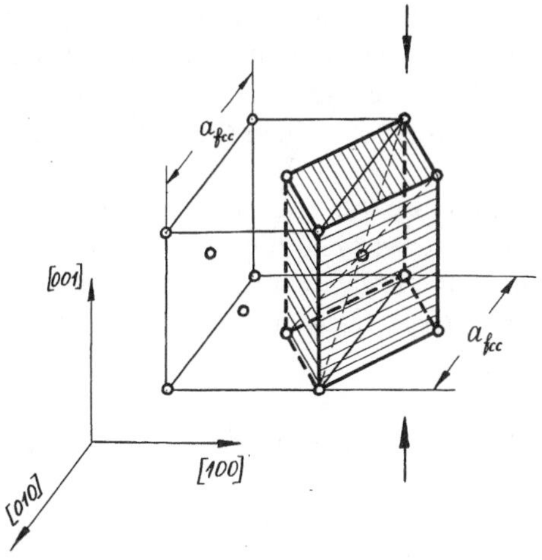

Figure 56. A schematic of the fcc → bcc Bain distortion. The heavy line and shading show a component of the fcc lattice that is transformed into a bcc lattice unit cell. Arrows show the tetragonal distortion direction.

*A glide-mirror plane of the space group is represented by a mirror plane in the point-group symmetry of a crystal.

lattice are transformed into (112) planes of the bcc martensite lattice instead of (011) martensite phase planes which are formed from the (011) parent phase planes in the cubic-to-tetragonal phase transition discussed earlier. For that reason fcc → bcc transformations give rise to bcc martensite twin-related domains with (112) twinning planes.

In the bcc → fcc transformation the (011) reflection plane of the parent phase is transformed into the (111) plane of the fcc phase, and twin-related fcc phase domains are fitted together along the (111) twinning planes.

To sum, the domain mechanism of heterogeneous transformations is not the only mechanism that may lead to invariant plane macroscopic deformations. One may also consider the plastic deformation mechanisms resulting in the invariant plane strain when combined with a homogeneous crystal lattice rearrangement. These may involve slip, dislocation glide, stacking faults, and so forth. Accommodation mechanisms of this type are seemingly realized in, for example, the formation of the so-called dislocated martensite in iron-based alloys. In that case, however, because real plastic deformation modes are involved, we cannot expect the martensitic transformation to be fully reversible.

6.5. EXAMPLE OF CRYSTALLOGRAPHIC THEORY OF MARTENSITE TRANSFORMATIONS FOR CUBIC-TO-TETRAGONAL CRYSTAL LATTICE REARRANGEMENT

The technique described in the preceding sections will now be applied to the cubic-to-tetragonal crystal lattice rearrangement. The plane of stress-free conjugation of two different structure domains of tetragonal martensite and the rigid body rotation $\hat{\mathbf{R}}$ that provides continuous transformation of the crystal lattice of one domain into that of another through the conjugation plane will be determined.

As was mentioned in Section 6.4, the cubic-to-tetragonal crystal lattice rearrangement may be realized by homogeneous strain of three types leading to different orientations of the tetragonal axes, along the [100], [010], and [001] parent phase directions, respectively. These types of strain are described by the matrices

$$\hat{\mathbf{B}}_1=\begin{pmatrix}\eta_3 & 0 & 0\\ 0 & \eta_1 & 0\\ 0 & 0 & \eta_1\end{pmatrix},\quad \hat{\mathbf{B}}_2=\begin{pmatrix}\eta_1 & 0 & 0\\ 0 & \eta_3 & 0\\ 0 & 0 & \eta_1\end{pmatrix},\quad \hat{\mathbf{B}}_3=\begin{pmatrix}\eta_1 & 0 & 0\\ 0 & \eta_1 & 0\\ 0 & 0 & \eta_3\end{pmatrix} \tag{6.5.1}$$

The representations (6.5.1) refer to the Cartesian basis with the axes x, y, and z coinciding with the [100], [010], and [001] parent cubic phase directions, respectively. The quantities, η_3 and η_1, are the length ratios between the martensite and parent phase straight-line segments collinear with and perpendicular to the tetragonal distortion axis, respectively. Since the matrices $\hat{\mathbf{B}}_1$ and $\hat{\mathbf{B}}_2$ in (6.5.1) are related by reflection in the $(1\bar{1}0)$ cubic phase mirror plane, the vector $\mathbf{m}_0$ normal to that plane is

$$\mathbf{m}_0=\left(\frac{1}{\sqrt{2}},\frac{\bar{1}}{\sqrt{2}},0\right) \tag{6.5.2}$$

The vector $\mathbf{m}_0$ is transformed by the rearrangements $\hat{\mathbf{B}}_1$ and $\hat{\mathbf{B}}_2$ into $\mathbf{m}_0'$ and $\mathbf{m}_0''$, respectively [see Eq. (6.4.9)]:

$$\begin{aligned}(\mathbf{m}_0')_i&=(\hat{\mathbf{B}}_1^{-1})_{ji}(\mathbf{m}_0)_j\\(\mathbf{m}_0'')_i&=(\hat{\mathbf{B}}_2^{-1})_{ji}(\mathbf{m}_0)_j\end{aligned} \tag{6.5.3}$$

where according to (6.5.1)

$$\hat{\mathbf{B}}_1^{-1}=\begin{pmatrix}\frac{1}{\eta_3}&0&0\\0&\frac{1}{\eta_1}&0\\0&0&\frac{1}{\eta_1}\end{pmatrix},\quad \hat{\mathbf{B}}_2^{-1}=\begin{pmatrix}\frac{1}{\eta_1}&0&0\\0&\frac{1}{\eta_3}&0\\0&0&\frac{1}{\eta_1}\end{pmatrix}$$

Substitution of the matrix representations $\hat{\mathbf{B}}_1^{-1}$ and $\hat{\mathbf{B}}_2^{-1}$ and the components of the unit vector $\mathbf{m}_0$ from (6.5.2) into (6.5.3) yields

$$\mathbf{m}_0'=\left(\frac{1}{\eta_3\sqrt{2}},-\frac{1}{\eta_1\sqrt{2}},0\right),\quad \mathbf{m}_0''=\left(\frac{1}{\eta_1\sqrt{2}},-\frac{1}{\eta_3\sqrt{2}},0\right) \tag{6.5.4}$$

$$\frac{\mathbf{m}_0'}{|\mathbf{m}_0'|}=\left(\frac{\eta_1}{\sqrt{\eta_1^2+\eta_3^2}},-\frac{\eta_3}{\sqrt{\eta_1^2+\eta_3^2}},0\right),\quad \frac{\mathbf{m}_0''}{|\mathbf{m}_0''|}=\left(\frac{\eta_3}{\sqrt{\eta_1^2+\eta_3^2}},-\frac{\eta_1}{\sqrt{\eta_1^2+\eta_3^2}},0\right)$$

The angle by which domains of the second type (those formed by the transformation $\hat{\mathbf{B}}_2$) should be rotated to provide crystal lattice continuity along the interdomain boundary plane $(1\bar{1}0)$ is determined from Eqs. (6.4.10) and (6.5.4)

$$\cos\phi=\frac{(\mathbf{m}_0''\mathbf{m}_0')}{|\mathbf{m}_0''|\cdot|\mathbf{m}_0'|}=\frac{2\eta_1\eta_3}{\eta_1^2+\eta_3^2} \tag{6.5.5}$$

According to Eqs. (6.4.12) and (6.5.4) the unit vector $\mathbf{p}$ determining the axis of rigid body rotation $\hat{\mathbf{R}}$ which brings the second domain of the tetragonal phase in coincidence with the first one is

$$\mathbf{p}=\frac{1}{|\sin\phi|}\left(\frac{\mathbf{m}_0''}{|\mathbf{m}_0''|}\times\frac{\mathbf{m}_0'}{|\mathbf{m}_0'|}\right)=\frac{1}{|\sin\phi|}\left(0,0,\frac{\eta_1^2-\eta_3^2}{\eta_1^2+\eta_3^2}\right) \tag{6.5.6}$$

It follows from (6.5.5) that

$$|\sin\phi|=\sqrt{1-\cos^2\phi}=\sqrt{1-\left(\frac{2\eta_1\eta_3}{\eta_1^2+\eta_3^2}\right)^2}=\frac{|\eta_1^2-\eta_3^2|}{\eta_1^2+\eta_3^2} \tag{6.5.7}$$

Substituting (6.5.7) into (6.5.6) gives

$$\mathbf{p}=\left(0,0,\frac{\eta_1^2-\eta_3^2}{|\eta_1^2-\eta_3^2|}\right)=\begin{cases}(0,0,1) & \text{if } \eta_1^2>\eta_3^2\\(0,0,\bar{1}) & \text{if } \eta_1^2<\eta_3^2\end{cases} \tag{6.5.8}$$

The rotation direction is derived from the direction of **p** by the right-hand screw rule. According to Eq. (1.4.9) the rigid body rotation matrix $\hat{\mathbf{R}}$ is*

$$(\hat{\mathbf{R}})_{ij}=\delta_{ij}\cos\phi+p_ip_j(1-\cos\phi)-\delta_{ijk}p_k|\sin\phi| \tag{6.5.9}$$

Substituting Eqs. (6.5.5), (6.5.6), and (6.5.7) into Eq. (6.5.9) yields the matrix

$$(\hat{\mathbf{R}})_{ij}=\begin{pmatrix} \dfrac{2\eta_1\eta_3}{\eta_1^2+\eta_3^2} & -\dfrac{\eta_1^2-\eta_3^2}{\eta_1^2+\eta_3^2} & 0 \\ \dfrac{\eta_1^2-\eta_3^2}{\eta_1^2+\eta_3^2} & \dfrac{2\eta_1\eta_3}{\eta_1^2+\eta_3^2} & 0 \\ 0 & 0 & 1 \end{pmatrix} \tag{6.5.10}$$

Since the domain of the first type corresponding to the transformation matrix $\hat{\mathbf{B}}_1$ is not subjected to rotation whereas the domain of the second type formed by the transformation $\hat{\mathbf{B}}_2$ undergoes rotation (6.5.10), Eq. (6.3.3) may be written in the form

$$\langle\hat{\mathbf{A}}\rangle=x\hat{\mathbf{B}}_1+(1-x)\hat{\mathbf{R}}\hat{\mathbf{B}}_2 \tag{6.5.11}$$

where x is the volume fraction of domains of the first type.

The multiplication of the matrix (6.5.10) by the matrix $\hat{\mathbf{B}}_2$ from (6.5.1) yields

$$\hat{\mathbf{R}}\hat{\mathbf{B}}_2=\begin{pmatrix} \dfrac{2\eta_1^2\eta_3}{\eta_1^2+\eta_3^2} & -\dfrac{\eta_1^2-\eta_3^2}{\eta_1^2+\eta_3^2}\eta_3 & 0 \\ \dfrac{\eta_1^2-\eta_3^2}{\eta_1^2+\eta_3^2}\eta_1 & \dfrac{2\eta_1\eta_3^2}{\eta_1^2+\eta_3^2} & 0 \\ 0 & 0 & \eta_1 \end{pmatrix} \tag{6.5.12}$$

Substituting (6.5.12) and the matrix $\hat{\mathbf{B}}_1$ from (6.5.1) into (6.5.11) gives

$$\langle\hat{\mathbf{A}}\rangle=\begin{pmatrix} \eta_3(1-y) & \eta_3 y & 0 \\ -\eta_1 y & \eta_1(1+y) & 0 \\ 0 & 0 & \eta_1 \end{pmatrix} \tag{6.5.13}$$

where

$$y=\frac{-\eta_1^2+\eta_3^2}{\eta_1^2+\eta_3^2}(1-x) \tag{6.5.14}$$

As in Section 6.3 we should next determine the Hermitian matrix $\langle\hat{\mathbf{A}}\rangle^+\langle\hat{\mathbf{A}}\rangle$ and its eigenvalues and eigenvectors. It follows from (6.5.13) that the product $\langle\hat{\mathbf{A}}\rangle^+\langle\hat{\mathbf{A}}\rangle$ is

$$\langle\hat{\mathbf{A}}\rangle^+\langle\hat{\mathbf{A}}\rangle=\begin{pmatrix} \eta_3^2(1-2y+2y^2) & -2\eta_1\eta_3 y^2 & 0 \\ -2\eta_1\eta_3 y^2 & \eta_1^2(1+2y+2y^2) & 0 \\ 0 & 0 & \eta_1^2 \end{pmatrix} \tag{6.5.15}$$

*The minus sign in the third term of Eq. (6.5.9) is associated with the back rotation from **m**″ to **m**′ which is taken to restore the continuity.

The eigenvalues of the matrix (6.5.15) are determined from the secular equation

$$\det \|(\langle\hat{\mathbf{A}}\rangle^{+}\langle\hat{\mathbf{A}}\rangle)_{ij}-\lambda^2\delta_{ij}\|=0 \tag{6.5.16}$$

where $\det\|\cdots\|$ is the determinant, δ_{ij} the Kronecker symbol, and λ^2 the eigenvalue. Eq. (6.5.16) is a cubic equation in the unknown λ^2. The determinant (6.5.16) can be calculated using the matrix elements (6.5.15). It is

$$[\eta_3^2(1-2y+2y^2)-\lambda^2][\eta_1^2(1+2y+2y^2)-\lambda^2][\eta_1^2-\lambda^2]-4\eta_1^2\eta_3^2y^4(\eta_1^2-\lambda^2)=0 \tag{6.5.17}$$

The first root of Eq. (6.5.17) is

$$\lambda_3^2=\eta_1^2 \tag{6.5.18a}$$

The corresponding eigenvector is

$$\mathbf{e}_3=(0,0,1) \tag{6.5.18b}$$

If $\lambda^2\neq\lambda_3^2=\eta_1^2$, Eq. (6.5.17) may be divided by the factor $\eta_1^2-\lambda^2$ to obtain a quadratic equation in λ^2 which can be solved to obtain λ_1^2 and λ_2^2:

$$[\eta_3^2(1-2y+2y^2)-\lambda^2][\eta_1^2(1+2y+2y^2)-\lambda^2]-4\eta_1^2\eta_3^2y^4=0 \tag{6.5.19}$$

Eq. (6.5.19) is reduced to the canonical form

$$\lambda^4-\lambda^2[(\eta_1^2+\eta_3^2)(1+2y^2)-2(\eta_3^2-\eta_1^2)y]+\eta_1^2\eta_3^2=0 \tag{6.5.20}$$

According to Section 6.3 [see also Eq. (1.7.11)] the operator $\langle\hat{\mathbf{A}}\rangle$ describes invariant plane strain if at least one of its eigenvalues is equal to unity. Let us put $\lambda^2=1$ in (6.5.20) and find the conditions that should be met to provide invariant plane strain. This drives us to the equation in the unknown y:

$$1-[(\eta_1^2+\eta_3^2)(1+2y^2)-2(\eta_3^2-\eta_1^2)y]+\eta_1^2\eta_3^2=0 \tag{6.5.21a}$$

or

$$2(\eta_1^2+\eta_3^2)y^2+2(\eta_1^2-\eta_3^2)y-1+\eta_1^2+\eta_3^2-\eta_1^2\eta_3^2=0 \tag{6.5.21b}$$

The solution of this quadratic equation, (6.5.21), yields

$$y_{1,2}^0=\frac{1}{2(\eta_1^2+\eta_3^2)}\left[-\eta_1^2+\eta_3^2\pm\sqrt{(\eta_1^2-\eta_3^2)^2+2(\eta_1^2+\eta_3^2)(1-\eta_1^2-\eta_3^2+\eta_1^2\eta_3^2)}\right] \tag{6.5.22}$$

The values $y_{1,2}^0$ provide the solution to Eq. (6.3.5). We may now find the values x^0 using (6.5.22) and (6.5.14):

$$x_{1,2}^0=\frac{1}{2}\pm\frac{1}{2}\sqrt{1+2\frac{\eta_1^2+\eta_3^2}{(\eta_1^2-\eta_3^2)^2}(1-\eta_1^2-\eta_3^2+\eta_1^2\eta_3^2)} \tag{6.5.23}$$

Both x^0 values in fact correspond physically to the same situation: since $x_1^0=1-x_2^0$, and x_1^0 is the volume fraction of domains of the first type, the choice of x_2^0 instead of x_1^0 is a mere permutation of domains of different types; domains of the first type are replaced with domains of the second type and vice versa.

We thus come to the conclusion that Eq. (6.5.20) has the eigenvalue $\lambda_2^2=1$ at $y=y_0$ where y_0 is determined by Eq. (6.5.22). The second eigenvalue λ_1^2 is found as the other solution to the secular equation (6.5.20) at $y=y_0$. For that purpose Eq. (6.5.20) is convenient to rewrite in the identical form

$$(\lambda^2-1)^2-(\lambda^2-1)[(\eta_1^2+\eta_3^2)(1+2y_0^2)-2(\eta_3^2-\eta_1^2)y_0]+2(\lambda^2-1)+1$$
$$-[(\eta_1^2+\eta_3^2)(1+2y_0^2)-2(\eta_3^2-\eta_1^2)y_0]+\eta_1^2\eta_3^2=0 \qquad (6.5.24)$$

According to (6.5.21a), the sum of the three last terms vanishes at $y=y_0$. We thus obtain

$$(\lambda^2-1)^2-(\lambda^2-1)[(\eta_1^2+\eta_3^2)(1+2y_0^2)-2(\eta_3^2-\eta_1^2)y_0-2]=0 \qquad (6.5.25)$$

If $\lambda^2\neq\lambda_2^2=1$, Eq. (6.5.25) may be divided by λ^2-1. This gives immediately the root $\lambda^2=\lambda_1^2$:

$$\lambda_1^2=(\eta_1^2+\eta_3^2)(1+2y_0^2)-2(\eta_3^2-\eta_1^2)y_0-1 \quad \text{or}$$
$$\lambda_1^2=2(\eta_1^2+\eta_3^2)y_0^2+2(\eta_1^2-\eta_3^2)y_0-1+\eta_1^2+\eta_3^2 \qquad (6.5.26)$$

Substituting (6.5.21b) into Eq. (6.5.26) yields eventually

$$\lambda_1^2=\eta_1^2\eta_3^2 \qquad (6.5.27)$$

The eigenvector $\mathbf{e}_1$ corresponding to the eigenvalue (6.5.27) is easy to find as

$$\mathbf{e}_1=\frac{1}{\sqrt{4\eta_3^2y_0^4+\eta_1^2(1+2y_0+2y_0^2-\eta_3^2)^2}}\left(\eta_1(1+2y_0+2y_0^2-\eta_3^2),\ 2\eta_3y_0^2,\ 0\right) \qquad (6.5.28)$$

The vector $\mathbf{n}$ normal to the habit plane which determines its orientation with respect to the parent phase crystal lattice and the macroscopical shear are found from Eqs. (1.7.13) and (1.7.16) as

$$\mathbf{n}=\sqrt{\frac{\lambda_3^2-1}{\lambda_3^2-\lambda_1^2}}\,\mathbf{e}_3+\sqrt{\frac{1-\lambda_1^2}{\lambda_3^2-\lambda_1^2}}\,\mathbf{e}_1$$
$$\varepsilon_0\boldsymbol{l}=(\lambda_3-\lambda_1)\left[\lambda_1\sqrt{\frac{\lambda_3^2-1}{\lambda_3^2-\lambda_1^2}}\,\mathbf{e}_3-\lambda_3\sqrt{\frac{1-\lambda_1^2}{\lambda_3^2-\lambda_1^2}}\,\mathbf{e}_1\right] \qquad (6.5.29)$$

if $\lambda_3^2>1$, $\lambda_1^2<1$. The ordering $\lambda_3>1>\lambda_1$ corresponds to the situation $\eta_1^2>1$ and $\eta_3^2<1$ as follows from the equalities $\lambda_1^2=\eta_1^2\eta_3^2$ and $\lambda_3^2=\eta_1^2$.* Substituting (6.5.18a) and (6.5.27) into (6.5.29) yields

$$\mathbf{n}=\sqrt{\frac{\eta_1^2-1}{\eta_1^2(1-\eta_3^2)}}\,\mathbf{e}_3+\sqrt{\frac{1-\eta_1^2\eta_3^2}{\eta_1^2(1-\eta_3^2)}}\,\mathbf{e}_1 \qquad (6.5.30)$$

$$\varepsilon_0\boldsymbol{l}=\eta_1(1-\eta_3)\left(\eta_3\sqrt{\frac{\eta_1^2-1}{1-\eta_3^2}}\,\mathbf{e}_3-\sqrt{\frac{1-\eta_1^2\eta_3^2}{1-\eta_3^2}}\,\mathbf{e}_1\right) \qquad (6.5.31)$$

where $\mathbf{e}_3$ is given by (6.5.18b), $\mathbf{e}_1$ and y_0 are expressed in terms of the crystal

*If, on the contrary, $\eta_1^2<1$ and $\eta_3^2>1$, the results obtained still hold except for permutation of the indexes at the eigenvalues λ_1 and λ_3 and their respective eigenvectors. In that case $\lambda_1^2=\eta_1^2$ and $\lambda_3^2=\eta_1^2\eta_3^2$. The eigenvector $\mathbf{e}_1$ is given by (6.5.18b) and $\mathbf{e}_3$ by (6.5.28).

lattice rearrangement parameters η_1 and η_3 by Eqs. (6.5.28) and (6.5.22). Eqs. (6.5.30) and (6.5.31) thus fully solve the problem of the determination of the habit plane orientation from the crystal lattice parameters of the parent and martensite phases. The macroscopic shape deformation

$$\langle\hat{\mathbf{A}}\rangle = x_0\hat{\mathbf{B}}_1 + (1-x_0)\hat{\mathbf{R}}\hat{\mathbf{B}}_2 \tag{6.5.32}$$

associated with the martensite transformation gives rise to a gap between the martensite and parent crystal lattices along the habit plane. To eliminate this gap, the martensite crystal should be rotated as rigid body. The combination of the shape deformation and rotation, $\hat{\mathbf{R}}_0\langle\hat{\mathbf{A}}\rangle$, provides crystal lattice continuity at the interphase boundary. In this case invariant plane strain may be written in the canonical dyadic form

$$\hat{\mathbf{R}}_0\langle\hat{\mathbf{A}}\rangle = \hat{\mathbf{I}} + \varepsilon_0 \boldsymbol{l} * \mathbf{n} \tag{6.5.33}$$

Since the matrix $\langle\hat{\mathbf{A}}\rangle$ and the vectors $\mathbf{n}$ and $\boldsymbol{l}$ are known, Eq. (6.5.33) defines the rigid body rotation matrix $\hat{\mathbf{R}}_0$. It follows from Eq. (6.5.33) that

$$\langle\hat{\mathbf{A}}\rangle = \hat{\mathbf{R}}_0^{-1}(\hat{\mathbf{I}} + \varepsilon_0 \boldsymbol{l} * \mathbf{n}) = \hat{\mathbf{R}}_0^{+}(\hat{\mathbf{I}} + \varepsilon_0 \boldsymbol{l} * \mathbf{n}) \tag{6.5.34}$$

as the identity $\hat{\mathbf{R}}_0^{-1} = \hat{\mathbf{R}}_0^{+}$ holds for any unitary matrix. Eq. (6.5.34) may be transformed into

$$\hat{\mathbf{R}}_0^{+} = \langle\hat{\mathbf{A}}\rangle(\hat{\mathbf{I}} + \varepsilon_0 \boldsymbol{l} * \mathbf{n})^{-1} \tag{6.5.35}$$

Since

$$(\hat{\mathbf{I}} + \varepsilon_0 \boldsymbol{l} * \mathbf{n})^{-1} = \hat{\mathbf{I}} - \frac{\varepsilon_0 \boldsymbol{l} * \mathbf{n}}{1 + \varepsilon_0(\boldsymbol{l}\mathbf{n})} \tag{6.5.36}$$

we have

$$\hat{\mathbf{R}}_0^{+} = \langle\hat{\mathbf{A}}\rangle\left(\hat{\mathbf{I}} - \frac{\varepsilon_0 \boldsymbol{l} * \mathbf{n}}{1 + \varepsilon_0(\boldsymbol{l}\mathbf{n})}\right)$$

and

$$\hat{\mathbf{R}}_0 = \left(\hat{\mathbf{I}} - \frac{\varepsilon_0 \boldsymbol{l} * \mathbf{n}}{1 + \varepsilon_0(\boldsymbol{l}\mathbf{n})}\right)^{+}\langle\hat{\mathbf{A}}\rangle^{+} = \left(\hat{\mathbf{I}} - \frac{\varepsilon_0 \mathbf{n} * \boldsymbol{l}}{1 + \varepsilon_0(\boldsymbol{l}\mathbf{n})}\right)\langle\hat{\mathbf{A}}\rangle^{+} \tag{6.5.37}$$

Substituting (6.5.32) into (6.5.33) gives

$$x_0\hat{\mathbf{R}}_0\hat{\mathbf{B}}_1 + (1-x_0)\hat{\mathbf{R}}_0\hat{\mathbf{R}}\hat{\mathbf{B}}_2 = \hat{\mathbf{I}} + \varepsilon_0 \boldsymbol{l} * \mathbf{n} \tag{6.5.38}$$

Comparison of Eqs. (6.3.3) and (6.5.32) shows that

$$\hat{\mathbf{A}}_1 = \hat{\mathbf{R}}_0\hat{\mathbf{B}}_1 \tag{6.5.39}$$

where $\hat{\mathbf{A}}_1$ is the crystal lattice rearrangement leading to domains of the first type. Domains of the second type are generated by the transformation

$$\hat{\mathbf{A}}_2 = \hat{\mathbf{R}}_0\hat{\mathbf{R}}\hat{\mathbf{B}}_2 \tag{6.5.40}$$

The matrices (6.5.39) and (6.5.40) describe orientational relationships between the parent and martensite phase crystal lattices. The transformation of the crystal lattice vectors is given by Eq. (1.4.2):

$$\mathbf{r}_M(1)=\hat{\mathbf{A}}_1\mathbf{r}_p=\hat{\mathbf{R}}_0\hat{\mathbf{B}}_1\mathbf{r}_p$$
$$\mathbf{r}_M(2)=\hat{\mathbf{A}}_2\mathbf{r}_p=\hat{\mathbf{R}}_0\hat{\mathbf{R}}\hat{\mathbf{B}}_2\mathbf{r}_p \tag{6.5.41}$$

where $\mathbf{r}_p$ is a parent phase vector and $\mathbf{r}_M(1)$ and $\mathbf{r}_M(2)$ are the corresponding vectors within martensite domains of the first and second type, respectively. Eq. (6.5.41) thus describes orientation relationships between the crystal lattice directions. The orientation relationships between the crystal lattice planes may be obtained from Eq. (1.5.5) in the form

$$\mathbf{H}_M(1)=\mathbf{H}_p\hat{\mathbf{A}}_1^{-1}=\mathbf{H}_p(\hat{\mathbf{R}}_0\hat{\mathbf{B}}_1)^{-1}$$
$$\mathbf{H}_M(2)=\mathbf{H}_p\hat{\mathbf{A}}_2^{-1}=\mathbf{H}_p(\hat{\mathbf{R}}_0\hat{\mathbf{R}}\hat{\mathbf{B}}_2)^{-1} \tag{6.5.42}$$

where $\mathbf{H}_p$ is a reciprocal lattice vector of the parent phase, $\mathbf{H}_M(1)$ and $\mathbf{H}_M(2)$ are the corresponding reciprocal lattice vectors in martensite domains of the first and second type, respectively, related to the parent phase basis.

In Section 6.6. a Fe-31.0 Ni alloy will be treated as an example in calculating martensite crystal morphology arising in the fcc → bcc crystal lattice rearrangement. As mentioned in Section 6.4, all that has been said of the cubic-to-tetragonal phase martensite transformation remains valid for the fcc → bcc and bcc → fcc transformations since these crystal lattice rearrangements are described by tetragonal Bain distortions (see Fig. 56).

All equations of the crystallographic theory just discussed may be simplified if the martensitic transformation results in only a small tetragonal distortion. The coefficients η_1 and η_3 may then be written in the form

$$\eta_3=1-\varepsilon_{33}^0$$
$$\eta_1=1+\varepsilon_{11}^0 \tag{6.5.43}$$

where ε_{33}^0 and ε_{11}^0 are small positive values characterizing the components of the tetragonal strain in the directions collinear with and normal to the tetragonal distortion axis, respectively. Since ε_{33}^0 and ε_{11}^0 are small quantities, the equations of the theory may be expanded in power series of ε_{11}^0 and ε_{33}^0 and truncated to leave the first nonvanishing terms only. It follows from (6.5.43) that the eigenvalues (6.5.18a) and (6.5.27) may be written in the form

$$\lambda_3=\eta_1=1+\varepsilon_{11}^0$$
$$\lambda_1=\eta_1\eta_3=(1+\varepsilon_{11}^0)(1-\varepsilon_{33}^0)\approx 1+\varepsilon_{11}^0-\varepsilon_{33}^0 \tag{6.5.44}$$

Since the calculations have been performed for the case $\lambda_1<1<\lambda_3$, Eqs. (6.5.44) imply that

$$\varepsilon_{11}^0-\varepsilon_{33}^0<0 \quad \text{and} \quad \varepsilon_{11}^0>0 \quad \text{or}$$
$$\varepsilon_{33}^0>\varepsilon_{11}^0>0 \tag{6.5.45}$$

Substituting Eqs. (6.5.43) and (6.5.44) into (6.5.29) yields the following approximations for the habit plane normal $\mathbf{n}$ and macroscopic shear $\varepsilon_0\boldsymbol{l}$:

$$\mathbf{n}=\sqrt{\frac{(1+\varepsilon_{11}^0)^2-1}{(1+\varepsilon_{11}^0)^2-(1+\varepsilon_{11}^0-\varepsilon_{33}^0)^2}}\,\mathbf{e}_3+\sqrt{\frac{1-(1+\varepsilon_{11}^0-\varepsilon_{33}^0)^2}{(1+\varepsilon_{11}^0)^2-(1+\varepsilon_{11}^0-\varepsilon_{33}^0)^2}}\,\mathbf{e}_1$$

$$\approx\sqrt{\frac{\varepsilon_{11}^0}{\varepsilon_{33}^0}}\,\mathbf{e}_3+\sqrt{\frac{\varepsilon_{33}^0-\varepsilon_{11}^0}{\varepsilon_{33}^0}}\,\mathbf{e}_1 \tag{6.5.46a}$$

$$\varepsilon_0\mathbf{l}=\varepsilon_{33}^0\left[(1+\varepsilon_{11}^0-\varepsilon_{33}^0)\sqrt{\frac{(1+\varepsilon_{11}^0)^2-1}{(1+\varepsilon_{11}^0)^2-(1+\varepsilon_{11}^0-\varepsilon_{33}^0)^2}}\,\mathbf{e}_3\right.$$

$$\left.-(1+\varepsilon_{11}^0)\sqrt{\frac{1-(1+\varepsilon_{11}^0-\varepsilon_{33}^0)^2}{(1+\varepsilon_{11}^0)^2-(1+\varepsilon_{11}^0-\varepsilon_{33}^0)^2}}\,\mathbf{e}_1\right]$$

$$\approx\varepsilon_{33}^0\left[\sqrt{\frac{\varepsilon_{11}^0}{\varepsilon_{33}^0}}\,\mathbf{e}_3-\sqrt{\frac{\varepsilon_{33}^0-\varepsilon_{11}^0}{\varepsilon_{33}^0}}\,\mathbf{e}_1\right] \tag{6.5.46b}$$

Eq. (6.5.22) gives at small ε_{11}^0 and ε_{33}^0

$$y_0\approx\frac{-\varepsilon_{11}^0-\varepsilon_{33}^0+|\varepsilon_{11}-\varepsilon_{33}|}{2}=\frac{-\varepsilon_{11}-\varepsilon_{33}-(\varepsilon_{33}-\varepsilon_{11})}{2}=-\varepsilon_{11} \tag{6.5.47}$$

since $\varepsilon_{33}^0>\varepsilon_{11}^0$.

Taking into consideration Eq. (6.5.47), and passing to the limit in Eq. (6.5.28), we obtain

$$\mathbf{e}_1=(1,0,0) \tag{6.5.48}$$

Application of Eq. (6.5.18b) gives $\mathbf{e}_3=(0,\ 0,\ 1)$. Substituting this result and (6.5.48) into (6.5.46) yields eventually

$$\mathbf{n}=\left(\sqrt{\frac{\varepsilon_{33}^0-\varepsilon_{11}^0}{\varepsilon_{33}^0}},\ 0,\ \sqrt{\frac{\varepsilon_{11}^0}{\varepsilon_{33}^0}}\right)$$

$$\varepsilon_0\mathbf{l}=\varepsilon_{33}^0\left(\sqrt{\frac{\varepsilon_{33}^0-\varepsilon_{11}^0}{\varepsilon_{33}^0}},\ 0,\ -\sqrt{\frac{\varepsilon_{11}^0}{\varepsilon_{33}^0}}\right)$$

$$\varepsilon_0=\varepsilon_{33}^0 \tag{6.5.49}$$

6.6. MARTENSITE CRYSTAL MORPHOLOGY IN THE CASE OF FCC → BCC CRYSTAL LATTICE REARRANGEMENT: NUMERICAL EXAMPLE

To exemplify the application of the crystallographic theory described in the preceding section, we shall carry out numerical calculations of a martensite transformation involving an fcc → bcc crystal lattice rearrangement.

By definition, the parameters η_1 and η_3 characterizing an fcc → bcc crystal lattice rearrangement in the Bain distortion are given by

$$\eta_1=\frac{a_{\text{bcc}}}{a_{\text{fcc}}}\sqrt{2},\quad \eta_3=\frac{a_{\text{bcc}}}{a_{\text{fcc}}}$$

Since for the Fe-31.0 Ni alloy (93)

$$a_{\text{fcc}}=3.591 \text{ Å}$$
$$a_{\text{bcc}}=2.875 \text{ Å}$$

we have

$$\eta_1=\sqrt{2}\,\frac{2.875}{3.591}=1.1322$$

$$\eta_3=\frac{2.875}{3.591}=0.8006 \tag{6.6.1}$$

Substituting these quantities into (6.5.22) and (6.5.23) yields

$$y_0=0.12884 \qquad x_0=0.614 \tag{6.6.2}$$

Using the numerical values cited in (6.5.28), we may find the eigenvector $\mathbf{e}_1$

$$\mathbf{e}_1=(0.9814,\ 0.1919,\ 0) \tag{6.6.3}$$

Since according to (6.5.18b),

$$\mathbf{e}_3=(0,\ 0,\ 1)$$

the habit plane orientation and macroscopic shear may now be calculated from Eqs. (6.5.30) and (6.5.31):

$$\mathbf{n}=(0.61095,\ 0.11949,\ 0.78251) \tag{6.6.4a}$$

$$\varepsilon_0 \boldsymbol{l}=(-0.15638,\ -0.0305,\ 0.16014) \tag{6.6.4b}$$

$$\varepsilon_0(\boldsymbol{l}\mathbf{n})=0.02627 \tag{6.6.4c}$$

Using the values (6.6.1) and $y=y_0=0.128766$, the matrix (6.5.13) can be constructed:

$$\langle\hat{\mathbf{A}}\rangle=\begin{pmatrix} 0.9037 & 0.1031 & 0 \\ -0.1458 & 0.9864 & 0 \\ 0 & 0 & 1.1322 \end{pmatrix} \tag{6.6.5}$$

The matrix (6.5.36) is obtained from (6.6.4) as follows:

$$\hat{\mathbf{I}}-\frac{\varepsilon_0 \boldsymbol{l} * \mathbf{n}}{1+\varepsilon_0(\boldsymbol{l}\mathbf{n})}=\begin{pmatrix} 1.0932 & 0.0175 & 0.1192 \\ 0.0175 & 1.0033 & 0.0224 \\ -0.0930 & -0.0175 & 0.8810 \end{pmatrix} \tag{6.6.6}$$

Substitution of (6.6.5) and (6.6.6) into (6.5.37) then gives the rotation matrix $\hat{\mathbf{R}}_0$:

$$\hat{\mathbf{R}}_0=\begin{pmatrix} 1.0932 & 0.0175 & 0.1192 \\ 0.0175 & 1.0033 & 0.0224 \\ -0.0930 & -0.0175 & 0.8810 \end{pmatrix}\begin{pmatrix} 0.9036 & -0.1458 & 0 \\ 0.1031 & 0.9864 & 0 \\ 0 & 0 & 1.1322 \end{pmatrix}$$

$$=\begin{pmatrix} 0.9897 & -0.1421 & 0.1274 \\ 0.1192 & 0.9871 & 0.0254 \\ -0.0858 & -0.0037 & 0.9974 \end{pmatrix} \tag{6.6.7}$$

The matrix multiplication (6.5.39) yields the matrix $\hat{\mathbf{A}}_1$:

$$\hat{\mathbf{A}}_1=\begin{pmatrix} 0.9897 & -0.1421 & 0.1274 \\ 0.1192 & 0.9871 & 0.0254 \\ -0.0858 & -0.0037 & 0.9974 \end{pmatrix}\begin{pmatrix} 0.8006 & 0 & 0 \\ 0 & 1.1322 & 0 \\ 0 & 0 & 0.8006 \end{pmatrix}$$

$$=\begin{pmatrix} 0.7924 & -0.1609 & 0.1020 \\ 0.0955 & 1.1176 & 0.0203 \\ -0.0687 & -0.0042 & 0.7986 \end{pmatrix} \tag{6.6.8}$$

On the assumption of the Bain crystal lattice correspondence for the matrix $\hat{\mathbf{B}}_1$,

$$[100]_{\text{fcc}}\|[100]_{\text{bcc}}$$
$$[010]_{\text{fcc}}\|[011]_{\text{bcc}}$$
$$[001]_{\text{fcc}}\|[01\bar{1}]_{\text{bcc}} \tag{6.6.9}$$

we find that the fcc lattice translation $[\frac{1}{2}\frac{\bar{1}}{2}0]_{\text{f}}$ is transformed into the bcc lattice translation $[\frac{1}{2}\frac{\bar{1}}{2}\frac{\bar{1}}{2}]_{\text{b}}$.

The bcc translation vector $\mathbf{T}_{[\frac{1}{2}\frac{\bar{1}}{2}\frac{\bar{1}}{2}]_{\text{b}}}$ formed in the crystal lattice rearrangement $\hat{\mathbf{A}}_1$ from the fcc translation vector $\mathbf{T}_{[\frac{1}{2}\frac{\bar{1}}{2}0]_{\text{f}}}$ is determined from Eq. (6.5.41)

$$\mathbf{T}_{[\frac{1}{2}\frac{\bar{1}}{2}\frac{\bar{1}}{2}]_{\text{b}}}=\hat{\mathbf{R}}_0\hat{\mathbf{B}}_1\mathbf{T}_{[\frac{1}{2}\frac{\bar{1}}{2}0]_{\text{f}}}=\begin{pmatrix} 0.7924 & -0.1609 & 0.1020 \\ 0.0955 & 1.1176 & 0.0203 \\ -0.0687 & -0.0042 & 0.7986 \end{pmatrix}\begin{pmatrix} \frac{1}{2}a_{\text{fcc}} \\ \frac{\bar{1}}{2}a_{\text{fcc}} \\ 0 \end{pmatrix}$$

$$=\tfrac{1}{2}a_{\text{fcc}}(0.9531,\ -1.0219,\ -0.0455) \tag{6.6.10}$$

Eq. (6.6.10) gives the bcc lattice translation $\mathbf{T}_{[\frac{1}{2}\frac{\bar{1}}{2}\frac{\bar{1}}{2}]_{\text{b}}}$ in the basis related to the fcc parent phase.

The disorientation angle, α, made by the $[1\bar{1}\bar{1}]_{\text{bcc}}$ and $[1\bar{1}0]_{\text{fcc}}$ directions in the bcc and fcc lattices, respectively, is

$$\cos\alpha=\frac{\mathbf{T}_{[\frac{1}{2}\frac{\bar{1}}{2}\frac{\bar{1}}{2}]_{\text{b}}}\mathbf{T}_{[\frac{1}{2}\frac{\bar{1}}{2}0]_{\text{f}}}}{|\mathbf{T}_{[\frac{1}{2}\frac{\bar{1}}{2}\frac{\bar{1}}{2}]_{\text{b}}}||\mathbf{T}_{[\frac{1}{2}\frac{\bar{1}}{2}0]_{\text{f}}}|}=\begin{pmatrix} 0.6820 \\ -0.7312 \\ -0.0326 \end{pmatrix}\begin{pmatrix} 1/\sqrt{2} \\ -1/\sqrt{2} \\ 0 \end{pmatrix}=0.9993$$

$$\alpha=2.09^{\circ}$$

Therefore the orientation relation

$$[1\bar{1}\bar{1}]_{\text{bcc}}\|[1\bar{1}0]_{\text{fcc}}$$

is realized to within 2°.

The orientation relations between the fcc and bcc crystal lattice planes are determined by the matrix $\hat{\mathbf{A}}_1^{-1}$ inverse to $\hat{\mathbf{A}}_1$ (see Section 1.5). Proceeding from the matrix $\hat{\mathbf{A}}_1$ given by (6.6.8), we can calculate the matrix $\hat{\mathbf{A}}_i^{-1}$:

$$\hat{\mathbf{A}}_1^{-1}=\begin{pmatrix} 1.2268 & 0.1759 & -0.1611 \\ -0.1067 & 0.8793 & -0.0087 \\ 0.1050 & 0.0197 & 1.2383 \end{pmatrix} \tag{6.6.11}$$

The Bain distortion $\hat{\mathbf{B}}_1$ characterized by the relations (6.6.9) transforms the

$(111)_{fcc}$ fcc lattice plane into the $(110)_{bcc}$ bcc lattice plane. The reciprocal lattice vector $\mathbf{H}_{(110)_b}$ corresponding to the $(110)_{bcc}$ crystal lattice plane is formed from the parent reciprocal lattice vector $\mathbf{H}_{(111)_f}$ corresponding to the plane $(111)_{fcc}$. It is determined from Eq. (6.5.42):

$$\mathbf{H}_{(110)_b}=(\hat{\mathbf{A}}_1^{-1})^{+}\mathbf{H}_{(111)_f}=\begin{pmatrix} 1.2268 & -0.1067 & 0.1050 \\ 0.1759 & 0.8793 & 0.0197 \\ -0.1611 & -0.0087 & 1.2383 \end{pmatrix}\begin{pmatrix} 1 \\ 1 \\ 1 \end{pmatrix}\frac{1}{a_{fcc}}$$

$$=\frac{1}{a_{fcc}}\begin{pmatrix} 1.2251 \\ 1.0751 \\ 1.0684 \end{pmatrix} \tag{6.6.12}$$

where the matrix $\hat{\mathbf{A}}_1^{-1}$ is given by Eq. (6.6.11). In (6.6.12) the vector $\mathbf{H}_{(110)_b}$ is written in the basis related to the fcc lattice. The disorientation angle, ϕ, made by the $(110)_{bcc}$ and $(111)_{fcc}$ planes is found from Eq. (1.5.6)

$$\cos\phi=\frac{\mathbf{H}_{(110)b}}{H_{(110)_b}}\cdot\frac{\mathbf{H}_{(111)f}}{H_{(111)f}}=\begin{pmatrix} 0.6286 \\ 0.5516 \\ 0.5482 \end{pmatrix}\begin{pmatrix} 1/\sqrt{3} \\ 1/\sqrt{3} \\ 1/\sqrt{3} \end{pmatrix}=0.998$$

$$\phi=3.62^\circ$$

We thus find these planes to be parallel,

$$(111)_{fcc}\|(110)_{bcc}$$

to within 3.6°.

The orientational relationship of martensite domains of the second type (formed by the transformation $\hat{\mathbf{A}}_2=\hat{\mathbf{R}}_0\hat{\mathbf{R}}\hat{\mathbf{B}}_2$) is twin related to the orientational relationship determined for domains of the first type. Such orientational relationships are known as Kurdjumov-Sacks orientational relationships.

In sum, the crystallographic theory of martensite crystals formulated here, which is based on alternating Bain distortions $\hat{\mathbf{B}}_1$ and $\hat{\mathbf{B}}_2$ with alternating directions of tetragonal axes, shows a good agreement with the experimental data on thin-plate internally twinned martensite crystals. The theory affords accurate predictions of habit plane orientations, magnitude and direction of macroscopic shear, and orientational relationships. The excellent review by Wayman (94) provides a detailed discussion of these problems.

6.7. SLIP MODEL OF FORMATION OF LATH MARTENSITE IN FERROUS ALLOYS

There are two basic morphologies of bulk martensite in ferrous alloys. The first one is formed by internally twinned thin-plate martensite crystals whose habit is near the $\{295\}_{fcc}$ plane. This morphology is found in Fe-Ni-C (95), Fe-7Al-2C (96) steel and nonferrous alloys, such as Au-Cd, Cu-Zn, Ni-Al. The second type of morphology is associated with the so-called dislocated lath martensite found in low-carbon steel (less than about 0.4 percent C by weight) and iron-nickel steel (97, 98).

The habit plane of lath martensite in quenched Fe at 0.2 percent C by weight and Fe at 0.6 percent C by weight steel was found to be $\{557\}_{\text{fcc}}$ (98). The $(111)_{\text{fcc}}$ habit was observed in lath martensite in Fe-20Ni-5Mn alloys (99).

In order to explain the formation of twin-free laths of martensite, an assumption must be made that the deformation modes combine with the Bain distortion to produce invariant plane strain without generating transformation twins. The simplest crystallographic mechanism meeting this requirement is the slip shear. We shall discuss in this section crystallographic mechanism based on $(011)_{\text{bcc}}$ slip modes of martensite crystals.

Let the fcc → bcc crystal lattice rearrangement be the Bain distortion $\mathbf{B}_3$ [matrix (6.5.1)] and the Bain strain compensated by the $\{011\}_{\text{bcc}}$ slip mechanism. Assume the two slip modes shown in Fig. 57 operate:

$$\begin{aligned}&[0\bar{1}1](011)_{\text{bcc}}\\&[011](0\bar{1}1)_{\text{bcc}}\end{aligned} \tag{6.7.1}$$

These modes transform a macroscopic vector $\mathbf{r}_b$ of the martensite crystal into the vector

$$\mathbf{r}'_b = \mathbf{r}_b + \mathbf{T}(\mathbf{r}_b)$$

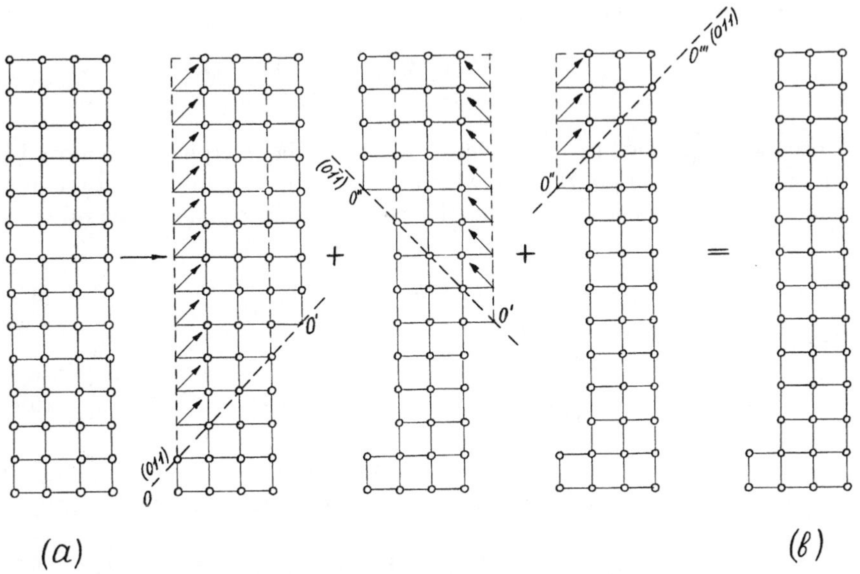

Figure 57. Schematic drawing of the shape change of a bcc lattice caused by the two-slip mechanism. 00′, 0′0″, and 0″0‴ are traces of the slip planes (011) and (0$\bar{1}$1). Arrows indicate the slip directions. It is shown how a bcc lattice component formed upon the Bain distortion, the drawing (*a*), is transformed into the component (*b*) as a result of the three slips.

where $\mathbf{T}(\mathbf{r}_b)$ is the total displacement generated by the slip:

$$\mathbf{r}_b \rightarrow \mathbf{r}'_b = \mathbf{r}_b + \mathbf{T}(\mathbf{r}_b) \tag{6.7.2}$$

The schematic in Fig. 57 shows that the total displacement $\mathbf{T}(\mathbf{r}_b)$ is

$$\mathbf{T}(\mathbf{r}_b) = \pm \mathbf{T}_{[011]_b} \times n_1 \pm \mathbf{T}_{[0\bar{1}1]_b} \times n_2 \tag{6.7.3}$$

where n_1 and n_2 are the numbers of the slip planes $(0\bar{1}1)_{bcc}$ and $(011)_{bcc}$, respectively, intersecting with the macroscopic vector $\mathbf{r}_b$. The sign $\pm$ in Eq. (6.7.3) depends on the choice of the direction of the slip translation vectors, either $[011]_b$ and $[0\bar{1}1]_b$ or $-[011]_b$ and $-[0\bar{1}1]_b$. We shall take the sign to be plus.

The integer n_1 may also be expressed as the ratio of the total number of $(0\bar{1}1)_{bcc}$ planes intersected by the vector $\mathbf{r}_b$ [which is equal to $(\mathbf{H}_{0\bar{1}1)}\mathbf{r}_b)$] to the average number, m_1, of $(0\bar{1}1)_{bcc}$ planes lying in between the nearest $(0\bar{1}1)_{bcc}$ slip planes:

$$n_1 = \frac{(\mathbf{H}_{(0\bar{1}1)_b}\mathbf{r}_b)}{m_1} \tag{6.7.4a}$$

A similar relation may be written for n_2:

$$n_2 = \frac{(\mathbf{H}_{(011)_b}\mathbf{r}_b)}{m_2} \tag{6.7.4b}$$

where m_2 is the average number of $(011)_{bcc}$ planes between the nearest $(011)_{bcc}$ slip planes.

Substituting (6.7.4) into (6.7.3) gives

$$\mathbf{T}(\mathbf{r}_b) = \mathbf{T}_{[011]_b}\frac{(\mathbf{H}_{(0\bar{1}1)_b}\mathbf{r}_b)}{m_1} + \mathbf{T}_{[0\bar{1}1]_b}\frac{(\mathbf{H}_{(011)_b}\mathbf{r}_b)}{m_2} \tag{6.7.5}$$

Since the step formed by the $[0\bar{1}1](011)_{bcc}$ slip at the interphase is removed by the subsequent $[011](0\bar{1}1)_{bcc}$ slip which starts from the step point at the interphase, the numbers m_1 and m_2 may be assumed to be equal to each other (see Fig. 57)

$$m_1 = m_2 = m \tag{6.7.6}$$

Substituting (6.7.6) into (6.7.5) yields

$$\mathbf{T}(\mathbf{r}_b) = \frac{1}{m}[\mathbf{T}_{[011]_b} * \mathbf{H}_{(0\bar{1}1)_b} + \mathbf{T}_{[0\bar{1}1]_b} * \mathbf{H}_{(011)_b}]\mathbf{r}_b \tag{6.7.7}$$

where the symbol $*$ designates the dyadic multiplication of vectors. Taking into consideration (6.7.7), the transformation (6.7.2) may be written in the form

$$\mathbf{r}'_b = \mathbf{r}_b + \mathbf{T}(\mathbf{r}_b) = \left[\hat{\mathbf{I}} + \frac{1}{m}(\mathbf{T}_{[011]_b} * \mathbf{H}_{(0\bar{1}1)_b} + \mathbf{T}_{[0\bar{1}1]_b} * \mathbf{H}_{(011)_b})\right]\mathbf{r}_b \tag{6.7.8}$$

Since

$$\mathbf{r}_b = \hat{\mathbf{B}}_3\mathbf{r}_f \tag{6.7.9}$$

where $\mathbf{r}_f$ is the fcc lattice vector undergoing transformation into $\mathbf{r}_b$ under the Bain distortion $\hat{\mathbf{B}}_3$, Eq. (6.7.9) becomes

$$\mathbf{r}_b' = \left[\hat{\mathbf{I}} + \frac{1}{m}(\mathbf{T}_{[011]_b} * \mathbf{H}_{(0\bar{1}1)_b} + \mathbf{T}_{[0\bar{1}1]_b} * \mathbf{H}_{(011)_b})\right]\hat{\mathbf{B}}_3\mathbf{r}_f \qquad (6.7.10)$$

The Bain distortion is characterized by the following relations between the fcc and bcc lattice vectors

$$\begin{aligned} &[100]_{bcc} \to [\tfrac{1}{2}\tfrac{1}{\underline{2}}0]_{fcc} \\ &[010]_{bcc} \to [\tfrac{1}{2}\tfrac{\bar{1}}{2}0]_{fcc} \\ &[001]_{bcc} \to [001]_{fcc} \end{aligned} \qquad (6.7.11)$$

Hence

$$\begin{aligned} &[011]_{bcc} \to [\tfrac{1}{\underline{2}}\tfrac{\bar{1}}{2}1]_{fcc} \qquad (011)_{bcc} \to (1\bar{1}1)_{fcc} \\ &[0\bar{1}1]_{bcc} \to [\tfrac{\bar{1}}{2}\tfrac{1}{2}1]_{fcc} \qquad (0\bar{1}1)_{bcc} \to (\bar{1}11)_{fcc} \end{aligned} \qquad (6.7.12)$$

The relations (6.7.12) between the bcc and fcc translations and the lattice planes may be written in another form, using the fcc basis:

$$\mathbf{T}_{[011]_b} = \hat{\mathbf{B}}_3\mathbf{T}_{[\frac{1}{2}\frac{\bar{1}}{2}1]_f} = \begin{pmatrix} \eta_1 & 0 & 0 \\ 0 & \eta_1 & 0 \\ 0 & 0 & \eta_3 \end{pmatrix}\begin{pmatrix} \frac{1}{2} \\ -\frac{1}{2} \\ 1 \end{pmatrix} a_{fcc} = a_{fcc}(\tfrac{1}{2}\eta_1, -\tfrac{1}{2}\eta_1, \eta_3) \qquad (6.7.13a)$$

$$\mathbf{T}_{[0\bar{1}1]_b} = \hat{\mathbf{B}}_3\mathbf{T}_{[\frac{\bar{1}}{2}\frac{1}{2}1]_f} = \begin{pmatrix} \eta_1 & 0 & 0 \\ 0 & \eta_1 & 0 \\ 0 & 0 & \eta_3 \end{pmatrix}\begin{pmatrix} -\frac{1}{2} \\ \frac{1}{2} \\ 1 \end{pmatrix} a_{fcc} = a_{fcc}(-\tfrac{1}{2}\eta_1, \tfrac{1}{2}\eta_1, \eta_3)$$

$$\mathbf{H}_{(011)_b} = \mathbf{H}_{(1\bar{1}1)_f}\hat{\mathbf{B}}_3^{-1}, \quad \mathbf{H}_{(0\bar{1}1)_b} = \mathbf{H}_{(\bar{1}11)_f}\hat{\mathbf{B}}_3^{-1} \qquad (6.7.13b)$$

where

$$\mathbf{H}_{(1\bar{1}1)_f} = \frac{1}{a_{fcc}}(1, \bar{1}, 1), \quad \mathbf{H}_{(\bar{1}11)_f} = \frac{1}{a_{fcc}}(\bar{1}, 1, 1)$$

Substituting (6.7.13b) into (6.7.10) yields

$$\mathbf{r}_b' = \left[\hat{\mathbf{B}}_3 + \frac{1}{m}(\mathbf{T}_{[011]_b} * \mathbf{H}_{(\bar{1}11)_f} + \mathbf{T}_{[0\bar{1}1]_b} * \mathbf{H}_{(1\bar{1}1)_f})\right]\mathbf{r}_f \qquad (6.7.14)$$

where the components of the vectors $\mathbf{T}_{[011]_b}$ and $\mathbf{T}_{[0\bar{1}1]_b}$ are given by Eq. (6.7.13a).

It follows from (6.7.14) that

$$\mathbf{r}_b' = \hat{\mathbf{A}}\mathbf{r}_f \qquad (6.7.15)$$

where

$$\hat{\mathbf{A}} = \hat{\mathbf{B}}_3 + \frac{1}{m}(\mathbf{T}_{[011]_b} * \mathbf{H}_{(\bar{1}11)_f} + \mathbf{T}_{[0\bar{1}1]_b} * \mathbf{H}_{(1\bar{1}1)_f}) \qquad (6.7.16)$$

is the operator of shape deformation. Eqs. (6.7.13a) and (6.7.16) make it possible to construct a matrix representation $\hat{\mathbf{A}}$ in the conventional fcc crystal lattice basis

$$\hat{\mathbf{A}}=\begin{pmatrix}\eta_1 & 0 & 0\\ 0 & \eta_1 & 0\\ 0 & 0 & \eta_3\end{pmatrix}+\frac{1}{m}\left[\begin{pmatrix}\frac{1}{2}\eta_1\\ -\frac{1}{2}\eta_1\\ \eta_3\end{pmatrix}*\begin{pmatrix}\bar{1}\\ 1\\ 1\end{pmatrix}+\begin{pmatrix}-\frac{1}{2}\eta_1\\ \frac{1}{2}\eta_1\\ \eta_3\end{pmatrix}*\begin{pmatrix}1\\ \bar{1}\\ 1\end{pmatrix}\right]$$

$$=\begin{pmatrix}\eta_1\left(1-\frac{1}{m}\right) & \frac{1}{m}\eta_1 & 0\\ \frac{1}{m}\eta_1 & \eta_1\left(1-\frac{1}{m}\right) & 0\\ 0 & 0 & \eta_3\left(1+\frac{2}{m}\right)\end{pmatrix} \tag{6.7.17}$$

The matrix $\hat{\mathbf{A}}$ in (6.7.17) is a Hermitian matrix and therefore does not include rotations. This circumstance make it possible to determine its eigenvalues and eigenvectors directly. These are

$$\begin{aligned}\lambda_3&=\eta_1 & \mathbf{e}_3&=\left(\frac{1}{\sqrt{2}},\frac{1}{\sqrt{2}},0\right)\\ \lambda_2&=\eta_1\left(1-\frac{2}{m}\right) & \mathbf{e}_2&=\left(\frac{1}{\sqrt{2}},\frac{\bar{1}}{\sqrt{2}},0\right)\\ \lambda_1&=\eta_3\left(1+\frac{2}{m}\right) & \mathbf{e}_1&=(0,0,1)\end{aligned} \tag{6.7.18}$$

According to Eq. (6.3.5), the matrix $\hat{\mathbf{A}}$ describes an invariant plane strain if one of its eigenvalues can be made equal to unity. Since the eigenvalue λ_2 depends on the slip parameter, m [Eq. (6.7.18)], we may set

$$\lambda_2=\eta_1\left(1-\frac{2}{m}\right)=1 \tag{6.7.19}$$

Eq. (6.7.19) determines the average distance, $m=m_0$, between the nearest $(011)_{bcc}$ slip planes as a condition for invariant plane strain:

$$m_0=\frac{2\eta_1}{\eta_1-1} \tag{6.7.20}$$

The use of the eigenvalues and eigenvectors (6.7.18) at $m=m_0=2\eta_1/(\eta_1-1)$ in Eq. (1.7.13) and (1.7.16) furnishes the habit plane normal $\mathbf{n}$ and macroscopic shear vector $\varepsilon_0\boldsymbol{l}$:

$$\mathbf{n}=\sqrt{\frac{(\eta_1^2-1)\eta_1^2}{\eta_1^4-\eta_3^2(2\eta_1-1)^2}}\left(\frac{1}{\sqrt{2}},\frac{1}{\sqrt{2}},0\right)+\sqrt{\frac{\eta_1^2-\eta_3^2(2\eta_1-1)^2}{\eta_1^4-\eta_3^2(2\eta_1-1)^2}}(0,0,1)$$

or

$$\mathbf{n}=\left(\sqrt{\frac{\eta_1^2(\eta_1^2-1)}{2[\eta_1^4-\eta_3^2(2\eta_1-1)^2]}},\sqrt{\frac{\eta_1^2(\eta_1^2-1)}{2[\eta_1^4-\eta_3^2(2\eta_1-1)^2]}},\sqrt{\frac{\eta_1^2-\eta_3^2(2\eta_1-1)^2}{\eta_1^4-\eta_3^2(2\eta_1-1)^2}}\right) \tag{6.7.21}$$

$$\varepsilon_0 \boldsymbol{l} = \frac{\eta_1^2 - \eta_3(2\eta_1 - 1)}{\eta_1} \left[\frac{\eta_3(2\eta_1 - 1)}{\eta_1} \sqrt{\frac{(\eta_1^2 - 1)\eta_1^2}{\eta_1^4 - \eta_3^2(2\eta_1 - 1)^2}} \left(\frac{1}{\sqrt{2}}, \frac{1}{\sqrt{2}}, 0 \right) - \eta_1 \sqrt{\frac{\eta_1^2 - \eta_3^2(2\eta_1 - 1)^2}{\eta_1^4 - \eta_3^2(2\eta_1 - 1)^2}} (0, 0, 1) \right]$$

or

$$\varepsilon_0 \boldsymbol{l} = \frac{\eta_1^2 - \eta_3(2\eta_1 - 1)}{\eta_1} \left(\sqrt{\frac{(\eta_1^2 - 1)\eta_3^2(2\eta_1 - 1)^2}{2[\eta_1^4 - \eta_3^2(2\eta_1 - 1)^2]}}, \sqrt{\frac{(\eta_1^2 - 1)\eta_3^2(2\eta_1 - 1)^2}{2[\eta_1^4 - \eta_3^2(2\eta_1 - 1)^2]}}, - \sqrt{\frac{\eta_1^2[\eta_1^2 - \eta_3^2(2\eta_1 - 1)^2]}{\eta_1^4 - \eta_3^2(2\eta_1 - 1)^2}} \right) \tag{6.7.22}$$

According to Eq. (1.7.4), the martensitic transformation and slip produce the macroscopic deformation that is an invariant plane strain

$$\hat{\mathbf{A}}_{\text{inv}} = \hat{\mathbf{I}} + \varepsilon_0 \boldsymbol{l} * \mathbf{n} \tag{6.7.23}$$

where $\mathbf{n}$ and $\boldsymbol{l}$ are given by Eqs. (6.7.21) and (6.7.22), respectively. It follows from Eq. (1.8.2) that

$$(\hat{\mathbf{A}}_{\text{inv}}^{-1})^+ = \hat{\mathbf{I}} - \varepsilon_0 \frac{\mathbf{n} * \boldsymbol{l}}{1 + \varepsilon_0(\mathbf{n}\boldsymbol{l})} \tag{6.7.24}$$

With the crystal lattice parameters of bcc and fcc iron,

$$a_{\text{bcc}} = 2.86 \text{ Å}$$
$$a_{\text{fcc}} = 3.56 \text{ Å}$$

we have

$$\eta_1 = \sqrt{2} \frac{a_{\text{bcc}}}{a_{\text{fcc}}} = 1.1361 \qquad \eta_3 = \frac{a_{\text{bcc}}}{a_{\text{fcc}}} = 0.8033 \tag{6.7.25}$$

Substitution of these numerical values into (6.7.20), (6.7.21), and (6.7.22) yields

$$\begin{aligned} m &= m_0 = 16.7 \\ \mathbf{n} &= (0.5494, 0.5494, 0.6295) \\ \varepsilon_0 \boldsymbol{l} &= 0.2365(0.4942, 0.4942, -0.7151) \\ &= (0.1169, 0.1169, -0.1692) \end{aligned} \tag{6.7.26}$$

The vector $\mathbf{n}$ in Eq. (6.7.26) corresponds to the $(778)_{\text{fcc}}$ habit plane which is close to the parent phase $(111)_{\text{fcc}}$ plane. The deviation of the habit plane from $(111)_{\text{fcc}}$ depends on the angle between the vector $\mathbf{n}$ and the vector normal to the $(111)_{\text{fcc}}$ plane equal to 3.7°.

The transformations of a crystal lattice vector and plane can be determined from Eqs. (1.4.2), (1.8.3), (6.7.23), and (6.7.24):

$$\mathbf{r}_b = \hat{\mathbf{A}}_{\text{inv}} \mathbf{r}_f = \mathbf{r}_f + \varepsilon_0 \boldsymbol{l}(\mathbf{n}\mathbf{r}_f) \tag{6.7.27}$$

$$\mathbf{H}_b = (\hat{\mathbf{A}}_{\text{inv}}^{-1})^+ \mathbf{H}_f = \mathbf{H}_f - \frac{\varepsilon_0 \mathbf{n}(\boldsymbol{l}\mathbf{H}_f)}{1 + \varepsilon_0(\boldsymbol{l}\mathbf{n})} \tag{6.7.28}$$

After the transformation the $(111)_{fcc}$ plane becomes the $(101)_{bcc}$ one, and the $[0\bar{1}1]$ fcc lattice direction gives the $[\bar{1}11]_{bcc}$ direction of the bcc lattice [in the particular case described by the relations (6.7.11)]. These planes and directions will be used to calculate the orientational relationships. In so doing, Eqs. (6.7.28) and (6.7.27) with the vectors $\varepsilon_0 \boldsymbol{l}$ and $\mathbf{n}$ given by (6.7.26) should be applied to the reciprocal lattice vector $\mathbf{H}_{(111)_f} = 1/a_{fcc}\,(1, 1, 1)$ and fcc lattice translation $\mathbf{T}_{[0\bar{1}1]_f}$, respectively:

$$\mathbf{H}_{(101)_b} = \mathbf{H}_{(111)_f} - \frac{\varepsilon_0 \mathbf{n}(\boldsymbol{l}\mathbf{H}_{(111)f})}{1 + \varepsilon_0(\boldsymbol{l}\mathbf{n})} = \frac{1}{a_{fcc}}(0.9655, 0.9655, 0.9602)_f$$

$$\mathbf{T}_{[\bar{1}11]_b} = \mathbf{T}_{[0\bar{1}1]_f} + \varepsilon_0 \boldsymbol{l}(\mathbf{n}\mathbf{T}_{[0\bar{1}1]_f}) = a_{fcc}(0.0093,\ -0.9906,\ 0.9864)_f$$

The scalar products

$$\cos \mathscr{S} = \frac{\mathbf{H}_{(101)_b}\mathbf{H}_{(111)_f}}{|\mathbf{H}_{(101)_b}||\mathbf{H}_{(111)_f}|} = 0.999995$$

$$\cos \alpha = \frac{\mathbf{T}_{[\bar{1}11]_f}\mathbf{T}_{[0\bar{1}1]_f}}{|\mathbf{T}_{[\bar{1}11]_b}||\mathbf{T}_{[0\bar{1}1]_f}|} = 0.999997$$

give the cosines of the angle $\mathscr{S}$ between the $(111)_{fcc}$ and $(101)_{bcc}$ planes and angle α between the $[0\bar{1}1]_f$ and $[\bar{1}11]_b$ directions. The angle values are

$$\mathscr{S} = 0.1726^\circ \quad \text{and} \quad \alpha = 0.14^\circ$$

The orientation relation between the fcc parent phase and bcc martensite phase is thus

$(111)_{fcc}$ is parallel to $(101)_{bcc}$ (to within 0.17°).
$[0\bar{1}1]_{fcc}$ is parallel to $[\bar{1}11]_{bcc}$ (to within 0.14°).

The conclusions that can be drawn from these results are as follows: the slip model of the fcc → bcc crystal lattice rearrangement yields the twin-free martensite crystal with the $(111)_{fcc}$ habit plane (the deviation of the habit plane from the $(111)_{fcc}$ one is of about 3.8°) which show Kurdjumov-Sacks orientation relations. These results are in good agreement with the experimental data on lath martensite. The habit planes of lath martensite in quenched Fe-0.2%C and Fe-0.6%C steel were determined to be $(557)_{fcc}$ (98). The $(111)_{fcc}$ habit plane of isolated laths in low-carbon steel was observed by Thomas and Rao (100). According to Wakasa and Wayman the habit plane in both isolated laths and laths comprising packets in Fe-20.0Ni-5.0Mn steel usually depart from $(111)_{fcc}$ by 4.5°, the magnitude of the departure ranging from 2.5 to 8° (according to the calculation results just cited, the habit and $(111)_{fcc}$ planes make an angle of 3.7°). The Kurdjumov-Sacks orientation relations were also observed.

It seems now pertinent to discuss the conditions that might make slip the major mechanism. As is known, decrease of stacking fault energy of an fcc parent phase (austenite) results in a tendency for slip, twin, and stacking fault instability of the crystal associated with the preferred motion of perfect (slip) and partial (twinning, stacking fault formation) dislocations along $(111)_{fcc}$ planes of austenite. As follows from relations between austenite and martensite directions, slip along $(111)_{fcc}$ planes of austenite is equivalent to slip along $(011)_{bcc}$ martensite planes. For that reason the $(011)_{bcc}$ slip mechanism may become the major mechanism in the case of low stacking fault energy when slip is facilitated. It also proves to be energetically favored. Perfect dislocations crossing a martensite crystal leave a perfect bcc lattice behind whereas the alternative mechanism considered in Section 6.5 involves the formation of $(112)_{bcc}$ twin boundaries intersecting crystal body and therefore additional energy expenditures. In the case of high stacking fault energies (Fe-30.0Ni-C, Fe-Al-C, Fe_3Pt alloys) the motion of dislocation arrays along the $(111)_{fcc}$ austenite [$(101)_{bcc}$ martensite] planes is hindered, and the mechanism resulting in formation of thin-plate twinned martensite crystals is "switched on." The same should be expected in high-carbon steel where the C-atomic pinning dislocations make the dislocation motion across a martensite crystal impossible. In these cases the slip mechanism proves to be inhibited, and martensitic transformations proceed without involving plastic deformation modes. In particular, C atoms cannot affect directly the martensitic transformation that proceeds by the mechanism described in Section 6.5, since the Bain distortion is perfectly homogeneous and cannot be treated in terms of a dislocation array.

These considerations make it possible to draw certain qualitative conclusions on the temperature and composition as factors determining the morphology of a martensitic crystal. In fact an alloying element can produce effects of two types: it may decrease the M_s temperature or change the stacking fault energy. If a martensite transformation occurs at a low M_s temperature when the dislocation motion is hindered, plastic deformation modes cannot be involved and a purely Bain crystal lattice rearrangement resulting in a thin-plate $(112)_{bcc}$ twinned morphology may be expected. On the other hand, an alloying element that does not affect the M_s temperature considerably but decreases the stacking fault energy of austenite should promote the slip mechanism of the $\gamma \rightarrow \alpha$ rearrangement and therefore facilitate the formation of lath crystals. As is known, elements that stimulate the formation of hcp phases in ferrous alloys decrease the stacking fault energy of austenite and thus should facilitate the formation of lath martensite.*

Mn, Cr, Re, Co, and other elements may serve as examples. On the other hand, such elements as Ni and Al do not form hcp phases when dissolved in Fe and increase the stacking fault energy of austenite [the increase becomes even more significant if the alloying element gives rise to long-range ordering, as observed in Fe_3Pt (101) and Fe-9.7Al-1.4C (102)]. To attain the desired M_s temperature and stacking fault energy values, combinations of alloying elements of different types can be utilized. This is the method for controlling morphology of martensite crystals.

*The fcc → hcp transition may be treated in terms of a stacking fault instability when the sequence of close-packed $(111)_{fcc}$ planes of the fcc lattice, ABCABCABC . . . , is transformed into the sequence ABABAB . . . of the hcp structure. This transformation occurs by regular faulting of the fcc lattice.

The experimental data confirm the conclusions concerning the dependence of martensite morphology on the stacking fault energy. Breedis showed the transition from the lath to thin-plate morphology to occur with increase of stacking fault energy (103). Ansell, Carr, and Strife carried out a detailed experimental study of this effect in Fe-Ni-C and Fe-Ni-Cr-C steel (104).

Finally, it should be emphasized that the presence of numerous dislocations observed in lath martensite is easy to understand within the framework of the slip mechanism. We make a natural assumption that some part of the dislocations involved in the $\langle 011\rangle(011)_{bcc}$ slip cannot pass from one end on the martensite crystal to the other without being halted by crystal lattice defects or the stress field and remain "frozen" in the martensite crystal.

6.8. CRYSTAL LATTICE ABNORMALITIES OF IRON-CARBON MARTENSITE

X-ray studies of carbon steel martensite have led to the conclusion that the martensite is a supersaturated interstitial solid solution of carbon in an α iron. The carbon atoms occupy interstitial sites between the nearest iron atoms in the $[001]_b$ direction (105–107), which results in the tetragonality of the martensite lattice. This conclusion has been confirmed by the crystallographic analysis of the diffusionless martensite transformation. If the $\gamma \rightarrow \alpha$ rearrangement is described by a uniform deformation [for example, the Bain (108) or Kurdjumov-Sachs (72) deformation], and if the carbon atoms occupy octahedral interstitial sites in the γ-lattice, all carbon atoms fall into octahedral sites lying parallel to one of the three $\langle 001\rangle$ directions of the body-centered lattice. In the body-centered lattice of α iron there are three body-centered sublattices of octahedral interstitial sites displaced by $[\frac{1}{2}, 0, 0]$, $[0, \frac{1}{2}, 0]$, and $[0, 0, \frac{1}{2}]$ relative to the iron lattice. Octahedral interstitial sites of the first sublattice, O_x, lie between iron atoms in the $[100]_b$ directions. The octahedral interstitial sites of the second, O_y, and the third, O_z, sublattices correspondingly lie between iron atoms in the $[010]_b$ and $[001]_b$ directions. Octahedral interstices have the tetragonal symmetry. Therefore an interstitial atom placed in these interstices produces a tetragonal distortion of the host iron lattice. The tetragonal axes are parallel to the $[100]_b$, $[010]_b$, and $[001]_b$ directions when C atoms occupy O_x, O_y, and O_z octahedral sites, respectively. The tensors of the bcc lattice expansion concentration coefficients are

$$u_{ij}(1)=\begin{pmatrix} u_{33} & 0 & 0 \\ 0 & u_{11} & 0 \\ 0 & 0 & u_{11} \end{pmatrix}, \quad u_{ij}(2)=\begin{pmatrix} u_{11} & 0 & 0 \\ 0 & u_{33} & 0 \\ 0 & 0 & u_{11} \end{pmatrix}$$

$$u_{ij}(3)=\begin{pmatrix} u_{11} & 0 & 0 \\ 0 & u_{11} & 0 \\ 0 & 0 & u_{33} \end{pmatrix} \tag{6.8.1}$$

The numbers in parentheses, 1, 2, and 3, designate the tensors corresponding to the situations where the interstitial atoms occupy the O_x, O_y, and O_z sublattices of octahedral interstices, respectively. The tensors (6.8.1) are written in the Cartesian coordinates related to the $[100]_b$, $[010]_b$, and $[001]_b$ axes of the bcc host lattice. If the interstitial atoms occupy the O_x, O_y, and O_z sublattices simultaneously, the resulting crystal lattice distortion is given by the superposition

$$\bar{u}_{ij}=u_{ij}(1)n_1+u_{ij}(2)n_2+u_{ij}(3)n_3 \tag{6.8.2}$$

where n_1, n_2, and n_3 are the fractions of the O_x, O_y, and O_z interstices, respectively, occupied by interstitial atoms. The values n_1, n_2, and n_3 meet the conservation condition:

$$n_1+n_2+n_3=n \tag{6.8.3}$$

where $n=N_C/N$, N_C is the number of carbon atoms and N is the number of host atoms. Substituting Eqs. (6.8.1) into (6.8.2) yields

$$\bar{u}_{ij}=\begin{pmatrix} u_{11}(n_2+n_3)+u_{33}n_1 & 0 & 0 \\ 0 & u_{11}(n_3+n_1)+u_{33}n_2 & 0 \\ 0 & 0 & u_{11}(n_1+n_2)+u_{33}n_3 \end{pmatrix} \tag{6.8.4}$$

The orthorhombic distortion (6.8.4) transforms the bcc host lattice unit translations, [100], [010], and [001], into the orthorhombic lattice translations a, b, and c, respectively (109):

$$\begin{aligned} a&=a_0[1+u_{11}(n_2+n_3)+u_{33}n_1] \\ b&=a_0[1+u_{11}(n_3+n_1)+u_{33}n_2] \\ c&=a_0[1+u_{11}(n_1+n_2)+u_{33}n_3] \end{aligned} \tag{6.8.5}$$

where a_0 is the crystal lattice parameter of the bcc Fe host lattice, a, b, c are the crystal lattice parameters of the orthorhombic interstitial solution. If all carbon atoms occupy the O_z interstices, for example,

$$n_1=n_2=0, \quad n_3=n \tag{6.8.6}$$

The distribution (6.8.6) will be referred to as the completely ordered distribution.

Substituting (6.8.6) into (6.8.5) yields

$$\begin{aligned} a&=b=a_0(1+u_{11}n) \\ c&=a_0(1+u_{33}n) \end{aligned} \tag{6.8.7}$$

For martensite of plain carbon steel (110)

$$u_{11}=-0.01, \quad u_{33}=0.86 \tag{6.8.8}$$

If a part of carbon atoms are distributed over O_x and O_y sites with the occupation probabilities n_1 and n_2, respectively, we have a partially disordered state.

The problem of locating C-atoms is important because it is closely related to the mechanism of the crystal lattice rearrangement.

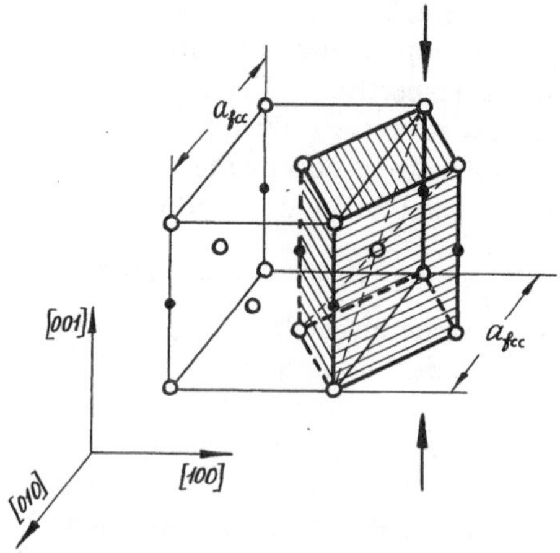

Figure 58. Positions of carbon atoms after the Bain crystal lattice rearrangement: ○ = Fe, • = C. The bcc unit cell is shaded. Arrows show the Bain contraction.

For example, the $\gamma \rightarrow \alpha$ rearrangement described by the Bain distortion (6.5.1) places all the carbon atoms into octahedral interstices of only one type, say, the O_z interstices if the Bain axis is parallel to the $[001]_b$ direction. This is illustrated by Fig. 58.

Lyssak et al. (111–113) and Alshevsky and Kurdjumov (114) made some new important discoveries that contribute to our understanding of the crystallographic mechanism of martensitic transformations in steel. They found that freshly formed martensite in manganese steel (κ' martensite) had an unusual orthorhombic lattice with the crystal lattice parameters a and c larger and smaller, respectively, than the corresponding parameters of "normal" carbon steel martensite at a given carbon content. The abnormal lattice thus features a lower c/a ratio than the "normal" one.

Because heating steel to room temperature causes an increase in the axial ratio, c/a (though not to the "normal" value), and eliminates the orthorhombic distortion, clearly the reduced axial ratio and slight orthorhombicity of the as-quenched κ' martensite are caused by partial disordering of carbon atoms in the crystal lattice rearrangement. The fact that the axial ratio increases spontaneously may be explained if we assume that C-atoms in as-quenched κ' martensite occupy O_z, O_x, and O_y sites, and diffusion of carbon atoms results in their transfer from O_x and O_y to O_z sites.

These data provide convincing evidence of a distortion mechanism more complex than the pure Bain type involved in the $\gamma \rightarrow \alpha$ rearrangement. The formation of κ' martensite is not in fact an exceptional phenomenon: κ' martensite was also observed in rhenium and chromium steels and in some nickel steels (115, 116). It was also found to appear in Fe-8Ni-1.75C in cooling from the liquid nitrogen to liquid helium temperature (116). Finally, κ' martensite was observed even in plain carbon steel containing 1.6 to 1.8 percent carbon (117).

To account for the crystal lattice abnormalities in martensite, Lyssak and Nikolin (112) assumed that the $\gamma \rightarrow \alpha$ rearrangement occurred through an intermediate hcp lattice of ε martensite. According to them, the $\gamma \rightarrow \varepsilon \rightarrow \alpha$ crystal lattice rearrangement shifts carbon atoms to tetrahedral interstices, causing an orthorhombic distortion. A similar idea was advocated by Fujita (118) who supposed that 50 percent of all carbon atoms of as-quenched martensite occupied tetrahedral interstices. Recovery of tetragonality at room temperature was ascribed to transfer of carbon atoms from tetrahedral to octahedral interstices. This mechanism, however, fails to explain some observed results.

According to (113, 115) the crystal lattice parameters of κ' martensite, $a_{\kappa'}$, $b_{\kappa'}$, $c_{\kappa'}$, undergo considerable changes on transforming to the usual tetragonal martensite. But the product of the crystal lattice parameters (the unit cell volume)

$$a_{\kappa'} \cdot b_{\kappa'} \cdot c_{\kappa'}$$

as well as the sum of the parameters

$$b_{\kappa'} + c_{\kappa'}$$

remain constant:

$$a_{\kappa'} \cdot b_{\kappa'} \cdot c_{\kappa'} = a_\alpha \cdot a_\alpha \cdot c_\alpha \tag{6.8.10}$$

and

$$b_{\kappa'} + c_{\kappa'} = a_\alpha + c_\alpha \tag{6.8.11}$$

These relations are easy to check by substituting the numerical parameter values. We have for as-quenched κ' martensite in Fe-6.0Mn-1.0C (119):

$$a_{\kappa'} = 2.866\ \text{Å}, \quad b_{\kappa'} = 2.882\ \text{Å}, \quad c_{\kappa'} = 2.955\ \text{Å} \tag{6.8.12a}$$

At room temperature (after the transformation of κ' martensite to tetragonal α martensite) the crystal lattice parameters become

$$a_\alpha = b_\alpha = 2.864\ \text{Å}, \quad c_\alpha = 2.975\ \text{Å} \tag{6.8.12b}$$

Testing these values in Eq. (6.8.10)

$$\frac{a_{\kappa'} \cdot b_{\kappa'} \cdot c_{\kappa'}}{a_\alpha \cdot a_\alpha \cdot c_\alpha} = \frac{2.866 \cdot 2.882 \cdot 2.955}{2.864 \cdot 2.864 \cdot 2.975} = 1.0002$$

we find that the unit cell volumes of κ' and α martensites are in fact the same to within 0.02 percent, i.e. to within the experimental error of the determination of crystal lattice parameters.

Next we obtain for (6.8.11)

$$b_{\kappa'} + c_{\kappa'} = 2.882 + 2.955 = 5.837\ \text{Å}$$
$$a_\alpha + c_\alpha = 2.864 + 2.975 = 5.839\ \text{Å}$$

The ratio of the two values obtained is equal to

$$\frac{b_{\kappa'} + c_{\kappa'}}{a_\alpha + c_\alpha} = \frac{5.837}{5.839} = 0.99965 = 1 - 0.00035$$

The sums are thus equal to each other to within 0.03 percent, again to within the measurement errors.

The empirical relations (6.8.10) and (6.8.11) cannot be explained within the model of tetrahedral interstices being occupied by carbon atoms. There is no reason to expect that volume expansion caused by insertion of carbon atoms into tetrahedral interstices may be the same as that caused by carbon atoms in octahedral sites because octahedral and tetrahedral interstices differ in their size. Neither is there a plausible reason for the relation (6.8.11) to hold.

If, however, we assume, after Khachaturyan, Rumynina and Kurdjumov (120), that carbon atoms in κ' martensite occupy only the octahedral interstices of the two types, for example, O_y and O_z, relations (6.8.10) and (6.8.11) can be readily explained. In this case

$$n_1 = 0 \tag{6.8.13}$$

and Eq. (6.8.3) may be written as

$$n_3 = n - n_2 \tag{6.8.14}$$

Substituting Eqs. (6.8.13) and (6.8.14) into (6.8.5) yields

$$\begin{aligned} a_{\kappa'} &= a_0(1 + u_{11} n) \\ b_{\kappa'} &= a_0[1 + u_{11} n + (u_{33} - u_{11}) n_2] \\ c_{\kappa'} &= a_0[1 + u_{33} n - (u_{33} - u_{11}) n_2] \end{aligned} \tag{6.8.15}$$

Combining the second and third equations in (6.8.15), we obtain

$$b_{\kappa'} + c_{\kappa'} = a_0[1 + (u_{11} + u_{33}) n] = \text{constant} \tag{6.8.16}$$

Eq. (6.8.16) proves relation (6.8.11) because its right-hand side depends on the total carbon concentration rather than the distribution of carbon atoms between the O_y and O_z sublattices characterized by the value n_2.

Relation (6.8.10) is also easy to obtain from the hypothesis that carbon atoms occupy only octahedral interstices. Multiplication of all three Eqs. (6.8.15) and truncation of all higher than first-order in u_{11} and u_{33} terms yields

$$a_{\kappa'} \cdot b_{\kappa'} \cdot c_{\kappa'} = a_0^3[1 + (2u_{11} + u_{33}) n] = \text{constant} \tag{6.8.17}$$

Eq. (6.8.17) proves relation (6.8.10) because, as in (6.8.16), its right-hand side only depends on the total carbon content, n, and is independent of the distribution of carbon atoms between the O_y and O_z sublattices given by n_2.

It follows that we can explain crystal lattice abnormalities observed in as-quenched κ' martensite with the assumption that carbon atoms occupy octahedral O_y and O_z interstices. It now remains to determine the crystallographic mechanism of the $\gamma \rightarrow \kappa'$ rearrangement responsible for the distribution of carbon atoms over those two sublattices.

As mentioned above, it cannot be a mere Bain distortion. According to Roitburd and Khachaturyan (121) and Kurdjumov and Khachaturyan (122, 123), this mechanism involves the Bain distortion combined with $[011](01\bar{1})_b$ transformation twins in the bcc host lattice. The twins do not violate the bcc

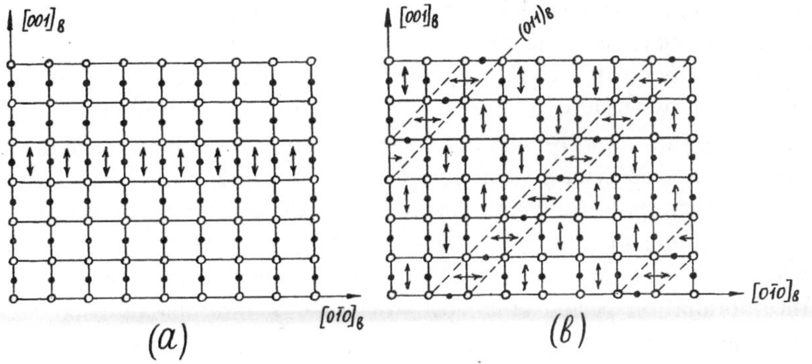

Figure 59. An (100) plane of martensite with $[0\bar{1}1](011)$ twins: ○=Fe, •=C. Arrows indicate the directions of tetragonal distortions caused by carbon atoms in the matrix and twin layers. (*a*) Untwinned martensite; (*b*) twinned martensite.

host lattice perfection because the $(011)_b$ twin plane is a belateral symmetry plane of the bcc lattice. However, they change C-atom positions radically. As shown in Fig. 59, $[0\bar{1}1](011)_b$ twins transfer carbon atoms from O_z to O_y sites. The smallest twin thickness is equal to one interplanar distance. In this limiting case twinning is equivalent to the $[0\bar{1}1](011)_b$ slip. It thus follows that the $[0\bar{1}1](011)_b$ slip, as well as the $[0\bar{1}1](011)_b$ twin, shifts C-atoms from O_z to O_y octahedral sites.

It should be noted that the slip may be interpreted as boundary-to-boundary glide of perfect dislocations with the $[0\bar{1}1]$ Burgers vector in the $(011)_b$ plane.

It was shown in Section 6.7 that the Bain distortion combined with the $[0\bar{1}1](011)_b$ slip mode can explain the formation of dislocated "lath" martensite having the $(111)_{fcc}$ habit.

It follows from Eq. (6.7.26) that the $(011)_b$ and $(0\bar{1}1)_b$ slip planes are regularly spaced and separated by $m_0 \approx 16$ host lattice planes. Therefore each slip system transfers $1/m_0 = \frac{1}{16}$ of all carbon atoms from O_z to O_y sites. Both slip systems, $[0\bar{1}1](011)_b$ and $[011](0\bar{1}1)_b$, transfer $2(1/m_0) = \frac{1}{8}$ of all carbon atoms:

$$n_2 = \tfrac{1}{8}n = 0.125n \qquad (6.8.18)$$

Eq. (6.8.18) shows that 12.5 percent of all carbon atoms occupy O_y interstices and that 87.5 percent of carbon atoms remain on O_z sites. These numerical data are in agreement with the neutron diffraction data on as-quenched martensite in Fe-8.0Ni-1.5C steel reported by Entin, Somenkov, and Shil'shtein (124) who employed isotopic substitution (Fe^{57} with a low scattering amplitude and Ni^{62} with a negative scattering amplitude) to increase the contribution to scattering from carbon.

The calculations performed in Section 6.7 are based on the assumption that the $\gamma \rightarrow \alpha$ rearrangement is a combination of the Bain distortion and two slip modes, $[0\bar{1}1](011)_b$ and $[011](0\bar{1}1)_b$. Similar calculations were carried out in the work (120) where the Bain distortion was combined with a single-slip mode, $[0\bar{1}1](011)_b$, to explain the structure characteristics of κ' martensite in Fe-6.0Mn-1.0C steel.

The calculation results (120) show that a macroscopic shape deformation becomes an invariant plane strain if the separation between the nearest $(011)_b$ slip planes is equal to 5-6 $(011)_b$ interplanar distances ($m_0 \approx 5$-6). Martensite crystals are shown to have a near $(225)_f$ habit. This is in agreement with the data reported by Dunne and Bowles (125) who determined the martensite habit and by Oshima and Wayman (126) and Oshima, Azuma, and Fujita (127) who observed extra spots on martensite electron diffraction patterns. The extra spots were found on the $\langle 011 \rangle$ directions on both sides of the fundamental reflections. The distance between the extra and fundamental spots corresponded to the modulation period equal to 5 or 6 $(011)_b$ interplanar distances. The diffraction effects are in agreement with the calculated value $m_0 \approx 5$-6 (120). In fact the model described in the work cited locates all O_y carbon atoms on the $(011)_b$ slip planes. The latter are periodically spaced, the spacing being of about 5-6 interplanar distances. Therefore carbon atoms are distributed within the $(011)_b$ slip planes with the same periodicity. This "long-period" distribution of carbon atoms gives rise to superlattice reflection arrays along the $[011]_b^*$ direction in the reciprocal lattice space. The observed extra spots may be assigned to these superlattice reflections.

It may be noted in passing that extra spots observed in the works (126, 127) furnish one more quantitative evidence in favor of the distribution of carbon atoms over octahedral interstices in κ' martensite. In fact, if C atoms of κ' martensite occurring in Fe-6.0Mn-1.0C steel occupy O_y sites only in every sixth $(011)_b$ plane (every slip plane) (120), we have

$$n_2 = \tfrac{1}{6}n, \quad n_3 = n - \tfrac{1}{6}n = \tfrac{5}{6}n \tag{6.8.19}$$

The crystal lattice parameters of room-temperature α martensite (all carbon atoms occupy O_z interstices) correspond to

$$n_1 = n_2 = 0, \quad n_3 = n \tag{6.8.20}$$

Substituting (6.8.20) into (6.8.5) yields

$$\begin{aligned} a_\alpha &= b_\alpha = a_0(1 + u_{11}n) \\ c_\alpha &= a_0(1 + u_{33}n) \end{aligned} \tag{6.8.21}$$

It follows from (6.8.21) that at small values of $u_{11}n$ and $u_{33}n$ the tetragonality ratio is given by

$$\frac{c_\alpha}{a_\alpha} = \frac{a_0(1 + u_{33}n)}{a_0(1 + u_{11}n)} \approx 1 + (u_{33} - u_{11})n \tag{6.8.22}$$

Substituting Eqs. (6.8.22) and (6.8.19) into (6.8.15), we obtain the relations between the crystal lattice parameters of α and κ' martensite:

$$\begin{aligned} a_{\kappa'} &= a_\alpha, \quad b_{\kappa'} = a_\alpha + \left(\frac{c_\alpha}{a_\alpha} - 1\right) a_0 \frac{1}{6} \\ c_{\kappa'} &= c_\alpha - \left(\frac{c_\alpha}{a_\alpha} - 1\right) a_0 \frac{1}{6} \end{aligned} \tag{6.8.23}$$

where a_0 is the crystal lattice parameter of pure αFe. Using the observed crystal lattice parameters of α martensite (6.8.12b) in Eq. (6.8.23), we can calculate the crystal lattice parameters of κ' martensite and compare them with the experimental values (6.8.12a). Coincidence of the two sets would lend a support to the assumption made at the beginning of this section about the distribution of carbon atoms over octahedral interstices. Substitution of the numerical values into Eq. (6.8.23) gives

$$a_{\kappa'} = 2.864 \text{ Å}$$

$$b_{\kappa'} = a_\alpha + \left(\frac{c_\alpha}{a_\alpha} - 1\right) a_0 \frac{1}{6} = 2.864 + \left(\frac{2.975}{2.864} - 1\right) \frac{2.865}{6} = 2.882 \text{ Å}$$

$$c_{\kappa'} = c_\alpha - \left(\frac{c_\alpha}{a_\alpha} - 1\right) a_0 \frac{1}{6} = 2.975 - \left(\frac{2.975}{2.864} - 1\right) \frac{2.865}{6} = 2.956 \text{ Å} \tag{6.8.24}$$

The calculated values (6.8.24) are in a striking agreement with the observed ones [see (6.8.12a)]. As a final note, traces of extended planar defects in $\{011\}$ planes of martensite that may be interpreted as $\{011\}_\mathrm{b}$ twins or slip bands have been observed in several electron microscopic studies (128–131).

In sum, the discovery of crystal lattice abnormalities in as-quenched martensite phases of certain low stacking fault energy steels provides new information about the crystallographic mechanisms of martensitic transformations. The mechanism of the $\gamma \rightarrow \alpha$ rearrangement involving both $(011)_\mathrm{b}$ slip modes and the conventional Bain distortion seems to give the best explanation for the observed structure characteristics of martensite crystals in low stacking fault energy steels.

7

ELASTIC STRAIN CAUSED BY CRYSTAL LATTICE REARRANGEMENT

7.1. INTRODUCTION

It has already been pointed out that solid phase transformations involve appearance of isolated "islands" of the new phase inside the parent phase matrix which grow, interact with each other, and eventually form a complex multiphase substructure. The shapes, volume fractions, crystallographic orientations, internal surfaces, and mutual arrangement of the phase transformation products, all that is understood under alloy morphology, significantly affects mechanical, electric, magnetic, and other properties of practical importance.

X-ray and especially electron microscopic studies have revealed the existence of various types of alloy morphology. Depending on the crystallography and kinetics of the phase transformation, the shape of inclusions may vary from spherical and polyhedral to plate- and needlelike. In some cases, the shapes of inclusions were reported to change even during the phase transformation. The most intriguing characteristic of multiphase morphologies is, however, the mutual arrangement of the new phase particles. In many systems regular arrays of new phase particles (modulated and "tweed" structures) are observed.

The detailed morphology of multiphase crystals cannot be explained by the classic thermodynamics of phase transformations based on consideration of bulk chemical free energy and interphase energy terms only. Indeed, the equilibrium shape of a new phase is conventionally related to the interphase energy effect. The shape is determined from the condition that the surface energy at a given inclusion volume be minimal [see, e.g. (80)]. This approach results in the conclusion that a new phase particle should be a polyhedron whose facets are determined by anisotropy of the specific surface energy coefficient. This approach provides an explanation of equilibrium faceting of an isolated new phase particle, but it cannot but fail to describe many cases of inclusions inside a parent phase. For that reason the interphase energy approach cannot be applied to describe the most interesting and numerous cases of plate- and needlelike inclusions.* The key factor for understanding the mechanisms of formation of heterogeneous crystal structures is the elastic strain energy effect arising from a mismatch of crystal lattices of coexisting phases along interphase boundaries. Unlike chemical free energy which only depends on the volume of a new phase particle, elastic strain energy also depends on the particle shape. This is similar to the case of ferroelectric and ferromagnetic crystals. Apart from shape mutual arrangement and orientation relations between the phase transformation products contribute to elastic strain energy stored in strain field which thus controls all alloy morphology features.

It seems worthwhile to consider certain points in more detail. A crystal lattice rearrangement gives rise to a new phase crystal lattice different from the parent one. The two lattices must be adjusted to each other along the interphase. If their adjustment proceeds by elastic displacements of atoms from their regular positions, we call it coherent conjugation. In this case all parent phase crystal lattice planes are transformed continuously into new phase lattice planes. Coherent conjugation involves maximum elastic displacements and therefore generates maximum elastic strain in an alloy. If a mismatch of the two phases is removed partially by dislocations, a semicoherent conjugation occurs. Semicoherent conjugation decreases elastic strain, though at the cost of increase of interphase energy associated with dislocation cores. It is generally believed that in the early stage of the transformation, when the basic features of alloy morphology are formed, coherent conjugation is realized.

The strain energy induced by a phase transformation depends strongly on the shape of coherent inclusions. The mismatch between adjacent boundary planes of the inclusion and the parent phase depends on the orientation of these planes. The mismatch varies from one boundary plane to another as well as the elastic strain field generated by this mismatch. Any variation of the inclusion shape affects the "spectrum" of the orientations of boundary planes and therefore results in change the elastic strain field.

Strain field and the amount of energy stored in it also depend on the mutual arrangement of inclusions in an alloy. In general, each inclusion generates an anisotropic strain field. If inclusions are located with respect to each other so that their strain fields are canceled out, the overall strain energy decreases considerably. This shows that the strain energy must depend on the relative inclusion positions. Therefore, the crystallography of a phase transformation that determines the mismatch of crystal lattices has a definite relation to strain energy and multiphase alloy morphology.

The first studies of the elastic strain problem were conducted in 1957 and

*Their formation cannot be explained by the phase transformation kinetics because their morphology remains the same during the whole transformation process.

1959 by Eshelby whose works have since become a classic (132, 133). In (132, 133) the elastic strain energy induced by a coherent ellipsoidal inclusion in an ellastically isotropic medium was calculated on the assumption that both phases had the same elastic moduli. The general theory of strain energy was proposed in 1966 by Khachaturyan (134); the exact equation for strain energy and the Fourier transform of elastic displacements in an arbitrary two-phase coherent mixture (a mixture formed in an arbitrary crystal lattice rearrangement and characterized by arbitrary shape and mutual locations of inclusions and by elastic anisotropy) were derived. In this work the suggestion was first made that strain energy calculations may also have other interesting applications. It was shown that minimizing strain energy makes it possible to determine the morphology of coherent precipitates (habit and orientation relations). The technique involved calculating habit plane orientations that depend on the axial ratio for tetragonal precipitates in a cubic parent phase. The same idea was formulated independently by Roitburd who considered a two-dimensional inclusion model (135) with different moduli.

In 1969 Khachaturyan and Shatalov extended the theory (134) to arbitrary coherent mixtures comprising inclusions formed by crystal lattice rearrangements of different types (136). It should, however, be noted that the generality was attained at the cost of the assumption that all the phases making up the mixture had equal elastic moduli.

Various cases of isolated inclusions of a certain shape having the same elastic moduli as the parent phase matrix were discussed in a number of papers. Sass, Mura, and Cohen expressed in 1967 the displacement field associated with a cuboidal precipitate in the form of a Fourier series (137). Favier in 1969 proposed the exact solution for inclusions having the shape of a rectangular parallelepiped (138). Sankaran and Laird in 1976 calculated the deformation field associated with a square plate inclusion (139). All these calculations have been carried out for elastic strain field in the limiting case of isotropic elasticity and homogeneous moduli. In 1978 Lee and Johnson evaluated the strain energy of a coherent cuboidal precipitate in an anisotropic matrix (140).

The next step in the development of the theory was made in 1977 by Lee, Barnett, and Aaronson (141). These authors obtained the exact solution of the strain energy problem for an ellipsoidal coherent inclusion in an anisotropic medium with inclusion and medium moduli differing from each other. Unfortunately, the exact close solution cannot be generalized to inclusions of other shapes nor to mixture of several types of inclusions. For that reason and to provide the possibility of analyzing multiphase coherent mixtures, the theory (136) based on the assumption of uniform moduli will be applied below. The cases where the differences in the moduli are of significance will be discussed separately.

7.2. STRAIN ENERGY OF MULTIPHASE ALLOYS

Following the line of reasoning pursued by Khachaturyan and Shatalov (136), let us consider an arbitrary coherent multiphase mixture of new phase particles formed in crystal lattice rearrangements of ν types. It is assumed that:

1. The parent and transformed phase particles have the same elastic moduli (the homogeneous moduli approximation).
2. The average length dimension of an inclusion and the average distance between the nearest inclusions are small compared with the typical length dimensions of the crystal.
3. External boundaries of the multiphase mixture are stress-free.

Let us introduce the stress-free strain tensors:

$$\varepsilon_{ij}^0(1),\ \varepsilon_{ij}^0(2),\ \ldots,\ \varepsilon_{ij}^0(p),\ \ldots,\ \varepsilon_{ij}^0(\nu) \tag{7.2.1}$$

which describe the macroscopic shape deformation of the parent phase caused by crystal lattice rearrangements of the 1st, 2nd, ... pth, ..., νth kinds associated with the phase transformation in the stress-free state.

Let the transformation to the multiphase mixture involve six successive steps (see the diagram in Fig. 60):

Step 1. Choose a group of small clusters in the parent phase crystal, and cut the clusters out of it. If the clusters are large enough to ignore the surface effects, the energy of the system remains unaffected in this step.

Step 2. Let each cluster be transformed to a new phase corresponding to one of the crystal lattice rearrangements under stress-free conditions. The clusters are thus transformed into a set of new phase inclusions. The associated expansion of clusters involves no strain energy change since clusters are stress-free. The expansion is described by strain tensors $\varepsilon_{ij}^0(p)$ ($p=1, 2, \ldots, \nu$) associated with the macroscopic shape deformation in the stress-free state.

Step 3. Let surface traction be applied to each cluster to restore the shape it had before the transformation. The restoration of the shape of the pth kind particle requires the homogeneous strain

$$\varepsilon_{ij}=-\varepsilon_{ij}^0(p) \tag{7.2.2}$$

According to Hooke's law the internal stress induced by surface traction is

$$\sigma_{ij}=\lambda_{ijkl}(-\varepsilon_{ij}^0(p)) \tag{7.2.3}$$

where λ_{ijkl} is the elastic modulus tensor. Since the final homogeneous strain within a pth kind particle is determined by Eq. (7.2.2) the mechanical energy changes by

$$\Delta E_{\text{self}}(p)=\tfrac{1}{2}v_p\lambda_{ijkl}\varepsilon_{ij}^0(p)\varepsilon_{kl}^0(p) \tag{7.2.4}$$

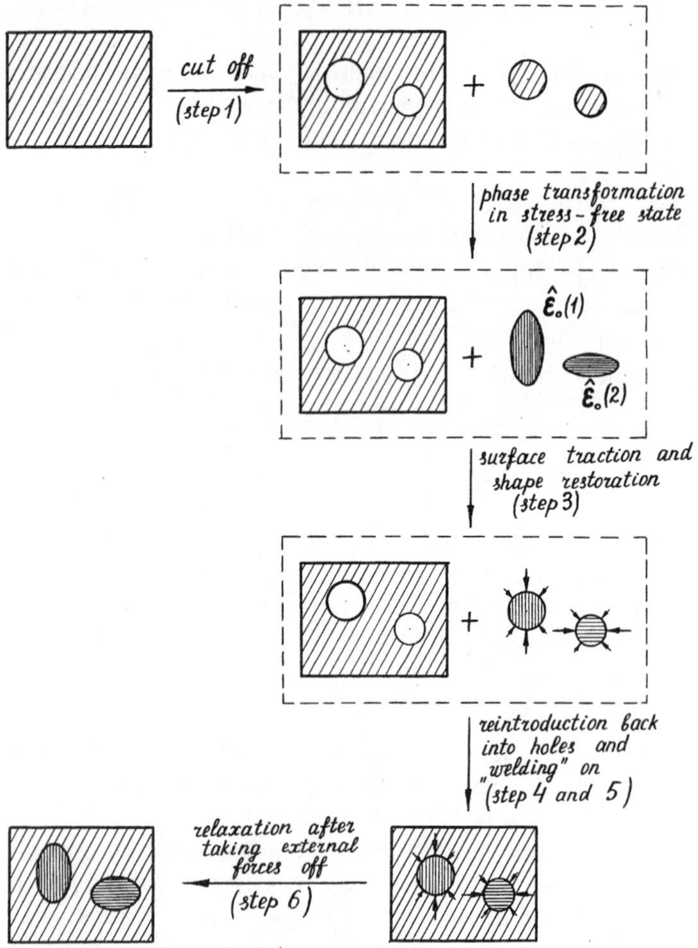

Figure 60. Successive steps to form coherent new phase inclusions.

where v_p is the volume of a particle of the pth kind. The total energy change of all pth kind particles is

$$
\begin{aligned}
N_p \Delta E_{\text{self}}(p) &= \tfrac{1}{2} N_p v_p \lambda_{ijkl} \varepsilon_{ij}^0(p) \varepsilon_{kl}^0(p) \\
&= \tfrac{1}{2} V_p \lambda_{ijkl} \varepsilon_{ij}^0(p) \varepsilon_{kl}^0(p)
\end{aligned}
\tag{7.2.5}
$$

where N_p is the number of all pth kind particles, $V_p = N_p v_p$ is the total volume of pth kind particles. Finally, the total energy change associated with the shape restoration of all particles is the sum of the energy changes (7.2.5):

$$
\Delta E_3 = \tfrac{1}{2} \sum_p V_p \lambda_{ijkl} \varepsilon_{ij}^0(p) \varepsilon_{kl}^0(p) \tag{7.2.6}
$$

Step 4. Let the particles ($p = 1, 2, \ldots, \nu$) be reintroduced back into their holes left in the parent crystal after their removal. Since after Step 3 each inclusion just fits into the space from which it was removed, the insertion does not give rise to any energy change:

$$\Delta E_4 = 0$$

Step 5. Let us "weld on" all introduced particles to the parent phase.

Step 6. Finally, let the particles relax by introducing the equilibrium elastic strain into the lattice. During relaxation each particle will initiate crystal lattice displacements in its vicinity. The displacement is opposed by elastic resistance of the lattices. The associated relaxation energy per unit volume may therefore be presented as power series in the deformations $\varepsilon_{ij}(\mathbf{r})$. Truncation after the square term yields

$$f_{\text{relax}}(\mathbf{r}) = -\sigma_{ij}^0(\mathbf{r})\varepsilon_{ij} + \tfrac{1}{2}\lambda_{ijkl}\varepsilon_{ij}\varepsilon_{kl} \tag{7.2.7}$$

where $\sigma_{ij}^0(\mathbf{r})$ and λ_{ijkl} are the first- and second-order expansion coefficients. The total relaxation energy is given as integral of (7.2.7) over the system volume V

$$\Delta E_{\text{relax}} = \int_{(V)} \left[-\sigma_{ij}^0(\mathbf{r})\varepsilon_{ij} + \tfrac{1}{2}\lambda_{ijkl}\varepsilon_{ij}\varepsilon_{kl}\right] dV \tag{7.2.8}$$

Since the elastic relaxation is a spontaneous process,

$$\Delta E_{\text{relax}} \leqslant 0 \tag{7.2.9}$$

Physically, the effect of relaxation is to remove a part of the elastic distortion energy introduced in Step 3 when transformed clusters were deformed to fit properly into the parent phase body.

Summation of the contributions to strain energy from all steps gives the total elastic strain energy:

$$E_{\text{elast}} = \Delta E_3 + \Delta E_{\text{relax}} = \tfrac{1}{2}\sum_p V_p \lambda_{ijkl}\varepsilon_{ij}^0(p)\varepsilon_{kl}^0(p) + \int_{(V)} \left[-\sigma_{ij}^0(\mathbf{r})\varepsilon_{ij} + \tfrac{1}{2}\lambda_{ijkl}\varepsilon_{ij}\varepsilon_{kl}\right] dV \tag{7.2.10}$$

Linear terms in ε_{ij} appear in (7.2.7), (7.2.8), and (7.2.10) because a multi-connected heterogeneous medium is being considered. Unlike the usual case of homogeneous media, the systems under consideration are strained in the stress-free state. In other words, if $\sigma_{ij}(\mathbf{r}) \equiv 0$, we have $\varepsilon_{ij}(\mathbf{r}) \not\equiv 0$ (σ_{ij} is the stress tensor).

Let the strain tensor of a stress-free state ($\sigma_{ij} \equiv 0$) be $\varepsilon_{ij}^0(r)$. Using the notation (7.2.1), we have

$$\varepsilon_{ij}^0(\mathbf{r}) = \begin{cases} \varepsilon_{ij}^0(1) & \text{if } \mathbf{r} \text{ is inside any first-type particle} \\ \cdots\cdots & \cdots\cdots \\ \varepsilon_{ij}^0(p) & \text{if } \mathbf{r} \text{ is inside any } p\text{th-type particle} \\ \cdots\cdots & \cdots\cdots \\ \varepsilon_{ij}^0(\nu) & \text{if } \mathbf{r} \text{ is inside any } \nu\text{th-type particle} \\ 0 & \text{otherwise} \end{cases} \tag{7.2.11}$$

The condition (7.2.11) may be written in the condensed form:

$$\varepsilon_{ij}^0(\mathbf{r}) = \sum_{p=1}^{v} \tilde{\theta}_p(\mathbf{r})\varepsilon_{ij}^0(p) \qquad (7.2.12)$$

where $\tilde{\theta}_p(\mathbf{r})$ is the shape function of particles of type p equal to unity inside a particle and to zero outside it. The shape function $\tilde{\theta}_p(\mathbf{r})$ is in general a multi-connected function and may describe an arbitrary set of particles of the pth type.

According to the usual relation of elasticity (142) the elastic stress is related to the elastic strain as

$$\sigma_{ij}(\mathbf{r}) = \frac{\delta E_{\text{elast}}}{\delta \varepsilon_{ij}(\mathbf{r})} = -\sigma_{ij}^0(\mathbf{r}) + \lambda_{ijkl}\varepsilon_{kl} \qquad (7.2.13)$$

The strain, $\varepsilon_{ij}^0(\mathbf{r})$, corresponding to the stress-free state may be found from Eq. (7.2.13) by setting $\sigma_{ij}(\mathbf{r}) \equiv 0$:

$$-\sigma_{ij}^0(\mathbf{r}) + \lambda_{ijkl}\varepsilon_{kl}^0(\mathbf{r}) \equiv 0 \qquad (7.2.14)$$

Substituting (7.2.12) into (7.2.14) yields the definition of the function $\sigma_{ij}^0(\mathbf{r})$:

$$\sigma_{ij}^0(\mathbf{r}) = \sum_{p=1}^{v} \sigma_{ij}^0(p)\tilde{\theta}_p(\mathbf{r}) \qquad (7.2.15)$$

where

$$\sigma_{ij}^0(p) = \lambda_{ijkl}\varepsilon_{kl}^0(p) \qquad (7.2.16)$$

Substituting (7.2.15) into (7.2.8) gives

$$\Delta E_{\text{relax}} = \int_{(V)} dV \left[-\sum_{p=1}^{v} \sigma_{ij}^0(p)\tilde{\theta}_p(\mathbf{r})\varepsilon_{ij} + \tfrac{1}{2}\lambda_{ijkl}\varepsilon_{ij}\varepsilon_{kl} \right] \qquad (7.2.17)$$

7.2.1. Evaluation of the Relaxation Energy

We begin by writing the elastic strain field, ε_{ij}, as the sum of homogeneous and heterogeneous strains:

$$\varepsilon_{ij}(\mathbf{r}) = \bar{\varepsilon}_{ij} + \delta\varepsilon_{ij}(\mathbf{r}) \qquad (7.2.18)$$

where the homogeneous strain $\bar{\varepsilon}_{ij}$ is defined so that

$$\int_{(V)} \delta\varepsilon_{ij}(\mathbf{r})\, dV = 0 \qquad (7.2.19)$$

The quantity, $\bar{\varepsilon}_{ij}$, is then the uniform macroscopic strain determining the macroscopic shape deformation of the crystal as a whole produced by internal stress due to the presence of new phase particles. The heterogeneous strain field, $\delta\varepsilon_{ij}(\mathbf{r})$, is chosen such that it has no macroscopic effects: it does not affect the macroscopic crystal shape. The validity of this assumption depends on the validity of the starting Assumption (1).

Substituting (7.2.18) into (7.2.17) and the use of the condition (7.2.19) yield two relaxation energy terms

$$\Delta E_{\text{relax}} = E_{\text{relax}}^{\text{hom}} + E_{\text{relax}}^{\text{heter}} \tag{7.2.20}$$

where

$$E_{\text{relax}}^{\text{hom}} = -\sum_{p=1}^{v} V_p \sigma_{ij}^0(p) \bar{\varepsilon}_{ij} + \frac{V}{2} \lambda_{ijkl} \bar{\varepsilon}_{ij} \bar{\varepsilon}_{kl} \tag{7.2.20a}$$

$$E_{\text{relax}}^{\text{heter}} = \int_{(V)} dV \left[-\sum_{p=1}^{v} \sigma_{ij}^0(p) \Delta\tilde{\theta}_p(\mathbf{r}) \delta\varepsilon_{ij} + \tfrac{1}{2} \lambda_{ijkl} \delta\varepsilon_{ij} \delta\varepsilon_{kl} \right] \tag{7.2.20b}$$

$$\Delta\tilde{\theta}_p(\mathbf{r}) = \tilde{\theta}_p(\mathbf{r}) - \frac{V_p}{V}$$

The relation

$$\int \tilde{\theta}_p(\mathbf{r})\, dV = V_p \tag{7.2.21}$$

following from the definition of $\tilde{\theta}_p(\mathbf{r})$ was employed to derive Eqs. (7.2.20).

The relaxation energy value depends on two sets of internal parameters: the components of the homogeneous strain, $\bar{\varepsilon}_{ij}$, and the local displacements, $\mathbf{u}(\mathbf{r})$, which are related to the heterogeneous strain by the usual elasticity relation:

$$\delta\varepsilon_{ij} = \frac{1}{2}\left(\frac{\partial u_i}{\partial r_j} + \frac{\partial u_j}{\partial r_i} \right) \tag{7.2.22}$$

Mechanical equilibrium is attained when variations of homogeneous strain and local displacements $\mathbf{u}(\mathbf{r})$ reduce the relaxation energy value to minimum. The equilibrium conditions thus give two sets of equations:

$$\frac{\partial \Delta E_{\text{relax}}}{\partial \bar{\varepsilon}_{ij}} = 0 \tag{7.2.23a}$$

$$\frac{\partial \Delta E_{\text{relax}}}{\partial u_i(\mathbf{r})} = 0 \tag{7.2.23b}$$

which are the necessary conditions for the minimum of the relaxation energy (7.2.20).

1. Calculation of the Strain Energy Associated with Homogeneous Strain Relaxation

Substituting (7.2.20) into (7.2.23a) gives

$$-\sum_{p=1}^{v} V_p \sigma_{ij}^0(p) + V \lambda_{ijkl} \bar{\varepsilon}_{kl} = 0$$

or (7.2.24)

$$\lambda_{ijkl} \bar{\varepsilon}_{kl} = \sum_{p=1}^{v} w_p \sigma_{ij}^0(p)$$

where $w_p = V_p/V$ is the volume fraction of particles of type p. Substituting (7.2.16) into (7.2.24) yields

$$\lambda_{ijkl}\bar{\varepsilon}_{kl} = \sum_{p=1}^{v} \lambda_{ijkl}\varepsilon_{kl}^{0}(p)w_p \tag{7.2.25}$$

To obtain this parameter in the simple form, Eq. (7.2.25) may be solved for $\bar{\varepsilon}_{kl}$:

$$\bar{\varepsilon}_{ij} = \sum_{p=1}^{v} \varepsilon_{ij}^{0}(p)w_p \tag{7.2.26}$$

Eq. (7.2.26) shows that relaxation does involve homogeneous strain. It also shows that expansion of the crystal caused by coherent new phase particles is proportional to the volume fractions of the phases comprising the multiphase system.

After substituting the uniform relaxation strain $\bar{\varepsilon}_{ij}$ given by Eq. (7.2.26), the expression for the homogeneous relaxation energy (7.2.20a) may be simplified to

$$E_{\text{relax}}^{\text{hom}} = -\frac{V}{2}\sum_{p=1}^{v}\sum_{q=1}^{v} \lambda_{ijkl}\varepsilon_{ij}^{0}(p)\varepsilon_{kl}^{0}(q)w_p w_q \tag{7.2.27}$$

2. Calculation of the Strain Energy Associated with Heterogeneous Strain Relaxation

According to Eq. (7.2.22), the heterogeneous strain $\delta\varepsilon_{ij}(\mathbf{r})$ is related to local displacements $\mathbf{u}(\mathbf{r})$. The equilibrium local displacement field may be found from the minimum conditions (7.2.23b). To calculate the first variational derivative of ΔE_{relax} in $\mathbf{u}(\mathbf{r})$, we must first calculate the first variation of ΔE_{relax}.

It follows from (7.2.22) and from the symmetry properties of the tensors involved,

$$\sigma_{ij}^{0}(p) = \sigma_{ji}^{0}(p), \quad \lambda_{ijkl} = \lambda_{klij} = \lambda_{jikl},$$

that the first variation of ΔE_{relax} may be written as

$$\delta\Delta E_{\text{relax}} = \int_{(V)} \left[-\sum_{p=1}^{v} \sigma_{ij}^{0}(p)\Delta\tilde{\theta}_p(\mathbf{r})\frac{\partial \delta u_i}{\partial r_j} + \lambda_{ijkl}\frac{\partial u_k}{\partial r_l}\frac{\partial \delta u_i}{\partial r_j}\right] dV \tag{7.2.28}$$

Since the variation $\delta\mathbf{u}(\mathbf{r})$ vanishes at the body surface, the Gauss theorem may be applied to Eq. (7.2.28) to reduce it to

$$\delta\Delta E_{\text{relax}} = \int_{(V)} \left[\sum_{p=1}^{v} \sigma_{ij}^{0}(p)\frac{\partial \Delta\tilde{\theta}_p(\mathbf{r})}{\partial r_j} - \lambda_{ijkl}\frac{\partial^2 u_k}{\partial r_j \partial r_l}\right] \delta u_i(\mathbf{r})\, dV \tag{7.2.29}$$

It follows from Eq. (7.2.29) that

$$\frac{\delta \Delta E_{\text{relax}}}{\delta u_i(\mathbf{r})} = \sum_{p=1}^{v} \sigma_{ij}^{0}(p)\frac{\partial \Delta\tilde{\theta}_p(\mathbf{r})}{\partial r_j} - \lambda_{ijkl}\frac{\partial^2 u_k}{\partial r_j \partial r_l} \tag{7.2.30}$$

Substituting (7.2.30) into (7.2.23b) results in the equilibrium equation for local displacements:

$$\lambda_{ijkl}\frac{\partial^2 u_k}{\partial r_j \partial r_l} = \sum_{p=1}^{v} \sigma_{ij}^{0}(p)\frac{\partial \Delta\tilde{\theta}_p(\mathbf{r})}{\partial r_j} \tag{7.2.31}$$

Assumption (1), which is in fact equivalent to the assumption of macroscopic homogeneity of the body, and the definition of the heterogeneous strain (7.2.18) leads to the conclusion that local displacements, $\mathbf{u}(\mathbf{r})$, vanish at the body surface. The latter is the boundary condition for Eq. (7.2.31).

Multiplication of (7.2.31) by $\exp(-i\mathbf{kr})$, integration over the body volume, and application of the Gauss theorem at the boundary condition $\mathbf{u}(\mathbf{r}_s)=0$ yield the equilibrium equation for the Fourier transforms:

$$\lambda_{ijkl}k_jk_kv_l(\mathbf{k}) = -i\sum_{p=1}^{\nu}\sigma_{ij}^0(p)k_j\Delta\theta_p(\mathbf{k}) \tag{7.2.32}$$

where

$$\mathbf{v}(\mathbf{k}) = \int_{(V)} dV\cdot\mathbf{u}(\mathbf{r})e^{-i\mathbf{kr}} \tag{7.2.33a}$$

$$\Delta\theta_p(\mathbf{k}) = \int_{(V)} dV\cdot\Delta\tilde{\theta}_p(\mathbf{r})e^{-i\mathbf{kr}} \tag{7.2.33b}$$

are the Fourier transforms of the functions $\mathbf{u}(\mathbf{r})$ and $\tilde{\theta}_p(\mathbf{r})$.

The solution to Eq. (7.2.32) gives for $\mathbf{v}(\mathbf{k})$

$$v_i(\mathbf{k}) = -i\sum_{p=1}^{\nu} G_{ij}(\mathbf{k})\sigma_{jk}^0(p)k_k\Delta\theta_p(\mathbf{k}) \tag{7.2.34}$$

This equation may be written in the symbolic form

$$\mathbf{v}(\mathbf{k}) = -i\sum_{p=1}^{\nu}\hat{\mathbf{G}}(\mathbf{k})\hat{\sigma}^0(p)\mathbf{k}\Delta\theta_p(\mathbf{k}) \tag{7.2.35}$$

where $G_{ij}(\mathbf{k})$ and $\sigma_{ij}^0(p)$ are the matrix representations of the operators $\hat{\mathbf{G}}(\mathbf{k})$ and $\hat{\sigma}^0(p)$. The matrix $G_{ij}(\mathbf{k})$ is the inverse tensor to $(\hat{\mathbf{G}}^{-1})_{ij}=\lambda_{iklj}k_kk_l$ [i.e., $G_{il}(\mathbf{k})(\hat{\mathbf{G}}^{-1}(\mathbf{k}))_{lj}=\delta_{ij}$]. The matrix $G_{ij}(\mathbf{k})$ is the Fourier transform of the Green function of anisotropic elasticity.

By definition, $(\hat{\mathbf{G}}^{-1}(\mathbf{k}))_{ij}=k^2\lambda_{iklj}n_kn_l$ where $\mathbf{n}=\mathbf{k}/k$ is a unit vector. The Green function $G_{ij}(\mathbf{k})$ inverse to $(\hat{\mathbf{G}}^{-1}(\mathbf{k}))_{ij}$ may therefore be written in the form

$$G_{ij}(\mathbf{k}) = \frac{1}{k^2}\Omega_{ij}(\mathbf{n}) \tag{7.2.36a}$$

where the tensor $\Omega_{ij}(\mathbf{n})$ is inverse to $\Omega_{ij}^{-1}(\mathbf{n})=\lambda_{iklj}n_kn_l$:

$$\Omega_{ik}(\mathbf{n})\Omega_{kj}^{-1}(\mathbf{n}) = \delta_{ij} \tag{7.2.36b}$$

The heterogeneous relaxation energy (7.2.20b) may be written in terms of the Fourier transforms of local displacements. Using the definition (7.2.22), one may write Eq. (7.2.20b) through the displacements $\mathbf{u}(\mathbf{r})$:

$$E_{\text{relax}}^{\text{heter}} = \int_{(V)} dV\left[-\sum_{p=1}^{\nu}\sigma_{ij}^0(p)\Delta\tilde{\theta}_p(\mathbf{r})\frac{\partial u_i}{\partial r_j} + \tfrac{1}{2}\lambda_{ijkl}\frac{\partial u_i}{\partial r_j}\frac{\partial u_k}{\partial r_l}\right] \tag{7.2.37}$$

Substitution of the back Fourier transforms

$$\mathbf{u}(\mathbf{r}) = \iiint_{-\infty}^{\infty} \frac{d^3k}{(2\pi)^3} \mathbf{v}(\mathbf{k}) e^{i\mathbf{kr}}$$

$$\Delta\tilde{\theta}_p(\mathbf{r}) = \iiint_{-\infty}^{\infty} \frac{d^3k}{(2\pi)^3} \Delta\theta_p(\mathbf{k}) e^{i\mathbf{kr}} \tag{7.2.38}$$

into (7.2.37), integration over $\mathbf{r}$, and application of the relation

$$\int_{(V)} dV \cdot e^{i(\mathbf{k}+\mathbf{k}')\mathbf{r}} = (2\pi)^3 \cdot \delta(\mathbf{k}+\mathbf{k}') \tag{7.2.39}$$

where $\delta(\mathbf{k})$ is the Dirak delta function yields

$$E_{\text{relax}}^{\text{heter}} = \int \frac{d^3k}{(2\pi)^3} \left[-i \sum_{p=1}^{v} \left(\sigma_{ij}^0(p) \Delta\theta_p(\mathbf{k}) k_j v_i^*(\mathbf{k}) + \tfrac{1}{2} \lambda_{ijkl} k_j k_l v_i(\mathbf{k}) v_k^*(\mathbf{k}) \right) \right] \tag{7.2.40}$$

The integrand may be rewritten in the symbolic form

$$E_{\text{relax}}^{\text{heter}} = \int \frac{d^3k}{(2\pi)^3} \left[-i \sum_{p=1}^{v} \left((\mathbf{k}\hat{\sigma}^0(p)\mathbf{v}^*(\mathbf{k}))\Delta\theta_p(\mathbf{k}) + (\mathbf{v}(\mathbf{k})\hat{\mathbf{G}}^{-1}(\mathbf{k})\mathbf{v}^*(\mathbf{k})) \right) \right] \tag{7.2.41}$$

where

$$(\hat{\mathbf{G}}^{-1})_{ij} = \lambda_{iklj} k_k k_l \tag{7.2.42}$$

Using the solution (7.2.35) for the local displacement field in Eq. (7.2.41), one obtains for the heterogeneous relaxation strain energy

$$E_{\text{relax}}^{\text{heter}} = \int \frac{d^3k}{(2\pi)^3} \sum_{p,q} \left[-(\mathbf{k}\hat{\sigma}^0(p)\hat{\mathbf{G}}(\mathbf{k})\hat{\sigma}^0(q)\mathbf{k}) \Delta\theta_p(\mathbf{k}) \Delta\theta_q^*(\mathbf{k}) \right.$$
$$\left. + \frac{1}{2} (\mathbf{k}\hat{\sigma}^0(p)\hat{\mathbf{G}}(\mathbf{k})\hat{\mathbf{G}}^{-1}(\mathbf{k})\hat{\mathbf{G}}(\mathbf{k})\hat{\sigma}^0(q)\mathbf{k}) \Delta\theta_p(\mathbf{k}) \Delta\theta_q^*(\mathbf{k}) \right] \tag{7.2.43}$$

or, as $\hat{\mathbf{G}}\hat{\mathbf{G}}^{-1} = \hat{\mathbf{I}}$, the simplier equation

$$E_{\text{relax}}^{\text{heter}} = -\frac{1}{2} \sum_{p,q} \int \frac{d^3k}{(2\pi)^3} (\mathbf{k}\hat{\sigma}^0(p)\hat{\mathbf{G}}(\mathbf{k})\hat{\sigma}^0(\mathbf{q})\mathbf{k}) \Delta\theta_p(\mathbf{k}) \Delta\theta_q^*(\mathbf{k}) \tag{7.2.44}$$

It follows from (7.2.44), (7.2.27), and (7.2.20) that the total relaxation energy is

$$\Delta E_{\text{relax}} = -\frac{V}{2} \sum_{p,q} \lambda_{ijkl} \varepsilon_{ij}^0(p) \varepsilon_{kl}^0(q) w_p w_q$$
$$-\frac{1}{2} \sum_{p,q} \int \frac{d^3k}{(2\pi)^3} (\mathbf{k}\hat{\sigma}^0(p)\hat{\mathbf{G}}(\mathbf{k})\hat{\sigma}^0(\mathbf{q})\mathbf{k}) \Delta\theta_p(\mathbf{k}) \Delta\theta_q^*(\mathbf{k}) \tag{7.2.45}$$

Combining Eqs. (7.2.45) and (7.2.6) yields the total elastic strain energy contribution [see Eq. (7.2.10)]:

$$E_{\text{elast}} = \frac{1}{2}\sum_p V_p\lambda_{ijkl}\varepsilon_{ij}^0(p)\varepsilon_{kl}^0(p) - \frac{V}{2}\sum_{p,q}\lambda_{ijkl}\varepsilon_{ij}^0(p)\varepsilon_{kl}^0(q)w_p w_q$$
$$-\frac{1}{2}\sum_{p,q}\int\frac{d^3k}{(2\pi)^3}(\mathbf{k}\hat{\sigma}(p)\hat{\mathbf{G}}(\mathbf{k})\hat{\sigma}(q)\mathbf{k})\Delta\theta_p(\mathbf{k})\Delta\theta_q^*(\mathbf{k}) \qquad (7..2.46)$$

Eq. (7.2.46) is simplified significantly if the identity

$$\int\frac{d^3k}{(2\pi)^3}\Delta\theta_p(\mathbf{k})\Delta\theta_q^*(\mathbf{k}) = \delta_{pq}V_p - \frac{V_pV_q}{V} \qquad (7.2.47)$$

is taken into account. The identity (7.2.47) may be obtained from the Parseval theorem:

$$I_{pq} = \int\frac{d^3k}{(2\pi)^3}\Delta\theta_p(\mathbf{k})\Delta\theta_q^*(\mathbf{k}) = \int_{(V)}\Delta\tilde{\theta}_p(\mathbf{r})\Delta\tilde{\theta}_q(\mathbf{r})\,dV \qquad (7.2.48)$$

Using the definitions $\Delta\tilde{\theta}_p(\mathbf{r}) = \tilde{\theta}_p(\mathbf{r}) - V_p/V = \tilde{\theta}_p(\mathbf{r}) - w_p$ and $\int\theta_p(\mathbf{r})dV = V_p$, we may rewrite Eq. (7.2.48) as follows:

$$I_{pq} = \int_{(V)}(\tilde{\theta}_p(\mathbf{r}) - w_p)(\tilde{\theta}_q(\mathbf{r}) - w_q)dV = \int_{(V)}\tilde{\theta}_p(\mathbf{r})\tilde{\theta}_q(\mathbf{r})dV - w_pw_qV \qquad (7.2.49)$$

Since a point $\mathbf{r}$ cannot belong to two particles of different types simultaneously, the integral in (7.2.49) is only nonzero if $p=q$. Bearing in mind the identity $(\tilde{\theta}_p(\mathbf{r}))^2 = \tilde{\theta}_p(\mathbf{r})$ that follows from the definition of $\tilde{\theta}_p(\mathbf{r})$ and Eq. (7.2.21), we may rewrite Eq. (7.2.49):

$$I_{pq} = \delta_{pq}V_p - w_pw_qV = \delta_{pq}V_p - \frac{V_pV_q}{V} \qquad (7.2.50)$$

The relation (7.2.50) proves the identity (7.2.47).

We may apply Eq. (7.2.36a) to rewrite (7.2.46) in the form

$$E_{\text{elast}} = \frac{V}{2}\sum_p\lambda_{ijkl}\varepsilon_{ij}^0(p)\varepsilon_{kl}^0(p)w_p - \frac{1}{2}V\sum_{p,q}\lambda_{ijkl}\varepsilon_{ij}^0(p)\varepsilon_{kl}^0(q)w_pw_q$$
$$-\frac{1}{2}\sum_{p,q}\int\frac{d^3k}{(2\pi)^3}\mathbf{n}\hat{\sigma}^0(p)\hat{\mathbf{\Omega}}(\mathbf{n})\hat{\sigma}^0(q)\mathbf{n}\Delta\theta_p(\mathbf{k})\Delta\theta_q^*(\mathbf{k}) \qquad (7.2.51)$$

where $\mathbf{n} = \mathbf{k}/k$. Since

$$\Delta\theta_p(\mathbf{k}) = \int dV\left[\tilde{\theta}_p(\mathbf{r}) - \frac{V_p}{V}\right]e^{i\mathbf{kr}} = \begin{cases}\theta_p(\mathbf{k}) & \text{if } \mathbf{k}\neq 0\\ 0 & \text{otherwise}\end{cases}$$

Eq. (7.2.51) may be also represented as

$$E_{\text{elast}} = \frac{1}{2}V\sum_p\lambda_{ijkl}\varepsilon_{ij}^0(p)\varepsilon_{kl}^0(p)w_p - \frac{1}{2}V\sum_{p,q}\lambda_{ijkl}\varepsilon_{ij}^0(p)\varepsilon_{kl}^0(q)w_pw_q$$
$$-\frac{1}{2}\sum_{p,q}\int\frac{d^3k}{(2\pi)^3}\mathbf{n}\hat{\sigma}^0(p)\hat{\mathbf{\Omega}}(\mathbf{n})\hat{\sigma}^0(q)\mathbf{n}\theta_p(\mathbf{k})\theta_q^*(\mathbf{k}) \qquad (7.2.52)$$

where sign $⨍$ has the meaning that a volume $(2\pi)^3/V$ about $k=0$ is to be excluded from the integration. When V is large, this exclusion defines the "principle value" of the integral.

Using the identity (7.2.47) in Eq. (7.2.51), we have

$$E_{\text{elast}}=\frac{1}{2}\sum_{p,q}\int\frac{d^3k}{(2\pi)^3}\left[\lambda_{ijkl}\varepsilon^0_{ij}(p)\varepsilon^0_{kl}(q)-(\mathbf{n}\hat{\sigma}^0(p)\hat{\mathbf{\Omega}}(\mathbf{n})\hat{\sigma}^0(q)\mathbf{n})\right]$$
$$\times\Delta\theta_p(\mathbf{k})\Delta\theta^*_q(\mathbf{k}) \qquad (7.2.53a)$$

or

$$E_{\text{elast}}=\frac{1}{2}\sum_{p,q}⨍\frac{d^3k}{(2\pi)^3}\left[\lambda_{ijkl}\varepsilon^0_{ij}(p)\varepsilon^0_{kl}(q)-(\mathbf{n}\hat{\sigma}^0(p)\hat{\mathbf{\Omega}}(\mathbf{n})\hat{\sigma}^0(q)\mathbf{n})\right]\theta_p(\mathbf{k})\theta^*_q(\mathbf{k}) \qquad (7.2.53b)$$

It is relevant to mention here that our analysis has been carried out for the case of a finite volume body. The analysis proved possible because the absence of external forces applied to boundaries was used as the boundary condition. The removal of external forces leads to additional strain usually called the image force effect. The image force effect brings on the homogeneous relaxation strain (7.2.26) which may be detected by x-ray measurements of lattice parameters and is considered in the first two terms of the expression for elastic strain energy (7.2.52).

Since the shape function $\tilde{\theta}_p(\mathbf{r})$ may be a multiconnected function, it can describe an isolated particle as well as a set of arbitrarily distributed particles. For this reason Eqs. (7.2.52) and (7.2.53) may be applied to describe both the strain energy of an isolated new phase particle and of an arbitrary system of new phase particles. In the latter case Eq. (7.2.52) includes terms that depend on the mutual arrangement of the particles and describe strain-induced interactions. Lastly, Eq. (7.2.52) does not impose any restrictions on the shape and mutual arrangement of these particles. All the information on the details of the substructure of the heterogeneous crystal is contained in the Fourier transforms of the shape function $\tilde{\theta}_p(\mathbf{r})$.

7.3. STRAIN-INDUCED INTERACTIONS BETWEEN COHERENT NEW PHASE INCLUSIONS

As mentioned in Section 7.2, interference of strain fields generated by different coherent inclusions results in strain-induced interactions that play the major part in such processes as sympatatic nucleation and strain-induced coarsening. Strain-induced pairwise interaction energies may be evaluated as suggested by Khachaturyan and Shatalov (136). The overall shape function $\tilde{\theta}_p(\mathbf{r})$ of a set of inclusions may be written as the sum of the shape functions of individual inclusions:

$$\tilde{\theta}_p(\mathbf{r})=\sum_{\alpha=1}^{N_p}\tilde{\theta}_p(\alpha,\mathbf{r}-\mathbf{R}_\alpha) \qquad (7.3.1)$$

where $\tilde{\theta}_p(\alpha, \mathbf{r}-\mathbf{R}_\alpha)$ is the shape function of the αth particle of the type p, the vector $\mathbf{R}_\alpha$ determines the position of its center of gravity, N_p is the number of particles.

The Fourier transform of the shape function (7.3.1) is

$$\theta_p(\mathbf{k}) = \sum_{\alpha=1}^{N_p} \theta_p(\alpha, \mathbf{k}) e^{-i\mathbf{k}\mathbf{R}_\alpha} \tag{7.3.2}$$

Substituting Eq. (7.3.2) into (7.2.52) yields

$$E_{\text{elast}} = -\frac{V}{2}\sum_{p,q} \lambda_{ijkl}\varepsilon^0_{ij}(p)\varepsilon^0_{kl}(q) w_p w_q + \frac{1}{2}\sum_{p,\alpha,q,\beta} \iiint_{-\infty}^{\infty} \frac{d^3k}{(2\pi)^3} [\lambda_{ijkl}\varepsilon^0_{ij}(p)\varepsilon^0_{kl}(q)$$
$$-(\mathbf{n}\hat{\sigma}^0(p)\hat{\mathbf{\Omega}}(\mathbf{n})\hat{\sigma}^0(q)\mathbf{n})]\theta_p(\alpha, \mathbf{k})\theta^*_q(\beta, \mathbf{k}) e^{-i\mathbf{k}(\mathbf{R}_\alpha - \mathbf{R}_\beta)} \tag{7.3.3}$$

where $w_p = \sum_\alpha v_\alpha(p)/V$ is the volume fraction of all inclusions of the pth type, $v_\alpha(p)$ is the volume of the αth inclusion of the pth type, V the total volume of the crystal. Separating the terms with $\alpha=\beta$ from those with $\alpha \neq \beta$ in (7.3.3), and using the identity

$$\int \frac{d^3k}{(2\pi)^3} |\theta_p(\alpha, \mathbf{k})|^2 = v_\alpha(p) \tag{7.3.4}$$

(the latter is valid within macroscopic accuracy $v_\alpha(p)/V \to 0$) which gives the relation

$$\tfrac{1}{2}\sum_p \lambda_{ijkl}\varepsilon^0_{ij}(p)\varepsilon^0_{kl}(p) V_p = \tfrac{1}{2}\sum_{p,\alpha} \lambda_{ijkl}\varepsilon^0_{ij}(p)\varepsilon^0_{kl}(p) v_\alpha(p)$$
$$= \frac{1}{2}\sum_{p,\alpha} \int \lambda_{ijkl}\varepsilon^0_{ij}(p)\varepsilon^0_{kl}(p) |\theta_p(\alpha, \mathbf{k})|^2 \frac{d^3k}{(2\pi)^3} \tag{7.3.5}$$

we have

$$E_{\text{elast}} = \frac{1}{2}\sum_{p,\alpha} \int \frac{d^3k}{(2\pi)^3} [\lambda_{ijkl}\varepsilon^0_{ij}(p)\varepsilon^0_{kl}(p) - (\mathbf{n}\hat{\sigma}^0(p)\hat{\mathbf{\Omega}}(\mathbf{n})\hat{\sigma}^0(p)\mathbf{n})] |\theta_p(\alpha, \mathbf{k})|^2$$
$$-\tfrac{1}{2}V \sum_{p,q} \lambda_{ijkl}\varepsilon^0_{ij}(p)\varepsilon^0_{kl}(q) w_p w_q$$
$$-\tfrac{1}{2}\sum_{p,\alpha}\sum_{\substack{q,\beta \\ (\alpha\neq\beta)}} \int (\mathbf{n}\hat{\sigma}^0(p)\hat{\mathbf{\Omega}}(\mathbf{n})\hat{\sigma}^0(q)\mathbf{n})\theta_p(\alpha, \mathbf{k})\theta^*_q(\beta, \mathbf{k}) e^{-i\mathbf{k}(\mathbf{R}_\alpha - \mathbf{R}_\beta)} \tag{7.3.6}$$

It will be shown in Section 8.1 that the term

$$E_p(\alpha) = \frac{1}{2}\int \frac{d^3k}{(2\pi)^3} [\lambda_{ijkl}\varepsilon^0_{ij}(p)\varepsilon^0_{kl}(p) - \mathbf{n}\hat{\sigma}^0(p)\hat{\mathbf{\Omega}}(\mathbf{n})\hat{\sigma}^0(p)\mathbf{n}] |\theta_p(\alpha, \mathbf{k})|^2 \tag{7.3.7}$$

gives the strain energy of a single αth inclusion of the pth type in an infinite anisotropic medium. Substituting (7.3.5) into (7.3.6) gives

$$E_{\text{elast}} = \sum_{\alpha, p} E_p(\alpha) - \frac{V}{2} \sum_{p,q} \lambda_{ijkl} \varepsilon_{ij}^0(p) \varepsilon_{kl}^0(q) w_p w_q$$

$$- \frac{1}{2} \sum_{p,\alpha} \sum_{\substack{q,\beta \\ (\alpha \neq \beta)}} \iiint_{-\infty}^{\infty} \frac{d^3k}{(2\pi)^3} (\mathbf{n}\hat{\sigma}^0(p)\hat{\mathbf{\Omega}}(\mathbf{n})\hat{\sigma}^0(q)\mathbf{n}) \theta_p(\alpha, \mathbf{k}) \theta_q^*(\beta, \mathbf{k}) e^{-i\mathbf{k}(\mathbf{R}_\alpha - \mathbf{R}_\beta)} \quad (7.3.8)$$

Eq. (7.3.8) may be rewritten in the form

$$E_{\text{elast}} = E_{\text{self}} + E_{\text{inter}} \quad (7.3.9)$$

where

$$E_{\text{self}} = \sum_{p,\alpha} E_p(\alpha) \quad (7.3.10)$$

is the sum of self-energies of all inclusions (the self-energy of an inclusion is the strain energy necessary to insert a single inclusion particle into an infinite elastic medium),

$$E_{\text{inter}} = -\frac{V}{2} \sum_{p,q} \lambda_{ijkl} \varepsilon_{ij}^0(p) \varepsilon_{kl}^0(q) w_p w_q + \frac{1}{2} \sum_{p,\alpha} \sum_{\substack{q,\beta \\ (\alpha \neq \beta)}} W_{pq}(\mathbf{R}_\alpha - \mathbf{R}_\beta) \quad (7.3.11)$$

is the strain-induced interaction energy,

$$W_{pq}(\mathbf{R}_\alpha - \mathbf{R}_\beta) = - \iiint_{-\infty}^{\infty} \frac{d^3k}{(2\pi)^3} (\mathbf{n}\hat{\sigma}^0(p)\hat{\mathbf{\Omega}}(\mathbf{n})\hat{\sigma}^0(q)\mathbf{n}) \theta_p(\alpha, \mathbf{k}) \theta_q^*(\beta, \mathbf{k}) e^{-i\mathbf{k}(\mathbf{R}_\alpha - \mathbf{R}_\beta)} \quad (7.3.12)$$

is the strain-induced interaction energy between the αth and βth inclusions of the pth and qth types located at the points $\mathbf{R}_\alpha$ and $\mathbf{R}_\beta$, respectively. It follows from (7.3.12) that

$$\begin{aligned} V_{pq}(\alpha, \beta, \mathbf{k}) &= -(\mathbf{n}\hat{\sigma}^0(p)\hat{\mathbf{\Omega}}(\mathbf{n})\hat{\sigma}^0(q)\mathbf{n}) \theta_p(\alpha, \mathbf{k}) \theta_q^*(\beta, \mathbf{k}) \\ &= -n_i \sigma_{ij}^0(p) \Omega_{jk}(\mathbf{n}) \sigma_{kl}^0(q) n_l \theta_p(\alpha, \mathbf{k}) \theta_q^*(\beta, \mathbf{k}) \end{aligned} \quad (7.3.13)$$

is the Fourier transform of the pairwise strain-induced interaction energies of inclusions. Eq. (7.3.12) shows that strain-induced pairwise interaction energies depend on the crystallography of the phase transformation and shapes of interacting particles. Inclusion shapes are described by the Fourier transforms of the shape functions $\tilde{\theta}_p(\alpha, \mathbf{r})$ and $\tilde{\theta}_q(\beta, \mathbf{r})$, whereas the crystallography of the phase transformation and the elastic properties of the phases are buried in the term $\mathbf{n}\hat{\sigma}^0(p)\hat{\mathbf{\Omega}}(\mathbf{n})\hat{\sigma}^0(q)\mathbf{n}$.

It need be emphasized that the first term of the interaction (7.3.11) is configurationally independent image force induced interaction associated with the "stress-free" boundary condition. Interactions of this type depend on the volume fractions of the phase constituents rather than on the mutual positions of inclusions. As for the second term in Eq. (7.3.11), it gives configurationally dependent strain-induced interactions affected by the shape of inclusions and their mutual arrangement.

8

MORPHOLOGY OF SINGLE COHERENT INCLUSION

8.1. STRAIN ENERGY AND SHAPE OF SINGLE COHERENT INCLUSION WITHIN INFINITE MATRIX

As has been shown by Khachaturyan the shape and orientation of a new phase coherent inclusion in an anisotropic crystal can be determined from analysis of the strain energy (134).

The equation for the strain energy of an isolated coherent inclusion within an infinite anisotropic matrix may be obtained from (7.2.53) as a particular case. This may readily be done if:

1. The phase transition is assumed to involve one type of crystal lattice rearangements only.
2. The shape function entering Eq. (7.2.53b) describes a simply connected region of an inclusion of a finite volume.

Eq. (7.2.53) is then simplified:*

$$E=\frac{1}{2}\iiint_{-\infty}^{\infty}\frac{d^3k}{(2\pi)^3}\,B(\mathbf{n})|\theta(\mathbf{k})|^2 \tag{8.1.1}$$

where $\theta(\mathbf{k})$ is the Fourier transform of the shape function of an arbitrarily shaped coherent inclusion,

$$B(\mathbf{n})=\lambda_{ijkl}\varepsilon_{ij}^{0}\varepsilon_{kl}^{0}-n_i\sigma_{ij}^{0}\Omega_{jl}(\mathbf{n})\sigma_{lm}^{0}n_m \tag{8.1.2}$$

*The difference in Eq. (8.1.1) between the "principle value" of the integral and the usual one can be ignored since the infinite matrix is considered.

is the function of the direction $\mathbf{n}=\mathbf{k}/k$, ε_{ij}^0 is an arbitrary stress-free transformation strain

$$\sigma_{ij}^0=\lambda_{ijkl}\varepsilon_{kl}^0 \tag{8.1.2a}$$

$\Omega_{jl}(\mathbf{n})$ is the inverse tensor to

$$\Omega_{ij}^{-1}(\mathbf{n})=\lambda_{iklj}n_k n_l \tag{8.1.3}$$

By definition,

$$B(\mathbf{n})\geqslant 0 \tag{8.1.4}$$

It should be stressed that all information on the elastic properties of the system and crystallography of the phase transformation is contained in the term $B(\mathbf{n})$ in the integrand of Eq. (8.1.1), while information on the shape and volume of inclusions is included in the term $|\theta(\mathbf{k})|^2$.

The expression for the Fourier transform of elastic displacements (7.2.35) is also simplified. It becomes

$$\mathbf{v}(\mathbf{k})=-i\hat{\mathbf{G}}(\mathbf{k})\hat{\sigma}^0\mathbf{k}\theta(\mathbf{k})$$

or, taking into account relation (7.2.36a),

$$\mathbf{v}(\mathbf{k})=-i\frac{1}{k^2}\hat{\mathbf{\Omega}}\left(\frac{\mathbf{k}}{k}\right)\hat{\sigma}^0\mathbf{k}\theta(\mathbf{k}) \tag{8.1.5}$$

The description of crystal lattice plane rearrangements formulated in Section 1.5 was based on the reciprocal lattice concept borrowed from the diffraction theory. We shall now apply the terminology of the diffraction theory, once more noting that the function $|\theta(\mathbf{k})|^2$ which appears in (8.1.1) and describes the effect of the inclusion shape on the strain energy in that equation is at the same time the Laue interference function [see, e.g. (143)]. The Laue interference function

$$|\theta(\mathbf{k})|^2=\left|\int_{(V)} dV e^{-i\mathbf{kr}}\right|^2 \tag{8.1.6}$$

describes broadening of Laue x-ray reflections caused by finiteness of the crystal whose scattering is measured ($\mathbf{k}$ is the distance from the reciprocal lattice point in the reciprocal space in this case).

Thus the Laue interference function for an inclusion having the shape of a rectangular parallelepiped may be obtained from (8.1.6) if integration is carried out over the parallelepiped volume. The result is

$$|\theta(\mathbf{k})|^2=64V^2\frac{\sin^2\frac{1}{2}k_xL_1}{(k_xL_1)^2}\frac{\sin^2\frac{1}{2}k_yL_2}{(k_yL_2)^2}\frac{\sin^2\frac{1}{2}k_zL_3}{(k_zL_3)^2} \tag{8.1.7}$$

where L_1, L_2, L_3 are the parallelepiped dimensions, $\mathbf{k}=(k_x, k_y, k_z)$ are the projections of $\mathbf{k}$ on the edges L_1, L_2, L_3.

The Laue interference function for an ellipsoidal inclusion is

$$|\theta(\mathbf{k})|^2=V^2\left|3\frac{\sin\phi(\mathbf{k})-\phi(\mathbf{k})\cos\phi(\mathbf{k})}{[\phi(\mathbf{k})]^3}\right|^2 \tag{8.1.8}$$

where

$$\phi(\mathbf{k})=\sqrt{L_{ij}k_ik_j}=\sqrt{(\mathbf{k}\hat{\mathbf{L}}\mathbf{k})} \tag{8.1.8a}$$

L_{ij} is the tensor inverse to $(\hat{\mathbf{L}}^{-1})_{ij}$ that determines the standard form of the equation for the ellipsoid surface

$$(\hat{\mathbf{L}}^{-1})_{ij}r_ir_j=1 \tag{8.1.9}$$

The eigenvalues of the tensor L_{ij} are squares of the ellipsoid semiaxes, a_1^2, a_2^2, a_3^2.

It should be noted that Eq. (8.1.9) is applicable to a large variety of inclusions, such as spherical inclusions ($L_{ij}=R^2\delta_{ij}$ where $a_1=a_2=a_3=R$ is the sphere radius), platelike inclusions ($a_1 \sim a_2 \gg a_3$), and needlelike ones ($a_1 \sim a_2 \ll a_3$), where a_1, a_2, a_3 are the semiaxes.

The ellipsoidal model is of interest from the standpoint of the elasticity theory. As shown by Eshelby for the isotropic elasticity case, and by Valpole (144) and Willis (145) for anisotropic elasticity, elastic strain inside an ellipsoidal inclusion (eigenstrain) is always homogeneous.

Since we have at our disposal the closed form of the total strain energy generated by a coherent inclusion of a new phase having an arbitrary shape, it is natural to raise the question, What is the shape that minimizes strain energy at a given inclusion volume V? The correct answer to this question would enable us to predict habit planes and orientations of new phase precipitates.

The solution to the problem may be obtained by analyzing the general properties of the integrand in (8.1.1), such as positiveness of $|\theta(\mathbf{k})|^2$ and dependence of $B(\mathbf{n})=B(\mathbf{k}/k)$ on the direction of $\mathbf{k}$ rather than its absolute value.

Since $B(\mathbf{n})\geqslant 0$ and $|\theta(\mathbf{k})|^2>0$, we have

$$E=\frac{1}{2}\int\frac{d^3k}{(2\pi)^3}B(\mathbf{n})|\theta(\mathbf{k})|^2\geqslant\frac{1}{2}\min B(\mathbf{n})\int\frac{d^3k}{(2\pi)^3}|\theta(\mathbf{k})|^2 \tag{8.1.10}$$

where $\min B(\mathbf{n})$ is the minimum value of the function $B(\mathbf{n})$. It follows from Eq. (7.2.47) that

$$\int\frac{d^3k}{(2\pi)^3}|\theta(\mathbf{k})|^2\equiv V \tag{8.1.11}$$

Substitution of the identity (8.1.11) into the right-hand side of inequality (8.1.10) yields

$$E=\frac{1}{2}\int\frac{d^3k}{(2\pi)^3}B(\mathbf{n})|\theta(\mathbf{k})|^2\geqslant\frac{1}{2}(\min B(\mathbf{n}))V \tag{8.1.12}$$

The right-hand side of inequality (8.1.12) gives the lower limit for the strain energy values (8.1.1) at the given volume. As will be shown below, the lower limit may be attained with platelike inclusions whose aspect ratio approaches zero and whose habit plane is normal to the vector $\mathbf{n}$ corresponding to the minimum of $B(\mathbf{n})$.

We shall now introduce the unit vector $\mathbf{n}_0$ such that the function $B(\mathbf{n})$ has its minimum at $\mathbf{n}=\mathbf{n}_0$:

$$B(\mathbf{n}_0) = \min B(\mathbf{n}) \tag{8.1.13}$$

It is easy to see that (8.1.12) becomes an equality if the function $|\theta(\mathbf{k})|^2$ on the right-hand side only differs from zero in an infinitely thin and infinitely long rod in the $\mathbf{k}$-space directed along the unit vector $\mathbf{n}_0$.

The function $|\theta(\mathbf{k})|^2$ behaves so if the inclusion described by the function $\tilde{\theta}(\mathbf{r})$ is an infinite platelet of infinitesimal thickness normal to the unit vector $\mathbf{n}_0$. This is in fact a well-known result of the diffraction theory: scattering from a platelike crystal is characterized by conversion of diffraction spots into rods in the reciprocal space, the direction of the rods being normal to the habit of the crystal [see, e.g. (143)].

We arrive at the following conclusion:

the minimum strain energy at a given inclusion volume is attained if the inclusion is "rolled out" to give an infinite platelet of infinitesimal thickness whose habit is normal to the vector $\mathbf{n}_0$ minimizing the function $B(\mathbf{n})$.

The strain energy of such an inclusion is proportional to its volume V:

$$E_{\text{bulk}} = \min \mathrm{E} = \tfrac{1}{2}[\min B(\mathbf{n})]V = \tfrac{1}{2}B(\mathbf{n}_0)V \tag{8.1.14}$$

It should, however, be remembered that an "unrolling" of a coherent inclusion into an infinitesimal thickness sheet minimizing the strain energy can never take place in actual systems because of a competing effect, an unlimited increase of the interphase surface energy. Competition between the strain and surface energies (bulk chemical free energy depends on the inclusion volume only and remains unaffected by shape changes) leads to inclusions of various shapes observed by electron microscopy.

Qualitatively, it is quite clear that platelike inclusions should be expected in cases of low interphase energies and considerable mismatch between the parent phase and inclusion crystal lattices. On the other hand, with large interphase energies and small crystal lattice mismatch, spherical and polyhedral inclusions should predominate.

8.1.1. Bulk Energy Associated with an Invariant Plane Strain Transformation

We shall now demonstrate that the basic concept of the phenomenological theory of martensitic transformations (1, 2), the requirement that the invariant plane must also be the habit plane of a martensitic crystal, follows directly from the condition of the strain energy minimum. To prove this, let us substitute the invariant plane distortion (1.7.3)

$$u^0_{ij} = \varepsilon_0 l_i n^0_j \tag{8.1.15}$$

into Eq. (8.1.2) for $B(\mathbf{n})$ to show that the minimum of the resultant function $B(\mathbf{n})$ is zero and occurs at $\mathbf{n} = \mathbf{n}_0$ which enters Eq. (8.1.15) and defines the direction normal to the invariant plane. The distortion (8.1.15) rather than its symmetrical part (stress-free deformation ε^0_{ij}) may be used in (8.1.2) since the antisymmetrical part of the distortion related to rigid body rotations does not affect strain energy in linear elasticity.

Substituting (8.1.15) into (7.2.16) yields

$$\sigma_{ij}^{0}=\varepsilon_0\lambda_{ijsp}l_s n_p^0 \tag{8.1.16}$$

Using Eq. (8.1.16) in Eq. (8.1.2), we obtain

$$B(\mathbf{n})=\varepsilon_0^2[\lambda_{ijlm}l_i n_j^0 l_l n_m^0-n_i(\lambda_{ijsp}l_s n_p^0)\Omega_{jl}(\mathbf{n})(\lambda_{lmqr}l_q n_r^0)n_m] \tag{8.1.17}$$

We shall now substitute $\mathbf{n}=\mathbf{n}_0$ into (8.1.17) to prove that $B(\mathbf{n})$ reaches its minimum at $\mathbf{n}=\mathbf{n}_0$:

$$B(\mathbf{n}_0)=\varepsilon_0^2[(\lambda_{ijlm}n_j^0 n_m^0)l_i l_l-(\lambda_{ijsp}n_i^0 n_p^0)l_s\Omega_{jl}(\mathbf{n}_0)(\lambda_{lmqr}n_r^0 n_m^0)l_q] \tag{8.1.18}$$

Since the elastic modulus symmetry gives

$$\lambda_{ijlm}=\lambda_{lmij}=\lambda_{jilm}=\lambda_{ijml}$$

we have from Eq. (8.1.3)

$$\begin{aligned}\lambda_{ijlm}n_j^0 n_m^0&=\lambda_{ijml}n_j^0 n_m^0=(\hat{\mathbf{\Omega}}^{-1}(\mathbf{n}_0))_{il}\\ \lambda_{ijsp}n_i^0 n_p^0&=\lambda_{spij}n_p^0 n_i^0=(\hat{\mathbf{\Omega}}^{-1}(\mathbf{n}_0))_{sj}\\ \lambda_{lmqr}n_r^0 n_m^0&=\lambda_{lmrq}n_m^0 n_r^0=(\hat{\mathbf{\Omega}}^{-1}(\mathbf{n}_0))_{lq}\end{aligned} \tag{8.1.19}$$

Substituting (8.1.19) into (8.1.18) yields

$$B(\mathbf{n}_0)=\varepsilon_0^2[(\hat{\mathbf{\Omega}}^{-1}(\mathbf{n}_0))_{il}l_i l_l-l_s(\hat{\mathbf{\Omega}}^{-1}(\mathbf{n}_0))_{sj}(\hat{\mathbf{\Omega}}^{-1}(\mathbf{n}_0))_{jl}(\hat{\mathbf{\Omega}}^{-1}(\mathbf{n}_0))_{lq}l_q] \tag{8.1.20}$$

or in the symbolic form

$$B(\mathbf{n}_0)=\varepsilon_0^2[\boldsymbol{l}\hat{\mathbf{\Omega}}^{-1}(\mathbf{n}_0)\boldsymbol{l}-\boldsymbol{l}(\hat{\mathbf{\Omega}}^{-1}(\mathbf{n}_0)\hat{\mathbf{\Omega}}(\mathbf{n}_0)\hat{\mathbf{\Omega}}^{-1}(\mathbf{n}_0))\boldsymbol{l}] \tag{8.1.21}$$

Since, by definition (7.2.36b),

$$\hat{\mathbf{\Omega}}(\mathbf{n})\hat{\mathbf{\Omega}}^{-1}(\mathbf{n})=\hat{\mathbf{I}}$$

where $\hat{\mathbf{I}}$ is an identity operator, we have

$$B(\mathbf{n}_0)=0 \tag{8.1.22}$$

Bearing in mind that $B(\mathbf{n})\geqslant 0$, we conclude that the vector $\mathbf{n}_0$ provides the absolute minimum of the function $B(\mathbf{n})$. In other words, in the case of invariant plane strain (8.1.15) being a stress-free transformation strain, Eq. (8.1.14) reads

$$E_{\text{bulk}}=\tfrac{1}{2}[\min B(\mathbf{n})]V=\tfrac{1}{2}B(\mathbf{n}_0)V=0 \tag{8.1.23}$$

It thus follows from the minimum strain energy condition that the vector normal to the invariant plane is also normal to the habit plane; in other words, the invariant plane always coincides with the habit plane.

This conclusion first obtained in (148) throws a bridge between the phenomenological theory of the martensite crystal habit based on purely geometrical considerations (1, 2) and the theory of optimum shapes of coherent inclusions based on consideration of strain energy as depending on inclusion shapes and orientations.

8.1.2. Bulk and Edge Energies of a Platelike Inclusion.

We shall proceed with platelike inclusions by writing the strain energy (8.1.1) in the form

$$E=\frac{1}{2}\int\frac{d^3k}{(2\pi)^3}\,[B(\mathbf{n}_0)+\Delta B(\mathbf{n})]|\theta(\mathbf{k})|^2$$

$$=\frac{1}{2}B(\mathbf{n}_0)\int\frac{d^3k}{(2\pi)^3}|\theta(\mathbf{k})|^2+\frac{1}{2}\int\frac{d^3k}{(2\pi)^3}\Delta B(\mathbf{n})|\theta(\mathbf{k})|^2 \qquad (8.1.24)$$

where

$$B(\mathbf{n}_0)=\min B(\mathbf{n})$$

and, by definition,

$$\Delta B(\mathbf{n})=B(\mathbf{n})-B(\mathbf{n}_0)>0 \qquad (8.1.25)$$

Taking into consideration (8.1.11), we may simplify Eq. (8.1.24):

$$E=\tfrac{1}{2}B(\mathbf{n}_0)V+E_{\text{edge}} \qquad (8.1.26)$$

where

$$E_{\text{edge}}=\frac{1}{2}\int\frac{d^3k}{(2\pi)^3}\Delta B(\mathbf{n})|\theta(\mathbf{k})|^2 \qquad (8.1.27)$$

Comparison of Eqs. (8.1.26) and (8.1.14) shows that the first term of (8.1.26) describes the strain energy of an infinite plate of infinitesimal thickness (the limit $D\to 0$ at a constant volume V where D is the thickness of the inclusion). The second term in (8.1.26) is therefore the energy correction associated with finite plate thickness. This term is positive since, by definition (8.1.25), $\Delta B(\mathbf{n})\geqslant 0$ and $|\theta(\mathbf{k})|^2>0$.

Now consider the origin of the first term in (8.1.26). We shall demonstrate that this term arises from homogeneous strain responsible for returning transformed phase habit plane atoms to their initial positions (the positions they occupied before the phase transition). The process is driven by external forces acting upon inclusion edges in the habit plane. The force along the direction $\mathbf{n}_0$ normal to the habit plane is zero:

$$\sigma_{ij}n_j^0=0 \qquad (8.1.28)$$

where σ_{ij} is stress associated with homogeneous strain produced by the forces applied. The platelike inclusion is therefore in a constrained stressed state. In this state all crystal lattice translations localized in the habit plane of the inclusion are identical to those at the habit plane of the parent phase. This identity is the direct consequence of coherent adjustment of the phases along the habit plane. The strain resulting in coincidence of the crystal lattices along some plane is an invariant plane strain and may be written in the form (1.7.3):*

$$u_{ij}=S_i n_j^0 \tag{8.1.29}$$

where $\mathbf{n}_0$ is the vector normal to the habit plane, $\mathbf{S}_i$ is the shear vector.

The invariant plane strain (8.1.29) inside a constrained inclusion does not in general coincide with the stress-free transformation strain ε_{ij}^0 and therefore involves an additional elastic strain, ε_{ij}^{el}, which may be found as difference between the total strain,

$$\tfrac{1}{2}(u_{ij}+u_{ji})=\tfrac{1}{2}(S_i n_j^0+S_j n_i^0),$$

and stress-free transformation strain ε_{ij}^0:

$$\varepsilon_{ij}^{el}=\tfrac{1}{2}(u_{ij}+u_{ji})-\varepsilon_{ij}^0=\tfrac{1}{2}(S_i n_j^0+S_j n_i^0)-\varepsilon_{ij}^0 \tag{8.1.30}$$

The stress induced by the elastic strain (8.1.30) is determined by Hooke's law

$$\begin{aligned}\sigma_{ij}&=\lambda_{ijkl}\varepsilon_{kl}^{el}=\tfrac{1}{2}\lambda_{ijkl}(S_k n_l^0+S_l n_k^0)-\sigma_{ij}^0\\&=\lambda_{ijkl}S_l n_k^0-\sigma_{ij}^0\end{aligned} \tag{8.1.31}$$

where σ_{ij}^0 is defined by the equation

$$\sigma_{ij}^0=\lambda_{ijkl}\varepsilon_{kl}^0 \tag{8.1.32}$$

The symmetry of the tensor $\lambda_{ijkl}(\lambda_{ijkl}=\lambda_{ijlk})$ was considered in deriving Eq. (8.1.31). Substituting (8.1.31) into (8.1.28) yields

$$\lambda_{ijkl}n_j^0 n_k^0 S_l=\sigma_{ij}^0 n_j^0$$

or taking into account (8.1.3)

$$(\hat{\mathbf{\Omega}}^{-1}(\mathbf{n}_0))_{il}S_l=\sigma_{ij}^0 n_j^0 \tag{8.1.33}$$

The operator form of Eq. (8.1.33) is

$$\hat{\mathbf{\Omega}}^{-1}(\mathbf{n}_0)\mathbf{S}=\hat{\sigma}^0\mathbf{n}_0 \tag{8.1.33a}$$

The solution of Eq. (8.1.33a) using the operator $\hat{\mathbf{\Omega}}(\mathbf{n}_0)$, which is the inverse of $\hat{\mathbf{\Omega}}^{-1}(\mathbf{n}_0)$, yields

$$\mathbf{S}=\hat{\mathbf{\Omega}}(\mathbf{n}_0)\hat{\sigma}^0\mathbf{n}_0 \tag{8.1.34}$$

Assume now the elastic strain to be fully localized within the inclusion. The elastic strain energy can then be written

$$E_{\text{bulk}}=\tfrac{1}{2}V\lambda_{ijkl}\varepsilon_{ij}^{el}\varepsilon_{kl}^{el} \tag{8.1.35}$$

where ε_{ij}^{el} is given by Eq. (8.1.30). Substituting Eq. (8.1.30) into (8.1.35) and

*This strain includes the elastic strain contribution and should be thus distinguished from the stress-free invariant plane strain (8.1.15).

taking into consideration of the symmetry of the tensor λ_{ijkl} yields

$$\begin{aligned} E_{\text{bulk}} &= \tfrac{1}{2}V\lambda_{ijkl}(S_i n_j^0 - \varepsilon_{ij}^0)(S_l n_k^0 - \varepsilon_{kl}^0) \\ &= \tfrac{1}{2}V\lambda_{ijkl}\varepsilon_{ij}^0\varepsilon_{kl}^0 + \tfrac{1}{2}V\lambda_{ijkl}n_j^0 n_k^0 S_i S_l - V\lambda_{ijkl}\varepsilon_{ij}^0 n_k^0 S_l \end{aligned} \tag{8.1.36}$$

Using the definitions (8.1.3) and (8.1.32), we obtain

$$E_{\text{bulk}} = \tfrac{1}{2}V\lambda_{ijkl}\varepsilon_{ij}^0\varepsilon_{kl}^0 + \tfrac{1}{2}V(\mathbf{S}\hat{\mathbf{\Omega}}^{-1}(\mathbf{n}_0)\mathbf{S}) - V(\mathbf{n}_0\hat{\sigma}^0\mathbf{S}) \tag{8.1.37}$$

The application of Eqs. (8.1.34) in (8.1.37) and the relation

$$\hat{\mathbf{\Omega}}(\mathbf{n}_0)\hat{\sigma}^0\mathbf{n}_0 = \mathbf{n}_0(\hat{\mathbf{\Omega}}(\mathbf{n}_0)\hat{\sigma}^0)^+ = \mathbf{n}_0\hat{\sigma}^{0+}\hat{\mathbf{\Omega}}^+(\mathbf{n}_0)$$

result in

$$\begin{aligned} E_{\text{bulk}} &= \tfrac{1}{2}V\lambda_{ijkl}\varepsilon_{ij}^0\varepsilon_{kl}^0 + \tfrac{1}{2}V(\mathbf{n}_0\hat{\sigma}^{0+}\hat{\mathbf{\Omega}}^+(\mathbf{n}_0)\hat{\mathbf{\Omega}}^{-1}(\mathbf{n}_0)\hat{\mathbf{\Omega}}(\mathbf{n}_0)\hat{\sigma}^0\mathbf{n}_0) \\ &\quad - V(\mathbf{n}_0\hat{\sigma}^0\hat{\mathbf{\Omega}}(\mathbf{n}_0)\hat{\sigma}^0\mathbf{n}_0) \end{aligned} \tag{8.1.38}$$

Since $\hat{\mathbf{\Omega}}^{-1}\hat{\mathbf{\Omega}} = \hat{\mathbf{I}}$ and $\hat{\sigma}^0$ and $\hat{\mathbf{\Omega}}(\mathbf{n}_0)$ are Hermitian matrices ($\hat{\sigma}^{0+} = \hat{\sigma}^0$, $\hat{\mathbf{\Omega}}^+(\mathbf{n}_0) = \hat{\mathbf{\Omega}}(\mathbf{n}_0)$) Eq. (8.1.38) may be simplified

$$E_{\text{bulk}} = \tfrac{1}{2}V[\lambda_{ijkl}\varepsilon_{ij}^0\varepsilon_{kl}^0 - (\mathbf{n}_0\hat{\sigma}^0\hat{\mathbf{\Omega}}(\mathbf{n}_0)\hat{\sigma}^0\mathbf{n}_0)] \tag{8.1.39}$$

The expression (8.1.39) fully coincides with the first term of Eq. (8.1.26) since, by definition (8.1.2),

$$B(\mathbf{n}_0) = \lambda_{ijkl}\varepsilon_{ij}^0\varepsilon_{kl}^0 - (\mathbf{n}_0\hat{\sigma}^0\hat{\mathbf{\Omega}}(\mathbf{n}_0)\hat{\sigma}^0\mathbf{n}_0)$$

Since the first term of (8.1.26) coincides completely with the elastic strain energy (8.1.39) and the correction E_{edge} in (8.1.26) is asymptotically small for thin inclusions, we arrive at the conclusion that the state of a thin coherent platelike inclusion must meet the basic conditions in the derivation of Eq. (8.1.39):

1. The elastic strain is homogeneous and localized within the inclusion.
2. The elastic strain provides coincidence of the constrained crystal lattice of the inclusion and undistorted stress-free crystal lattice of the parent phase along the habit plane or, in other words, the total strain inside the inclusion which is a combination of elastic strain, and stress-free transformation strain is always an invariant plane strain.

There is one point to be emphasized. *Since elastic strain is localized within a coherent inclusion whereas the parent phase is stress-free and remains undistorted, the basic assumption of the theory just described—that of identical elastic moduli of the parent and transformed phases—proves to be unnecessary. All the results obtained for a platelike coherent inclusion also hold for different moduli since the function* $B(\mathbf{n})$ *in* Eq. (8.1.26) *includes only the elastic modulus of the inclusion.*

The second term of Eq. (8.1.26) makes a clear physical sense. It may be interpreted as the energy of a dislocation loop in an anisotropic parent crystal that envelops the inclusion in the habit plane. This may be proved as follows.

According to Eq. (8.1.25), the function $\Delta B(\mathbf{n})$ assumes the minimum value equal to zero at $\mathbf{n}=\mathbf{n}_0$ and increases if $\mathbf{n}$ deviates from $\mathbf{n}_0$. The integration over $\mathbf{k}$ in (8.1.27) gives a result that includes contributions from various directions of $\mathbf{n}$, also those deviating from $\mathbf{n}_0$. On the other hand, since the inclusion is a plate normal to $\mathbf{n}_0$, and its thickness is equal to D and length to L, the function $|\theta(\mathbf{k})|^2$ differs from zero only within a thin and extended rod in the $\mathbf{k}$ space which emerges from the origin, $\mathbf{k}=0$, in the direction defined by $\mathbf{n}_0$. The typical length of this rod is $2\pi/D$; its typical thickness is $2\pi/L$ (Fig. 61). The integration in (8.1.32) is therefore over the rod in the $\mathbf{k}$-space as shown in Fig. 61. This is the reason why the typical deviation $\delta\mathbf{n}$ of the vector $\mathbf{n}=\mathbf{k}/k$ from the direction of the rod, $\mathbf{n}_0$, is of the order of $\delta n \approx (2\pi/L)/(2\pi/D)=D/L$. We may now estimate the typical value of the energy, E_{edge}. It is given by

$$E_{\text{edge}} \sim B(\mathbf{n}_0)V\frac{D}{L} \sim \lambda\varepsilon_0^2 V\frac{D}{L} \tag{8.1.40}$$

where λ and ε_0 are the typical modulus and the typical stress-free transformation strain, respectively, and $\lambda\varepsilon_0^2$ the typical specific strain energy.

Taking into account that $V \sim L^2 D$, we may rewrite relation (8.1.40) in the

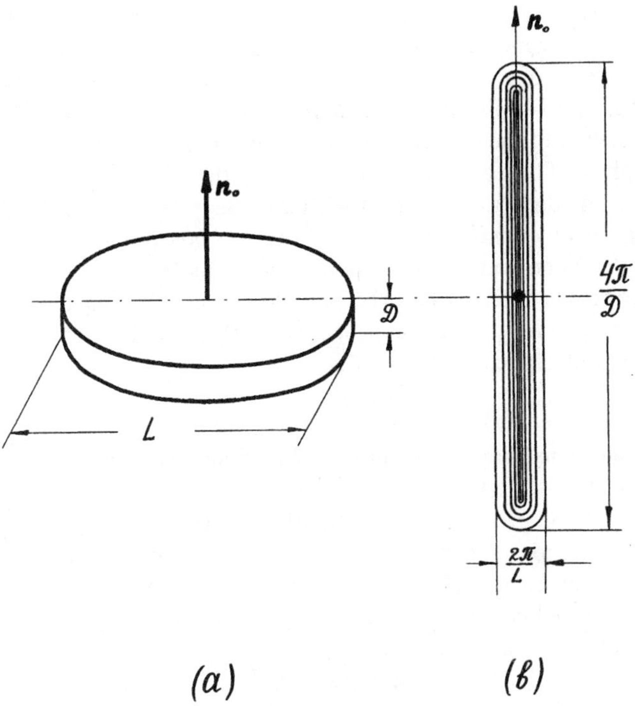

Figure 61. Scheme of a platelike inclusion (*a*) and corresponding "intensity" distribution $|\theta(\mathbf{k})|^2$ in the $\mathbf{k}$-space (*b*).

form

$$E_{\text{edge}} \sim \lambda(\varepsilon_0 D)^2 L \tag{8.1.41}$$

The accurate calculation carried out in Section 8.8 will show that the value of E_{edge} is in fact proportional to the plate perimeter P in the habit plane

$$E_{\text{edge}} \sim \lambda(\varepsilon_0 D)^2 P \tag{8.1.42}$$

Eqs. (8.1.41) and (8.1.42) do not contradict each other since $P \sim L$. The energy E_{edge} is proportional to the perimeter length since it is associated with the crystal lattice mismatch between the matrix and the inclusion along the edges of the platelike inclusion. It follows from Eq. (8.1.42) that the energy correction E_{edge} can be interpreted as "string" energy with the line tension coefficient equal to $\lambda(\varepsilon_0 D)^2$. The alternative interpretation attributes the energy E_{edge} with the energy of some dislocation loop with the Burger vector $b \sim \varepsilon_0 D$ enveloping the platelike inclusion in the habit plane along the habit plane perimeter.

It is of interest that accurate calculations of E_{edge} for the inclusion whose thickness is the interplanar distance of the habit plane and stress-free distortion is given by the dyadic product of the type (8.1.15) (149) give the value equal to the dislocation loop energy calculated in the anisotropic elasticity approximation (see Section 8.8). Really this is not a surprising result since a dislocation loop is in fact a segment of an extra plane and may therefore be interpreted as a coherent platelike inclusion whose thickness is equal to the interplanar distance.

The strain field generated by the edges of a platelike inclusion which can be treated as the field of some effective dislocation loop is a long-range one. In the case of an asymptotically thin inclusion ($D/L \to 0$) the energy of this field as well as the energy of the dislocation loop is determined by the elastic moduli of the parent phase. The reason is that the contribution from the strain field energy associated with edges and localized within an asymptotically thin inclusion is also asymptotically small. This conclusion depends on the geometry of the problem rather than on the elastic moduli of the inclusion and matrix. We can therefore conclude that in the case of different matrix and inclusion elastic moduli, the energy E_{edge} is still described by Eq. (8.1.27) where the elastic moduli of the matrix only appear.

8.1.3. Equilibrium Shape of a Coherent Platelike Inclusion

Returning to the problem of the inclusion shape, we will proceed from the idea of competition between the interphase (surface) energy and shape-dependent strain energy E_{edge}. The surface energy of a platelike inclusion accurate to the higher order terms in D/L is given by

$$E_S \sim \gamma_s \frac{V}{D} \sim \gamma_s L^2 \tag{8.1.43}$$

where γ_s is the specific interphase energy (interphase energy per area unit).

Combining relations (8.1.40) and (8.1.43), we find that the overall shape-dependent free energy is

$$E = E_{\text{edge}} + E_S \approx \lambda \varepsilon_0^2 V \frac{D}{L} + \gamma_s \frac{V}{D} \tag{8.1.44}$$

Since $L \sim \sqrt{V/D}$ for a platelike inclusion, Eq. (8.1.44) may be rewritten in the form:

$$E \approx \lambda \varepsilon_0^2 \sqrt{V} D^{\frac{3}{2}} + \gamma_s V D^{-1} \tag{8.1.45}$$

Minimization of E with respect to D at a constant volume V,

$$\frac{dE}{dD} = 0$$

yields

$$D \simeq \left[\left(\frac{\gamma_s}{\lambda \varepsilon_0^2} \right)^2 V \right]^{\frac{1}{5}} \tag{8.1.46}$$

and

$$L \simeq \sqrt{\frac{V}{D}} \simeq \left(\frac{\lambda \varepsilon_0^2}{\gamma_s} V^2 \right)^{\frac{1}{5}} \tag{8.1.47}$$

Dividing (8.1.46) by (8.1.47), we obtain

$$\frac{D}{L} \simeq \left(\frac{\gamma_s}{\lambda \varepsilon_0^2} \frac{1}{V^{\frac{1}{3}}} \right)^{\frac{3}{5}} \tag{8.1.48}$$

Eq. (8.1.48) may be simplified by introducing the material constant

$$r_0 = \frac{\gamma_s}{\lambda \varepsilon_0^2} \tag{8.1.49}$$

having the length dimension.

Substituting (8.1.49) into (8.1.48) yields

$$\frac{D}{L} \simeq \left(\frac{r_0}{V^{\frac{1}{3}}} \right)^{\frac{3}{5}} = \left(\frac{r_0^3}{V} \right)^{\frac{1}{5}} \tag{8.1.50}$$

Eq. (8.1.50) makes it possible to find the equilibrium aspect ratio. Platelike inclusions with small aspect ratios, $D/L \ll 1$, occur if

$$\frac{r_0}{V^{\frac{1}{3}}} \ll 1 \quad \text{or} \quad \frac{\gamma_s}{\lambda \varepsilon_0^2 V^{\frac{1}{3}}} \ll 1 \tag{8.1.51}$$

However, if

$$\frac{\gamma_s}{\lambda \varepsilon_0^2 V^{\frac{1}{3}}} = \frac{r_0}{V^{\frac{1}{3}}} \sim 1 \tag{8.1.52}$$

the relation (8.1.50) predicts an equiaxial shape of inclusions.

It thus follows from Eqs. (8.1.50) *and* (8.1.49) *that a platelike inclusion is formed if the specific surface energy, γ_S, is small, the mismatch, ε_0, between the inclusion and parent phase crystal lattices is large, and the inclusion volume, V, is also large.* This conclusion is equivalent to the requirement that the strain energy effect be far more manifest than the surface effect. A violation of this condition leads to the violation of the inequality (8.1.51) and hence the formation of equiaxial inclusions.

One more situation that deserves attention is the case of accidental degeneracy of the function $B(\mathbf{n})$ with respect to $\mathbf{n}$, such as the case when the function $B(\mathbf{n})$ has the same value for two crystallographically nonequivalent vectors $\mathbf{n}_0$ and $\mathbf{n}_1$. This is the case of two platelike precipitates with the habit planes normal to $\mathbf{n}_0$ and $\mathbf{n}_1$, respectively, having the same strain energies (8.1.14). This ambiguity may be removed if one takes into consideration that the strain energy (8.1.14) is in fact the zero aspect ratio limit attained when the interphase energy contribution may be neglected. If, however, the latter should be included the choice between the two habit plane variants (habit planes normal to $\mathbf{n}_0$ or $\mathbf{n}_1$) is straightforward; the plane providing the lower interphase energy is to be preferred.

The situation is somewhat more complex if $B(\mathbf{n}_1)$ is slightly larger than $B(\mathbf{n}_0)$ while the interphase energy of the habit plane perpendicular to $\mathbf{n}_1$ is lower. The elastic strain and interphase energies then compete with each other, the former term as given by Eq. (8.1.14) favoring the habit plane normal to $\mathbf{n}_0$ and the latter favoring the plane normal to $\mathbf{n}_1$. In actual systems the choice of the habit will be determined by relative contributions from the strain and interphase energies to the total inclusion free energy.

As shown above, the contribution from the interphase energy depends on the equilibrium aspect ratio being the lower the smaller the latter quantity becomes. The habit plane orientation will therefore be determined by the vector $\mathbf{n}_0$ corresponding to the absolute minimum of the strain energy (8.1.1.14) if the equilibrium aspect ratio is asymptotically small and the interphase energy contribution may be ignored. According to the inequality (8.1.51) this is the case of a large new phase particle volume which is typical for late stages of the decomposition. On the other hand, the aspect ratio may be such that the habit plane will be oriented perpendicular to $\mathbf{n}_1$ to minimize the interphase energy. This is the case of comparatively small new phase particles formed in early stages of the decomposition. Increase of a new phase particle volume during coarsening may thus result in a change of the habit plane orientation. We shall consider the conditions that may cause a change in the habit plane orientation during coarsening.

The total free energy of a platelike inclusion whose habit plane is perpendicular to the vector $\mathbf{n}$ includes the bulk chemical free energy, the bulk strain energy (8.1.14), and the interphase energy

$$E(\mathbf{n}) = [(f_1 - f_0) + \tfrac{1}{2}B(\mathbf{n})]V + 2\gamma_s(\mathbf{n})L^2 \tag{8.1.53}$$

where f_1 and f_0 are the specific chemical free energies of the new and parent phases, respectively, $\gamma_s(\mathbf{n})$ is the specific interphase energy coefficient related to the boundary plane normal to $\mathbf{n}$, $2L^2$ is the interphase area, L the size of the plate in the habit plane.

The change of the habit plane orientation from $\mathbf{n}_0$ to $\mathbf{n}_1$ results in the free energy change

$$\Delta E = \tfrac{1}{2}\Delta B(\mathbf{n}_1)V + 2\Delta\gamma_S L^2 \tag{8.1.54}$$

where

$$\Delta B(\mathbf{n}_1) = B(\mathbf{n}_1) - \min B(\mathbf{n}) = B(\mathbf{n}_1) - B(\mathbf{n}_0)$$
$$\Delta\gamma_s = \gamma_s(\mathbf{n}_1) - \gamma_s(\mathbf{n}_0)$$

By definition,

$$\Delta B(\mathbf{n}_1) > 0 \quad \text{and} \quad \Delta\gamma_s < 0 \tag{8.1.55}$$

The habit plane orientation normal to the vector $\mathbf{n}_1$ which does not correspond to the minimum of the strain energy, $\frac{1}{2}B(\mathbf{n}_1)V$, will be favored if

$$\Delta E = \tfrac{1}{2}\Delta B(\mathbf{n}_1)V + 2\Delta\gamma_s L^2 \leqslant 0 \tag{8.1.56}$$

Since $V = DL^2$, the inequality (8.1.56) is reduced to

$$\frac{1}{2}\Delta B(\mathbf{n}_1)D + 2\Delta\gamma_s \leqslant 0 \quad \text{or} \quad D \leqslant \frac{4|\Delta\gamma_s|}{\Delta B(\mathbf{n}_1)} \tag{8.1.57}$$

In other words, the lower interphase energy habit will be preferred for thin inclusions formed in early stages of the decomposition. Substitution of the relation (8.1.46) into (8.1.57) yields the condition

$$\frac{\Delta B(\mathbf{n}_1)}{4|\Delta\gamma_s|}\left(\frac{\gamma_s(\mathbf{n}_1)}{\lambda\varepsilon_0^2}\right)^{\frac{2}{3}} V^{\frac{1}{3}} \leqslant 1 \tag{8.1.58}$$

for the habit plane to be normal to $\mathbf{n}_1$ rather than $\mathbf{n}_0$. The inequality (8.1.58) shows that we may in fact expect the lower interphase energy habit in the beginning of the transformation when the precipitate volume V is not large enough for the condition (8.1.58) to be violated.

Finally, it should be remembered that the parent phase plane of the highest symmetry may be expected to have the lowest interphase energy since atomic adjustment of the different crystal lattices along this plane does not require the formation of ledges. On the other hand, a coherent adjustment of crystal lattices along an irrational habit plane implies that ledges will form which in turn generate local displacements and thus interphase energy.

Therefore in certain cases the habit planes of precipitate particles formed early in the transformation process may be expected to coincide with high-symmetry planes. Later in the process the habit plane may change its orientation to minimize the strain energy.

8.2. ELLIPSOIDAL INCLUSION IN ANISOTROPIC PARENT PHASE: HOMOGENEOUS MODULUS CASE

The equation for the elastic strain energy of an ellipsoidal coherent inclusion having the same modulus as the parent phase follows directly from the general case (8.1.1). To obtain it, we must substitute the Fourier transform of the ellipsoidal shape function (8.1.8) into (8.1.1). The result is

$$E_{\text{ellips}} = \frac{1}{2} V^2 \iiint_{-\infty}^{\infty} \frac{d^3k}{(2\pi)^3} B(\mathbf{n}) \left[3 \frac{\sin \phi(\mathbf{k}) - \phi(\mathbf{k}) \cos \phi(\mathbf{k})}{[\phi(\mathbf{k})]^3} \right]^2 \tag{8.2.1}$$

where

$$\mathbf{n} = \frac{\mathbf{k}}{k} \quad \text{and} \quad \phi(\mathbf{k}) = \sqrt{(\mathbf{k}\hat{\mathbf{L}}\mathbf{k})} \tag{8.2.2}$$

The matrix $\hat{\mathbf{L}}$ is determined by Eq. (8.1.9) which describes the shape of the ellipsoidal inclusion (the eigenvalues of the Hermitian matrix $\hat{\mathbf{L}}$ are the squared lengths of the ellipsoid semiaxes; the eigenvectors are the directions of the principal semiaxes).

The Fourier transform of the elastic displacements $\mathbf{u}(\mathbf{r})$ caused by a coherent ellipsoidal inclusion is given by Eq. (8.1.5)

$$\mathbf{v}(\mathbf{k}) = -i \frac{1}{k} \hat{\mathbf{\Omega}}(\mathbf{n}) \hat{\sigma}^0 \mathbf{n} \theta(\mathbf{k}) \tag{8.2.3}$$

where $\theta(\mathbf{k})$ is given by (8.1.8).

The back Fourier transform of Eq. (8.2.3) yields the displacement field

$$\mathbf{u}(\mathbf{r}) = -i \iiint_{-\infty}^{\infty} \frac{d^3k}{(2\pi)^3} \frac{1}{k} \hat{\mathbf{\Omega}}(\mathbf{n}) \hat{\sigma}^0 \mathbf{n} \theta(\mathbf{k}) e^{i\mathbf{k}\mathbf{r}} \tag{8.2.4a}$$

or, in the suffix form,

$$u_i(\mathbf{r}) = -i \iiint_{-\infty}^{\infty} \frac{d^3k}{(2\pi)^3} \frac{1}{k} \Omega_{ij}(\mathbf{n}) \sigma^0_{jk} n_k \theta(\mathbf{k}) e^{i\mathbf{k}\mathbf{r}} \tag{8.2.4b}$$

The curvilinear distortion $u_{ij}(\mathbf{r}) = \partial u_i(\mathbf{r})/\partial r_j$ can be calculated by taking the coordinate derivative of (8.2.4b):

$$\begin{aligned} u_{ij}(\mathbf{r}) &= -i \frac{\partial}{\partial r_j} \int \frac{d^3k}{(2\pi)^3} \frac{1}{k} \Omega_{ik}(\mathbf{n}) \sigma^0_{kl} n_l \theta(\mathbf{k}) e^{i\mathbf{k}\mathbf{r}} \\ &= \int \frac{d^3k}{(2\pi)^3} n_j \Omega_{ik}(\mathbf{n}) \sigma^0_{kl} n_l \theta(\mathbf{k}) e^{i\mathbf{k}\mathbf{r}} \end{aligned} \tag{8.2.5a}$$

or, in the symbolic form,

$$\hat{\mathbf{u}}(\mathbf{r})=\int \frac{d^3k}{(2\pi)^3}\,\hat{\mathbf{\Omega}}(\mathbf{n})\hat{\sigma}^0\mathbf{n}*\mathbf{n}\theta(\mathbf{k})e^{i\mathbf{kr}} \tag{8.2.5b}$$

where $*$ means the dyadic multiplication of vectors. Substitution of the Fourier transform of the ellipsoid shape function (8.1.8) into (8.2.5a) yields

$$u_{ij}(\mathbf{r})=V\iiint_{-\infty}^{\infty}\frac{d^3k}{(2\pi)^3}\,n_j\Omega_{ik}(\mathbf{n})\sigma^0_{kl}n_l\left[3\,\frac{\sin\phi(\mathbf{k})-\phi(\mathbf{k})\cos\phi(\mathbf{k})}{(\phi(\mathbf{k}))^3}\right]e^{i\mathbf{kr}} \tag{8.2.6}$$

We must first consider the simplest case of a spherical inclusion of the radius R.

By definition of the tensor L_{ij} [see comments to Eq. (8.1.9)], we have for a spherical inclusion

$$L_{ij}=R^2\delta_{ij} \tag{8.2.7}$$

Substituting (8.2.7) into (8.2.2) yields

$$\phi(\mathbf{k})=\sqrt{k_iL_{ij}k_j}=\sqrt{k_iR^2\delta_{ij}k_j}=Rk \tag{8.2.8}$$

With this in mind we obtain from (8.2.6)

$$u_{ij}(\mathbf{r})=V\iiint_{-\infty}^{\infty}\frac{d^3k}{(2\pi)^3}\,n_j\Omega_{ik}(\mathbf{n})\sigma^0_{kl}n_l\left[3\,\frac{\sin kR-kR\cos kR}{(kR)^3}\right]\cos kr \tag{8.2.9}$$

The exponential $\exp(i\mathbf{kr})$ in (8.2.9) is replaced by $\cos\mathbf{kr}$ since the remaining part of the integrand is an even function of $\mathbf{k}$.

The volume element in the $\mathbf{k}$-space is

$$d^3k=k^2dk\,dO_{\mathbf{n}} \tag{8.2.10}$$

where $dO_{\mathbf{n}}$ is the solid angle element in the direction $\mathbf{n}=\mathbf{k}/k$.

Using Eq. (8.2.10) and carrying out the integration in (8.2.9) followed by the integration over the solid angle $O_{\mathbf{n}}$, we obtain

$$u_{ij}(\mathbf{r})=\oint\frac{dO_{\mathbf{n}}}{(2\pi)^3}\,\tilde{u}_{ij}(\mathbf{n})V\int_0^{\infty}k^2dk\left[3\,\frac{\sin kR-kR\cos kR}{(kR)^3}\right]\cos k(\mathbf{nr}) \tag{8.2.11a}$$

where

$$\tilde{u}_{ij}(\mathbf{n})=n_j\Omega_{ik}(\mathbf{n})\sigma^0_{kl}n_l$$

$$V=\frac{4\pi}{3}R^3 \tag{8.2.11b}$$

The direct evaluation of the integral over $\mathbf{k}$ gives

$$I(r)=\frac{4\pi}{3}R^3\int_0^{\infty}k^2dk\left[3\,\frac{\sin kR-kR\cos kR}{(kR)^3}\right]\cos k(\mathbf{nr})=2\pi^2 \tag{8.2.12}$$

if

$$\mathbf{nr} \leqslant R \tag{8.2.13}$$

When the point $\mathbf{r}$ lies within the sphere of the radius R, the inequality (8.2.13) is certainly fulfilled. Substituting Eq. (8.2.12) into (8.2.11a) gives

$$u_{ij}(\mathbf{r}) = \oint \frac{dO_{\mathbf{n}}}{4\pi} \tilde{u}_{ij}(\mathbf{n}) = \langle n_j \Omega_{ik}(\mathbf{n}) \sigma^0_{kl} n_l \rangle_{\mathbf{n}} = \text{const.} \tag{8.2.14}$$

where the symbol $\langle \cdots \rangle_{\mathbf{n}}$ implies averaging over all directions of the unit vector $\mathbf{n}$: $\langle \cdots \rangle_{\mathbf{n}} = 1/4\pi \oint dO_{\mathbf{n}}(\cdots)$.

Eq. (8.2.14) thus shows that curvilinear distortion inside a constrained coherent spherical inclusion is homogeneous and determined by the constant (8.2.14).

The latter conclusion is easy to extend to the more complex case of an ellipsoidal inclusion in an anisotropic medium. To do this, we should carry out homogeneous deformation of space using the Hermitian deformation operator $\hat{\mathbf{A}}$:

$$\mathbf{r}' = \hat{\mathbf{A}}\mathbf{r} \tag{8.2.15}$$

which converts the ellipsoidal inclusion into a spherical one of the same volume. The latter requires that the determinant of the matrix $\hat{\mathbf{A}}$ be equal to unity

$$\det \|\hat{\mathbf{A}}\| = 1 \tag{8.2.16}$$

The deformation will change the function $\tilde{u}_{ij}(\mathbf{n})$ in (8.2.11b) which, however, will still depend on the direction $\mathbf{n}$ of the wave vector $\mathbf{k}$ rather than its absolute value. In this case the calculation of the distortion tensor, $u_{ij}(\mathbf{r})$, within an ellipsoidal inclusion is reduced to the calculation of the distortion within a spherical inclusion considered above. Naturally, we must obtain the same result, the distortion tensor within an ellipsoidal inclusion should be constant.

A detailed calculation of the distortion within an inclusion is given below. The equation for the surface of an ellipsoidal inclusion is given by (8.1.9):

$$L_{ij}^{-1} r_i r_j = 1 \tag{8.2.17}$$

The homogeneous "deformation" of space (8.2.15) makes it possible to rewrite Eq. (8.2.17) in the form

$$(\hat{\mathbf{A}}^{+} \hat{\mathbf{L}}^{-1} \hat{\mathbf{A}})_{ij} r'_i r'_j = 1 \tag{8.2.18}$$

where

$$\hat{\mathbf{A}}^{+} = \hat{\mathbf{A}} \tag{8.2.19}$$

according to the definition (8.2.15).

An ellipsoidal inclusion is transformed into a spherical one if

$$\hat{\mathbf{A}} \hat{\mathbf{L}}^{-1} \hat{\mathbf{A}} = \frac{1}{R^2} \hat{\mathbf{I}} \tag{8.2.20}$$

where R is the radius of the sphere having the same volume. The condition for the latter reads

$$\hat{\mathbf{A}}=\frac{1}{R}\hat{\mathbf{L}}^{\frac{1}{2}} \tag{8.2.21}$$

(according to their definitions, the matrices $\hat{\mathbf{L}}$, $\hat{\mathbf{L}}^{-1}$, and $\hat{\mathbf{L}}^{\frac{1}{2}}$ commute with each other). It follows from Eq. (8.2.21) that

$$\hat{\mathbf{A}}^{-1}=R\hat{\mathbf{L}}^{-\frac{1}{2}} \tag{8.2.22}$$

Let us change of variables in the integral (8.2.6):

$$\mathbf{k}=\hat{\mathbf{A}}^{-1}\mathbf{k}'$$

Taking into consideration Eq. (8.2.22), we can rewrite the latter relation as

$$\mathbf{k}=R\hat{\mathbf{L}}^{-\frac{1}{2}}\mathbf{k}' \tag{8.2.23}$$

Substituting Eq. (8.2.23) into (8.2.2) and making use of the relation $\hat{\mathbf{L}}^{-\frac{1}{2}}\hat{\mathbf{L}}\hat{\mathbf{L}}^{-\frac{1}{2}}=\hat{\mathbf{I}}$, we have

$$\phi(\mathbf{k})=\sqrt{(\mathbf{k}\hat{\mathbf{L}}\mathbf{k})}=\sqrt{\mathbf{k}'R^2\hat{\mathbf{L}}^{-\frac{1}{2}}\hat{\mathbf{L}}\hat{\mathbf{L}}^{-\frac{1}{2}}\mathbf{k}'}=Rk'$$

Therefore the change of variables (8.2.23) in Eq. (8.2.6) with $\tilde{u}_{ij}(\mathbf{n})$ given by Eq. (8.2.11b) leads to

$$u_{ij}(\mathbf{r})=\det\|\hat{\mathbf{A}}^{-1}\|\int\frac{d^3k'}{(2\pi)^3}\tilde{u}_{ij}(\mathbf{n}(\mathbf{n}'))\frac{4\pi}{3}R^3\left[3\frac{\sin k'R-k'R\cos k'R}{(k'R)^3}\right]e^{i\mathbf{k}'\mathbf{r}'} \tag{8.2.25}$$

where $\mathbf{r}'=\hat{\mathbf{A}}\mathbf{r}$, $\mathbf{kr}=\mathbf{k}'\mathbf{r}'$, $\mathbf{n}'=\mathbf{k}'/k'$

$$\mathbf{n}=\mathbf{n}(\mathbf{n}')=\frac{\mathbf{k}}{k}=\frac{R\hat{\mathbf{L}}^{-\frac{1}{2}}\mathbf{k}'}{|R\hat{\mathbf{L}}^{-\frac{1}{2}}\mathbf{k}'|}=\frac{\mathbf{L}^{-\frac{1}{2}}\mathbf{n}'}{\sqrt{(\mathbf{n}'\mathbf{L}^{-1}\mathbf{n}')}} \tag{8.2.26}$$

Since the space deformation $\hat{\mathbf{A}}$ does not affect the inclusion volume,

$$\det\|\hat{\mathbf{A}}^{-1}\|=1$$

The integral (8.2.25) coincides with (8.2.9) and may therefore be reduced to (8.2.14)

$$u_{ij}(\mathbf{r})=\oint\frac{dO_{\mathbf{n}'}}{4\pi}\tilde{u}_{ij}(\mathbf{n}(\mathbf{n}'))=\langle\tilde{u}_{ij}(\mathbf{n}(\mathbf{n}'))\rangle_{\mathbf{n}'}=u^*_{ij}=\text{const.} \tag{8.2.27}$$

if the vector $\mathbf{r}'$ is within the sphere of the radius R or, which is the same, the point $\mathbf{r}$ lies within the ellipsoid.

The symbol $\langle\cdots\rangle_{\mathbf{n}'}$ implies averaging over all directions of the vector $\mathbf{n}'$ (integration over the unit sphere surface).

Substituting Eq. (8.2.11b) into (8.2.27) yields

$$u^*_{ij}=\oint\frac{dO_{\mathbf{n}'}}{4\pi}[n_j\Omega_{ik}(\mathbf{n})\sigma^0_{kl}n_l]_{\mathbf{n}=\mathbf{n}(\mathbf{n}')}=S_{ijkl}\sigma^0_{kl} \tag{8.2.28}$$

where

$$S_{ijkl}=\oint \frac{dO_{\mathbf{n}'}}{4\pi}\left[n_j\Omega_{ik}(\mathbf{n})n_l\right]_{\mathbf{n}=\mathbf{n}(\mathbf{n}')}=\langle n_j\Omega_{ik}(\mathbf{n})n_l\rangle_{\mathbf{n}'} \tag{8.2.29}$$

Eq. (8.2.27) proves the conclusion on homogeneous distortion inside an ellipsoidal inclusion made above. This result has been obtained by several authors using another formalism (Eshelby's method) (144, 145).

A similar technique may be applied to derive the expression for the strain energy of a coherent ellipsoidal inclusion in the form of the average over all directions $\mathbf{n}$. To do this, we may introduce a change of variables (8.2.23) in Eq. (8.2.1). With (8.2.8) we obtain

$$E_{\text{ellips}}=\frac{1}{2}V^2\int\frac{d^3k'}{(2\pi)^3}B(\mathbf{n}(\mathbf{n}'))\left[3\frac{\sin k'R-k'R\cos k'R}{(k'R)^3}\right]^2 \tag{8.2.30}$$

The integration over k' followed by integration over the solid angle $O_{\mathbf{n}'}$ yields

$$E_{\text{ellips}}=\tfrac{1}{2}V^2\oint dO_{\mathbf{n}'}B(\mathbf{n}(\mathbf{n}'))\int_0^\infty\frac{(k')^2dk'}{(2\pi)^3}\left[3\frac{\sin k'R-k'R\cos k'R}{(k'R)^3}\right]^2 \tag{8.2.31}$$

For a spherical particle of the radius R Eq. (8.1.11) gives

$$V\equiv\int\frac{d^3k}{(2\pi)^3}|\theta(\mathbf{k})|^2=\oint dO_{\mathbf{n}}\int_0^\infty\frac{k^2dk}{(2\pi)^3}V\left|3\frac{\sin kR-kR\cos kR}{(kR)^3}\right|^2$$

$$=4\pi V\int_0^\infty\frac{k^2dk}{(2\pi)^3}\left|3\frac{\sin kR-kR\cos kR}{(kR)^3}\right|^2 \tag{8.2.32}$$

It follows from this equation that

$$\int_0^\infty\frac{k^2dk}{(2\pi)^3}\left|3\frac{\sin kR-kR\cos kR}{(kR)^3}\right|^2=\frac{1}{4\pi} \tag{8.2.33}$$

Substituting (8.2.33) into (8.2.31) yields

$$E_{\text{ellips}}=\frac{1}{2}V^2\oint\frac{dO_{\mathbf{n}'}}{4\pi}B(\mathbf{n}(\mathbf{n}'))=\frac{1}{2}V^2\langle B(\mathbf{n}(\mathbf{n}'))\rangle_{\mathbf{n}'} \tag{8.2.34}$$

The calculations carried out in this section show that ellipsoidal inclusions differ from inclusions of all other shapes in that they allow to obtain a comparatively simple solution because an ellipsoid can be reduced to a sphere of the same volume by homogeneous space deformation which is something that cannot be done with inclusions of other shapes.

8.3. LIMIT TRANSITION TO ESHELBY'S THEORY OF ELLIPSOIDAL INCLUSIONS IN ISOTROPIC MATRICES

The theory of multiphase coherent mixtures by Khachaturyan and Shatalov (136, 148) described in the preceding sections includes all the results of Eshelby's theory (132, 133) of ellipsoidal inclusions in isotropic matrices. To pass from general theory to the particular case, Eqs. (8.1.1) and (8.1.5) for the strain energy and elastic displacements should be simplified as follows (150). A single ellipsoidal inclusion within an infinite matrix must be considered under the assumption of isotropic homogeneous elastic moduli. This was done in part in Section 8.2 where inclusions of ellipsoidal shape were used as a limiting case. This approach led to Eq. (8.2.28) for the homogeneous distortion insidê an inclusion and Eq. (8.2.34) for the strain energy. Eqs. (8.2.28) and (8.2.34) are valid for an anisotropic medium. To simplify the solutions further, we must apply the isotropic medium approximation.

The elastic modulus tensor of an isotropic medium has the following non-vanishing components

$$\lambda_{1111}=\lambda_{2222}=\lambda_{3333}=c_{11}=2\mu\frac{1-\sigma_1}{1-2\sigma_1}$$

$$\lambda_{1122}=\lambda_{1133}=\lambda_{2233}=c_{12}=2\mu\frac{\sigma_1}{1-2\sigma_1}$$

$$\lambda_{1212}=\lambda_{1313}=\lambda_{2323}=c_{44}=\mu \tag{8.3.1}$$

where μ is the shear modulus, σ_1 the Poisson's ratio, c_{11}, c_{12}, c_{44} are Voigt's designations of elastic constants.

The invariant form of the elastic modulus tensor whose components are given by (8.3.1) is

$$\lambda_{ijkl}=2\mu\frac{\sigma_1}{1-2\sigma_1}\delta_{ij}\delta_{kl}+\mu(\delta_{ik}\delta_{jl}+\delta_{il}\delta_{jk}) \tag{8.3.2}$$

According to Eqs. (8.1.3) and (8.3.2) the tensor $\hat{\mathbf{\Omega}}^{-1}(\mathbf{n})$ may be written for the isotropic case

$$\Omega_{ij}^{-1}(\mathbf{n})=\lambda_{iklj}n_k n_l=2\mu\frac{\sigma_1}{1-2\sigma_1}n_i n_j+\mu(n_i n_j+\delta_{ij})$$

or

$$\Omega_{ij}^{-1}(\mathbf{n})=\frac{\mu}{1-2\sigma_1}n_i n_j+\mu\delta_{ij} \tag{8.3.3}$$

The inverse tensor $\Omega_{ij}(\mathbf{n})$ may be determined from (7.2.36b). It is given as

$$\Omega_{ij}(\mathbf{n})=\frac{\delta_{ij}}{\mu}-\frac{1}{2\mu(1-\sigma_1)}n_i n_j \tag{8.3.4}$$

Substituting the tensor (8.3.4) into Eq. (8.2.29) yields

$$S_{ijkl}=\langle n_j\Omega_{ik}(\mathbf{n})n_l\rangle_{\mathbf{n}'}=\frac{\delta_{ik}}{\mu}\langle n_j n_l\rangle_{\mathbf{n}'}-\frac{1}{2\mu(1-\sigma_1)}\langle n_i n_j n_k n_l\rangle_{\mathbf{n}'} \tag{8.3.5}$$

where $\langle\cdots\rangle_{\mathbf{n}'}$ denotes averaging over all directions $\mathbf{n}'$:

$$\langle\cdots\rangle_{\mathbf{n}'}=\oint\frac{dO_{\mathbf{n}'}}{4\pi}(\cdots) \tag{8.3.6}$$

Without the loss of generality the directions of the Cartesian coordinate axes may be chosen along the principal axes of the ellipsoidal inclusion (along the principal directions of the tensors L_{ij} and L_{ij}^{-1}).

By definition of $\hat{\mathbf{L}}^{-1}$ [see Eq. (8.1.9)], we then have

$$\hat{\mathbf{L}}^{-1}\mathbf{e}_1=\frac{1}{a_1^2}\mathbf{e}_1,\quad \hat{\mathbf{L}}^{-1}\mathbf{e}_2=\frac{1}{a_2^2}\mathbf{e}_2,\quad \hat{\mathbf{L}}^{-1}\mathbf{e}_3=\frac{1}{a_3^2}\mathbf{e}_3 \tag{8.3.7}$$

where $\mathbf{e}_1, \mathbf{e}_2, \mathbf{e}_3$ are the unit vectors along the Cartesian axes that are the principal directions of the ellipsoidal inclusion and a_1, a_2, a_3 are the ellipsoid semiaxes. Similar relations are valid for $\hat{\mathbf{L}}^{-\frac{1}{2}}$:

$$\hat{\mathbf{L}}^{-\frac{1}{2}}\mathbf{e}_1=\frac{1}{a_1}\mathbf{e}_1,\quad \hat{\mathbf{L}}^{-\frac{1}{2}}\mathbf{e}_2=\frac{1}{a_2}\mathbf{e}_2,\quad \hat{\mathbf{L}}^{-\frac{1}{2}}\mathbf{e}_3=\frac{1}{a_3}\mathbf{e}_3 \tag{8.3.8}$$

The invariant representation of the vectors $\mathbf{n}$ and $\mathbf{n}'$ written in Cartesian coordinates are

$$\begin{aligned}\mathbf{n}&=(n_1, n_2, n_3)=n_1\mathbf{e}_1+n_2\mathbf{e}_2+n_3\mathbf{e}_3\\ \mathbf{n}'&=(n_1', n_2', n_3')=n_1'\mathbf{e}_1+n_2'\mathbf{e}_2+n_3'\mathbf{e}_3\end{aligned} \tag{8.3.9}$$

where n_1, n_2, n_3 and n_1', n_2', n_3' are the coordinates of the vectors $\mathbf{n}$ and $\mathbf{n}'$, respectively.

Substitution of Eqs. (8.3.9) into (8.2.26) and the use of Eqs. (8.3.7) and (8.3.8) lead to

$$\mathbf{n}=(n_1, n_2, n_3)=\frac{n_1'}{a_1 g(\mathbf{n}')}\mathbf{e}_1+\frac{n_2'}{a_2 g(\mathbf{n}')}\mathbf{e}_2+\frac{n_3'}{a_3 g(\mathbf{n}')}\mathbf{e}_3$$

or

$$n_1=\frac{n_1'}{a_1 g(\mathbf{n}')},\quad n_2=\frac{n_2'}{a_2 g(\mathbf{n}')},\quad n_3=\frac{n_3'}{a_3 g(\mathbf{n}')} \tag{8.3.10}$$

where

$$g(\mathbf{n}')=\sqrt{\left(\frac{n_1'}{a_1}\right)^2+\left(\frac{n_2'}{a_2}\right)^2+\left(\frac{n_3'}{a_3}\right)^2} \tag{8.3.11}$$

Using the components of the vector $\mathbf{n}$ given by (8.3.10) and the definition (8.3.6), we obtain

$$\langle n_i n_j\rangle_{\mathbf{n}'}=\oint\frac{dO_{\mathbf{n}'}}{4\pi}\frac{n_i'}{a_i}\frac{n_j'}{a_j}\frac{1}{g^2(\mathbf{n}')} \tag{8.3.12a}$$

$$\langle n_i n_j n_k n_l\rangle_{\mathbf{n}'}=\oint\frac{dO_{\mathbf{n}'}}{4\pi}\frac{n_i'}{a_i}\frac{n_j'}{a_j}\frac{n_k'}{a_k}\frac{n_l'}{a_l}\frac{1}{g^4(\mathbf{n}')} \tag{8.3.12b}$$

where $i, j, k, l=1, 2, 3$, and the integration is over the surface of the sphere of unit radius.

It is easy to see that the integral (8.3.12a) vanishes if $i\neq j$. The same is true of the integral (8.3.12b) if at least one of the indexes i, j, k, l differs from the remaining three indexes. With this in mind, one has only to find three types of nonzero average values:

$$\langle n_i^2\rangle,\quad \langle n_i^4\rangle,\quad \text{and}\quad \langle n_i^2 n_j^2\rangle \tag{8.3.13}$$

Since $\mathbf{n}$ is a unit vector, that is, $\sum_{i=1}^{i=3} n_i^2=1$, the average values (8.3.13) are not independent and are related by

$$\langle n_1^2\rangle_{\mathbf{n}'}+\langle n_2^2\rangle_{\mathbf{n}'}+\langle n_3^2\rangle_{\mathbf{n}'}=\sum_{i=1}^{3}\langle n_i^2\rangle_{\mathbf{n}'}=\left\langle \sum_{i=1}^{3} n_i^2\right\rangle_{\mathbf{n}'}=1 \tag{8.3.14a}$$

$$\sum_{j=1}^{3}\langle n_i^2 n_j^2\rangle_{\mathbf{n}'}=\left\langle n_i^2\sum_{j=1}^{3} n_j^2\right\rangle_{\mathbf{n}'}=\langle n_i^2\rangle_{\mathbf{n}'} \tag{8.3.14b}$$

and

$$\left\langle\left(\sum_{i=1}^{3} n_i^2\right)^2\right\rangle_{\mathbf{n}'}=\left\langle\sum_{i,j} n_i^2 n_j^2\right\rangle_{\mathbf{n}'}=\sum_{i,j}\langle n_i^2 n_j^2\rangle_{\mathbf{n}'}=1$$

or

$$\sum_{i=1}^{3}\langle n_i^4\rangle_{\mathbf{n}'}+2(\langle n_1^2 n_2^2\rangle_{\mathbf{n}'}+\langle n_1^2 n_3^2\rangle_{\mathbf{n}'}+\langle n_2^2 n_3^2\rangle_{\mathbf{n}'})=1 \tag{8.3.14c}$$

where, according to the definition (8.3.12b),

$$\langle n_i^2 n_j^2\rangle_{\mathbf{n}'}=\oint\frac{dO_{\mathbf{n}'}}{4\pi}\left(\frac{n_i'}{a_i}\right)^2\left(\frac{n_j'}{a_j}\right)^2\frac{1}{\left[\left(\dfrac{n_1'}{a_1}\right)^2+\left(\dfrac{n_2'}{a_2}\right)^2+\left(\dfrac{n_3'}{a_3}\right)^2\right]^2} \tag{8.3.14d}$$

The constant eigenstrain ε_{ij}^* within the ellipsoidal inclusion may be found by the symmetrization of Eq. (8.2.28), which gives

$$\begin{aligned}\varepsilon_{ij}{}^*&=\tfrac{1}{2}(u_{ij}^*+u_{ji}^*)=\tfrac{1}{2}(S_{ijkl}+S_{jikl})\sigma_{kl}^0\\&=\tfrac{1}{2}(S_{ijkl}+S_{jikl})\lambda_{klmn}\varepsilon_{mn}^0\end{aligned} \tag{8.3.15}$$

or

$$\varepsilon_{ij}^*=s_{ijmn}\varepsilon_{mn}^0 \tag{8.3.16}$$

where

$$s_{ijmn}=\tfrac{1}{2}(S_{ijkl}+S_{jikl})\lambda_{klmn} \tag{8.3.17}$$

is Eshelby's tensor which relates the eigenstrain ε_{ij}^* and the stress-free transformation strain ε_{ij}^0.

Substituting Eq. (8.3.2) into (8.3.15) yields

$$s_{ijmn}=\frac{2\mu\sigma_1}{1-2\sigma_i}\frac{S_{ijkk}+S_{jikk}}{2}\delta_{mn}+2\mu\frac{S_{ijmn}+S_{jimn}+S_{ijnm}+S_{jinm}}{4} \tag{8.3.18}$$

The nonzero components of the matrix s_{ijmn} are found from Eqs. (8.3.5) and (8.3.17) to be

$$s_{iiii}=\frac{2-\sigma_1}{1-\sigma_1}\langle n_i^2\rangle_{\mathbf{n}'}-\frac{1}{1-\sigma_1}\langle n_i^4\rangle_{\mathbf{n}'}$$

$$s_{iijj}=\frac{\sigma_1}{1-\sigma_1}\langle n_i^2\rangle_{\mathbf{n}'}-\frac{1}{1-\sigma_1}\langle n_i^2 n_j^2\rangle_{\mathbf{n}'},\quad i\neq j$$

$$s_{ijij}=\frac{1}{2}(\langle n_i^2\rangle_{\mathbf{n}'}+\langle n_j^2\rangle_{\mathbf{n}'})-\frac{1}{1-\sigma_1}\langle n_i^2 n_j^2\rangle_{\mathbf{n}'},\quad i\neq j \tag{8.3.19}$$

Let us introduce the definition

$$\langle n_i^2\rangle_{\mathbf{n}'}=\oint\frac{dO_{\mathbf{n}'}}{4\pi}\left(\frac{n_i'}{a_i}\right)^2\left[\left(\frac{n_1'}{a_1}\right)^2+\left(\frac{n_2'}{a_2}\right)^2+\left(\frac{n_3'}{a_3}\right)^2\right]^{-1}=\frac{1}{4\pi}I_i \tag{8.3.20}$$

where the indexes i, j, k form a cyclic sequence: (i, j, k), (k, i, j), (j, k, i).

The coefficients, I_i, defined by (8.3.20) coincide with the corresponding coefficients of the original Eshelby's paper (132, 133). This will be shown explicitly for the coefficient

$$\langle n_3^2\rangle_{\mathbf{n}'}=\frac{1}{4\pi}I_3=\oint\frac{dO_{\mathbf{n}'}}{4\pi}\left(\frac{n_3'}{a_3}\right)^2\left[\left(\frac{n_1'}{a_1}\right)^2+\left(\frac{n_2'}{a_2}\right)^2+\left(\frac{n_3'}{a_3}\right)^2\right]^{-1} \tag{8.3.21}$$

In the spherical coordinate system we have

$$\mathbf{n}'=(\cos\phi\sin\theta,\ \sin\phi\sin\theta,\ \cos\theta) \tag{8.3.22}$$

and

$$dO_{\mathbf{n}'}=\sin\theta d\theta d\phi \tag{8.3.23}$$

Substituting Eqs. (8.3.22) and (8.3.23) into (8.3.21) yields

$$\langle n_3^2\rangle_{\mathbf{n}'}=\int_0^{\pi}\tfrac{1}{2}\sin\theta d\theta\int_0^{2\pi}\frac{d\phi}{2\pi}\frac{\cos^2\theta}{a_3^2}\left[\frac{\cos^2\phi\sin^2\theta}{a_1^2}+\frac{\sin^2\phi\sin^2\theta}{a_2^2}+\frac{\cos^2\theta}{a_3^2}\right]^{-1}$$

$$=\int_0^{\pi}\frac{\sin\theta d\theta}{2}\frac{\cos^2\theta}{a_3^2}\Phi(\theta) \tag{8.3.24}$$

where

$$\Phi(\theta)=\int_0^{2\pi}\frac{d\phi}{2\pi}\left[\frac{\cos^2\phi\sin^2\theta}{a_1^2}+\frac{\sin^2\phi\sin^2\theta}{a_2^2}+\frac{\cos^2\theta}{a_3^2}\right]^{-1}$$

$$=\int_0^{2\pi}\frac{d\phi}{2\pi}\left[\frac{\cos^2\theta}{a_3^2}+\frac{\sin^2\theta}{2}\left(\frac{1}{a_3^2}+\frac{1}{a_1^2}\right)+\frac{\sin^2\theta}{2}\left(\frac{1}{a_1^2}-\frac{1}{a_2^2}\right)\cos 2\phi\right]^{-1}$$

$$=\left[\left(\frac{\cos^2\theta}{a_3^2}+\frac{\sin^2\theta}{a_2^2}\right)\left(\frac{\cos^2\theta}{a_3^2}+\frac{\sin^2\theta}{a_1^2}\right)\right]^{-\frac{1}{2}} \tag{8.3.25}$$

[see Eq. 3.645 in (151)].

Substituting Eq. (8.3.25) into (8.3.24) gives

$$\langle n_3^2\rangle_{n'} = \int_0^{\pi} \frac{\sin\theta\cos^2\theta}{2a_3^2}\,d\theta\left[\left(\frac{\cos^2\theta}{a_3^2}+\frac{\sin^2\theta}{a_2^2}\right)\left(\frac{\cos^2\theta}{a_3^2}+\frac{\sin^2\theta}{a_1^2}\right)\right]^{-\frac{1}{2}}$$

$$=\tfrac{1}{2}a_1a_2\int_0^{\pi}\sin\theta[(a_2^2+a_3^2tg^2\theta)(a_1^2+a_3^2tg^2\theta)]^{-\frac{1}{2}}d\theta \qquad (8.3.26)$$

Using the new variable,

$$u=a_3^2tg^2\theta$$

the integral (8.3.26) may be transformed into

$$\langle n_3^2\rangle_{n'}=\tfrac{1}{2}a_1a_2a_3\int_0^{\infty}\frac{du}{(a_3^2+u)\Delta(u)} \qquad (8.3.27)$$

where $\Delta(u)=(a_3^2+u)^{\frac{1}{2}}(a_2^2+u)^{\frac{1}{2}}(a_1^2+u)^{\frac{1}{2}}$

Comparison of Eqs. (8.3.21) and (8.3.27) gives

$$I_3=2\pi a_1a_2a_3\int_0^{\infty}\frac{du}{(a_3^2+u)\Delta(u)} \qquad (8.3.28)$$

Similar calculations for $\langle n_1^2\rangle_{n'}$ and $\langle n_2^2\rangle_{n'}$ show that the constants I_1, I_2, I_3 defined by Eq. (8.3.20) may be written in the general form

$$I_i=2\pi a_1a_2a_3\int_0^{\infty}\frac{du}{(a_i^2+u)\Delta(u)} \qquad (8.3.29)$$

where $i=1, 2, 3$. It is easy to see that these constants coincide with Eshelby's constants I_a, I_b, I_c (133).

The following relations may be obtained by differentiating (8.3.20) with respect to a_i and a_j and from the definitions (8.3.14d):

$$\langle n_i^2n_j^2\rangle_{n'}=\begin{cases}\langle n_i^2\rangle_{n'}+\tfrac{1}{2}a_i\dfrac{\partial}{\partial a_i}\langle n_i^2\rangle_{n'} & \text{if } i=j\\[2ex] \tfrac{1}{2}a_j\dfrac{\partial}{\partial a_j}\langle n_i^2\rangle_{n'} & \text{if } i\neq j\end{cases} \qquad (8.3.30)$$

Let us introduce new functions defined by

$$\frac{3a_j^2}{4\pi}I_{ij}=\begin{cases}\langle n_i^2\rangle_{n'}-2\langle n_i^2n_j^2\rangle_{n'} & \text{if } i\neq j\\ 3\langle n_i^2\rangle_{n'}-2\langle n_i^4\rangle_{n'} & \text{if } i=j\end{cases} \qquad (8.3.31)$$

and show that these functions coincide with Eshelby's functions of the type I_{ab} and I_{aa}.

Substituting (8.3.30) into the definitions (8.3.31) transforms them into

$$\frac{3a_j^2I_{ij}}{4\pi}=\langle n_i^2\rangle_{n'}-a_j\frac{\partial}{\partial a_j}\langle n_i^2\rangle_{n'} \qquad (8.3.32)$$

With (8.3.20), (8.3.32) becomes

$$\frac{3I_{ij}}{4\pi}=-\frac{\partial}{\partial a_j}\frac{\langle n_i^2\rangle_{\mathbf{n}'}}{a_j}=-\frac{1}{4\pi}\frac{\partial}{\partial a_j}\left(\frac{I_i}{a_j}\right) \tag{8.3.33}$$

Eq. (8.3.33) holds irrespective of whether $i=j$ or $i\neq j$. Using the representation (8.3.29) in Eq. (8.3.33), we obtain

$$3I_{ij}=-\frac{\partial}{\partial a_j}\left[2\pi a_i a_k\int_0^\infty\frac{du}{(a_i^2+u)^{\frac{3}{2}}(a_j^2+u)^{\frac{1}{2}}(a_k^2+u)^{\frac{1}{2}}}\right]$$

$$=2\pi a_i a_j a_k\int_0^\infty\frac{du}{(a_i^2+u)^{\frac{3}{2}}(a_j^2+u)^{\frac{3}{2}}(a_k^2+u)}$$

or

$$I_{ij}=\frac{2\pi}{3}a_1a_2a_3\int_0^\infty\frac{du}{(a_i^2+u)(a_j^2+u)\Delta(u)} \tag{8.3.34}$$

if $i\neq j$.

A similar treatment of Eqs. (8.3.29) and (8.3.33) gives for $i=j$

$$3I_{ii}=-\frac{\partial}{\partial a_i}\left[2\pi a_j a_k\int_0^\infty\frac{du}{(a_i^2+u)^{\frac{3}{2}}(a_j^2+u)^{\frac{1}{2}}(a_k^2+u)^{\frac{1}{2}}}\right]$$

$$=3\cdot 2\pi a_i a_j a_k\int_0^\infty\frac{du}{(a_i^2+u)^{\frac{5}{2}}(a_j^2+u)^{\frac{1}{2}}(a_k^2+u)^{\frac{1}{2}}}$$

or

$$I_{ii}=2\pi a_1a_2a_3\int_0^\infty\frac{du}{(a_i^2+u)^2\Delta(u)} \tag{8.3.35}$$

Comparison of the constants (8.3.34) and (8.3.35) with the constants of the types I_{ab} and I_{aa} from Eshelby's paper shows them to be fully identical to each other.

It now remains to demonstrate that the tensor s_{ijkl} which relates the eigenstrain within an ellipsoidal inclusion to the stress-free deformation strain ε_{ij}^0 [see Eq. (8.3.16)] coincides with the tensor derived by Eshelby.

Substituting Eqs. (8.3.31) and (8.3.20) into (8.3.19) yields

$$s_{iiii}=\frac{2-\sigma_1}{1-\sigma_1}\langle n_i^2\rangle_{\mathbf{n}'}-\frac{1}{1-\sigma_1}\langle n_i^4\rangle_{\mathbf{n}'}=\frac{1-2\sigma_1}{2(1-\sigma_1)}\langle n_i^2\rangle_{\mathbf{n}'}$$

$$+\frac{1}{2(1-\sigma_1)}(3\langle n_i^2\rangle_{\mathbf{n}'}-2\langle n_i^4\rangle_{\mathbf{n}'})=\frac{1-2\sigma_1}{8\pi(1-\sigma_1)}I_i+\frac{3a_i^2I_{ii}}{8\pi(1-\sigma_1)}$$

or

$$s_{iiii}=\frac{1-2\sigma_1}{8\pi(1-\sigma_1)}I_i+\frac{3a_i^2I_{ii}}{8\pi(1-\sigma_1)} \tag{8.3.36a}$$

$$s_{iijj} = \frac{\sigma_1}{1-\sigma_1}\langle n_i^2\rangle_{\mathbf{n}'} - \frac{1}{1-\sigma_1}\langle n_i^2 n_j^2\rangle_{\mathbf{n}'} = -\frac{1-2\sigma_1}{2(1-\sigma_1)}\langle n_i^2\rangle_{\mathbf{n}'}$$

$$+\frac{1}{2(1-\sigma_1)}(\langle n_i^2\rangle_{\mathbf{n}'} - 2\langle n_i^2 n_j^2\rangle_{\mathbf{n}'}) = -\frac{1-2\sigma_1}{8\pi(1-\sigma_1)}I_i + \frac{3a_j^2}{8\pi(1-\sigma_1)}I_{ij}$$

or

$$s_{iijj} = -\frac{1-2\sigma_1}{8\pi(1-\sigma_1)}I_i + \frac{3a_j^2}{8\pi(1-\sigma_1)}I_{ij} \tag{8.3.36b}$$

$$s_{ijij} = \frac{1}{2}(\langle n_i^2\rangle_{\mathbf{n}'} + \langle n_j^2\rangle_{\mathbf{n}'}) - \frac{1}{1-\sigma_1}\langle n_i^2 n_j^2\rangle_{\mathbf{n}'} = \frac{1}{2}\left[1 - \frac{1}{2(1-\sigma_1)}\right](\langle n_i^2\rangle_{\mathbf{n}'} + \langle n_j^2\rangle_{\mathbf{n}'})$$

$$-\frac{1}{4(1-\sigma_1)}(\langle n_i^2\rangle_{\mathbf{n}'} - 2\langle n_i^2 n_j^2\rangle_{\mathbf{n}'}) - \frac{1}{4(1-\sigma_1)}(\langle n_j^2\rangle_{\mathbf{n}'} - 2\langle n_i^2 n_j^2\rangle_{\mathbf{n}'})$$

$$= \frac{1-2\sigma_1}{8\pi(1-\sigma_1)}\frac{I_i + I_j}{2} - \frac{3}{8\pi(1-\sigma_1)}I_{ij}\frac{a_i^2 + a_j^2}{2} \tag{8.3.36c}$$

or

$$s_{ijij} = \frac{1-2\sigma_1}{8\pi(1-\sigma_1)}\frac{I_i + I_j}{2} - \frac{3}{8\pi(1-\sigma_1)}I_{ij}\frac{a_i^2 + a_j^2}{2}$$

Comparison of the components of the tensor s_{ijkl} (8.3.36) with the tensor components calculated by Eshelby (133) shows the two tensors to be in fact identical to each other.

We have thus demonstrated that the theory formulated in Section 8.2 includes Eshelby's theory of ellipsoidal coherent inclusions in isotropic media.

8.4. ELLIPSOIDAL INCLUSION IN ANISOTROPIC PARENT PHASE: THE CASE OF DIFFERENT MODULI

As shown by Lee, Barnett, and Aaronson (141), the strain energy problem for an ellipsoidal coherent inclusion in an infinite anisotropic matrix also has an analytical solution in the case of an inhomogeneous system, where the inclusion and matrix differ by their elastic moduli. In principle the possibility of solving the problem is based on the fact that constrained strain within a coherent ellipsoidal inclusion is homogeneous. The solution to the problem will be given below in terms of the Fourier method (134).

Consider an inhomogeneous system comprising an ellipsoidal coherent inclusion in an anisotropic crystal. We assume that the elastic constants of the inclusion differ from those of the matrix. In the case of linear elasticity the relaxation strain energy may be written in the form (7.2.17):

$$\Delta E_{\text{relax}} = \int [-\sigma_{ij}^0 \tilde{\theta}(\mathbf{r})\varepsilon_{ij} + \tfrac{1}{2}\lambda_{ijkl}(\mathbf{r})\varepsilon_{ij}\varepsilon_{kl}]dV \tag{8.4.1}$$

where the index p is omitted, since we are considering a crystal lattice transformation involving rearrangements of only one type, $\tilde{\theta}(\mathbf{r})$ is the shape function of the ellipsoidal inclusion,

$$\sigma_{ij}^0 = \lambda_{ijkl}^{(\text{incl})} \varepsilon_{kl}^0 \tag{8.4.2}$$

$\lambda_{ijkl}^{\text{incl}}$ are the inclusion elastic modulus tensor elements, ε_{ij}^0 is an arbitrary stress-free transformation strain describing the crystal lattice rearrangement. The elastic moduli of the system $\lambda_{ijkl}(\mathbf{r})$ are coordinate dependent and may be written as

$$\lambda_{ijkl}(\mathbf{r}) = \lambda_{ijkl}^0 + \Delta\lambda_{ijkl}\tilde{\theta}(\mathbf{r}) \tag{8.4.3}$$

where λ_{ijkl}^0 are the elastic modulus tensor elements of the parent phase, $\Delta\lambda_{ijkl} = \lambda_{ijkl}^{(\text{incl})} - \lambda_{ijkl}^0$.

The elastic stress is related to the strain by the usual elasticity relation

$$\sigma_{ij}(\mathbf{r}) = \frac{\delta \Delta E_{\text{relax}}}{\delta \varepsilon_{ij}(\mathbf{r})}$$

from which it follows that

$$\sigma_{ij}(\mathbf{r}) = -\sigma_{ij}^0\tilde{\theta}(\mathbf{r}) + \lambda_{ijkl}(\mathbf{r})\varepsilon_{kl} \tag{8.4.4}$$

Substituting the definition (8.4.3) into (8.4.4) yields

$$\sigma_{ij}(\mathbf{r}) = -\sigma_{ij}^0\tilde{\theta}(\mathbf{r}) + \Delta\lambda_{ijkl}\tilde{\theta}(\mathbf{r})\varepsilon_{kl} + \lambda_{ijkl}^0\varepsilon_{kl} \tag{8.4.5}$$

Assume that the strain is constant within the inclusion, $\varepsilon_{ij} = \varepsilon_{ij}^*$, if the point, $\mathbf{r}$, is within the inclusion. This assumption enables one to calculate the elastic strain in a closed form. If the strain within the inclusion thus determined proves to be homogeneous, the elastic problem will be solved, since the solution of the linear elasticity equations is unique. The corresponding calculations for an ellipsoidal inclusion are carried out below.

The second term in (8.4.5) vanishes outside the inclusion because of the factor $\tilde{\theta}(\mathbf{r})$, and thus the strain ε_{kl} may be replaced by the constant ε_{kl}^* in this term. Eq. (8.4.5) then becomes

$$\sigma_{ij}(\mathbf{r}) = -\bar{\sigma}_{ij}\tilde{\theta}(\mathbf{r}) + \lambda_{ijkl}^0\varepsilon_{kl} \tag{8.4.6}$$

where

$$\bar{\sigma}_{ij} = \sigma_{ij}^0 - \Delta\lambda_{ijkl}\varepsilon_{kl}^* \tag{8.4.7}$$

is a constant.

According to the elasticity theory the equation of the elastic equilibrium is

$$\frac{\partial \sigma_{ij}}{\partial r_j} = 0 \tag{8.4.8}$$

Substituting Eq. (8.4.6) into (8.4.8) yields

$$\lambda_{ijkl}^0 \frac{\partial \varepsilon_{kl}}{\partial r_j} = -\bar{\sigma}_{ij} \frac{\partial}{\partial r_j}\tilde{\theta}(\mathbf{r}) \tag{8.4.9}$$

Referring to the definition of the elastic strain

$$\varepsilon_{kl}=\frac{1}{2}\left(\frac{\partial u_k}{\partial r_l}+\frac{\partial u_l}{\partial r_k}\right) \tag{8.4.10}$$

we obtain from (8.4.9)

$$\lambda^0_{ijkl}\frac{\partial^2 u_l}{\partial r_j \partial r_k}=-\bar{\sigma}_{ij}\frac{\partial}{\partial r_j}\tilde{\theta}(\mathbf{r}) \tag{8.4.11}$$

Eq. (8.4.11) coincides with (7.2.31) to within the differences in the definitions of the constants λ^0_{ijkl} and $\bar{\sigma}_{ij}$. The latter equation describes the case of homogeneous systems, and all the results obtained in Section 8.2 remain valid if the tensors λ_{ijkl}, $\Omega_{ij}(\mathbf{n})$, and σ^0_{ij} are replaced with the tensors λ^0_{ijkl}, $\Omega^0_{ij}(\mathbf{n})$ and $\bar{\sigma}_{ij}$, respectively, where $\Omega^0_{ij}(\mathbf{n})$ is the inverse tensor to $(\hat{\Omega}_0^{-1})_{ij}=\lambda^0_{iklj}n_k n_l$:

$$\begin{aligned}\lambda_{ijkl} &\rightarrow \lambda^0_{ijkl}\\ \Omega_{ij}(\mathbf{n}) &\rightarrow \Omega^0_{ij}(\mathbf{n})\\ \sigma^0_{ij} &\rightarrow \bar{\sigma}_{ij}\end{aligned} \tag{8.4.12}$$

For instance, the Fourier transform of the displacement field may be obtained from Eq. (8.2.3) in the form

$$\mathbf{v}(\mathbf{k})=-i\frac{1}{k}\hat{\Omega}^0(\mathbf{n})\hat{\bar{\sigma}}\mathbf{n}\theta(\mathbf{k}) \tag{8.4.13}$$

As shown in Section 8.2, the field of the type (8.4.13) produces a constant strain within an ellipsoidal inclusion consistent with the starting assumption of the theory. Referring to (8.4.12), the constrained distortion [see Eq. (8.2.27)] may be written

$$u^*_{ij}=\oint\frac{dO_{\mathbf{n}'}}{4\pi}\left[n_j\Omega^0_{ik}(\mathbf{n})\bar{\sigma}_{kl}n_l\right]_{\mathbf{n}=\mathbf{n}(\mathbf{n}')} \tag{8.4.14}$$

where according to Eqs. (8.2.26)

$$\mathbf{n}=\mathbf{n}(\mathbf{n}')=\frac{\hat{\mathbf{L}}^{-\frac{1}{2}}\mathbf{n}'}{\sqrt{(\mathbf{n}'\hat{\mathbf{L}}^{-1}\mathbf{n}')}}$$

The symmetrization of Eq. (8.4.14) gives

$$\varepsilon^*_{ij}=\tfrac{1}{2}(u^*_{ij}+u^*_{ji})=\oint\frac{dO_{\mathbf{n}'}}{4\pi}\frac{1}{2}\left[n_j\Omega^0_{ik}(\mathbf{n})\bar{\sigma}_{kl}n_l+n_i\Omega^0_{jk}(\mathbf{n})\bar{\sigma}_{kl}n_l\right]_{\mathbf{n}=\mathbf{n}(\mathbf{n}')} \tag{8.4.15}$$

Eq. (8.4.15) may be rewritten in the dense form

$$\varepsilon^*_{ij}=S_{ijkl}\bar{\sigma}_{kl} \tag{8.4.16}$$

where

$$S_{ijkl}=\oint\frac{dO_{\mathbf{n}'}}{4\pi}\frac{1}{2}\left[n_j\Omega^0_{ik}(\mathbf{n})n_l+n_i\Omega^0_{jk}(\mathbf{n})n_l\right]_{\mathbf{n}=\mathbf{n}(\mathbf{n}')} \tag{8.4.17}$$

is a constant forth-rank tensor depending on the elastic moduli λ^0_{ijkl} of the matrix and the shape of the ellipsoidal inclusion but independent of the stress-free transformation strain ε_{ij}^0.

Substituting Eq. (8.4.7) into (8.4.16) results in the linear equation

$$\varepsilon_{ij}^* = S_{ijkl}(\lambda_{klmn}^0 \varepsilon_{mn}^0 - \Delta\lambda_{klmn}\varepsilon_{mn}^*) \tag{8.4.18}$$

Since the components of the tensor S_{ijkl} are constants depending on the shape tensor L_{ij} and elastic constants only [see Eq. (8.4.17)], Eq. (8.4.18) provides the possibility of finding the constant strain tensor ε_{ij}^* within the ellipsoidal inclusion.

It thus follows that Eq. (8.4.17) together with Eq. (8.4.18) fully determine all the parameters of Eq. (8.4.13) describing elastic displacements and thus furnish the final solution to the problem. For instance, the strain energy of the problem may readily be obtained if it is written in terms of the Fourier transform of the displacement field (8.4.13). The procedure has already been exemplified for the case of homogeneous moduli in Section 7.2 where the solution (7.2.44) has been obtained. With the index p removed (we now consider transformations involving rearrangements of only one type), and with the substitution (8.4.12) necessary to pass from homogeneous to inhomogeneous elastic moduli, Eq. (7.2.44) reads

$$E_{\text{relax}}^{\text{heter}} = -\frac{1}{2}\int \frac{d^3k}{(2\pi)^3} (\mathbf{n}\hat{\bar{\sigma}}\hat{\boldsymbol{\Omega}}^0(\mathbf{n})\hat{\bar{\sigma}}\mathbf{n})|\theta(\mathbf{k})|^2 \tag{8.4.19}$$

The same transformations as lead from Eq. (8.2.8) to (8.2.34) may be applied to reduce Eq. (8.4.19) to the integral over the surface of the sphere of the unit radius

$$E_{\text{relax}}^{\text{heter}} = -\frac{1}{2}V\oint \frac{dO_{\mathbf{n}'}}{4\pi}[(\mathbf{n}\hat{\bar{\sigma}}\hat{\boldsymbol{\Omega}}^0(\mathbf{n})\hat{\bar{\sigma}}\mathbf{n})]_{\mathbf{n}=\mathbf{n}(\mathbf{n}')} \tag{8.4.20}$$

The suffix form of Eq. (8.4.20) is

$$E_{\text{relax}}^{\text{heter}} = -\frac{1}{2}V\oint \frac{dO_{\mathbf{n}'}}{4\pi}[n_i\Omega_{jk}^0(\mathbf{n})n_l]_{\mathbf{n}=\mathbf{n}(\mathbf{n}')}\bar{\sigma}_{ij}\bar{\sigma}_{kl}$$

or using the notation (8.4.17)

$$E_{\text{relax}}^{\text{heter}} = -\frac{1}{2}VS_{ijkl}\bar{\sigma}_{ij}\bar{\sigma}_{kl} \tag{8.4.21}$$

The total elastic strain energy is given by the sum of the elastic strain energy (7.2.6)

$$\Delta E_3 = \frac{1}{2}V\lambda_{ijkl}^{(\text{incl})}\varepsilon_{ij}^0\varepsilon_{kl}^0 \tag{8.4.22}$$

associated with the restoration of the initial inclusion shapes after the stress-free transformation (Step 3 in Section 7.2) and the relaxation energy (8.4.21) associated with loosening of the strain ΔE_{relax} (Step 6 in 7.2). The total strain energy of an ellipsoidal inclusion whose elastic moduli differ from those of the matrix is thus given by

$$E_{\text{ellips}} = \Delta E_3 + E_{\text{relax}}^{\text{heter}} = \frac{1}{2} V[\lambda_{ijkl}^{\text{incl}} \, \varepsilon_{ij}^0 \varepsilon_{kl}^0 - S_{ijkl} \bar{\sigma}_{ij} \bar{\sigma}_{kl}] \tag{8.4.23}$$

where $\bar{\sigma}_{ij}$ are given by Eq. (8.4.7) and ε_{ij}^* by Eq. (8.4.18).

8.5. CRYSTAL LATTICE PARAMETERS AND ORIENTATION RELATIONS OF COHERENT CONSTRAINED PLATELIKE NEW PHASE PARTICLES

The crystal lattice parameters and orientation relations are the primary experimental data which are comparatively easy to obtain from x-ray patterns of two-phase alloys.

A coherent new phase particle in a constrained state has a crystal lattice different from that of the particle in the stress-free state (e.g., the stress-free state may be realized by cutting off the particle from the matrix). The difference between the lattices corresponding to constrained and stress-free states is due to elastic strain caused by crystal lattice mismatch along interphase boundaries.

It was shown in Section 8.1 that a coherent new phase inclusion has a platelike equilibrium shape if the magnitude of mismatch is large and interphase energy small. Any coherent platelike inclusion undergoes homogeneous elastic strain to restore the habit plane to the state it had before the transformation occurred, to provide the exact coincidence of the habit planes of the new phase particle and the undistorted matrix. Since the combination of the stress-free transformation strain, ε_{ij}^0, and the elastic strain within the inclusion results in the coincidence of the crystal habit planes before and after the transformation, the total strain within the inclusion is by definition an invariant plane strain, the habit plane being the invariant plane. According to (1.7.3) an invariant plane strain is described by the dyadic product of two vectors one of which, $\mathbf{n}_0$, is normal to the invariant plane and the other one is the shear vector $\mathbf{S} = \varepsilon \boldsymbol{l}$. The calculations in Section 8.1.2 fully confirm the latter statement. According to Eq. (8.1.29) the homogeneous strain within a platelike coherent inclusion may be represented as follows

$$u_{ij}^* = S_i(\mathbf{n}_0) n_j^0 \tag{8.5.1}$$

where $\mathbf{S}(\mathbf{n}) = \varepsilon \boldsymbol{l}$ is the shear vector given by (8.1.34)

$$\mathbf{S}(\mathbf{n}_0) = \hat{\mathbf{\Omega}}(\mathbf{n}_0) \hat{\sigma}^0 \mathbf{n}_0 \tag{8.5.2}$$

Eqs. (8.5.1) and (8.5.2) are fairly general. They do not require equality of the matrix and inclusion anisotropic elastic moduli. If these are different, the shear vector $\mathbf{S}(\mathbf{n}_0)$ should be calculated from the elastic moduli of the inclusion.

We shall demonstrate below that the theory (134) formulated in Section 8.1 also leads to Eq. (8.5.1). The application of the back Fourier transformation to the Fourier transform (8.1.5) gives for the displacements

$$\mathbf{u}(\mathbf{r}) = -i \iiint_{-\infty}^{\infty} \frac{\mathbf{S}(\mathbf{n})}{k} \theta(\mathbf{k}) e^{i\mathbf{kr}} \frac{d^3k}{(2\pi)^3} \tag{8.5.3}$$

where

$$\mathbf{S}(\mathbf{n}) = \hat{\mathbf{\Omega}}(\mathbf{n})\hat{\sigma}^0\mathbf{n}, \quad \mathbf{n} = \frac{\mathbf{k}}{k} \tag{8.5.4}$$

The coordinate derivative of Eq. (8.5.3) yields the distortion tensor

$$u_{ij}(\mathbf{r}) = \frac{\partial u_i(\mathbf{r})}{\partial r_j} = \int S_i(\mathbf{n}) n_j \theta(\mathbf{k}) e^{i\mathbf{kr}} \frac{d^3k}{(2\pi)^3} \tag{8.5.5}$$

In the case of a platelike inclusion, the Fourier transform, $\theta(\mathbf{k})$, of the shape function only takes nonzero values within the thin and extended rod in the **k**-space emerging from the origin point $\mathbf{k}=0$ in the direction $\mathbf{n}_0$ (see Section 8.1). The thickness and the length of the rod are $2\pi/D$ and $2\pi/L$ where D and L are the thickness and the length of the inclusion, respectively ($D/L \ll 1$). For that reason the integration in (8.5.5) is only over this rod. We may therefore rewrite Eq. (8.5.5) in the form

$$u_{ij}(\mathbf{r}) = S_i(\mathbf{n}_0) n_j^0 \int \frac{d^3k}{(2\pi)^3} \theta(\mathbf{k}) e^{i\mathbf{kr}} \tag{8.5.6}$$

with accuracy of the order $D/L \ll 1$.

The integral in Eq. (8.5.6) is the back Fourier transform of the function $\theta(\mathbf{k})$ and hence is equal to $\tilde{\theta}(\mathbf{r})$:

$$u_{ij}(\mathbf{r}) = S_i(\mathbf{n}_0) n_j^0 \tilde{\theta}(\mathbf{r}) = \begin{cases} u_{ij}^* = S_i(\mathbf{n}_0) n_j^0 & \text{if } \mathbf{r} \text{ within inclusion} \\ 0 & \text{otherwise} \end{cases} \tag{8.5.7}$$

Eq. (8.5.7) coincides with Eqs. (8.5.1) derived on the assumption that coherent fitting together of a platelike inclusion and matrix results in homogeneous invariant plane strain concentrated within the inclusion and vanishing outside it.

The matrix of invariant plane strain is given by (1.7.4) where according to (8.5.1)

$$\varepsilon \mathbf{l} = \mathbf{S}(\mathbf{n}_0) \tag{8.5.8}$$

that is,

$$\hat{\mathbf{A}} = \hat{\mathbf{I}} + \mathbf{S}(\mathbf{n}_0) * \mathbf{n}_0 \tag{8.5.9}$$

Any crystal lattice translation **r** of the parent phase is transformed into $\mathbf{r}'$ after the phase transition

$$\mathbf{r}' = \hat{\mathbf{A}}\mathbf{r}(\hat{\mathbf{I}} + \mathbf{S}(\mathbf{n}_0) * \mathbf{n}_0)\mathbf{r} = \mathbf{r} + \mathbf{S}(\mathbf{n}_0)(\mathbf{n}_0\mathbf{r}) \tag{8.5.10}$$

Eq. (8.5.10) determines the orientation relations between the crystal lattices of the inclusion and parent phase. It also shows that all the crystal lattice translations in the habit plane remain unaffected by the crystal lattice rearrangement.

For example, if the new phase unit cell is chosen such that two of its base vectors lie in the habit plane these two new phase parameters are exactly equal to the corresponding translations of the parent phase.

As for the crystal lattice planes their transformations may be described by the transformations of the parent phase reciprocal lattice in the crystal lattice rearrangement (see Section 1.5). It follows from Eqs. (1.8.3) and (8.5.8) that an invariant plane strain within a platelike coherent inclusion results in the reciprocal lattice transformation given by

$$\mathbf{H}'=\mathbf{H}-\mathbf{n}_0\frac{(\mathbf{S}(\mathbf{n}_0)\mathbf{H})}{1+(\mathbf{n}_0\mathbf{S}(\mathbf{n}_0))} \tag{8.5.11}$$

where $\mathbf{H}'$ is the reciprocal lattice vector of the new phase formed from the reciprocal lattice vector of the parent phase, $\mathbf{H}$, after the transformation.

Eq. (8.5.11) may be rewritten in the form

$$\Delta\mathbf{H}=\mathbf{H}'-\mathbf{H}=-\mathbf{n}_0\frac{(\mathbf{S}(\mathbf{n}_0)\mathbf{H})}{1-(\mathbf{n}_0\mathbf{S}(\mathbf{n}_0))} \tag{8.5.12}$$

Eq. (8.5.12) shows that the formation of a platelike coherent new phase precipitate results in splitting of the parent phase reflections in the single-crystal diffraction pattern. This splitting occurs along the direction $\mathbf{n}_0$ normal to the habit plane of the precipitate. Equation (8.5.12) also shows that all parent phase reflections making up the reciprocal lattice plane normal to the vector $\mathbf{S}(\mathbf{n}_0)$ that passes through the zero reciprocal lattice point do not undergo splitting at all (the scalar product of these reflections $\mathbf{HS}(\mathbf{n}_0)$ is equal to zero). Figure 62 shows a typical single crystal diffraction pattern from a single crystal containing platelike coherent inclusions.

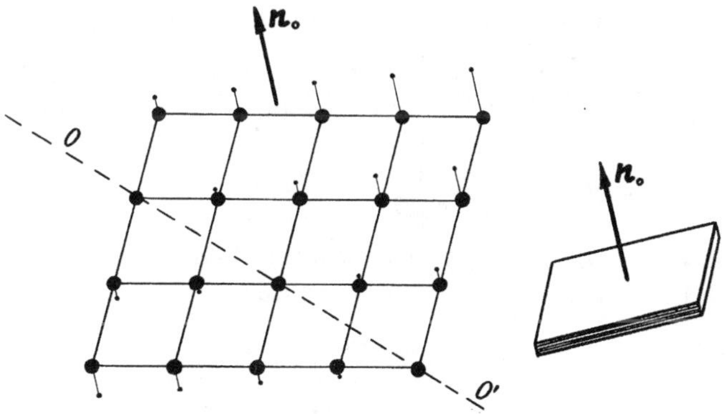

Figure 62. Scheme of the diffraction pattern from coherent platelike new phase precipitates: ● a parent phase diffraction spot; • a precipitate phase diffraction spot. 00′ is the trace of a plane in the reciprocal lattice of the parent phase normal to the shear vector $\boldsymbol{l}$ which contains the reciprocal lattice origin; $\mathbf{n}_0$ is a unit vector normal to the habit.

A detailed analysis of the splittings makes it possible to calculate the vector $\mathbf{S}(\mathbf{n}_0)$ and, using (8.5.2), to determine information on the elastic moduli of new phase particles. In many cases, when new phase single crystals are not available, this may be the only means to determine the single crystal elastic moduli of the new phase. In the next chapter we show that the results obtained in this section provide the explanation of some experimental data on decomposed alloys.

8.6. HABIT PLANE AND ORIENTATION RELATIONS OF TETRAGONAL PRECIPITATES IN CUBIC PARENT PHASES

To exemplify the analysis of actual situations, we shall now consider the case of a tetragonal phase inclusion in a cubic parent phase matrix. In that we shall follow Wen, Kostlan, et al. (152).

According to Eq. (8.1.13), the determination of the habit plane orientation of a precipitate is reduced to the minimization of the function $B(\mathbf{n})$ defined by (8.1.2) with respect to the unit vector $\mathbf{n}$. The nonzero components of the elastic modulus of a cubic crystal λ_{ijkl} are

$$\begin{aligned} &\lambda_{1111}=\lambda_{2222}=\lambda_{3333}=c_{11} \\ &\lambda_{1122}=\lambda_{1133}=\lambda_{2233}=c_{12} \\ &\lambda_{1212}=\lambda_{1313}=\lambda_{2323}=c_{44} \end{aligned} \tag{8.6.1}$$

Using the constants (8.6.1) and the symmetry relations for the tensor λ_{ijkl} in Eq. (8.1.3), we may write the components of the tensor $\Omega_{ij}^{-1}(\mathbf{n})$ in the form

$$\begin{aligned} &\Omega_{ii}^{-1}(\mathbf{n})=c_{12}+(c_{11}-c_{12})n_i^2 \\ &\Omega_{ij}^{-1}(\mathbf{n})=(c_{11}+c_{44})n_in_j \quad \text{if } i\neq j \end{aligned} \tag{8.6.2}$$

Eq. (8.6.2) and the identity

$$\hat{\mathbf{\Omega}}(\mathbf{n})\hat{\mathbf{\Omega}}^{-1}(\mathbf{n})=\hat{\mathbf{I}}$$

lead to the equations for the components of the inverse tensor

$$\begin{aligned} &\Omega_{ii}(\mathbf{n})=\frac{c_{44}+(c_{11}-c_{44})(n_j^2+n_k^2)+\xi(c_{11}+c_{12})n_j^2n_k^2}{c_{44}D(\mathbf{n})} \\ &\Omega_{ij}(\mathbf{n})=-\frac{(c_{12}+c_{44})(1+\xi n_k^2)}{c_{44}D(\mathbf{n})}n_in_j \end{aligned} \tag{8.6.3}$$

where the indexes i, j, k form a cyclic sequence [summation over the repeated indexes in Eqs. (8.6.2) and (8.6.3) is not implied],

$$\xi=\frac{c_{11}-c_{12}-2c_{44}}{c_{44}} \tag{8.6.4a}$$

is the elastic anisotropy parameter and

$$\begin{aligned} D(\mathbf{n})=&c_{11}+\xi(c_{11}+c_{12})(n_1^2n_2^2+n_1^2n_3^2+n_2^2n_3^2) \\ &+\xi^2(c_{11}+2c_{12}+c_{44})n_1^2n_2^2n_3^2 \end{aligned} \tag{8.6.4b}$$

If the crystal lattice rearrangement is from a cubic to a tetragonal phase the stress-free transformation strain is given by the tensor

$$\varepsilon_{ij}^0=\begin{pmatrix}\varepsilon_{11}^0 & 0 & 0\\ 0 & \varepsilon_{11}^0 & 0\\ 0 & 0 & \varepsilon_{33}^0\end{pmatrix} \tag{8.6.5}$$

which is referred to the Cartesian coordinate axes [100], [010], [001] of the cubic parent phase.

Here

$$\varepsilon_{11}^0=\frac{a-a_0}{a_0}, \quad \varepsilon_{33}^0=\frac{c-a_0}{a_0} \tag{8.6.6}$$

where c and a are the crystal lattice parameters of the tetragonal phase, a_0 is the crystal lattice parameter of the cubic phase. The fcc → bcc and bcc → fcc crystal lattice rearrangements are also determined by the tensor (8.6.6) where

$$\varepsilon_{11}^0=\frac{a_b}{a_f}\sqrt{2}-1, \quad \varepsilon_{33}^0=\frac{a_b}{a_f}-1 \tag{8.6.7}$$

for the fcc → bcc transition and

$$\varepsilon_{11}^0=\frac{a_f}{a_b\sqrt{2}}-1, \quad \varepsilon_{33}^0=\frac{a_f}{a_b}-1 \tag{8.6.8}$$

for the bcc → fcc transition (a_f and a_b are the crystal lattice parameters of the fcc and bcc phases, respectively). Substitution of the tensor ε_{ij}^0 from (8.6.5) into (8.1.2) yields

$$B(\mathbf{n})=\bar{B}-\frac{(\sigma_{33}^0)^2}{c_{11}}\psi(\mathbf{n}) \tag{8.6.9}$$

where

$$\psi(\mathbf{n})=[\alpha_1^2+An_3^2(1-n_3^2)-(\alpha_1^2-1)n_3^4+\xi Bn_1^2n_2^2n_3^2+2\xi\alpha_1^2n_1^2n_2^2]$$
$$\times\left[1+\xi\frac{c_{11}+c_{12}}{c_{11}}(n_1^2n_2^2+n_1^2n_3^2+n_2^2n_3^2)+\frac{c_{11}+2c_{12}+c_{44}}{c_{11}}\xi^2n_1^2n_2^2n_3^2\right]^{-1} \tag{8.6.10}$$

$$\alpha_1=\frac{\sigma_{11}^0}{\sigma_{33}^0}=\frac{(c_{11}+c_{12})t_1+c_{12}}{2c_{12}t_1+c_{11}} \tag{8.6.11a}$$

$$t_1=\frac{\varepsilon_{11}^0}{\varepsilon_{33}^0} \tag{8.6.11b}$$

$$A=\frac{c_{11}}{c_{44}}+\alpha_1^2\left(\frac{c_{11}}{c_{44}}-2\right)-2\alpha_1\left(\frac{c_{12}}{c_{44}}+1\right) \tag{8.6.11c}$$

$$B=\frac{c_{11}+c_{12}}{c_{44}}+2\alpha_1^2\frac{c_{11}-c_{44}}{c_{44}}-4\alpha_1\frac{c_{12}+c_{44}}{c_{44}} \tag{8.6.11d}$$

$$\sigma^0_{ij}=\lambda_{ijkl}\varepsilon^0_{kl}=\begin{pmatrix}\sigma^0_{11} & 0 & 0\\ 0 & \sigma^0_{11} & 0\\ 0 & 0 & \sigma^0_{33}\end{pmatrix} \tag{8.6.11e}$$

$$\begin{aligned}\sigma^0_{11}&=(c_{11}+c_{12})\varepsilon^0_{11}+c_{12}\varepsilon^0_{33}\\ \sigma^0_{33}&=2c_{12}\varepsilon^0_{11}+c_{11}\varepsilon^0_{33}\end{aligned} \tag{8.6.11f}$$

$$\bar{B}=\lambda_{ijkl}\varepsilon^0_{ij}\varepsilon^0_{kl}=\varepsilon^0_{ij}\sigma^0_{kl}=2\varepsilon^0_{11}\sigma^0_{11}+\varepsilon^0_{33}\sigma^0_{33} \tag{8.6.11g}$$

Since $\bar{B}$ in Eq. (8.6.9) is a constant, $B(\mathbf{n})$ reaches its minimum when the function $\psi(\mathbf{n})$ has the maximum value.

In the spherical coordinate system the vector $\mathbf{n}$ may be written

$$\mathbf{n}=(n_1, n_2, n_3)=(\cos\phi\,\sin\theta, \sin\phi\,\sin\theta, \cos\theta) \tag{8.6.12}$$

where θ is the angle made by the vector $\mathbf{n}$ and the [001] direction (z-axis), ϕ is the angle between the projection of the vector $\mathbf{n}$ on the (001) plane and the direction [100] (x-axis).

Substituting (8.6.12) into Eq. (8.6.10) yields

$$\psi(\mathbf{n})=\psi(\theta, \phi)=\frac{A_0(\theta)+\xi B_0(\theta)\sin^2 2\phi}{A_1(\theta)+\xi B_1(\theta)\sin^2 2\phi} \tag{8.6.13}$$

where

$$A_0(\theta)=\alpha_1^2+A\cos^2\theta(1-\cos^2\theta)-(\alpha_1^2-1)\cos^4\theta$$

$$B_0(\theta)=\tfrac{1}{4}B\cos^2\theta\sin^4\theta+\tfrac{1}{2}\xi\alpha_1^2\sin^4\theta$$

$$A_1(\theta)=1+\xi\frac{c_{11}+c_{12}}{c_{11}}\cos^2\theta(1-\cos^2\theta)$$

$$B_1(\theta)=\frac{c_{11}+c_{12}}{4c_{11}}\sin^4\theta+\frac{c_{11}+2c_{12}+c_{44}}{4c_{11}}\xi\cos^2\theta\sin^4\theta$$

The values of θ_0 and ϕ_0 at which the function $\psi(\theta, \phi)$ given by (8.6.13) reaches its absolute maximum determine the vector $\mathbf{n}_0$ and thus the habit plane orientation (see Section 8.1).

The problem of finding θ_0 and ϕ_0 is actually reduced to solving the set of two transcendental equations,

$$\frac{\partial\psi(\theta, \phi)}{\partial\theta}=0, \quad \frac{\partial\psi(\theta, \phi)}{\partial\phi}=0 \tag{8.6.15}$$

in two unknowns, θ and ϕ. This problem was solved by Wen, Kostlan, et al. (152).

Roughly, the line of reasoning pursued in (152) is as follows. The function

$$\psi(x)=\frac{A_0+\xi B_0 x}{A_1+\xi B_1 x} \tag{8.6.16}$$

where $x=\sin^2 2\phi$ is a monotonic function since the sign of its first derivative

$$\frac{d\psi(x)}{dx}=\frac{A_1B_0-A_0B_1}{(A_1+\xi B_1 x)^2}\xi \tag{8.6.17}$$

is independent of x and is only determined by the sign of the numerator

$$[A_1(\theta)B_0(\theta)-A_0(\theta)B_1(\theta)]\xi \tag{8.6.18}$$

in (8.6.17). Since $x=\sin^2 2\phi$ and therefore ranges from 0 to 1, the maximum of $\psi(x)$ which is a monotonic function of x should occur at one of the limiting x values, that is, at 0 if $(A_1B_0-A_0B_1)\xi<0$ or 1 if $(A_1B_0-A_0B_1)\xi>0$. In the first case $x=\sin^2 2\phi=0$; the solution is $\phi=\phi_0=0$. In the second case $x=\sin^2 2\phi=1$; it is $\phi=\phi_0=\pi/4$. As has been proved in (152), the inequality

$$A_1(\theta)B_0(\theta)-A_0(\theta)B_1(\theta)>0$$

holds irrespective of both θ and ϕ with any values of elastic moduli. The sign of the anisotropy parameter, ξ, therefore solely determines the position of the maximum of $\psi(\theta, \phi)$ with respect to ϕ:

1. If $\xi<0$, the product (8.6.18) is negative, and $\psi(\theta, \phi)$ reaches its maximum at $\phi=0$.
2. If $\xi>0$, the product (8.6.18) is positive, and the maximum occurs at $\phi=\pi/4$.

In the special case of ξ equal to zero (elastic isotropy) the maximum of $\psi(\theta, \phi)$ is degenerate with respect to ϕ.

The normal, $\mathbf{n}_0$, to the habit plane is thus given by

$$\mathbf{n}_0=\begin{cases}(\sin\theta, \quad 0, \quad \cos\theta) & \text{if } \xi<0\\ \left(\dfrac{\sin\theta}{\sqrt{2}}, \dfrac{\sin\theta}{\sqrt{2}}, \cos\theta\right) & \text{if } \xi>0\end{cases} \tag{8.6.19}$$

and the Miller indexes of the habit plane are

$$(hkl)_{\text{hab}}=\begin{cases}(hol) & \text{if } \xi<0\\ (hhl) & \text{if } \xi>0\end{cases} \tag{8.6.20}$$

Eqs. (8.6.19) and (8.6.20) show that the problem of determining the habit plane orientation is solved by maximizing (8.6.10) with respect to only one variable, θ, either at $\phi=0$ or at $\phi=\pi/4$, that is, by maximizing the functions

$$\psi(\theta, 0)=\frac{A_0(\theta)}{A_1(\theta)} \quad \text{or} \quad \psi\left(\theta, \frac{\pi}{4}\right)=\frac{A_0(\theta)+\xi B_0(\theta)}{A_1(\theta)+\xi B_1(\theta)}$$

respectively.

According to (152) the maximizing yields

$$\mathbf{n}_0=(\sin\theta_0, 0, \cos\theta_0) \quad \text{if } \xi<0$$

where

$$\cos^2\theta_0=\begin{cases}0 & \text{if } -\infty<t_1<-\dfrac{c_{11}+c_{12}}{c_{12}} \quad\text{and}\quad 1\leqslant t_1\leqslant\infty\\ 1+\dfrac{c_{11}+2c_{12}}{c_{11}+c_{12}}\dfrac{t_1}{1-t_1} & \text{if } -\dfrac{c_{11}}{c_{12}}-1<t_1<0\\ 1 & \text{if } 0\leqslant t_1<1\end{cases} \tag{8.6.21}$$

and

$$\mathbf{n}_0=\left(\frac{1}{\sqrt{2}}\sin\theta_0, \frac{1}{\sqrt{2}}\sin\theta_0, \cos\theta_0\right) \quad \text{if } \xi>0$$

where

$$\cos^2\theta_0=\begin{cases}0 & \text{if } -\infty<t_1<t_1^0 \text{ and } t_3^0<t_1<\infty\\ 1-2\dfrac{(\xi+2)(c_{11}+2c_{12})t_1}{\xi(c_{11}+2c_{12})(2t_1-1)+4(c_{11}+c_{12})(t_1-1)} & \text{if } t_1^0\leqslant t_1<0\\ 1 & \text{if } 0\leqslant t_1\leqslant t_2^0\\ \dfrac{\xi(c_{11}+2c_{12})+4c_{11}(1-t_1)}{\xi(c_{11}+2c_{12})(1+2t_1)} & \text{if } t_2^0<t_1\leqslant t_3^0\end{cases} \tag{8.6.22}$$

$$t_1^0=-\frac{c_{11}+c_{12}}{c_{12}}-\xi\frac{c_{11}+2c_{12}}{4c_{12}}$$

$$t_2^0=\frac{2c_{11}}{2c_{11}+\xi(c_{11}+2c_{12})}$$

$$t_3^0=1+\xi\frac{c_{11}+2c_{12}}{4c_{11}}$$

$$t_1^0<t_2^0<t_3^0$$

The orientation relations between the precipitate and parent phase lattices are determined by Eqs. (8.5.10) and (8.5.12) which include the vector $\mathbf{n}_0$ normal to the habit plane given by Eqs. (8.6.21) and (8.6.22) and the shear vector $\mathbf{S}(\mathbf{n}_0)$ given by (8.5.2).

Substituting the components of the vector σ_{ij}^0 and $\Omega_{ij}(\mathbf{n}_0)$ from Eqs. (8.6.11f) and (8.6.3) into Eq. (8.5.2) yields

$$\mathbf{S}(\mathbf{n}_0)=(\mathbf{S}_1(\mathbf{n}_0), S_1(\mathbf{n}_0), S_3(\mathbf{n}_0)) \tag{8.6.23}$$

if $\xi > 0$ where

$$S_1(\mathbf{n}_0)=\frac{\sigma_{33}^0}{D_1(\theta)}\frac{\sin\theta}{\sqrt{2}}\left[\frac{(c_{11}+c_{12})t_1+c_{12}}{c_{11}+2c_{12}t_1}\left(\frac{c_{11}}{c_{44}}+\frac{c_{11}+c_{12}}{2c_{44}}\sin^2\theta\right.\right.$$

$$\left.\left.+\xi\frac{c_{11}-c_{44}}{2c_{44}}\cos^2\theta\sin^2\theta\right)+\frac{c_{12}+c_{44}}{c_{44}}\left(1+\frac{1}{2}\xi\sin^2\theta\right)\cos^2\theta\right] \quad (8.6.24a)$$

$$S_3(\mathbf{n}_0)=\frac{\sigma_{33}^0\cos\theta}{D_1(\theta)}\left[-2\frac{(c_{11}+c_{12})t_1+c_{12}}{c_{11}+2c_{12}t_1}\frac{c_{12}+c_{44}}{c_{44}}\left(1+\frac{1}{2}\xi\sin^2\theta\right)\frac{\sin^2\theta}{4}\right.$$

$$\left.+\frac{c_{11}}{c_{44}}-\frac{c_{11}-c_{44}}{c_{44}}\cos^2\theta+\xi\frac{c_{11}+c_{12}}{c_{44}}\frac{\sin^4\theta}{4}\right] \quad (8.6.24b)$$

$$D_1(\theta)=c_{11}+\xi(c_{11}+c_{12})\left[\cos^2\theta(1-\cos^2\theta)+\frac{\sin^4\theta}{4}\right]$$

$$+\frac{1}{4}\xi^2(c_{11}+2c_{12}+c_{44})\cos^2\theta\sin^4\theta$$

and

$$\mathbf{S}(\mathbf{n}_0)=(S_1(\mathbf{n}_0),\,0,\,S_3(\mathbf{n}_0)) \quad (8.6.25)$$

if $\xi < 0$ where

$$S_1(\mathbf{n}_0)=\frac{\sigma_{33}^0}{D_2(\theta)}\sin\theta\left[\frac{(c_{11}+c_{12}t_1)+c_{12}}{c_{11}+2c_{12}t_1}\left(1+\frac{c_{11}-c_{44}}{c_{44}}\cos^2\theta\right)-\frac{c_{12}+c_{44}}{c_{44}}\cos^2\theta\right] \quad (8.6.26a)$$

$$S_3(\mathbf{n}_0)=\frac{\sigma_{33}^0}{D_2(\theta)}\cos\theta\left[-\frac{(c_{11}+c_{12}t_1)+c_{12}}{c_{11}+2c_{12}t_1}\frac{c_{12}+c_{44}}{c_{44}}\sin^2\theta+1-\frac{c_{11}-c_{44}}{c_{44}}\cos^2\theta\right] \quad (8.6.26b)$$

$$D_2(\theta)=c_{11}+\xi(c_{11}+c_{12})\sin^2\theta\cos^2\theta$$

An interesting observation is that the $B(\mathbf{n})$ minimum is provided if the habit plane contains the invariant strain line and the lowest modulus direction (153). The observation was made for some cases of cubic-tetragonal and cubic-orthorhombic transformations. It seems however that the invariant line hyposesis is not always applicable. For example, it fails to explain the case $\xi>0$ and $t_2^0<t_1<t_3^0$ (see Eq. (8.6.22)).

8.7. EQUILIBRIUM SHAPE OF COHERENT INCLUSION

The term platelike inclusion is applied to any new phase precipitate whose thickness is far smaller than other size dimensions. Such precipitates may be formed as disks, plane polygons, rectangulars, "laths", and so forth. The analysis of equilibrium inclusion shapes carried out in Section 8.1 should therefore be extended to include more accurate treatment of competition between strain and interphase interactions. This analysis was made by Khachaturyan and Hairapetyan (154).

Following (154), we assume that the equilibrium shape of a precipitate particle is determined by the condition that the sum of elastic and interphase energies be minimal at a given precipitate volume V. If the thickness of a platelike inclusion D is constant, the volume conservation condition is reduced to the condition of a constant habit plane surface area $S=V/D$. The condition that the latter quantity be constant implies that shape variations leave the interphase energy, $E_s=2\gamma_s S=2\gamma_s V/D$ where γ_s is the interphase energy coefficient, unaffected. It should, however, be observed that the energy, $E_s=2\gamma_s V/D$, may only be taken for the total interphase energy to the extent that the contributions to the surface energy from crystal edges which are certainly shape dependent mav be neglected. The crystal edge interphase energy is given by

$$E_s^{\text{edge}}=D\oint_P \gamma_s(\mathbf{m})dl_{\mathbf{m}} \tag{8.7.1}$$

where $dl_{\mathbf{m}}$ is a linear element of the contour P enveloping the inclusion in the habit plane, $ds=Ddl$ the surface element of the inclusion edge, the vector $\mathbf{m}$ is the unit vector normal to the edge surface element $Ddl_{\mathbf{m}}$. It lies in the habit plane and is perpendicular to the linear element $dl_{\mathbf{m}}$. The quantity $\gamma_s(\mathbf{m})$ is the interphase energy coefficient of the edge boundary plane normal to the vector $\mathbf{m}$. The integration in (8.7.1) is over the contour P.

It follows from Eq. (8.7.1) that the interphase energy of inclusion edges is of the order

$$E_s^{\text{edge}}\approx\gamma_s DP$$

where the product DP is the area of the inclusion edge surface. The perimeter P value is of the order of the typical inclusion length L. The interphase energy of an inclusion edge is only a small fraction of the total interphase energy if the aspect D/L is far less than unity:

$$\frac{E_s^{\text{edge}}}{E_s+E_s^{\text{edge}}}\approx\frac{\gamma_s DP}{2\gamma_s L^2+\gamma_s DP}\approx\frac{\gamma_s DP}{2\gamma_s L^2}\frac{1}{1+\gamma_s DP/(2\gamma_s L^2)}\approx\frac{D}{2L}\frac{1}{1+D/2L}\approx\frac{D}{2L}\ll 1 \tag{8.7.2}$$

where P is set approximately equal to L. This fraction may therefore be neglected. The case when the interphase energy E_s^{edge} plays an important part in the variational procedure will be discussed in the end of this section. The problem of determining the equilibrium shape of a platelike inclusion in the habit plane is thus reduced to the variational problem of determining the elastic strain energy minimum at a constant surface area $S=V/D$.

The shape dependent part of the elastic strain energy (8.1.27) may be written in the form

$$E_{\text{edge}}=\frac{1}{2}\iiint_{-\infty}^{\infty}\frac{d^3k}{(2\pi)^3}\,\Delta B(\mathbf{n})|\theta(\mathbf{k})|^2$$

$$=\frac{1}{2}\int_{-\infty}^{\infty}\frac{dk_z}{2\pi}\frac{4\sin^2\frac{1}{2}k_zD}{k_z^2}\iint_{-\infty}^{\infty}\frac{d^2\tau}{(2\pi)^2}\,\Delta B\left(\frac{\mathbf{k}}{k}\right)\int_{(S)}d^2\rho\int_{(S)}d^2\rho' e^{-i\tau(\rho-\rho')} \tag{8.7.3}$$

where the Fourier transform of the shape function is given by

$$\theta(\mathbf{k})=\frac{2\sin\frac{1}{2}k_zD}{k_z}\int_{(S)}d^2\rho e^{-i\tau\rho} \tag{8.7.4}$$

$\mathbf{k}=(\boldsymbol{\tau}, k_z)$, $\boldsymbol{\tau}=(k_x, k_y)$ is the projection of the vector $\mathbf{k}$ onto the habit plane, k_z is the projection of the vector $\mathbf{k}$ on the direction $\mathbf{n}_0$ normal to the habit, $\boldsymbol{\rho}$ is the projection of the vector $\mathbf{r}$ on the habit. If the Cartesian axes are chosen such that z is collinear with $\mathbf{n}_0$, and x and y lie in the habit plane, we have $\mathbf{r}=(x, y, z)$, $\boldsymbol{\rho}=(x, y)$. The integration in Eq. (8.7.4) is over the area S of the habit plane. This area is limited by the curve $y=y(x)$ having two branches, $y=y_+(x)$ and $y=y_-(x)$ (see Fig. 63).

As shown in Appendix 3, at $D/L \to 0$, the integral (8.7.3) value approaches [see Eq. (A.3.22)]

$$E_{\text{edge}}=\oint \delta(\mathbf{m})dl_{\mathbf{m}} \tag{8.7.5}$$

where

$$\delta(\mathbf{m}) \approx \left(\frac{D^2}{4\pi}\ln\frac{L}{D}\right)\beta_{ij}(\mathbf{n}_0)m_im_j \tag{8.7.6}$$

The vector $\mathbf{m}$ is the unit vector lying in the habit plane and normal to the line element $dl_{\mathbf{m}}$ of the closed perimeter contour P described by the curve $y=y(x)$,

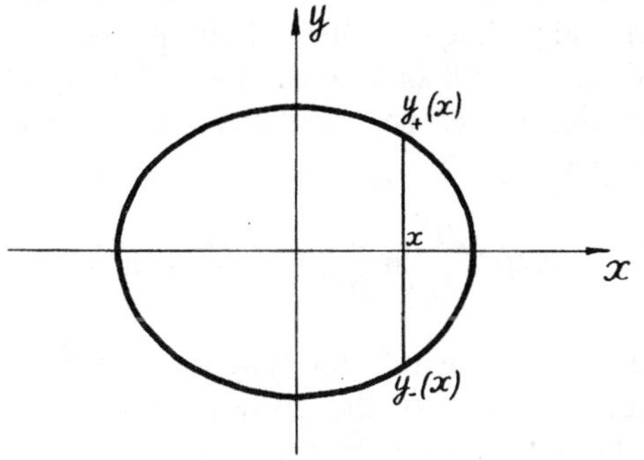

Figure 63. The scheme of a contour $y=y(x)$ describing the shape of a platelike precipitate in the habit plane.

$\beta_{ij}(\mathbf{n}_0)$ is the second-rank tensor,

$$\beta_{ij}(\mathbf{n}_0)=\frac{1}{2}\begin{pmatrix} \dfrac{\partial^2 \Delta B(\mathbf{n})}{\partial n_x^2} & \dfrac{\partial^2 \Delta B(\mathbf{n})}{\partial n_x \partial n_y} \\ \\ \dfrac{\partial^2 \Delta B(\mathbf{n})}{\partial n_x \partial n_y} & \dfrac{\partial^2 \Delta B(\mathbf{n})}{\partial n_y^2} \end{pmatrix}_{\mathbf{n}=\mathbf{n}_0} \tag{8.7.7}$$

If the Cartesian axes x and y are along the principal axes of the tensor $\beta_{ij}(\mathbf{n}_0)$ defined by Eq. (8.7.7), Eq. (8.7.6) may be simplified and written as

$$\delta(\mathbf{m})=\frac{D^2}{4\pi}\ln\frac{L}{D}(\beta_1 m_x^2+\beta_2 m_y^2) \tag{8.7.8}$$

where the coefficients β_1 and β_2 are the eigenvalues of the tensor $\beta_{ij}(\mathbf{n}_0)$ (the corresponding eigenvectors are directed along x and y, respectively). The sum $\beta_1 m_x^2+\beta_2 m_y^2$ is the diagonal representation of the quadratic form $\beta_{ij}(\mathbf{n}_0)m_i m_j$. Since, by definition, the function $\Delta B(\mathbf{n})$ reaches its minimum at $\mathbf{n}=\mathbf{n}_0$, the first nonvanishing term in the Taylor expansion of $\Delta B(\mathbf{n})$ is

$$\Delta B(\mathbf{n})\simeq\beta_{ij}(\mathbf{n}_0)\delta n_i \delta n_j+\cdots \tag{8.7.9}$$

where $\delta\mathbf{n}$ is a small deviation from the direction $\mathbf{n}_0$. Since $\Delta B(\mathbf{n})$ is positive irrespective of $\delta\mathbf{n}$, the quadratic form (8.7.9) must be positive definite. Hence all the eigenvalues of the matrix $\beta_{ij}(\mathbf{n}_0)$ are positive. Thus

$$\beta_1>0 \quad \text{and} \quad \beta_2>0$$

For definiteness, let

$$\beta_1>\beta_2$$

(the case $\beta_2>\beta_1$ may be obtained by mere permutation of the axes x and y).

Formally, the function $\delta(\mathbf{m})$ may be treated as line tension coefficient referred to the normal to $\mathbf{m}$ straight line lying in the habit plane. By the definition (8.7.7) the components of the tensor $\beta_{ij}(\mathbf{n}_0)$ have values of the order of $\lambda\varepsilon_0^2$ where λ is a typical elastic modulus value. The coefficient $\delta(\mathbf{m})$ in (8.7.8) may therefore roughly be estimated as

$$\delta(\mathbf{m})\sim\frac{D^2}{4\pi}\ln\frac{L}{D}\lambda\varepsilon_0^2\approx\frac{\lambda(D\varepsilon_0)^2}{4\pi}\ln\frac{L}{D} \tag{8.7.10}$$

With this in mind, we may identify the function $\delta(\mathbf{m})$ with the linear tension coefficient of a linear element of the dislocation loop perpendicular to the vector $\mathbf{m}$ with the Burgers vector of the order of $\varepsilon_0 D$.

Since $\mathbf{m}=(m_x, m_y)$ is the unit vector normal to the linear element $dl_{\mathbf{m}}$ of the contour $y=y(x)$, the components of this vector and of the element $dl_{\mathbf{m}}$ may be expressed in terms of the derivative dy/dx:

$$m_x = -\frac{dy/dx}{\sqrt{1+(dy/dx)^2}}$$

$$m_y = \frac{1}{\sqrt{1+(dy/dx)^2}}$$

$$dl = \sqrt{1+(dy/dx)^2}\, dx \tag{8.7.11}$$

The use of (8.7.11) in (8.7.8) and substitution of the result into the integral (8.7.5) along the contour $y=y(x)$ yield

$$E_{\text{edge}} = \oint \delta(m_x, m_y)dl_m = \frac{D^2}{4\pi}\ln\frac{L}{D}\left\{\beta_1 \oint \frac{(dy/dx)^2}{1+(dy/dx)^2}\sqrt{1+\left(\frac{dy}{dx}\right)^2}\, dx\right.$$

$$\left. +\beta_2 \oint \frac{1}{1+(dy/dx)^2}\sqrt{1+\left(\frac{dy}{dx}\right)^2}\, dx\right\}$$

$$= \frac{D^2}{4\pi}\ln\frac{L}{D}\left\{\beta_1 \oint \frac{(dy/dx)^2}{\sqrt{1+(dy/dx)^2}}\, dx + \beta_2 \oint \frac{dx}{\sqrt{1+(dy/dx)^2}}\right\} \tag{8.7.12}$$

The latter equation makes it possible to formulate the variational problem. The equilibrium shape of an inclusion in the habit plane is determined from the equation

$$\frac{D^2}{4\pi}\ln\frac{L}{D}\left\{\beta_1 \int \frac{(dy_+/dx)^2\, dx}{\sqrt{1+(dy_+/dx)^2}} + \beta_2 \int \frac{dx}{\sqrt{1+(dy_+/dx)^2}}\right.$$

$$\left. +\beta_1 \int \frac{(dy_-/dx)^2\, dx}{\sqrt{1+(dy_-/dx)^2}} + \beta_2 \int \frac{dx}{\sqrt{1+(dy_-/dx)^2}}\right\} = \min \tag{8.7.13}$$

under the additional condition

$$\int y_+(x)dx - \int y_-(x)dx = S = \text{constant} \tag{8.7.14}$$

where the integral along the contour $y=y(x)$ is divided into two integrals along the branches $y=y_+(x)$ and $y=y_-(x)$.

The problem of determining the equilibrium inclusion shape in the habit plane is thus a two-dimensional analogue of the problem of the equilibrium faceting of a crystal [see, e.g. (80)].

Since the analytical expression for the line tension coefficient (8.7.8) is known, the equilibrium precipitate shape can actually be determined. The minimization of the strain energy (8.7.12) under the condition (8.7.14) can be carried out by the Lagrange method of undetermined multipliers. In fact one may introduce the generating functional

$$\Phi=\frac{D^2}{4\pi}\ln\frac{L}{D}\left\{\beta_1\int\frac{p_+^2}{\sqrt{1+p_+^2}}dx+\beta_2\int\frac{dx}{\sqrt{1+p_+^2}}+\beta_1\int\frac{p_-^2}{\sqrt{1+p_-^2}}dx\right.$$
$$\left.+\beta_2\int\frac{dx}{\sqrt{1+p_-^2}}\right\}-\mu\frac{D^2}{4\pi}\ln\frac{L}{D}\left(\int y_+dx-\int y_-dx\right) \tag{8.7.15}$$

where $p_+=dy_+/dx$, $p_-=dy_-/dx$, μ is the undetermined multiplier.

The necessary minimum condition for the functional (8.7.15) is

$$\delta\Phi=0 \tag{8.7.16}$$

where $\delta\Phi$ is the first variation of the functional (8.7.15) with respect to y. The evaluation of the first variation reduces Eq. (8.7.16) to

$$\int\left\{-\frac{d}{dx}\frac{p}{dp_+}\left(\frac{\beta_1p_+^2+\beta_2}{\sqrt{1+p_+^2}}\right)-\mu\right\}\delta y_+dx+\int\left\{-\frac{d}{dx}\frac{d}{dp_-}\left(\frac{\beta_1p_-^2+\beta_2}{\sqrt{1+p_-^2}}\right)+\mu\right\}\delta y_-dx=0 \tag{8.7.17}$$

Since the variations $\delta y_+(x)$ and $\delta y_-(x)$ are infinitesimal arbitrary quantities, Eq. (8.7.17) requires that

$$-\frac{d}{dx}\frac{d}{dp_+}\left(\frac{\beta_1p_+^2+\beta_2}{\sqrt{1+p_+^2}}\right)-\mu=0$$
$$-\frac{d}{dx}\frac{d}{dp_-}\left(\frac{\beta_1p_-^2+\beta_2}{\sqrt{1+p_-^2}}\right)+\mu=0 \tag{8.7.18}$$

Eqs. (8.7.18) may be rewritten in the condensed form

$$-\frac{d}{dx}\frac{d}{dp_\pm}\left(\frac{\beta_1p_\pm^2+\beta_2}{\sqrt{1+p_\pm^2}}\right)=\pm\mu \tag{8.7.19}$$

The solution to Eq. (8.7.19) describes the equilibrium shape of a platelike inclusion in the habit plane. Introducing the anisotropy parameter

$$\alpha=\frac{\beta_1-\beta_2}{\beta_1} \tag{8.7.20}$$

we may rewrite Eq. (8.7.19) in yet another form

$$-\frac{d}{dx}\frac{d}{dp_\pm}\left(\sqrt{p_\pm^2+1}-\frac{\alpha}{\sqrt{p_\pm^2+1}}\right)=\pm\mu \tag{8.7.21}$$

Since $\beta_1>\beta_2$, we have

$$0\leqslant\alpha\leqslant 1 \tag{8.7.22}$$

The integration of Eq. (8.7.21) over x yields

$$\frac{d}{dp}\left(\sqrt{p^2+1}-\frac{\alpha}{\sqrt{p^2+1}}\right)=\mp\mu x$$

or

$$\frac{p}{\sqrt{p^2+1}}+\frac{\alpha p}{(p^2+1)^{\frac{3}{2}}}=\mp\mu x \tag{8.7.23}$$

Since

$$\frac{d}{dx}=\frac{dy}{dx}\frac{d}{dy}=p\frac{d}{dy}$$

Eq. (8.7.21) may also be rewritten

$$-p\frac{d}{dy}\frac{d}{dp}\left(\sqrt{p^2+1}-\frac{\alpha}{\sqrt{p^2+1}}\right)=\pm\mu \tag{8.7.24}$$

Substitution of the relation

$$\frac{d}{dy}=\frac{dp}{dy}\frac{d}{dp}$$

into (8.7.24) gives

$$-p\frac{dp}{dy}\frac{d^2}{dp^2}\left(\sqrt{p^2+1}-\frac{\alpha}{\sqrt{p^2+1}}\right)=\pm\mu$$

or (8.7.25)

$$-p\frac{d^2}{dp^2}\left(\sqrt{p^2+1}-\frac{\alpha}{\sqrt{p^2+1}}\right)dp=\pm\mu dy$$

The integration of Eq. (8.7.25) yields

$$\frac{1-2\alpha}{\sqrt{1+p^2}}+\frac{\alpha}{(1+p^2)^{\frac{3}{2}}}=\pm\mu y_{\pm} \tag{8.7.26}$$

It should be noted that the integration constants in (8.7.23) and (8.7.26) are omitted since they describe mere rigid body translations of a precipitate particle along the x and y directions, respectively.

Eqs. (8.7.23) and (8.7.26) are the parametric equations describing the two branches, $y=y_+(x)$ and $y=y_-(x)$, of the curve $y=y(x)$ which determines the equilibrium shape of an inclusion in the habit plane at a given value of the anisotropy parameter α of the line tension coefficient. The Lagrange multiplier μ plays the part of a scale factor. Its variation results in the isogonal shape transformation which may be treated as variation of the length scale.

With $\alpha=0$, the parameteric equations (8.7.23) and (8.7.26) become

$$\frac{p}{\sqrt{p^2+1}}=\mp\mu x \qquad \frac{1}{\sqrt{1+p^2}}=\pm\mu y \tag{8.7.27}$$

By squaring and then summing these equations, we obtain

$$x^2+y^2=\frac{1}{\mu^2} \tag{8.7.28}$$

which is the equation describing a circle. The Lagrange multiplier μ plays the role of the inverse radius in (8.7.28). This result shows that, with the zero anisotropy of the linear tension coefficient, the inclusion must have the circular disk shape.

An analytical expression for the contour $y = y(x)$ may also be obtained in the case $\alpha = 0.5$. Eqs. (8.7.23) and (8.7.26) then give

$$\begin{aligned} p(1+p^2)^{-\frac{1}{2}} + \tfrac{1}{2}p(1+p^2)^{-\frac{3}{2}} &= \mp\mu x \\ (1+p^2)^{-\frac{3}{2}} &= \pm 2\mu y \end{aligned} \tag{8.7.29}$$

The parameter p can be excluded to obtain

$$[2+(2\mu y)^{\frac{2}{3}}][1-(2\mu y)^{\frac{2}{3}}]^{\frac{1}{2}} = \mp 2\mu x \tag{8.7.30}$$

The function (8.7.30) describing the inclusion shape in the case $\alpha = 0.5$ is plotted in Fig. 64. The inclusion is oval-shaped in the habit plane (the particle has rounded edges).

If α exceeds 0.5, the functions $y = y(p)$ and $x = x(p)$ obtained from Eqs. (8.7.23) and (8.7.26) describes two different curves. These are shown in Fig. 65a and b for $\alpha = 0.9$. The contour in Fig. 65c describes the shape corresponding to the strain energy minimum whereas that shown in Fig. 65d corresponds to the strain energy maximum [it should be remembered that the original variational equation (8.7.18) is based on the extremum condition, $\delta\Phi = 0$, rather than on the minimum condition, and both strain energy maximum and minimum satisfy that equation]. Of course actual phase transformations go in the direction of

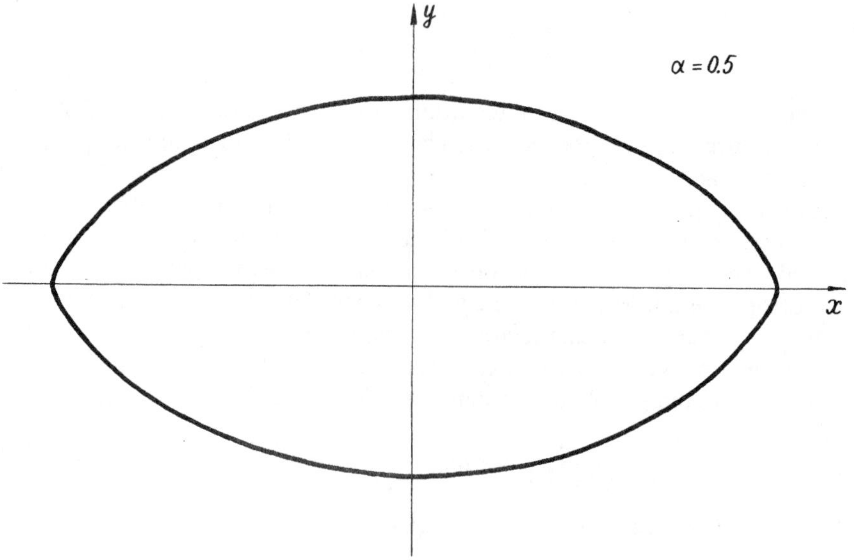

Figure 64. The calculated oval shape of the habit plane of a platelike inclusion at $\alpha = 0.5$.

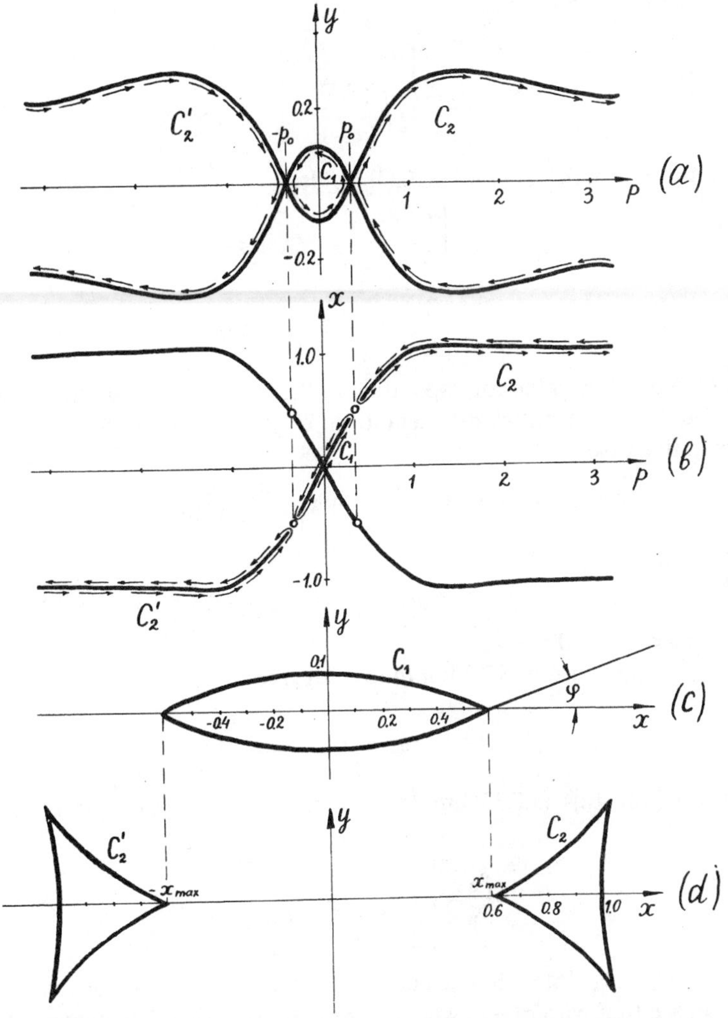

Figure 65. Formation of a coherent platelike inclusion with sharp ends ($\alpha=0.9$). (*a*) Dependence. $y=y(p)$; (*b*) dependence. $x=x(p)$; (*c*) the shape of the inclusion. $y=y(x)$, in the habit plane corresponding to the minimum elastic energy. It is described by curve C_1. (*d*) The shape of the inclusion in the habit plane corresponding to the maximum elastic energy. It is described by curves C_2 and C_2'.

the strain energy minimum, and actual inclusion shapes are described by the contour in Fig. 65c.

The parameter p_0 value corresponding to the half-length x_0 of the equilibrium inclusion can be found from the half-length definition $y(x_0)=0$ in the form

$$\mu y(p_0)=\frac{1-2\alpha}{\sqrt{1+p_0^2}}+\frac{\alpha}{(1+p_0^2)^{\frac{3}{2}}}=0 \tag{8.7.31}$$

The solution of (8.7.31) is

$$p_0=\begin{cases}\infty & \text{if } \alpha\leqslant 0.5\\ \sqrt{\dfrac{1-\alpha}{2\alpha-1}} & \text{if } \alpha\geqslant 0.5\end{cases} \tag{8.7.32}$$

Substituting Eq. (8.7.32) into (8.7.23) yields

$$x_0=\begin{cases}\dfrac{1}{\mu} & \text{if } \alpha\leqslant 0.5\\ \dfrac{2}{\mu}\sqrt{\alpha(1-\alpha)} & \text{if } \alpha\geqslant 0.5\end{cases} \tag{8.7.33}$$

The parameter p_1 value corresponding to the inclusion half-width can be found likewise, from the half-width definition $x(y_0)=0$. The application of (8.7.23) gives in this case

$$x(p_1)=\frac{p_1}{\sqrt{1+p_0^2}}+\alpha\frac{p_1}{(1+p_1^2)^{\frac{3}{2}}}=0 \tag{8.7.34}$$

The solution to (8.7.34) is

$$p_1=0 \tag{8.7.35}$$

for both $\alpha\leqslant 0.5$ and $\alpha\geqslant 0.5$.

Substituting p_1 from (8.7.35) into (8.7.26) gives

$$y_0=\left|\frac{1}{\mu}(1-\alpha)\right| \tag{8.7.36}$$

It follows from Eqs. (8.7.33) and (8.7.36) that the width-to-length ratio y_0/x_0 is

$$\frac{y_0}{x_0}=\begin{cases}1-\alpha & \text{if } \alpha\leqslant 0.5\\ \dfrac{1}{2}\sqrt{\dfrac{1-\alpha}{\alpha}} & \text{if } \alpha\geqslant 0.5\end{cases} \tag{8.7.37}$$

The ratio y_0/x_0 is plotted in Fig. 66a with respect to the anisotropy parameter α. One can see that with $\alpha\to 1$, which is the extreme case of strong anisotropy of the line tension coefficient (8.7.6), the ratio y_0/x_0 approaches zero, i.e. the inclusion has a needlelike (lath) shape when $x_0 \gg y_0 \gg D$.* It should be noted that really needlelike shape requires that $x_0 \gg y_0 \sim D$. "Sharpness" of inclusion ends is determined by the angle ϕ between the tangent to the contour $y=y(x)$ at $x=x_0$ and the axis x. Since $\tan\phi$ is given by the first derivative $(dy/dx)_{x=x_0}$, it may be found from Eq. (8.7.32):

$$\tan\phi=\left(\frac{dy}{dx}\right)_{x=x_0}=p_0=\begin{cases}\infty & \text{if } \alpha\leqslant 0.5\\ \sqrt{\dfrac{1-\alpha}{2\alpha-1}} & \text{if } \alpha>0.5\end{cases}$$

*It should, however, be remembered that the theory is inapplicable when $y_0 \approx D$.

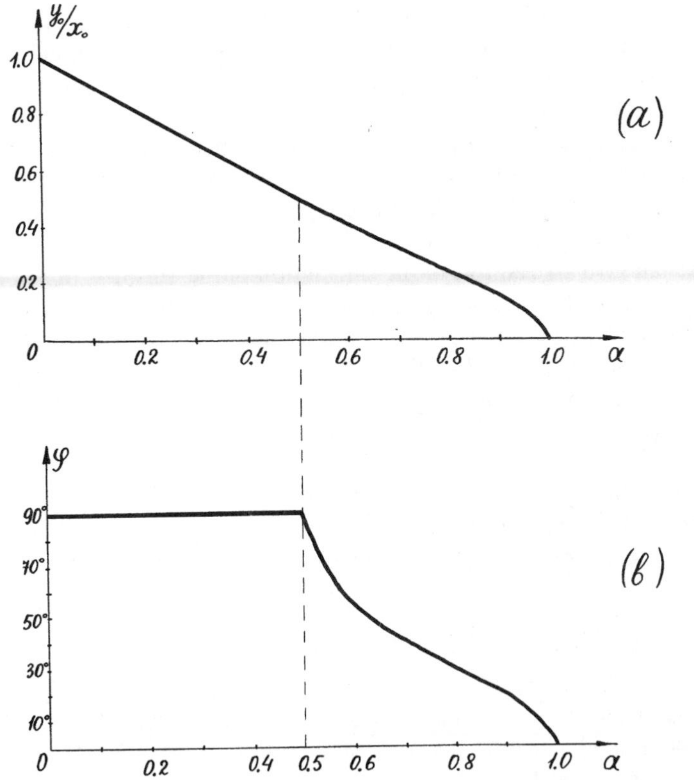

Figure 66. (*a*) The dependence of the width-to-length ratio, y_0/x_0, of a platelike coherent inclusion on the anisotropy parameter α. (*b*) "Sharpness" of inclusion ends, ϕ, as a function of the anisotropy parameter, α. (See Fig. 65c).

if the definition $dy/dx=p$ is recalled. We thus have

$$\phi=\begin{cases}\dfrac{\pi}{2} & \text{if } \alpha\leqslant 0.5\\[2ex] \arctan\sqrt{\dfrac{1-\alpha}{2\alpha-1}} & \text{if } \alpha>0.5\end{cases} \tag{8.7.38}$$

The plot of ϕ with relation to the anisotropy parameter α is given in Fig. 66b.

Figure 66b shows that a platelike inclusion makes an oval in the habit plane ($\phi=\pi/2$) if $\alpha\leqslant 0.5$. If $\alpha>0.5$, the inclusion has sharp ends along the x-axis. Eqs. (8.7.23) and (8.7.26), however, become invalid in the vicinity of sharp ends. In this case a more rigorous calculation of the quantity, $\delta(\mathbf{m})$, is needed. The result should be an oval with the curvature radius of the order of D.

We will now return to the problem of the edge surface energy (8.7.1) contribution to the strain energy. According to what has been said here, this contribution becomes significant when the anisotropy parameter α vanishes and the line tension coefficient $\delta(\mathbf{m})$ turns isotropic [the case of a diagonal $\beta_{ij}(\mathbf{n}_0)$ tensor]. In particular, this may occur when the normal to the habit plane, $\mathbf{n}_0$, is collinear with a three-, four-, or sixfold symmetry axis and the stress-free transformation strain tensor ε_{ij}^0 is invariant under rotations about that axis. It is then necessary to take into consideration anisotropy of the edge surface energy. However in principle, the theory remains the same. The edge surface energy anisotropy only results in the replacement of the coefficient $\delta(\mathbf{m})$ with $\delta(\mathbf{m})+D\gamma_s(\mathbf{m})$ in all the equations.

We shall give an example of the calculation of the tensor $\beta_{ij}(\mathbf{n}_0)$ in Eq. (8.7.9) that determines the anisotropy coefficient α. We shall do it by expanding the function $\Delta B(\mathbf{n})$ in (8.7.9) in powers of $\delta\mathbf{n}$ where $\delta\mathbf{n}$ is the projection of the deviation vector $\Delta\mathbf{n}=\mathbf{n}-\mathbf{n}_0$ on the habit plane. The invariant form of the vector $\mathbf{n}$ presented in terms of the vector $\delta\mathbf{n}$ is

$$\mathbf{n}=\delta\mathbf{n}+(1-\tfrac{1}{2}(\delta\mathbf{n})^2)\mathbf{n}_0 \tag{8.7.39}$$

where $\mathbf{n}_0$ is the unit vector normal to the habit plane. If the Cartesian coordinates are chosen such that the z-axis is normal to the habit plane (collinear with $\mathbf{n}_0$) whereas the x- and y-axes lie in the habit plane, Eq. (8.7.39) may be rewritten in the form

$$\mathbf{n}=(\delta n_x, \delta n_y, 1-\tfrac{1}{2}(\delta\mathbf{n})^2) \tag{8.7.40}$$

where δn_x and δn_y are the x and y components of $\delta\mathbf{n}$, respectively.

Let us consider a tetragonal phase inclusion with the (001) habit. Substitution of (8.7.40) into (8.6.9) and expansion of the result into the Taylor series in $\delta\mathbf{n}$ up to the first nonvanishing term yields

$$B(\mathbf{n})=\bar{B}-\frac{(\sigma_{33}^0)^2}{c_{11}}\psi(\mathbf{n})\approx B(\mathbf{n}_0)-\frac{(\sigma_{33}^0)^2}{c_{11}}\left(A+2\alpha_1^2-2-\xi\frac{c_{11}+c_{12}}{c_{11}}\right)(\delta\mathbf{n})^2 \tag{8.7.41}$$

where $B(\mathbf{n}_0)=\bar{B}-(\sigma_{33}^0)^2/c_{11}$.

Eq. (8.7.41) may be rewritten

$$\Delta B(\mathbf{n})=B(\mathbf{n})-B(\mathbf{n}_0)=\beta(\delta\mathbf{n})^2=\beta\delta n_x^2+\beta\delta n_y^2 \tag{8.7.42}$$

where

$$\beta_1=\beta_2=\beta=-\frac{(\sigma_{33}^0)^2}{c_{11}}\left[A+2\alpha_1^2-2-\frac{\xi(c_{11}+c_{12})}{c_{11}}\right] \tag{8.7.43}$$

Substituting Eqs. (8.6.11f) and (8.6.11c) into (8.7.43) yields

$$\beta=-\frac{(2c_{12}\varepsilon_{11}^0+c_{11}\varepsilon_{33}^0)^2}{c_{11}}\left[\frac{c_{11}(1+\alpha_1^2)-2c_{12}\alpha_1-2(\alpha_1+1)c_{44}}{c_{44}}-\xi\frac{c_{11}+c_{12}}{c_{11}}\right] \tag{8.7.44}$$

where

$$\alpha_1 = \frac{(c_{11}+c_{12})\varepsilon_{11}^0 + c_{12}\varepsilon_{33}^0}{2c_{12}\varepsilon_{11}^0 + c_{11}\varepsilon_{33}^0}$$

$$\xi = \frac{c_{11}-c_{12}-2c_{44}}{c_{44}}$$

Since $\beta_1 = \beta_2$ [Eq. (8.7.43)], the anisotropy parameter α vanishes, and the (001) precipitate must have a disklike shape. It is easy to pass to the limiting case of a cubic precipitate. To do this, we must put $\varepsilon_{11}^0 = \varepsilon_{33}^0 = \varepsilon_0$ in Eq. (8.7.44). The result is

$$\beta = -\frac{(c_{11}+2c_{12})^2}{c_{11}^2}(\varepsilon_0)^2 \xi (c_{11}-c_{12}) \tag{8.7.45}$$

since $\alpha_1 = 1$ and $[c_{11}(1+\alpha_1^2) - 2c_{12}\alpha_1 - 2(\alpha_1+1)c_{44}]/c_{44} = 2\xi$ at $\varepsilon_{11}^0 = \varepsilon_{33}^0$. The ξ value must be negative because the (001) habit with the cubic inclusion within a cubic matrix is realized when $\xi < 0$.

As shown in Section 8.1, the minimization of the sum of the strain and interphase energies yields the equilibrium aspect ratio (8.1.50) at a given inclusion volume V. We shall now show, following Morris, Khachaturyan and Wen (150), that the dependence of the equilibrium aspect ratio on the inclusion volume provides the possibility to determine the interphase energy. We shall consider a tetragonal platelike precipitate with the (001) habit.

It follows from Eq. (8.7.43) that in this case $\beta_1 = \beta_2$, and therefore the line tension coefficient (8.7.8)

$$\delta(\mathbf{m}) = \beta \frac{D^2}{4\pi} \ln \frac{2R}{D} = \text{constant} \tag{8.7.46}$$

is isotropic.

As shown above, this leads to a disklike precipitate. The integration in Eq. (8.7.5), along the circle with $\delta(\mathbf{m})$ given by (8.7.42), yields

$$E_{\text{edge}} = \frac{1}{2}\beta D^2 R \ln \frac{2R}{D} \tag{8.7.47}$$

where R is the disk radius. The interphase energy of the disklike precipitate is

$$E_s = 2\pi R^2 \gamma_s \tag{8.7.48}$$

where γ_s is the interphase specific energy related to the habit plane.

Combining Eqs. (8.7.47) and (8.7.48), we get the total energy

$$E_{\text{total}} = \frac{1}{2}\beta D^2 R \ln \frac{2R}{D} + 2\pi R^2 \gamma_s \tag{8.7.49}$$

Using the definition

$$K = \frac{D}{2R} \tag{8.7.50}$$

of the aspect ratio of an inclusion, we may rewrite Eq. (8.7.49) in the form

$$E_{\text{total}} = -\frac{1}{4}\beta\frac{D^3}{K}\ln K + \frac{\gamma_s \pi D^2}{2K^2} \tag{8.7.51}$$

It follows from the volume definition

$$V = \pi R^2 D$$

and Eq. (8.7.50) that

$$V = \frac{\pi D^3}{4K^2}$$

and

$$D = \left(\frac{4V}{\pi}\right)^{\frac{1}{3}} K^{\frac{2}{3}} \tag{8.7.52}$$

Substituting Eq. (8.7.52) into (8.7.49) gives

$$E_{\text{total}} = V\left[-\frac{\beta}{\pi}K\ln K + \gamma_s\left(\frac{2\pi}{V}\right)^{\frac{1}{3}} K^{-\frac{2}{3}}\right] \tag{8.7.53}$$

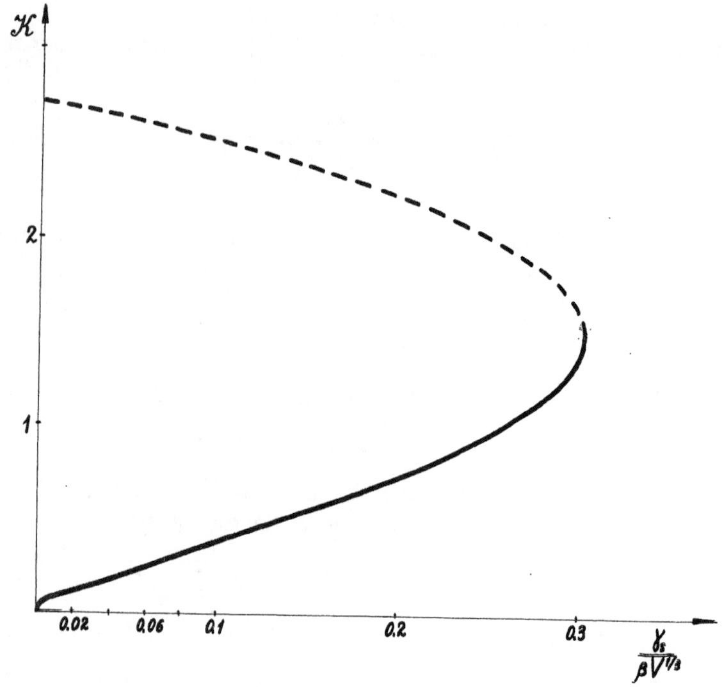

Figure 67. The dependence of the equilibrium aspect ratio, K, on the dimensionless parameter, $\gamma_s/\beta V^{1/3}$ (solid line).

The minimum condition $dE_{\text{total}}/dK=0$ results in the equilibrium aspect ratio condition

$$K^{\frac{5}{3}} \ln \frac{K}{e} = -\left(\frac{16\pi^4}{27}\right)^{\frac{1}{3}} \frac{\gamma_s}{\beta V^{\frac{1}{3}}} \tag{8.7.54}$$

where e is the base of natural logarithm. The dependence of the equilibrium aspect ratio on the dimensionless parameter $\gamma_s/\beta V^{\frac{1}{3}}$ given by Eq. (8.7.54) is plotted in Fig. 67. Using the plot $K=K(\gamma_s/\beta V^{\frac{1}{3}})$ from Fig. 67, one can find the interphase energy coefficient γ_s if the inclusion volume and aspect ratio are known (e.g., from an electron microscopic experiment).

8.8. EQUILIBRIUM SHAPE OF INCLUSION CHARACTERIZED BY INVARIANT PLANE TRANSFORMATION STRAIN

In this section we consider the important case when the stress-free transformation strain is an invariant plane strain:

$$u_{ij}=\varepsilon_0 l_i n_j^0 \tag{8.8.1}$$

where $\boldsymbol{l}$ and $\mathbf{n}_0$ are the unit vectors along the shear direction and the direction normal to the invariant plane, respectively. We shall follow the line of reasoning suggested by Khachaturyan and Rumynina (149). It was shown in Section 8.1 that in this case a new phase inclusion has a platelike shape, its habit plane being normal to the invariant plane. According to Eq. (8.1.26) the strain energy of a platelike inclusion is given by two terms:

$$E=E_{\text{bulk}}+E_{\text{edge}} \tag{8.8.2}$$

where $E_{\text{bulk}}=\frac{1}{2}B(\mathbf{n}_0)V$, with $\mathbf{n}_0$ being the unit vector normal to the habit plane, E_{edge} the strain energy described by Eqs. (8.7.5) and (8.7.6), and V the inclusion volume.

The first term in (8.8.2) which is proportional to the inclusion volume renormalizes the bulk chemical free energy whereas the second one is proportional to the inclusion perimeter and describes the edge effect. As follows from Eq. (8.1.23), the volume term vanishes in the case under consideration [when the stress-free strain is described by Eq. (8.8.1)]. The total strain energy is thus given by the term E_{edge}.

It is of interest that the solution to this problem may also be applied to determine the strain energy associated with a plane dislocation loop. In fact a dislocation loop, which is a fragment of an extra plane, can always be treated as an extremely thin platelike coherent inclusion whose thickness is equal to the interplanar distance, whose perimeter contour coincides with the dislocation loop itself and the vector $\varepsilon_0\boldsymbol{l}$ multiplied by the interplanar distance is the Burgers vector $\mathbf{b}$.

To calculate the shape of a platelike inclusion in the habit plane, one should determine the components of the tensor $\beta_{ij}(\mathbf{n}_0)$ in Eq. (8.7.6). Eq. (8.7.9) defines these components as coefficients of the quadratic terms in the Taylor expansion of $\Delta B(\mathbf{n})$. The unit vector, $\mathbf{n}$, departing slightly from the normal to the habit plane, $\mathbf{n}_0$, can be written in the invariant form (8.7.39)

$$\mathbf{n}=\mathbf{n}_0+\delta\mathbf{n}-\tfrac{1}{2}(\delta\mathbf{n})^2\mathbf{n}_0 \tag{8.8.3}$$

where $\delta\mathbf{n}$ is the projection of a small deviation, $\mathbf{n}-\mathbf{n}_0$, onto the habit plane. The function $B(\mathbf{n})$ from (8.1.2) can be presented in the symbolic form

$$B(\mathbf{n})=\bar{B}-(\mathbf{n}\hat{\sigma}^0\hat{\Omega}\hat{\sigma}^0\mathbf{n}) \tag{8.8.4}$$

where $\bar{B}=\lambda_{ijkl}\varepsilon^0_{ij}\varepsilon^0_{kl}=\text{constant}$. It follows from Eq. (8.8.4) that the expansion (8.7.9) up to the quadratic terms can be written as follows:

$$\begin{aligned}\Delta B(\mathbf{n})&=\delta^2B(\mathbf{n})+\cdots=-\delta^2(\mathbf{n}\hat{\sigma}^0\hat{\Omega}(\mathbf{n})\hat{\sigma}^0\mathbf{n})+\cdots\\ &=-[(\delta^2\mathbf{n}\hat{\sigma}^0\hat{\Omega}(\mathbf{n}_0)\hat{\sigma}^0\mathbf{n}_0)+(\mathbf{n}_0\hat{\sigma}^0\hat{\Omega}(\mathbf{n}_0)\hat{\sigma}^0\delta^2\mathbf{n})+(\delta\mathbf{n}\hat{\sigma}^0\hat{\Omega}(\mathbf{n}_0)\hat{\sigma}^0\delta\mathbf{n})\\ &\quad+(\delta\mathbf{n}\hat{\sigma}^0\delta\hat{\Omega}(\mathbf{n}_0)\hat{\sigma}^0\mathbf{n}_0)+(\mathbf{n}_0\hat{\sigma}^0\delta\hat{\Omega}(\mathbf{n}_0)\hat{\sigma}^0\delta\mathbf{n})+(\mathbf{n}_0\hat{\sigma}^0\delta^2\hat{\Omega}(n_0)\hat{\sigma}^0\mathbf{n}_0)]\end{aligned} \tag{8.8.5}$$

where δ^2 is the second variation symbol.

To determine the variations $\delta\hat{\Omega}$ and $\delta^2\hat{\Omega}$, we may use the Taylor expansions:

$$\hat{\Omega}(\mathbf{n})=\hat{\Omega}(\mathbf{n}_0)+\delta\hat{\Omega}(\mathbf{n})+\delta^2\hat{\Omega}(\mathbf{n})+\cdots \tag{8.8.6a}$$

$$\hat{\Omega}^{-1}(\mathbf{n})=\hat{\Omega}^{-1}(\mathbf{n}_0)+\delta\hat{\Omega}^{-1}(\mathbf{n})+\delta^2\hat{\Omega}^{-1}(\mathbf{n})+\cdots \tag{8.8.6b}$$

and the equality

$$\hat{\Omega}(\mathbf{n})\hat{\Omega}^{-1}(\mathbf{n})=\hat{\mathbf{I}} \tag{8.8.7}$$

It follows from Eq. (8.8.7) that

$$\hat{\Omega}(\mathbf{n})=[\hat{\Omega}^{-1}]^{-1} \tag{8.8.8}$$

Substituting Eq. (8.8.6b) into (8.8.8) gives

$$\begin{aligned}\hat{\Omega}(\mathbf{n})&=[\hat{\Omega}^{-1}(\mathbf{n}_0)+\delta\hat{\Omega}^{-1}(\mathbf{n})+\delta^2\hat{\Omega}^{-1}(\mathbf{n})+\cdots]^{-1}\\ &=[\hat{\Omega}^{-1}(\mathbf{n}_0)(\hat{\mathbf{I}}+\hat{\Omega}(\mathbf{n}_0)\delta\hat{\Omega}^{-1}(\mathbf{n})+\hat{\Omega}(\mathbf{n}_0)\delta^2\hat{\Omega}^{-1}(\mathbf{n})+\cdots)]^{-1}\end{aligned} \tag{8.8.9}$$

Taking into consideration the operator identity,*

$$[\hat{\mathbf{A}}\hat{\mathbf{B}}]^{-1}=\hat{\mathbf{B}}^{-1}\hat{\mathbf{A}}^{-1}$$

we can rewrite Eq. (8.8.9) in the form

$$\hat{\Omega}(\mathbf{n})=[\hat{\mathbf{I}}+\hat{\Omega}(\mathbf{n}_0)\delta\hat{\Omega}^{-1}(\mathbf{n})+\hat{\Omega}(\mathbf{n}_0)\delta^2\hat{\Omega}^{-1}(\mathbf{n})+\cdots]^{-1}\hat{\Omega}(\mathbf{n}_0) \tag{8.8.10}$$

Expanding the first term in Eq. (8.8.10) in the power series of $\hat{\Omega}(\mathbf{n}_0)\delta\hat{\Omega}^{-1}(\mathbf{n})$, $\hat{\Omega}(\mathbf{n}_0)\delta^2\hat{\Omega}^{-1}(\mathbf{n}),\ldots$, and taking into consideration that the values $\hat{\Omega}(\mathbf{n}_0)\delta\hat{\Omega}^{-1}(\mathbf{n})$, $\hat{\Omega}(\mathbf{n}_0)\delta^2\hat{\Omega}^{-1}(\mathbf{n}),\ldots$ are the first-, second-, and so on, order of magnitude quanti-

*To prove this identity, it will suffice to show that the identity $(\hat{\mathbf{A}}\hat{\mathbf{B}})^{-1}(\hat{\mathbf{A}}\hat{\mathbf{B}})=\hat{\mathbf{I}}$ remains valid after substituting $\hat{\mathbf{B}}^{-1}\hat{\mathbf{A}}^{-1}$ for $(\hat{\mathbf{A}}\hat{\mathbf{B}})^{-1}$. In fact $\hat{\mathbf{B}}^{-1}\hat{\mathbf{A}}^{-1}\hat{\mathbf{A}}\hat{\mathbf{B}}=\hat{\mathbf{B}}^{-1}(\hat{\mathbf{A}}^{-1}\hat{\mathbf{A}})\hat{\mathbf{B}}=\hat{\mathbf{B}}^{-1}\hat{\mathbf{B}}=\hat{\mathbf{I}}$ since $\hat{\mathbf{B}}^{-1}\hat{\mathbf{B}}=\hat{\mathbf{I}}$ and $\hat{\mathbf{A}}^{-1}\hat{\mathbf{A}}=\hat{\mathbf{I}}$ by definition.

ties with respect to $\delta\mathbf{n}$, we obtain

$$\hat{\mathbf{\Omega}}(\mathbf{n})=[\hat{\mathbf{I}}-\hat{\mathbf{\Omega}}(\mathbf{n}_0)\delta\hat{\mathbf{\Omega}}^{-1}(\mathbf{n})+(\hat{\mathbf{\Omega}}(\mathbf{n}_0)\delta\hat{\mathbf{\Omega}}^{-1}(\mathbf{n}))^2-\hat{\mathbf{\Omega}}(\mathbf{n}_0)\delta^2\hat{\mathbf{\Omega}}^{-1}(\mathbf{n})+\cdots]\hat{\mathbf{\Omega}}(\mathbf{n}_0)$$

or

$$\hat{\mathbf{\Omega}}(\mathbf{n})=\hat{\mathbf{\Omega}}(\mathbf{n}_0)-\hat{\mathbf{\Omega}}(\mathbf{n}_0)\delta\hat{\mathbf{\Omega}}^{-1}(\mathbf{n})\hat{\mathbf{\Omega}}(\mathbf{n}_0)+(\hat{\mathbf{\Omega}}(\mathbf{n}_0)\delta\hat{\mathbf{\Omega}}^{-1}(\mathbf{n}))^2\hat{\mathbf{\Omega}}(\mathbf{n}_0)$$
$$-\hat{\mathbf{\Omega}}(\mathbf{n}_0)\delta^2\hat{\mathbf{\Omega}}^{-1}(\mathbf{n})\hat{\mathbf{\Omega}}(\mathbf{n}_0)+\cdots \tag{8.8.11}$$

Comparison of the expansions (8.8.11) and (8.8.6a) shows that

$$\delta\hat{\mathbf{\Omega}}(\mathbf{n})=-\hat{\mathbf{\Omega}}(\mathbf{n}_0)\delta\hat{\mathbf{\Omega}}^{-1}(\mathbf{n})\hat{\mathbf{\Omega}}(\mathbf{n}_0) \tag{8.8.12}$$

$$\delta^2\hat{\mathbf{\Omega}}(\mathbf{n})=(\hat{\mathbf{\Omega}}(\mathbf{n}_0)\delta\hat{\mathbf{\Omega}}^{-1}(\mathbf{n}))^2\hat{\mathbf{\Omega}}(\mathbf{n}_0)-\hat{\mathbf{\Omega}}(\mathbf{n}_0)\delta^2\hat{\mathbf{\Omega}}^{-1}(\mathbf{n})\hat{\mathbf{\Omega}}(\mathbf{n}_0) \tag{8.8.13}$$

Substituting Eq. (8.8.1) into the definition

$$\sigma_{ij}^0=\lambda_{ijkl}u_{kl}^0$$

and of σ_{ij}^0 into the product $\hat{\mathbf{\Omega}}(\mathbf{n}_0)\hat{\sigma}^0\mathbf{n}_0$ gives

$$\hat{\mathbf{\Omega}}(\mathbf{n}_0)\hat{\sigma}^0\mathbf{n}_0=\Omega_{ij}(\mathbf{n}_0)\sigma_{jl}^0n_l^0=\varepsilon_0\Omega_{ij}(\mathbf{n}_0)\lambda_{jlsp}l_sn_p^0n_l^0$$
$$=\varepsilon_0\Omega_{ij}(\mathbf{n}_0)\Omega_{js}^{-1}(\mathbf{n}_0)l_s=\varepsilon_0l_i \tag{8.8.14}$$

that is,

$$\hat{\mathbf{\Omega}}(\mathbf{n}_0)\hat{\sigma}^0\mathbf{n}_0=\varepsilon_0\boldsymbol{l}$$

since

$$\lambda_{jlsp}n_p^0n_l^0=\lambda_{jlps}n_p^0n_l^0=\Omega_{js}^{-1}(\mathbf{n}_0)$$

Using Eqs. (8.8.12) to (8.8.14) and the relation $\delta^2\mathbf{n}=-\frac{1}{2}(\delta\mathbf{n})^2\mathbf{n}_0$ [see Eq. (8.8.3)], we may rewrite Eq. (8.8.5) as

$$\Delta B(\mathbf{n})=(\mathbf{n}_0\hat{\sigma}^0\hat{\mathbf{\Omega}}(\mathbf{n}_0)\hat{\sigma}^0\mathbf{n}_0)(\delta\mathbf{n})^2-(\delta\mathbf{n}\hat{\sigma}^0\hat{\mathbf{\Omega}}(\mathbf{n}_0)\hat{\sigma}^0\delta\mathbf{n})$$
$$+\varepsilon_0(\delta\mathbf{n}\hat{\sigma}^0\hat{\mathbf{\Omega}}(\mathbf{n}_0)\delta\hat{\mathbf{\Omega}}^{-1}\boldsymbol{l})+\varepsilon_0(\boldsymbol{l}\delta\hat{\mathbf{\Omega}}^{-1}\hat{\mathbf{\Omega}}(\mathbf{n}_0)\hat{\sigma}^0\delta\mathbf{n})$$
$$-\varepsilon_0^2(\boldsymbol{l}\delta\hat{\mathbf{\Omega}}^{-1}\hat{\mathbf{\Omega}}(\mathbf{n}_0)\delta\hat{\mathbf{\Omega}}^{-1}\boldsymbol{l})+\varepsilon_0^2(\boldsymbol{l}\delta^2\hat{\mathbf{\Omega}}^{-1}\boldsymbol{l}) \tag{8.8.15}$$

With the definition

$$\delta\hat{\mathbf{\Omega}}^{-1}\boldsymbol{l}=\frac{1}{\varepsilon_0}(\hat{\sigma}^0+\hat{\sigma}_1^+)\delta\mathbf{n} \tag{8.8.16}$$

where

$$(\hat{\sigma}_1)_{ij}=\sigma_{ij}^{(1)}=\varepsilon_0\lambda_{ilmj}l_ln_m^0 \tag{8.8.17}$$

Eq. (8.8.15) may be written

$$\Delta B(\mathbf{n})=[\varepsilon_0^2(\boldsymbol{l}\hat{\mathbf{\Omega}}^{-1}(\mathbf{n}_0)\boldsymbol{l})(\delta\mathbf{n})^2-(\delta\mathbf{n}\hat{\sigma}_1\hat{\mathbf{\Omega}}(\mathbf{n}_0)\hat{\sigma}_1^+\delta\mathbf{n})$$
$$+\varepsilon_0^2(\boldsymbol{l}\delta^2\hat{\mathbf{\Omega}}^{-1}(\mathbf{n})\boldsymbol{l})]+\cdots \tag{8.8.18}$$

Since

$$(\delta^2\hat{\mathbf{\Omega}}^{-1}(\mathbf{n}))_{ij}=\lambda_{ilmj}\delta n_l\delta n_m-(\hat{\mathbf{\Omega}}^{-1}(\mathbf{n}_0))_{ij}(\delta\mathbf{n})^2$$

we have

$$\Delta B(\mathbf{n}) = \varepsilon_0^2(\delta\mathbf{n}\hat{\boldsymbol{\Omega}}^{-1}(\boldsymbol{l})\delta\mathbf{n}) - (\delta\mathbf{n}\hat{\sigma}_1\hat{\boldsymbol{\Omega}}(\mathbf{n}_0)\hat{\sigma}_1^+\delta\mathbf{n}) \tag{8.8.19a}$$

or in the suffix form

$$\Delta B(\mathbf{n}) = \varepsilon_0^2[\Omega_{ij}^{-1}(\boldsymbol{l}) - (\hat{\sigma}_1\hat{\boldsymbol{\Omega}}(\mathbf{n}_0)\hat{\sigma}_1^+)_{ij}]\delta n_i \delta n_j + \cdots \tag{8.8.19b}$$

Comparison of Eqs. (8.7.9) and (8.8.19b) yields

$$\hat{\boldsymbol{\beta}}(\mathbf{n}_0) = \varepsilon_0^2\hat{\boldsymbol{\Omega}}^{-1}(\boldsymbol{l}) - \hat{\sigma}_1\hat{\boldsymbol{\Omega}}(\mathbf{n}_0)\hat{\sigma}_1^+ \tag{8.8.20a}$$

or

$$\beta_{ij}(\mathbf{n}_0) = \varepsilon_0^2\Omega_{ij}^{-1}(\boldsymbol{l}) - (\hat{\sigma}_1\hat{\boldsymbol{\Omega}}(\mathbf{n}_0)\hat{\sigma}_1^+)_{ij} \tag{8.8.20b}$$

It is easy to see that the vector $\mathbf{n}_0$ (the vector normal to the habit plane) is an eigenvector of the operator (8.8.20a). The corresponding eigenvalue is equal to zero. Since $\hat{\boldsymbol{\beta}}(\mathbf{n}_0)$ is a Hermitian matrix, two other eigenvectors, $\mathbf{e}_1$ and $\mathbf{e}_2$, corresponding to the positive eigenvalues β_1 and β_2 are orthogonal to each other and to the vector $\mathbf{n}_0$ and therefore lie in the habit plane.

Thus the problem of determining the shape of a coherent inclusion in the habit plane is reduced to (1) the diagonalization of the matrix $\hat{\boldsymbol{\beta}}(\mathbf{n}_0)$ in Eq. (8.8.20a), (2) the determination of the anisotropy parameter, α, and (3) the calculation of the shape of the inclusion from the parametric equations (8.7.23) and (8.7.26).

Using Eq. (8.8.20a) in Eq. (8.7.6), and Eq. (8.7.6) in Eq. (8.7.5), we have finally for the strain energy of an inclusion

$$E_{\text{edge}} = \frac{D^2}{4\pi}\varepsilon_0^2 \ln\frac{L}{D}\oint_P [\mathbf{m}(\hat{\boldsymbol{\Omega}}^{-1}(\boldsymbol{l}) - \hat{\sigma}_1\hat{\boldsymbol{\Omega}}(\mathbf{n}_0)\hat{\sigma}_1^+)\mathbf{m}]dl_{\mathrm{m}} \tag{8.8.21}$$

Bearing in mind that with a dislocation loop, $D = d$ where d is the interplanar distance and $d\varepsilon_0 = b$ is the Burgers vector modulus, we can see that Eq. (8.8.21) coincides completely with the expression for the dislocation energy in an anisotropic elastic medium (155, 156). Representation (8.8.20b) of the tensor $\beta_{ij}(\mathbf{n}_0)$ was also reported in (91).

8.9. SHAPE OF FERROMAGNETIC PRECIPITATES

In many technologically important alloys the decomposition into ferromagnetic and paramagnetic phases is used to improve the magnetic properties. It is of interest to elucidate the effect of the magnetostatic energy on the shape of precipitates as has been done before with the elastic strain energy. This will enable us to demonstrate the profound analogy between the elastic and magnetostatic energies of multiphase alloys. The technique of the **k**-space analysis that has been developed earlier may be as efficient in the case of systems of ferromagnetic particles as it is in the case of coherent precipitates.

Magnetostatic energy may be represented as the sum of interacting magnetic dipoles:

$$E_{\text{mag}}=\frac{1}{2}\iint d^3r\, d^3r'\, m_i(\mathbf{r})\left(\frac{\delta_{ij}}{|\mathbf{r}-\mathbf{r}'|^3}-\frac{3(r_i-r_i')(r_j-r_j')}{|\mathbf{r}-\mathbf{r}'|^5}\right)m_j(\mathbf{r}') \tag{8.9.1}$$

where $\mathbf{m}(\mathbf{r})$ is the magnetization density at the point $\mathbf{r}$. The integration in (8.9.1) is over the whole crystal body. Making use of the Fourier representations

$$\mathbf{m}(\mathbf{r})=\int\frac{d^3k}{(2\pi)^3}\mathbf{M}(\mathbf{k})e^{i\mathbf{kr}}$$

$$\frac{\delta_{ij}}{r^3}-3\frac{r_ir_j}{r^5}=4\pi\int\frac{d^3k}{(2\pi)^3}\frac{k_ik_j}{k^2}e^{i\mathbf{kr}} \tag{8.9.2}$$

in Eq. (8.9.1) we have*

$$E_{\text{mag}}=2\pi\int\frac{d^3k}{(2\pi)^3}\frac{|\mathbf{M}(\mathbf{k})\mathbf{k}|^2}{k^2} \tag{8.9.3}$$

Consider an arbitrary system of ferromagnetic particles with several directions of magnetization designated by the index p. In this case the spatial distribution of magnetization will be given by the equation

$$\mathbf{m}(\mathbf{r})=M_0\sum_p\mathbf{e}(p)\tilde{\theta}_p(\mathbf{r}) \tag{8.9.4}$$

[compare with Eq. (7.2.15)] where $\tilde{\theta}_p(\mathbf{r})$ is the shape function that describes the distribution of ferromagnetic phase domains with the magnetization direction parallel to the unit vector $\mathbf{e}(p)$. The magnetic vectorial dipole $M_0\mathbf{e}(p)$ plays the same role in Eq. (8.9.4) as the tensor elastic dipole $\hat{\sigma}^0(p)$ does in Eq. (7.2.15). The Fourier transform of Eq. (8.9.4) yields

$$M(\mathbf{k})=M_0\sum_p\mathbf{e}(p)\theta_p(\mathbf{k}) \tag{8.9.5}$$

Substituting Eq. (8.9.5) into (8.9.3), we obtain

$$E_{\text{mag}}=2\pi M_0^2\sum_{p,q}\int\frac{d^3k}{(2\pi)^3}B_{\text{mag}}^{pq}(\mathbf{n})\theta_p(\mathbf{k})\theta_q^*(\mathbf{k}) \tag{8.9.6}$$

where

$$\mathbf{n}=\frac{\mathbf{k}}{k}$$

$$B_{\text{mag}}^{pq}(\mathbf{n})=(\mathbf{e}(p)\mathbf{n})(\mathbf{e}(q)\mathbf{n}) \tag{8.9.7}$$

[compare with Eq. (7.2.53b)]. Since the shape functions $\tilde{\theta}_p(\mathbf{r})$ may in general describe multiply connected regions occupied by the ferromagnetic phase, Eq. (8.9.6) as well as Eq. (7.2.53b) is applicable to an arbitrary set of ferromagnetic or monodomain particles in a paramagnetic matrix. The profound analogy

*For simplicity, the magnetic susceptibility is assumed to be equal to unity.

between the elastic and magnetostatic energies arises from the fact that both the function (8.9.7) and the corresponding function in (7.2.53b) depend on the direction **n** of the wave vector **k** rather than on the absolute value of **k**.

The representation

$$\tilde{\theta}_p(\mathbf{r})=\sum_{\alpha} \tilde{\theta}_p^0(\alpha, \mathbf{r}-\mathbf{R}_\alpha) \tag{8.9.8}$$

describes the shape function $\tilde{\theta}_p(\mathbf{r})$ as a sum of the shape functions $\tilde{\theta}_p^0(\alpha, \mathbf{r}-\mathbf{R}_\alpha)$ where $\tilde{\theta}_p^0(\alpha, \mathbf{r}-\mathbf{R}_\alpha)$ is the single simply connected particle α at the position $\mathbf{R}_\alpha$ with the magnetization direction p. The Fourier transform of Eq. (8.9.8) yields

$$\theta_p(\mathbf{k})=\sum_{p} \theta_p^0(\alpha, \mathbf{k})e^{i\mathbf{k}\mathbf{R}_\alpha} \tag{8.9.9}$$

Substituting Eq. (8.9.9) into (8.9.6), we have

$$E_{\text{mag}}=2\pi M_0^2 \sum_{p,q} \sum_{\alpha,\beta} e^{i\mathbf{k}(\mathbf{R}_\alpha-\mathbf{R}_\beta)} \int \frac{d^3k}{(2\pi)^3} B_{\text{mag}}^{pq}(\mathbf{n})\theta_p^0(\alpha, \mathbf{k})\theta_q^{0*}(\beta, \mathbf{k}) \tag{8.9.10}$$

It follows from Eq. (8.9.10) that the Fourier transform of the magnetostatic interaction energy of two arbitrary magnetic particles is

$$V_{pq}^{\alpha\beta}(\mathbf{k})=2\pi M_0^2 B_{\text{mag}}^{pq}(\mathbf{n})\theta_p^0(\alpha, \mathbf{k})\theta_q^{0*}(\beta, \mathbf{k}) \tag{8.9.11}$$

The pairwise interaction energy $W_{pq}^{\alpha\beta}(\mathbf{R}_\alpha-\mathbf{R}_\beta)$ is given by the back Fourier transform of (8.9.11):

$$W_{pq}^{\alpha\beta}(\mathbf{R}_\alpha-\mathbf{R}_\beta)=2\pi M_0^2 \int \frac{d^3k}{(2\pi)^3} B_{\text{mag}}^{pq}(\mathbf{n})\theta_p^0(\alpha, \mathbf{k})\theta_q^{0*}(\beta, \mathbf{k})e^{i\mathbf{k}(\mathbf{R}_\alpha-\mathbf{R}_\beta)} \tag{8.9.12}$$

[compare with Eq. (7.3.12)].

Let us consider a single ferromagnetic precipitate within a paramagnetic matrix. In this case Eq. (8.9.10) is reduced to

$$E_{\text{mag}}=2\pi M_0^2 \int \frac{d^3k}{(2\pi)^3} B_{\text{mag}}(\mathbf{n})|\theta(\mathbf{k})|^2 \tag{8.9.13}$$

where

$$B_{\text{mag}}(\mathbf{n})=(\mathbf{en})^2 \tag{8.9.14}$$

and $\theta(\mathbf{k})$ is the Fourier transform of the shape function of a single precipitate, **e** is the direction of magnetization.

One may readily see that the function $B_{\text{mag}}(\mathbf{n})$ is analogous to the function $B(\mathbf{n})$ given by Eq. (8.1.2). The difference is that the function $B_{\text{mag}}(\mathbf{n})=(\mathbf{ne})^2$ is degenerate with respect to all the vectors **n** belonging to a cone whose axis is parallel to the magnetization direction **e**. The minimum value, $B_{\text{mag}}(\mathbf{n})=0$, is assumed when the vectors **n** lie in the plane normal to **e**. This degeneration gives rise to the basic difference between particle shapes minimizing the elastic and magnetostatic energies.

As shown in Section 8.1, the minimum value is, as a rule, taken by $B(\mathbf{n})$ on several discrete orientations $\mathbf{n}_0$ which may be brought into coincidence with each other by the symmetry operations of the precipitate phase. The latter circumstance results in the formation of platelike precipitates with the habit normal to $\mathbf{n}_0$. The degeneration of min $B_{mag}(\mathbf{n})$ with respect to all the vectors $\mathbf{n}$ belonging to the plane normal to the magnetization axis $\mathbf{e}$ leads to a different result: to a rodlike shape of precipitates, the rod axis being directed along $\mathbf{e}$.

It follows from Eq. (8.9.13) that the minimal magnetostatic energy, E_{mag}, is zero if the function $|\theta(\mathbf{k})|^2$ describing the shape of the precipitate assumes non-zero values only within the plane in the $\mathbf{k}$-space crossing the point $\mathbf{k}=0$ and directed perpendicular to the axis $\mathbf{e}$ [in this case the integration in Eq. (8.9.13) is in effect taken over $\mathbf{k}$ normal to $\mathbf{e}$ when $\mathbf{ne}=\mathbf{ke}/k=0$]. Such a function $|\theta(\mathbf{k})|^2$ arises when the precipitate volume, V, is stretched into an infinitely thin and infinitely long rod whose axis is parallel to the magnetization direction $\mathbf{e}$. Formation of such an infinite rod cannot, however, be realized since the interphase energy in this case would also tend to infinity. Competition between the magnetostatic and interphase energies provides an optimal shape which is a compromise between the two energies. According to Eq. (8.9.14) the function $B_{mag}(\mathbf{n})$ assumes its minimal value equal to zero at $\mathbf{n}$ lying in the plane normal to $\mathbf{e}$ and increases if $\mathbf{n}$ deviates from that plane. The integration over $\mathbf{k}$ in (8.9.13) makes contributions only from directions of $\mathbf{n}$, deviating from the plane normal to $\mathbf{e}$. If, on the other hand, the precipitate is a rod directed along $\mathbf{e}$ and its radius is equal to R_0 and half-length to H_0, the function $|\theta(\mathbf{k})|^2$ differs from zero only within a thin and extended disk in the $\mathbf{k}$-space normal to $\mathbf{e}$. The typical thickness of this disk is $2\pi/H_0$; its typical radius is $2\pi/R_0$. Therefore the integration in (8.9.13) is carried out over the disk in the $\mathbf{k}$-space. In the integration the typical deviation of the vector $\mathbf{n}$ from the plane normal to $\mathbf{e}$, $\delta\mathbf{n}$, where $(\mathbf{ne})^2=0$, is of the order of

$$\delta n \sim \frac{2\pi/H_0}{2\pi/R_0} = \frac{R_0}{H_0}$$

In the case of a rodlike shape,

$$\frac{R_0}{H_0} \ll 1 \tag{8.9.15}$$

The typical value of the magnetostatic energy is thus

$$E_{mag} \sim \tfrac{1}{2} M_0^2 V \frac{R_0}{H_0} \tag{8.9.16}$$

where $M_0^2 V$ is the typical magnetostatic energy. Taking into account that $V \sim R_0^2 H_0$, we may rewrite Eq. (8.9.16) as

$$E_{mag} \sim \tfrac{1}{2} M_0^2 R_0^3 \tag{8.9.17}$$

This estimate may be exemplified by the following calculations. Let the rod have a shape of an extended circular cylinder, whose axis is parallel to $\mathbf{e}$. The

Fourier transform of the shape function of such a cylinder is

$$\theta(\mathbf{k}) = \int_{(V)} e^{-i\mathbf{kr}} d^3r = 4\pi R_0^2 H_0 \frac{\sin k_z H_0}{k_z H_0} \frac{J_1(\tau R_0)}{\tau R_0} \tag{8.9.18}$$

where $J_1(x)$ is the Bessel function of the first order; the z-axis of the Cartesian coordinate system is parallel to the cylinder axis direction $\mathbf{e}$. The value k_z is the projection of the $\mathbf{k}$ vector on the axis $\mathbf{e}$; $\boldsymbol{\tau} = (k_x, k_y)$ is the projection of the $\mathbf{k}$ vector on the x, y coordinate plane. In this case

$$B_{\text{mag}}(\mathbf{n}) = (\mathbf{ne})^2 = \frac{(\mathbf{ke})^2}{k^2} = \frac{k_z^2}{k_z^2 + \tau^2} \tag{8.9.19}$$

Substituting (8.9.19) and (8.9.18) to Eq. (8.9.13) yields

$$E_{\text{mag}} = 2\pi M_0^2 \iiint_{-\infty}^{\infty} \frac{k_z^2}{k_z^2 + \tau^2} (4\pi R_0^2 H_0)^2 \frac{\sin^2 k_z H_0}{(k_z H_0)^2} \left(\frac{J_1(\tau R_0)}{\tau R_0} \right)^2 \frac{dk_z}{2\pi} \frac{d^2\tau}{(2\pi)^2} \tag{8.9.20}$$

Integrating (8.9.20) over k_z, we obtain

$$E_{\text{mag}} = \frac{1}{2} M_0^2 4\pi \left[\frac{4}{3\pi} - \int_0^\infty \frac{J_1^2(x)}{x^2} e^{-2x(H_0/R_0)} dx \right] R_0^3 \tag{8.9.21}$$

Since in the case of a long rod $H_0/R_0 \gg 1$, the integral in Eq. (8.9.21) can readily be estimated by term-by-term integration of the Taylor expansion of the Bessel function in x. The result is

$$E_{\text{mag}} = 2\pi M_0^2 R_0^3 \left(\frac{4}{3\pi} - \frac{1}{8} \frac{R_0}{H_0} + 0\left(\frac{R_0}{H_0} \right) \right) \tag{8.9.22}$$

Eq. (8.9.22) is in agreement with the estimate (8.9.17). To estimate the equilibrium shape of a ferromagnetic precipitate, we shall consider the sum of the magnetostatic energy (8.9.22) and the interphase energy

$$E_s = \gamma_s 4\pi R_0 H_0 \tag{8.9.23}$$

where $4\pi R_0^2 H_0$ is the lateral surface area of the rod (the area of the rod bases, $2\pi R_0^2$, may be neglected), γ_s is the specific interphase energy.

In the case of $R_0/H_0 \ll 1$, this sum is

$$E = \frac{8 M_0^2 R_0^3}{3} + \gamma_s 4\pi R_0 H_0 \tag{8.9.24}$$

Introducing the aspect ratio

$$K = \frac{H_0}{R_0}$$

we may represent Eq. (8.9.24) in terms of the aspect ratio K and volume V:

$$E = \frac{4}{3\pi} M_0^2 V K^{-1} + 2\gamma_s (2\pi V^2 K)^{\frac{1}{3}} \tag{8.9.25}$$

[compare Eq. (8.9.25) with Eq. (8.7.53)]. Minimizing Eq. (8.9.25) with respect to K at a given volume, V, we obtain the equation for the equilibrium aspect ratio

$$K_{eq}=\frac{V^{\frac{1}{3}}}{2\pi}\left(\frac{4M_0^2}{\gamma_s}\right)^{\frac{3}{4}} \tag{8.9.26}$$

Eq. (8.9.26) shows that rodlike shapes (the case $K_{eq} \to H_0/R_0 \gg 1$) are realized for large precipitates (the volume V is large) far from the Curie point (large magnetization M_0) and small interphase energy coefficient, γ_s.

As follows from Eq. (8.9.26), the equilibrium aspect ratio depends on the precipitate volume V. This results in the change of the morphology of the precipitate with coarsening. At the early stage of aging when the volume V is small the equiaxial shape may be expected. Coarsening will lead to an increase in the volume and therefore elongation of precipitates in accordance with Eq. (8.9.26).

In many real alloys the shape of ferromagnetic particles is also affected by elastic strain. The elastic strain effect can readily be taken into account if one combines Eqs. (8.1.1) and (8.9.13). The resulting equation,

$$E=\frac{1}{2}\int\frac{d^3k}{(2\pi)^3}\,B_{eff}(\mathbf{n})|\theta(\mathbf{k})|^2 \tag{8.9.27a}$$

where

$$B_{eff}(\mathbf{n})=B(\mathbf{n})+4\pi M_0^2(\mathbf{ne})^2 \tag{8.9.27b}$$

has the same form as Eqs. (8.1.1) and (8.9.13). This means that the treatment applied to analyze the morphology of a single precipitate may as well be applied to analyze the shape controlled by both effects, the elastic strain and magnetization.

It should be pointed out that Eq. (8.9.6) enables us to obtain a compact equation for the demagnetization factor of an arbitrary set of ferromagnetic particles:

$$E_{mag}=\tfrac{1}{2}M_0^2 V N_{demag} \tag{8.9.28}$$

where

$$N_{demag}=4\pi\sum_{pq}\frac{1}{V}\int\frac{d^3k}{(2\pi)^3}\,(\mathbf{ne}(p))(\mathbf{ne}(q))\theta_p(\mathbf{k})\theta_q^*(\mathbf{k})$$

is the demagnitization factor. In the case of a single monodomain ferromagnetic particle, the equation for the demagnitization factor is simplified. It will be

$$N_{demag}=4\pi\frac{1}{V}\iiint_{-\infty}^{\infty}\frac{d^3k}{(2\pi)^3}\,(\mathbf{ne})^2|\theta(\mathbf{k})|^2 \tag{8.9.29}$$

Eq. (8.9.29) is applicable to any shape of a ferromagnetic particle, the shape being determined by the function $|\theta(\mathbf{k})|^2$.

An example of the calculation of the integral (8.9.29) in the case of an ellipsoidal shape particle may be found in Section 8.3. If the magnetization direction **e** is parallel to the principal ellipsoid axis and is chosen to be the z-axis, Eq. (8.9.29) becomes

$$N_{\text{demag}} = \frac{4\pi}{V} \iiint_{-\infty}^{\infty} n_z^2 |\theta(\mathbf{k})|^2 \frac{d^3k}{(2\pi)^3} \tag{8.9.30}$$

where $\theta(\mathbf{k})$ is given by Eq. (8.1.8). As has been shown in the end of Section 8.2, Eq. (8.9.30) may be rewritten in the form

$$N_{\text{demag}} = 4\pi \langle n_z^2 \rangle_{\mathbf{n}'} \tag{8.9.31}$$

where $\langle n_z^2 \rangle_{\mathbf{n}'}$ is defined by Eq. (8.3.21). As follows from Eq. (8.3.27), Eq. (8.9.31) may be represented as

$$N_{\text{demag}} = 2\pi a_1 a_2 a_3 \int_0^{\infty} \frac{du}{(a_3^2 + u)\Delta(u)} \tag{8.9.32}$$

The above-considered effect of magnetization on the shape of ferromagnetic particles is the basis for thermomagnetic treatment when the precipitation reaction occurs in a permanent magnetic field. In this case rodlike precipitates whose axis is parallel to the magnetic field direction arise. Such a structure is widely used, for instance, in ALNICO alloys to produce high coersive materials for permanent magnets. To choose the best temperature range for the thermomagnetic treatment, one should know the temperature dependence of the interphase energy γ_s and magnetization M_0^2 entering Eq. (8.9.26). The calculation of the interphase energy in Problem 1 to Section 4.6 (see Fig. 44) shows that the interphase energy coefficient γ_s vanishes when the temperature tends to the top of the miscibility gap. In this connection one may see from Eq. (8.9.26) that the formation of the most elongated particles may be expected within the range near the top of the miscibility gap if the precipitate phase is ferromagnetic.

Note that the parallel rodlike precipitates formed in thermomagnetic treatment may also be treated as parallel magnetic dipoles. Since parallel dipoles repel each other, the magnetostatic interaction inhibits coarsening with the coalescence of neighboring rodlike precipitates. This repulsion also gives rise to the formation of an ordered distribution of rods in the plane normal to the rod axis.

Perhaps the reader has already noticed that the foregoing consideration strongly resembles that applied to bubble domain configurations in ferromagnetic films. The formation of regularly spaced quasi-periodic distributions of bubble domains is in fact physically the same phenomenon as the emergence of quasi-periodic distributions of parallel rodlike precipitates of the ferromagnetic phase formed in isothermal aging in a permanent magnetic field. Finally, the calculation technique formulated here is good for both ferromagnetic particles in a paramagnetic matrix and domain structures in ferromagnetic crystals.

8.10. RODLIKE PRECIPITATES

As was shown in Section 8.1, the elastic strain energy approaches its minimum value when a precipitate is "rolled" into a platelet whose habit is normal to the vector $\mathbf{n}_0$ minimizing the function $B(\mathbf{n})$. The interphase energy increase hinders this process and provides a certain equilibrium value of the aspect ratio [see Eq. (8.7.54)]. An important question here is whether the elastic energy minimization will result in a rodlike precipitate. As will be shown, the answer is in the positive. In some special cases rodlike precipitates prove to be more stable than platelike ones. One of such cases has been considered in Section 8.9. It was shown there that a precipitate has a rodlike shape if the minimum of the function $B(\mathbf{n})$ is degenerate with respect to any $\mathbf{n}$ lying in a plane. The problem of a rodlike precipitate is therefore reduced to finding the phase transformation crystallography (the stress-free transformation strain ε_{ij}^0) and the elastic anisotropy that would ensure the desirable degeneration of the function $B(\mathbf{n})$ with respect to $\mathbf{n}$. Since $\mathbf{n}$ enters $B(\mathbf{n})$ only through the tensor $\Omega_{ij}(\mathbf{n})$ and vector $\sigma_{ij}^0 n_j$ [see Eq. (8.1.2)], the degeneration of $B(\mathbf{n})$ may only be provided if $\hat{\Omega}(\mathbf{n})$ and $\hat{\sigma}^0\mathbf{n}$ are spherically or cylindrically isotropic. Spherical isotropy of $\hat{\Omega}(\mathbf{n})$ may be expected when decomposition occurs in cubic alloys based on an almost elastically isotropic solvents such as Al, Nb, Mo, W, and so on. Cylindrical isotropy of $\hat{\Omega}(\mathbf{n})$ is always realized in hexagonal phases because it is associated with the crystal lattice symmetry and not with the individual physical properties of the alloy (142). The symmetry axis in this case coincides with the sixfold axis of the hexagonal phase. As for the vector $\hat{\sigma}^0\mathbf{n}$, it has the cylindrical symmetry $C_{\infty h}$ in the cases of cubic → tetragonal phase transformations (the symmetry axis is parallel to one of the ⟨100⟩ matrix lattice directions) or in the cases of the cubic → trigonal and cubic → hexagonal phase transformations (the symmetry axis is parallel to one of the ⟨111⟩ matrix lattice directions). The cylindrical symmetry of $\hat{\sigma}^0\mathbf{n}$ also occurs in the case of decomposition of a hexagonal alloy into two hexagonal phases.

Summing up the foregoing consideration, one can say that the function $B(\mathbf{n})$ may be cylindrically degenerate with respect to $\mathbf{n}$* in the cases of the cubic → tetragonal and cubic → trigonal phase transitions if the transformation occurs in an alloy based on almost elastically isotropic cubic solvents (Al, Nb, Mo, W, and so on), and in the cases of the cubic → hexagonal and hexagonal → hexagonal phase transformations.

The cylindrical degeneration of $B(\mathbf{n})$ is not sufficient, however, for a rodlike precipitate to be formed. The latter occurs if $B(\mathbf{n})$ also assumes its minimal value at $\mathbf{n}$ perpendicular to the cylinder axis. In this case the desirable degeneration takes place with respect to any $\mathbf{n}$ belonging to the plane normal to the cylinder axis.

The latter requirement for $B(\mathbf{n})$ puts certain quantitative constraints on the

*The spherical degeneration of $B(\mathbf{n})$ in fact means that $B(\mathbf{n})$ is a constant. This case is of no interest because the elastic strain energy does not then depend on the shape of the particle.

stress-free transformation strain and crystal lattice anisotropy. The chances, however, of the absolute minimum of $B(\mathbf{n})$ occurring at $\mathbf{n}$ normal to the cylinder axis are very high since the $C_{\infty h}$ point group symmetry of $B(\mathbf{n})$ ensures its extremum for any $\mathbf{n}$ normal to the symmetry axis.

If $B(\mathbf{n})$ has a cylindrical symmetry with respect to an axis directed along the direction $\mathbf{e}$, the function $B(\mathbf{n})$ will in fact depend on the scalar product $\mathbf{ne}$;

$$B(\mathbf{n}) = B(\mathbf{ne})^2$$

In the case we are interested in, when $B(\mathbf{n})$ assumes its absolute minimum at $\mathbf{n}$ normal to the symmetry axis direction $\mathbf{e}$, at $\mathbf{ne}=0$, one can expand $B(\mathbf{n})$ in a power series of $\mathbf{ne}$:

$$B(\mathbf{n}) = B(\mathbf{ne}) = \min B(\mathbf{n}) + \beta(\mathbf{ne})^2 + \cdots \tag{8.10.1}$$

Truncation of the higher than second-order terms yields

$$\Delta B(\mathbf{n}) = B(\mathbf{n}) - \min B(\mathbf{n}) = \beta(\mathbf{ne})^2 \tag{8.10.2}$$

Substituting (8.10.2) to (8.1.27), we obtain

$$E_{\text{edge}} = \frac{1}{2}\beta \int (\mathbf{ne})^2 |\theta(\mathbf{k})|^2 \frac{d^3k}{(2\pi)^3} \tag{8.10.3}$$

Eq. (8.10.3) for the elastic energy has the same form as Eq. (8.9.13) describing the magnetostatic energy of a ferromagnetic particle, and all the conclusions drawn in Section 8.9 are thus equally applicable to the present case. In other words, we can state that in the cases of cylindrical symmetry of the function $B(\mathbf{n})$, when this function assumes its minimum value at all $\mathbf{n}$ normal to the symmetry axis, the elastic strain effect results in the formation of rodlike precipitates, the rod axis coinciding with the symmetry axis direction $\mathbf{e}$. Referring to the correspondence between Eq. (8.10.3) and (8.9.13), we substitute the coefficient β for $4\pi M_0^2$ and rewrite Eq. (8.9.26) as follows:

$$K_{\text{eq}} = \frac{V^{\frac{1}{4}}}{2\pi}\left(\frac{\beta}{\pi\gamma_s}\right)^{\frac{3}{4}}$$

where K_{eq} is the equilibrium aspect ratio (length to radius ratio). As in the case of ferromagnetic particles the most elongated precipitates may be expected within the range near the top of the miscibility gap where γ_s tends to zero.

The closest analogy to the rodlike particles formed in cooling under a permanent magnetic field may be expected when a hexagonal matrix decomposes into two hexagonal phases. Since the hexagonal lattice is uniaxial, all rodlike particles should be parallel to the hexagonal axis. They produce a structure that strongly resembles the one obtained by the thermomagnetic treatment of ALNICO alloys. These precipitates may be treated as parallel "elastic" dipoles because their interaction has the same form as repulsive interaction of a pair of dipoles considered in Section 8.9. It prohibits coalescence of neighboring rodlike precipitates with coarsening and ensures ordered distributions of parallel precipitates.

For instance, decomposition carried out in thin (001) hexagonal films seems to result in the formation of ordered distributions of rodlike precipitates resembling distributions produced by bubble domains.

It will be shown below that the crystal lattice of a coherent rodlike precipitate, as well as the crystal lattice of a platelike one, is homogeneously strained. The strain (8.5.5) may be represented as follows:

$$u_{ij}(\mathbf{r})=\int n_i S_j(\mathbf{n})\theta(\tau, k_z)e^{i\tau\rho+ik_z z}\frac{d^3k}{(2\pi)^3} \tag{8.10.4}$$

where $\theta(\mathbf{k})=\theta(\boldsymbol{\tau}, k_z)$ is determined by Eq. (8.9.18), $\mathbf{kr}=\mathbf{r}\boldsymbol{\rho}+k_z z$, the z-axis is parallel to the rod axis vector $\mathbf{e}$, $\mathbf{r}=(\boldsymbol{\rho}, z)$, $\boldsymbol{\rho}=(x, y)$, (x, y, z) are the coordinates of the vector $\mathbf{r}$.

As mentioned earlier, the function $\theta(\boldsymbol{\tau}, k_z)$ describing a rodlike precipitate assumes nonzero values only within a thin disk in the $\mathbf{k}$-space which is normal to the rod axis vector $\mathbf{e}$. The integration in Eq. (8.10.4) is then indeed carried out over the plane in the $\mathbf{k}$-space normal to $\mathbf{e}$. In this connection one may rewrite Eq. (8.10.4) in the form

$$u_{ij}(\mathbf{r})=\int m_i S_j(\mathbf{m})\theta(\tau, k_z)e^{i\tau\rho+ik_z z}\frac{d^3k}{(2\pi)^3} \tag{8.10.5}$$

where $\mathbf{m}=\boldsymbol{\tau}/\tau$ is the unit vector parallel to $\boldsymbol{\tau}$ and situated in the plane normal to $\mathbf{e}$. The representation (8.10.5) is fulfilled within an accuracy of a small ratio $R_0/H_0 \ll 1$.

Since $\mathbf{S}(\mathbf{m})=\hat{\boldsymbol{\Omega}}(\mathbf{m})\hat{\sigma}^0\mathbf{m}$ [see Eq. (8.5.4)] is assumed to be cylindrically isotropic, we obtain

$$\mathbf{S}(\mathbf{m})=\kappa_0\mathbf{m} \tag{8.10.6}$$

where κ_0 is a constant. Substituting Eq. (8.10.6) into (8.10.5) yields

$$u_{ij}(\mathbf{r})=\kappa_0\int_{-\infty}^{\infty}\frac{dk_z}{2\pi}e^{ik_z z}\iint_{-\infty}^{\infty}\frac{d^2\tau}{(2\pi)^2}\theta(\tau, k_z)m_i m_j e^{i\tau\rho} \tag{8.10.7}$$

where $d^3k=dk_z d^2\tau=dk_z\tau d\tau d\phi$ is a volume element in the $\mathbf{k}$-space. Let us consider the integral

$$I_3=\iint_{-\infty}^{\infty}\frac{d^2\tau}{(2\pi)^2}\theta(\tau, k_z)m_i m_j e^{i\tau\rho}$$

It may be rewritten as

$$I_3=\int_0^{\infty}\frac{\tau d\tau}{2\pi}\theta(\tau, k_z)\int_0^{2\pi}\frac{d\phi}{2\pi}m_i m_j e^{i\tau\rho} \tag{8.10.8}$$

Since

$$\int_0^{2\pi} \frac{d\phi}{2\pi} m_i m_j e^{i\tau\rho} = \frac{1}{2}(\delta_{ij} - e_i e_j) \int_0^{2\pi} \frac{d\phi}{2\pi} e^{i\tau\rho}$$

we may represent Eq. (8.10.8) in the form

$$\begin{aligned} I_3 &= \frac{1}{2}(\delta_{ij} - e_i e_j) \int_0^{\infty} \frac{\tau d\tau}{2\pi} \int_0^{2\pi} \frac{d\phi}{2\pi} \theta(\tau, k_z) e^{i\tau\rho} \\ &= \frac{1}{2}(\delta_{ij} - e_i e_j) \int \frac{d^2\tau}{(2\pi)^2} \theta(\tau, k_z) e^{i\tau\rho} \end{aligned} \tag{8.10.9}$$

Substituting Eq. (8.10.9) into (8.10.7), we obtain

$$\begin{aligned} u_{ij}(\mathbf{r}) &= \frac{1}{2} \kappa_0 (\delta_{ij} - e_i e_j) \int_{-\infty}^{\infty} \frac{dk_z}{2\pi} \int \frac{d^2\tau}{(2\pi)^2} \theta(\tau, k_z) e^{i\tau\rho + ik_z z} \\ &= \frac{1}{2} \kappa_0 (\delta_{ij} - e_i e_j) \int \frac{d^3k}{(2\pi)^3} \theta(\mathbf{k}) e^{i\mathbf{kr}} \end{aligned} \tag{8.10.10}$$

The integral in the right-hand side of Eq. (8.10.10) is the back Fourier transform for the shape function $\tilde{\theta}(\mathbf{r})$. Taking this into account, we obtain

$$\begin{aligned} u_{ij}(\mathbf{r}) &= \frac{1}{2} \kappa_0 (\delta_{ij} - e_i e_j) \tilde{\theta}(\mathbf{r}) \\ &= u_{ij}^* \tilde{\theta}(\mathbf{r}) = \begin{cases} u_{ij}^* & \text{within the particle} \\ 0 & \text{otherwise} \end{cases} \end{aligned} \tag{8.10.11}$$

where

$$u_{ij}^* = \tfrac{1}{2}\kappa_0 (\delta_{ij} - e_i e_j) = \tfrac{1}{2}\kappa_0 \begin{pmatrix} 1 & 0 & 0 \\ 0 & 1 & 0 \\ 0 & 0 & 0 \end{pmatrix} \tag{8.10.12}$$

The tensor representation (8.10.12) is related to the Cartesian coordinate system where the z-axis is parallel to $\mathbf{e}$.

It follows from Eqs. (8.10.11) and (8.10.12) that homogeneous elastic strain is concentrated only within the rodlike precipitate. The strain (8.10.12) provides the precise coincidence of any crystal lattice vector of a precipitate parallel to the rod axis and the corresponding crystal lattice vector of the undistorted matrix. Therefore the crystal lattice parameter of the precipitate chosen to be parallel to the precipitate axis should be precisely equal to the corresponding crystal lattice parameter of the matrix.

As shown above, in the case of the cubic → tetragonal phase transformation the cylindrical axis $\mathbf{e}$ should be parallel to one of the directions ⟨100⟩ of the cubic matrix. Therefore, if rodlike precipitates arise, they should be directed along the ⟨100⟩ axes. In this situation the crystal lattice parameter c of a precipitate parallel to the rod direction should be precisely equal to the parameter a_0 of the undistorted matrix.

In the case of the cubic → tetragonal transformation in an isotropic alloy [see Eqs. (8.3.2) and (8.3.4)], we obtain

$$\hat{\sigma}^0\mathbf{m}=\frac{2\mu}{1-2\sigma_1}(\varepsilon_{11}^0+\sigma_1\varepsilon_{33}^0)\mathbf{m}$$

$$\mathbf{S}(\mathbf{m})=\hat{\boldsymbol{\Omega}}(\mathbf{m})\hat{\sigma}^0\mathbf{m}=\left(\frac{\hat{\mathbf{I}}}{\mu}-\frac{1}{2\mu(1-\sigma_1)}\mathbf{m}*\mathbf{m}\right)$$

$$\times\frac{2\mu}{1-2\sigma_1}(\varepsilon_{11}^0+\sigma_1\varepsilon_{33}^0)\mathbf{m}=(\varepsilon_{11}^0+\sigma_1\varepsilon_{33}^0)\mathbf{m} \tag{8.10.13}$$

where σ_1 is the Poisson ratio.

Comparing (8.10.13) and (8.10.6), we have

$$\kappa_0=\frac{\varepsilon_{11}^0+\sigma_1\varepsilon_{33}^0}{1-\sigma_1} \tag{8.10.14}$$

In accordance with the theoretical predictions rodlike precipitates parallel to the ⟨100⟩ directions seem to be observed in Al-based alloys, Al-Cu-Mg (146), Al-Mg-Si and Al-Mg-Ge (147). The reported crystal lattice parameters of tetragonal rodlike precipitates in Al-Cu-Mg (146)

$$a=b=5.5\ \text{Å},\quad c=a_{\text{Al}}^0=4.04\ \text{Å}$$

really demonstrate the precise coincidence of the parameter c of the constrained precipitate with the parameter $a_{\text{Al}}^0=4.04$ Å of the fcc aluminium matrix predicted by the foregoing theoretical analysis.

It is of interest that the x-ray studies of ω-phase precipitates in Ti-, Zr-, and Hf-based alloys (bcc → hcp transformation) have shown that diffuse maxima form the {111} "planes" in the reciprocal lattice of the matrix. These "planes" can be interpreted as being a result of scattering by thin, extended rodlike precipitates of the ω phase. Such an assumption is also in agreement with the above theoretical analysis since the precipitate phase is hexagonal (this ensures the cylindrical elastic isotropy) and the direction of the c-axis is parallel to one of the {111} directions of the bcc matrix. As shown above, in this situation the formation of rodlike precipitates parallel to the hexagonal axis [parallel to the ⟨111⟩ directions of the bcc matrix] should be expected.

9

HABIT PLANE AND ORIENTATION RELATIONS IN PRECIPITATES: COMPARISON WITH EXPERIMENTAL DATA

The theory in Chapter 8 is applicable to a single precipitate in an infinite parent phase lattice formed by homogeneous transformation strain. The theory therefore applies to alloys occurring in two-phase fields of T-c equilibrium diagrams near the solvus. According to the lever rule the total volume of the new phase must be small in this case and the composition of the parent phase remains nearly constant during the transformation. This transformation is a decomposition reaction controlled by the diffusion mass transport and resulting in the formation of small new phase coherent precipitates in the first stage of the process. The habit plane theory given in Section 8.6 applies to this stage of the decomposition directly. The next stage is a coarsening process that develops by growth of large and dissolution of small precipitates. Coarsening is controlled by two factors, namely, by (1) a decrease in the elastic strain energy which depends on the shape, size, and separations among precipitates and (2) a relaxation of the interphase energy. We shall refer to the former process as strain-induced coarsening. This will be discussed in detail later.

Strain-induced coarsening occurs because the chemical potential is not constant along the precipitate boundary. Its variation is caused by the strain field generated by all the other precipitates. Strain-induced variation of the chemical potential along the boundary results in the diffusion transport of the new phase substance along the boundary to insure the equilization of the chemical potential. This leads to shape changes and translational motion of a precipitate, both providing relaxation of the strain energy (translational motion results from transport of the new phase substance from one side of a precipitate to another one). It may seem surprising that, despite its fundamental importance in explaining the structure of the majority of multiphase alloys, strain-induced coarsening has not been given enough theoretical consideration.

Interphase energy decrease is the second factor controlling coarsening. Interphase energy decreases during coarsening because the equilibrium compositions of the parent and new phases depend on the shape and size of precipitates. This is a manifestation of the surface tension effect. As a consequence precipitates of different sizes differ in their stabilities (the smaller the particle the lower its stability). This is the origin of the driving force of the coarsening process. It will be shown in Chapter 11 that strain-induced coarsening results in the formation of thin-plate multidomain particles that look like "sandwiches" composed of adjacent twin-related new phase platelets. The habit of the thin-plate particles is determined by the strain energy minimum condition. The thin-plate morphology is in many respects analogous to that of martensitic crystals because both morphologies are controlled by the same strain-energy minimum condition. The habit of multidomain new phase particles is substantially different from habits of single homogeneous precipitates. The calculation of the habit of multidomain particles formed in coarsening will be exemplified in Chapter 11.

The calculated and observed habit plane orientations and orientation relations should therefore be compared only with caution. One should be certain about the stage of the decomposition, whether the observed platelike inclusion is a single coherent new phase particle that may be treated in terms of the theory given in Section 8.6 or whether it is already a heterogeneous packet of precipitates formed in coarsening. These two cases are comparatively easy to distinguish. For example, strain-induced coarsening of tetragonal precipitates in a cubic matrix will be shown in Section 11.4 to result in thin-plate packets of twin-related monodomain precipitates with the (101) habit and alternating directions of the tetragonal axes. The diffraction pattern from such a packet exhibits splittings of precipitate reflections along the [101] direction of the reciprocal lattice of the parent phase.

On the other hand, it was shown in Section 8.6 that the (101) habit of a homogeneous tetragonal precipitate, which is a particular case of the (h0l) habit, may only arise with unrealistically accidental numerical values of the stress-free transformation tensor components and elastic constants and therefore can hardly be expected to occur.

We shall consider several cases where the theoretical predictions of the habit, orientation relations, and crystal lattice parameters of precipitates in the constrained state may be compared with the characteristics determined by the electron microscopic and x-ray diffraction techniques.

The only parameter that describes the crystallography of the cubic → tetragonal phase transformation is the tetragonality ratio:

$$t_1 = \frac{\varepsilon_{11}^0}{\varepsilon_{33}^0} = \frac{a - a_0}{c - a_0}$$

where a and c are the crystal lattice parameters of the tetragonal phase, and a_0 is the crystal lattice parameter of the parent cubic phase.

The theoretical analysis in Section 8.6 shows that the habit plane orientation can be presented in terms of only one parameter, t_1, if the elastic constants of the precipitate are known. The calculations of the habit plane orientation of tetragonal precipitates in Fe and V- based alloys were made using Eqs. (8.6.21) and (8.6.22). The results are given in Fig. 68. The following precipitate elastic constants were used:

$$c_{11}=2.42\times10^{12}\ \text{dyne/cm}^2,\ c_{12}=1.46\times10^{12}\ \text{dyne/cm}^2$$
$$c_{44}=1.12\times10^{12}\ \text{dyne/cm}^2\ \text{for}\ \alpha\text{-Fe based alloys}$$
$$c_{11}=2.28\times10^{12}\ \text{dyne/cm}^2,\ c_{12}=1.19\times10^{12}\ \text{dyne/cm}^2$$
$$c_{44}=0.426\times10^{12}\ \text{dyne/cm}^2\ \text{for V-based alloys}$$

The plots $n_3=\cos\theta_0$ as opposed to t_1 in Fig. 68 may be used to predict the precipitate habit orientations because the parameter t_1 can be found from

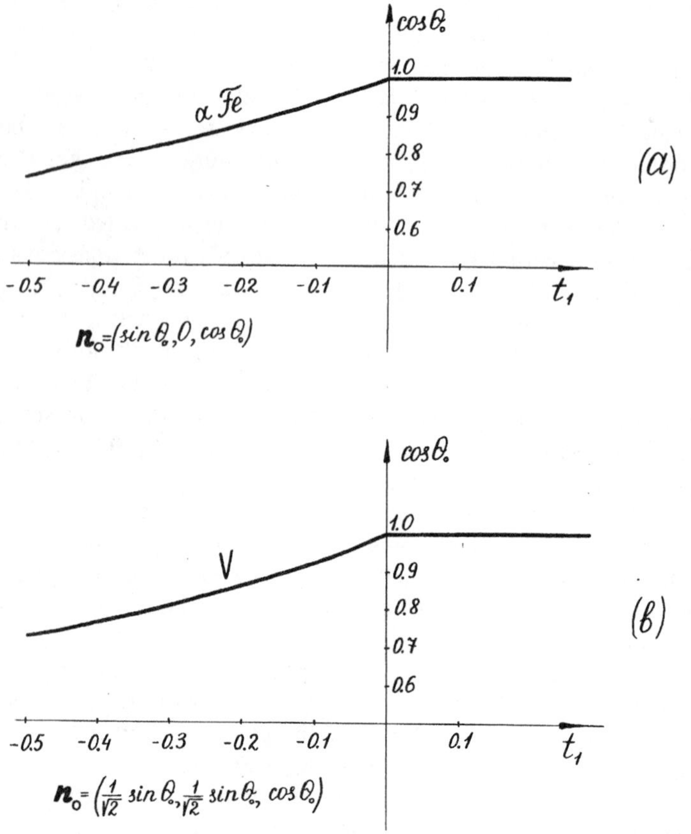

Figure 68. The calculated habit plane orientation of a tetragonal precipitate as a function of the tetragonality ratio, t_1, (*a*) a Fe-based alloy; (*b*) a V-based alloy.

independent measurements of crystal lattice parameters [see Eqs. (8.6.6), (8.6.7), (8.6.8), and (8.6.11b)]. As shown in Section 8.6, the vector $\mathbf{n}_0$ normal to the habit plane is either

$$\mathbf{n}_0=(\sqrt{1-n_3^2}, 0, n_3)$$

if the anisotropy parameter (8.6.4a) $\xi<0$ or

$$\mathbf{n}_0=\left(\sqrt{\frac{1-n_3^2}{2}}, \sqrt{\frac{1-n_3^2}{2}}, n_3\right)$$

if $\xi>0$. Here $n_3=\cos\theta_0$. The $n_3=\cos\theta_0$ value is determined by Eq. (8.6.21) if $\xi<0$ and (8.6.22) if $\xi>0$. With ξ being less than 0, the maximum of the function $\psi(\mathbf{n})=\psi(\theta, \phi)$ occurs at $\phi=0$, whereas with $\xi>0$, the maximum is at $\phi=\pi/4$. The typical plot of $\psi(\theta, 0)$ and $\psi(\theta, \pi/4)$ as opposed to θ for the case of α-Fe, $t_1=-0.06$, is shown in Fig. 69.

It should, however, be remembered that the use of n_3 and t_1 plots given in Fig. 68 to predict the habit plane orientation has two limitations. The first limitation is described by the inequality (8.1.51) which holds with small aspect ratios, D/L. Its origin is the neglect of interphase energy contributions to the total strain energy inherent in the theoretical consideration carried out in Section 8.6. The second limitation is due to the fact that the plots in Figs. 68a and b were calculated with the elastic constants of pure Fe and V, although the precipitates in question are alloys where Fe and V are solvent components. Their elastic constants may therefore differ from those used to calculate the plots. This is important because, as shown in Section 8.1, habit plane orientations

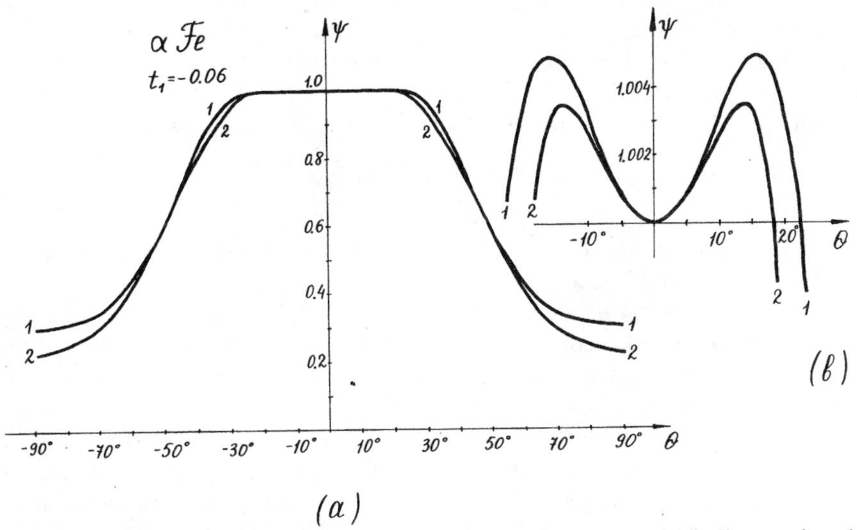

Figure 69. (*a*) The plot, $\psi(\theta, \phi)$ with respect to θ in the case of α-Fe ($t_1=-0.06$). (*b*) is an enlarged treatment of the diagram depicted in (*a*). 1. $\psi(\theta, 0)$ with respect to θ, 2. $\psi(\theta, \pi/4)$ with respect to θ.

of thin platelike precipitates are determined by the elastic constants of the precipitate rather than the parent phase.

The limitations mentioned grow still more significant if there are two or more orientations of the precipitate habit that provide almost the same bulk strain energy (the case of degeneracy of the strain energy). The preferred habit plane in this case is that characterized by the lower interphase energy. The choice of the habit plane also requires refinement based on more accurate values for the precipitate elastic constants.

9.1. MORPHOLOGY AND CRYSTAL LATTICE CORRESPONDENCE OF NITRIDE PRECIPITATES IN IRON-NITROGEN MARTENSITE

We shall demonstrate how the elastic strain theory formulated in Section 8.6 can be applied to determine the morphology and crystal lattice correspondence of nitride precipitates in the iron-nitrogen martensite. This will be done according to Morris, Hong, Wedge, and Khachaturyan (157).

As is known, quenching of Fe-N as well as Fe-C austenites results in the formation of the tetragonal martensitic phase. The crystal lattice parameters of iron-nitrogen martensite as well as those of iron-carbon martensite show a linear dependence on the concentration of interstitial atoms (158). However, the decomposition reaction that occurs in tempered Fe-N martensite differs from the decomposition of Fe-C martensite. It leads to the formation of ordered nitride precipitates, namely, the bcc-based ordered nitride $\alpha''(Fe_8N)$ further transformed into another fcc-based nitride, $\gamma'(Fe_4N)$ (159).

According to Jack (159) the structure of the α'' phase may be interpreted as tetragonal interstitial superlattice in the bcc host lattice of α-Fe, the N atoms making up a double-period bcc lattice within the O_z sublattice of octahedral interstices, with $a=b\approx c\approx 2a_0$ where a_0 is the bcc crystal lattice parameter of the α-Fe bcc host lattice.

The crystal lattice parameters of the α'' phase are as follows:

$$a_{\alpha''}=2a_0=2\times 2.86=5.72\ \text{Å}$$
$$c_{\alpha''}=6.292\ \text{Å}\approx 2a_0 \tag{9.1.1}$$

It should be noted that the spacing $a_{\alpha''}$ of the tetragonal α'' phase is exactly equal to twice the crystal lattice parameter of the parent phase (α-Fe).

It thus follows that the crystal lattice parameters of the α'' phase reported by Jack (159) characterize constrained precipitates because the coincidence of the crystal lattice parameters is itself evidence of an elastic strain adjustment of the crystal lattice of the α'' phase to the crystal lattice of α-Fe. For that reason the crystal lattice parameters of the α'' phase (9.1.1) cannot be employed to determine the stress-free transformation strain ε_{ij}^0. To find ε_{ij}^0, the crystal lattice parameters of the single-phase ordered α'' solid solution have been measured by Suyazov, Usikov, and Mogutnov (160) for an Fe-8.56N alloy (N/Fe$=0.0936$). By definition, this single phase alloy is in the stress-free state. The experimental crystal lattice parameters

$$a_{\alpha''} = 5.692 \text{ Å}$$
$$c_{\alpha''} = 6.180 \text{ Å} \tag{9.1.2}$$

corresponding to the stress-free state differ from the parameters (9.1.1) of constrained precipitates of the α'' phase.

Using Eq. (8.6.6) and the parameters (9.1.2), the stress-free transformation strain can be calculated:

$$\varepsilon_{11}^0\left(\frac{9.36}{100}\right) = \frac{a_{\alpha''} - 2a_0}{2a_0} = \frac{5.692 - 2 \times 2.86}{2 \times 2.86} = -0.004895$$

$$\varepsilon_{33}^0\left(\frac{9.36}{100}\right) = \frac{c_{\alpha''} - 2a_0}{2a_0} = \frac{6.180 - 2 \times 2.86}{2 \times 2.86} = 0.080419$$

$$t_1 = \frac{\varepsilon_{11}^0\left(\frac{9.36}{100}\right)}{\varepsilon_{33}^0\left(\frac{9.36}{100}\right)} = -0.0608 \tag{9.1.3}$$

The quantity N/Fe$=9.36/100$ is the nitrogen to iron ratio corresponding to the composition of the alloy under consideration, 8.56 atomic percent N.

The linear dependence of the stress-free strain on the composition may be applied to determine the stress-free transformation strain of the stoichiometric α'' phase (Fe_8N) from the numerical values (9.1.3):

$$\varepsilon_{11}^0 = \varepsilon_{11}^0\left(\frac{12.5}{100}\right) = \varepsilon_{11}^0\left(\frac{9.36}{100}\right)\frac{12.5}{9.36} = -0.004895 \times \frac{12.5}{9.36} = -0.006537$$

$$\varepsilon_{33}^0 = \varepsilon_{33}^0\left(\frac{12.5}{100}\right) = \varepsilon_{33}^0\left(\frac{9.36}{100}\right)\frac{12.5}{9.36} = 0.080419 \times \frac{12.5}{9.36} = 0.107397 \tag{9.1.4}$$

where 12.5 is the N/Fe ratio corresponding to the stoichiometry, Fe_8N. To determine the habit of the α'' phase precipitate, we must consider the plot $\psi(\theta, 0)$ vs. θ at $t_1 = -0.06$ depicted in Fig. 69. This figure shows that the maximum of the $\psi(\theta, 0)$ with relation to θ features a plateau ranging from $-17°$ to $17°$ around the point $\theta = 0$ (around the direction [001]). The function is nearly constant in this range: its maximum, 1.005, falls at $\theta = \pm 16.2°$, while, at $\theta = 0$, its value is equal to 1. We thus have a degeneracy with all habit plane orientations from $-17°$ to $17°$ resulting in almost the same bulk strain energy (8.1.14). In this situation we have to consider the interphase energy contribution and choose the habit corresponding to the lowest interphase energy value.

The interphase energy between the α'' phase and the bcc α-Fe matrix arises from N-N interactions that lead to ordering of interstitial atoms over the O_z

octahedral interstices of α-Fe. Ordering producing the α'' phase occurs because nearest and next-nearest N-N interaction energies are positive (the N-N inter-actions are repulsive). In the case of attractive interactions decomposition would occur.

The removal of N-atoms on the bcc host lattice side of the (001) plane to create the $(001)_{\alpha''}/(001)_{\alpha}$ coherent interphase boundary between the α'' phase, and the α-Fe matrix breaks repulsive positive bonds between nearest and next-nearest N-atoms separated by the $(001)_{\alpha}$ plane. This decreases the interphase energy contribution to the free energy of the system. The larger the number of repulsive bonds broken, the larger the decrease of the free energy. In this respect the $(001)_{\alpha''}/(001)_{\alpha}$ boundary is favored since its formation involves the dissociation of the maximum number of repulsive bonds and hence provides the lowest interphase energy.

Returning to the choice of the habit plane orientation from the range $-17°$ to $17°$ where the strain energy is almost constant, we conclude that the $(001)_{\alpha}$ habit plane providing the lowest interphase energy ($\theta=0$) should be given preference. This agrees with the electron microscopic observations of the $\{001\}$ habit of α'' precipitates.

A departure of the elastic constants of the α'' phase from those of α-Fe might also remove the degeneracy. The calculation of the orientation relations and crystal lattice parameters of the α'' phase can be made on the basis of Eqs. (8.5.10) and (8.5.12) where $\mathbf{n}_0=(0, 0, 1)$, and $\mathbf{S}(\mathbf{n}_0)$ is given by Eq. (8.5.2).

It follows from Eq. (8.6.5) that, with the cubic-to-tetragonal crystal lattice rearrangement, the matrix $\hat{\sigma}^0$ in (8.6.11e) has the form

$$\hat{\sigma}^0=\begin{pmatrix} \sigma_{11}^0 & 0 & 0 \\ 0 & \sigma_{11}^0 & 0 \\ 0 & 0 & \sigma_{33}^0 \end{pmatrix} \tag{9.1.5}$$

where

$$\begin{aligned} \sigma_{11}^0&=(c_{11}+c_{12})\varepsilon_{11}^0+c_{12}\varepsilon_{33}^0 \\ \sigma_{33}^0&=2c_{12}\varepsilon_{11}^0+c_{11}\varepsilon_{33}^0 \end{aligned} \tag{9.1.6}$$

The diagonal form of the matrix (9.1.5) corresponds to the Cartesian basis whose axes coincide with the [100], [010], and [001] directions of the bcc α-Fe parent lattice.

With σ_{ij}^0 given by (9.1.5), the vector $\mathbf{S}(\mathbf{n}_0)$ is reduced to the simplest form

$$S_i(\mathbf{n}_0)=\Omega_{i3}(\mathbf{n})\sigma_{33}^0 \tag{9.1.7}$$

where $\mathbf{n}_0=(0, 0, 1)$. It follows from Eq. (8.6.3) that all off-diagonal components of the tensor $\Omega_{ij}(\mathbf{n}_0)$ vanish at $\mathbf{n}_0=(0, 0, 1)$. Eq. (9.1.7) may then be rewritten in the form

$$\mathbf{S}(\mathbf{n}_0)=((0, 0, \Omega_{33}(\mathbf{n}_0))\sigma_{33}^0)_{\mathbf{n}_0=(0,0,1)} \tag{9.1.8}$$

The use of Eqs. (8.6.3) and (9.1.6) in (9.1.8) yields

$$\mathbf{S}(\mathbf{n}_0)=\left(0,\,0,\,\frac{\sigma_{33}^0}{c_{11}}\right)=\left(0,\,0,\,\frac{c_{11}\varepsilon_{33}^0+2c_{12}\varepsilon_{11}^0}{c_{11}}\right)$$

$$=\left(0,\,0,\,\varepsilon_{33}^0+2\frac{c_{12}}{c_{11}}\varepsilon_{11}^0\right) \qquad (9.1.9)$$

Substituting the numerical values (9.1.3) and the α-Fe elastic constant values, $c_{11}=2.42\times10^{12}$ dyne/cm² and $c_{12}=1.46\times10^{12}$ dyne/cm², into (9.1.9), we obtain

$$\mathbf{S}(\mathbf{n}_0)=(0,\,0,\,0.099512) \qquad (9.1.10)$$

Substitution of the vector $\mathbf{S}(\mathbf{n}_0)$ into (8.5.10) provides a means for calculating the crystal lattice parameters of a constrained α″ phase precipitate from the crystal lattice parameters of the α″ phase in the stress-free state.

As the $\alpha\rightarrow\alpha''$ crystal lattice rearrangement involves the transformations

$$\mathbf{r}_1=(2a_0,\,0,\,0)_\alpha\rightarrow(a_{\alpha''},\,0,\,0)_{\alpha''}=\mathbf{r}'_1$$
$$\mathbf{r}_2=(0,\,2a_0,\,0)_\alpha\rightarrow(0,\,a_{\alpha''},\,0)_{\alpha''}=\mathbf{r}'_2$$
$$\mathbf{r}_3=(0,\,0,\,2a_0)_\alpha\rightarrow(0,\,0,\,c_{\alpha''})_{\alpha''}=\mathbf{r}'_3$$

substitution of (9.1.10) and

$$\mathbf{r}_1=(2a_0,\,0,\,0),\quad \mathbf{r}_2=(0,\,2a_0,\,0),\quad \mathbf{r}_3=(0,\,0,\,2a_0)$$

into Eq. (8.5.10) gives

$$\mathbf{r}_1=\mathbf{r}'_1,\quad \mathbf{r}_2=\mathbf{r}'_2$$

and hence

$$a_{\alpha''}=2a_0=2\times2.86=5.72\text{ Å} \qquad (9.1.11a)$$

and

$$c_{\alpha''}=|\mathbf{r}'_3|=2a_0+0.099512\times2a_0=2\times2.86\times1.099512=6.29\text{ Å} \quad (9.1.11b)$$

The calculated crystal lattice parameters (9.1.11) are in excellent agreement with the x-ray spacings (9.1.1) reported by Jack (159). Indeed, the agreement seems convincing because the calculation does not involve any fitting variable. All the numerical values used were obtained in independent measurements.

Eq. (8.5.10) with $\mathbf{S}(\mathbf{n}_0)$ given by (9.1.10) yields

$$\mathbf{r}_1=\mathbf{r}'_1,\quad \mathbf{r}_2=\mathbf{r}'_2,\quad \text{and } \mathbf{r}_3\parallel\mathbf{r}'_3$$

We thus have the Bain orientation relations

$$[100]_\alpha\parallel[100]_{\alpha''},\quad [010]_\alpha\parallel[010]_{\alpha''},\quad [001]_\alpha\parallel[001]_{\alpha''}$$

This is also in agreement with the electron diffraction data (161).

It should be stressed that the exact equality of the crystal lattice parameters, $2a_0=a_{\alpha''}$, is not an accidental coincidence. This equality originates from elastic strain relaxation of the crystal lattice and confirms the general statements made in Section 8.5.

We shall now calculate the habit of a γ' phase precipitate and compare the result with the experimental data. The phase is an ordered fcc-based iron nitride with the stoichiometric composition Fe_4N (159). The crystal lattice parameter of the γ' phase near its stability limit is

$$a_{\gamma'} = 3.791\ \text{Å}$$

whereas the crystal lattice parameter of the bcc matrix is

$$a_0 = 2.860\ \text{Å}$$

Using the definition (8.6.8) and the numerical values of the crystal lattice parameters we can calculate the components of the stress-free transformation strain, ε_{ij}^0, for the $\alpha \to \gamma'$ crystal lattice rearrangement:

$$\begin{aligned}
\varepsilon_{11}^0 &= \frac{a_{\gamma'}}{a_0\sqrt{2}} - 1 = \frac{3.791}{2.86\sqrt{2}} - 1 = -0.06271 \\
\varepsilon_{33}^0 &= \frac{a_{\gamma'}}{a_0} - 1 = \frac{3.791}{2.86} - 1 = 0.32552 \qquad (9.1.12) \\
t_1 &= -0.1926
\end{aligned}$$

With the numerical value of t_1, (9.1.12), we can find the habit plane orientation from the n_3 with relation to the t_1 plot (Fig. 68a) to obtain the habit plane normal

$$\mathbf{n}_0 = (0.484,\ 0,\ 0.875)_{\text{bcc}}$$

which is in agreement with the $(102)_{\text{bcc}}$ habit of γ' phase precipitates observed in (161).

9.2. MORPHOLOGY OF PRECIPITATES IN Nb-O INTERSTITIAL SOLUTION

Nb-O interstitial bcc solution provides another example that may be treated in terms of the theory given in Section 8.6. As shown by van Landuyt, Gevers, and Amelinkx (162), the decomposition of Nb-O alloys produces Nb_2O suboxide platelets with the $(103)_{\text{bcc}}$ habit. The suboxide, Nb_2O, forms by transfer of O-atoms to the sole O_z sublattice of octahedral interstices in the bcc host lattice Nb followed by segregation and ordering of O-atoms over the sublattice sites. The occupancy of the O_z sublattice corresponding to the Nb_2O stoichiometry causes a slight pseudotetragonal distortion of the bcc crystal lattice of the magnitude depending on the composition ratio O/Nb$=\bar{n}$. On the assumption of a linear concentration dependence of the Nb host lattice crystal lattice parameters, the tetragonality factor $t_1 = \varepsilon_{11}^0/\varepsilon_{33}^0$ of the precipitate may be written in terms of the concentration coefficients, u_{11} and u_{33}, of the bcc crystal lattice expansion effected by interstitial atoms in the O_z interstices:

$$t_1 = \frac{\varepsilon_{11}^0}{\varepsilon_{33}^0} = \frac{u_{11}\bar{n}}{u_{33}\bar{n}} = \frac{u_{11}}{u_{33}}$$

According to the estimate made in (163)

$$t_1 = -0.12 \tag{9.2.1}$$

The elastic constants of the precipitate are not known. We assume that the elastic constants are unaffected by the insertion of O-atoms in the bcc host lattice of Nb. The elastic constants of Nb_2O precipitates are thus set equal to those of pure Nb:

$$\begin{aligned} c_{11} &= 2.46 \times 10^{12}\ \text{dyne/cm}^2 \\ c_{12} &= 1.34 \times 10^{12}\ \text{dyne/cm}^2 \\ c_{44} &= 0.287 \times 10^{12}\ \text{dyne/cm}^2 \end{aligned} \tag{9.2.2}$$

Note that crystals characterized by the elastic constant values (9.2.2) show near isotropic elasticity. Actually, the deviation from elastic isotropy is measured by the departure of the ratio $(c_{11} - c_{12})/2c_{44}$ from unity. This ratio is equal to 0.42 for α-Fe, 0.64 for Ta, 1.95 for Nb, 1.29 for Mo and 1.28 for V. It appears that Nb features close to isotropic elasticity. The effect of isotropy is illustrated in Fig. 70 where the plots of $\psi(\theta, 0)$ and $\psi(\theta, \pi/4)$ with respect to θ almost coincide with each other.

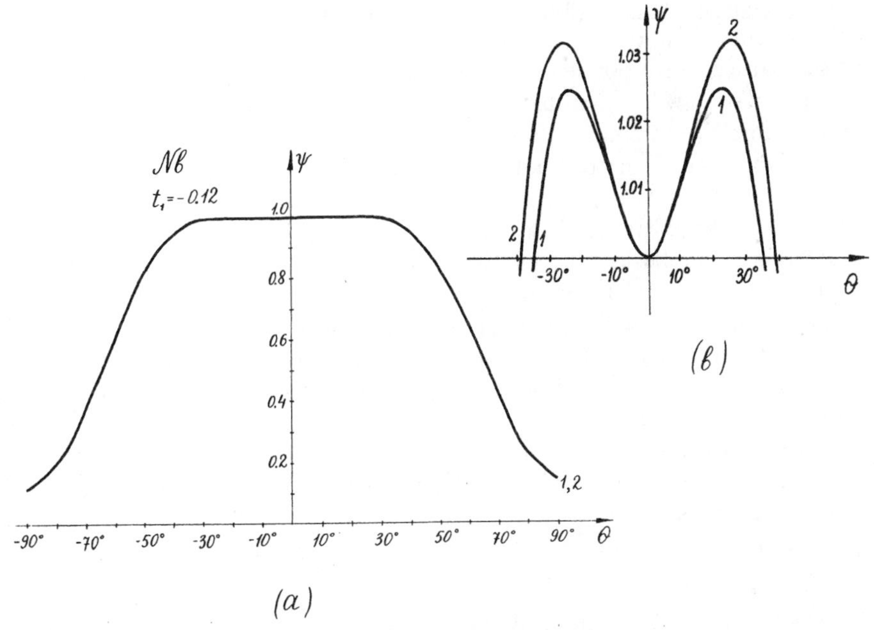

Figure 70. (*a*) The plot $\psi(\theta, \phi)$ with respect to θ in the case of Nb ($t_1 = -0.12$). (*b*) is an enlarged treatment of the diagram depicted in (*a*). 1. $\psi(\theta, 0)$ with respect to θ, 2. $\psi(\theta, \pi/4)$ with respect to θ.

The curves $\psi(\theta, 0)$ and $\psi(\theta, \pi/4)$ with respect to θ reach their maxima, 1.0245 and 1.0321, at almost the same point ($\theta = 23.07°$). The first maximum corresponds to the unit vector

$$\mathbf{n}_0 = (0.39, 0, 0.920) \tag{9.2.3}$$

and the second one to

$$\mathbf{n}_0 = (0.277, 0.277, 0.920) \tag{9.2.4}$$

The rational plane whose normal is closest to the direction (9.2.3) is the $(103)_{bcc}$ plane (the deviation of about 5°), and the rational plane with a near (9.2.4) normal is the $(113)_{bcc}$ plane (the deviation of about 2°). With the difference between the two maxima being about 0.7 percent, we again have the degeneracy case. The degeneracy is now due to almost isotropic elasticity of the crystal in question. The choice between the two habits, $(103)_{bcc}$ and $(113)_{bcc}$, should be based on consideration of the interphase energy terms corresponding to these boundaries, and of the sign of the deviation of the elastic anisotropy ratio, $(c_{11} - c_{12})/2c_{44}$, from unity. The information we have on the interphase energy and, even more so, the elastic constants is far from sufficient. In fact the introduction of about 33 atomic percent O into the Nb bcc lattice may change the sign of elastic anisotropy (the sign of the deviation of the $(c_{11} - c_{12})/2c_{44}$ ratio from unity). This lack of information makes the final choice between the two variants of the habit impossible. Nevertheless, one of the calculated planes, $(103)_{bcc}$, is in agreement with the electron microscopic observations (162).

The theory described in Section 8.6 provides the possibility to calculate the crystal lattice correspondence between the precipitate and bcc Nb matrix. This calculation will be omitted because there is no sufficient experimental data on the subject to be compared with.

Note that very thin coherent precipitates with the $(103)_{bcc}$ habit were also observed in Nb-C (164), V-C (165) and Mo-C bcc interstitial solutions (166) which, like Nb-O, have near isotropic elasticity.

It has been emphasized in (164, 165) that these coherent precipitates seemingly have intermediate bcc-based structures closely related to those of the matrices. Aging of the precipitates yields hexagonal metal carbides, V_2C, Mo_2C, and Nb_2C. The crystal structures of the carbides are close to the Nb_2O-type bcc-based superlattices (167). The former may be obtained from the latter by diffusionless bcc → hcp host lattice rearrangement which does not affect the positions of ordered interstitial atoms. In this connection we may assume the metastable carbides observed in Nb-C, V-C, and Mo-C solutions to have the same structure as that of Nb_2O suboxide.* In this regard, the crystallography of the precipitation reactions in Nb-C, V-C, and Mo-C alloys should be the

*Ta_2O suboxide (23, 109) and $V_2H(V_2D)$ hydride (168–170) have the same structure. It will be shown later that this structure is a typical consequence of strain-induced ordering of interstitial atoms over octahedral sites of the bcc host lattice (171).

same.* The elastic characteristics of these solutions are also close to each other (the solutions are almost isotropic elastically). If the parameter t_1 of these alloys has the value near $t_1 = -0.1$, the mathematical treatment that led to the $(103)_{bcc}$ habit for Nb_2O precipitates also applies to Nb_2C, V_2C, and Mo_2C and gives the same habit variants.

9.3. MORPHOLOGY OF β-PHASE PRECIPITATES IN V-H ALLOYS

According to Westlake the precipitation in V-H alloys results in the formation of β-phase platelets on the $\{227\}$ planes (172). The electron microscopic study confirmed this conclusion (173). The authors (173) reported the β-phase to form comparatively large platelike precipitates with small aspect ratios ($D/L \ll 1$) in the bcc host lattice of V. Since the precipitates have a homogeneous crystal lattice, they may be thought to grow individually rather than by the strain-induced coarsening mechanism involving agglomeration of individual particles to large plates composed of twin-related lamelae of the new phase.

If a homogeneous inclusion is coherent, it must be in a constrained state, and its habit plane orientation must be determined by the strain energy minimum condition. Unlike those in the Fe-N and Nb-O alloys considered above, new phase precipitates in the V-H alloys are sufficiently [of about 0.5 μ (173)] thick, and one may be certain that the interphase energy cannot affect the habit plane orientation determined by the strain energy minimum condition in this case [see inequality (8.1.58)]. The strain energy minimization approach differs from the approach suggested by Bowles, Muddle, and Wayman to explain the (227) habit of β phase precipitates in vanadium hydride (174). The authors (174) proceed from the hypothesis that the macroscopic stress-free shape deformation is reduced to the invariant plane strain by means of a combination of the stress-free transformation strain and extra strain produced by the $\langle 11\bar{2}\rangle(111)_{bcc}$ dislocation glide across the β-phase crystal. Their analysis is fully within the frame of the conventional crystallographic theory of the martensitic transformation (see Section 6.5).

We shall demonstrate, following Wen, Kostlan, Hong, Khachaturyan, and Morris (152), that all morphological characteristics of β-phase precipitates observed can also be explained by the assumption that the precipitates are coherent homogeneous inclusions. The two approaches do not contradict each other. In fact coherent precipitates are always in a constrained state and therefore cannot be considered stable. A transition to the stable state requires relaxation of internal stress within precipitate particles. Usually, this involves

*The fact that the Nb_2O and Ta_2O suboxides and the V_2H hydride, unlike V_2C, Nb_2C, and Mo_2C, do not form hexagonal modifications is very instructive. The O and H atoms are far smaller than C. Therefore the elastic distortion produced by C-atoms is strong enough to cause bcc host lattice instability with respect to the bcc $\rightarrow$ hcp crystal lattice rearrangement, whereas with O and H it is not.

transition to semicoherent interphase boundaries, the formation of a regular array of misfit dislocations. The dislocation structure providing the stress-free state is just the structure considered by Bowles, Muddle, and Wayman to explain the observed morphology of β hydride. The two approaches thus describe different (coherent and semicoherent) stages of the decomposition reaction.

The calculation of the coherent precipitate morphology will be done as before. Hydrogen atoms of the β-phase are ordered and occupy half of the O_z octahedral interstices in a regular way. The bcc host lattice is slightly distorted from the original cubic symmetry because of the ordered arrangement of H-atoms. Since H-atoms occupy only the O_z octahedral interstices; the crystal lattice of the β-phase is pseudotetragonal. A slight departure from pure tetragonal distortion is due to the fact that ordering of H-atoms within the O_z sublattice corresponds to monoclynic rather than tetragonal symmetry.

The crystal lattice parameters at room temperature are $a=3.002$ Å, $c=3.311$ Å (175). The stoichiometric composition of the β-phase corresponds to the formula V_2H. Its atomic structure was first determined by Somenkov et al. (168). As mentioned in Section 9.2, it is the same as the Nb_2O and Ta_2O superstructures (162, 23).

The vanadium matrix lattice has the spacing

$$a_0=3.032\ \text{Å}$$

The stress-free transformation strain and the tetragonality factor t_1 are therefore determined by the relations

$$\varepsilon_{11}^0=\frac{a-a_0}{a_0}=\frac{3.002-3.032}{3.032}=-0.0099$$

$$\varepsilon_{33}^0=\frac{c-a_0}{a_0}=\frac{3.311-3.032}{3.032}=0.0890$$

$$t_1=\frac{\varepsilon_{11}^0}{\varepsilon_{33}^0}=-0.111 \qquad (9.3.1)$$

The ε_{11}^0 and ε_{33}^0 values are small because of a comparatively small "size" of H-atoms. Since the introduction of H-atoms into the bcc host lattice of V has only little effect on the V host lattice parameters, we may assume that H-atoms do not affect the elastic constants as well. We shall therefore set the elastic constants of the β-phase equal to those of pure vanadium:

$$c_{11}=2.28\times10^{12}\ \text{dyne/cm}^2$$
$$c_{12}=1.19\times10^{12}\ \text{dyne/cm}^2$$
$$c_{44}=0.426\times10^{12}\ \text{dyne/cm}^2 \qquad (9.3.2)$$

The habit plane orientation corresponding to the tetragonality factor $t_1=-0.11$ is determined from Fig. 68b where $n_3=\cos\theta_0$ is plotted against t_1.

The habit plane normal providing the maximum of $\psi(\mathbf{n})$ equal to 1.02441 is

$$\mathbf{n}_0=(0.277, 0.277, 0.920) \tag{9.3.3}$$

The habit plane normal observed in (173) is

$$\mathbf{n}_{\text{exp}}=(0.293, 0.236, 0.926) \tag{9.3.4}$$

[The unit vector normal to the $(227)_{\text{bcc}}$ plane is $\mathbf{n}(227)=(0.265, 0.265, 0.927)$.] The deviation of the $\mathbf{n}_0$ vector calculated, (9.3.3), from that observed, (9.3.4), is about 0.9°.

The agreement between the calculated and observed habits may be considered a very good one since the calculations involve no fitting variable. All the parameter values used (the elastic constants and crystal lattice parameters of the phases) are taken from independent measurements.

The orientation relations and crystal lattice parameters of the β-phase in a constrained state may be determined from Eqs. (8.5.10) an (8.5.12) with $\mathbf{n}_0$ and $\mathbf{S}(\mathbf{n}_0)$ given by Eqs. (9.3.3) and (8.5.2).

Numerical calculations give for the vector $\mathbf{S}(\mathbf{n}_0)$ [see (8.5.2)]:

$$\mathbf{S}(\mathbf{n}_0)=\varepsilon_0 \boldsymbol{l} \tag{9.3.5}$$

where

$$\varepsilon_0=1.889\,\frac{\sigma_{33}^0}{c_{11}}=1.889\,\frac{2c_{12}\varepsilon_{11}^0+c_{11}\varepsilon_{33}^0}{c_{11}}=1.889\left(\varepsilon_{33}^0+\frac{2c_{12}}{c_{11}}\,\varepsilon_{11}^0\right) \tag{9.3.6}$$

$$\boldsymbol{l}=(-0.278,\ -0.278, 0.919) \tag{9.3.7}$$

is the unit vector in the shear direction.

Substitution of the numerical values (9.3.1) and (9.3.2) into (9.3.6) yields

$$\varepsilon_0=0.07827 \tag{9.3.8}$$

The ε_0 value measured in (173) is equal to

$$0.07864\pm0.00987 \tag{9.3.9}$$

This is in excellent agreement with the calculation result (9.3.8). The vector $\boldsymbol{l}$ observed in (173) is

$$\boldsymbol{l}_{\text{exp}}=(-0.2603,\ -0.2707, 0.9267) \tag{9.3.10}$$

The deviation of the calculated shear direction, (9.3.7), from the observed direction, (9.3.10), is of 2°. We thus find that the vector $\boldsymbol{l}$ calculated from Eq. (8.5.2) also well fits the observed one.

The crystal lattice of the β-phase is closely related to the bcc matrix lattice, and the crystal lattice correspondence between them is as follows:

$$\begin{aligned}&[100]_{\text{V}} \rightarrow [100]_{\beta}\\ &[010]_{\text{V}} \rightarrow [010]_{\beta}\\ &[001]_{\text{V}} \rightarrow [001]_{\beta}\end{aligned} \tag{9.3.11}$$

The crystal lattice parameters of the β-phase in a constrained state may be calculated from Eq. (8.5.10)

$$\mathbf{r}'_\beta = \mathbf{r}_V + \mathbf{S}(\mathbf{n}_0)(\mathbf{n}_0\mathbf{r}_V) \tag{9.3.12}$$

which describes the bcc to β-phase crystal lattice rearrangement. Here, $\mathbf{r}_V$ and $\mathbf{r}'_\beta$ are related reference vectors of the V bcc matrix lattice and β-phase lattice, respectively.

Let the bcc host lattice translations be

$$\mathbf{r}_V = \begin{cases} (a_0, 0, 0) \\ (0, a_0, 0) \\ (0, 0, a_0) \end{cases} \tag{9.3.13}$$

This determines the unit cell of the bcc matrix lattice. According to the crystal lattice correspondence (9.3.11), substitution of $\mathbf{n}_0$ from (9.3.3), $\mathbf{S}(\mathbf{n}_0)$ from (9.3.5), and $\mathbf{r}_V$ from (9.3.13) into Eq. (9.3.12) furnishes the crystal lattice parameters of the β-phase:

$$\begin{aligned}
\mathbf{a}_1(\beta) &= (a_0, 0, 0) + 0.07827(-0.278, -0.278, 0.919)a_0 0.277 \\
&= a_0(0.994, -0.006, 0.0199) \\
\mathbf{a}_2(\beta) &= (0, a_0, 0) + 0.07827(-0.278, -0.278, 0.919)a_0 0.277 \\
&= a_0(-0.006, 0.994, 0.0199) \\
\mathbf{a}_3(\beta) &= (0, 0, a_0) + 0.07827(-0.278, -0.278, 0.919)a_0 0.920 \\
&= a_0(-0.020, -0.020, 1.066)
\end{aligned} \tag{9.3.14}$$

where $\mathbf{a}_1(\beta)$, $\mathbf{a}_2(\beta)$ and $\mathbf{a}_3(\beta)$ are three crystal lattice translation vectors of the β-phase given in the Cartesian basis built on the [100], [010], and [001] vanadium matrix directions.

It follows from Eqs. (9.3.14) that

$$|\mathbf{a}_1(\beta)| = |\mathbf{a}_2(\beta)| = a = a_0\, 0.994 = 3.0138 \text{ Å}$$
$$|\mathbf{a}_3(\beta)| = c = a_0\, 1.0663 = 3.2332 \text{ Å}$$

These parameters differ from the parameters $a = 3.002$ Å and $c = 3.311$ Å reported for the stress-free β-phase because they really have been obtained for the hydride in a constrained state.

Since the reciprocal lattice of the constrained-state hydride is transformed according to Eq. (8.5.11),

$$\mathbf{H}_\beta = \mathbf{H}_V - \mathbf{n}_0 \frac{(\mathbf{S}(\mathbf{n}_0)\mathbf{H}_V)}{1 + (\mathbf{S}(\mathbf{n}_0)\mathbf{n}_0)} \tag{9.3.15}$$

It follows from Eq. (9.3.15) that any crystal lattice plane of the bcc matrix whose reciprocal lattice vector is normal to the vector $\boldsymbol{l}$ retains its orientation and interplanar distance in the bcc → β-phase crystal lattice rearrangement. In particular, we have for the $(1\bar{1}0)_{bcc}$ plane:

$$(1\bar{1}0)_V \| (1\bar{1}0)_\beta$$

9.4. MORPHOLOGY OF COHERENT PRECIPITATES OF CUBIC PHASE IN CUBIC MATRIX

The decomposition of a cubic phase into a two-phase mixture of cubic phases is characterized by the simplest crystallography because the crystal lattice rearrangement is described by the stress-free transformation strain

$$\varepsilon_{ij}^{0}=\varepsilon_0\delta_{ij} \tag{9.4.1}$$

where δ_{ij} is the Kronecker delta symbol,

$$\varepsilon_0=\frac{a_p-a_0}{a_0} \tag{9.4.2}$$

is the parameter describing the crystal lattice mismatch between the two adjacent phases, and a_p and a_0 are the crystal lattice parameters of the cubic phase precipitate and matrix, respectively. If the Vegard law holds, the mismatch parameter ε_0 can be written in terms of the concentration coefficient of crystal lattice expansion. For a binary alloy it is given by

$$\varepsilon_0=\frac{da}{adn}(n_p-n_0) \tag{9.4.3}$$

where n_p and n_0 are the atomic fractions of the solute in the precipitate and matrix, respectively, and da/adn is the concentration coefficient of linear expansion.

If one substitutes the stress-free transformation strain (9.4.1) into Eq. (8.1.1) and (8.1.2) and employs the definition (8.6.3) and (8.6.4), the strain energy of an arbitrary inclusion is obtained in the form

$$E=\frac{1}{2}\int B(\mathbf{n})|\theta(\mathbf{k})|^2\frac{d^3k}{(2\pi)^3} \tag{9.4.4}$$

where

$$B(\mathbf{n})=(c_{11}+2c_{12})\varepsilon_0^2\left[3-(c_{11}+2c_{12})\right.$$
$$\left.\times\frac{1+2\xi(n_1^2n_2^2+n_1^2n_3^2+n_2^2n_3^2)+3\xi^2n_1^2n_2^2n_3^2}{c_{11}+\xi(c_{11}+c_{12})(n_1^2n_2^2+n_1^2n_3^2+n_2^2n_3^2)+\xi^2(c_{11}+2c_{12}+c_{44})n_1^2n_2^2n_3^2}\right]$$
$$\mathbf{n}=(n_1, n_2, n_3) \tag{9.4.5}$$

The bulk strain energy $E_{\text{bulk}}(\mathbf{n})$ of a plate-shaped precipitate whose habit is normal to the unit vector $\mathbf{n}$ may be found from Eq. (9.4.4) and is described by Eq. (8.1.14)

$$E_{\text{bulk}}=\tfrac{1}{2}VB(\mathbf{n}) \tag{9.4.6}$$

where $B(\mathbf{n})$ is given by Eq. (9.4.5).

It is worthy of note that $B(\mathbf{n})$ coincides with the effective modulus $Y(\mathbf{n})$ in Cahn's theory of spinodal decomposition for the symmetry directions $\langle 100\rangle$,

$\langle 110\rangle$, and $\langle 111\rangle$ and would coincide for all other directions **n** if the calculations in (43) were more accurate.

As shown at the end of Section 8.6, the bulk energy (9.4.6) is minimized at **n**, directed either along $\langle 100\rangle$, if $\xi=(c_{11}-c_{12}-2c_{44})/c_{44}<0$, or along $\langle 111\rangle$, if $\xi>0$. The first case thus corresponds to the $\{100\}$ habit whereas the latter one corresponds to the $\{111\}$ habit.

According to Section 8.7 the equilibrium shape of a precipitate is determined by competition between the interphase chemical energy and the edge strain energy, the equilibrium shape of a cubic phase precipitate in a cubic matrix being that of a platelike circular disk.

For such a precipitate the equilibrium aspect ratio $K=D/2R$ at a given volume V can be found from the transcendental equation (8.7.54) and Eq. (8.7.45) (see Fig. 67):

$$K^{\frac{5}{3}}\ln\frac{K}{e}=-A \tag{9.4.7}$$

where R and D are the precipitate particle radius and thickness, respectively,

$$A=\left(\frac{16\pi^4}{27}\right)^{\frac{1}{3}}\frac{c_{11}^2}{(c_{11}+2c_{12})^2|\xi|}\frac{\gamma_s}{(c_{11}-c_{12})\varepsilon_0^2V^{\frac{1}{3}}}$$

or

$$A=3.864\frac{c_{11}^2}{(c_{11}+2c_{12})^2}\frac{1}{|\xi|}\frac{\gamma_s}{(c_{11}-c_{12})\varepsilon_0^2V^{\frac{1}{3}}} \tag{9.4.8}$$

Let us introduce the material constant r_0

$$r_0=\frac{c_{11}^2}{(c_{11}+2c_{12})^2|\xi|}\frac{\gamma_s}{(c_{11}-c_{12})\varepsilon_0^2} \tag{9.4.9}$$

having the length dimension. The dimensionless parameter A in Eq. (9.4.8) may then be written in the form

$$A=\frac{3.864r_0}{V^{\frac{1}{3}}} \tag{9.4.10}$$

This eventually determines the equilibrium shape of the precipitate. All the approximations inherent in Eq. (9.4.7) are based on the assumption $K\ll 1$ (thin-plate approximation). However, to estimate the typical value of the dimensionless parameter (9.4.10) which determines the limit of the applicability of the theory, we must put K equal to unity. This K value corresponds to the equiaxial shape of the inclusion. Substituting $K=1$ into the left-hand side of Eq. (9.4.7), we obtain

$$A=A_0=1 \tag{9.4.11}$$

Since K as a function of A vanishes at $A=0$ and increases monotonically in the range $0<A<1$, the platelike shapes (those characterized by $K=D/2R\ll 1$) are formed if

$$\frac{3.864r_0}{V^{\frac{1}{3}}} \ll 1 \tag{9.4.12}$$

If the inequality (9.4.12) holds, and $\xi=(c_{11}-c_{12}-2c_{44})/c_{44}<0$, the platelike precipitate has the $\{100\}$ habit.

The $\{100\}$ habit should be observed for a large variety of cubic alloys where cubic phase precipitates arise because almost all cubic alloys known show elastic anisotropy with $\xi<0$. This is in complete agreement with the x-ray and electron microscopic observations. The $\{100\}$ habit was in fact reported for all cases of platelike cubic phase precipitates in cubic matrices [see, e.g. (176)].

Certain conclusions on the shapes of coherent precipitates can be made on the basis of the above consideration. If the mismatch parameter ε_0 is small, the inequality (9.4.12) may prove violated, and precipitates may have the equiaxial shape. The same effect may be observed if the precipitate volume V is small. The fact that volume V appears in the inequality (9.4.12) makes it possible to predict changes of precipitate shapes during the coarsening process. Indeed, if a new phase critical nucleus has a small volume and the inequality (9.4.12) does not hold, precipitates formed in the initial stage of the decomposition will have the equiaxial shape. Later in the process, as the precipitate volume increases during coarsening (provided the coherent nature of inclusions is retained), the inequality (9.4.12) may turn valid, and transition from the equiaxial to a platelike shape will be observed.

If the critical nucleus volume meets the inequality (9.4.12), all precipitates will have platelike shapes from the beginning of the aging process.

In sum, a platelike shape may be expected if the mismatch parameter ε_0 is large, the specific interphase energy small, and the precipitate volume V sufficiently large. Conversely, if the crystal lattice mismatch is small, the interphase energy large, and the precipitate particles small, the equiaxial shape is predicted.

We shall now estimate the typical distance r_0 for Ni-based substitutional alloys. Let us assume that the interphase energy γ_s is of the order of

$$\gamma_s \sim 10 \text{ erg/cm}^2$$

and the elastic constants coincide with those of pure Ni:

$$c_{11}=2.5\times10^{12} \text{ dyne/cm}^2$$
$$c_{12}=1.6\times10^{12} \text{ dyne/cm}^2$$
$$c_{44}=1.185\times10^{12} \text{ dyne/cm}^2$$

Substitution of these values into Eq. (9.4.9) yields

$$r_0=0.172\frac{10^{-3}}{\varepsilon_0^2}\text{Å} \tag{9.4.13}$$

With $\varepsilon_0 \approx 0.001$ (small crystal lattice mismatch), Eq. (9.4.13) gives

$$r_0 \approx 172 \text{ Å} \tag{9.4.14a}$$

Table 9.1

Alloy	Crystal lattice mismatch parameter	Shape of precipitates
Ni-Si	0.0024	Equiaxial shape
Ni-Cr-Ti-Al	0.0028	
Ni-Cu-Al	0.0025	
Ni-Al	0.0048	Equiaxial and later platelike
Ni-Mo-Si	0.0065	
Ni-Al-Ti	0.0076	Platelike
Ni-Ti	0.0085	
Ni-Cu-Si	0.01	
Cu-Ni-Fe	0.017	
Cu-Ni-Co	0.017	
Co-Ti	0.020	
Au-Pt	0.034	

If $\varepsilon_0 \approx 0.01$, we obtain

$$r_0 \approx 1.72\ \text{Å} \tag{9.4.14b}$$

The estimate (9.4.14) shows that, with a crystal lattice mismatch of 1 percent, platelike precipitate particles are formed from the very beginning of the process because $r_0 = 1.72$ Å is of the same order of magnitude as the crystal lattice parameter.

The experimental results seem to confirm the theoretical conclusions made. Table 9.1, quoted from (176), demonstrates how the morphology of precipitates depends on the crystal lattice mismatch parameter ε_0.

Alloys characterized by small ε_0 values do in fact contain precipitates having equiaxial shapes whereas alloys with larger ε_0 values have platelike shapes. The effect of shape variation discussed above as inference from the inequality (9.4.12) was observed experimentally in Ni-Al alloys (177, 178) characterized by intermediate mismatch parameter values. This effect was also reported for Ni-Al-Ti (179).

9.4.1. Strain-induced Tetragonality.

The comparison of theoretical predictions with experimental data would not be complete without a discussion of the phenomenon of strain-induced tetragonality. The effect directly follows from coherent fitting together of the cubic crystal lattices of a platelike precipitate and the matrix.

With $\varepsilon_{11}^0 = \varepsilon_{22}^0 = \varepsilon_{33}^0 = \varepsilon_0$, Eq. (9.1.9) gives

$$\mathbf{S}(\mathbf{n}_0) = \left[0, 0, \varepsilon_0\left(1 + 2\frac{c_{12}}{c_{11}}\right)\right] \tag{9.4.15}$$

if $\mathbf{n}_0 = (0, 0, 1)$ (the case of $\xi < 0$).

Application of Eq. (8.5.10) to the basis vectors $\mathbf{r}_1 = (a_0, 0, 0)$, $\mathbf{r}_2 = (0, a_0, 0)$, and $\mathbf{r}_3 = (0, 0, a_0)$ of the parent phase unit cell with $\mathbf{S}(\mathbf{n}_0)$ taken from (9.4.15) results in

$$\mathbf{r}'_1 = \mathbf{r}_1 = (a_0, 0, 0)$$

$$\mathbf{r}'_2 = \mathbf{r}_2 = (0, a_0, 0)$$

$$\mathbf{r}'_3 = \left[0, 0, a_0 + a_0\varepsilon_0\left(1 + 2\frac{c_{12}}{c_{11}}\right)\right] \tag{9.4.16}$$

The three vectors (9.4.16) form the basis of the unit cell of the precipitate crystal lattice in a constrained state. They are mutually orthogonal and $|\mathbf{r}'_2| = |\mathbf{r}'_1| = a_0$. The vectors $\mathbf{r}'_1$, $\mathbf{r}'_2$, and $\mathbf{r}'_3$ are thus the crystal lattice parameters of a slightly tetragonal lattice:

$$c = a_0\left(1 + \varepsilon_0 + 2\varepsilon_0\frac{c_{12}}{c_{11}}\right)$$

$$a = a_0$$

$$\frac{c}{a} \simeq 1 + \varepsilon_0 + 2\varepsilon_0\frac{c_{12}}{c_{11}} \tag{9.4.17}$$

The tetragonality (9.4.17) results from elastic "adjustment" of the precipitate crystal lattice to the parent phase lattice under the conditions that the precipitate is coherent and has a platelike shape. Moreover the exact coincidence of the parameter a with the parent phase lattice parameter a_0 and the slight tetragonality of the precipitate observed may be considered an indication that the precipitate is coherent, platelike shaped and has the $\{100\}$ habit.

In the later stages of aging, when interphase boundaries turn semicoherent because of the formation of misfit dislocations, the elastic accommodation strain within the precipitate particle relaxes, and the precipitate crystal lattice becomes cubic.

In general, the theoretical results obtained for an isolated coherent precipitate particle also hold with a semicoherent precipitate because the formation of misfit dislocations can be treated in terms of renormalization of the material constants of the system, namely, as a decrease of the mismatch parameter ε_0 and an increase of the interphase energy γ_s.

9.4.2. Strain-induced Tetragonality and Diffraction

The cubic-to-cubic crystal lattice rearrangement also results in the rearrangement of the reciprocal lattice and consequently affects the diffraction pattern.

The reciprocal lattice points corresponding to the diffraction spots of the matrix are transformed into those of the precipitate according to Eq. (8.5.11)

$$\mathbf{H}=\mathbf{H}_0-\mathbf{n}_0\frac{(\mathbf{S}(\mathbf{n}_0)\mathbf{H}_0)}{1+(\mathbf{n}_0\mathbf{S}(\mathbf{n}_0))} \tag{9.4.18}$$

or

$$\Delta\mathbf{H}=\mathbf{H}-\mathbf{H}_0=-\mathbf{n}_0\frac{(\mathbf{S}(\mathbf{n}_0)\mathbf{H}_0)}{1+\mathbf{n}_0\mathbf{S}(\mathbf{n}_0)} \tag{9.4.19}$$

where $\mathbf{H}_0$ and $\mathbf{H}$ are the reciprocal lattice vectors of the matrix and constrained precipitate, respectively, $\mathbf{n}_0=(0, 0, 1)$ is the vector normal to the (001) habit plane, $\mathbf{S}(\mathbf{n}_0)$ is given by Eq. (9.4.15), $\Delta\mathbf{H}$ is the shift of the precipitate diffraction spot $\mathbf{H}$ with respect to the matrix phase spot $\mathbf{H}_0$. Substitution of Eq. (9.4.15) and $\mathbf{n}_0=(0, 0, 1)$ into Eq. (9.4.19) yields

$$\Delta\mathbf{H}=\left(0,\ 0,\ -\frac{H_3^0}{a_0}\varepsilon_0\frac{1+2\dfrac{c_{12}}{c_{11}}}{1+\varepsilon_0\left(1+2\dfrac{c_{12}}{c_{11}}\right)}\right) \tag{9.4.20}$$

where (H_1^0, H_2^0, H_3^0) are the Miller indexes of the matrix reflection corresponding to the reciprocal lattice vector $\mathbf{H}_0=(H_1^0, H_2^0, H_3^0)/a_0$.

It follows from Eq. (9.4.20) that the formation of a precipitate affects the diffraction pattern of an alloy and results in splitting of the matrix reflections along the direction [001] normal to the habit plane, the separation between the matrix and precipitate reflections given by Eq. (9.4.19) being

$$|\Delta\mathbf{H}|=-\frac{\varepsilon_0}{a_0}\frac{1+2\dfrac{c_{12}}{c_{11}}}{1+\varepsilon_0\left(1+2\dfrac{c_{12}}{c_{11}}\right)}H_3^0 \tag{9.4.21}$$

The diffraction spots from thin-plate precipitates are broadened and have the shape of a rod in the reciprocal space normal to the habit plane (see Fig. 71). The length of the rod is of the order of $1/D$; its thickness is of about $1/2R$. The diffraction spots of the precipitate can be resolved if the half length of the spots, $1/D$, is less than the separation ΔH between the splitted precipitate and matrix diffraction spots. Otherwise, if

$$\Delta H<\frac{1}{D} \quad \text{or} \quad D\Delta H<1 \tag{9.4.22}$$

the precipitate spots cannot be observed, and the diffraction pattern from the two-phase alloy looks like as if it were taken from a homogeneous single-crystal alloy with dilation point defects.

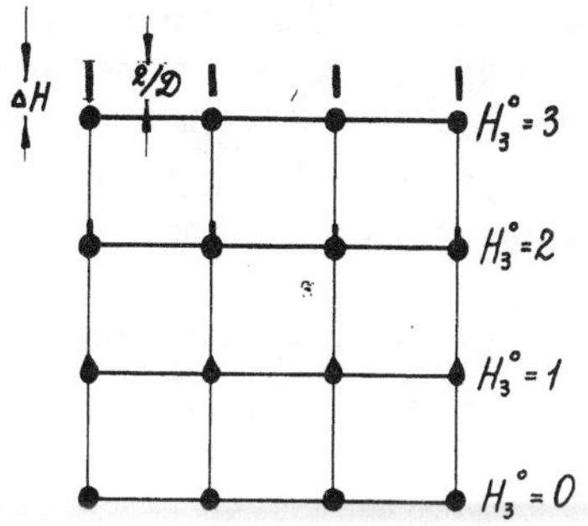

Figure 71. Resolution of diffraction spots formed by thin-plate coherent precipitates of a cubic phase in a cubic matrix. The index H_3^0 enumerates layer lines: ΔH is the shift of the diffraction spot of a coherent precipitate from that of the matrix, and $2/D$ is the length of a precipitate diffraction spot in the reciprocal lattice associated with the finite thickness of the precipitate.

Substituting (9.4.21) into (9.4.22) gives the relation

$$\frac{\varepsilon_0 D}{a_0} H_3^0 \frac{1+2\dfrac{c_{12}}{c_{11}}}{1+\varepsilon_0\left(1+2\dfrac{c_{12}}{c_{11}}\right)} < 1$$

which is the criterion showing when the precipitate spots cannot in principle be resolved. If on the other hand,

$$\frac{\varepsilon_0 D}{a_0} H_3^0 \frac{1+2\dfrac{c_{12}}{c_{11}}}{1+\varepsilon_0\left(1+2\dfrac{c_{12}}{c_{11}}\right)} > 1$$

the precipitate reflections can be observed.

The effect of strain-induced tetragonality in coherent plate-shaped precipitates was reported by several authors. For instance, Butler and Thomas observed this effect in Cu-Ni-Fe alloys (65). In agreement with the theoretical predictions, a slight tetragonality of precipitates manifested itself by splittings of the diffraction spots along the interphase normal [001] in the selected area diffraction pattern. The cubic symmetry was only regained when the interphase network of misfit dislocations was completed (when the coherent interphase was transformed into a semicoherent one). This result is also in agreement with the x-ray data reported by Hargreaves (181) who detected slight tetragonality in 50%Cu-35%Ni-15%Fe after 24 hrs at 800° C. The tetragonal phase was replaced by the cubic stress-free phase after 240 hrs at 800° C.

9.4.3. Sphere-to-cube Shape Transformation

To conclude this Section, we shall consider shape transformations that occur during coarsening. This effect was in part discussed already. It was shown that coarsening transforms equiaxial fine precipitate to platelike particles. We shall demonstrate that precipitates may remain equiaxial and still undergo shape transformations. We shall show that spherical precipitates formed in the early stage of strain-induced coarsening are transfigured to cuboidal particles during the growth process. Still later, cuboidal particles aggregate to form (001) plates. It will also be shown that the sphere-to-cube transformation conforms to the theory given below.

The strain energy of a cuboidal precipitate is given by Eq. (9.4.4) where $\theta(k)$ is the Fourier transform of the shape function of a cube:

$$\theta(\mathbf{k}) = \iiint_{-\infty}^{\infty} \tilde{\theta}(\mathbf{r}) e^{-i\mathbf{kr}} d^3 r = \int_{(V)} e^{-i\mathbf{kr}} d^3 r$$

$$= 8V \frac{\sin \frac{1}{2}k_x L}{k_x L} \frac{\sin \frac{1}{2}k_y L}{k_y L} \frac{\sin \frac{1}{2}k_z L}{k_z L} \tag{9.4.23}$$

where V is the precipitate volume, L is the cube edge dimension, (k_x, k_y, k_z) are the components of the vector $\mathbf{k}$ along the [100], [010], and [001] directions of the cubic parent phase, respectively.

The strain energy of a cuboidal precipitate is thus given by the integral (8.1.1)

$$E_{\text{cube}} = \frac{1}{2} V^2 \iiint_{-\infty}^{\infty} B\left(\frac{\mathbf{k}}{k}\right) \frac{\sin^2 \frac{1}{2}k_x L}{(\frac{1}{2}k_x L)^2} \frac{\sin^2 \frac{1}{2}k_y L}{(\frac{1}{2}k_y L)^2} \frac{\sin^2 \frac{1}{2}k_z L}{(\frac{1}{2}k_z L)^2} \frac{d^3 k}{(2\pi)^3} \tag{9.4.24}$$

The exact analytical integration of (9.4.24) is hardly possible. The only example of a numerical calculation of the integral (9.4.24) for a cuboid cubic-phase precipitate with the elastic constants of copper may be found in the work by Lee and Johnson (140). To obtain a more general solution for an arbitrary set of elastic constants, we can do nothing but sacrifice accuracy for a simplification of Eq. (9.4.24). Thus we may replace Eq. (9.4.5) which determines $B(\mathbf{n})$ with an approximate relation:

$$B(\mathbf{n}) \cong (c_{11} + 2c_{12})\varepsilon_0^2 \Bigg[2\frac{c_{11} - c_{12}}{c_{11}} - 4\frac{c_{11} + 2c_{12}}{c_{11}} \frac{\Delta}{c_{11} + c_{12} + 2c_{44}}$$

$$\times (n_1^2 n_2^2 + n_1^2 n_3^2 + n_2^2 n_3^2) - 54 \frac{c_{11} + 2c_{12}}{c_{11}} \frac{\Delta^2 n_1^2 n_2^2 n_3^2}{(c_{11} + c_{12} + 2c_{44})(c_{11} + 2c_{12} + 4c_{44})} \Bigg] \tag{9.4.25}$$

where

$$\Delta = c_{11} - c_{12} - 2c_{44}$$

Eq. (9.4.25) provides a fairly good approximation to the function $B(\mathbf{n})$. It assumes exactly the same values as Eq. (9.4.5) at $\mathbf{n}$ parallel to the [100], [110], and [111] symmetry directions and serves as a good extrapolation function at intermediate directions.

Substituting Eq. (9.4.25) into (9.4.24) with

$$n_1^2=\frac{k_x^2}{k_x^2+k_y^2+k_z^2},\quad n_2^2=\frac{k_y^2}{k_x^2+k_y^2+k_z^2},\quad n_3^2=\frac{k_z^2}{k_x^2+k_y^2+k_z^2}$$

and term-by-term integration of the resulting equation yields

$$E_{\text{cube}}=\frac{1}{2}(c_{11}+2c_{12})\varepsilon_0^2V\left[2\frac{c_{11}-c_{12}}{c_{11}}-12\frac{c_{11}+2c_{12}}{c_{11}}\frac{\Delta}{c_{11}+c_{12}+2c_{44}}I_1\right.$$
$$\left.-54\frac{c_{11}+2c_{12}}{c_{11}}\frac{\Delta^2 I_2}{(c_{11}+c_{12}+2c_{44})(c_{11}+2c_{12}+4c_{44})}\right] \tag{9.4.26}$$

where

$$I_1=\iiint_{-\infty}^{\infty}\frac{\sin^2x\,\sin^2y\,\sin^2z\,dxdydz}{(x^2+y^2+z^2)^2z^2\quad(2\pi)^3}=0.006931$$

$$I_2=\iiint_{-\infty}^{\infty}\frac{\sin^2x\,\sin^2y\,\sin^2z\,dxdydz}{(x^2+y^2+z^2)^3\quad(2\pi)^3}=0.000959 \tag{9.4.27}$$

In the case of a spherical precipitate the strain energy is given by Eq. (8.2.34) (with $\mathbf{n}'=\mathbf{n}$):

$$E_{\text{sphere}}=\tfrac{1}{2}V<B(\mathbf{n})> \tag{9.4.28}$$

where

$$\langle\cdots\rangle=\oint\frac{dO_{\mathbf{n}}}{4\pi}(\cdots) \tag{9.4.29}$$

is the average over all the directions $\mathbf{n}$, $dO_{\mathbf{n}}$ is the solid angle element normal to $\mathbf{n}$. Substituting (9.4.25) into (9.4.28) gives

$$E_{\text{sphere}}=\frac{1}{2}(c_{11}+2c_{12})\varepsilon_0^2V\left[2\frac{c_{11}-c_{12}}{c_{11}}-12\frac{c_{11}+2c_{12}}{c_{11}}\frac{\Delta}{c_{11}+c_{12}+2c_{44}}I_1^0\right.$$
$$\left.-54\frac{c_{11}+2c_{12}}{c_{11}}\frac{\Delta^2 I_2^0}{(c_{11}+c_{12}+2c_{44})(c_{11}+2c_{12}+4c_{44})}\right] \tag{9.4.30}$$

where

$$I_1^0=\langle n_1^2n_2^2\rangle=\oint\frac{dO_{\mathbf{n}}}{4\pi}n_1^2n_2^2=\frac{1}{15}$$

$$I_2^0=\langle n_1^2n_2^2n_3^2\rangle=\frac{1}{105} \tag{9.4.31}$$

Lastly, a thin-plate precipitate has the strain energy (8.1.14) where, according to Eq. (9.4.25),

$$\min B(\mathbf{n})=(B(\mathbf{n}_0))_{\mathbf{n}_0=(0,0,1)}=2\frac{c_{11}+2c_{12}}{c_{11}}(c_{11}-c_{12})\varepsilon_0^2$$

Using the latter expression in Eq. (8.1.14), we obtain

$$E_{\text{plate}}=\frac{1}{2}B(\mathbf{n}_0)V=\frac{c_{11}+2c_{12}}{c_{11}}(c_{11}-c_{12})\varepsilon_0^2 V \tag{9.4.32}$$

To estimate strain energies for spherical, cuboidal, and thinplate precipitates, we shall substitute the elastic constants of nickel,

$$c_{11}=2.5\times 10^{12},\quad c_{12}=1.6\times 10^{12};\quad c_{44}=1.185\times 10^{12}\ \text{dyne/cm}^2 \tag{9.4.33}$$

into Eqs. (9.4.26), (9.4.30), and (9.4.32).

The numerical calculations yield

$$\begin{aligned}E_{\text{sphere}}&=2.6381c_{44}\varepsilon_0^2 V\\ E_{\text{cube}}&=1.8352c_{44}\varepsilon_0^2 V\\ E_{\text{plate}}&=1.7316c_{44}\varepsilon_0^2 V\end{aligned} \tag{9.4.34}$$

Comparison of the strain energies (9.4.34) shows that the strain energy of a precipitate is strongly dependent on its shape. The major decrease (by 32 percent) in the strain energy occurs on going from a sphere to a cuboid particle. Further transformation of cuboid particles to platelets only provides a 5.5 percent lowering of the strain energy.

Below we shall demonstrate how competition between strain and interphase energies effects the transformation of spherical precipitates to cuboid ones. To find the critical precipitate volume at which the transformation becomes inevitable, we must compare the total free energies of spherical and cuboidal particles given as sums of the precipitate strain and interphase energies (the bulk chemical free energy does not depend on particle shape and may therefore be ommitted from consideration).

With the numerical values (9.4.27) the sum of the interphase energy, $6\gamma_s V^{\frac{2}{3}}$, and the strain energy (9.4.26) for a cuboidal particle is given by

$$F_{\text{cube}}=\frac{V}{2}(c_{11}+2c_{12})\varepsilon_0^2\left\{2\frac{c_{11}-c_{12}}{c_{11}}-0.0831\frac{c_{11}+2c_{12}}{c_{11}}\frac{\Delta}{c_{11}+c_{12}+2c_{44}}\right.$$
$$\left.-0.0517\frac{c_{11}+2c_{12}}{c_{11}}\frac{\Delta^2}{(c_{11}+c_{12}+2c_{44})(c_{11}+2c_{12}+4c_{44})}\right\}+6\gamma_s V^{\frac{2}{3}} \tag{9.4.35}$$

where $6V^{\frac{2}{3}}=6L^2=S$ is the total interphase area of the cuboidal precipitate, γ_s is the specific interphase free energy.

For a sphere, the interphase area is given by $\gamma_s(36\pi)^{\frac{1}{3}}V^{\frac{2}{3}}$. From this and from Eq. (9.4.30) for the strain energy and the numerical values (9.4.31), we obtain

$$F_{\text{sphere}}=\frac{V}{2}(c_{11}+2c_{12})\varepsilon_0^2\left\{2\frac{c_{11}-c_{12}}{c_{11}}-0.8\frac{c_{11}+2c_{12}}{c_{11}}\frac{\Delta}{c_{11}+c_{12}+2c_{44}}\right.$$

$$\left.-0.5142\frac{c_{11}+2c_{12}}{c_{11}}\frac{\Delta^2}{(c_{11}+c_{12}+2c_{44})(c_{11}+2c_{12}+4c_{44})}\right\}+\gamma_s(36\pi)^{\frac{1}{3}}V^{\frac{2}{3}} \tag{9.4.36}$$

where $(36\pi)^{\frac{1}{3}}V^{\frac{2}{3}}$ is the total surface of a sphere expressed via its volume.

The sphere-to-cuboid particle transformation occurs when

$$F_{\text{cube}}=F_{\text{sphere}} \tag{9.4.37}$$

Hence we obtain using Eqs. (9.4.35) and (9.4.36) in (9.4.37),

$$\frac{\gamma_s c_{11}(c_{11}+c_{12}+2c_{44})}{(c_{11}+2c_{12})^2|\Delta|\varepsilon_0^2V^{\frac{1}{3}}}\left[0.3584+0.2312\frac{\Delta}{c_{11}+2c_{12}+4c_{44}}\right]^{-1}=1$$

or

$$\frac{2c_{11}^2}{(c_{11}+2c_{12})^2|\xi|}\frac{\gamma_s}{(c_{11}-c_{12})\varepsilon_0^2V^{\frac{1}{3}}}\Lambda_0=1 \tag{9.4.38}$$

where

$$\Lambda_0=\frac{(c_{11}-c_{12})(c_{11}+c_{12}+2c_{44})}{2c_{11}c_{44}}\left[0.3584+0.2312\frac{c_{44}\xi}{c_{11}+2c_{12}+4c_{44}}\right]^{-1} \tag{9.4.39}$$

Using the definition (9.4.9) in Eq. (9.4.38), we have

$$\Lambda_0\frac{r_0}{V^{\frac{1}{3}}}=1 \tag{9.4.40}$$

[compare with Eq. (9.4.10) and (9.4.11).].

The calculation of the dimensionless parameter Λ_0 with the elastic constants (9.4.33) yields

$$\Lambda_0=3.016 \tag{9.4.41}$$

The calculation of the length parameter r_0 with $\gamma_s=10$ erg/cm^2 in (9.4.13) results in

$$r_0=\frac{1.72}{\varepsilon_0^2}\times10^{-4}\,\text{Å}$$

From this and Eq. (9.4.40), we may find the critical edge dimension of a cuboidal precipitate

$$L_c=\sqrt[3]{V_c}=3.016r_0=3.016\frac{1.72\times10^{-4}}{\varepsilon_0^2}\,\text{Å}$$

$$=\frac{5.187}{\varepsilon_0^2}\times10^{-4}\,\text{Å} \tag{9.4.42}$$

If $L > L_c = \sqrt[3]{V_c}$, cuboidal precipitates are formed. If on the other hand, $\sqrt[3]{V} < \sqrt[3]{V_c}$, spherical precipitates are favored.

For instance, with $\gamma_s = 10$ erg/cm^2, $\varepsilon_0 = 0.001$, the critical size L_c calculated from Eq. (9.4.42) is

$$L_c = \frac{5.187}{(0.001)^2} \times 10^{-4}\ \text{Å} = 518\ \text{Å}$$

In the case specified, spherical precipitates undergo transformation to cuboidal ones if they overgrow the critical size of 518 Å.

It is of interest to compare these theoretical predictions with the experimental data.

The electron microscopic study of the decomposition in Ni-Al alloys has shown that precipitates of the cubic γ' phase (Ni_3Al) are randomly distributed spherical particles that are transformed into cuboidal ones with coarsening (177). Still later, cuboidal precipitates aggregate to form platelets with the {100} habit (177). The electron micrographs of cuboidal precipitates in Cu-22at%Au alloy (187) are shown in Fig. 72.

To estimate the critical size L_c that determines the size ranges of cuboidal ($L > L_c$) and spherical precipitates ($L < L_c$) in Ni-Al alloys, we may use the numerical values $\gamma_s = 14.4$ erg/cm^2 and $\varepsilon_0 \simeq 0.005$ (177) and the elastic constants (9.4.33). Substitution of these parameter values into Eq. (9.4.9) yields

$$r_0 = 9.90\ \text{Å} \tag{9.4.43}$$

Using Eq. (9.4.40) with $\Lambda_0 = 3.016$, we obtain

$$L_c = \sqrt[3]{V_c} = \Lambda_0 r_0 = 3.016 \times 9.90 \simeq 30\ \text{Å} \tag{9.4.44}$$

According to the estimate (9.4.44) spherical precipitates are transformed into cuboidal ones when their size exceeds $L_c \sim 30$ Å. This conclusion is in a reasonable agreement with the electron microscopic observations.

In Ni-Si the misfit parameter ε_0 is nearly two times lower than with Ni-Al ($\varepsilon_0 \approx 0.0025$). The calculation with $\gamma_s = 10.5$ erg/cm^2 and $\varepsilon_0 = 0.0025$ (179) yields

$$r_0 = 28.9\ \text{Å} \quad \text{and} \quad L_c = \Lambda_0 r_0 = 3.016 \times 28.9 \simeq 87\ \text{Å} \tag{9.4.45}$$

The electron microscopic study (179) has shown that the decomposition of Ni-Si alloy results in the formation of spherical precipitates of diameters ranging from 50 to 80 Å. Further growth of precipitates results in their transformation to cuboidal particles (180). This result agrees with the theoretical prediction $L_c = 87$ Å for the critical size because the diameter of spherical precipitates observed was less than 87 Å.

The lowest misfit parameter ε_0 was found for Ni-Cr-Al ($\varepsilon_0 \approx 0.001$). The estimation of the critical size for the sphere-to-cuboid shape transformation with $\gamma_s = 10$ erg/cm^2 yields

$$L_c \approx 518\ \text{Å}$$

This is also in agreement with the electron microscopic observation of spherical precipitates having diameters of the order of 10^3 Å (179).

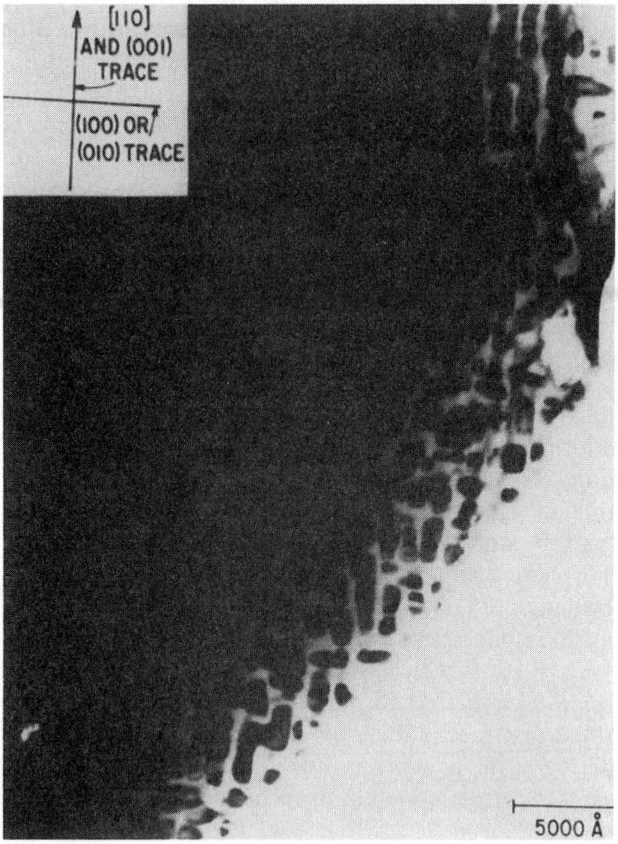

Figure 72. Cuboidal precipitates in the Cu-22.0 Au alloy (187). (Courtesy of M. J. Marcinkovski and L. Zwell.)

Reconfiguration of a spherical precipitate into a cuboidal one can be also utilized to determine the interphase energy coefficient γ_s. Indeed, if the critical diameter of a spherical precipitate is known from electron microscopic observations, the value r_0 and thus the interphase energy coefficient γ_s can be calculated from Eqs. (9.4.40) and (9.4.9), respectively.

9.5. GP ZONES IN SOLID SOLUTIONS: Al-Cu ALLOYS

X-ray and electron microscopic studies have shown that aging of some supersaturated solid solutions results in the formation of so-called Guinier-Preston zones (GP zones) which further develop into metastable or stable precipitate phases (182). GP zones are small segregations formed by atomic redistribution over crystal lattice sites of the homogeneous solid solution. They may have either equiaxial (Cu-Co, Al-Zn, Al-Ag, etc.) or platelike (Al-Cu) shapes. The equiaxial shape corresponds to a small difference between the atomic radii of the solute and solvent atoms (less than about 3 percent). If on the other hand, the difference is comparatively large, GP zones have platelike shapes. For instance, in the case of Al-Cu alloys where GP zones are a few interplanar distances thick, the difference in the atomic radii amounts to about 12 percent. According to Guinier (182) GP zones cannot be regarded as new phase precipitates because they do not have well-defined boundaries and their structure is continuously transformed into the parent phase structure. It has, however, been shown in Section 8.5 that the crystal lattice planes parallel to the habit plane of a constrained coherent platelike precipitate and undistorted parent phase are geometrically identical to each other and that a continuous transformation of one crystal lattice into another one without a precipitate boundary is a feature of coherent fitting. Guinier's argument therefore provides no basis for treating GP zones as formations different from thin coherent new phase precipitates. The x-ray studies (183–186) and the computer simulation carried out by Morris and Khachaturyan (64) (see Section 5.5) confirm this conclusion and demonstrate that the GP zone stage of aging may well be understood in terms of metastable *T-c* phase diagrams (see Fig. 50c). Following Wen et al., we shall discuss the formation of GP zones in Al-Cu alloys (152).

The theoretical results obtained in Section 8.1 may be applied to analyze the morphology of GP zones which will be treated as coherent metastable phase precipitates in a cubic matrix formed in isomorphic decomposition. In this case metastable phase precipitates also have cubic symmetry in the stress-free state. The spacing of the new phase, however, differs from that of the matrix. The stress-free transformation strain in such systems is a pure dilation:

$$\varepsilon_{ij}^0 = \varepsilon_0 \delta_{ij} = \frac{da}{adn}(n_p - n_0)\delta_{ij} \tag{9.5.1}$$

where $\varepsilon_0 = (da/adn)(n_p - n_0)$ is the concentration coefficient of crystal lattice expansion, n, the atomic fraction of solute atoms, n_p and n_0 are the atomic fractions of the solute in the precipitate and matrix, respectively, δ_{ij} is the Kronecker symbol.

To evaluate the elastic strain energy of a cubic precipitate in a cubic matrix, we may use Eq. (9.4.4). If $\xi < 0$, the function $B(\mathbf{n})$ in (9.4.4) assumes its minimum value:

$$B(\mathbf{n}_0) = (c_{11} + 2c_{12})\varepsilon_0^2 \left(3 - \frac{c_{11} + 2c_{12}}{c_{11}}\right)$$

at $\mathbf{n} = \mathbf{n}_0 = \langle 100 \rangle$. The conclusion follows that the minimum strain energy (8.1.14) is associated with a platelike precipitate having the $\{100\}$ habit (see Section 8.1).

To determine the orientation relations and crystal lattice parameters of constrained cubic precipitates, one should employ Eq. (9.4.17). According to Eqs. (9.4.17) a coherent platelike cubic phase inclusion with the (001) habit must be slightly tetragonal, the relations between the crystal lattice parameters being

$$a_p = a_0$$

$$c_p = a_0\left(1+\varepsilon_0+2\varepsilon_0\frac{c_{12}}{c_{11}}\right)$$

$$\frac{c_p}{a_p} = 1+\varepsilon_0+2\varepsilon_0\frac{c_{12}}{c_{11}} \tag{9.5.2}$$

where a_p and c_p are the crystal lattice parameters of the constrained precipitate.

According to the definition of the stress-free strain ε_0,

$$a_0(1+\varepsilon_0) = a_p^0 \tag{9.5.3}$$

where a_p^0 is the spacing of the cubic lattice of the precipitate in the stress-free state. Substituting Eq. (9.5.3) into (9.5.2) gives

$$a_p = a_0$$

$$c_p = a_p^0 + 2\frac{c_{12}}{c_{11}}(a_p^0 - a_0) \tag{9.5.4}$$

Eqs. (9.5.4) and (9.5.2) solve the problem of independent determination of crystal lattice parameters of constrained precipitates from the crystal lattice parameters, a_0 and a_p^0, of the phases in the stress-free state.

The orientation relation

$$\{100\}_p \| \{100\}_0 \tag{9.5.5}$$

follows directly from the relations (9.4.16).

As shown in Section 9.4 [the relation (9.4.12)], spherical zones may be expected at small strain magnitudes, ε_0, whereas platelike zones should occur at large ε_0. This conclusion is in a good agreement with the fact that GP zones are spherical in Al-Zn and Al-Ag and platelike in Al-Cu alloys.

Let us assume that the elastic constants of the Al-based matrix of Al-Cu alloys are close to those of pure Al:

$$c_{11} = 1.068 \times 10^{12} \text{ dyne/cm}^2$$

$$c_{12} = 0.607 \times 10^{12} \text{ dyne/cm}^2$$

$$c_{44} = 0.282 \times 10^{12} \text{ dyne/cm}^2 \tag{9.5.6}$$

Since $\xi = -0.365 < 0$, the function $B(\mathbf{n})$ given by Eq. (9.4.5) has its minimum at $\mathbf{n}_0 = \{100\}$. Coherent precipitates should therefore have the $\{100\}$ habit.

The relation

$$\frac{da}{adn} \approx \frac{a_{\text{Cu}} - a_{\text{Al}}}{a_{\text{Al}}} = \frac{3.605 - 4.04}{4.04} = -0.1 \tag{9.5.7}$$

where $a_{\text{Cu}} = 3.605$ Å and $a_{\text{Al}} = 4.04$ Å may be used to estimate the stress-free strain ε_0.

According to Gerold (188), the best fit between the calculated and observed x-ray diffuse scattering from Al-4at%Cu is obtained on the assumption that a GP zone is a sole (001) plane containing Cu-atoms only. On both of its sides, the adjacent fcc Al matrix (001) planes occupied with Al are displaced toward the GP zone center by $u=0.2$ Å (see Fig. 73). Within this model the inward displacements of the (001) precipitate boundaries are given by

$$u=u_{33}^{*}\frac{a_{\mathrm{Al}}}{4}=\left(1+2\frac{c_{12}}{c_{11}}\right)\varepsilon_0\frac{a_{\mathrm{Al}}}{4} \tag{9.5.8}$$

where $a_{\mathrm{Al}}/4$ is the distance between the precipitate centrum and its boundary. The displacements (9.5.8) are just the displacements of the nearest to the GP zone Al (001) planes toward the Cu (001) plane in the middle,

$$u_{33}^{*}=\frac{c_p-a_0}{a_0}=\varepsilon_0\left(1+2\frac{c_{12}}{c_{11}}\right)$$

[see Eq. (9.5.2)]. Substitution of the numerical values from (9.5.6) and (9.5.7) into (9.5.8) yields

$$u=\left(1+\frac{2\times 0.607}{1.068}\right)(-0.1)\frac{4.04}{4}=-0.216\text{ Å} \tag{9.5.9}$$

in a very good agreement with the experimental value, $u=-0.2$ Å, obtained from the x-ray measurements (189).

To explain the appearance of solute atom segregations whose thickness is equal to one or a few interplanar distances the microscopic kinetic theory of decomposition based on consideration of crystal lattice structures of solid solutions must be applied. The microscopic approach was applied by Morris and Khachaturyan (64) who analyzed spinodal decomposition within the framework of the atomic model described in Section 5.5. The computer simulation of spinodal decomposition carried out in (64) has demonstrated that

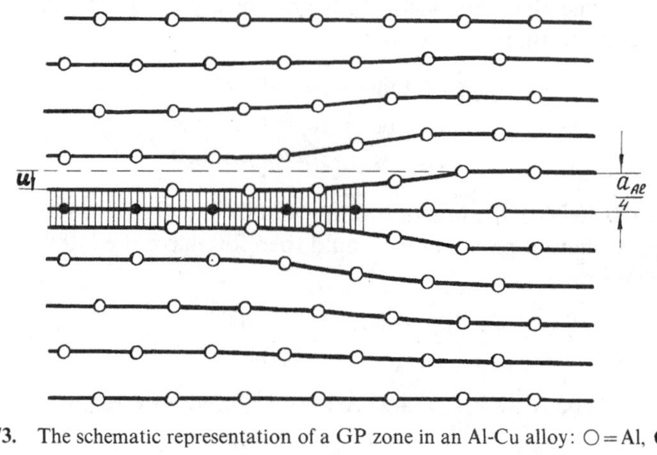

Figure 73. The schematic representation of a GP zone in an Al-Cu alloy: ○ = Al, ● = Cu.

precipitates one or several planes thick may actually be formed in alloys near the solubility limit (Fig. 50c). The detailed results of the computer simulation are cited in Section 5.5.

Aging above 100° C results in the dissolution of GP zones and appearance of platelets of the θ'' metastable phase along $\{100\}$ planes. The thicknesses and diameters of the platelets are of about 20 to 400 Å, respectively. The orientation relation between the crystal lattices of the parent and θ'' phase is

$$\{100\}_{\theta''} \| \{100\}_0 \tag{9.5.10}$$

The θ'' phase has a lattice similar to that of the fcc matrix. In a first approximation, it may be treated as an Al-based substitutional superlattice with $a_{\theta''} \approx a_{\text{Al}}$ and $c_{\theta''} \approx 2a_{\text{Al}}$ (190).

The x-ray measurements showed [see for instance (176)] that the θ'' phase is simple tetragonal with

$$c = 7.7 \text{ Å}$$

$$a = 4.04 \text{ Å} \tag{9.5.11}$$

The habit plane orientation, orientation relations, and crystal lattice parameters of constrained precipitates of the θ'' phase will fit the theoretical predictions. The (001) habit follows from Eq. (9.4.5) since the function $B(\mathbf{n})$ assumes its minimum value at $\mathbf{n} = \mathbf{n}_0 = (0, 0, 1)$.

The orientation relation (9.5.5) obtained from the theoretical consideration agrees with the observed orientation relation (9.5.10).

Lastly, the crystal lattice parameters of the θ'' phase also fit the theoretical predictions (9.5.2). In fact, since the habit plane of the θ'' precipitate is (001), the crystal lattice parameters of the θ'' phase, $a_{\theta''}$, and matrix, a_{Al}, lying in the (001) plane should be the same (this is true for any crystal lattice translation of the θ'' and parent phases in the habit plane).

With the θ'' phase composition $n \approx \frac{1}{4}$ fitting the stoichiometry Al_3Cu, the crystal lattice parameter $c_{\theta''}$ is [see Eq. (9.5.2)]

$$c_{\theta''} = 2a_{\text{Al}}\left[1 + \left(1 + 2\frac{c_{12}}{c_{11}}\right)\frac{da}{adn}\, n_p\right]$$

$$= 2 \times 4.04\left[1 + \left(1 + 2\frac{0.607}{1.068}\right)(-0.1)0.25\right] = 7.65 \text{ Å} \tag{9.5.12}$$

The calculated value, $c_{\theta''} = 7.65$ Å, well compares with that obtained experimentally, $c_{\theta''} = 7.70$ Å (the exact coincidence is attained with n_{Cu} set equal to 0.22 instead of 0.25 as in our calculations).

The intermediate θ' phase that succeeds the θ'' phase in the course of aging has the face-centered tetragonal lattice (fct) with the crystal lattice parameters (191):

$$a_{\theta'} = a_{\text{Al}} = 4.04 \text{ Å}$$

$$c_{\theta'} = 5.8 \text{ Å} \tag{9.5.13}$$

As with θ'' phase precipitates the coincidence of the crystal lattice parameters of the θ' phase, $a_{\theta'}$, and matrix, a_{Al}, shows θ' phase precipitates to be platelike coherent inclusions with the $\{001\}$ habit. This is in agreement with electron microscopic observations.

Precipitates of the η' metastable hexagonal phase in Al-Zn-Mg alloys provide one more evidence for the validity of the theoretical claim that a platelike coherent precipitate and a matrix should have the same crystal lattice parameters in the habit plane. Low-temperature aging was shown to result in the precipitation of the hexagonal η' phase with the crystal lattice parameters (192):

$$a_{\eta'} = 4.96\ \text{Å}, \quad c_{\eta'} = 8.68\ \text{Å}$$

The Al-based matrix has the parameter

$$a_0 = 4.054\ \text{Å}$$

The $[\frac{\bar{1}}{2}\frac{\bar{1}}{2}1]_0$ and $[\frac{\bar{1}}{2}1\frac{\bar{1}}{2}]_0$ translations of the fcc matrix lying in the $(111)_0$ plane are equal to

$$T([\tfrac{\bar{1}}{2}\tfrac{\bar{1}}{2}1]) = T([\tfrac{\bar{1}}{2}1\tfrac{\bar{1}}{2}]) = a_0\sqrt{\frac{3}{2}} = 4.05\sqrt{\frac{3}{2}} = 4.96\ \text{Å}$$

The translations $\mathbf{T}([\frac{\bar{1}}{2}\frac{\bar{1}}{2}1])$ and $\mathbf{T}([\frac{\bar{1}}{2}1\frac{\bar{1}}{2}])$ in the (111) matrix plane making an angle of 60° are thus exactly equal to the hcp η' phase translations $[100]_{\eta'}$ and $[010]_{\eta'}$ in the $(001)_{\mathrm{hcp}}$ plane ($a_{\eta'} = 4.96$ Å) which also make an angle of 60°. This coincidence of the crystal lattice parameters may be understood if η' phase precipitates are assumed to be coherent platelets with a $\{111\}_0$ habit that corresponds to the $(001)_{\mathrm{hcp}}$ η' phase plane.

9.6. EQUILIBRIUM SHAPE OF MARTENSITIC "LATHS"

As was shown in Section 6.7, the crystallographic mechanism of the $\gamma \to \alpha$ rearrangement which is the combination of the Bain distortion and the $[0\bar{1}1](011)_b$ slip leads to the formation of twin-free martensitic plates with the Kurdjumov-Sacks orientation relations and the habit near the $(111)_A$ austenite plane. The mechanism described in Section 6.7 seems to provide an explanation of the main morphological characteristics of lath martensite except for elongation of martensite lath-shaped planes along the $\langle 110\rangle_A$ direction of the austenite lattice. We shall now apply the theory given in Section 8.8 to calculate the shape of martensitic crystals in the habit plane and to see if the $\gamma \to \alpha$ rearrangement mechanism suggested can account for the formation of laths elongated along $\langle 110\rangle_A$, as determined by a number of authors [see, e.g. (99)].

It was found in Section 8.1 that with heterogeneous moduli, the energy E_{edge} which eventually determines the inclusion shape in the habit plane depends on the matrix elastic moduli. Therefore we must calculate the edge strain energy E_{edge} in Eq. (8.7.5) using the elastic moduli of the austenite parent phase. The elastic moduli of austenite were only determined for Fe-Ni alloys in a limited composition range, from about 30 to 50 percent Ni. Lath martensite was observed, in particular, in Fe-20Ni-5Mn (99) and Fe up to 28 percent Ni (97, 98) alloys. Unfortunately, no data on the elastic constants of those alloys is available. The nearest in the composition austenite phase whose elastic constants are known is Fe-31.5Ni (194). The elastic constant values for this alloy are

$$c_{11}=1.404\times 10^{12}\ \text{dyne/cm}^2$$
$$c_{12}=0.84\times 10^{12}\ \text{dyne/cm}^2$$
$$c_{44}=1.121\times 10^{12}\ \text{dyne/cm}^2 \qquad (9.6.1)$$

According to Section 8.8, to calculate the shape of an inclusion in the habit plane, we must (1) calculate the tensor $\beta_{ij}(\mathbf{n}_0)$ in Eq. (8.8.20b), (2) find its eigenvalues β_1 and β_2 and eigenvectors $\mathbf{e}_1$ and $\mathbf{e}_2$ which determine the directions of the x and y-axes, (3) evaluate the anisotropy parameter α in Eq. (8.7.20) and, finally, (4) determine the inclusion shape in the habit plane, $y=y(x)$, from the parametric equations (8.7.23) and (8.7.26).

To realize this, we shall employ the calculated vector $\mathbf{n}_0$ normal to the habit plane, the vector $\boldsymbol{l}$, and the strain ε_0 given by Eqs. (6.7.26):

$$\mathbf{n}_0=(0.5494,\ 0.5494,\ 0.6295) \qquad (9.6.2a)$$

$$\boldsymbol{l}=(0.494,\ 0.494,\ -0.715)$$
$$\varepsilon_0=0.2365 \qquad (9.6.2b)$$

The nonzero components of the elastic modulus tensor λ_{ijkl} for a cubic crystal written in terms of the elastic constants c_{11}, c_{12}, and c_{44} are given by Eq. (8.6.1). Substituting the nonzero components of the tensor λ_{ijkl} taken from (8.6.1) and the components of the tensor u^0_{ij} from (8.8.1) into Eq. (8.8.17), we obtain

$$\sigma^{(1)}_{11}=\sigma^{(1)}_{22}=c_{44}\varepsilon_0\left[\frac{c_{11}}{c_{44}}\,l_1n^0_1+l_2n^0_2+l_3n^0_3\right]$$

$$\sigma^{(1)}_{33}=c_{44}\varepsilon_0\left[\frac{c_{11}}{c_{44}}\,l_3n^0_3+l_1n^0_1+l_2n^0_2\right]$$

$$\sigma^{(1)}_{ij}=c_{44}\varepsilon_0\left[\frac{c_{12}}{c_{44}}\,l_in^0_j+l_jn^0_i\right] \quad \text{if } i\neq j \qquad (9.6.3)$$

where $\mathbf{n}_0=(n^0_1, n^0_2, n^0_3)$, $\boldsymbol{l}=(l_1, l_2, l_3)$. Using the numerical values (9.6.1) and (9.6.2) in Eq. (9.6.3), we find

$$\hat{\sigma}_1=c_{44}\varepsilon_0\begin{pmatrix}0.1611 & 0.4747 & -0.1599\\ 0.4747 & 0.1611 & -0.1599\\ 0.0167 & 0.0167 & -0.0207\end{pmatrix} \qquad (9.6.4)$$

$$(\hat{\sigma}^+_1)_{ij}=\sigma^{(1)}_{ji} \qquad (9.6.5)$$

The components of the tensor $\Omega_{ij}(\mathbf{n})$ for a cubic crystal are given by Eq. (8.6.3).

Substitution of the numerical values (9.6.2a) and (9.6.1) into (8.6.3) yields

$$\hat{\Omega}(\mathbf{n}_0)=\frac{1}{c_{44}}\begin{pmatrix} 0.8176 & -0.2147 & -0.3315 \\ -0.2147 & 0.8176 & -0.3315 \\ -0.3315 & -0.3315 & 0.8792 \end{pmatrix} \tag{9.6.6}$$

The components of the tensor $\Omega_{ij}^{-1}(\mathbf{n})$ are given by Eq. (8.6.2). Substituting the vector $\boldsymbol{l}$ from (9.6.2a) for $\mathbf{n}$, and using the numerical values (9.6.1), we may rewrite Eq. (8.6.2) in the form

$$\hat{\Omega}^{-1}(\boldsymbol{l})=c_{44}\begin{pmatrix} 0.8717 & 0.4883 & -0.7067 \\ 0.4883 & 0.8717 & -0.7067 \\ -0.7067 & -0.7067 & 1.0061 \end{pmatrix} \tag{9.6.7}$$

It follows from Eq. (8.8.20a) that

$$\hat{\beta}(\mathbf{n}_0)=\varepsilon_0^2\hat{\Omega}^{-1}(\boldsymbol{l})-\hat{\sigma}_1\hat{\Omega}(\mathbf{n}_0)\hat{\sigma}_1^+ \tag{9.6.8}$$

As shown in Section 8.8, the vector $\mathbf{n}_0$ is always an eigenvector of the matrix (9.6.8) corresponding to the zero eigenvalue. It is easy to show that, with $\mathbf{n}_0$, $\boldsymbol{l}$, and ε_0 given by (9.6.2), the vector

$$\mathbf{e}_2=\left(\frac{1}{\sqrt{2}}, \frac{\bar{1}}{\sqrt{2}}, 0\right) \tag{9.6.9}$$

is also an eigenvector of the tensor $\hat{\beta}(\mathbf{n}_0)$. Actually, the direct calculations show that the vector (9.6.9) is an eigenvector of the matrices (9.6.4), (9.6.5), (9.6.6), and (9.6.7):

$$\hat{\sigma}_1\mathbf{e}_2=-0.3136c_{44}\varepsilon_0\mathbf{e}_2, \quad \hat{\sigma}_1^+\mathbf{e}_2=-0.3136c_{44}\varepsilon_0\mathbf{e}_2$$

$$\hat{\Omega}\mathbf{e}_2=\frac{1}{c_{44}}1.0323\mathbf{e}_2, \quad \hat{\Omega}^{-1}\mathbf{e}_2=c_{44}0.3834\mathbf{e}_2 \tag{9.6.10}$$

Since the vector $\mathbf{e}_2$ is the eigenvector of all the operators entering the expression for the operator $\hat{\beta}(\mathbf{n}_0)$ in Eq. (9.6.8), it is also an eigenvector of the operator $\hat{\beta}(\mathbf{n}_0)$

$$\hat{\beta}(\mathbf{n}_0)\mathbf{e}_2=\beta_2\mathbf{e}_2 \tag{9.6.11}$$

where the quantity β_2 may be determined by substituting Eq. (9.6.10) into Eq. (9.6.8):

$$\hat{\beta}(\mathbf{n}_0)\mathbf{e}_2=c_{44}\varepsilon_0^2[0.3834-(-0.3136)1.0323(-0.3136)]=0.2819c_{44}\varepsilon_0^2$$

Hence

$$\beta_2=0.2819c_{44}\varepsilon_0^2 \tag{9.6.12}$$

Since $\hat{\beta}(\mathbf{n}_0)$ is a Hermitian operator, its eigenvectors should be mutually orthogonal. As two of them [$\mathbf{n}_0$ and $\mathbf{e}_2$, see Eqs. (9.6.2a) and (9.6.9)] are known, the third one, $\mathbf{e}_1$, is easy to determine. It is

$$\mathbf{e}_1=(-0.4451, \ -0.4451, 0.7769) \tag{9.6.13}$$

Since

$$\hat{\boldsymbol{\beta}}(\mathbf{n}_0)\mathbf{e}_1 = \beta_1 \mathbf{e}_1$$
$$\beta_1 = (\mathbf{e}_1 \hat{\boldsymbol{\beta}}(\mathbf{n}_0)\mathbf{e}_1) = \varepsilon_0^2(\mathbf{e}_1 \hat{\boldsymbol{\Omega}}^{-1}(\boldsymbol{l})\mathbf{e}_1) - (\mathbf{e}_1 \hat{\sigma}_1 \hat{\boldsymbol{\Omega}}(\mathbf{n}_0)\hat{\sigma}_1^{+} \mathbf{e}_1) \tag{9.6.14}$$

Let us introduce the vector

$$\mathbf{F} = \hat{\sigma}_1^{+} \mathbf{e}_1 \tag{9.6.15}$$

which may be calculated from Eqs. (9.6.4) and (9.6.13) to be

$$\mathbf{F} = (-0.2699, -0.2699, 0.1262) \tag{9.6.16}$$

Eq. (9.6.14) can be written as

$$\beta_1 = \varepsilon_0^2(\mathbf{e}_1 \hat{\boldsymbol{\Omega}}^{-1}(\boldsymbol{l})\mathbf{e}_1) - (\mathbf{F}\hat{\boldsymbol{\Omega}}(\mathbf{n}_0)\mathbf{F}) \tag{9.6.17}$$

Substitution of Eqs. (9.6.7), (9.6.6), (9.6.13), and (9.6.16) into the convolution (9.6.17) yields

$$\beta_1 = c_{44}\varepsilon_0^2(2.1239 - 0.1471) = c_{44}\varepsilon_0^2 1.9768 \tag{9.6.18}$$

It follows from Eqs. (9.6.12) and (9.6.18) that the anisotropy parameter (8.7.20) is equal to

$$\alpha = \frac{\beta_1 - \beta_2}{\beta_2} = \frac{1.9768 - 0.2819}{1.9768} = 0.8574 \tag{9.6.19}$$

According to Section 8.7, platelike inclusions are elongated in the direction of the eigenvector $\mathbf{e}_2$ which corresponds to the lowest eigenvalue. In the case under consideration it is the eigenvector $\mathbf{e}_2 = (1/\sqrt{2}, \bar{1}/\sqrt{2}, 0)$. The theory suggested thus predicts martensite laths to be elongated along the $[1\bar{1}0]_A$ direction. This fully agrees with the electron microscopic observations (99).

According to Section 8.7, the width-direction of an inclusion coincides with the direction of the eigenvector $\mathbf{e}_1$ corresponding to the largest eigenvalue β_1. It follows from Eq. (9.6.13) that this direction makes an angle of 3.8° with the direction $[\bar{1}\bar{1}2]_A$.

Substitution of the calculated anisotropy parameter value, $\alpha = 0.8574$, into Eqs. (8.7.23) and (8.7.26) yields

$$x = \pm[p(p^2+1)^{-\frac{1}{2}} + 0.8574p(p^2+1)^{-\frac{3}{2}}]$$
$$y = \pm[0.8574(p^2+1)^{-\frac{3}{2}} - 0.7148(p^2+1)^{-\frac{1}{2}}] \tag{9.6.20}$$

(the scale factor μ is set to be equal to unity). The function $y = y(x)$ given by (9.6.20) in the parametric form is plotted in Fig. 74. This function describes the equilibrium shape of a martensitic lath in the habit plane that is consistent with the mechanism described in Section 6.7 for the $\gamma \to \alpha$ transformation. For simplicity, a 3.8° departure of the calculated habit from the $(111)_A$ plane is not shown in the Fig. 74.

The theory described thus furnishes the correct predictions of the $(111)_A$ habit, $[1\bar{1}0]_A$ longitudinal direction, and orientation relations of lath martensite.

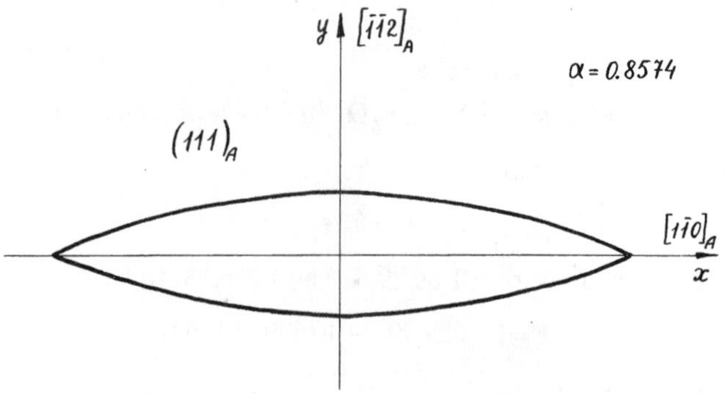

Figure 74. The calculated equilibrium shape of a martensitic lath in the $(111)_A$ habit plane ($\alpha=0.8574$).

There is, however, a descrepancy in the width-to-length ratio. The theory predicts this ratio to be equal to 0.2, whereas according to the authors of the electron microscopic study (99) the width-to-length ratio value is at least ten times less than 0.2. Two reasons may account for this descrepancy. First, as shown above, the elongated morphology is associated with the small value of $\beta_2=0.2819c_{44}\varepsilon_0^2$ which is about seven times less than the value of $\beta_1=1.977c_{44}\varepsilon_0^2$. Unlike β_1, β_2 is very sensitive to variations of elastic constants. Nevertheless, the elastic constants of Fe-Ni martensite are known to be highly dependent on the alloy composition (194). In fact the use of the elastic constants measured for the Fe-31.5Ni austenite (the invar range) in the calculations of lath martensite observed in the Fe-20Ni austenite may seriously affect the β_2 value and be the reason of the difference between the predicted and observed width-to-length ratio values.

The descrepancy may also be due to an ambiguity inherent in the interpretation of electron microscopic images. For instance, the elongated lath observed might actually be an alignment of a few laths not resolved in the electron micrograph.

10

STRAIN-INDUCED COARSENING IN COHERENT ALLOYS CONSISTING OF TWO CUBIC-SYMMETRY PHASES

In the previous chapter we analyzed the elastic strain effect on the morphology of a single coherent new phase inclusion in a parent phase. These results are applicable when the density of inclusions is small and the typical separation between new phase particles far exceeds the typical particle dimensions. This is the condition enabling the independence of shape, orientation, and location of an inclusion on the combined strain field generated by all the other inclusions present in the crystal. If on the other hand, the typical distance between inclusions is commensurate with inclusion dimensions (the case of a high density of inclusions), the mutual influence of inclusions grows substantial and cannot be ignored. Interference of strain fields generated by various inclusions results in strain-induced interactions (see Section 7.3). These interactions generate a tendency for rearranging new phase particles to produce a certain spatial pattern such that the strain fields from individual particles be cancelled out to provide strain relaxation. Because long-range strain-induced interactions are highly dependent on shapes, orientations, and mutual arrangement of inclusions, they control evolution of the morphology of a two-phase coherent mixture with time to provide strain relaxation. This process will be called strain-induced coarsening. As mentioned in the introduction to Chapter 9, strain-induced coarsening results in translational motion of inclusions as well as transformations of inclusion shapes by diffusion transport of atoms from one side of an inclusion to its another side ("uphill" diffusion). The driving force of strain-induced coarsening far exceeds that of usual coarsening because it is determined by strain-induced interactions proportional to the new phase volume, whereas with usual coarsening the driving force of the process is proportional to the interphase surface area. For that reason, in most cases of interest strain-induced coarsening governs morphology transformations in a two-phase alloy during annealing.

We shall consider examples of two-phase coherent mixture morphologies that provide minimization of the strain energy and therefore arise in strain-induced coarsening. The first example is a coherent two-phase mixture of cubic symmetry phases with different crystal lattice parameters. In this case the strain energy minimum corresponds to so-called modulated structures actually observed in many alloys of this type. Second, a coherent mixture of cubic and tetragonal phases will be considered. Strain energy minimization now leads to formation of twinned plates with martensitic morphology. These structures were also observed in many decomposed alloys.

10.1. MODULATED STRUCTURE IN COHERENT MIXTURE OF TWO CUBIC-SYMMETRY PHASES

In 1940 Bradley (195) and in 1943 Daniel and Lipson (196) discovered side-bands in the x-ray powder diffraction patterns taken from Cu-Ni-Fe alloys. Later the side-band phenomenon in x-ray diffraction was studied by Hargreaves (181), Biederman and Kneller (197), Hillert et al. (198), and others. The effect was interpreted as resulting from x-ray diffraction on periodic modulations of the concentration profiles with a period many times larger than the crystal lattice parameter. Since then, a wealth of experimental data on one-, two-, and three-dimensional modulated structures has been collected. Modulated structures were observed in many systems by both x-ray diffraction and electron microscopy. Their existence was reported for Au-Ni (199), Au-Pt (200), Fe-Ni-Al (201), Fe-Be (202–204), Al-Ni (205, 177), Cu-Ti (206, 207), Cu-Ni-Co (197, 208), ALNICO (209–211), and so on.

The studies cited have demonstrated the modulated structures to result in the appearance of satellites along the $\langle 100 \rangle$ directions in the reciprocal lattice in the vicinity of the fundamental matrix reflections.

As often happens, it was shown later that the modulated structures in fact are pseudoperiodic rather than periodic. For instance, pseudoperiodic basket-weave structures were observed in Cu-Ni-Fe by Livak and Thomas (212) and in nonstoichiometric β-CuZn by Kubo and Wayman (35). Although the structures observed were not periodic, they gave satellite reflections near the fundamental spots.

The start on the problem of modulated structures in cubic alloys was made by Cahn who studied the initial stage of the decomposition of a cubic solution into two cubic phases (42, 43). He has demonstrated that the spinodal instability generates a concentration wave whose wavelength is determined by competition between thermodynamic factors favoring long-wave concentration fluctuations and kinetic factors that act to the contrary and suppress long-range diffusion prerequisite for the development of long concentration waves (see Section 5.3). As shown by Rioja and Laughlin (67), Cahn's mechanism is in fact observed in the classic Al-Cu system where the side-band phenomenon appears in the initial stage of the decomposition preceding the formation of GP zones. Cahn's spinodal mechanism, however, fails to explain all the variety of modulated structures. The first reason for that is that Cahn's explanation is based on the assumption that the deviations of the concentrations from their mean value are only small. This assumption makes it possible to linearize the nonlinear kinetic equation (5.1.19) and neglect "anharmonic interactions" between concentration waves. Naturally, the approximation of small concentration profile variations is only valid at the initial stages of the decomposition. Further descriptions of decomposition must take into consideration the nonlinear terms of the kinetic equation (5.1.19).

According to the results of the computer simulation carried out by Morris and Khachaturyan (see Section 5.4) who obtained the exact numerical solution of the nonlinear equation (5.1.19), the linear stage of the spinodal decomposition described by Cahn is limited in time by the condition

$$\frac{tD}{a^2} \lesssim 0.01$$

where t is time, D the diffusivity, a the crystal lattice parameter. The later stages of decomposition should be treated in terms of nonlinear theory. They result in the formation of a fine mixture of concentration heterogeneities with near equilibrium compositions. After the equilibrium compositions are attained, the rate of the decomposition drops, and the process goes into its coarsening stage. The appearance of satellites reasonably well resolved was calculated to occur at $t \gg a^2/D$, outside the time range of the applicability of the linear theory.

There is also important experimental evidence for the formation of some modulated structures in strain-induced coarsening rather than in the spinodal decomposition. According to Ardell and Nicholson (177) and Hornbogen and Roth (205), the initial stage of the decomposition in Ni-Al produces randomly distributed cuboid precipitate particles. Pseudoperiodic arrays of particles only arise later in the course of coarsening.

The mechanism explaining the formation of modulated structures was suggested by Khachaturyan (213–215). It involves the strain-induced coarsening discussed here. The problem solved in (213–215) amounts to the determination of the concentration distribution that corresponds to the free energy minimum. Unlike the conventional thermodynamic theory of phase transformations, this theory includes strain energy contributions to the free energy of the system.

It also takes into account the fact that the decomposition occurs within the parent phase matrix. The solution of the free energy minimization problem within this approach predicts the formation of a domain structure rather than, like with the conventional decomposition thermodynamics, the appearance of the bicrystal composed by two equilibrium phases. This is a qualitatively new result. The domains (referred to as concentration domains below) are similar to magnetic and ferroelectric domains. They form modulated structures whose morphologies coincide with those observed in electron microscopic studies. To understand the mechanism suggested, one should recall that cubic new phase precipitates have platelike shapes and the $\{100\}$ habit if the elastic anisotropy, $\xi=(c_{11}-c_{12}-2c_{44})/c_{44}$, is negative, as is usually the case (see Section 9.4). Let us consider some volume V within the crystal where the decomposition has taken place. The volume chosen is surrounded by parent phase matrix and includes alternating platelets of two equilibrium phases with the (001) habit (see Fig. 75). In what follows, this volume will be called a complex. If the crystal lattice parameter with respect to composition dependence deviates from linearity insignificantly, any atomic redistribution within the complex, including the formation of the complex itself, will not have a noticeable effect on the volume, V, and shape of the complex. Therefore one may assume the macroscopic shape deformation to be zero. In that case the decomposition does not spoil fitting of the complex into its hole in the matrix and therefore does not generate a strain field. The latter conclusion is valid, however, only if internal heterogeneity of the complex comprising alternating platelets of two equilibrium phases may be neglected. This neglect is justified at large distances from the complex boundaries. By "large distance" we mean

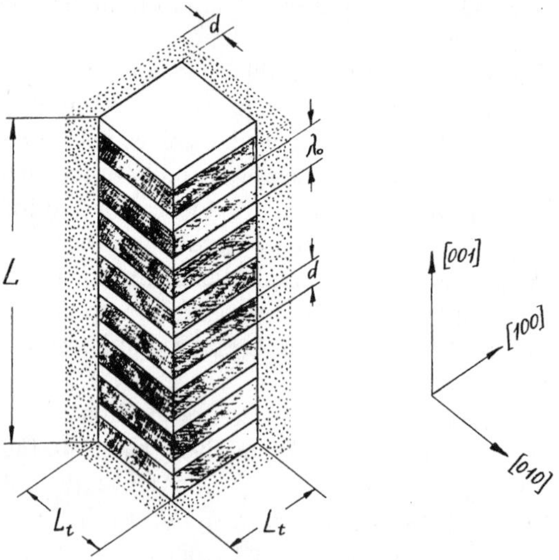

Figure 75. The scheme of a one-dimensional modulated structure comprising coherent platelets of two equilibrium phases with the (001) habit. The phases are designated by black and white layers. Shadow approximately shows the surface layer where the elastic strain is concentrated.

a distance significantly greater than the typical heterogeneity dimension, the platelet thickness, d. The long-range elastic strain field thus vanishes, and the remaining short-range strain is concentrated in the vicinity of the complex boundary within a thin surface layer enveloping the edge boundaries of platelets (see Fig. 75). The surface layer thickness is of the order of d, and its volume is of the order of dLP where L is the linear dimension of the complex in the [001] direction, P the platelet perimeter. The typical specific strain energy is of the order

$$e_{\text{str}} \approx \lambda\varepsilon_0^2(c_1-c_2)^2 \tag{10.1.1}$$

where c_1 and c_2 are the compositions of two coexisting equilibrium phases, $\varepsilon_0(c_1-c_2)$ is the crystal lattice mismatch, λ the typical modulus.

The characteristic strain energy stored in the surface layer is the product of the specific free energy e_{str} and the surface layer volume, dLP:

$$E_{\text{edge}} \simeq e_{\text{str}}\, dLP \simeq \lambda\varepsilon_0^2(c_1-c_2)^2 dLP \tag{10.1.2}$$

This relation shows that the strain energy E_{edge} decreases linearly with the typical platelet thickness. The layer thickness cannot, however, decrease infinitely. Indeed a decrease of the platelet thickness d at a constant complex volume V will result in an increase in the number of platelets comprising the complex and therefore in the number of platelet interphase boundaries. The result will be an increase in the interphase energy. The latter quantity is of the order of

$$E_s \approx \gamma_s S \frac{L}{d} \tag{10.1.3}$$

where γ_s is the interphase specific energy, S the area of the interphase between adjacent platelets. One can see that $E_s \to \infty$ if $d \to 0$.

The total energy is

$$E_{\text{total}} = E_{\text{edge}} + E_s \approx \lambda\varepsilon_0^2(c_1-c_2)^2 dLP + \gamma_s \frac{V}{d} \tag{10.1.4}$$

where $LS = V$. The minimization of the total energy with respect to the thickness d gives the equation

$$\frac{dE_{\text{total}}}{dd} \approx \lambda\varepsilon_0^2(c_1-c_2)^2 LP - \frac{\gamma_s V}{d^2} = 0 \tag{10.1.5}$$

the solution of which is

$$d \approx \sqrt{r_0 \frac{V}{PL}} \approx \sqrt{r_0 L_c} \tag{10.1.6}$$

Eq. (10.1.6) determines the equilibrium platelet thickness. Here

$$r_0 = \frac{\gamma_s}{\lambda\varepsilon_0^2(c_1-c_2)^2} \tag{10.1.7}$$

is the material constant having the length dimension, $L_c = V/PL$ is the typical dimension of the complex in the (001) plane.

Because L_c is a constant, the typical platelet thickness is also a constant. It thus follows that a complex is a sandwich made up by periodically alternating platelets of equilibrium phases having the same thickness. On the whole this structure is in agreement with one-dimensional modulated structures observed by electron microscopy.

It is noteworthy that the relation

$$d \sim \sqrt{L_c}$$

coincides with the relation describing the thickness of ferromagnetic domains as depending on the crystal size [see, e.g. (216)]. This is not accidental. The point is that new phase platelets may be interpreted as domains of a certain kind, specifically, elastic concentration domains. In fact there is a strong similarity between the elastic concentration and magnetic domains:

1. The equilibrium state of an uniaxial ferromagnetic crystal is doubly degenerate with respect to two magnetization directions ("upward" and "downward"). This is the reason why ferromagnetic domains are formed. The same is true of a two-phase alloy. Indeed the equilibrium state of an alloy in a two-phase field of an equilibrium phase diagram is also doubly degenerate with respect to two compositions describing coexisting equilibrium phases because the thermodynamic potential, $\Omega = f(c) - \mu c$, has two equal minima in this field at these very compositions (here μ is the chemical potential).

2. The domain structure of a ferromagnetic crystal is known to be more stable than the monodomain homogeneous state [see, e.g. (216)]. The reason for that is that the formation of domains results in the vanishing of the long-range magnetic field and the localization of the residual magnetic field (scattering field) within a thin surface layer of the crystal, exactly like the formation of concentration domains within a complex results in localization of long-range strain field within a thin layer enveloping the complex.

3. The energy of the Bloch walls plays the same part in the formation of magnetic domains as the interphase energy between adjacent platelets of equilibrium phases does in the formation of concentration domains. The only difference in principle between the concentration and magnetic domains is that in the latter case the role of a complex is played by the crystal as a whole.

In general, a volume of the complex is not constant: during the decomposition the complex grows by absorbing new portions of the matrix. The assumption of a constant complex volume introduced to prove the higher stability of a domain structure is justified when the complex growth is limited either by crystal lattice defects or by collisions with other growing complexes. The latter is analogous to what occurs during grain formation in solidification processes. Figure 76 provides an example of a concentration domain structure formed in collisions of growing complexes.

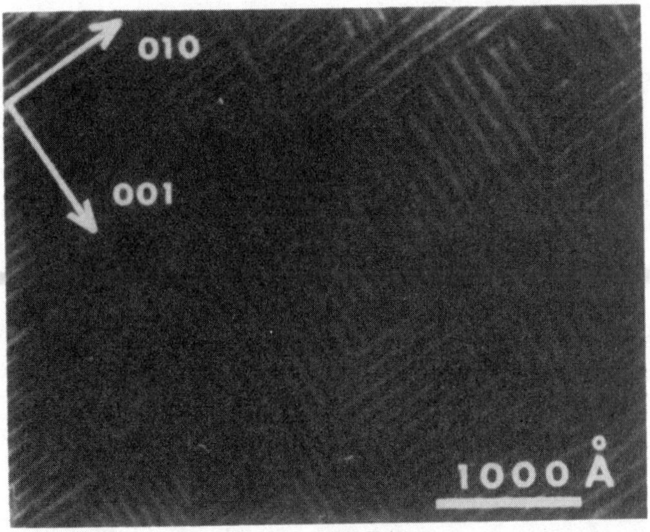

Figure 76. The electron microscopic image of complexes in β-CuZn consisting of one-dimensional modulated structures with the different modulation directions (35). (Courtesy of C. M. Wayman, H. Kubo, and I. Cornelis.)

10.2. STRAIN ENERGY OF CONCENTRATION HETEROGENEITY IN CUBIC SOLID SOLUTIONS

We shall describe in this section a continuous theory of the strain energy of a concentration heterogeneity within an infinite body of a cubic solid solution. We shall proceed from the following assumptions:

1. The concentration dependence of the crystal lattice parameter of the solution follows Vegard's law:

$$a(\bar{c}) = a_0(1 + \varepsilon_0 \bar{c}) \tag{10.2.1}$$

within the range of concentration variations due to heterogeneities. Here $\bar{c}$ is the atomic fraction of the solute element,

$$\varepsilon_0 = \frac{da}{a d\bar{c}} \tag{10.2.2}$$

is the concentration expansion coefficient, $a(\bar{c})$ and a_0 are the crystal lattice parameters of the solid solution of the concentration $\bar{c}$ and of the reference solution, respectively.

2. Elastic moduli are not affected by the composition.

It follows from Eq. (10.2.1) that

$$\varepsilon_{ij}^{0}(\mathbf{r}) = \varepsilon_0(\bar{c}(\mathbf{r}) - \bar{c})\delta_{ij} \tag{10.2.3}$$

where $\varepsilon_{ij}^{0}(\mathbf{r})$ is the local stress-free transformation strain at the point $\mathbf{r}$. This quantity characterizes the crystal lattice mismatch between a matrix and a heterogeneity of the local composition $\bar{c}(\mathbf{r})$. The stress-free transformation strain of a new phase inclusion in the matrix is described by Eq. (7.2.12) which, in the case of a cubic inclusion in a cubic matrix, has the form

$$\varepsilon_{ij}^{0}(\mathbf{r}) = \varepsilon_0 \delta_{ij} \tilde{\theta}(\mathbf{r}) \tag{10.2.4}$$

As shown above, the stress-free strain (10.2.4) leads to the strain energy given by Eq. (9.4.4).

It is easy to see that Eqs. (10.2.4) and (10.2.3) are similar to each other: Eq. (10.2.4) may be obtained from (10.2.3) by mere substitution

$$\tilde{\theta}(\mathbf{r}) \rightarrow \bar{c}(\mathbf{r}) - \bar{c} \tag{10.2.5}$$

This substitution transforms Eq. (9.4.4) for the strain energy into

$$E = \frac{1}{2} \int B(\mathbf{n}) |\tilde{c}(\mathbf{k})|^2 \frac{d^3k}{(2\pi)^3} \tag{10.2.6}$$

since

$$\theta(\mathbf{k}) \rightarrow \tilde{c}(\mathbf{k}) \tag{10.2.7}$$

where

$$\tilde{c}(\mathbf{k}) = \int (\bar{c}(\mathbf{r}) - \bar{c}) e^{-i\mathbf{k}\mathbf{r}} d^3r \tag{10.2.8}$$

and

$$B(\mathbf{n}) = (c_{11} + 2c_{12})\varepsilon_0^2 \left[3 - \frac{c_{11} + 2c_{12}}{c_{11}} \right.$$
$$\left. \times \frac{1 + 2\xi(n_1^2 n_2^2 + n_1^2 n_3^2 + n_2^2 n_3^2) + 3\xi^2 n_1^2 n_2^2 n_3^2}{1 + \xi\left(1 + \frac{c_{12}}{c_{11}}\right)(n_1^2 n_2^2 + n_1^2 n_3^2 + n_2^2 n_3^2) + \xi^2 \left(1 + 2\frac{c_{12}}{c_{11}} + \frac{c_{44}}{c_{12}}\right) n_1^2 n_2^2 n_3^2} \right] \tag{10.2.9}$$

One should remember that the total strain energy (10.2.6) includes two effects: configurationally dependent strain-induced interactions among concentration heterogeneities determined by the mutual arrangement of the heterogeneities and configurationally independent self-energy having no bearance on the interactions and therefore independent of the mutual arrangement of the heterogeneities. The physical meaning these two effects make and the problem of separating the corresponding contributions in Eq. (10.2.6) will be discussed later in terms of the microscopic theory of strain-induced interactions.

Although in addition to the strain-induced interaction energy Eq. (10.2.6) includes the self-energy, it can be applied to determine heterogeneous states because the self-energy only renormalizes the reference state but does not affect the morphology of heterogeneities.

Eq. (10.2.6) can always be written as the sum of two terms

$$E = E_{\text{bulk}} + E_{\text{edge}} \tag{10.2.10}$$

where

$$E_{\text{bulk}} = \frac{1}{2}\min B(\mathbf{n}) \int |\tilde{c}(\mathbf{k})|^2 \frac{d^3k}{(2\pi)^3} = \frac{1}{2}\min B(\mathbf{n}) \int (\Delta c(\mathbf{r}))^2 d^3r \tag{10.2.11}$$

$$E_{\text{edge}} = \frac{1}{2}\int \Delta B(\mathbf{n})|\tilde{c}(\mathbf{k})|^2 \frac{d^3k}{(2\pi)^3} \tag{10.2.12}$$

$$\Delta c(\mathbf{r}) = \bar{c}(\mathbf{r}) - \bar{c} \tag{10.2.13}$$

$$\Delta B(\mathbf{n}) = B(\mathbf{n}) - \min B(\mathbf{n}) \tag{10.2.14}$$

and min $B(\mathbf{n})$ is the minimum value of $B(\mathbf{n})$.

The identity

$$\iiint_{-\infty}^{\infty} |\tilde{c}(\mathbf{k})|^2 \frac{d^3k}{(2\pi)^3} \equiv \iiint_{-\infty}^{\infty} (\Delta c(\mathbf{r}))^2 d^3r$$

is used in Eq. (10.2.11).

Since according to the definition (10.2.14) $\Delta B(n) \geqslant 0$, the integrand in (10.2.12) is a nonnegative quantity,

$$E_{\text{edge}} \geqslant 0 \tag{10.2.15}$$

With this in mind, one comes to the conclusion that E_{bulk} given by Eq. (10.2.11) is the minimal value of the strain energy (10.2.10) attained asymptotically at $E_{\text{edge}} \to 0$.

Combining the free energy of the reference state

$$F^0_{\text{ref}} = \int f_{\text{ref}}(\bar{c}(\mathbf{r}))d^3r \tag{10.2.16}$$

and the strain energy (10.2.10), we obtain

$$F = F_{\text{bulk}} + E_{\text{edge}} \tag{10.2.17}$$

where $f_{\text{ref}}(\bar{c}(\mathbf{r}))$ is the specific chemical free energy of the reference state. According to Eq. (10.2.11)

$$F_{\text{bulk}} = F^0_{\text{ref}} + E_{\text{bulk}} = \int f(\bar{c}(\mathbf{r}))d^3r \tag{10.2.18}$$

where

$$f(\bar{c}(\mathbf{r})) = f_{\text{ref}}(\bar{c}(\mathbf{r})) + \frac{1}{2}\min B(\mathbf{n})(\bar{c}(\mathbf{r}) - \bar{c})^2 \tag{10.2.19}$$

is the specific free energy of the solution.

The gradient terms proportional to $(\nabla\tilde{c})^2$ are ommitted from Eq. (10.2.18) because we are considering large-scale two-phase heterogeneities where the gradient terms may be included in the interphase energy.

Let us assume the decomposition reaction occurs within a rectangular parallelepiped-shaped complex. Its internal structure can be determined by the minimization of the sum of the chemical, strain, and interphase energies under the condition of the conservation of the complex volume V.

It follows from Eq. (10.2.17) and the inequality (10.2.15) that the minimum value of the free energy F_{bulk} (10.2.18) is the lower limit for the total free energy $F(F \geqslant F_{\text{bulk}})$. Comparison of Eqs. (10.2.17) and (10.2.12) shows that $F \to F_{\text{bulk}}$ if only $|\tilde{c}(\mathbf{k})|^2$ does not vanish at those directions $\mathbf{n}=\mathbf{n}_0$ of the vector $\mathbf{k}$ that provide the minimization of $B(\mathbf{n})$ in (10.2.9) or, according to the definition (10.2.14), reduce $\Delta B(\mathbf{n})$ to zero. In other words

$$B(\mathbf{n}_0)=\min B(\mathbf{n}) \tag{10.2.20}$$

The situation here is analogous to that in Section 8.1 where habit plane orientation was determined.

It follows from Eq. (10.2.9) that at $\xi=(c_{11}-c_{12}-2c_{44})/c_{44}<0$, which is typical for almost all the cubic alloys known, $B(\mathbf{n})$ assumes its minimal value equal to

$$B(\mathbf{n}_0)=\min B(\mathbf{n})=(c_{11}+2c_{12})\varepsilon_0^2\left(3-\frac{c_{11}+2c_{12}}{c_{11}}\right)$$
$$=\frac{2(c_{11}+2c_{12})}{c_{11}}(c_{11}-c_{12}) \tag{10.2.21}$$

at $\mathbf{n}=\mathbf{n}_0=\langle 100\rangle$.

In the case of $\xi<0$, the free energy F asymptotically approaches $F_{\text{bulk}}(F \to F_{\text{bulk}}, E_{\text{edge}} \to 0)$ if the function $|\tilde{c}(\mathbf{k})|^2$ vanishes outside thin and long rods extended along the $\langle 100\rangle$ cubic axes and intersecting each other at the reciprocal lattice origin. This requirement is met if $\tilde{c}(\mathbf{r})$ within the complex may be given by one of the relations

$$\tilde{c}(\mathbf{r})=\bar{c}+c_1^{[001]}(z) \tag{10.2.22}$$

$$\tilde{c}(\mathbf{r})=\bar{c}+c_2^{[100]}(x)+c_2^{[010]}(y) \tag{10.2.23}$$

$$\tilde{c}(\mathbf{r})=\bar{c}+c_3^{[100]}(x)+c_3^{[010]}(y)+c_3^{[001]}(z) \tag{10.2.24}$$

where x, y, z are the coordinates of the vector $\mathbf{r}$ along the [100], [010], and [001] directions, respectively. Outside the complex

$$\tilde{c}(\mathbf{r})\equiv\bar{c}$$

Eq. (10.2.22) describes a one-dimensional concentration modulation in the direction [001], and Eqs. (10.2.23) and (10.2.24) describe a superposition of one-dimensional modulations along the [100], [010], and [001] directions.

If d is the typical modulation length, and L_t is the typical size of the complex in the transversal direction normal to the modulation axis, the thickness and length of the corresponding rods in the **k**-space will be of the order of $2\pi/L_t$ and $2\pi/d$, respectively. In this case the requirement that the rod be thin and long is reduced to the inequality

$$d/L_t \ll 1 \tag{10.2.25}$$

From (10.2.25) it is easy to see that the minimum of the free energy (10.2.17) coincides with the minimum of the free energy F_{bulk} [Eq. (10.2.18)] to within $d/L_t \ll 1$ (the order of magnitude of E_{edge}).

In its turn the minimization of F_{bulk} is carried out by the conventional procedure (see Section 4.1) and yields the usual result: the minimum of the free energy F_{bulk} is realized if the alloy is decomposed into two phases whose equilibrium compositions, c_1^0 and c_2^0, are determined as the coordinates of two tangent points on the tangent to the free energy curve, as opposed to the $\bar{c}$ curve, $f(\bar{c})$, given by Eq. (10.2.19) (see Fig. 37).

Because the specific free energy (10.2.19) involves the strain energy, the equilibrium compositions c_1^0 and c_2^0 of the coexisting phases correspond to the coherent *T-c* diagram and differ from the compositions given by the stable *T-c* diagram.

The introduction of the strain energy into the thermodynamics of the decomposition immediately gives a nontrivial result. The morphology of a two-phase coherent mixture that affects the free energy becomes the additional thermodynamic parameter to be determined from the free energy minimum condition. In particular, we have been able to show, from purely thermodynamic considerations, that the free energy is minimized only if the morphology of the complex is described by Eqs. (10.2.22) to (10.2.24). The formation of a two-phase mixture of an arbitrary morphology proves impossible.

We shall now establish the correspondence between heterogeneities of the types (10.2.22) to (10.2.24) and elastic strains generated by these heterogeneities.

The Fourier transform of the strain $\varepsilon_{ij}(\mathbf{r})$

$$\tilde{\varepsilon}_{ij}(\mathbf{k})=\int \varepsilon_{ij}(\mathbf{r})e^{-i\mathbf{kr}}d^3r=\frac{1}{2}\int\left(\frac{\partial u_i}{\partial r_j}+\frac{\partial u_j}{\partial r_i}\right)e^{-i\mathbf{kr}}d^3r$$

where $\mathbf{u}(\mathbf{r})$ is the elastic displacement can be rewritten as

$$\tilde{\varepsilon}_{ij}(\mathbf{k})=\tfrac{1}{2}i(k_j v_i(\mathbf{k})+k_i v_j(\mathbf{k})) \tag{10.2.26}$$

where

$$\mathbf{v}(\mathbf{k})=\int \mathbf{u}(\mathbf{r})e^{-i\mathbf{kr}}d^3r \tag{10.2.27}$$

is the Fourier transform of the displacement, $\mathbf{u}(\mathbf{r})$.

Using the correspondence (10.2.7) in Eq. (8.1.5), we obtain the relation

$$v_i(\mathbf{k})=-\frac{i}{k}\Omega_{ij}(\mathbf{n})n_j(c_{11}+2c_{12})\varepsilon_0\tilde{c}(\mathbf{k}) \tag{10.2.28}$$

where $\Omega_{ij}(\mathbf{n})$ is given by Eq. (8.6.3), $\sigma^0=(c_{11}+2c_{12})\varepsilon_0$.

Substituting Eq. (10.2.28) into Eq. (10.2.26) gives

$$\tilde{\varepsilon}_{ij}(\mathbf{k})=\tfrac{1}{2}(c_{11}+2c_{12})\varepsilon_0[n_i\Omega_{jl}(\mathbf{n})n_l+n_j\Omega_{il}(\mathbf{n})n_l]\tilde{c}(\mathbf{k}) \qquad (10.2.29)$$

The back Fourier transform of (10.2.29) yields

$$\varepsilon_{ij}(\mathbf{r})=\int \tilde{\varepsilon}_{ij}(\mathbf{k})e^{i\mathbf{kr}}\frac{d^3k}{(2\pi)^3}=\frac{1}{2}(c_{11}+2c_{12})\varepsilon_0 \times \int [n_i\Omega_{jl}(\mathbf{n})n_l+n_j\Omega_{il}(\mathbf{n})n_l]\tilde{c}(\mathbf{k})e^{i\mathbf{kr}}\frac{d^3k}{(2\pi)^3} \qquad (10.2.30)$$

With the one-dimensional modulation (10.2.22) the function $\tilde{c}(\mathbf{k})$ in the integrand in (10.2.30) vanishes outside the direction $\mathbf{k}\|[001]$. Therefore Eq. (10.2.30) can be rewritten in the form

$$\varepsilon_{ij}(\mathbf{r})=\frac{1}{2}(c_{11}+2c_{12})\varepsilon_0[n_i\Omega_{jl}(\mathbf{n})n_l+n_j\Omega_{il}(\mathbf{n})n_l]_{\mathbf{n}=[001]}\int \tilde{c}(\mathbf{k})e^{i\mathbf{kr}}\frac{d^3k}{(2\pi)^3} \qquad (10.2.31)$$

Using the back Fourier transform

$$\int \tilde{c}(\mathbf{k})e^{i\mathbf{kr}}\frac{d^3k}{(2\pi)^3}=\bar{c}(\mathbf{r})-\bar{c}$$

with the one-dimensional distribution (10.2.22), we obtain

$$\varepsilon_{ij}(\mathbf{r})=\tfrac{1}{2}(c_{11}+2c_{12})\varepsilon_0[n_i\Omega_{jl}(\mathbf{n})n_l+n_j\Omega_{il}(\mathbf{n})n_l]_{\mathbf{n}=[001]}c_1^{[001]}(z) \qquad (10.2.32)$$

Eq. (10.2.32) holds with accuracy to $d/L \ll 1$. Employing Eqs. (8.6.3) for the components $\Omega_{jl}(\mathbf{n})$ in a cubic crystal, we can rewrite Eq. (10.2.32) eventually as

$$\varepsilon_{ij}(\mathbf{r})=\left(1+2\frac{c_{12}}{c_{11}}\right)\varepsilon_0\begin{pmatrix}0&0&0\\0&0&0\\0&0&1\end{pmatrix}c_1^{[001]}(z) \qquad (10.2.33)$$

In the case of the two-dimensional distribution (10.2.23) the function $\tilde{c}(\mathbf{k})$ does not vanish only within the two rods in the $\mathbf{k}$-space. One of these is parallel to the axis [100] and the other one to the axis [010]. The corresponding equation for $\varepsilon_{ij}(\mathbf{r})$ is

$$\varepsilon_{ij}(\mathbf{r})=\left(1+2\frac{c_{12}}{c_{11}}\right)\varepsilon_0\left[\begin{pmatrix}1&0&0\\0&0&0\\0&0&0\end{pmatrix}c_2^{[100]}(x) + \begin{pmatrix}0&0&0\\0&1&0\\0&0&0\end{pmatrix}c_2^{[010]}(y)\right] \qquad (10.2.34)$$

10.3. ONE-DIMENSIONAL MODULATED STRUCTURES

As shown in Section 10.2, the free energy F becomes minimal if the complex is transformed into a sandwichlike structure comprising alternating (001) platelets of the equilibrium phases (see Fig. 75). The compositions of these phases, c_1^0 and c_2^0, are determined by the tangent to the specific free energy curve, $f(\bar{c})$. The total volumes of the first and second phases are equal to $\gamma_1 V$ and $(1-\gamma_1)V$ where the equilibrium fracture of the first phase, γ_1, is determined by the lever rule

$$\gamma_1 = \frac{c_2^0 - \bar{c}}{c_2^0 - c_1^0} \tag{10.3.1}$$

Since the free energy F_{bulk} in Eq. (10.2.18) depends solely on the ratio of volumes of the coexisting phases, to determine the mutual arrangement of the platelets and their thicknesses one must investigate the strain energy E_{edge} which makes a small contribution to F.

Let us calculate the value of E_{edge} for a periodic square wave concentration modulation (10.2.22). By definition, the function $\Delta c(\mathbf{r}) = \bar{c}(\mathbf{r}) - \bar{c}$ vanishes outside the complex. This makes it possible to rewrite Eq. (10.2.8) in the form

$$\tilde{c}(\mathbf{k}) = \int_{(V)} (\bar{c}(\mathbf{r}) - \bar{c}) e^{-i\mathbf{k}\mathbf{r}} d^3r = \int_{(V)} c_1^{[001]}(z) e^{-i\mathbf{k}\mathbf{r}} d^3r \tag{10.3.2}$$

where the integration is over the volume of the complex and where the definition (10.2.22) is used.

Since the complex is assumed to be a cylinder with the [001] axis, all the cross-sections of the complex by (001) planes are alike. Eq. (10.3.2) may therefore be rewritten as

$$\tilde{c}(\mathbf{k}) = \int_{-\frac{1}{2}L}^{\frac{1}{2}L} c_1^{[001]}(z) e^{-ik_z z} S(\boldsymbol{\tau}) dz \tag{10.3.3}$$

where L is the length of the complex in the [001] direction, $\mathbf{r} = (\boldsymbol{\rho}, z)$, $\boldsymbol{\rho} = (x, y)$, are the coordinates of the vector $\mathbf{r} = (x, y, z)$, $\mathbf{k} = (\boldsymbol{\tau}, k_z)$, $\boldsymbol{\tau} = (k_x, k_y)$ are the coordinates of the vector $\mathbf{k} = (k_x, k_y, k_z)$,

$$S(\boldsymbol{\tau}) = \int_{(S)} d^2\rho \, e^{-i\boldsymbol{\tau}\boldsymbol{\rho}} \tag{10.3.4}$$

[The integration in (10.3.4) is over the cross section S formed by intersecting the complex with a (001) plane.]

Let us assume that $c_1^{[001]}(z)$ is a periodic function of z within the complex and that its spacing is equal to λ_0. Outside the complex, $c_1^{[001]}(z) \equiv 0$ by definition. Thus, if

$$z = \xi + n\lambda_0 \tag{10.3.5}$$

where $0 \leqslant \xi \leqslant \lambda_0$ and n is an integer within the range $-\frac{1}{2}L/\lambda_0 \leqslant n \leqslant \frac{1}{2}L/\lambda_0$, we have the relation

$$c_1^{[001]}(z) = c_1^{[001]}(\xi + n\lambda_0) = c_1^{[001]}(\xi) \tag{10.3.6}$$

Using Eqs. (10.3.5) and (10.3.6), Eq. (10.3.3) may be rewritten as follows:

$$\tilde{c}(k) = S(\boldsymbol{\tau}) \int_{-L/2}^{L/2} c_1^{[001]}(z) e^{-ik_z z} dz$$
$$= S(\boldsymbol{\tau}) \sum_{n=-L/(2\lambda_0)}^{L/(2\lambda_0)} \int_0^{\lambda_0} c_1^{[001]}(\xi) e^{-ik_z(\xi + n\lambda_0)} d\xi$$
$$= S(\boldsymbol{\tau}) A(k_z) \sum_{n=-L/(2\lambda_0)}^{L/(2\lambda_0)} e^{-ik_z \lambda_0 n} \tag{10.3.7}$$

where

$$A(k_z) = \int_0^{\lambda_0} c_1^{[001]}(\xi) e^{-ik_z \xi} d\xi \tag{10.3.8}$$

Since

$$\sum_{n=-L/(2\lambda_0)}^{n=L/(2\lambda_0)} e^{-ik_z\lambda_0 n} = \frac{\sin \frac{1}{2}k_z L}{\sin \frac{1}{2}k_z \lambda_0} \tag{10.3.9}$$

Eq. (10.3.7) becomes

$$\tilde{c}(\mathbf{k}) = S(\boldsymbol{\tau}) A(k_z) \frac{\sin \frac{1}{2}k_z L}{\sin \frac{1}{2}k_z \lambda_0} \tag{10.3.10}$$

Substituting Eq. (10.3.10) into Eq. (10.2.12) yields

$$E_{\text{edge}} = \frac{1}{2} \int \Delta B\left(\frac{k_z}{k}, \frac{\boldsymbol{\tau}}{k}\right) |S(\boldsymbol{\tau})|^2 |A(k_z)|^2 \frac{\sin^2 \frac{1}{2}k_z L}{\sin^2 \frac{1}{2}k_z \lambda_0} \frac{dk_z}{2\pi} \frac{d^2\tau}{(2\pi)^2} \tag{10.3.11}$$

where

$$k = \sqrt{k_z^2 + \tau^2} \tag{10.3.12}$$

The periodic function of k_z,

$$\frac{\sin^2 \frac{1}{2}k_z L}{\sin^2 \frac{1}{2}k_z \lambda_0}$$

has very sharp maxima at the "reciprocal lattice points"

$$k_z(h) = \frac{2\pi}{\lambda_0} h, \quad h = 0, \pm 1, \pm 2, \ldots \pm \infty \tag{10.3.13}$$

The half-width of these maxima is $\Delta k_z \sim 2\pi/L$; the maxima value is $L^2/\lambda_0^2 (L/\lambda_0 \gg 1)$. On the other hand, the other functions in the integrand in (10.3.11) are smooth. These characteristics of the integrand make it possible to rewrite Eq. (10.3.11) in the form

$$E_{\text{edge}} \cong \frac{1}{2} \sum_{h=-\infty}^{\infty} \left| A\left(\frac{2\pi h}{\lambda_0}\right) \right|^2 \int_{-\pi/\lambda_0}^{\pi/\lambda_0} \frac{\sin^2 \frac{1}{2}k_z L}{\sin^2 \frac{1}{2}k_z \lambda_0} \frac{dk_z}{2\pi}$$
$$\times \iint_{-\infty}^{\infty} |S(\boldsymbol{\tau})|^2 \Delta B\left(\frac{2\pi h/\lambda_0}{k(h)}, \frac{\boldsymbol{\tau}}{k(h)}\right) \frac{d^2\tau}{(2\pi)^2} \tag{10.3.14}$$

where

$$k(h)=\sqrt{\left(\frac{2\pi}{\lambda_0}\right)^2+\tau^2}$$

Eq. (10.3.14) is valid, with accuracy to the ratio $\lambda_0/L \ll 1$.
Since

$$\int_{-\pi/\lambda_0}^{\pi/\lambda_0} \frac{\sin^2 \frac{1}{2}k_zL}{\sin^2 \frac{1}{2}k_z\lambda_0}\frac{dk_z}{2\pi}=\frac{L}{\lambda_0^2} \tag{10.3.15}$$

Eq. (10.3.14) may be rewritten as

$$E_{\text{edge}}=\frac{1}{2}\frac{L}{\lambda_0^2}\sum_{h=-\infty}^{\infty}\left|A\left(\frac{2\pi h}{\lambda_0}\right)\right|^2\iint_{-\infty}^{\infty}|S(\boldsymbol{\tau})|^2\Delta B\left(\frac{2\pi h/\lambda_0}{k(h)},\frac{\boldsymbol{\tau}}{k(h)}\right)\frac{d^2\tau}{(2\pi)^2} \tag{10.3.16}$$

As shown in Appendix 3 [see Eqs. (A.3.18) and (A.3.19)],

$$\iint_{-\infty}^{\infty}|S(\boldsymbol{\tau})|^2\Delta B\left(\frac{k_z}{k},\frac{\boldsymbol{\tau}}{k}\right)\frac{d^2\tau}{(2\pi)^2}\cong I_1(k_z)+I_2(k_z) \tag{10.3.17}$$

where

$$I_1(k_z)=\frac{\beta_2}{2|k_z|}\oint\frac{dl}{1+(dy/dx)^2} \tag{10.3.18}$$

$$I_2(k_z)=\frac{\beta_1}{2|k_z|}\oint\left(\frac{dy}{dx}\right)^2\frac{1}{1+(dy/dx)^2}\,dl \tag{10.3.19}$$

β_1 and β_2 are the eigenvalues of the matrix $\beta_{ij}(\mathbf{n}_0)$ corresponding to the eigenvectors lying in the plane normal to the vector $\mathbf{n}_0$ minimizing $B(\mathbf{n})$ [see Eq. (8.7.9)], $y=y(x)$ is the curve describing the shape of the complex in the (001) cross section. The integration in Eqs. (10.3.18) and (10.3.19) is over the $y=y(x)$ contour in the (x, y) plane. Eq. (10.3.17) is valid for modulated structures if

$$\frac{\lambda_0}{L_t}\ll 1 \tag{10.3.20}$$

where L_t is the typical lateral size of the complex in the direction normal to the [001] axis.

Substituting (10.3.18) and (10.3.19) into (10.3.17) at $k_z=(2\pi/\lambda_0)h$ yields

$$\iint_{-\infty}^{\infty}|S(\boldsymbol{\tau})|^2\Delta B\left(\frac{2\pi h/\lambda_0}{k(h)},\frac{\boldsymbol{\tau}}{k(h)}\right)\frac{d^2\tau}{(2\pi)^2}\cong\frac{\lambda_0}{4\pi|h|}\oint\frac{\beta_2+\beta_1(dy/dx)^2}{1+(dy/dx)^2}\,dl \tag{10.3.21}$$

Substituting Eq. (10.3.21) into (10.3.16) results in

$$E_{\text{edge}}=\frac{1}{8\pi}\frac{L}{\lambda_0}\oint\frac{\beta_2+\beta_1(dy/dx)^2}{1+(dy/dx)^2}\,dl\sum_{h=-\infty}^{\infty}\frac{1}{|h|}\left|A\left(\frac{2\pi h}{\lambda_0}\right)\right|^2 \tag{10.3.22}$$

According to Eq. (8.7.45), in the case of a coherent mixture of cubic phases,

$$\beta=\beta_1=\beta_2=\left(\frac{c_{11}+2c_{12}}{c_{11}}\right)^2|\xi|(c_{11}-c_{12})\varepsilon_0^2 \tag{10.3.23}$$

Using this equation in Eq. (10.3.22), we obtain

$$E_{\text{edge}}=\frac{\text{LP}\beta}{8\pi\lambda_0}\sum_{h=-\infty}^{\infty}\frac{1}{|h|}\left|A\left(\frac{2\pi h}{\lambda_0}\right)\right|^2 \tag{10.3.24}$$

where $\oint dl=P$ is the perimeter of the (001) cross section of the complex.

The Fourier coefficients $A(2\pi h/\lambda_0)$ can be calculated from Eq. (10.3.8) for the square wave periodic concentration profile depicted in Fig. 77. The result is

$$A\left(\frac{2\pi h}{\lambda_0}\right)=\int_0^{\lambda_0}c_1^{[001]}(\xi)e^{-i(2\pi h/\lambda_0)\xi}d\xi=(c_1^0-\bar{c})\int_0^{\gamma_1\lambda_0}e^{-i(2\pi h/\lambda_0)\xi}d\xi$$

$$+(c_2^0-\bar{c})\int_{\gamma_1\lambda_0}^{\lambda_0}e^{-i(2\pi h/\lambda_0)\xi}d\xi$$

$$=\begin{cases}(c_1^0-c_2^0)e^{-i\pi\gamma_1 h}\dfrac{\sin\pi h\gamma_1}{\pi h/\lambda_0} & \text{if } h\neq 0\\ 0 & \text{if } h=0 \quad *\end{cases} \tag{10.3.25}$$

Substituting Eq. (10.3.25) into Eq. (10.3.24) results in

$$E_{\text{edge}}=\frac{\text{LP}\beta}{8\pi^3}(c_1^0-c_2^0)^2\lambda_0\sum_{h=-\infty}^{\infty}{}'\frac{\sin^2\pi h\gamma_1}{|h|^3} \tag{10.3.26}$$

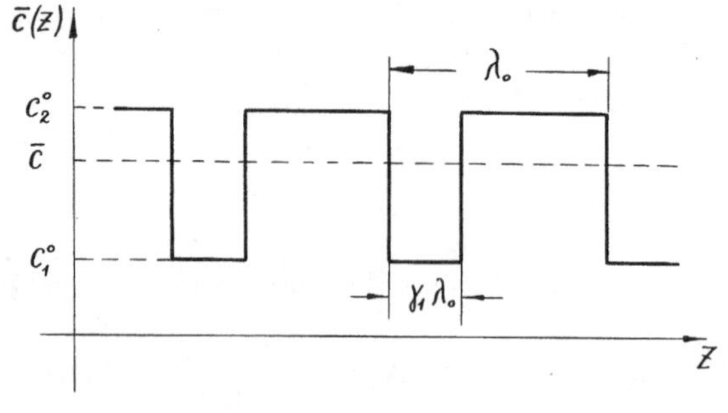

Figure 77. The square-wave periodic concentration profile related to a one-dimensional modulated structure.

*The conservation condition for solute atoms used in this equation, $c_1^0\gamma_1\lambda_0+c_2^0(1-\gamma_1)\lambda_0=\bar{c}$, follows from the lever rule (10.3.1).

where prime implies the exclusion of the term corresponding to $h=0$. Eq. (10.3.26) may also be written in the form

$$E_{\text{edge}}=\frac{S_L\beta\alpha(\gamma_1)}{8\pi^3}(c_1^0-c_2^0)^2\lambda_0 \tag{10.3.27}$$

where $S_L=LP$ is the area of the lateral boundary of the complex,

$$\alpha(\gamma_1)=2\sum_{h=1}^{\infty}\frac{\sin^2\pi h\gamma_1}{|h|^3} \tag{10.3.28}$$

The plot $\alpha(\gamma_1)$ with relation to γ_1 is given in Fig. 78. Using Eq. (10.3.23) in (10.3.27), we may rewrite Eq. (10.3.27) as

$$E_{\text{edge}}=S_L\frac{\alpha(\gamma_1)}{8\pi^3}\left(\frac{c_{11}+2c_{12}}{c_{11}}\right)^2|\xi|(c_{11}-c_{12})\varepsilon_0^2(c_2^0-c_1^0)^2\lambda_0 \tag{10.3.29}$$

It follows from Eq. (10.3.29) that the quantity

$$\frac{E_{\text{edge}}}{S_L}=\frac{\alpha(\gamma_1)}{8\pi^3}\left(\frac{c_{11}+2c_{12}}{c_{11}}\right)^2|\xi|(c_{11}-c_{12})\varepsilon_0^2(c_2^0-c_1^0)^2\lambda_0 \tag{10.3.30}$$

plays the part of the specific surface energy related to the lateral boundary of the complex. The appearance of the strain energy (10.3.29) proportional to the lateral boundary area has been already discussed in Section 10.1. It is associated with the strain localized near the edges of platelets of the two phases comprising the complex. This strain field is an analogue of the magnetic scattering field in ferromagnetic domains.

The only parameter of the modulated structure that remains to be determined is the period, λ_0. The equilibrium value of λ_0 corresponds to a compromise

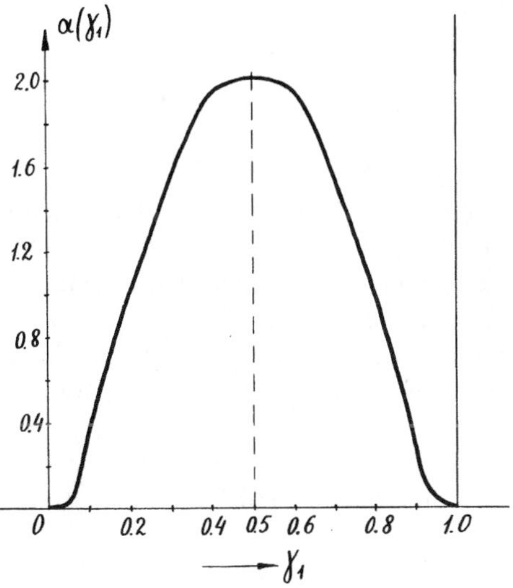

Figure 78. The calculated plot $\alpha(\gamma_1)$ with respect to γ_1.

between the strain energy (10.3.29) which vanishes at $\lambda_0 \to 0$ and the interphase energy which increases infinitely at $\lambda_0 \to 0$ (the zero λ_0 value corresponds to the unlimited increase of the number of interphase boundaries between platelets of the two phases making up the complex).

To calculate the period, λ_0, the strain energy (10.3.29) should be combined with the interphase energy

$$E_s = 2\gamma_{(001)} S \frac{L}{\lambda_0} = \frac{2\gamma_{(001)}}{\lambda_0} V \tag{10.3.31}$$

where $\gamma_{(001)}$ is the interphase energy coefficient corresponding to the interphase boundary along the (001) plane, $2L/\lambda_0$ is the total number of interphase boundaries in the complex, S is the area of each boundary. The total energy may thus be written as

$$\Delta F = E_{\text{edge}} + E_s = S_L \frac{\alpha(\gamma_1)}{8\pi^3} \left(\frac{c_{11} + 2c_{12}}{c_{11}} \right)^2 |\xi| (c_{11} - c_{12}) \varepsilon_0^2 (c_2^0 - c_1^0)^2 \lambda_0 + \frac{2\gamma_{(001)} V}{\lambda_0} \tag{10.3.32}$$

The minimum condition, $d\Delta F / d\lambda_0 = 0$, yields the equilibrium period value

$$\lambda_0 = \sqrt{r_1 L_t} \tag{10.3.33}$$

where

$$r_1 = \frac{16\pi^3 \gamma_{(001)}}{(c_{11} - c_{12}) \varepsilon_0^2 (c_2^0 - c_1^0)^2 |\xi| \alpha(\gamma_1) \left(\dfrac{c_{11} + 2c_{12}}{c_{11}} \right)^2} \tag{10.3.34}$$

is the material constant with the length dimension [compare with the constant r_0 in Eqs. (8.1.49) and (9.4.9)],

$$L_t = \frac{V}{S_L} = \frac{S}{P} \tag{10.3.35}$$

is the typical lateral size of the complex.

Eq. (10.3.33) demonstrates the square root dependence of the heterogeneity length λ_0 on the size of the complex. As mentioned in Section 10.1, a similar relation holds with ferromagnetics where the magnetic domain size shows a square root dependence on the size of the crystal.

Finally, it should be noted that the one-dimensional structure obtained (see Fig. 75) insures the absolute minimization of the free energy if all interphase boundaries are coherent.

In conclusion, a few points on the applicability of the present theory should be mentioned. According to Eqs. (10.3.20) and (10.3.33) the theory is valid provided the inequality

$$\sqrt{\frac{r_1}{L_t}} \ll 1 \tag{10.3.36}$$

holds. Violation of the inequality (10.3.36) indicates that the interphase energy becomes commensurate with the bulk strain energy responsible for "rolling" precipitates into thin and extended platelets rather than with the strain energy E_{edge}. In that case the mechanism of the formation of modulated structures described above ceases to operate. The inequality (10.3.36) thus provides the possibility of predicting the formation of microstructures with a well-defined regularity. The pseudoperiodic distributions of precipitates will be the nearer to the perfectly periodic distribution, the stronger the inequality (10.3.36), the smaller the characteristic length, r_1 [see Eq. (10.3.34)]. We thus obtain the following conditions for the existence of regular modulated structures:

1. The coexisting equilibrium phases should have crystal lattice parameters that differ greatly from each other (a large ε_0 value).
2. The solid solution should be characterized by a large elastic anisotropy value (large ξ value).
3. The interphase specific energy should be sufficiently small.

As regards the last point, the best conditions for the formation of periodic structures are realized near spinodal curves where the value of $\gamma_{(001)}$ is small (see Fig. 44).

Violation of any of these conditions leads to a disordering of the regular arrangement and deep reconstruction of the morphology of the two-phase mixture. For instance, when the typical precipitate dimension in the initial stage of the decomposition is sufficiently small ($r_1/L_t \ll 1$), one should except the formation of randomly distributed precipitates. Further growth of precipitates may bring about a rearrangement to a regular modulated structure [an increase of L_t may improve the inequality (10.3.36)].

We now shall consider the limiting case when the modulated structure period λ_0 is too small for a treatment in terms of macroscopic thermodynamics.

With the modulated structure period given by Eq. (10.3.33), the equilibrium thickness, d, of a platelet of one of the coexisting phases comprising the complex (see Fig. 75) is

$$d=\gamma_1\lambda_0=\gamma_1\sqrt{r_1 L_t} \tag{10.3.37}$$

where the volume fraction of the phase, γ_1, is determined by the lever rule (10.3.1) and therefore depends on the composition of the alloy rather than its structure.

According to Eq. (10.3.37) the d value drops with a decrease in the size of the complex, L_t, and the constant r_1 (small interphase energy $\gamma_{(001)}$, large crystal lattice mismatch ε_0, and elastic anisotropy $|\xi|$). An obvious question is, To what extent may the thickness, d, decrease without violating the validity of the macroscopic approximation? Indeed, if the platelet becomes too thin, it can no longer be interpreted in such macroscopic terms as specific bulk free energy and specific interphase energy and cannot thus be considered as a new phase particle. This is so when the thickness, d, becomes commensurate with the interphase boundary thickness. We may therefore conclude that Eq. (10.3.37) holds until d becomes less than the interphase boundary thickness. Further decrease of d is blocked because the "interphase" energy associated with the gradient terms in Eq. (4.2.3) would become of the same order of magnitude as the "bulk" energy in that case. This would result in a significant increase of the free energy.

Therefore, if d given by Eq. (10.3.37) becomes commensurate with the interphase boundary thickness, d ,

$$d \approx d_{\text{interph}}$$

we have

$$\lambda_0^* \approx \frac{d_{\text{interph}}}{\gamma_1} \tag{10.3.38}$$

where λ_0^* is a limit value of the period λ_0. Using Eq. (10.3.38), we may rewrite Eq. (10.3.33)

$$\lambda_0 = \begin{cases} \sqrt{r_1 L_t} & \text{if } \dfrac{d_{\text{interph}}}{\gamma_1 \sqrt{r_1 L_t}} \ll 1 \\[2ex] \lambda_0^* \approx \dfrac{d_{\text{interph}}}{\gamma_1} & \text{otherwise} \end{cases} \tag{10.3.39}$$

The value d_{interph} is a thermodynamic function of the composition and temperature plotted in Fig. 46. It follows from Eq. (10.3.39) that coarsening (increase of L_t) does not affect the period λ_0 of the modulated structure at the first stage of the decomposition. The period increases later, when $d_{\text{interph}}/(\gamma_1\sqrt{r_1 L_t}) \ll 1$. The dependence of λ_0 on size of the complex L_t (or aging time) is plotted schematically in Fig. 79. It is noteworthy that the limit period λ_0^* becomes larger, the closer the decomposition is to the spinodal curve.

The calculations carried out above enable us to obtain a semiquantitative estimate of the smallest possible value of the modulated structure period, λ_0, in terms of the interphase boundary thickness. An accurate calculation should, however, be based on the minimization of the sum of the strain energy E_{edge} [Eq. (10.3.29)] and the free energy of a one-dimensional distribution (4.5.12) with respect to the period λ_0 under the conditions (4.5.9).

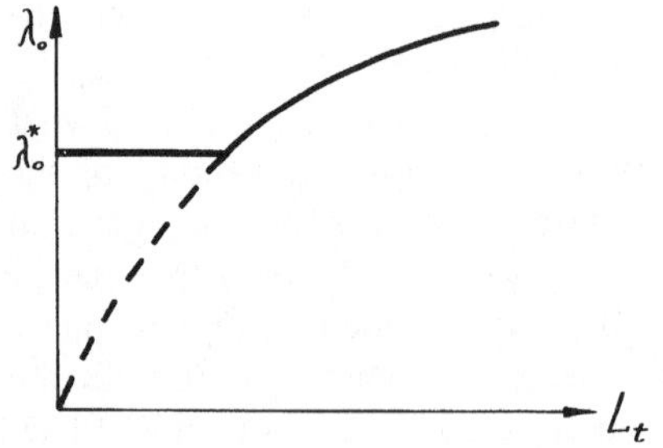

Figure 79. The typical dependence of the modulated structure period λ_0 on the size of the complex L_t.

10.4 TWO-DIMENSIONAL MODULATED STRUCTURES

As shown in Section 10.2, apart from one-dimensional distributions (10.2.22) there are two-dimensional concentration distributions (10.2.23) that provide the minimization of the strain energy.

The two-dimensional concentration distribution (10.2.23) like the one-dimensional distribution (10.2.22) is mainly described by the bulk free energy F_{bulk} in (10.2.18). The small correction E_{edge} in (10.2.17) characterizes the surface energy of the complex generated by elastic strain. Its value is of the order of $d/L \ll 1$ of E_{bulk} where d is the typical heterogeneity length, E_{bulk} is the bulk strain energy (10.2.11).

For this reason the equilibrium compositions and volumes of the phases making up the complex should be determined from the condition of the minimum E_{bulk} value. The necessary condition for E_{bulk} to be minimal at a given number of solute atoms is

$$\frac{\partial f(\bar{c})}{\partial \bar{c}} = \mu \tag{10.4.1}$$

where $f(\bar{c})$ is given by Eq. (10.2.19) and μ is an undetermined Lagrange coefficient playing the role of the chemical potential. Eq. (10.4.1) determines the equilibrium concentrations of the phases formed in the decomposition of the complex. In a general case it has three solutions (see Fig. 80)

$$c_1(\mu), \quad c_2(\mu), \quad \text{and} \quad c_3(\mu) \tag{10.4.2}$$

The minimum number of different values assumed by the two-dimensional distribution (10.2.23) is equal to three. This is the case where each of the functions, $c_2^{[100]}(x)$ and $c_2^{[010]}(y)$, may only take two different values, $c_1(\mu)/2$ and $c_2(\mu)/2$. The function $\bar{c}(\mathbf{r})$ is then equal to $c_1(\mu)$, $c_2(\mu)$, or $\frac{1}{2}[c_1(\mu)+c_2(\mu)]$. The equality

$$c_3(\mu) = \tfrac{1}{2}[c_1(\mu)+c_2(\mu)] \tag{10.4.3}$$

where $c_3(\mu)$ is the third solution of Eq. (10.4.1) is the equation determining μ.

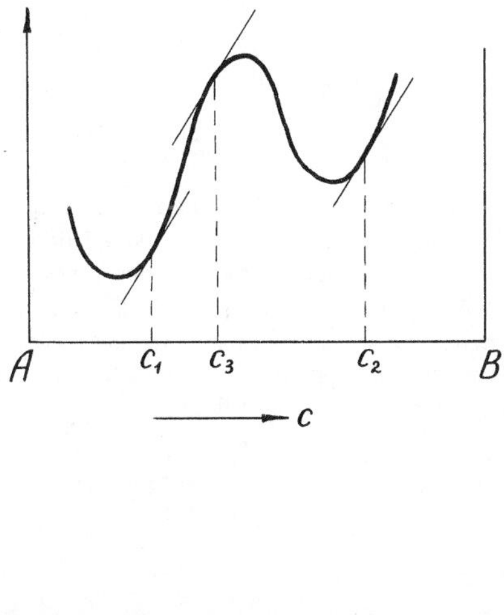

Figure 80. The graphical solution of Eq. (10.4.1).

We shall use γ_1 to denote the fraction of the volume V of the complex where the function $c_2^{[100]}(x)$ has the value $(c_1-\bar{c})/2$, and γ_2 will be used to denote the fraction of the volume of the complex where the function $c_2^{[010]}(y)$ has the value $(c_1-\bar{c})/2$. According to Fig. 81 the volume of the phase having the composition c_1 will be $\gamma_1\gamma_2 V$, the volume of the phase of the composition c_2 will be $(1-\gamma_1)(1-\gamma_2)V$, and the volume of the phase of the intermediate composition $\frac{1}{2}(c_1+c_2)$ will be $(\gamma_1+\gamma_2-2\gamma_1\gamma_2)V$. The condition of the conservation of the number of solute atoms reads

$$c_1\gamma_1\gamma_2+c_2(1-\gamma_1)(1-\gamma_2)+\tfrac{1}{2}(c_1+c_2)(\gamma_1+\gamma_2-2\gamma_1\gamma_2)=\bar{c} \tag{10.4.4}$$

or

$$\gamma_1+\gamma_2=2\,\frac{c_2-\bar{c}}{c_2-c_1} \tag{10.4.5}$$

and the free energy (10.2.18) may be written

$$F_{\text{bulk}}=V\left[\,f(c_1)\gamma_1\gamma_2+f(c_2)(1-\gamma_1)(1-\gamma_2)+f\left(\frac{c_1+c_2}{2}\right)(\gamma_1+\gamma_2-2\gamma_1\gamma_2)\right] \tag{10.4.6}$$

Minimization of (10.4.6) with respect to γ_1 and γ_2, taking into consideration of (10.4.5), yields

$$\gamma_1=\gamma_2=\frac{c_2-\bar{c}}{c_2-c_1} \tag{10.4.7}$$

The limitations imposed by this result on the form of the functions $c_2^{[100]}(x)$ and $c_2^{[010]}(y)$ are still insufficient to determine the function $\bar{c}(\mathbf{r})$ unambiguously: there is an infinite set of functions of the type (10.2.23) that fit these limitations and with respect to which the free energy F_{bulk} is degenerate. For that reason the determination of the function $\bar{c}(\mathbf{r})$ requires, as in the case of the one-dimensional distribution, the minimization of the additional terms not included in the bulk free energy thus far. These additional terms are the elastic strain energy E_{edge} proportional to the lateral boundary area of the complex and the interphase energy E_s of the interphase boundaries within the complex.

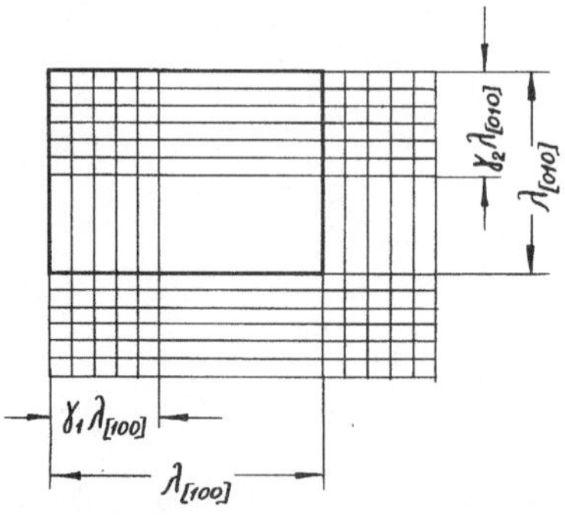

Figure 81. An unit cell of a two-dimensional modulated structure in the cross section (001). Horizontal and vertical hatchings show the one-dimensional modulation along the [010] and [100] directions, respectively: $\lambda_{[010]}$ and $\lambda_{[100]}$ are the periods of the one-dimensional square-wave modulations. The heavy line indicates a unit cell.

It follows from Eq. (10.2.23) that the strain energy E_{edge} in (10.2.12) for the two-dimensional distribution, which is a mere superposition of two one-dimensional distributions, may be separated into the sum of two quantities. As the two-dimensional distribution (10.2.23) is the superposition of the one-dimensional distributions $c_2^{[100]}(x)$ and $c_2^{[010]}(y)$, the strain energy E_{edge} in (10.2.12) may be written as the sum

$$E_{\text{edge}} = E_{\text{edge}}^{[100]} + E_{\text{edge}}^{[010]} \tag{10.4.8}$$

where the terms $E_{\text{edge}}^{[100]}$ and $E_{\text{edge}}^{[010]}$ refer to the one-dimensional distributions $c_2^{[100]}(x)$ and $c_2^{[010]}(y)$, respectively. The problem of the calculation of $E_{\text{edge}}^{[100]}$ and $E_{\text{edge}}^{[010]}$ has in fact already been solved in Section 10.3. These values are given by the equations

$$E_{\text{edge}}^{[100]} = S_{[100]} \frac{\alpha(\gamma_1)}{32\pi^3} \left(\frac{c_{11}+2c_{12}}{c_{11}} \right)^2 |\xi|(c_{11}-c_{12})\varepsilon_0^2(c_2-c_1)^2 \lambda_{[100]} \tag{10.4.9}$$

$$E_{\text{edge}}^{[010]} = S_{[010]} \frac{\alpha(\gamma_1)}{32\pi^3} \left(\frac{c_{11}+2c_{12}}{c_{11}} \right)^2 |\xi|(c_{11}-c_{12})\varepsilon_0^2(c_2-c_1)^2 \lambda_{[010]} \tag{10.4.10}$$

where $S_{[100]}$, $S_{[010]}$, $\lambda_{[100]}$, and $\lambda_{[010]}$ are the lateral areas of the complex and the periods of the one-dimensional distributions $c_2^{[100]}(x)$ and $c_2^{[010]}(y)$ along directions [100] and [010], respectively. Eqs. (10.4.9) and (10.4.10) follow from Eq. (10.3.29) after the substitutions

$$c_1^0 \to \tfrac{1}{2}c_1$$
$$c_2^0 \to \tfrac{1}{2}c_2$$

For brevity, let us consider the case where the interphase specific free energy, $\gamma_{(001)}$, is proportional to the square of the difference between the compositions of the coexisting phases c_1 and c_2:

$$\gamma_{(001)} = \xi(c_1 - c_2)^2 \tag{10.4.11}$$

Eq. (10.4.11) is valid for a stepwise concentration profile realized in the case of pairwise interactions. It is easy to see, from assumption (10.4.11), that the interphase energy E_s as well as the strain energy E_{edge} may be written as the sum of interphase energies corresponding to two one-dimensional distributions:

$$E_s = \frac{\gamma_{(001)}}{2\lambda_{[100]}} V + \frac{\gamma_{(001)}}{2\lambda_{[010]}} V \tag{10.4.12}$$

where $\gamma_{(001)}$ is the interphase energy corresponding to the concentration jump, $c_2 - c_1$, on the interphase boundary. It follows from Eqs (10.4.9), (10.4.10), and (10.4.12) that the periods $\lambda_{[100]}$ and $\lambda_{[010]}$ can be determined independently by minimizing the sum of Eqs. (10.4.9), (10.4.10), and (10.4.12) with respect to these parameters. A procedure fully similar to that applied to the one-dimensional problem yields

$$\lambda_{[100]} = \sqrt{r_1 L_{[100]}}, \quad \lambda_{[010]} = \sqrt{r_1 L_{[010]}}$$

where $L_{[100]}=S_{(100)}/P_{(100)}$ and $L_{[010]}=S_{(010)}/P_{(010)}$ are the characteristic linear dimensions of the complex in the cross sections made by the (100) and (010) planes, respectively [compare with Eq. (10.3.35)], $S_{(100)}$, $S_{(010)}$ and $P_{(100)}$, $P_{(010)}$ are the areas and perimeters of those cross sections, respectively,

$$r_1=\frac{16\pi^3\gamma_{(001)}}{(c_{11}-c_{12})\varepsilon_0^2(c_1-c_2)^2|\xi|\alpha(\gamma_1)[(c_{11}+2c_{12})/c_{11}]^2} \tag{10.4.13}$$

is the material constant having the length dimension.

From symmetry considerations the (100) and (010) planes are in every respect equivalent to each other. Therefore one may expect the approximate equality $L_{[100]}\approx L_{[010]}$ to hold in the majority of cases. This and Eq. (10.4.12) give

$$\lambda_{[100]}=\lambda_{[010]}=\lambda_0 \tag{10.4.14}$$

The structure of the complex is thus a periodic modulated structure consisting of rods along the [001] direction (Fig. 82). There are three types of rods, those with the compositions c_1 and c_2 near equilibrium compositions and rods of the composition $(c_1+c_2)/2$ near the solid solution composition. This modulated structure is in fact a three-phase coherent mixture.

The (001) cross section of the complex exhibits a two-dimensional square periodic lattice, whose sites are formed by squares of two phases of the compositions c_1 and c_2 and rectangulars of the composition $(c_1+c_2)/2$ connecting these squares. The basic translations of this lattice are along the [100] and [010] directions (see Fig. 82).

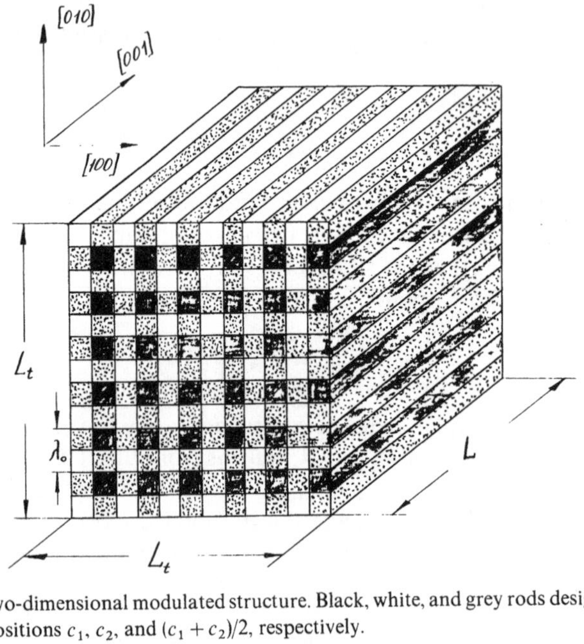

Figure 82. Two-dimensional modulated structure. Black, white, and grey rods designate the phases with the compositions c_1, c_2, and $(c_1+c_2)/2$, respectively.

The (100) or (010) cross sections of the complex show periodically distributed rods of the phases with the compositions c_1, c_2, and $(c_1+c_2)/2$ extended along the direction [001] (Fig. 82).

The free energy minimum of the two-dimensional distribution found for the class of functions (10.2.23) corresponds to a three-phase complex and does not therefore insure the absolute minimum of the free energy: one of the structure components of the complex has the composition $(c_1+c_2)/2$ close to the composition of the undecomposed matrix. The absolute minimum corresponds, as shown in Section 10.3, to a one-dimensional two-phase distribution.

The distribution shown in Fig. 82 for this reason is metastable, stable with respect to small composition variations. To make sure that the optimal two-dimensional distribution obtained is metastable, one must investigate variations of the free energy of this distribution under infinitesimal variations of the concentration that remove the function $\bar{c}(\mathbf{r})$ from the class of functions (10.2.23). At the same time it is sufficient to analyze only the most "dangerous" variations which are, on the one hand, associated with the minimum increase of the sum of the strain and interphase energies and, on the other hand, with the maximum decrease of the bulk chemical free energy. As one can see from Fig. 80, the composition $c_3=(c_1+c_2)/2$ corresponds to the convex part of the chemical free energy curve, and thus rods with this composition find themselves below the chemical spinodal. Metastability of these rods is provided by the strain and interphase energy contributions which inhibit decomposition. Since the minimum strain energy increase is produced by the formation of thin platelets with the (001) habit, and the maximum decrease of the bulk free energy occurs in the decomposition of the rods having the intermediate composition $(c_1+c_2)/2$, close or identical to the composition of the initial matrix, the most "dangerous" composition variations in a two-dimensional complex will be associated with the decomposition of those rods into alternating platelets, with the (001) habit having near equilibrium compositions c_1 and c_2.

The transformation of a rod with the composition $(c_1+c_2)/2$ into a periodic two-phase sandwich comprising alternating platelets with the compositions c_1 and c_2 results in the decrease of the bulk free energy by

$$\Delta F_{\text{bulk}}^{\text{rod}}=\left[f(c_1)\gamma_3+f(c_2)(1-\gamma_3)-f\left(\frac{c_1+c_2}{2}\right)\right]v \tag{10.4.15}$$

where v is the rod volume, γ_3 the volume fraction of the first phase of the composition c_1. According to the lever rule

$$\gamma_3=\frac{c_2-\bar{c}}{c_2-c_1}=\frac{c_2-\frac{1}{2}(c_1+c_2)}{c_2-c_1}=\frac{1}{2} \tag{10.4.16}$$

Substituting Eq. (10.4.16) into (10.4.15) yields

$$\Delta F_{\text{bulk}}^{\text{rod}}=\frac{1}{2}\left[f(c_1)+f(c_2)-2f\left(\frac{c_1+c_2}{2}\right)\right]v \tag{10.4.17}$$

The sum of the strain and interphase energies (10.3.32) changes upon the decomposition of the rod into a one-dimensional modulated structure by

$$\Delta F_{\text{rod}} = S_{\text{rod}} \frac{\alpha(\frac{1}{2})}{8\pi^3} \left(\frac{c_{11}+2c_{12}}{c_{11}} \right)^2 |\xi| (c_{11}-c_{12}) \varepsilon_0^2 (c_2-c_1)^2 \lambda_0 + 2\frac{\gamma_{(001)} v}{\lambda_0} \tag{10.4.18}$$

where S_{rod} is the lateral surface area of the rod (according to Section 10.3 this is just a one-dimensional structure that provides the minimum increase of strain and interphase energies).

Minimization of Eq. (10.4.18) gives the equilibrium period $\lambda_{[001]}$:

$$\lambda_{[001]} = \sqrt{r_1 l_{\text{rod}}} \tag{10.4.19}$$

where r_1 is given by Eq. (10.3.34) at $\gamma_1 = \frac{1}{2}$ and

$$l_{\text{rod}} = \frac{v}{S_{\text{rod}}} \tag{10.4.20}$$

is the typical thickness of the rod. Substituting Eq. (10.4.19) into (10.4.18) transforms the latter equation into

$$\Delta F_{\text{rod}} = \frac{4\gamma_{(001)} v}{\lambda_{[001]}} = \frac{4\gamma_{(001)} v}{\sqrt{r_1 l_{\text{rod}}}} \tag{10.4.21}$$

The total free energy change is given by the sum of Eqs. (10.4.21) and (10.4.17). From (10.4.19) we have

$$\Delta F_{\text{total}} = \left[\Delta f(c) + \frac{4\gamma_{(001)}}{\sqrt{r_1 l_{\text{rod}}}} \right] v \tag{10.4.22}$$

where

$$\Delta f(c) = \frac{1}{2} \left[f(c_1) + f(c_2) - 2f\left(\frac{c_1+c_2}{2} \right) \right] < 0 \tag{10.4.23}$$

since, by definition, $f(c)$ is convex at $c = (c_1+c_2)/2$. A complex having the structure shown in Fig. 82 thus proves metastable if the total free energy change of the formation of a one-dimensional modulated structure within the rod of the composition $(c_1+c_2)/2$ is positive. This condition is realised if

$$\frac{1}{2} \left[f(c_1) + f(c_2) - 2f\left(\frac{c_1+c_2}{2} \right) \right] + \frac{4\gamma_{(001)}}{\sqrt{r_1 l_{\text{rod}}}} > 0 \tag{10.4.24}$$

The condition (10.4.24) holds for a small supercooling when the specific free energy of a two-phase mixture (the final state) is close to the specific free energy of the matrix (the initial state). It also holds for short coarsening times when the size of the complex and therefore the thickness of the rods, l_{rod}, are sufficiently small. In other cases the two-dimensional structure depicted in Fig. 82 proves unstable with respect to a secondary decomposition of rods of the composition $(c_1+c_2)/2$ into sandwiches of alternating platelets of the compositions c_1 and c_2.

A secondary decomposition may thus result in the formation of a two-phase system from a three-phase one. As shown at the beginning of this section, the two-phase state insures the minimum of the bulk free energy F_{bulk}. A three-dimensional modulated structure generated in a secondary decomposition is shown in Fig. 83.

To sum up, the two-dimensional rodlike structure depicted in Fig. 82 is metastable if the following requirements are met simultaneously:

1. The supercooling into the two-phase field of the equilibrium diagram is small.
2. Complexes have comparatively small length dimensions (aging time is comparatively small).
3. The crystal lattice expansion concentration coefficient is rather large.

The latter condition may, for instance, be provided by alloying a solid solution with elements having considerably larger atomic radii than the solvent atoms.

Finally, we should note that the three-dimensional structure of the type (10.2.24), which is a superposition of three square concentration waves, cannot generally insure minimization of the total free energy. The reason for that is that the minimal number of phases corresponding to the distribution (10.2.24) is

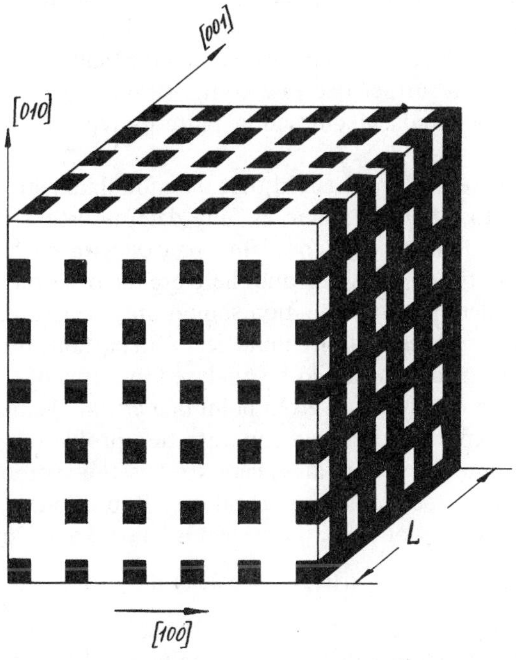

Figure 83. Three-dimensional modulated structure formed at the secondary decomposition of grey rods in the two-dimensional structure depicted in Fig. 82. Black and white regions designate the phases with the compositions c_1 and c_2, respectively.

four, whereas the necessary condition for the coexistence of phases in equilibrium with each other (10.4.1) does not allow that number to exceed three.

One more point should be mentioned. As with the one-dimensional modulated structure, the lower limit of the period of the two-dimensional modulated structure shown in Fig. 82 is given by the value λ_0^* (see Fig. 79). The dependence of the period on the size of the complex is similar to that plotted in Fig. 79.

10.5. FORMATION OF BASKETLIKE MODULATED STRUCTURES IN ORDERING OF MUTUAL ARRANGEMENT OF PRECIPITATES

It was shown in Sections 10.2 to 10.4 that any spatial rearrangement of the substructure of a two-phase coherent mixture of cubic phases which increases the intensity function $|\tilde{c}(\mathbf{k})|^2$ along the "soft" $\langle 100\rangle$ direction in the **k**-space decreases the strain energy of the system. Therefore, to predict evolution of the system during strain-induced coarsening, we should look for the morphology of a two-phase alloy that would increase the function $|\tilde{c}(\mathbf{k})|^2$ along the direction $\langle 100\rangle$ at the expense of all other directions. The simplest way to achieve this is changes of precipitate shapes. After further shape changes become ineffective, the process develops by other transformations of the system morphology. These transformations occur at the higher hierarchical level, involving changes in the mutual positions of precipitates resulting in the modulated structures. Changes of the mutual particle arrangement are also controlled by the $|c(\mathbf{k})|^2$ function which, as before, should increase along the "soft" $\langle 100\rangle$ direction in the **k**-space to provide a further decrease of the strain energy. This leads to the appearance of side-bands in diffraction patterns reported by numerous investigators.

It should, however, be emphasized that the modulated structures described in Sections 10.2 to 10.4 are treated as fully deterministic structures, as fully ordered configurations of two coexisting phases with each point **r** related unambiguously to the composition and therefore the positions of precipitates.

However, this deterministic situation should only be treated as a particular case. The more general description should deal with partially ordered modulated structures. These structures should be analyzed in terms of preferential probabilities to find a precipitate at a given point of a crystal. Within this approach it is more convenient to use a mean concentration profile $\langle\bar{c}(\mathbf{r})\rangle$ averaged over various distributions of precipitates (here $\langle\cdots\rangle$ is the symbol for averaging) rather than specific locations of precipitates. If precipitates are randomly distributed over the crystal body, the system is "disordered" and

$$\langle\bar{c}(\mathbf{r})\rangle=\bar{c}=\text{constant}$$

where $\bar{c}$ is the composition of the alloy. If on the contrary, $\langle\bar{c}(\mathbf{r})\rangle$ depends on the coordinate **r**, the spatial distribution of precipitates exhibits long-range order.

This situation is analogous to atomic ordering. Actually, a superstructure in a fully ordered alloy is described by a fully deterministic atomic distribution where each crystal lattice site is occupied by an atom of a definite type. At finite temperatures, however, only partial ordering occurs, and the crystal lattice sites may only be ascribed certain probabilities, $n(\mathbf{r}) = \langle \tilde{c}(\mathbf{r}) \rangle$, that they will be occupied with an atom of a definite type (see Section 3.1).

It was shown in Section 10.2 that the strain energy of a system of precipitates is determined by the Fourier transform $|\tilde{c}(\mathbf{k})|^2$. According to Eq. (10.2.6), the average strain energy may be written as

$$\langle E \rangle = \frac{1}{2} \int B(\mathbf{n}) \langle |\tilde{c}(\mathbf{k})|^2 \rangle \frac{d^3k}{(2\pi)^3} \tag{10.5.1}$$

With accuracy to macroscopically small fluctuations, this energy coincides with the strain energy of a particular realization of the distribution of precipitates.

Let us write $\tilde{c}(\mathbf{k})$ as

$$\tilde{c}(\mathbf{k}) = \langle \tilde{c}(\mathbf{k}) \rangle + \delta\tilde{c}(\mathbf{k}) \tag{10.5.2}$$

where

$$\langle \tilde{c}(\mathbf{k}) \rangle = \int (\langle \tilde{c}(\mathbf{r}) \rangle - \bar{c}) e^{-i\mathbf{kr}} d^3r \tag{10.5.3}$$

$$\delta\tilde{c}(\mathbf{k}) = \int (\tilde{c}(\mathbf{r}) - \langle \tilde{c}(\mathbf{r}) \rangle) e^{-i\mathbf{kr}} d^3r \tag{10.5.4}$$

Substituting Eq. (10.5.2) into the expression for $\langle |\tilde{c}(\mathbf{k})|^2 \rangle$ gives

$$\langle |\tilde{c}(\mathbf{k})|^2 \rangle = |\langle \tilde{c}(\mathbf{k}) \rangle|^2 + \langle |\delta\tilde{c}(\mathbf{k})|^2 \rangle \tag{10.5.5}$$

[The crossterms of the type $\langle \tilde{c}(\mathbf{k}) \rangle \langle \delta\tilde{c}(\mathbf{k}) \rangle$ vanish because by the definition (10.5.2), $\langle \delta\tilde{c}(\mathbf{k}) \rangle = 0$.]

The first term in Eq. (10.5.5) describes the coherent "intensity" of a partially ordered distribution of precipitates; the second one describes the "intensity" of fluctuations in this distribution.

The first term is an analogue of the coherent scattering intensity related to superlattice reflections [see the second term in Eq. (1.3.36)], and the second one may be put in correspondence with short-range order diffuses scattering (1.3.37) in the case of atomic ordering.

Substituting Eq. (10.5.5) into Eq. (10.5.1) yields

$$\langle E \rangle = \frac{1}{2} \int B(\mathbf{n}) |\langle \tilde{c}(\mathbf{k}) \rangle|^2 \frac{d^3k}{(2\pi)^3} + \frac{1}{2} \int B(\mathbf{n}) \langle |\delta\tilde{c}(\mathbf{k})|^2 \rangle \frac{d^3k}{(2\pi)^3} \tag{10.5.6}$$

Neglecting fluctuations in the distribution of precipitates, we may ommit the second term in Eq. (10.5.6). This is equivalent to the mean-field approximation widely applied in Chapter 3 to describe the order-disorder phenomenon in alloys. In the mean-field approximation

$$\langle E \rangle \approx \frac{1}{2} \int B(\mathbf{n}) |\langle \tilde{c}(\mathbf{k}) \rangle|^2 \frac{d^3k}{(2\pi)^3} \tag{10.5.7}$$

The mean-field approximation (10.5.7), as well as that used in the theory of an ordering well applies to the case of a high degree of ordering when the fluctuation term $\langle|\delta\tilde{c}(\mathbf{k})|^2\rangle$ is small.

We shall assume below that the mean-field equation (10.5.7) holds. This assumption provides the possibility to reduce the calculation of the strain energy to the calculations already made in Sections 9.4, 10.3, and 10.4. For instance, if new phase precipitates preferentially occupy the (001) platelike layer and are randomly distributed within this layer, the function

$$\Delta c(\mathbf{r}) = \langle\tilde{c}(\mathbf{r})\rangle - \bar{c}$$

describes the system heterogeneity. It is constant within the layer and vanishes outside it. The problem of the strain energy is reduced in this case to the problem solved in Section 9.4 for an isolated platelike precipitate. It thus follows that the preferential segregation of precipitate particles into the (001) plane decreases the strain energy somewhat like "rolling" a precipitate particle into a thin sheet.

The latter effect results in the directional ordering of the mutual arrangement of precipitates insuring the segregation of precipitates into the (001) plane. This may be interpreted as correlation of precipitate locations along certain directions.

If new phase precipitates form a regular array in the crystal matrix characterized by the concentration profile $\langle\tilde{c}(\mathbf{r})\rangle$ which meets the conditions (10.2.22) to (10.2.24), the strain energy of the heterogeneous structure also decreases. In this case the strain energy may be determined as if for the deterministic (fully ordered) modulated structures discussed in Sections 10.3 and 10.4.

For instance, applying Eq. (10.2.22) to the mean concentration profile,

$$\langle\tilde{c}(\mathbf{r})\rangle = \bar{c} + c_1^{[001]}(z) \tag{10.5.8}$$

where $c_1^{[001]}(z)$ is the square concentration wave in the direction of the [001] axis (z-axis). The amplitude of this wave is an analogue of the long-range order parameter of concentration waves describing ordering of atoms. The ordering given by Eq. (10.5.8) corresponds to the preferential occupation of periodically spaced (001) layers (within these layers precipitates are distributed randomly). A structure of this type is shown by the schematic in Fig. 84. It is implied that all the regions free from the first equilibrium phase are filled with the second phase and parent matrix.

The situation with the three-dimensional distribution (10.2.24) is very much the same. We have

$$\langle\tilde{c}(\mathbf{r})\rangle = \bar{c} + c_3^{[100]}(x) + c_3^{[010]}(y) + c_3^{[001]}(z) \tag{10.5.9}$$

where $c_3^{[100]}(x)$, $c_3^{[010]}(y)$, and $c_3^{[001]}(z)$ are the interpenetrating square concentration waves in the [100], [010], and [001] directions. The functions $c_3^{[100]}(x)$, $c_3^{[010]}(y)$, $c_3^{[001]}(z)$ are assumed to describe one-dimensional distributions of platelike precipitates with the (100), (010), and (001) habits if the inequality (9.4.12) responsible for the platelike shape holds. Precipitates of each kind are randomly distributed within the respective one-dimensional periodically spaced

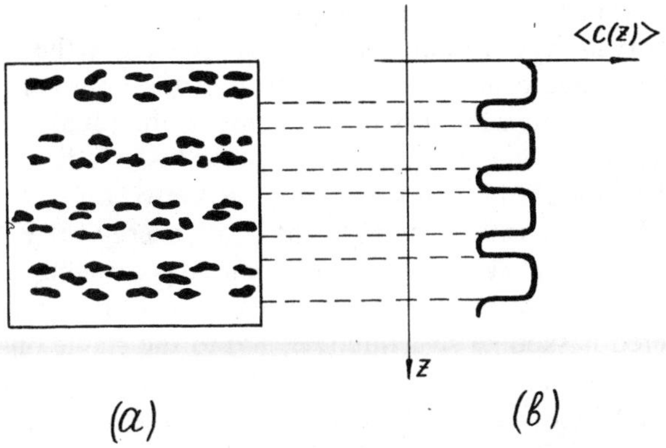

Figure 84. Schematic drawing of a "partially ordered" one-dimensional modulated structure. Precipitates preferentially occupy alternating periodically spaced (001) layers. (*a*) The spatial distribution of precipitates; (*b*) the mean-concentration profile corresponding to the distribution (*a*).

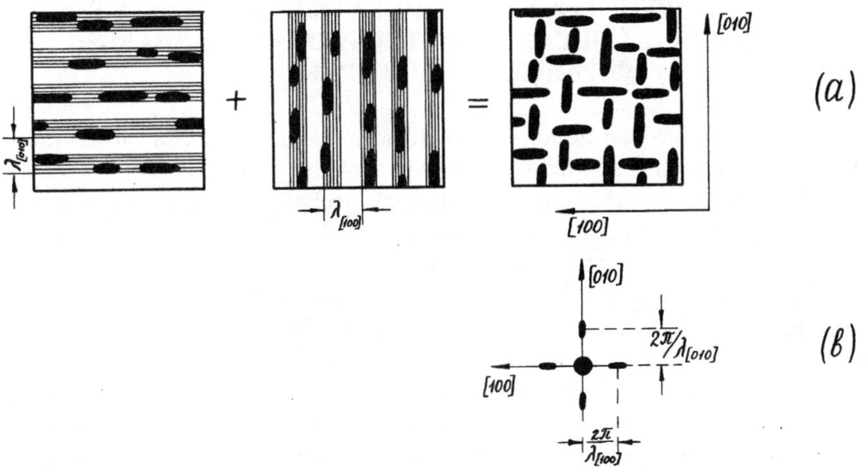

Figure 85. (*a*) Schematic drawing of a preferential distribution of precipitates in a "partially ordered" two-dimensional modulated structure which is a superposition of two "partially ordered" one-dimensional modulated structures. (*b*) The diffraction pattern from the structure in (*a*).

layer systems, as shown in Fig. 85. Precipitates with the (100) habit form the $c_3^{[100]}(x)$ mean concentration profile; those with the (010) and (001) habits form the $c_3^{[010]}(y)$ and $c_3^{[001]}(z)$ mean concentration profiles, respectively. For instance, the superposition of two one-dimensional distributions of these types yields the modulated structure exemplified in Fig. 85.

All "partially ordered" modulated structures described minimize the strain energy because they provide the redistribution of the coherent "intensity"

$|\langle\tilde{c}(\mathbf{k})\rangle|^2$ to ensure nonzero values along the "soft" $\langle 100\rangle$ directions only. The residual intensity in the other parts of the **k**-space related to the diffuse intensity $\langle|\delta\tilde{c}(\mathbf{k})|^2\rangle$ (short-range order) is neglected within the approximation (10.5.7). As for the coherent intensity, it is proportional to the squared moduli of the Fourier transforms $|\tilde{c}_{[100]}(\mathbf{k})|^2$, $|\tilde{c}_{[010]}(\mathbf{k})|^2$, $|\tilde{c}_{[001]}(\mathbf{k})|^2$ which do not vanish along the [100], [010], and [001] "soft" directions, respectively.

The larger these intensities, the lower the strain energy value. Starting from this, we arrive at the conclusion that strain-induced coarsening is a sequence of transformations from the initial random distribution of precipitates through a partially ordered basketlike structure (Fig. 85) to the stable one-dimensional structure (Fig. 75). The two-dimensional modulated metastable structure depicted in Fig. 82 may also be formed during the coarsening process.

It should be noted that the computer simulation of kinetics of strain-induced coarsening in a pseudo-two-dimensional coherent mixture of two cubic phases carried out by Wen, Morris, and Khachaturyan (234) seems to confirm the conclusions made above (see Fig. 109 in Section 12.5).

10.6. MORPHOLOGY OF MODULATED STRUCTURES IN TWO-PHASE COHERENT MIXTURES OF CUBIC PHASES: COMPARISON WITH EXPERIMENT

During the past two decades, a wealth of experimental data on the morphologies of modulated structures in two-phase cubic alloys have been collected due to progress in the electron microscopic technique. These data seem to provide material sufficient for verification of the theoretical mechanisms of formation of modulated structures.

The first, and somewhat general, conclusion that can be drawn from electron microscopy is that modulated structures cannot be described by single sine waves and are not in fact periodic, as assumed by Daniel and Lipson (196) to explain the side-band effect in x-ray diffraction patterns. For instance, satellite reflections in the vicinity of the fundamental spots were observed for nonperiodic, though regular, basket-weave structures by Livak and Thomas (212), Butler and Thomas (65) in Cu-Ni-Fe, Enami et al. (37) in Ni-Al, Kubo and Wayman in β-CuZn (35) (see Fig. 86). This effect can be accounted for if the spectrum of separation distances between precipitates has a sharp maximum at a certain distance that corresponds to the intersatellite spacing in the diffraction pattern (if the Patterson function of the system of precipitate particles has a sharp maximum corresponding to some separation distance).

The observation that satellites fall on the $\langle 100\rangle$ directions in the vicinity of the fundamental reflections in all cases studied is also easy to explain. It has been shown in Section 10.5 that satellite intensities are proportional to the quantity $|\langle\tilde{c}(\mathbf{k})\rangle|^2$. The appearance of satellites on the $\langle 100\rangle$ directions of the reciprocal lattice therefore shows the maxima of the function $|\langle\tilde{c}(\mathbf{k})\rangle|^2$ to fall on those directions. This fully agrees with the strain energy minimum condition

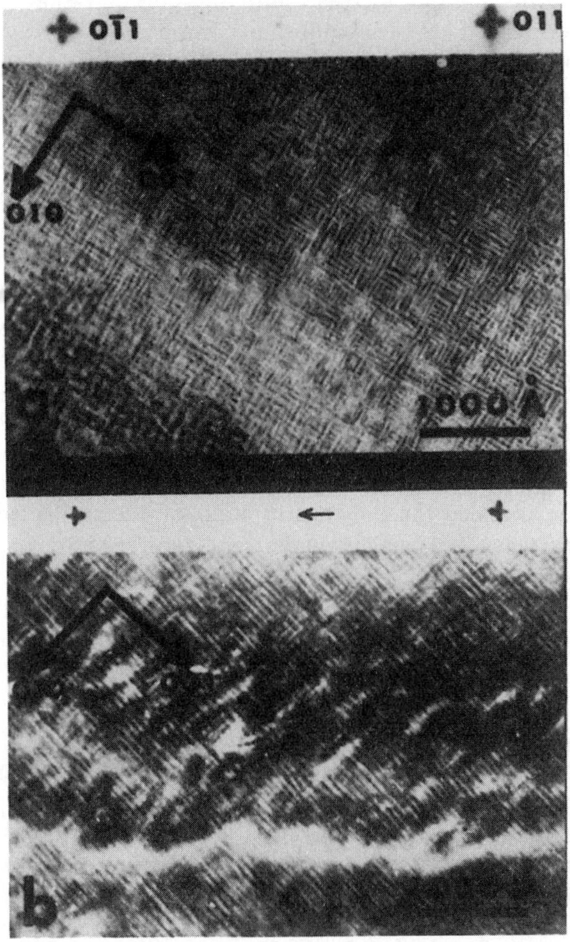

Figure 86. Electron micrographs of the basket-weave structure in β-CuZn. The bands at the top of these pictures show the side-band effect near the $(0\bar{1}1)$ and (011) diffraction spots on the diffraction pattern. (Courtesy of C. M. Wayman, H. Kubo, and I. Cornelis.)

which requires that the strain energy (10.5.7) assume its minimum value when the function $|\langle\tilde{c}(\mathbf{k})\rangle|^2$ vanishes everywhere except at the "soft" $\langle 100 \rangle$ directions.

The second conclusion concerns Cahn's theory of spinodal decomposition (42, 43). This theory describes the initial stage of aging under the spinodal curve as characterized by a sine wave concentration fluctuation that manifests itself by a side-band diffraction effect along the "soft" cubic directions. Computer simulation results prove that this process occurs, but only during a period of time commensurate with the time of an elementary diffusion event (see Section 5.4). Further in the process, as the concentration fluctuations grow, they develop a square concentration profile and thus can no longer be described by a linear equation. The fluctuations increase until the concentration profile maxima and minima correspond to the equilibrium compositions (see Figs. 50). According to the computer simulation results described in Section 5.4, the nonlinear stage of the decomposition is a comparatively fast process. It takes about 0.1 of elementary diffusion event times. The succeeding stage is the coarsening process which goes at a far lower rate to produce the modulated structures.

These theoretical conclusions are in agreement with the experimental results. According to Butler and Thomas, the development of the initial composition fluctuations in 51.5Cu-33.5Ni-15.0Fe is in fact a very rapid process (65). These fluctuations fit a square wave concentration profile better than a sine wave profile. They have large amplitudes (over 82 percent of the composition change occurs during 1 min at 625° C whereas structure changes are still noticeable for at least 40 to 200 hrs). Livak and Thomas (212) reported the decomposition of the asymmetrical 64-Cu-27Ni-9Fe alloy to occur approximately as described in Sections 9.4 and 10.2 to 10.5. According to these authors specimens that annealed for up to 10 hrs still contained a partially decomposed matrix. Their microstructure revealed the presence of cuboidal inclusions (see Section 9.4) and rodlike particles. The particles underwent transformation to $\{100\}$ platelets in about 100 hrs.

An important point in connection with this discussion of experimental results is that the effect of strain-induced interactions between new phase particles on the two-phase alloy morphology will be stronger the lower the ratio r_1/L_t in Eq. (10.3.36) (the larger the crystal lattice mismatch, the smaller the interphase energy $\gamma_{(001)}$, the larger the complex size L_t). In fact the microstructure formed in spinodal decomposition of Cu-Ni-Fe alloys, which are characterized by a small crystal lattice mismatch, consists of rodlike blocks separated by a diffuse interphase at the early stage of aging. In the later stages of the decomposition this microstructure developed into a distinct sandwichlike structure (65). In the case of β-CuZn alloys, which are characterized by a very large crystal lattice mismatch (ε_0 of about 0.068), well-defined structures composed of regular arrays of platelike particles with the $\{100\}$ habit were observed from the very beginning of the decomposition.

Electron microscopic studies proved the existence of stable one-dimensional sandwichlike modulated domain structures in cubic alloys (see Sections 10.1 and 10.3). Stable one-dimensional structures of the type shown in Figs. 76 and 87 were in fact observed by Biederman and Kneller in Cu-Ni-Fe (197), by Warlimont in Fe-Si (217), by Kúbo and Wayman in nonstoichiometric β-CuZn (35). An especially perfect one-dimensional modulated structure was observed in β-CuZn alloys (35) where the crystal lattice mismatch is large enough for the contribution from the strain energy to the thermodynamics and kinetics of the phase transformation to be significant. Figure 76 reproduces an electron microscopic image of a group of complexes of various orientations with respect to the crystallographic axes taken from a β-CuZn alloy.

Among two-dimensional modulated structures, the morphology of high coercive ALNICO alloys showing the best magnetic properties has seemingly

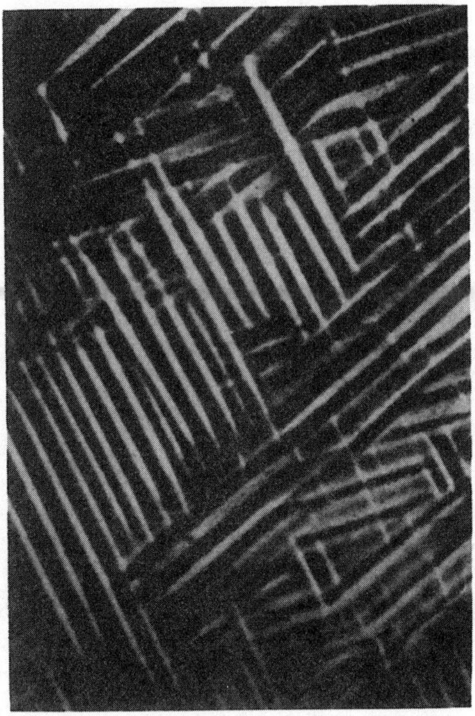

Figure 87. Electron micrograph of complexes of one-dimensional modulated structures in a Fe-Si alloy (217). ×35,000. (Courtesy of H. Warlimont.)

been studied in more detail than any other structure of this type. We shall now compare the results obtained in these studies with the theoretical predictions of Section 10.4.

According to De Vos (209), the two-dimensional modulated structure of ALNICO alloys is formed by periodically distributed rodlike newphase precipitates elongated in the [001] direction (see Fig. 88). In the cross section of the modulated structure by the (001) plane, the array of rodlike precipitates looks like a periodic quadratic lattice with the basic translations along the [100] and [010] axes (Fig. 88a). This perfect modulated structure is formed by annealing in a permanent magnetic field directed along the [001] axis. The magnetic field stabilizes the two-dimensional modulated structure and provides the formation of complexes with rodlike precipitates directed along the [001] axis only. This seems to be the major reason why strain-induced coarsening under magnetic field results in the formation of large-scale and therefore fairly perfect modulated structures. It is also of interest that the two-dimensional structure displayed in Fig. 88 provides the highest coercive force: rodlike shapes of the magnetic phase precipitates inhibit alternating magnetization.

Comparison of the electron microscopic photographs corresponding to the (001) and (100) planes (Fig. 88a and b) with the (001) and (100) cross sections of

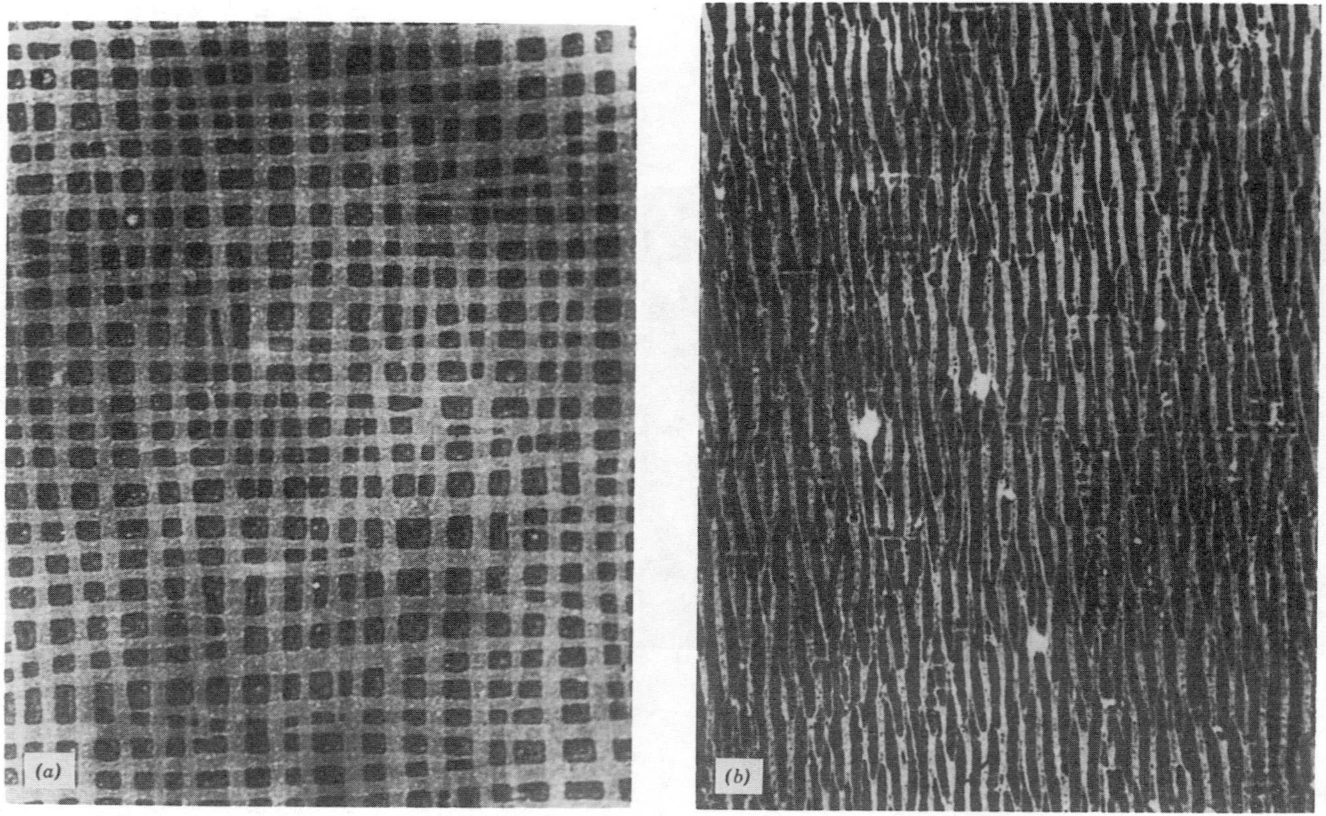

Figure 88. Electron micrographs of the two-dimensional modulated structure of a single crystal of ALNICO 8 after isothermal heat treatment in magnetic field (209). (*a*) (001) plane perpendicular to the field; (*b*) (100) plane parallel to the field. ×50,000. (Courtesy of K. J. de Vos.)

the complex schematically drawn in Fig. 82 substantiates the validity of the theory described in Section 10.3. Electron microscopic observations clearly demonstrate that the formation of the modulated structure occurs in strain-induced coarsening of randomly distributed precipitates, in agreement with the theoretical predictions of Section 10.5 (see Fig. 89).

The application of the transmission electron microscopy makes it possible to determine the structure of ALNICO alloys in even greater detail (210, 218, 219). Figure 90 reproduces the bright-field electron microscopic images of the two-dimensional modulated structure in the (100) and (001) planes. It shows distinctly that the two-dimensional pseudoperiodic structure comprises three phases. Two phases differ strongly in their compositions (white and black fields in Fig. 90a), and the third one is of intermediate composition (grey fields linking "white" phase sites of the two-dimensional quadratic lattice, compare with Fig. 82). As shown in Section 10.3, the composition of the third phase is $(c_1 + c_2)/2$ where c_1 and c_2 are the compositions of the "white" and "black" phases. The three-phase structures were observed in two-dimensional modulated structures

Figure 89. Electron micrograph of the two-phase structure of a single crystal of Ticonal after thermal treatment at 1240°C for 20 min and at 900°C for $\frac{1}{2}$ hr (209). ×50,000. (Courtesy of K. J. de Vos.)

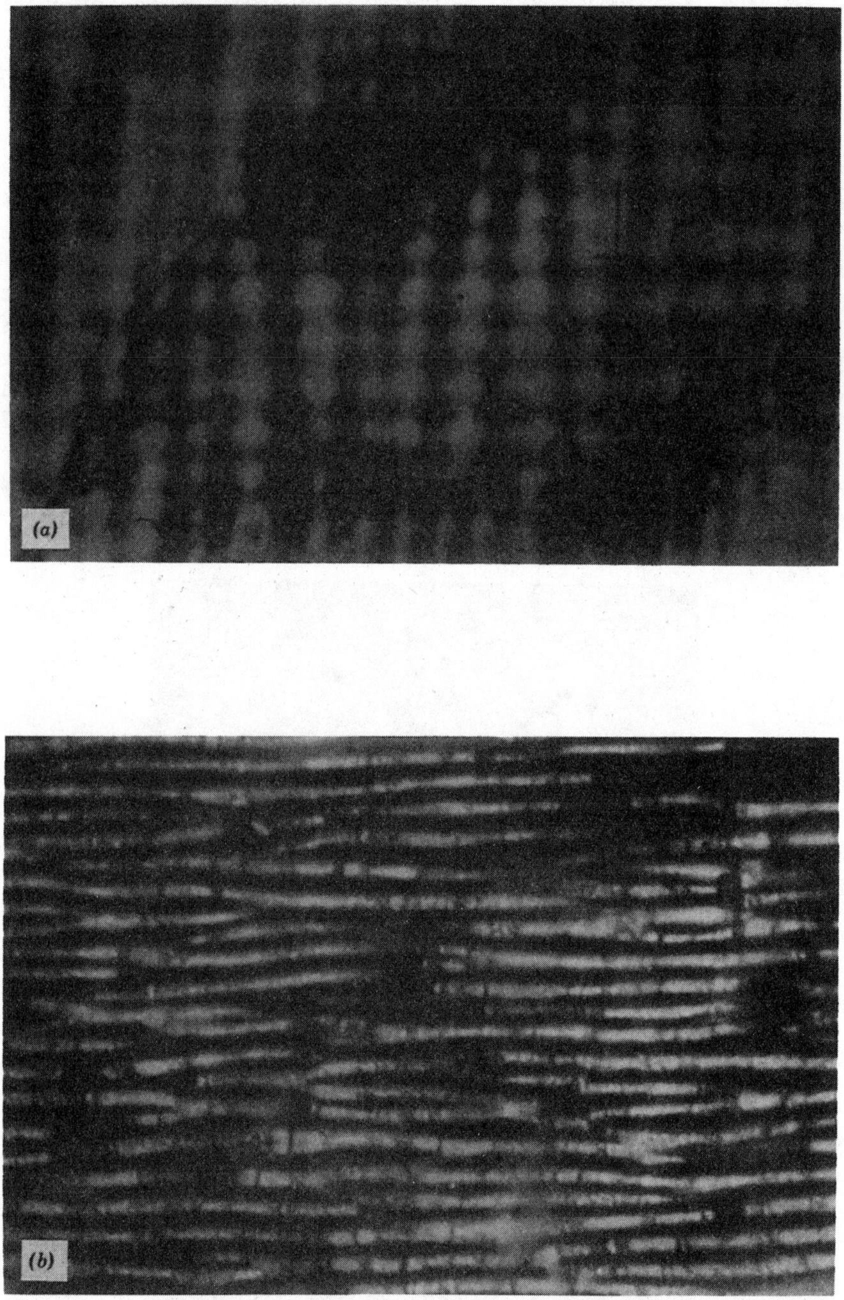

Figure 90. Bright-field electron micrographs of the two-dimensional modulated structure in a thin foil of ALNICO after thermomagnetic treatment. (*a*) The (001) thin foil orientation is normal to the magnetic field; (*b*) the (100) thin foil orientation is parallel to the magnetic field. ×60,000. (Courtesy of E. G. Knizhnik.)

by several authors, notably by de Vos (209). The appearance of the third-phase reflections was reported by Sumin and et al. (220). Their location with respect to the matrix phase reflections is in excellent agreement with the theoretical predictions (221).

The magnetic measurements also demonstrate the existence of the third phase with a Curie temperature between the Curie temperatures of the magnetic and almost nonmagnetic phases (222). The three-phase structure is an important argument in favor of the theory given in Section 10.4. It confirms its basic statement that two-dimensional structures are superpositions of one-dimensional distributions.

The strain-induced stabilization of the parent phase (third phase) in a two-dimensional modulated structure is in fact a manifestation of a thermoelastic equilibrium between the parent and new phases similar to the thermoelastic equilibrium observed in martensitic alloys. Cooling of the decomposed alloy also shifts the thermoelastic equilibrium on the side of the new phases, exactly as was the case with martensitic alloys. Bearing this in mind, we may conclude that cyclic heating and cooling of a two-dimensional modulated structure results in the appearance and disappearance of the third intermediate (parent) phase, as in martensitic structures. This was observed in experiments with ALNICO alloys carried out by Povolotsky (222) and Sumin et al. (220). Cooling of two-dimensional structures was reported to result in the absorption of the third phase rods by the first two phases without affecting the two-dimensional nature of the modulated structure.

As mentioned in Section 10.4, further coarsening of two-dimensional struc-

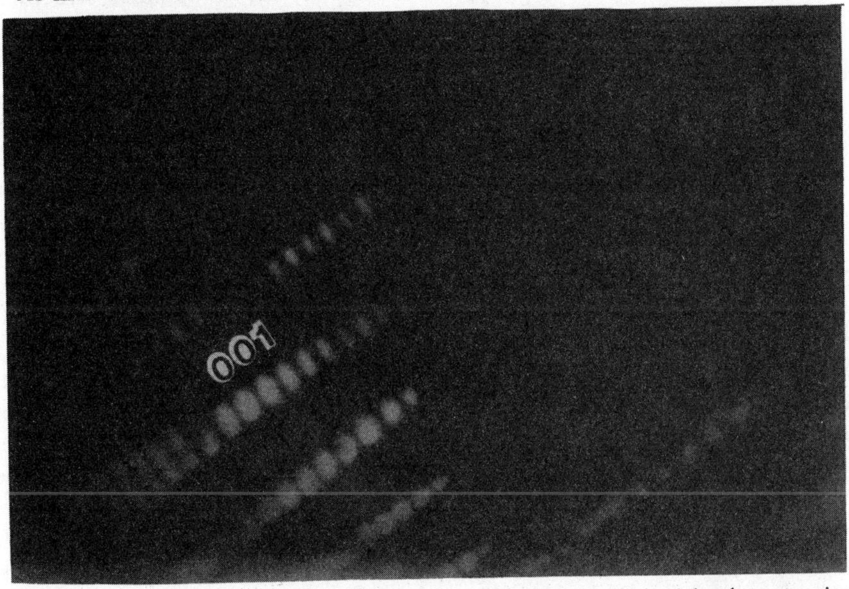

Figure 91. Bright-field electron micrographs from the two-dimensional modulated structure in a thin foil of ALNICO after secondary decomposition. The (100) thin foil orientation is parallel to the magnetic field. (Courtesy of E. G. Knizhnik.)

tures leads to a secondary decomposition and a formation of the three-dimensional two-phase modulated structures shown in Fig. 83. Electron microscopic studies of the late stages of the decomposition of ALNICO alloys confirm the occurrence of secondary decomposition (211, 219). Figure 91 demonstrates that intermediate phase rods of the two-dimensional structure undergo secondary decomposition in the later stages of aging to give sandwiches of alternating (001) platelets of two equilibrium phases [compare Fig. 91 with the (100) cross section in Fig. 83). All platelets have approximately the same thickness. This is in agreement with the value $\gamma_3 = \frac{1}{2}$ theoretically predicted in Eq. (10.4.16).

At the end of Section 10.3 the necessary condition given for the formation of a well-defined modulated structure is a large crystal lattice mismatch between two equilibrium phases. This suggests a new possibility for controlling the formation of modulated structures. Adding an alloying component with an atomic radius considerably exceeding that of the solvent should lead to an increase in crystal lattice mismatch because of the higher solubility of the alloying element in one of the coexisting phases. Such an increase would favor the modulated structure. Titanium may serve as example: added to ALNICO makes the two-dimensional structure more elaborate and thus improves the magnetic properties of the alloy.

Finally, the electron microscopic studies by Butler and Thomas (65), Kubo and Wayman (35), and Bouchard and Thomas (223) also confirm the existence of basketlike modulated structures derived from the strain energy minimum condition in Section 10.5. The typical example of such modulated structures observed in Cu-Ni-Fe (65), Cu-Mn-Al (223), and β-CuZn (35) is shown in Fig. 86. In agreement with theoretical predictions, these structures give rise to satellites in the "soft" $\langle 100 \rangle$ directions around the fundamental spots in the diffraction patterns. This is also in agreement with the results of the computer simulation of strain-induced coarsening which, as it will be shown in Chapter 12, predict the formation of metastable basketlike structures that give side-band reflections (see Fig. 111).

In comparing the theoretical results of Sections 10.2 through 10.5 with the electron microscopic data, one should note that the theory of modulated structures described in those sections is essentially that of thermodynamics. It excludes the problem concerning the sequence of structures preceding the stable one-dimensional sandwichlike distribution. We cannot, for instance, infer from that theory whether metastable two- and three-dimensional structures should arise through intermediate states or whether the stable one-dimensional structure may be formed directly. Moreover there is no guarantee that the structures discussed are the only metastable states possible. One cannot rule out the formation of other metastable structures exhibiting more complex morphologies. All these structures should, however, fit the strain energy minimum condition, and therefore the satellite reflections should be along the "soft" $\langle 100 \rangle$ directions.

Yet, despite its thermodynamic nature, certain conclusions about alloy morphologies may be made from consideration of strain energy theory. It seems probable that a random distribution of precipitates in the as-quenched state is transformed in time into a partially ordered basketlike structure depicted in Fig. 85 that develops eventually into a stable sandwichlike one-dimensional modulated structure (see Fig. 75). Under certain kinetic conditions we may also expect the formation of the two- and three-dimensional distributions described in Sections 10.4 and 10.5.

Electron microscopic studies (223) confirm these conclusions. Figure 92 shows as an example the evolution of modulated structures in Cu-Mn-Al (223). The initial stage of the decomposition gives a random distribution of equiaxial precipitates. This structure undergoes further transformations to give a three-dimensional basketlike structure and later the one-dimensional modulated structure. If the crystal lattice mismatch is large, the modulated structure may be formed in one step.

It should be remembered that the driving force of strain-induced coarsening is practically reduced to zero upon the formation of one of the modulated structures described in Sections 10.2 through 10.5. That is the reason why strain-induced rearrangement of random distributions to modulated structures is a far faster process than further transformations of first-formed partially ordered arrays.

As shown at the end of Section 10.3, under certain conditions the modulated structure period λ_0 may remain constant at the early stage of aging, its magnitude being a thermodynamic function of temperature and composition. This effect was, in particular, observed by Livak and Thomas [see Figs. 1 and 2 in (212)]. These authors also reported a very interesting related phenomenon. They found the modulated structure period to remain constant until the equilibrium magnetic Curie temperature was attained. This finding is easy to explain in terms of the mechanism advanced in the end of Section 10.3 to account for the constancy of the modulated structure period in the early stage of aging.

In fact it is easy to see that the equilibrium Curie temperature is attained when concentration heterogeneities formed in the decomposition can be regarded as macroscopic new phase precipitates. According to Eq. (10.3.39) an increase of the period λ_0 occurs when the thickness of heterogeneities becomes sufficiently large for heterogeneities to be considered macroscopic new phase precipitates. Therefore, both effects, the attainment of the equilibrium Curie temperature and growth of the modulated structure period should occur simultaneously.

After that stage the modulated structure period increases monotonically, which manifests itself by a decrease in spacings between the satellites and fundamental reflections in diffraction patterns during aging. An explanation of this phenomenon is rather simple. It involves dissolution of "extra planes" of the macrostructure formed by new phase platelets or rods and, accordingly, growth of the other planes. By varying the distance between the extra planes undergoing dissolution, we may practically monotonically vary the modulated structure period. This is analogous to the dislocation climb. Such "dislocations" were seemingly detected in the works (209, 219).

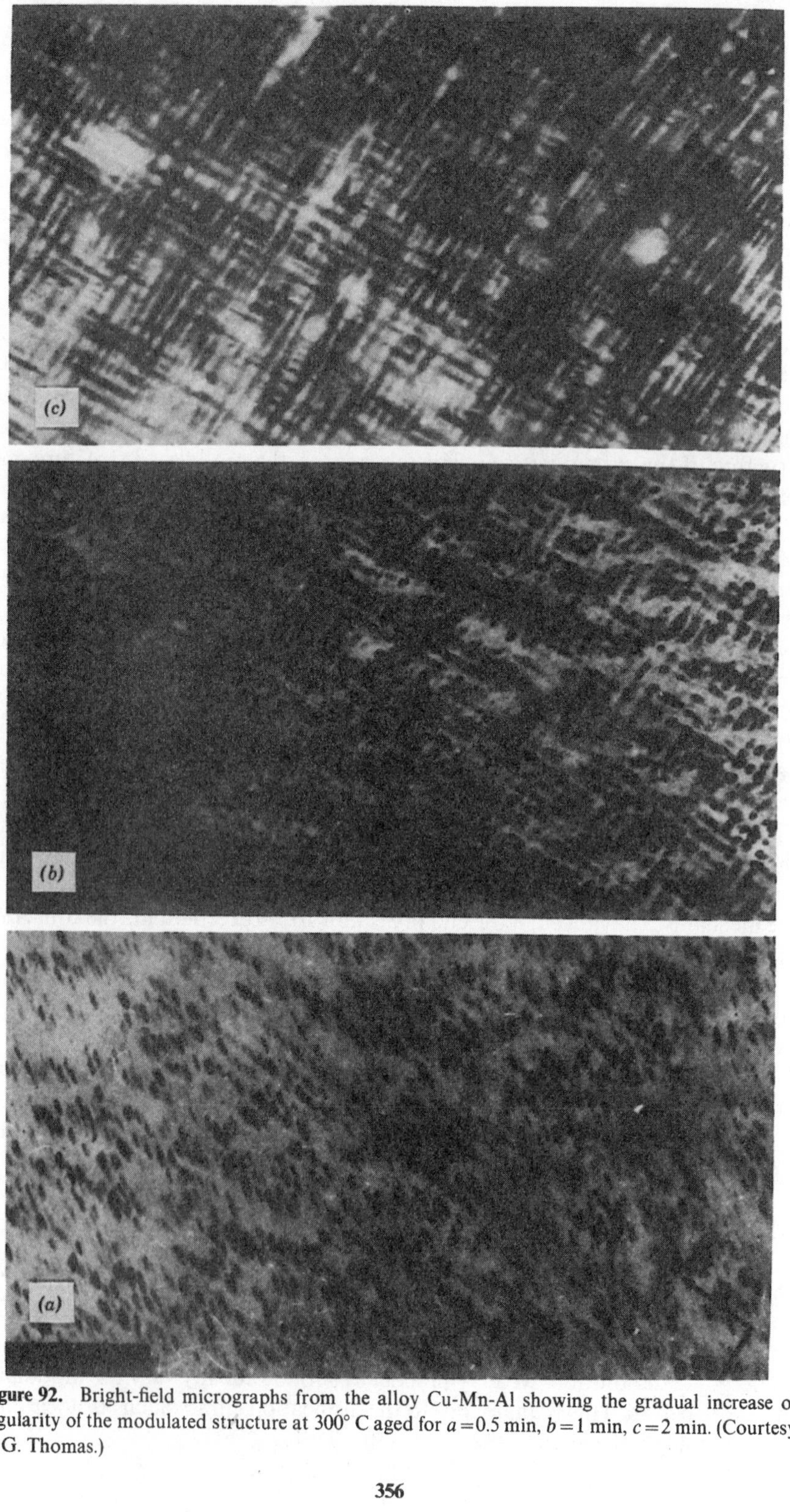

Figure 92. Bright-field micrographs from the alloy Cu-Mn-Al showing the gradual increase of regularity of the modulated structure at 300° C aged for $a = 0.5$ min, $b = 1$ min, $c = 2$ min. (Courtesy of G. Thomas.)

When the modulated structure grows coarse enough, another mechanism of strain relaxation becomes operative. This mechanism involves disarrangement of coherent interphases by formation of misfit dislocations. The replacement of coherent interphases with semicoherent interphases reduces the bulk strain energy but increases the effective interphase energy by the contribution from the dislocation cores. The loss of coherency thus leads to a total energy decrease only if new phase particles forming a modulated structure are large enough for the strain energy decrease proportional to the particle volume to outweight the interphase energy increase proportional to the interphase area. A detailed electron microscopic study of formation of semicoherent interphases in the latest stages of aging was undertaken, for example, in (65).

A small point may be mentioned in passing. The disappearance of the slight tetragonality of constrained phases composing the modulated structures is associated with strain relaxation in the transition to a semicoherent interphase. The removal of the tetragonality caused by the loss of coherency was reported in a number of papers (180, 208, 201, 224, 65).

Returning to the coherent stage of the decomposition, we should dwell on the effect of the volume fractions of coexisting phases on the morphology of two-phase coherent mixtures. According to the experimental data on Cu-Ni-Fe reported by Livak and Thomas (212), modulated structures with well-defined periodicity are formed, seemingly, when the volume fractions of the coexisting phases are approximately the same, when the alloy composition corresponds to the middle of the miscibility gap. This conclusion is in agreement with the theoretical predictions. Indeed the material constant r_1 having the length dimension, previously defined by Eq. (10.3.34), may be written as

$$r_1 = \frac{16\pi^3 \gamma_{(001)}}{(c_{11}-c_{12})\varepsilon_0^2(c_2^0-c_1^0)^2|\zeta|\alpha(\gamma_1)[(c_{11}+2c_{12})/c_{11}]^2} \tag{10.6.1}$$

where γ_1 is the volume fraction of the first phase, c_1^0 and c_2^0 are the compositions of the first and second phases, respectively.

According to the lever rule,

$$\gamma_1 = \frac{c_2^0 - \bar{c}}{c_2^0 - c_1^0} \tag{10.6.2}$$

where $\bar{c}$ is the alloy composition. It follows from Fig. 78 that $\alpha(\gamma_1)$ approaches zero at $\gamma_1 \to 0$, and $\alpha(\gamma_1)$ reaches its maximum at $\gamma_1 = \frac{1}{2}$, when both phases have the same volume fractions. Therefore r_1 given by (10.6.1) increases infinitely when γ_1 approaches zero and, conversely, becomes minimal at $\gamma_1 = \frac{1}{2}$. According to (10.6.2), the first situation is realized with solutions near the solubility limits, and the second one corresponds to a solution in the middle of the miscibility gap (a symmetrical case).

The criterion for the formation of a modulated structure [see Eq. (10.3.36)] reads

$$\sqrt{\frac{r_1}{L_t}} \ll 1 \tag{10.6.3}$$

Hence, with r_1 approaching infinity at $\gamma_1 \to 0$, the inequality (10.6.3) may prove violated. It thus follows that modulated structures cannot be formed in alloys near the solvus curve where $\bar{c} \approx c_1$ and, consequently, $r_1 \to \infty$. Instead, individual precipitate particles of the first phase appear. Strain-induced coarsening causes segregation of the particles into platelike (001) layers. This phenomenon was in fact observed in a Ni-Al alloy by Ardell and Nicholson (177).

10.7. STRAIN-INDUCED TETRAGONALITY OF CUBIC PHASES IN MODULATED STRUCTURES: Cu-Ni-Fe AND ALNICO ALLOYS

X-ray studies of heterogeneous cubic alloys that show the side-band effect in the diffraction patterns have demonstrated the succession of the decomposition stages to be as follows. First comes the satellite stage, then the stage of coexistence of slightly tetragonal phases, and, eventually, the equilibrium stage (the formation of semicoherent mixtures of two cubic phases). According to various authors, phases composing modulated structures owe their tetragonality to elastic strain generated by a coherent fitting together of coexisting cubic phases which have different crystal lattice parameters (180, 208, 201, 224, 65). A highly convincing argument in favor of this assumption has been found by Linetskii (225) and Lifshitz, Linetskii, and Milyaev (224) who studied ALNICO alloys to show that the observed tetragonality is associated with elastic strain and cannot be interpreted as an intrinsic property of the phases. According to Linetskii, the tetragonal β_2 phase turned cubic after the electrolytic extraction (225). Since the tetragonality is a strain-induced phenomenon resulting from coherent conjugation of cubic phases, the tetragonality ratios and orientation relations of the phases are determined by the modulated structure morphology.

The theoretical results cited in Section 10.2 make it possible to calculate the homogeneous strain within the phases forming a modulated structure and thus to predict the strain-induced tetragonality ratios and orientation relations of the phases.

According to Eq. (10.2.33), the alternating new phase platelets composing one-dimensional modulated structures (see Fig. 75) have tetragonal crystal lattices. The tetragonal distortion is given by

$$\varepsilon_{ij}(\mathbf{r}) = \frac{c_{11}+2c_{12}}{c_{11}}\varepsilon_0 \begin{pmatrix} 0 & 0 & 0 \\ 0 & 0 & 0 \\ 0 & 0 & 1 \end{pmatrix} c_1^{[001]}(z) \tag{10.7.1}$$

As shown in Section 10.3,

$$c(\mathbf{r}) = \bar{c} + c_1^{[001]}(z) = \begin{cases} c_1^0 & \text{if the coordinate } z \text{ of the point } \mathbf{r} \\ & \text{is within a first-phase platelet} \\ c_2^0 & \text{if the coordinate } z \text{ of the point } \mathbf{r} \\ & \text{is within a second-phase platelet} \end{cases} \tag{10.7.2}$$

Using (10.7.2) in Eq. (10.7.1), we find that the tetragonal distortions of the two phases comprising the modulated structure are

$$\varepsilon_{33}(\mathbf{r})=\begin{cases}\dfrac{c_{11}+2c_{12}}{c_{11}}\varepsilon_0(c_1^0-\bar{c}) & \text{for first-phase platelets}\\[2ex] \dfrac{c_{11}+2c_{12}}{c_{11}}\varepsilon_0(c_2^0-\bar{c}) & \text{for second-phase platelets}\end{cases} \tag{10.7.3}$$

Let the crystal lattice parameter of the stress-free cubic matrix phase be a. We can then calculate the slight strain-induced tetragonality of the two co-existing phases from Eq. (10.7.3):

$$c(1)=a(1+\varepsilon_{33}(z))=a\left[1+\frac{c_{11}+2c_{12}}{c_{11}}\varepsilon_0(c_1^0-\bar{c})\right]$$

$$a(1)=a(1+\varepsilon_{11}(z))=a \tag{10.7.4}$$

for the first phase and

$$c(2)=a\left[1+\frac{c_{11}+2c_{12}}{c_{11}}\varepsilon_0(c_2^0-\bar{c})\right]$$

$$a(2)=a \tag{10.7.5}$$

for the second one where $c(1)$, $a(1)$ and $c(2)$, $a(2)$ are the crystal lattice parameters of the first and second phases, respectively. It should be stressed that, as follows from Eqs. (10.7.4) and (10.7.5), the crystal lattice parameters $a(1)$ and $a(2)$ of the coexisting phases are equal to each other

$$a(1)=a(2)=a \tag{10.7.6}$$

The tetragonal axes of both phases are directed along [001].

Eqs. (10.7.4) and (10.7.5) may be simplified if the stress-free strains $\varepsilon_0(c_2^0-\bar{c})$ and $\varepsilon_0(c_1^0-\bar{c})$ related to the tetragonal phases are expressed in terms of the crystal lattice parameters of the stress-free cubic phases involved in the modulated structure. Let the crystal lattice parameters of the first- and second-cubic phases in the stress-free state be

$$a_0(1) \quad \text{and} \quad a_0(2)$$

According to the definition of the stress-free transformation strain, we then have

$$\varepsilon_0(c_1^0-\bar{c})=\frac{a_0(1)-a}{a}$$

$$\varepsilon_0(c_2^0-\bar{c})=\frac{a_0(2)-a}{a} \tag{10.7.7}$$

Excluding the terms $\varepsilon_0(c_1^0-\bar{c})$ and $\varepsilon_0(c_2^0-\bar{c})$ from Eqs. (10.7.4) and (10.7.5), and using the definition (10.7.7), we obtain

$$c(1)=a+\frac{c_{11}+2c_{12}}{c_{11}}(a_0(1)-a)=a_0(1)+2\frac{c_{12}}{c_{11}}(a_0(1)-a)$$

$$c(2)=a+\frac{c_{11}+2c_{12}}{c_{11}}(a_0(2)-a)=a_0(2)+2\frac{c_{12}}{c_{11}}(a_0(2)-a) \tag{10.7.8}$$

for the crystal lattice parameters and

$$\frac{c(1)}{a(1)}=\frac{a+[(c_{11}+2c_{12})/c_{11}][a_0(1)-a]}{a}=1+\frac{c_{11}+2c_{12}}{c_{11}}\frac{a_0(1)-a}{a}$$

$$\frac{c(2)}{a(2)}=\frac{a+[(c_{11}+2c_{12})/c_{11}][a_0(2)-a]}{a}=1+\frac{c_{11}+2c_{12}}{c_{11}}\frac{a_0(2)-a}{a} \tag{10.7.9}$$

for the axial ratios.

The theoretical relations (10.7.8) and (10.7.9) may be compared with the x-ray diffraction data.

10.7.1. Modulated Structure in Cu-Ni-Fe Alloys

Geisler and Newkirk studied aging of Cu-Ni-Fe alloys and found that the "satellite" stage of the decomposition was succeeded by the stage characterized by the appearance of reflections from two slightly tetragonal phases (226). Measurements of the crystal lattice parameters of these tetragonal phases, $a(1)$ and $a(2)$, gave the same value equal to the parent phase parameter a:

$$a(1)=a(2)=a=3.56\ \text{Å}$$

This agrees with the theoretical predictions (10.7.6) for the one-dimensional modulated structure. The observed axial ratios were $c(1)/a(1)=0.98$ and $c(2)/a(2)=1.02$.

According to the experimental data, the first and second equilibrium phases are cubic in the stress-free state and have the crystal lattice parameters $a_0(1)=3.53$ Å and $a_0(2)=3.59$ Å, respectively. The experimental data on the constrained tetragonal phases, respective stress-free cubic phases and matrix in the stress-free state are as follows

$$\frac{c(1)}{a(1)}=0.98 \quad a(1)=3.56\ \text{Å}$$

$$\frac{c(2)}{a(2)}=1.02 \quad a(2)=3.56\ \text{Å} \tag{10.7.10}$$

and

$$a_0(1)=3.53\ \text{Å}$$
$$a_0(2)=3.59\ \text{Å}$$
$$a=3.56\ \text{Å} \tag{10.7.11}$$

respectively. Using the experimental data on the cubic phases in the stress-free state (10.7.11) and the Ni elastic constants, $c_{11}=1.684\times10^{12}$ dyne/cm^2, $c_{12}=1.214\times10^{12}$ dyne/cm^2 taken to represent the elastic constants of Cu-Ni-Fe alloys, we obtain from Eq. (10.7.9):

$$\frac{c(1)}{a(1)}=1+\frac{1.684+2\times 1.214}{1.684}\cdot\frac{3.53-3.56}{3.56}=0.9794$$

$$\frac{c(2)}{a(2)}=1+\frac{1.684+2\times 1.214}{1.684}\cdot\frac{3.59-3.56}{3.56}=1.0205 \qquad (10.7.12)$$

The calculated axial ratios (10.7.12) well compare with the observed ones, (10.7.10).

The procedure for calculating (10.7.12) does not involve any fitting variable (all the parameters used are taken from independent measurements), and the agreement between the theoretical, (10.7.12), and experimental, (10.7.10), results should be considered an excellent one.

10.7.2 Crystal Lattice Parameters of Two-dimensional Structures

The elastic strain in the two-dimensional modulated structure depicted in Fig. 82 is described by Eq. (10.2.34):

$$\varepsilon_{ij}(\mathbf{r})=\frac{c_{11}+2c_{12}}{c_{11}}\varepsilon_0\left\{\begin{pmatrix}1&0&0\\0&0&0\\0&0&0\end{pmatrix}c_2^{[100]}(x)+\begin{pmatrix}0&0&0\\0&1&0\\0&0&0\end{pmatrix}c_2^{[010]}(y)\right\}$$

(10.7.13)

As shown in Section 10.4, two-dimensional modulated structures consist of periodically packed rods extended along the [001] direction and having the compositions c_1, c_2, and $(c_1+c_2)/2$. The corresponding concentration profile is a superposition of two one-dimensional modulations $c_2^{[100]}(x)$ and $c_2^{[010]}(y)$ which assume the values

$$\begin{aligned}c_2^{[100]}(x)&=\tfrac{1}{2}(c_1-\bar{c})\\ c_2^{[010]}(y)&=\tfrac{1}{2}(c_1-\bar{c})\end{aligned} \qquad (10.7.14)$$

within rods of the first phase having the composition c_1, and

$$\begin{aligned}c_2^{[100]}(x)&=\tfrac{1}{2}(c_2-\bar{c})\\ c_2^{[010]}(y)&=\tfrac{1}{2}(c_2-\bar{c})\end{aligned} \qquad (10.7.15)$$

within rods of the second phase having the composition c_2. Within the intermediate phase rods of the composition $(c_1+c_2)/2$ the functions $c_2^{[100]}(x)$ and $c_2^{[010]}(y)$ have the values

$$\begin{aligned}c_2^{[100]}(x)&=\tfrac{1}{2}(c_1-\bar{c})\\ c_2^{[010]}(y)&=\tfrac{1}{2}(c_2-\bar{c})\end{aligned} \qquad (10.7.16a)$$

within intermediate phase rods of the first kind, and

$$\begin{aligned} c_2^{[100]}(x) &= \tfrac{1}{2}(c_2 - \bar{c}) \\ c_2^{[010]}(y) &= \tfrac{1}{2}(c_1 - \bar{c}) \end{aligned} \tag{10.7.16b}$$

within intermediate phase rods of the second kind. Substitution of the values (10.7.14) and (10.7.15) into Eq. (10.7.13) gives the homogeneous strain within the first- and second-phase rods:

$$\varepsilon_{ij}(1) = \frac{c_{11} + 2c_{12}}{2c_{11}} \varepsilon_0 (c_1 - \bar{c}) \begin{pmatrix} 1 & 0 & 0 \\ 0 & 1 & 0 \\ 0 & 0 & 0 \end{pmatrix} \tag{10.7.17a}$$

and

$$\varepsilon_{ij}(2) = \frac{c_{11} + 2c_{12}}{2c_{11}} \varepsilon_0 (c_2 - \bar{c}) \begin{pmatrix} 1 & 0 & 0 \\ 0 & 1 & 0 \\ 0 & 0 & 0 \end{pmatrix} \tag{10.7.17b}$$

respectively. The strain within intermediate phase rods may also be determined from Eq. (10.7.13) after substituting the values (10.7.16a) and (10.7.16b) into this equation:

$$\varepsilon'_{ij}(3) = \frac{c_{11} + 2c_{12}}{2c_{11}} \varepsilon_0 \begin{pmatrix} c_1 - \bar{c} & 0 & 0 \\ 0 & c_2 - \bar{c} & 0 \\ 0 & 0 & 0 \end{pmatrix} \tag{10.7.18}$$

and

$$\varepsilon''_{ij}(3) = \frac{c_{11} + 2c_{12}}{2c_{11}} \varepsilon_0 \begin{pmatrix} c_2 - \bar{c} & 0 & 0 \\ 0 & c_1 - \bar{c} & 0 \\ 0 & 0 & 0 \end{pmatrix} \tag{10.7.19}$$

respectively.

The conclusion follows that the elastic strain (10.7.17) makes the first phase slightly tetragonal. The corresponding crystal lattice parameters are

$$\begin{aligned} c(1) &= a(1 + \varepsilon_{33}) = a \\ a(1) &= a(1 + \varepsilon_{11}) = a\left[1 + \frac{c_{11} + 2c_{12}}{2c_{11}} \varepsilon_0 (c_1 - \bar{c})\right] \end{aligned} \tag{10.7.20}$$

Eq. (10.7.17) predicts the second phase to be also slightly tetragonal and to have the parameters

$$\begin{aligned} c(2) &= a(1 + \varepsilon_{33}) = a \\ a(2) &= a(1 + \varepsilon_{11}) = a\left[1 + \frac{c_{11} + 2c_{12}}{2c_{11}} \varepsilon_0 (c_2 - \bar{c})\right] \end{aligned} \tag{10.7.21}$$

We may see from Eqs. (10.7.18) and (10.7.19) that the homogeneous distortion within intermediate third-phase rods is orthorhombic rather than tetragonal which presupposes the formation of a slightly orthorhombic phase. The crystal lattice parameters of the strain-produced orthorhombic phase (10.7.18) are

$$c(3)=a\left[1+\frac{c_{11}+2c_{12}}{2c_{11}}\varepsilon_0(c_1-\bar{c})\right]$$

$$b(3)=a\left[1+\frac{c_{11}+2c_{12}}{2c_{11}}\varepsilon_0(c_2-\bar{c})\right]$$

$$a(3)=a \tag{10.7.22}$$

The distortion (10.7.19) leads to the same orthorhombic phase rotated by 90° about the [001] axis with respect to the phase generated by (10.7.18).

The parameters $\varepsilon_0(c_1-\bar{c})$ and $\varepsilon_0(c_2-\bar{c})$ can be excluded from Eqs. (10.7.20) through (10.7.22), exactly as with the one-dimensional structure, by applying the stress-free transformation strain definitions (10.7.7). This gives

$$c(1)=a$$

$$a(1)=a+\frac{c_{11}+2c_{12}}{2c_{11}}(a_0(1)-a)$$

$$\frac{c(1)}{a(1)}=\left[1+\frac{c_{11}+2c_{12}}{2c_{11}}\frac{a_0(1)-a}{a}\right]^{-1}\cong 1-\frac{c_{11}+2c_{12}}{2c_{11}}\frac{a_0(1)-a}{a} \tag{10.7.23}$$

for the first phase, and

$$c(2)=a$$

$$a(2)=a+\frac{c_{11}+2c_{12}}{2c_{11}}(a_0(2)-a)$$

$$\frac{c(2)}{a(2)}=\left[1+\frac{c_{11}+2c_{12}}{2c_{11}}\frac{a_0(2)-a}{a}\right]^{-1}\cong 1-\frac{c_{11}+2c_{12}}{2c_{11}}\frac{a_0(2)-a}{a} \tag{10.7.24}$$

for the second one. The tetragonal axes of both tetragonal phases are directed along the same [001] direction coinciding with the longitudinal axis of the rods.

The crystal lattice parameters of the intermediate orthorhombic phase can be obtained from (10.7.22). They may be written

$$c(3)=a+\frac{c_{11}+2c_{12}}{2c_{11}}(a_0(1)-a)$$

$$b(3)=a+\frac{c_{11}+2c_{12}}{2c_{11}}(a_0(2)-a)$$

$$a(3)=a \tag{10.7.25}$$

It follows from Eqs. (10.6.23) and (10.6.24) that the parameters c of the tetragonal phases are strictly equal to each other and to the cubic phase parameter a:

$$c(1)=c(2)=a \tag{10.7.26}$$

It should be noted that the third-phase reflections are rather difficult to detect in x-ray and electron diffraction patterns. The point is that the crystal lattice parameters of the orthorhombic intermediate phase $a(3)$, $b(3)$, $c(3)$ coincide with the parameters of the two tetragonal phases [see Eqs. (10.7.22), (10.7.20), and (10.7.21)]. This difficulty can be overcome, however, the third phase was observed in (220).

10.7.3. Modulated Structures in ALNICO Alloys

According to De Vos (209), the aging of ALNICO alloys in a permanent magnetic field results in modulated structures whose morphology coincides with that theoretically predicted in Section 10.4 (see Fig. 82). The results of the detailed single-crystal x-ray study of ALNICO alloys by Linetskii (225) and by Lifshitz, Linetskii, and Milyaev (224) may be used to compare the theoretical crystal lattice parameters with the experimental ones. Rod-shaped precipitates in the ALNICO modulated structures were shown to comprise two tetragonal phases, β and β_2 (224, 225) (the third orthorhombic phase is difficult to be detected by x-ray diffraction). The tetragonal axes of both phases are directed along the rod axes [001].

The experimental crystal lattice parameters of the tetragonal phases β and β_2 are as follows (225):

$$\begin{aligned} c_\beta &= c(1) = 2.880\ \text{Å} \\ a_\beta &= a(1) = 2.863\ \text{Å} \end{aligned} \tag{10.7.27a}$$

$$\begin{aligned} c_{\beta_2} &= c(2) = 2.880\ \text{Å} \\ a_{\beta_2} &= a(2) = 2.909\ \text{Å} \end{aligned} \tag{10.7.27b}$$

The crystal lattice parameter of the parent cubic phase in as-quenched state was found to be

$$a = 2.880\ \text{Å} \tag{10.7.27c}$$

The β_2 and β_2 phases were also shown to become cubic in the stress-free state. This was proved by extraction of these phases from the matrix (225). The experimental crystal lattice parameters of the stress-free cubic β and β_2 phases are

$$\begin{aligned} a_\beta^0 &= a_0(1) = 2.863\ \text{Å} \\ a_{\beta_2}^0 &= a_0(2) = 2.905\ \text{Å} \end{aligned} \tag{10.7.28}$$

The observed coincidence of the parameters, c, of both tetragonal phases with the cubic matrix phase spacing, a [see (10.7.27)],

$$c_\beta = c_{\beta_2} = a = 2.880\ \text{Å}$$

is in a complete agreement with the theory predictions (10.7.26). The strongest argument in favor of the theory of modulated structures is, however, the excellent quantitative agreement between the observed and calculated crystal lattice parameters, a_β and a_{β_2}, for both tetragonal phases.

With the elastic constants of ALNICO set equal to those of α-Fe,

$$c_{11}=2.335\times10^{12}\ \text{dyne/cm}^2$$
$$c_{12}=1.355\times10^{12}\ \text{dyne/cm}^2 \tag{10.7.29}$$

and, using the crystal lattice parameters of the corresponding cubic phases in the stress-free state (10.7.28), we can calculate the crystal lattice parameters of the phases in the constrained state from Eqs. (10.7.23) and (10.7.24).

According to (10.7.23), the β-phase should be tetragonal and have the crystal lattice parameters

$$a_\beta^{\text{calc}}=a(1)=a+\frac{c_{11}+2c_{12}}{2c_{11}}(a_\beta^0-a)=2.880+\frac{2.335+2\times1.355}{2\times2.335}(2.863-2.880)$$
$$=2.8616\ \text{Å}$$
$$c_\beta^{\text{calc}}=c(1)=a=2.880\ \text{Å} \tag{10.7.30a}$$

Similarly, Eq. (10.7.24) yields for the β_2-phase

$$a_{\beta_2}^{\text{calc}}=a(2)=a+\frac{c_{11}+2c_{12}}{2c_{11}}(a_{\beta_2}^0-a)=2.880+\frac{2.335+2\times1.355}{2\times2.335}(2.905-2.880)$$
$$=2.9070\ \text{Å}$$
$$c_\beta^{\text{calc}}=c(2)=a=2.880\ \text{Å} \tag{10.7.30b}$$

The calculated crystal lattice parameters of the slightly tetragonal phases β and β_2 coincide with the x-ray values (10.7.27) to within the measurement errors.

Returning to the problem of aging stages touched upon in the beginning section, we must admit that there is hardly a sharp distinction between the "satellite" stage and the stage of coexistence of two tetragonal phases. In certain cases shortwave x-ray diffraction from decomposed alloys exhibits satellites in the vicinity of low-order parent phase reflections whereas the higher-order reflections show splittings indicative of a slight tetragonality. This is convincing evidence of the difference between the two stages being of optical rather than structural origin. The optical conditions for the detection of the tetragonal phase reflections are easy to derive. We already know that satellites arise from coherent diffraction on quasiperiodic concentration modulations. They may be treated as Laue reflections from a long-period concentration profile. Conversely, violation of coherency between waves scattered by various elements of the modulated structure results in tetragonal splittings.

A modulated structure produces coherent diffraction if splittings of the parent and new phase reflections caused by new phase tetragonality are larger than separations between the parent phase reflections and nearest satellites in the reciprocal space. The latter quantity is equal to

$$\frac{\Delta k}{2\pi}=\frac{1}{\lambda_0} \tag{10.7.31}$$

where λ_0 is the modulated structure period. The splitting in the reciprocal space between the parent and tetragonal phase reflections is described by Eq. (1.5.5b):

$$\frac{1}{2\pi}\Delta\mathbf{k}_{\text{spl}}=\mathbf{H}(\hat{\mathbf{I}}+\hat{\varepsilon})^{-1}-\mathbf{H}\cong-\mathbf{H}\hat{\varepsilon}=-\hat{\varepsilon}\mathbf{H} \tag{10.7.32}$$

where $\hat{\varepsilon}$ is the homogeneous strain transforming the parent phase lattice to the tetragonal phase lattice. In the case of the cubic-to-tetragonal phase transformation

$$\hat{\varepsilon}=\begin{pmatrix}\varepsilon_{11} & 0 & 0\\ 0 & \varepsilon_{11} & 0\\ 0 & 0 & \varepsilon_{33}\end{pmatrix} \tag{10.7.33}$$

Substituting Eq. (10.7.33) into (10.7.32) yields

$$\frac{1}{2\pi}\Delta\mathbf{k}_{\text{spl}}=-\frac{1}{a}(H_1\varepsilon_{11},\ H_2\varepsilon_{11},\ H_3\varepsilon_{33}) \tag{10.7.34}$$

where $\mathbf{H}=(H_1, H_2, H_3)/a$, and (H_1, H_2, H_3) are the diffraction indexes of the reflection.

Since (10.7.31) describes the separation in the [001] direction, it should be compared with the z component of the vector $\Delta\mathbf{k}_{\text{spl}}/2\pi$ in Eq. (10.7.34)

$$\frac{1}{2\pi}\Delta k^z_{\text{spl}}=-\frac{\varepsilon_{33}}{a}H_3 \tag{10.7.35}$$

Hence the optical criterion of the observation of the "satellite" stage reads

$$\frac{1}{\lambda_0}\gg\frac{H_3|\varepsilon_{33}|}{a}\quad\text{or}\quad\frac{a}{|\varepsilon_{33}|\lambda_0H_3}\gg1 \tag{10.7.36}$$

where $|\varepsilon_{33}|\lambda_0$ may be interpreted as the change in the period of the modulated structure in the [001] direction caused by the tetragonal distortion.

By inverting the sense of the inequality (10.7.36), we obtain the condition for the observation of the two-phase stage

$$\frac{a}{|\varepsilon_{33}|\lambda_0H_3}<1 \tag{10.7.37}$$

The inequality (10.7.37) affords the criterion for the resolution of reflections from the tetragonal phase and the matrix. It follows from the inequalities (10.7.36) and (10.7.37) that the appearance of satellites near low-index reflections (H_3 is a small integer) and, simultaneously, the replacement of the higher index reflections with tetragonal multiplets (H_3 is a large integer) may in fact take place.

The conclusion follows that the breaking of diffraction coherency and the observation of diffraction reflections from the components of a modulated structure rather than from the modulated structure as a whole should be expected at large modulated structure periods ($\lambda_0/a > 1$), large tetragonal distortions in the modulation directions $|\varepsilon_{33}|\lambda_0$, and high diffraction indexes H_3.

In conclusion, it should be mentioned that the theory described in this section may prove to be efficient in identifying modulated structure morphologies from only the diffraction data. The basic points are as follows.

1. If a modulated structure gives rise to diffraction reflections from two tetragonal phases with the same crystal lattice parameter a but different parameter c values, one of which is larger and the other one smaller than the crystal lattice parameter of the homogeneous alloy, the modulated structure is one-dimensional. It is a sandwich of periodically alternating platelets of two equilibrium phases depicted in Fig. 75.

2. The observation of two tetragonal phases whose crystal lattice parameters c are equal to each other, whereas the parameters a are not, shows that the modulated structure is two-dimensional (see Fig. 82) and represents a periodic distribution of rod-shaped precipitates in three phases.

11

MORPHOLOGY OF COHERENT MIXTURE OF CUBIC AND TETRAGONAL PHASES CONTROLLED BY ELASTIC STRAIN EFFECT

11.1. STABLE CONFIGURATIONS IN COHERENT MIXTURE OF CUBIC AND NONCUBIC PHASES

The strain-induced morphology of a coherent mixture of two cubic phases has been considered in Chapter 10. A phase transformation leading to a coherent mixture of two cubic phases can be accomplished by the only crystal lattice rearrangement. It is a particular case of a more general problem of a coherent mixture of noncubic new phase inclusions in a cubic parent phase. As shown in Section 1.6, the same new phase crystal lattice may be generated by a number of crystal lattice rearrangements if the crystal symmetry of the new phase differs from that of the parent phase. All these rearrangements may be derived from one of them by applying all the symmetry operations making up the point symmetry group of the parent crystal [see Eq. (1.6.9)]. The new phase structure domains produced by the symmetry-related rearrangements differ in their orientations with respect to the parent phase lattice. Thus, as shown in Section 1.6, the formation of tetragonal phase inclusions in a cubic matrix may involve crystal lattice rearrangements of three types, tetragonal distortions along the [100], [010], and [001] directions of the parent cubic phase (see Fig. 4). Each of these generates a structure domain of its own. The same is true for the fcc → bcc and bcc → fcc crystal lattice rearrangements involving the Bain distortion (see Section 6.4).

The number of structure domains that may be formed in the cubic-to-orthorhombic phase crystal lattice rearrangement increases to 6 because each of the three possible orientations of the axis c of the orthorhombic phase may occur in combination with two different orientations of the axis b (we assume the usual orientation relations with the axes a, b, c parallel to the $\langle 100 \rangle$ cubic phase directions). For instance, with c parallel to the [001] direction, b may have one of two orientations, [100] or [010].

The minimum strain energy of a single coherent inclusion in a parent phase was shown in Section 8.1 [Eq. (8.1.14)] to be

$$E_{\text{bulk}} = \frac{1}{2} \min B(\mathbf{n}) V \tag{11.1.1}$$

where V is the inclusion volume, and $\min B(\mathbf{n})$ is positive if the stress-free transformation strain is not an invariant plane strain. If the transformation strain is an invariant plane strain, $\min B(\mathbf{n})$ reduces to zero, and so does the bulk strain energy (11.1.1). Because the transformation strain is, as a rule, not an invariant plane strain, the total strain energy of a set of noninteracting inclusions is positive and proportional to the new phase volume. Hence comes a natural question if the bulk strain energy may be decreased by constructing the two-phase morphology such that it would maximize the negative contribution of strain-induced interactions to the bulk strain energy, namely, by a special choice of the types of new phase structure domains and of their mutual arrangement. The answer is almost always positive. The decrease of the strain energy, however, depends on the stress-free transformation strains,

$$\varepsilon_{ij}^0(1), \varepsilon_{ij}^0(2), \ldots, \varepsilon_{ij}^0(p), \ldots \tag{11.1.2}$$

which result in the formation of distinct orientational variants (structure domains) of the new phase. Moreover in the majority of cases the bulk strain energy may be eliminated completely with the proper choice of the structure domains of the low-symmetry phase.

It has been shown in Section 8.1.1 that the bulk strain energy of an inclusion vanishes on the conditions that (1) the stress-free transformation strain producing the inclusion is a homogeneous invariant plane strain and (2) the inclusion has a platelike shape whose habit plane is parallel to the invariant plane.

These two conditions may be satisfied only in part because the transformation strains (11.1.2) are not invariant plane strains. We can construct a plate comprising alternating platelets of various structure domains (see Section 6.3). If the adjacent domains are fitted together along geometrically identical planes, the formation of the heterogeneous sandwichlike plate involves no internal stress (Fig. 93). We will now assume that (1) the thickness of platelets is chosen such that the macroscopic shape deformation in the stress-free state be an invariant plane strain and (2) the habit plane of the heterogeneous plate coincides with the invariant plane of the shape deformation.

The heterogeneous new phase plate thus constructed meets all the requirements that allow the bulk strain energy to vanish but with a single exception. The plate is not homogeneous, and its coherent conjugation with the parent

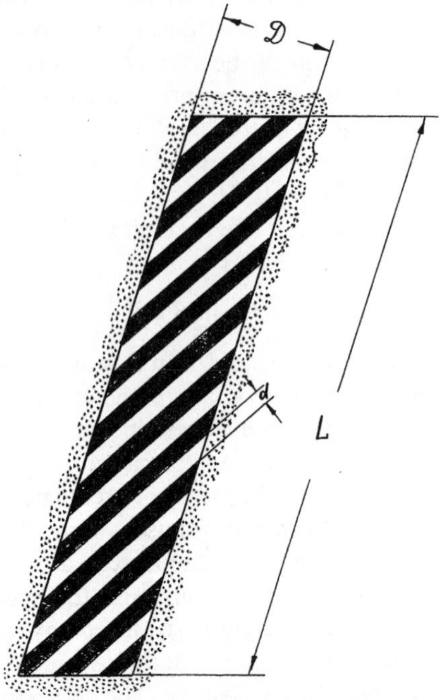

Figure 93. Schematic drawing of the thin-plate multidomain morphology of a tetragonal phase in a cubic phase matrix. Dark and white regions designate tetragonal phase domains with the [100] and [010] directions of the tetragonal axes. Interdomain boundaries are parallel to the (110) plane. Elastic strain is concentrated within the shaded layer regions near the interphase boundary. The typical thickness of this layer regions is d.

phase along the habit plane will generate local elastic strain concentrated within a thin layer in the vicinity of the interphase habit plane. This is true because a heterogeneous habit plane may be regarded only approximately as an invariant plane. The elastic strain caused by the heterogeneous structure of the habit plane can be neglected at large distances from the habit (at distances far exceeding the typical heterogeneity length). Since the typical heterogeneity length is of the same order of magnitude as the typical thickness of the structure domains, the elastic strain is concentrated near the habit plane within a layer of about the structure domain thickness (see Fig. 93). We may therefore estimate the elastic strain energy as

$$E_{\text{edge}} \approx \lambda \varepsilon_0^2 S d \tag{11.1.3}$$

where λ, ε_0, and d are the typical elastic modulus, transformation strain, and thickness of the structure domains, respectively, S is the interphase area.

The strain energy of the new phase conglomerate constructed thus proves to be proportional to the interphase surface area rather than to the new phase volume. We come to the conclusion that, with the proper morphology of the two-phase coherent mixture, the bulk strain energy may be lowered. This is similar to what was found for martensitic transformations (see Section 6.3). The conditions formulated above for reducing the bulk strain energy to zero are essentially the same as used in the crystallographic theories of martensitic transformations by Wechsler, Lieberman, and Read (1) and Bowles and Mackenzie (2).

It follows from Eq. (11.1.3) that the strain energy E_{edge} vanishes as the thickness of the structure domains approaches zero. The zero thickness cannot, however, be attained because at $d \to 0$ the number of interdomain boundaries and the corresponding surface energy term increase infinitely. The competition between the strain and surface energies results in an equilibrium thickness of structure domains. If the thickness of the plate is D, the surface energy coefficient is γ_s and its length and width is L, the combination of the surface energy

$$E_s \approx \gamma_s DL \frac{L}{d} = \frac{\gamma_s L^2 D}{d}$$

and the strain energy (11.1.3) gives

$$E_{\text{total}} \approx \lambda \varepsilon_0^2 L^2 d + \frac{\gamma_s L^2 D}{d} \tag{11.1.4}$$

The minimum condition

$$\frac{dE_{\text{total}}}{dd} = 0$$

yields the equilibrium domain thickness

$$d \approx \sqrt{r_0 D} \tag{11.1.5}$$

where

$$r_0 = \frac{\gamma_s}{\lambda \varepsilon_0^2}$$

is the material constant having the length dimension.

It is noteworthy that Eq. (11.1.5) resembles Eq. (10.1.6) which determines the thickness of concentration domains in a coherent mixture of two cubic phases. This is not an accidental coincidence. It reflects a strong analogy between the modulated structures considered in Chapter 10, martensitic crystals, coherent mixtures of cubic and noncubic phases, and magnetic and ferroelectric domains. The structure similarity between martensitic crystals and magnetic domains was discerned by Roitburd (227) [see also the reviews (253)]. As regards the theory of decomposed alloys, Eq. (11.1.5) was derived by Khachaturyan and Shatalov (148). In both studies cited, the strain energy was included in the form of (11.1.3).

We thus come to the conclusion that coherent new phase monodomain inclusions that are distinct orientational variants of the noncubic phase aggregate to form heterogeneous plates. The internal structure of plates is determined by the condition of a stress-free fitting together of the adjacent structure domains and by the requirement that the macroscopic shape deformation in the stress-free state be an invariant plane strain. The habit plane of the plate coincides with the invariant plane of the macroscopic shape deformation.

One more significant point should be mentioned. Regardless of the differences in the nature of phase transformations mentioned above, the final result of the transformations should be the same. The differences in the type of the phase transformation will affect the kinetics of the process rather than the microstructure of the final state. Therefore structures known from martensitic crystal crystallography will be formed eventually, structures comprising new phase crystals composed of alternating structure domains. Because adjacent structure domains are twin-related to ensure their stress-free fitting together, they are perceived as transformation twins. For instance, if the final martensite structure is directly formed in a diffusionless transformation, a phase transformation associated with the decomposition reaction occurs as a result of a succession of structure changes leading to a substantial reconstruction of the morphology of the two-phase coherent mixture. The reason for that is the necessity of long-distance diffusion. The structure evolution proceeds through a formation of single-phase precipitates distributed rather randomly, whose habit of course differs from that formed in the final stage. These single-domain precipitates grow and move in strain-induced coarsening and aggregate eventually to form multidomain new phase plates whose morphology is almost identical to the morphology of martensitic crystals. The structure similarity of the process under discussion to martensitic transformations is so close that even such a typically "martensitic" phenomenon as surface relief is observed at this stage. The examples of the martensitelike morphology of Ta-O alloys at the final stage of the decomposition may be found in (228, 229).

11.2. STRAIN ENERGY OF A TWO-PHASE ALLOY FORMED BY CUBIC AND TETRAGONAL PHASES

As mentioned in Section 11.1 the stable structure of a two-phase alloy composed of cubic and tetragonal phases does not depend on the nature of the phase transformation. The stable structure should be the same irrespective of whether the phase transformation is a decomposition involving long-range diffusion, ordering involving short-range diffusion (diffusion over distances of the order of interatomic separations), or a martensitic transformation involving no diffusion at all. The nature of the phase transformation determines the kinetics of the process rather than the final form of the microstructure. The remarkable fact in this connection is that the structure of a two-phase coherent mixture becomes an additional thermodynamic "parameter" if the strain energy is taken into consideration.

This "parameter" as well as all the other internal parameters of the system should be determined from the total free energy minimum condition. Indeed, as shown in Section 11.1, the strain energy that depends on the morphology of the two-phase structure is proportional to the noncubic phase volume. The Helmholtz free energy whose minimization determines the equilibrium state of the system is therefore also a function of the morphology of the two-phase state of the alloy. We thus arrive at the conclusion that the equilibrium morphology of a two-phase alloy as well as any other equilibrium property of the alloy can be found by the minimization of the free energy. A quantitative analysis of an equilibrium state of a two-phase coherent mixture composed of a tetragonal and cubic phase will now be given. We will roughly follow the line of reasoning suggested by Khachaturyan and Shatalov (148). The analysis is based on only one assumption that the difference between the elastic moduli of the cubic and tetragonal phases may be neglected.

Consider a platelike sandwich composed of alternating structure domains of the tetragonal phase coherently fitted to the cubic phase matrix. The stress-free transformation strain presented in the two types of structure domains is given by the tensors

$$\varepsilon_{ij}^0(1)=\begin{pmatrix}\varepsilon_{33}^0 & 0 & 0\\ 0 & \varepsilon_{11}^0 & 0\\ 0 & 0 & \varepsilon_{11}^0\end{pmatrix},\quad \varepsilon_{ij}^0(2)=\begin{pmatrix}\varepsilon_{11}^0 & 0 & 0\\ 0 & \varepsilon_{33}^0 & 0\\ 0 & 0 & \varepsilon_{11}^0\end{pmatrix} \tag{11.2.1}$$

for the first and second type of tetragonal phase structure domains, respectively, where

$$\varepsilon_{11}^0=\frac{a-a_0}{a_0},\quad \varepsilon_{33}^0=\frac{c-a_0}{a_0} \tag{11.2.2}$$

a, c are the crystal lattice parameters of the tetragonal phase, a_0 is the crystal lattice parameter of the cubic matrix in the stress-free state. The representation (11.2.1) of the tensors $\hat{\boldsymbol{\varepsilon}}^0(1)$ and $\hat{\boldsymbol{\varepsilon}}^0(2)$ corresponds to the Cartesian coordinate system whose axes are along the [100], [010], and [001] directions of the cubic phase matrix.

The total strain energy of an arbitrary coherent mixture of tetragonal phase domains of two types is given by Eq. (7.2.53b):

$$E_{\text{elast}}=\frac{1}{2}\sum_{p=1}^{2}\sum_{q=1}^{2}\int\frac{d^3k}{(2\pi)^3}\left[\sigma_{ij}^0(p)\varepsilon_{ij}^0(q)-\mathbf{n}\hat{\sigma}^0(p)\hat{\boldsymbol{\Omega}}(\mathbf{n})\hat{\sigma}^0(q)\mathbf{n}\right]\theta_p(\mathbf{k})\theta_q^*(\mathbf{k}) \tag{11.2.3}$$

where $p, q=1, 2$ are the indexes labeling the type of tetragonal phase domains, $\sigma_{ij}^0(p)=\lambda_{ijkl}\varepsilon_{kl}^0(p)$,

$$\hat{\sigma}^0(1)=\begin{pmatrix}\sigma_{33}^0 & 0 & 0\\ 0 & \sigma_{11}^0 & 0\\ 0 & 0 & \sigma_{11}^0\end{pmatrix},\quad \hat{\sigma}^0(2)=\begin{pmatrix}\sigma_{11}^0 & 0 & 0\\ 0 & \sigma_{33}^0 & 0\\ 0 & 0 & \sigma_{11}^0\end{pmatrix} \tag{11.2.4}$$

$$\begin{aligned}\sigma_{11}^0&=(c_{11}+c_{12})\varepsilon_{11}^0+c_{12}\varepsilon_{33}^0\\ \sigma_{33}^0&=2c_{12}\varepsilon_{11}^0+c_{11}\varepsilon_{33}^0\end{aligned} \tag{11.2.5}$$

$\mathbf{n}=\mathbf{k}/k$, $\Omega_{ij}(\mathbf{n})$ is the inverse Cristoffel tensor given by Eqs. (8.6.3),

$$\theta_p(\mathbf{k})=\int \tilde{\theta}_p(\mathbf{r})e^{-i\mathbf{kr}}d^3r \tag{11.2.6}$$

where $\tilde{\theta}_p(\mathbf{r})$ is the shape function equal to unity within structure domains of the pth type and to zero outside them.

Let us introduce the definition

$$\hat{\boldsymbol{\varepsilon}}^0(\mathbf{r})=\hat{\boldsymbol{\varepsilon}}^0(1)\tilde{\theta}_1(\mathbf{r})+\hat{\boldsymbol{\varepsilon}}^0(2)\tilde{\theta}_2(\mathbf{r}) \tag{11.2.7}$$

which describes the distribution of the stress-free transformation strain, and

$$\hat{\sigma}^0(\mathbf{r})=\hat{\sigma}^0(1)\tilde{\theta}_1(\mathbf{r})+\hat{\sigma}^0(2)\tilde{\theta}_2(\mathbf{r}) \tag{11.2.8}$$

which is related to the stress-free strain.

Using Eqs. (11.2.7) and (11.2.8) in Eq. (11.2.3), we obtain

$$E_{\text{elast}}=\frac{1}{2}\iiint_{-\infty}^{\infty}\frac{d^3k}{(2\pi)^3}\left[\sigma^0_{ij}(\mathbf{k})\varepsilon^0_{ij}(\mathbf{k})-\mathbf{n}\hat{\sigma}^0(\mathbf{k})\hat{\boldsymbol{\Omega}}(\mathbf{n})\hat{\sigma}^0(\mathbf{k})\mathbf{n}\right] \tag{11.2.9}$$

where $\sigma^0_{ij}(\mathbf{k})$ and $\varepsilon^0_{ij}(\mathbf{k})$ are Fourier transforms of the functions $\sigma^0_{ij}(\mathbf{r})$ and $\varepsilon^0_{ij}(\mathbf{r})$ given by Eqs. (11.2.7) and (11.2.8). We shall consider a crystal formed by agglomeration of all structure domains of the tetragonal phase coherently conjugated with the cubic matrix and shall find the morphology which minimizes the strain energy. Let us introduce the shape function $\tilde{\theta}(\mathbf{r})$ to describe a single-connected tetragonal phase crystal composed of structure domains of the first and second types. Such a multidomain crystal will be called a complex. The shape function of the complex, $\tilde{\theta}(\mathbf{r})$, is equal to unity when $\mathbf{r}$ is within the complex and zero when $\mathbf{r}$ is outside it.

Since the complex is composed of domains of the first and second types and does not contain cubic matrix, we have

$$\tilde{\theta}(\mathbf{r})=\tilde{\theta}_1(\mathbf{r})+\tilde{\theta}_2(\mathbf{r}) \tag{11.2.10}$$

Integration of Eq. (11.2.10) yields

$$V=V_1+V_2 \tag{11.2.11}$$

where V, V_1, V_2 are the volumes of the complex and of the domains of the first and second types, respectively. The Fourier expansion of Eq. (11.2.10) gives

$$\theta(\mathbf{k})=\theta_1(\mathbf{k})+\theta_2(\mathbf{k}) \tag{11.2.12}$$

where

$$\theta(\mathbf{k})=\int \tilde{\theta}(\mathbf{r})e^{-i\mathbf{kr}}d^3r \tag{11.2.13}$$

$$\theta_p(\mathbf{k})=\int \tilde{\theta}_p(\mathbf{r})e^{-i\mathbf{kr}}d^3r \tag{11.2.14}$$

Using the definition (11.2.10), the function $\tilde{\theta}_p(\mathbf{r})(p=1,2)$ may be written as the sum of its average over the complex and of variation $\Delta\tilde{\theta}_p(\mathbf{r})$

$$\tilde{\theta}_p(\mathbf{r})=\langle\tilde{\theta}_p(\mathbf{r})\rangle+\Delta\tilde{\theta}_p(\mathbf{r}) \tag{11.2.15}$$

Since

$$\langle\tilde{\theta}_p(\mathbf{r})\rangle=\frac{1}{V}\int_{(V)}\tilde{\theta}_p(\mathbf{r})d^3r=\frac{V_p}{V}\tilde{\theta}(\mathbf{r}) \tag{11.2.16}$$

Eq. (11.2.15) may be rewritten as

$$\tilde{\theta}_p(\mathbf{r})=\frac{V_p}{V}\tilde{\theta}(\mathbf{r})+\Delta\tilde{\theta}_p(\mathbf{r}) \tag{11.2.17}$$

The Fourier transform of Eq. (11.2.17) is

$$\theta_p(\mathbf{k})=\frac{V_p}{V}\theta(\mathbf{k})+\Delta\theta_p(\mathbf{k}) \tag{11.2.18}$$

where

$$\Delta\theta_p(\mathbf{k})=\int\Delta\tilde{\theta}_p(\mathbf{r})e^{-i\mathbf{kr}}d^3r \tag{11.2.19}$$

The summation over p in Eq. (11.2.18), taking into consideration Eqs. (11.2.11) and (11.2.12), gives

$$\theta(\mathbf{k})=\theta(\mathbf{k})+\Delta\theta_1(\mathbf{k})+\Delta\theta_2(\mathbf{k})$$

or

$$\Delta\theta_1(\mathbf{k})+\Delta\theta_2(\mathbf{k})=0 \tag{11.2.20}$$

Using Eqs. (11.2.18) and (11.2.20) in the Fourier transforms of Eqs. (11.2.7) and (11.2.8), we obtain

$$\hat{\varepsilon}^0(\mathbf{k})=\hat{\bar{\varepsilon}}_0\theta(\mathbf{k})+(\hat{\varepsilon}^0(1)-\hat{\varepsilon}^0(2))\Delta\theta_1(\mathbf{k}) \tag{11.2.21}$$

$$\hat{\sigma}^0(\mathbf{k})=\hat{\bar{\sigma}}_0\theta(\mathbf{k})+(\hat{\sigma}^0(1)-\hat{\sigma}^0(2))\Delta\theta_1(\mathbf{k}) \tag{11.2.22}$$

where

$$\hat{\bar{\varepsilon}}^0=\frac{V_1}{V}\hat{\varepsilon}^0(1)+\frac{V_2}{V}\hat{\varepsilon}^0(2)=\frac{V_1}{V}\hat{\varepsilon}^0(1)+\left(1-\frac{V_1}{V}\right)\hat{\varepsilon}^0(2) \tag{11.2.23}$$

is the stress-free transformation strain of the complex as a whole,

$$\hat{\bar{\sigma}}^0=\frac{V_1}{V}\hat{\sigma}^0(1)+\frac{V_2}{V}\hat{\sigma}^0(2) \tag{11.2.24}$$

According to the uncertainty relation the function $\theta(\mathbf{k})$ in Eq. (11.2.13) does not vanish within a small region in the $\mathbf{k}$-space near $\mathbf{k}=0$ having the volume

$$\Delta^3k\sim\frac{(2\pi)^3}{V}$$

On the other hand, the function $\Delta\theta_p(\mathbf{k})$ describing the heterogeneous structure of the complex vanishes in this region. In fact, as follows from the definition of $\Delta\theta_p(\mathbf{k})$

$$\Delta\theta_p(\mathbf{k}) = \int \Delta\tilde{\theta}_p(\mathbf{r}) e^{-i\mathbf{k}\mathbf{r}} d^3r \to \int \Delta\tilde{\theta}_p(\mathbf{r}) d^3r = 0 \tag{11.2.25}$$

when $\mathbf{k}$ approaches zero. The function $\Delta\theta_p(\mathbf{k})$ grows nonzero at distances exceeding $\sim 2\pi/d$ where d is the typical thickness of tetragonal domains. It follows that, if $d/L \ll 1$ where L is the typical length dimension of the complex (if the characteristic size of structure domains is far lower than that of the complex), the functions $\theta(\mathbf{k})$ and $\Delta\theta_p(\mathbf{k})$ are nonzero in different regions of the $\mathbf{k}$-space, and therefore

$$\Delta\theta_p(\mathbf{k})\theta(\mathbf{k}) \cong 0 \tag{11.2.26}$$

with accuracy to the ratio $d/L \ll 1$.

Bearing this in mind and substituting Eqs. (11.2.21) and (11.2.22) into (11.2.9), we obtain

$$E_{\text{elast}} = E_{\text{homog}} + E_{\text{heter}} \tag{11.2.27}$$

where

$$E_{\text{homog}} = \frac{1}{2} \int \frac{d^3k}{(2\pi)^3} (\overline{\sigma_{ij}^0}\, \overline{\varepsilon_{ij}^0} - \mathbf{n}\hat{\bar{\sigma}}^0 \hat{\Omega}(\mathbf{n}) \hat{\bar{\sigma}}^0 \mathbf{n}) |\theta(\mathbf{k})|^2 \tag{11.2.28}$$

and

$$E_{\text{heter}} = \frac{1}{2} \int \frac{d^3k}{(2\pi)^3} [\Delta\sigma_{ij}^0 \Delta\varepsilon_{ij}^0 - \mathbf{n}\Delta\hat{\sigma}^0 \hat{\Omega}(\mathbf{n}) \Delta\hat{\sigma}^0 \mathbf{n}] |\Delta\theta_1(\mathbf{k})|^2 \tag{11.2.29}$$

where

$$\Delta\hat{\varepsilon}^0 = \hat{\varepsilon}^0(1) - \hat{\varepsilon}^0(2) \tag{11.2.30a}$$

$$\Delta\hat{\varepsilon}^0 = \hat{\varepsilon}^0(1) - \hat{\varepsilon}^0(2) \tag{11.2.30b}$$

Comparison of Eqs. (11.2.28) and (8.1.1) shows that the first term in Eq. (11.2.27) has already been calculated. The strain energy E_{homog}in (11.2.27) is in fact the same as the strain energy of a single coherent homogeneous inclusion if its stress-free transformation strain ε_{ij}^0 is replaced by the mean value (11.2.23). The second term in Eq. (11.2.27), the quantity E_{heter}, describes the contribution associated with the internal heterogeneous structure of the complex which is made up of tetragonal domains of two kinds.

It will be shown below that the strain energy E_{heter} is proportional to the interphase area and therefore makes a considerable contribution to the interphase energy. The term E_{heter} arises from short-range elastic strain concentrated in the vicinity of the interphase boundary within a thin layer whose thickness is commensurate with the typical thickness of the tetragonal phase domains composing the complex. The total interphase energy which includes both the coherent interphase energy and the surface-dependent strain energy, E_{heter}, may be interpreted as interphase energy of a semicoherent boundary.

Eqs. (11.2.28) and (11.2.29) thus give a quantitative description of the parameters used in Section 11.1 to analyze the problem qualitatively.

11.3. MINIMIZATION OF "HOMOGENEOUS" STRAIN ENERGY AND EQUILIBRIUM STRUCTURE OF COHERENT MIXTURE OF CUBIC AND TETRAGONAL PHASES

As mentioned in Section 11.2, Eq. (11.2.28) for obtaining the strain energy, E_{homog}, may be applied to a homogeneous new phase coherent inclusion whose shape is described by the shape function, $\tilde{\theta}(\mathbf{r})$, and the homogeneous stress-free transformation strain by Eq. (11.2.23):

$$\hat{\bar{\varepsilon}}^0 = w\hat{\varepsilon}^0(1) + (1-w)\hat{\varepsilon}^0(2) \tag{11.3.1}$$

where $w = V_1/V$ is the volume fraction of tetragonal phase domains formed in the crystal lattice rearrangement $\hat{\varepsilon}^0(1)$.

In this connection we may use the theoretical results obtained in Section 8.1. It was shown there that the strain energy attains minimum when a single homogeneous coherent inclusion has the shape of a thin extended plate whose habit is perpendicular to the unit vector $\mathbf{n}_0$. The vector $\mathbf{n}_0$ is defined as the vector corresponding to the minimum of the function $B(\mathbf{n})$ given by Eq. (8.1.2):

$$B(\mathbf{n}_0) = \min B(\mathbf{n}) \tag{11.3.2}$$

It was also shown in Section 8.1 that the strain energy (11.2.28) can be written for a thin-plate inclusion as follows [see Eq. (8.1.26)]:

$$E_{\text{homog}} = \tfrac{1}{2}\min \bar{B}(\mathbf{n})V + \Delta E_{\text{homog}} \tag{11.3.3}$$

where, according to Eq. (11.2.28),

$$\bar{B}(\mathbf{n}) = \bar{\sigma}^0_{ij}\bar{\varepsilon}^0_{ij} - \mathbf{n}\hat{\bar{\sigma}}^0\hat{\Omega}(\mathbf{n})\hat{\bar{\sigma}}^0\mathbf{n} \geqslant 0 \tag{11.3.4}$$

$$\Delta E_{\text{homog}} = \frac{1}{2}\int \frac{d^3k}{(2\pi)^3}\,\Delta\bar{B}(\mathbf{n})|\theta(\mathbf{k})|^2 \tag{11.3.5}$$

and

$$\Delta\bar{B}(\mathbf{n}) = \bar{B}(\mathbf{n}) - \min \bar{B}(\mathbf{n}) \tag{11.3.6}$$

It was shown in Section 8.1 [see Eq. (8.1.40)] that the strain energy ΔE_{homog} having the form (11.3.5) is of the order of $\lambda\varepsilon_0^2V(D/L)$ where D is the plate thickness. L its characteristic length, λ the typical elastic modulus. Since $D/L \ll 1$,

$$\frac{\Delta E_{\text{homog}}}{\min \bar{B}(\mathbf{n})V} \sim \frac{D}{L} \ll 1 \tag{11.3.7}$$

this is, ΔE_{homog} is a small correction to the volume-dependent strain energy.

It is proportional to the perimeter of the plate rather than its volume.

Unlike the real transformation strains $\hat{\varepsilon}^0(1)$ and $\hat{\varepsilon}^0(2)$ which are material constants and cannot be changed the mean transformation strain, $\hat{\bar{\varepsilon}}^0$ in Eq. (11.3.1) is a variable internal thermodynamic parameter. It may be found from the strain energy minimization procedure.

According to the relation (11.3.7) the first (volume dependent) term in Eq. (11.3.3) makes the major contribution to the total strain energy. For that reason the strain energy minimization should be carried out by first minimizing the term:

$$E_{\text{bulk}} = \frac{1}{2} \min \bar{B}(\mathbf{n}) V$$

which is the first term in Eq. (11.3.3).

Since $\bar{B}(\mathbf{n}) \geqslant 0$, the minimal possible value of $\bar{B}(\mathbf{n})$ is zero. In certain cases the lower limit can in fact be attained. It follows from Eq. (8.1.23) that

$$\min \bar{B}(\mathbf{n}) = \bar{B}(\mathbf{n}_0) = 0$$

if the transformation strain $\hat{\bar{\varepsilon}}^0$ determining $\bar{B}(\mathbf{n})$ is an invariant plane strain:

$$\bar{\varepsilon}^0_{ij} = \frac{1}{2}\varepsilon_0(l_i n^0_j + l_j n^0_i) \tag{11.3.8a}$$

or

$$\hat{\bar{\varepsilon}}^0 = \frac{1}{2}\varepsilon_0(\boldsymbol{l} * \mathbf{n}_0 + \mathbf{n}_0 * \boldsymbol{l}) \tag{11.3.8b}$$

It thus follows that choosing the volume fraction, w, of tetragonal phase domains in Eq. (11.3.1) such that the mean transformation strain $\hat{\bar{\varepsilon}}^0$ is an invariant plane strain and therefore provide the fulfillment of the condition (11.3.8), we might in many cases reduce the first term in Eq. (11.3.3) to zero. We shall demonstrate below how the mean transformation strain, $\hat{\bar{\varepsilon}}^0$, can be represented in the form (11.3.8).

The matrix form of Eq. (11.3.1) is

$$\hat{\bar{\varepsilon}}^0 = w\hat{\varepsilon}^0(1) + (1-w)\hat{\varepsilon}^0(2) = \begin{pmatrix} w\varepsilon^0_{33} + (1-w)\varepsilon^0_{11} & 0 & 0 \\ 0 & w\varepsilon^0_{11} + (1-w)\varepsilon^0_{33} & 0 \\ 0 & 0 & \varepsilon^0_{11} \end{pmatrix} \tag{11.3.9}$$

The necessary (but not sufficient) condition for the strain (11.3.9) to be an invariant plane strain and thus to be represented in the form (11.3.8), is for one of the diagonal elements of the matrix (11.3.9), to vanish, for example,

$$w\varepsilon^0_{11} + (1-w)\varepsilon^0_{33} = 0 \tag{11.3.10}$$

This condition yields

$$w = w_0 = -\frac{\varepsilon^0_{33}}{\varepsilon^0_{11} - \varepsilon^0_{33}} \tag{11.3.11}$$

Eq. (11.3.10) requires that

$$\frac{\varepsilon_{11}^0}{\varepsilon_{11}^0-\varepsilon_{33}^0}<0 \tag{11.3.12}$$

(because of the natural physical condition $w_0>0$).

Substituting Eq. (11.3.11) into (11.3.9) yields

$$\hat{\boldsymbol{\varepsilon}}^0=\begin{pmatrix} \varepsilon_{11}^0+\varepsilon_{33}^0 & 0 & 0 \\ 0 & 0 & 0 \\ 0 & 0 & \varepsilon_{11}^0 \end{pmatrix} \tag{11.3.13}$$

It is easy to see that the strain (11.3.13) is an invariant plane strain when $\varepsilon_{11}^0+\varepsilon_{33}^0$ and ε_{11}^0 have opposite signs:

$$\frac{\varepsilon_{11}^0}{\varepsilon_{11}^0+\varepsilon_{33}^0}<0 \tag{11.3.14}$$

The set of inequalities (11.3.12) and (11.3.14) is reduced to the requirements

$$\begin{gathered} \varepsilon_{33}^0>0 \\ -\varepsilon_{33}^0<\varepsilon_{11}^0<0 \end{gathered} \tag{11.3.15}$$

or

$$\varepsilon_{33}^0<0 \qquad 0<\varepsilon_{11}^0<-\varepsilon_{33}^0 \tag{11.3.16}$$

The inequalities (11.3.15) and (11.3.16) determine the range of the crystallographic parameters ε_{11}^0 and ε_{33}^0 where the volume-dependent strain energy vanishes because the strain (11.3.13) can be written as the symmetric dyadic product (11.3.8) within this range. The vectors $\mathbf{n}_0$, $\boldsymbol{l}$ and the strain ε_0 in this case may be represented in terms of the parameters ε_{11}^0 and ε_{33}^0:

$$\mathbf{n}_0=\left(\sqrt{\frac{|\varepsilon_{11}^0+\varepsilon_{33}^0|}{|\varepsilon_{11}^0+\varepsilon_{33}^0|+|\varepsilon_{11}^0|}},\quad 0,\quad \sqrt{\frac{|\varepsilon_{11}^0|}{|\varepsilon_{11}^0+\varepsilon_{33}^0|+|\varepsilon_{11}^0|}}\right) \tag{11.3.17}$$

$$\boldsymbol{l}=\left(\sqrt{\frac{|\varepsilon_{11}^0+\varepsilon_{33}^0|}{|\varepsilon_{11}^0+\varepsilon_{33}^0|+|\varepsilon_{11}^0|}},\quad 0,\quad -\sqrt{\frac{|\varepsilon_{11}^0|}{|\varepsilon_{11}^0+\varepsilon_{33}^0|+|\varepsilon_{11}^0|}}\right) \tag{11.3.18}$$

$$\varepsilon_0=\pm(|\varepsilon_{11}^0+\varepsilon_{33}^0|+|\varepsilon_{11}^0|) \tag{11.3.19}$$

The positive and negative signs in Eq. (11.3.19) correspond to the situations described by the inequalities (11.3.15) and (11.3.16), respectively.

We thus arrive at the conclusion that the minimization of the bulk free energy results in the formation of thin multidomain crystals of the tetragonal phase. This morphology allows the complete vanishing of the bulk strain energy and hence the formation of the equilibrium morphology. The habit plane orientation, $\mathbf{n}_0$, the macroscopic shear direction, $\boldsymbol{l}$, and the shear value, ε_0, are determined by Eqs. (11.3.17) to (11.3.19). The equilibrium volume fraction parameter, w_0, is given by Eq. (11.3.11). All these quantities are expressed in terms of the stress-free transformation strains ε_{11}^0 and ε_{33}^0 which may be found from the crystal lattice parameters of the cubic and tetragonal phases [see Eq. (11.2.2)].

The equilibrium thin-plate morphology is fully identical to the morphologies predicted by the crystallographic theories by Wechsler, Lieberman, and Read (1) and Bowles and Mackenzie (2) for small transformation strains. Consider, for example, the case

$$\begin{aligned} \varepsilon_{11}^0 &= \tilde{\varepsilon}_{11}^0 \\ \varepsilon_{33}^0 &= -\tilde{\varepsilon}_{33}^0 \end{aligned} \tag{11.3.20}$$

where

$$\tilde{\varepsilon}_{33}^0 > \tilde{\varepsilon}_{11}^0 > 0 \tag{11.3.21}$$

[the relations (11.3.20) provide the transition to the designations used in Section 6.5].

Substitution of the definitions (11.3.20) into Eqs. (11.3.17), (11.3.18), and (11.3.19) and the use of the inequalities (11.3.21) yield

$$\begin{aligned} \mathbf{n} &= \left(\sqrt{\frac{\tilde{\varepsilon}_{33}^0 - \tilde{\varepsilon}_{11}^0}{\tilde{\varepsilon}_{33}^0}},\ 0,\ \sqrt{\frac{\tilde{\varepsilon}_{11}^0}{\tilde{\varepsilon}_{33}^0}}\right) \\ \boldsymbol{l} &= \left(\sqrt{\frac{\tilde{\varepsilon}_{33}^0 - \tilde{\varepsilon}_{11}^0}{\tilde{\varepsilon}_{33}^0}},\ 0,\ -\sqrt{\frac{\tilde{\varepsilon}_{11}^0}{\tilde{\varepsilon}_{33}^0}}\right) \\ \varepsilon_0 &= \tilde{\varepsilon}_{33}^0 \end{aligned} \tag{11.3.22}$$

It is easy to see that Eq. (11.3.22) which derives from the strain energy approach and Eq. (6.5.49) which follows from the crystallographic theories at small transformation strains are identical to each other.

It was shown in Section 8.7 that the strain energy (11.3.5) can be written as contour integral (8.7.12) over the perimeter, P, enveloping the habit plane of the thin multidomain plate of tetragonal phase:

$$\Delta E_{\text{homog}} = \frac{D^2}{4\pi} \ln \frac{L}{D} \oint \frac{\beta_1 + \beta_2 (dy/dx)^2}{1 + (dy/dx)^2}\, dl \tag{11.3.23}$$

where $y = y(x)$ gives the perimeter line in the habit plane. In the case of the invariant plane transformation strain $\hat{\boldsymbol{\varepsilon}}^0$, given by Eq. (11.3.8), the positive constants β_1 and β_2 in Eq. (11.3.23) are the eigenvalues of the tensor (8.8.20b):

$$\hat{\boldsymbol{\beta}}(\mathbf{n}_0) = \varepsilon_0^2 \hat{\boldsymbol{\Omega}}^{-1}(\boldsymbol{l}) - \hat{\sigma}_1 \hat{\boldsymbol{\Omega}}(\mathbf{n}_0) \hat{\sigma}_1^+ \tag{11.3.24}$$

The corresponding eigenvectors lie in the habit plane and determine the x- and y-axes [the z-axis is parallel to the normal to the habit plane $\mathbf{n}_0$; $\mathbf{n}_0$ is also an eigenvector of the tensor (11.3.24)].

The procedure for the minimization of the strain energy (11.3.23) is similar to that applied in Section 9.6 to "lath" martensite. It includes the calculation of the eigenvalues β_1 and β_2 of the tensor (11.3.24), the calculation of the anisotropy parameter α from Eq. (8.7.20), and, finally, the determination of the habit plane shape $y = y(x)$ from the parametric Eqs. (8.7.23) and (8.7.26). It is noteworthy that the strain energy ΔE_{homog}(11.3.23) is proportional to the perimeter length and can be interpreted in terms of the energy of some effective dislocation loop enveloping the multidomain plate along its perimeter with the Burgers vector equal to $\varepsilon_0 lD$.

We will now discuss the basic assumption that the elastic moduli of both phases are the same. It was shown in Section 8.1 that this assumption holds with thin-plate inclusions. We must merely use the elastic moduli of the inclusion to calculate the volume-dependent term of the strain energy, $\frac{1}{2}\min B(\mathbf{n})V$, and the elastic moduli of the matrix to calculate the elastic energy, E_{edge}, proportional to the perimeter length. But for one simplification, the same applies to multidomain plates. The simplification comes from the fact that the tetragonal phase elastic moduli are not needed in the calculations because the elastic strain within the plate responsible for the volume-dependent term in the strain energy vanishes. As for the strain energy, ΔE_{homog}, which can be interpreted as the strain energy of the effective dislocation loop, E_{edge} it describes long-range strain distributed mainly within the matrix. The strain energy (11.3.23) should therefore be calculated with the elastic moduli of the cubic matrix. Such calculations yield results accurate to $D/L \ll 1$.

Finally, it should be emphasized once more that the final equilibrium structure of a two-phase coherent alloy composed of a tetragonal phase and a cubic phase is the same regardless of the type of the phase transformation. In the decomposition a thin-plate multidomain tetragonal phase crystal is formed in successive transformations controlled by strain-induced coarsening because the solid state reaction of this type involves long-range diffusion. In martensitic transformations that do not involve diffusion the exact same thin-plate morphology is formed directly.

11.4. STRAIN ENERGY OF SEMI-COHERENT INTERPHASE AND EQUILIBRIUM DOMAIN STRUCTURE

The minimization of the volume-dependent strain energy carried out in the preceding section furnishes the equilibrium volume fraction of tetragonal phase domains, w_0, and the equilibrium thin-plate morphology. It, however, fails to provide information about the internal domain structure of a multidomain plate, including shape, size, and mutual arrangement of the domains comprising the plate. This information can be obtained by minimizing the "heterogeneous" strain energy (11.2.29) depending on the nonhomogeneous nature of the multidomain plate. The E_{heter} energy will be shown to be proportional to the interphase area: it represents the difference between the coherent and semicoherent interphase energies.

We now calculate the "heterogeneous" strain energy, E_{heter}. Eq. (11.2.29) may be rewritten in the form

$$E_{\text{heter}} = \frac{1}{2} \iiint_{-\infty}^{\infty} B_{\text{het}}(\mathbf{n}) |\Delta\theta_1(\mathbf{k})|^2 \frac{d^3k}{(2\pi)^3} \tag{11.4.1}$$

where

$$B_{\text{het}}(\mathbf{n}) = \Delta\sigma^0_{ij}\Delta\varepsilon^0_{ij} - \mathbf{n}\Delta\hat{\sigma}^0\hat{\Omega}(\mathbf{n})\Delta\hat{\sigma}^0\mathbf{n} \tag{11.4.2}$$

$$\Delta\hat{\sigma}^0 = \hat{\sigma}^0(1) - \hat{\sigma}^0(2) \tag{11.4.3a}$$

$$\Delta\hat{\boldsymbol{\varepsilon}}^0 = \hat{\boldsymbol{\varepsilon}}^0(1) - \hat{\boldsymbol{\varepsilon}}^0(2) \tag{11.4.3b}$$

$$\mathbf{n} = \frac{\mathbf{k}}{k}$$

Comparison of Eqs. (11.4.1) and (11.4.2) with Eqs. (8.1.1) and (8.1.2) shows E_{heter} to have the same mathematical structure as the strain energy of a coherent homogeneous inclusion. It may therefore be handled alike. The calculation details are as follows.

Eq. (11.4.1) may be rewritten in the form

$$E_{\text{heter}} = \frac{1}{2} \min B_{\text{het}}(\mathbf{n}) \int |\Delta\theta_1(\mathbf{k})|^2 \frac{d^3k}{(2\pi)^3} + \frac{1}{2}\int \Delta B_{\text{het}}(\mathbf{n}) |\Delta\theta_1(\mathbf{k})|^2 \frac{d^3k}{(2\pi)^3} \tag{11.4.4}$$

[compare with Eq. (8.1.24)] where

$$\Delta B_{\text{het}}(\mathbf{n}) = B_{\text{het}}(\mathbf{n}) - \min B_{\text{het}}(\mathbf{n}) \geqslant 0 \tag{11.4.5}$$

$\min B_{\text{het}}(\mathbf{n})$ is the minimum value of $B_{\text{het}}(\mathbf{n})$ corresponding to the vector $\mathbf{n} = \mathbf{m}$.

The integral in the first term of Eq. (11.4.4) may be written

$$\int |\Delta\theta_1(\mathbf{k})|^2 \frac{d^3k}{(2\pi)^3} \equiv \int (\Delta\tilde{\theta}_1(\mathbf{r}))^2 d^3r \tag{11.4.6}$$

Substitution of the definition (11.2.17) into the right-hand part of Eq. (11.4.6) yields

$$\int |\Delta\theta_1(\mathbf{k})|^2 \frac{d^3\mathbf{k}}{(2\pi)^3} \equiv \int (\tilde{\theta}_1(\mathbf{r}) - w\tilde{\theta}(\mathbf{r}))^2 d^3r$$

$$= \int [\tilde{\theta}_1^2(\mathbf{r}) - 2w\tilde{\theta}_1(\mathbf{r})\tilde{\theta}(\mathbf{r}) + w^2\tilde{\theta}^2(\mathbf{r})] d^3r \tag{11.4.7}$$

where $w = V_1/V$.

According to the definition of the shape functions $\tilde{\theta}_p(\mathbf{r})$ and $\tilde{\theta}(\mathbf{r})$ (they are equal to either zero or unity) the following identities hold:

$$\tilde{\theta}_1^2(\mathbf{r}) \equiv \tilde{\theta}_1(\mathbf{r})$$
$$\tilde{\theta}^2(\mathbf{r}) \equiv \tilde{\theta}(\mathbf{r})$$
$$\tilde{\theta}(\mathbf{r})\tilde{\theta}_1(\mathbf{r}) \equiv \tilde{\theta}_1(\mathbf{r}) \tag{11.4.8}$$

The latter identity holds because all the tetragonal phase domains described by the function $\tilde{\theta}_1(\mathbf{r})$ are within the complex described by the function $\tilde{\theta}(\mathbf{r})$. Using identities (11.4.8) in Eq. (11.4.7), we obtain

$$\int |\Delta\theta_1(\mathbf{k})|^2 \frac{d^3k}{(2\pi)^3} \equiv \int [\tilde{\theta}_1(\mathbf{r}) - 2w\tilde{\theta}_1(\mathbf{r}) + w^2\tilde{\theta}(\mathbf{r})]d^3r$$
$$= V_1 - 2wV_1 + w^2V = V(w - 2w^2 + w^2) = Vw(1-w)$$

or

$$\int |\Delta\theta_1(\mathbf{k})|^2 \frac{d^3k}{(2\pi)^3} \equiv Vw(1-w) \tag{11.4.9}$$

because, according to the definitions of the functions $\tilde{\theta}_1(\mathbf{r})$ and $\tilde{\theta}(\mathbf{r})$,

$$\int \tilde{\theta}_1(\mathbf{r})d^3r = V_1$$

$$\int \tilde{\theta}(\mathbf{r})d^3r = V$$

Substitution of identity (11.4.9) into Eq. (11.4.4) yields

$$E_{\text{heter}} = \frac{1}{2} \min B_{\text{het}}(\mathbf{n}) Vw(1-w) + \Delta E_{\text{heter}} \tag{11.4.10}$$

where

$$\Delta E_{\text{heter}} = \frac{1}{2} \int \Delta B_{\text{het}}(\mathbf{n}) |\Delta\theta_1(\mathbf{k})|^2 \frac{d^3k}{(2\pi)^3} \tag{11.4.11}$$

Comparison of Eqs. (11.4.10) and (11.3.3) shows these two equations to be similar in form. At the first sight the "heterogeneous" strain energy contains a term proportional to volume V. It will, however, be demonstrated that this term vanishes because $\min B_{\text{het}}(\mathbf{n}) = 0$.*

It follows from the definitions (11.4.3b) and (11.2.1) that the strain $\Delta\hat{\varepsilon}^0$ may

*The relation $\min B_{\text{het}}(\mathbf{n}) = 0$ was proved by Khachaturyan and Shatalov (148). Seven years later the same problem was analyzed in terms of isotropic elasticity by Mura, Mori, and Kato (230). They, however, overlooked the relation $\min B_{\text{het}}(\mathbf{n}) = 0$ and thus arrived at an erroneous conclusion that the heterogeneous internal structure of the martensitic plate results in the new effect disregarded previously—the appearance of a significant volume-dependent contribution to the strain energy that has the form of the first term in Eq. (11.4.10). They also missed the actual nonzero term of the strain energy given by (11.4.11).

be written

$$\Delta\hat{\boldsymbol{\varepsilon}}^0=\hat{\boldsymbol{\varepsilon}}^0(1)-\hat{\boldsymbol{\varepsilon}}^0(2)=(\varepsilon_{33}^0-\varepsilon_{11}^0)\begin{pmatrix}1 & 0 & 0\\ 0 & \bar{1} & 0\\ 0 & 0 & 0\end{pmatrix} \tag{11.4.12}$$

It is easy to see that the tensor (11.4.12) represents the invariant plane strain since it is a symmetrized dyadic product

$$\Delta\hat{\boldsymbol{\varepsilon}}^0=(\varepsilon_{33}^0-\varepsilon_{11}^0)(\mathbf{p}*\mathbf{m}+\mathbf{m}*\mathbf{p}) \tag{11.4.13}$$

of two vectors,

$$\mathbf{p}=\left(\frac{1}{\sqrt{2}},\frac{\bar{1}}{\sqrt{2}},0\right) \tag{11.4.14a}$$

and

$$\mathbf{m}=\left(\frac{1}{\sqrt{2}},\frac{1}{\sqrt{2}},0\right) \tag{11.4.14b}$$

Since $\Delta\hat{\boldsymbol{\varepsilon}}^0$ is a dyadic, we may use the theorem proved in Section 8.1.1. According to this theorem, min $B(\mathbf{n})=0$ if $B(\mathbf{n})$ defined by Eq. (8.1.2) depends on the strain ε_{ij}^0 which is a dyadic. The minimum value, zero, is attained at the argument, $\mathbf{n}$, value equal to one of the vectors entering the dyadic product. Applying this theorem to the function $B_{\text{het}}(\mathbf{n})$, (11.4.2) which has the form of $B(\mathbf{n})$ and depends on the dyadic (11.4.13), one may readily see that

$$\min B_{\text{het}}(\mathbf{n})=B_{\text{het}}(\mathbf{m})=B_{\text{het}}(\mathbf{p})=0 \tag{11.4.15}$$

Eq. (11.4.15) actually proves that the volume-dependent term in Eq. (11.4.10) for the strain energy vanishes, and thus

$$E_{\text{heter}}=\Delta E_{\text{heter}}=\frac{1}{2}\int \Delta B_{\text{het}}(\mathbf{n})|\Delta\theta_1(\mathbf{k})|^2\frac{d^3k}{(2\pi)^3} \tag{11.4.16}$$

It follows from Eqs. (11.4.15) that

$$B_{\text{het}}(\mathbf{p})=B_{\text{het}}(\mathbf{m})=0 \tag{11.4.17}$$

Bearing in mind Eqs. (11.4.17), we may find the domain morphology that minimizes the strain energy (11.4.16). It follows from Eq. (11.4.17) that E_{heter} in Eq. (11.4.16) approaches zero asymptotically when the function $|\Delta\theta_1(\mathbf{k})|^2$ only differs from zero within an infinitely thin and infinitely long rod in the **k**-space emerging from $\mathbf{k}=0$ and going in the direction of **m** or **p**. Such a favorable situation occurs if the function $\Delta\tilde{\theta}_1(\mathbf{r})$ [and $\tilde{\theta}_1(\mathbf{r})$] describes a domain structure comprising alternating thin platelets with either the (110) habit (normal to the vector **m**) or the $(1\bar{1}0)$ habit (normal to the vector **p**). Crystallographically, the two variants are equivalent. For definiteness, the platelets will be assumed to have the (110) habit. The corresponding thin-plate tetragonal crystal consisting of two types of domains with the (110) habit is given schematically in Fig. 94.

As shown in Section 8.7, the first nonvanishing term in the Taylor expansion

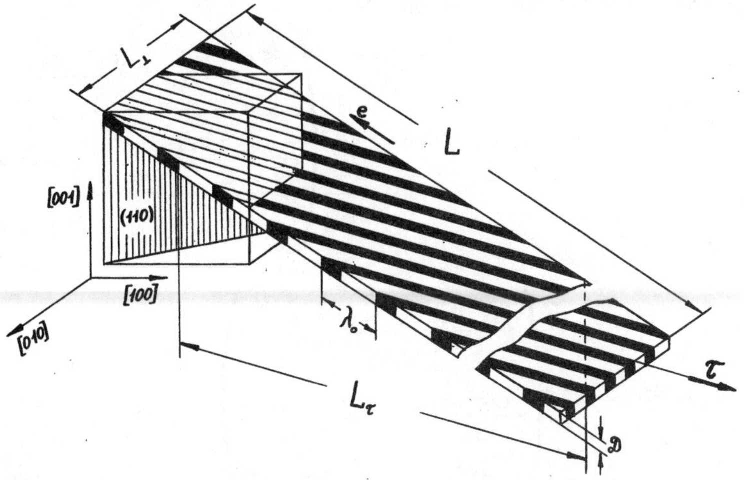

Figure 94. Schematic drawing of the thin-plate multidomain morphology minimizing elastic strain energy of tetragonal phase inclusions in a cubic matrix. L, $L_\perp$, and D are the length, width, and thickness, respectively, of the multidomain plate; black and white regions designate structural domains of the tetragonal phase whose tetragonal axis is parallel to the [100] and [010] directions, respectively; $\boldsymbol{\tau}$ is the unit vector parallel to the traces of (110) interdomain boundaries on the habit plane: L_τ is the length of "laths" formed by structural domains; and λ_0 is the modulation period.

of the function $\Delta B(\mathbf{n})=B(\mathbf{n})-\min B(\mathbf{n})$ has the form (8.7.9) where the tensor $\hat{\boldsymbol{\beta}}(\mathbf{n}_0)$ is given by Eq. (8.8.20). This is so if the stress-free transformation strain ε_{ij}^0 determining $\Delta B(\mathbf{n})$ is an invariant plane strain. Because the function $\Delta B_{\text{het}}(\mathbf{n})$ defined by Eqs. (11.4.5) and (11.4.2) depends on the invariant plane strain, $\Delta\hat{\boldsymbol{\varepsilon}}^0$ [Eq. (11.4.12)], the first nonvanishing term in the Taylor expansion of $\Delta B_{\text{het}}(\mathbf{n})$ is

$$\Delta B_{\text{het}}(\mathbf{n})=\beta_{ij}(\mathbf{m})\Delta n_i\Delta n_j=\Delta\mathbf{n}\hat{\boldsymbol{\beta}}(\mathbf{m})\Delta\mathbf{n} \tag{11.4.18}$$

where

$$\Delta\mathbf{n}=\mathbf{n}-\mathbf{m}$$

and $\hat{\boldsymbol{\beta}}(\mathbf{m})$ is determined by Eq. (8.8.20) which in the present case has the form

$$\hat{\boldsymbol{\beta}}(\mathbf{m})=(\varepsilon_{33}^0-\varepsilon_{11}^0)^2\hat{\boldsymbol{\Omega}}^{-1}(\mathbf{p})-\hat{\sigma}_1\hat{\boldsymbol{\Omega}}(\mathbf{m})\hat{\sigma}_1^+ \tag{11.4.19}$$

Using the Voigt designations,

$$\lambda_{1111}=\lambda_{2222}=\lambda_{3333}=c_{11}$$
$$\lambda_{1122}=\lambda_{1133}=\lambda_{2233}=c_{12}$$
$$\lambda_{1212}=\lambda_{1313}=\lambda_{2323}=c_{44}$$

and Eqs. (11.4.14), one may write the matrices $\hat{\sigma}_1$, $\hat{\sigma}_1^+$, and $\hat{\boldsymbol{\Omega}}^{-1}(\mathbf{p})$ [see the definitions (8.8.17) and (8.6.2)] as follows:

$$(\hat{\sigma}_1)_{ij} = (\varepsilon^0_{33} - \varepsilon^0_{11})\lambda_{ilkj} p_l m_k = (\varepsilon^0_{33} - \varepsilon^0_{11}) \begin{pmatrix} \frac{c_{11}-c_{44}}{2} & \frac{c_{12}-c_{44}}{2} & 0 \\ -\frac{c_{12}-c_{44}}{2} & -\frac{c_{11}-c_{44}}{2} & 0 \\ 0 & 0 & 0 \end{pmatrix} \tag{11.4.20}$$

$$(\hat{\sigma}_1^+)_{ij} = (\varepsilon^0_{33} - \varepsilon^0_{11}) \begin{pmatrix} \frac{c_{11}-c_{44}}{2} & -\frac{c_{12}-c_{44}}{2} & 0 \\ \frac{c_{12}-c_{44}}{2} & -\frac{c_{11}-c_{44}}{2} & 0 \\ 0 & 0 & 0 \end{pmatrix} \tag{11.4.21}$$

$$(\hat{\boldsymbol{\Omega}}^{-1}(\mathbf{p}))_{ij} = \lambda_{iklj} p_k p_l = \begin{pmatrix} \frac{c_{11}+c_{44}}{2} & -\frac{c_{12}+c_{44}}{2} & 0 \\ -\frac{c_{12}+c_{44}}{2} & \frac{c_{11}+c_{44}}{2} & 0 \\ 0 & 0 & c_{44} \end{pmatrix} \tag{11.4.22}$$

We shall demonstrate that the vectors

$$\mathbf{e}_1 = \mathbf{p} = \left(\frac{1}{\sqrt{2}}, \frac{\bar{1}}{\sqrt{2}}, 0\right) \quad \text{and} \quad \mathbf{e}_2 = (0, 0, 1)$$

are the eigenvectors of the matrix $\hat{\boldsymbol{\beta}}(\mathbf{m})$ in Eq. (11.4.19). Taking into consideration the representations (11.4.20), (11.4.21), and (11.4.22) of the matrices $\hat{\sigma}_1$, $\hat{\sigma}_1^+$, and $\hat{\boldsymbol{\Omega}}^{-1}(\mathbf{p})$, it is easy to see that

$$\hat{\sigma}_1^+ \mathbf{p} = (\varepsilon^0_{33} - \varepsilon^0_{11}) \frac{c_{11}+c_{12}-2c_{44}}{2} \mathbf{m} \tag{11.4.23}$$

$$\hat{\sigma}_1 \mathbf{m} = (\varepsilon^0_{33} - \varepsilon^0_{11}) \frac{c_{11}+c_{12}-2c_{44}}{2} \mathbf{p} \tag{11.4.24}$$

$$\hat{\boldsymbol{\Omega}}^{-1}(\mathbf{p})\mathbf{p} = \frac{c_{11}+c_{12}+2c_{44}}{2} \mathbf{p} \tag{11.4.25}$$

Using the definitions (8.6.3), we also obtain

$$\hat{\boldsymbol{\Omega}}(\mathbf{m})\mathbf{m} = \frac{2}{c_{11}+c_{12}+2c_{44}} \mathbf{m} \tag{11.4.26}$$

The application of the relations (1.4.23) through (11.4.26) yields

$$\hat{\boldsymbol{\beta}}(\mathbf{m})\mathbf{p}=(\varepsilon_{33}^0-\varepsilon_{11}^0)^2\hat{\boldsymbol{\Omega}}^{-1}(\mathbf{p})\mathbf{p}-\hat{\sigma}_1\hat{\boldsymbol{\Omega}}(\mathbf{m})\hat{\sigma}_1^+\mathbf{p}$$
$$=(\varepsilon_{33}^0-\varepsilon_{11}^0)^2\left[\frac{c_{11}+c_{12}+2c_{44}}{2}-\frac{(c_{11}+c_{12}-2c_{44})^2}{2(c_{11}+c_{12}+2c_{44})}\right]\mathbf{p}$$
$$=\frac{4(\varepsilon_{33}^0-\varepsilon_{11}^0)^2c_{44}(c_{11}+c_{12})}{c_{11}+c_{12}+2c_{44}}\mathbf{p}$$

or

$$\hat{\boldsymbol{\beta}}(\mathbf{m})\mathbf{p}=\frac{4c_{44}(c_{11}+c_{12})}{(c_{11}+c_{12}+2c_{44})}(\varepsilon_{33}^0-\varepsilon_{11}^0)^2\mathbf{p} \tag{11.4.27}$$

It follows from Eq. (11.4.27) that

$$\beta_1=4\frac{(c_{11}+c_{12})}{(c_{11}+c_{12}+2c_{44})}c_{44}(\varepsilon_{33}^0-\varepsilon_{11}^0)^2 \tag{11.4.28}$$

is the eigenvalue of the tensor $\hat{\boldsymbol{\beta}}(\mathbf{m})$ corresponding to the eigenvector

$$\mathbf{e}_1=\mathbf{p}=\left(\frac{1}{\sqrt{2}},\frac{\bar{1}}{\sqrt{2}},0\right).$$

Since the vector

$$\mathbf{e}_3=\mathbf{m}=\left(\frac{1}{\sqrt{2}},\frac{1}{\sqrt{2}},0\right)$$

is always an eigenvector of the Hermitian matrix $\hat{\boldsymbol{\beta}}(\mathbf{m})$ corresponding to the eigenvalue $\beta_3=0$ [see the discussion of Eq. (8.8.20)], the eigenvector $\mathbf{e}_2$ must be orthogonal to both

$$\mathbf{e}_1=\left(\frac{1}{\sqrt{2}},\frac{\bar{1}}{\sqrt{2}},0\right) \quad \text{and} \quad \mathbf{e}_3=\mathbf{m}=\left(\frac{1}{\sqrt{2}},\frac{1}{\sqrt{2}},0\right).$$

Hence

$$\mathbf{e}_2=(0, 0, 1) \tag{11.4.29}$$

We have from Eqs. (11.4.21) and (11.4.22)

$$\hat{\sigma}_1^+\mathbf{e}_2=0 \quad \text{and} \quad \hat{\boldsymbol{\Omega}}^{-1}(\mathbf{p})\mathbf{e}_2=c_{44}\mathbf{e}_2$$

and therefore

$$\hat{\boldsymbol{\beta}}(\mathbf{m})\mathbf{e}_2=(\varepsilon_{33}^0-\varepsilon_{11}^0)^2\hat{\boldsymbol{\Omega}}^{-1}(\mathbf{p})\mathbf{e}_2-\hat{\sigma}_1\hat{\boldsymbol{\Omega}}(\mathbf{m})\hat{\sigma}_1^+\mathbf{e}_2=c_{44}(\varepsilon_{33}^0-\varepsilon_{11}^0)^2\mathbf{e}_2$$

or

$$\hat{\boldsymbol{\beta}}(\mathbf{m})\mathbf{e}_2=c_{44}(\varepsilon_{33}^0-\varepsilon_{11}^0)^2\mathbf{e}_2 \tag{11.4.30}$$

It follows from Eq. (11.4.30) that the eigenvalue corresponding to the eigenvector $\mathbf{e}_2=(0, 0, 1)$ is

$$\beta_2=c_{44}(\varepsilon_{33}^0-\varepsilon_{11}^0)^2 \tag{11.4.31}$$

All the calculations described above have been carried out in the Cartesian coordinate system whose x-, y-, and z-axes are parallel to the [100], [010], and [001] directions of the cubic phase, respectively. It is more convenient, however, to use a Cartesian system whose axes are parallel to the $[1\bar{1}0]$, [001], and [110] cubic crystal directions because these axes are the principal axes of the tensor $\hat{\beta}(\mathbf{m})$. In terms of this new Cartesian system Eq. (11.4.18) becomes

$$\Delta B_{\text{het}}(\mathbf{n}) = \beta_1 \Delta n_1^2 + \beta_2 \Delta n_2^2 \tag{11.4.32}$$

where β_1 and β_2 are given by Eqs. (11.4.28) and (11.4.31).

Let us assume that the function $\Delta\theta_1(\mathbf{k})$ is related to the periodic distribution of the plate-shaped domains of the first type with the (110) habit. According to Fig. 94, this periodic distribution may be described as a one-dimensional modulated structure generated by successive translations of a (110) platelet of a domain of the first type by the unit translation vectors

$$\mathbf{T} = \lambda_0 \mathbf{e} \tag{11.4.33}$$

where λ_0 is the translation period in the unit vector $\mathbf{e}$ direction. As follows from Fig. 94, the vector $\mathbf{e}$ is orthogonal to the habit plane normal $\mathbf{n}_0$ of the complex and can therefore be written as

$$\mathbf{e} = (-n_3^0, 0, n_1^0) \tag{11.4.34}$$

in the Cartesian system related to the [100], [010], [001] directions of the cubic phase. According to Eq. (11.3.17)

$$n_1^0 = \sqrt{\frac{|\varepsilon_{11}^0 + \varepsilon_{33}^0|}{|\varepsilon_{11}^0 + \varepsilon_{33}^0| + |\varepsilon_{11}^0|}} \tag{11.4.35}$$

and

$$n_3^0 = \sqrt{\frac{|\varepsilon_{11}^0|}{|\varepsilon_{11}^0 + \varepsilon_{33}^0| + |\varepsilon_{11}^0|}}$$

are the components of the vector $\mathbf{n}_0$

$$\mathbf{n}_0 = (n_1^0, 0, n_3^0) \tag{11.4.36}$$

The transition to the new coordinate system related to the $[1\bar{1}0]$, [001], and [110] directions of the cubic matrix leads to a new coordinate representation of the vectors (11.4.34) and (11.4.36):

$$\mathbf{e} = \left(-\frac{1}{\sqrt{2}} n_3^0,\ n_1^0,\ -\frac{1}{\sqrt{2}} n_3^0 \right) \tag{11.4.37}$$

$$\mathbf{n}_0 = \left(\frac{1}{\sqrt{2}} n_1^0,\ n_3^0,\ \frac{1}{\sqrt{2}} n_1^0 \right) \tag{11.4.38}$$

Taking into consideration the equation

$$\Delta\theta_1(\mathbf{k})=\theta_1(\mathbf{k})-w\theta(\mathbf{k})\cong\begin{cases}0 & \text{when } |\mathbf{k}|\ll\dfrac{1}{L}\\ \theta_1(\mathbf{k}) & \text{when } |\mathbf{k}|\gg\dfrac{1}{L}\end{cases} \tag{11.4.39}$$

(L is the length of the complex) which follows from Eq. (11.2.25) and (11.2.18) and recalling what was said in discussing Eq. (11.2.25), we can rewrite Eq. (11.4.16) in the form

$$E_{\text{heter}}=\frac{1}{2}⨍\Delta B_{\text{het}}(\mathbf{n})|\theta_1(\mathbf{k})|^2\frac{d^3k}{(2\pi)^3} \tag{11.4.40}$$

The sign $⨍$ in Eq. (11.4.40) means that the region of the $\mathbf{k}$-space near $\mathbf{k}=0$ is excluded from the integration. The radius of the excluded volume is of the order of $\Delta k\sim 2\pi/L$. If the shape function of the generating platelet is $\tilde{\theta}_0(\mathbf{r})$, the shape function of the whole periodic distribution generated by this platelet is

$$\tilde{\theta}_1(\mathbf{r})=\sum_{n=-L/(2\lambda_0)}^{L/(2\lambda_0)}\tilde{\theta}_0(\mathbf{r}-n\lambda_0\mathbf{e}) \tag{11.4.41}$$

The Fourier transform of Eq. (11.4.41) is

$$\theta_1(\mathbf{k})=\theta_0(\mathbf{k})\sum_{n=-L/(2\lambda_0)}^{n=L/(2\lambda_0)}e^{-i(\mathbf{ke})\lambda_0 n}=\theta_0(\mathbf{k})\frac{\sin\frac{1}{2}(\mathbf{ke})L}{\sin\frac{1}{2}(\mathbf{ke})\lambda_0} \tag{11.4.42}$$

[compare with Eq. (10.3.10)].

Because the generating platelet is a noncircular cylinder disk with the (110) habit, the Fourier transform of its shape function

$$\theta_0(\mathbf{k})=\iiint_{-\infty}^{\infty}\tilde{\theta}_0(\mathbf{r})e^{-i\mathbf{kr}}d^3r=\int_{(v)}e^{-i\mathbf{kr}}d^3r \tag{11.4.43}$$

(v is the platelet volume) can always be represented in the form

$$\theta_0(\mathbf{k})=\frac{2\sin\frac{1}{2}k_z d}{k_z}S(k_x,k_y) \tag{11.4.44}$$

where d is the platelet thickness,

$$S(k_x,k_y)=\int_{S_{(110)}}\exp\left(-i(k_x x+k_y y)\right)dx\,dy \tag{11.4.45}$$

(k_x, k_y, k_z) and (x, y, z) are the projections of the vectors $\mathbf{k}$ and $\mathbf{r}$, respectively, on the coordinate axes $[1\bar{1}0]$, $[001]$, and $[110]$. The integration in Eq. (11.4.43) is over the generating platelet volume, and in Eq. (11.4.45) over the area $S_{(110)}$ of the platelet in the (110) plane.

Substituting Eq. (11.4.44) into (11.4.42) and (11.4.42) and (11.4.32) into (11.4.40) yields

$$E_{\text{heter}}=\frac{1}{2}\int\frac{\beta_1k_x^2+\beta_2k_y^2}{k_x^2+k_y^2+k_z^2}\,\frac{4\sin^2\frac{1}{2}k_zd}{k_z^2}\,|S(k_k,k_y)|^2\,\frac{\sin^2\frac{1}{2}(\mathbf{ke})L}{\sin^2\frac{1}{2}(\mathbf{ke})\lambda_0}\,\frac{d^3k}{(2\pi)^3} \tag{11.4.46}$$

where, according to the definition,

$$\Delta n_1^2=\frac{k_x^2}{k_x^2+k_y^2+k_z^2},\quad \Delta n_2^2=\frac{k_y^2}{k_x^2+k_y^2+k_z^2}$$

It is easy to see that the function $\sin^2\frac{1}{2}(\mathbf{ke})L/\sin^2\frac{1}{2}(\mathbf{ke})\lambda_0$ versus **ke** is periodic, with a period equal to $2\pi/\lambda_0$. It describes equidistant peaks located in the "reciprocal lattice points"

$$\mathbf{ke}=\frac{2\pi h}{\lambda_0},\quad h=0,\ \pm1,\ \pm2,\ \pm3,\ldots,\ \pm\infty \tag{11.4.47}$$

Substituting Eq. (11.4.37) into (11.4.47), we obtain

$$-\frac{1}{\sqrt{2}}k_xn_3^0+k_yn_1^0-\frac{1}{\sqrt{2}}k_zn_3^0=\frac{2\pi h}{\lambda_0} \tag{11.4.48}$$

The width of the peaks is of the order of $2\pi/L$, and the maximum value is $L/\lambda_0\gg1$. On the other hand, the other functions involved in the integrand are sufficiently smooth to be considered constant within the ranges near $\mathbf{ke}=2\pi h/\lambda_0$ corresponding to the peak positions. Taking this into account, we may rewrite Eq. (11.4.46):

$$E_{\text{heter}}=\frac{1}{2}\sum_{h=-\infty}^{\infty}{}'\iiint_{-\infty}^{\infty}\frac{d^3k}{(2\pi)^2}\,\frac{\beta_1k_x^2+\beta_2k_y^2}{k^2}\,\frac{4\sin^2\frac{1}{2}k_zd}{k_z^2}\,|S(k_x,k_y)|^2$$
$$\times\delta\left(\mathbf{ke}-\frac{2\pi h}{\lambda_0}\right)\int_{-\pi/\lambda_0}^{\pi/\lambda_0}\frac{d\xi}{2\pi}\,\frac{\sin^2\frac{1}{2}\xi L}{\sin^2\frac{1}{2}\xi\lambda_0} \tag{11.4.49}$$

Here priming implies that the term in the sum over h corresponding to $h=0$ is omitted. This is the case because the integration region in (11.4.46) near $\mathbf{k}=0$ is excluded. The function $\delta[\mathbf{ke}-(2\pi h/\lambda_0)]$ is the Dirak delta function.

Since

$$\int_{-\pi/\lambda_0}^{\pi/\lambda_0}\frac{d\xi}{2\pi}\,\frac{\sin^2\frac{1}{2}\xi L}{\sin^2\frac{1}{2}\xi\lambda_0}=\frac{L}{\lambda_0^2} \tag{11.4.50}$$

we can simplify Eq. (11.4.49) and write

$$E_{\text{heter}}=\frac{L}{2\lambda_0^2}\sum_{h=-\infty}^{\infty}{}'\int\frac{d^3k}{(2\pi)^2}\,\frac{\beta_1k_x^2+\beta_2k_y^2}{k_x^2+k_y^2+k_z^2}\,\frac{4\sin^2\frac{1}{2}k_zd}{k_z^2}\,|S(k_x,k_y)|^2\,\delta\left(\mathbf{ke}-\frac{2\pi h}{\lambda_0}\right) \tag{11.4.51}$$

Using the representation (11.4.48) in the argument of the delta-function in (11.4.51) and integrating over k_z, we obtain

$$E_{\text{heter}}=\frac{L}{2\lambda_0^2}\sum_{h=-\infty}^{\infty}{}' \iint_{-\infty}^{\infty}\frac{dk_x dk_y}{(2\pi)^2}\frac{\beta_1 k_x^2+\beta_2 k_y^2}{k_x^2+k_y^2+[2\pi h/\lambda_1+k_x-k_y(n_1^0\sqrt{2}/n_3^0)]^2}$$
$$\times\frac{4\sin^2\{\frac{1}{2}d[2\pi h/\lambda_1+k_x-k_y(n_1^0\sqrt{2}/n_3^0)]\}}{[2\pi h/\lambda_1+k_x-k_y(n_1^0\sqrt{2}/n_3^0)]^2}|S(k_x, k_y)|^2 \qquad (11.4.52)$$

where

$$\lambda_1=\frac{\lambda_0 n_3^0}{\sqrt{2}} \qquad (11.4.53)$$

is the separation between the nearest platelets of the same kind in the [110] direction.

Since $|S(k_x, k_y)|^2$ is nonzero if $k_x\sim 2\pi/D$ and $k_y\sim 2\pi/D$ where D is the thickness of the multidomain thin-plate crystal, the following approximation can be used

$$\frac{\beta_1 k_x^2+\beta_2 k_y^2}{k_x^2+k_y^2+[2\pi h/\lambda_1+k_x-k_y(n_1^0\sqrt{2}/n_3^0)]^2}\,\frac{4\sin^2\frac{1}{2}d[2\pi h/\lambda_1+k_x-k_y(n_1^0\sqrt{2}/n_3^0)]}{[2\pi h/\lambda_1+k_x-k_y(n_1^0\sqrt{2}/n_3^0)]^2}$$
$$\approx\frac{\beta_1 k_x^2+\beta_2 k_y^2}{k_x^2+k_y^2+(2\pi h/\lambda_1)^2}\frac{\lambda_1^2}{\pi^2}\frac{\sin^2\pi h(d/\lambda_1)}{h^2} \qquad (11.4.54)$$

The approximation (11.4.54) holds if

$$\frac{2\pi h}{\lambda_1}\gg k_x\sim k_y\sim\frac{2\pi}{D}$$

at any $|h|\geqslant 1$. This is the case when

$$\frac{\lambda_1}{D}\ll 1 \qquad (11.4.55)$$

namely, when the period of the domain distribution is far less than the thickness of the complex.

Substituting Eq. (11.4.54) into (11.4.52) yields

$$E_{\text{heter}}=\frac{1}{2}L\left(\frac{\lambda_1}{\lambda_0}\right)^2\sum_{h=-\infty}^{\infty}{}'\iint_{-\infty}^{\infty}\frac{dk_x dk_y}{(2\pi)^2}\frac{\beta_1 k_x^2+\beta_2 k_y^2}{k_x^2+k_y^2+(2\pi h/\lambda_1)^2}|S(k_x, k_y)|^2\frac{\sin^2\pi hw}{\pi^2 h^2} \qquad (11.4.56)$$

where $w=d/\lambda_1$ is the volume fraction of domains of the first type.

Let us introduce the definitions

$$I_1(h)=\iint_{-\infty}^{\infty}\frac{dk_x dk_y}{(2\pi)^2}\frac{\beta_2 k_y^2}{k_x^2+k_y^2+(2\pi h/\lambda_1)^2}|S(k_x, k_y)|^2 \qquad (11.4.57a)$$

$$I_2(h)=\iint_{-\infty}^{\infty}\frac{dk_x dk_y}{(2\pi)^2}\frac{\beta_1 k_x^2}{k_x^2+k_y^2+(2\pi h/\lambda_1)^2}|S(k_x, k_y)|^2 \tag{11.4.57b}$$

With these definitions, Eq. (11.4.56) may be rewritten in the form

$$E_{\text{heter}}=\frac{1}{2}L\left(\frac{\lambda_1}{\lambda_0}\right)^2\frac{1}{\pi^2}\sum_{h=-\infty}^{\infty}{}' \frac{\sin^2 \pi hw}{h^2}[I_1(h)+I_2(h)] \tag{11.4.58}$$

The integrals (11.4.57a) and (11.4.57b) are calculated in Appendix 3. They are equal to [see Eqs. (A.3.18) and (A.3.19)]

$$I_1(h)=\frac{\beta_2}{2}\frac{1}{|2\pi h/\lambda_1|}\oint\frac{dl}{1+(dy/dx)^2}$$

and

$$I_2(h)=\frac{\beta_1}{2}\frac{1}{|2\pi h/\lambda_1|}\oint\frac{dl\,(dy/dx)^2}{1+(dy/dx)^2}$$

where dl is the line element of the contour $y=y(x)$ enveloping a monodomain platelet in its (110) habit plane (see Fig. 94). The integrations in $I_1(h)$ and $I_2(h)$ are over the contour $y=y(x)$. Substituting the expressions for $I_1(h)$ and $I_2(h)$ into Eq. (11.4.58), we obtain

$$E_{\text{heter}}=\frac{1}{2}\lambda_1 L(n_3^0)^2\frac{1}{(2\pi)^3}\sum_{h=-\infty}^{\infty}{}' \frac{\sin^2 \pi hw}{|h|^3}\oint\frac{\beta_2+\beta_1(dy/dx)^2}{1+(dy/dx)^2}dl \tag{11.4.59}$$

where the relation $(\lambda_1/\lambda_0)^2=\frac{1}{2}(n_3^0)^2$ obtained from Eq. (11.4.53) is used. The integration in (11.4.59) is again over the contour $y=y(x)$ in the (110) habit plane of the platelet. The shape of the platelet in the (110) plane is that of a lath-like parallelogram. We can see from Fig. 94 that the longer side of the parallelogram is given by intersection of two planes, the (110) plane and the habit plane of the multidomain plate normal to the vector $\mathbf{n}_0$. The shorter side of the parallelogram is the intersection of the habit plane of the complex and the (010) plane. Its direction is parallel to the [001] direction of the cubic matrix.

Since the longer parallelogram side is the straight line obtained as intersection of two planes whose normals are

$$\mathbf{m}=\left(\frac{1}{\sqrt{2}}, \frac{1}{\sqrt{2}}, 0\right) \quad \text{and} \quad \mathbf{n}_0=(n_1^0, 0, n_3^0)$$

(the vector coordinates are here again related to the usual [100], [010], [001] basis), the direction of the intersection line is determined as the vector product of $\mathbf{m}$ and $\mathbf{n}_0$:

$$\boldsymbol{\tau}=\frac{\mathbf{m}\times\mathbf{n}_0}{|\mathbf{m}\times\mathbf{n}_0|}=\frac{1}{\sqrt{(n_1^0)^2+2(n_3^0)^2}}(n_3^0, -n_3^0, -n_1^0) \tag{11.4.60}$$

The unit vector $\boldsymbol{\tau}$ written in the new coordinate system whose x-, y-, and z-axes are parallel to the $[1\bar{1}0]$, [001], and [110] directions of the cubic matrix, respectively, is as follows

$$\boldsymbol{\tau}' = (\sqrt{2}n_3^0, -n_1^0, 0)\frac{1}{\sqrt{(n_1^0)^2 + 2(n_3^0)^2}} \tag{11.4.61}$$

The integration in Eq. (11.4.59) is actually over the parallelogram perimeter depicted in Fig. 95. The ratio of the contour integrals taken over the shorter and longer sides of the parallelogram is proportional to the ratio of the side lengths and, therefore, to the ratio D/L_τ (the parallelogram width is of the order of the thickness of the complex, D). Because

$$\frac{D}{L_\tau} \ll 1$$

we can restrict the integration in Eq. (11.4.59) to the longer sides of the parallelogram, AD and CB (see Fig. 95).

Since the segments AD and CB are parallel to the vector $\boldsymbol{\tau}'$ given by (11.4.61), they are described by the linear equations

$$y_+(x) = y_+(0) - \frac{n_1^0}{\sqrt{2}n_3^0}x$$

$$y_-(x) = y_-(0) - \frac{n_1^0}{\sqrt{2}n_3^0}x \tag{11.4.62}$$

respectively. It follows from (11.4.62) that

$$\frac{dy}{dx} = -\frac{n_1^0}{\sqrt{2}n_3^0} \tag{11.4.63}$$

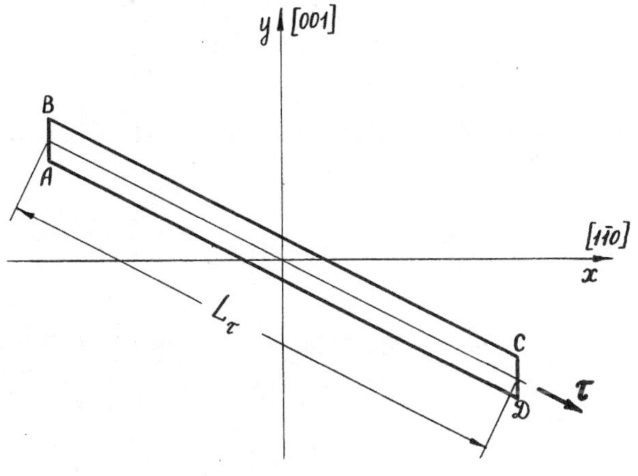

Figure 95. The (110) plane cross section of the thin-plate multidomain crystal showing the shape of interdomain boundaries; L_τ is the length of a parallelogram $ABCD$ describing the interdomain boundary [compare with Fig. 94].

Substitution of (11.4.63) into (11.4.59) yields

$$E_{\text{heter}}=\frac{1}{2}\lambda_1 L(n_3^0)^2\frac{1}{(2\pi)^3}\sum_{h=-\infty}^{\infty}{}' \frac{\sin^2 \pi h w}{|h|^3}\left[\frac{\beta_2+\frac{1}{2}\beta_1(n_1^0/n_3^0)^2}{1+\frac{1}{2}(n_1^0/n_3^0)^2}\right]2L_\tau \tag{11.4.64}$$

It follows from Fig. 94 that the width of the complex, $L_\perp$, in the direction [010] is $L_\tau\tau_y$ (by definition, τ_y is the projection of the unit vector, $\boldsymbol{\tau}$, on the axis [010] which represents the y-axis). According to the definition (11.4.60)

$$|\tau_y|=\frac{n_3^0}{\sqrt{(n_1^0)^2+2(n_3^0)^2}} \tag{11.4.65}$$

Therefore

$$L_\perp=L_\tau\tau_y=L_\tau\frac{n_3^0}{\sqrt{(n_1^0)^2+2(n_3^0)^2}}=\frac{L_\tau}{\sqrt{2+(n_1^0/n_3^0)^2}}$$

or

$$L_\tau=L_\perp\sqrt{2+\left(\frac{n_1^0}{n_3^0}\right)^2} \tag{11.4.66}$$

Using the definition (10.3.28) and Eq. (11.4.66) in Eq. (11.4.64), we obtain

$$E_{\text{heter}}=\frac{1}{2}\lambda_1 S\frac{\sqrt{2}}{(2\pi)^3}\alpha(w)\frac{\beta_2(n_3^0)^2+\beta_1\frac{1}{2}(n_1^0)^2}{\sqrt{1+\frac{1}{2}(n_1^0/n_3^0)^2}} \tag{11.4.67}$$

where $S_{\text{hab}}=2LL_\perp$ is the total area of two habit planes that are interphase boundaries. As follows from Eq. (11.4.35),

$$\left(\frac{n_1^0}{n_3^0}\right)^2=\left|\frac{\varepsilon_{11}^0+\varepsilon_{33}^0}{\varepsilon_{11}^0}\right| \tag{11.4.68}$$

Using Eqs. (11.4.68), (11.4.28), and (11.4.31) in Eq. (11.4.67), we obtain the final result

$$E_{\text{heter}}=\lambda_1 S_{\text{hab}}\frac{\sqrt{2}}{(2\pi)^3}\alpha(w)(\varepsilon_{33}^0-\varepsilon_{11}^0)^2 c_{44}$$

$$\times\frac{1}{2}\frac{|\varepsilon_{11}^0|+2|\varepsilon_{11}^0+\varepsilon_{33}^0|(c_{11}+c_{12})/(c_{11}+c_{12}+2c_{44})}{\sqrt{1+\frac{1}{2}(|\varepsilon_{11}^0+\varepsilon_{33}^0|/|\varepsilon_{11}^0|)}\sqrt{|\varepsilon_{11}^0+\varepsilon_{33}^0|+|\varepsilon_{11}^0|}} \tag{11.4.69}$$

Eq. (11.4.69) may be rewritten as

$$E_{\text{heter}}=S_{\text{hab}}\,\zeta_0 c_{44}(\varepsilon_{33}^0-\varepsilon_{11}^0)^2\lambda_1 \tag{11.4.70}$$

where

$$\zeta_0 = \frac{1}{2}\frac{\sqrt{2}}{(2\pi)^3}\alpha(w)\frac{|\varepsilon_{11}^0| + 2|\varepsilon_{11}^0 + \varepsilon_{33}^0|(c_{11}+c_{12})/(c_{11}+c_{12}+2c_{44})}{\sqrt{1+\frac{1}{2}(|\varepsilon_{11}^0+\varepsilon_{33}^0|/|\varepsilon_{11}^0|)}\sqrt{|\varepsilon_{11}^0+\varepsilon_{33}^0|+|\varepsilon_{11}^0|}} \tag{11.4.71}$$

is a dimensionless constant.

Since $\lambda_1 = d/w_0$ [w_0 is given by (11.3.11)], Eq. (11.4.70) is in a complete agreement with the qualitative estimation (11.1.3). In fact the characteristic modulus, λ, and the stress-free strain, ε_0, are, according to Eq. (11.4.70), equal to c_{44} and $\varepsilon_{33}^0 - \varepsilon_{11}^0$, respectively. The estimation (11.1.3) is obtained with accuracy the dimensionless constant, ζ_0/w_0.

Let us calculate the equilibrium period, λ_1, of a domain array consisting of a thin-plate complex. As shown in Section 11.1, the equilibrium period, λ_1, as well as the domain thickness, d, is determined by competition between the strain energy, E_{heter}, which approaches zero when $\lambda_1 \to 0$ and surface interdomain energy which increases infinitely when $\lambda_1 \to 0$. The surface energy includes contributions from all (110) interdomain boundaries separating the adjacent (110) twin-related tetragonal domains characterized by the stress-free strains, $\hat{\varepsilon}^0(1)$ and $\hat{\varepsilon}^0(2)$.

The total interdomain surface energy is equal to the interdomain area, S_p, between two adjacent domains (the area of the parallelogram depicted in Fig. 95) multiplied by the number of these boundaries, $2L/\lambda_0$. The total interdomain area is thus

$$2S_p\frac{L}{\lambda_0} \tag{11.4.72}$$

On the other hand, the total volume, V, of the complex may be represented as the "unit cell" volume of the periodic domain array multiplied by the number of all "unit cells", L/λ_0. Since the "unit cell" volume is $\lambda_1 S_p$ where λ_1 is the period in the [110] direction and S_p the area of the cross section of the complex by the (110) plane, we have

$$\lambda_1 S_p \frac{L}{\lambda_0} = V \tag{11.4.73}$$

Using Eq. (11.4.73) in (11.4.72), we obtain the total interdomain area in the form

$$2S_p\frac{L}{\lambda_0} = 2\frac{V}{\lambda_1} \tag{11.4.74}$$

The multiplication of the total interdomain surface area (11.4.74) by the surface energy coefficient $\gamma_{(110)}$ yields the total surface energy:

$$E_{\text{surface}} = \gamma_{(110)}\frac{2V}{\lambda_1} \tag{11.4.75}$$

Combining the total strain energy (11.4.70) and the interdomain surface energy

(11.4.75), we obtain

$$E_{\text{total}} = S_{\text{hab}}\, \zeta_0 c_{44}(\varepsilon_{33}^0 - \varepsilon_{11}^0)^2 \lambda_1 + 2\frac{\gamma_{(110)} V}{\lambda_1} \tag{11.4.76}$$

The minimization of Eq. (11.4.76) with respect to the period, λ_1, yields

$$\lambda_1 = \sqrt{\frac{\gamma_{(110)}}{\zeta_0 c_{44}(\varepsilon_{33}^0 - \varepsilon_{11}^0)^2} D} \tag{11.4.77}$$

where $D = 2V/S_{\text{hab}}$ is the thickness of the multidomain tetragonal phase plate. The thickness of tetragonal domains of the first type (with the [100] tetragonal axis) is

$$d_1 = w_0 \lambda_1$$

The thickness of domains of the second type (with the [010] tetragonal axis) is

$$d_2 = \lambda_1 - w_0 \lambda_1 = \lambda_1(1 - w_0)$$

As emphasized in Section 11.1, the relation of type of (11.4.77) is typical for domain structures of uniaxial magnetics and ferroelectrics because of the deep analogy between these phenomena.

Let us estimate the strain-induced interphase energy (11.4.69) for Fe-31Ni martensite. According to (194) in this alloy

$$\begin{aligned} c_{11} &= 1.404 \times 10^{12} \text{ dyne/cm}^2 \\ c_{12} &= 0.840 \times 10^{12} \text{ dyne/cm}^2 \\ c_{44} &= 1.121 \times 10^{12} \text{ dyne/cm}^2 \end{aligned} \tag{11.4.78a}$$

The crystal lattice parameter measurements (93) gave the values [see (6.6.1)]

$$\varepsilon_{11}^0 = \eta_1 - 1 = 0.1322, \quad \varepsilon_{33}^0 = \eta_3 - 1 = -0.1994 \tag{11.4.78b}$$

Substitution of the numerical parameter values (11.4.78b) into (11.3.11), (11.3.17), (11.3.18), and (11.3.19) gives

$$w_0 = 0.6010 \tag{11.4.79a}$$

$$\begin{aligned} \mathbf{n}_0 &= (0.5805, 0, 0.8142) \\ \boldsymbol{l} &= (0.5805, 0, -0.8142) \\ \varepsilon_0 &= 0.1994 \end{aligned} \tag{11.4.79b}$$

[compare (11.4.79a) with the value of $x_0 = 0.614$ obtained from Eq. (6.6.2) of the crystallographic theory].

It follows from the plot of $\alpha(\gamma_1)$ with relation to γ_1 (Fig. 78) that

$$\alpha(w_0) = \alpha(0.601) = 1.96 \tag{11.4.80}$$

Substituting (11:4.78) and (11.4.80) into (11.4.71), we obtain

$$\zeta_0 = 2.22 \times 10^{-3} \tag{11.4.81}$$

Maki and Wayman (231) reported the mean twin width in Fe-31Ni-0.23C martensite to be 102 Å, with the narrower region regarded as "twin" and the wider as "matrix." Within the theory formulated above, the terms "twin" and "matrix" should be referred to tetragonal domains of the second and first types, respectively. In fact the calculated thickness of domains of the first type is larger than that of domains of the second type:

$$d_1=\lambda_1 w_0=0.601\lambda_1 \tag{11.4.82a}$$

$$d_2=\lambda_1(1-w_0)=0.399\lambda_1 \tag{11.4.82b}$$

Assuming $d_2=102$ Å, we have

$$0.399\lambda_1=102\ \text{Å} \quad \text{and thus}$$

$$\lambda_1 \approx 255\ \text{Å}=2.55\times 10^{-6}\ \text{cm} \tag{11.4.83}$$

With the numerical values (11.4.83), (11.4.78), (11.4.81), and (11.4.70) we can estimate the strain energy contribution to the interphase energy:

$$\frac{E_{\text{heter}}}{S}=\zeta_0 c_{44}(\varepsilon_{33}^0-\varepsilon_{11}^0)^2\lambda_1 \approx 0.70\times 10^3\frac{\text{erg}}{\text{cm}^2} \tag{11.4.84}$$

It should be noted that Eq. (11.4.77) enables one to determine the interdomain (twin) specific surface energy if the domain thickness and the thickness of the multidomain tetragonal phase plate are known. For instance, using the electron micrograph of thin-plate martensite in Fe-30Ni-0.39C from (231), we can estimate the thickness D and period λ_1 (compare with Fig. 94). This gives

$$D \approx 1.650\times 10^{-4}\ \text{cm}$$

$$\lambda_1 \approx 220\ \text{Å}=2.2\times 10^{-6}\ \text{cm} \tag{11.4.85}$$

It follows from Eqs. (11.4.77) and (11.4.70) that

$$\frac{\lambda_1^2}{D}\zeta_0 c_{44}(\varepsilon_{33}^0-\varepsilon_{11}^0)^2=\gamma_{(110)} \tag{11.4.86}$$

Using the numerical data (11.4.85) and (11.4.81) and (11.4.78) in (11.4.86), we obtain

$$\gamma_{(110)} \approx 8\frac{\text{erg}}{\text{cm}^2} \tag{11.4.87}$$

In other words, the specific twin surface energy consists of about 1 percent of the strain-induced interphase energy (11.4.84).

The coherent specific cubic-tetragonal phase interphase energy seems to be of the same order of magnitude as the specific twin surface energy:

$$\gamma_{\text{interph}}^{\text{coherent}} \sim 10\frac{\text{erg}}{\text{cm}^2} \tag{11.4.88}$$

This value may also be estimated if we assume that the energy increase caused by the transformation of a bcc lattice unit cell into the fcc lattice fragment may be related to the new bcc → fcc interphase formed. This interphase energy is

$$\gamma_{\text{interph}}^{\text{coherent}} 6a_{\text{bcc}}^2 \tag{11.4.89}$$

where $6a_{\text{bcc}}^2$ is the surface area of the bcc lattice unit cell (a_{bcc} is the bcc lattice parameter). On the other hand, the total energy release in cooling-induced fcc-to-bcc transformation in Fe-31Ni amounts to 290 cal/mol (232). This means that the internal energy increase caused by the bcc → fcc transformation at low temperatures (at which the entropy contribution to the free energy may be neglected) is just equal to 290 cal/mole = 1218 J/mole = 1.218×10^{10} erg/mole. Since the bcc unit cell contains two atoms, the internal energy increase in the bcc unit cell transformation is

$$1.218 \times 10^{10} \times \frac{2}{6.02 \times 10^{23}} \text{ erg} = 0.404 \times 10^{-13} \text{ erg} \tag{11.4.90}$$

On the assumption that (11.4.89) is equal to (11.4.90), we obtain

$$6a_{\text{bcc}}^2 \gamma_{\text{interph}}^{\text{coherent}} = 0.404 \times 10^{-13} \text{ erg}$$

or

$$\gamma_{\text{interph}}^{\text{coheret}} = \frac{0.404 \times 10^{-13} \text{ erg}}{6 \times (2.86)^2 \times 10^{-16} \text{ cm}^2} = 8.23 \frac{\text{erg}}{\text{cm}^2} \tag{11.4.91}$$

This result compares well with the estimate (11.4.88). As is known, by following the same line of reasoning, one can correctly estimate the vacancy formation energy from the vacancy surface area and the sublimation energy per atom required to produce a vacancy.

It should be emphasized here that the "chemical" coherent interphase energy is about two orders of magnitude lower than that of a semicoherent boundary given by Eq. (11.4.70). Homogeneous martensitic crystals cannot be formed, however, because of the large volume-dependent strain energy contribution to the free energy. In actual systems the interphase boundary is seemingly always semicoherent, and its energy assumes the value of the order given by Eq. (11.4.70). The difference between coherent and semicoherent interphase energies should, however, be taken into account in nucleation theory.

A few points concerning the approximation employed in the strain energy calculations should be mentioned. Unlike the calculations of ΔE_{homog} carried out in Section 11.3, where the use of matrix elastic constants does not lead to large errors (the error is of the order of $D/L \ll 1$), the calculation of E_{heter} is affected substantially by differences between matrix and inclusion elastic constants. This is so because the short-range strain field is concentrated in the vicinity of the interphase boundary in both the matrix and inclusion. The calculations for Fe-Ni martensite described here can therefore only be regarded as an estimation.

11.5. STRAIN-INDUCED COARSENING IN COHERENT MIXTURE OF CUBIC AND TETRAGONAL PHASES

As already noted, the very fact that strain energy contributes to the free energy makes the two-phase morphology an additional "thermodynamic parameter." The free energy minimization with respect to this "parameter" leads to the equilibrium structure, a multidomain thin-plate of a tetragonal phase in a cubic matrix. With decomposition this equilibrium state may only be attained through strain-induced coarsening. In fact the as-quenched state is usually formed as a random distribution of fine monodomain precipitates of a tetragonal phase. The strain energy of the system is then proportional to the total volume of the tetragonal phase. Decrease of the strain energy is realized by both shape transformations of single precipitates and their mutual rearrangement. The latter process may be treated as partial ordering of the distribution of precipitates. As mentioned in Section 10.5, a similar process (formation of a basketlike structure) takes place during strain-induced coarsening of mixtures of two cubic phases. From this point of view the equilibrium state discussed in Sections 11.1 to 11.4 is a fully ordered state.

To account for partial ordering, we must replace the functions $\tilde{\theta}_1(\mathbf{r})$ and $\Delta\tilde{\theta}_1(\mathbf{r})$ with their mean values, $\langle\tilde{\theta}_1(\mathbf{r})\rangle$ and $\langle\Delta\tilde{\theta}_1(\mathbf{r})\rangle$, obtained by averaging over the whole crystal. According to the results obtained in Section 11.4, the heterogeneous strain energy tends to zero when the squared modulus of the Fourier transform $|\langle\Delta\theta_1(\mathbf{k})\rangle|^2$ only differs from zero at the reciprocal lattice points along the $\langle 110\rangle$ directions in the $\mathbf{k}$-space.

In particular, such is the case with a partially ordered one-dimensional distribution of precipitates within a "sandwich" composed of regularly spaced alternating (110) layers. Each layer is occupied preferentially by monodomain precipitates with the same tetragonal axis direction (such as [100]), whereas the layer nearest to it is occupied by precipitates with the twin-related direction of the tetragonal axis ([010] direction). A schematic of such a structure is given in Fig. 96.

The desirable distribution described by the function $|\langle\Delta\theta_1(\mathbf{k})\rangle|^2$ may also be provided by a three-dimensional distribution which is a superposition of one-dimensional layer systems depicted in Fig. 96 along some of the six $\langle 110\rangle$ directions. In this case the function $|\langle\Delta\theta_1(\mathbf{k})\rangle|^2$ will also differ from zero along chosen $\langle 110\rangle$ directions in the $\mathbf{k}$-space. Such distribution is in many respects similar to the "basket" structure formed in mixtures of two cubic phases (see Section 10.5).

It is noteworthy that partially ordered structures formed by regularly spaced $\{110\}$ layers filled with precipitates having preferentially the same tetragonal axis direction show a tendency to form $\langle 110\rangle$ precipitate arrays. It seems possible that the so-called "tweed" structure observed at the early stages of decomposition reactions involving cubic-to-tetragonal phase transformations are associated with the phenomenon described above. The tweed structure displays $\langle 110\rangle$ alignment on electron micrographs (Fig. 97). This conclusion seems to agree

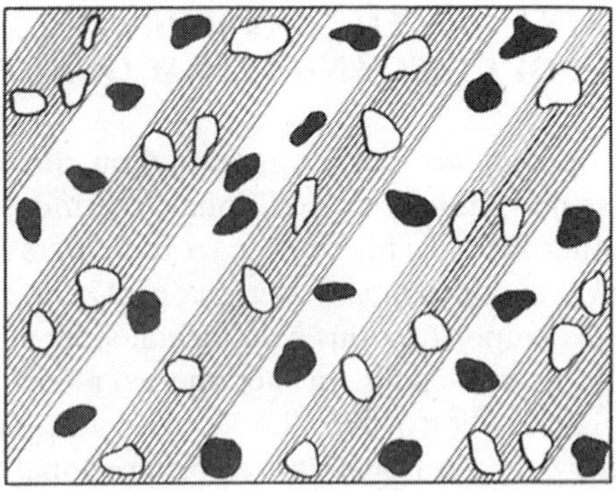

Figure 96. Schematic drawing of a "partially ordered" one-dimensional distribution of tetragonal phase precipitates within alternating regularly spaced (110) layers. Shaded layers are preferentially occupied by precipitates along the [100] direction of the tetragonal axis. All other layers are preferentially occupied by precipitates along the [010] direction of the tetragonal axis. Precipitates along the [100] and [010] directions of the tetragonal axis are rendered as white and black regions, respectively.

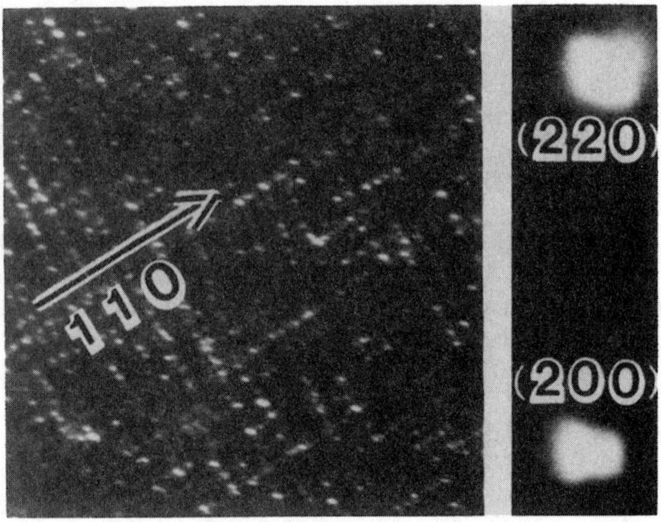

Figure 97. Dark-field electron micrograph of the tweed structure in a Cu-2.0% Be alloy (257). (Courtesy of R. J. Rioja and D. E. Laughlin.)

with the computer simulation results obtained by Wen et al. (233, 234) and described in Section 12.4.

The strain-induced coarsening process involves small-scale redistribution of precipitates which requires their movements over distances on the order of the mean separation length. Large-scale redistribution takes more time. It is therefore to be expected at the later stages of strain-induced coarsening. Strain energy decrease caused by large-scale morphology reconstruction is associated with changes in the function $|\langle\theta(\mathbf{k})\rangle|^2$ describing the shape of the complex as a whole and thus controls the shape of the complexes (the case of small **k**). The corresponding strain energy minimization has been described in Section 11.3. It leads to thin-plate shapes with small aspect ratios, D/L. Development of strain-induced coarsening should eventually yield thin-plate multidomain crystals depicted in Fig. 94. In this respect there is no difference between the martensitic transformation and decomposition.

11.6. MORPHOLOGY OF ALLOYS COMPOSED OF CUBIC AND TETRAGONAL PHASES: COMPARISON WITH EXPERIMENTAL OBSERVATIONS

The theoretical results described in the preceding sections bridge the strain energy approach and the phenomenological crystallographic theories (1, 2) of martensitic transformations. As demonstrated in Sections 8.1 and 11.3, the two approaches become fully equivalent in the particular case when the stress-free transformation strain is an invariant plane strain and the macroscopic shear is far less than unity. The reader may be referred to the excellent reviews by Wayman (94) and Christian (235) which demonstrate that the crystallographic theories give a very good description of martensite morphologies if plastic deformation modes are not involved into the transformation geometry. These problems have been discussed in part in Sections 6.5, 6.6, and 9.6. There is, however, a group of related phenomena that have not been studied as thoroughly as the martensitic transformation has been. Here belong decomposition reactions generating coherent tetragonal phase precipitates in a cubic phase matrix. According to what has been said in Sections 11.2 to 11.5, the decomposition should eventually lead to martensitelike morphologies if a coherent interphase is maintained. The basic difference between these two solid state reactions mainly pertains to kinetics. For instance, the final thin-plate multidomain structure of martensitic crystals is formed practically instantaneously, whereas with the decomposition this morphology is attained by successive transformations of shapes, sizes, orientations, and mutual arrangements of tetragonal phase monodomain precipitates. The origin of this difference is that structure transformations during the decomposition require long-range diffusion whereas martensitic transformations do not.

From the kinetic standpoint, ordering in stoichiometric alloys stands nearer to martensitic transformations than to decomposition. Both ordering and martensitic transformations proceed without concentration changes and therefore does not require long-range diffusion to provide different concentrations of the solute in the coexisting phases. Order-disorder transitions involve only short-range diffusion (within the superlattice unit cell volume) which allows long-range order and completion of the process.

Structure evolution in decomposition usually starts from a randomlike distribution of monodomain precipitates of a tetragonal phase. In the course of strain-induced coarsening, the "tweed" morphology develops (see Fig. 97). The reconstruction of the microstructure occurs through "uphill" diffusion which results in changes of precipitate shapes and precipitate motion to reduce the strain energy. According to the theoretical predictions of Section 11.5 confirmed by the computer simulation described in Section 12.4, precipitates form ⟨110⟩ regular arrays (see Fig. 96). The volume-dependent strain energy decreases in the formation of the "tweed" structure, but it is still nonzero. The volume-dependent elastic energy vanishes completely during coarsening of the tweed structure when monodomain tetragonal precipitates aggregate to produce a thin-plate multidomain tetragonal phase crystal. The resulting microstructure corresponds to the equilibrium state. The similarity of the morphology to that of martensitic crystals is indeed striking; even such purely martensitic structure characteristics as surface relief is also observed.

Experimental studies of all the stages of isothermal coarsening have found the rates of various processes involved incommensurable. If, for instance, a slow stage takes a reasonable duration of aging, it is very probable that fast stages will be missed, and vice versa: if a fast process takes a reasonable period of time, the detection of slow stages, far more prolonged, is hardly possible. This is the reason why various coarsening stages are usually observed under different temperature conditions.

A tweed structure was observed by Van Landyut experimentally in the decomposition of intersitial bcc-based Nb-O alloy (236). In agreement with the theoretical predictions coarsening results in the formation of groups of "sandwiches," each consisting of a small number of monodomain platelets of the tetragonal phase with the (110) habit and alternating directions of the tetragonal axis. Since neighboring platelets are twin-related about the (110) plane, each "sandwich" looks like an internally twinned tetragonal phase crystal (Fig. 98).* Further coarsening results in the formation of internally (110) twinned lenticular-shaped crystals (Fig. 99).

A similar sequence of microstructure transformations has been observed in substitutional solid solutions in transitions of fcc-disordered solutions into tetragonal-ordered phases of CuAuI, CoPt ($L1_0$ phases) and Ni_3V (DO_{22}). Electron microscopic studies of the early stages of the phase transformation in CuAuI (237, 238), CoPt (239), and Ni_3V (240) alloys performed at low temperatures revealed the presence of tweed microstructures. Coarsening of tweed structures during heating resolves the new phase particles into very dense regular arrays detectable in superlattice reflection images (237, 239, 240). Electron microscopic observations show the particles to be (001) thin plates arranged in a stair-step fashion, which roughly follows {110} traces in a (001)

*As a reminder to the reader, single monodomain precipitates of a new phase formed in the initial stage of aging of Nb-O alloy have the {310} habit (162) rather than the {110} one observed in lenticular new phase crystals (see Fig. 99), which are formed when monodomain precipitates aggregate into groups. The {310} habit may be explained in terms of the strain energy minimum condition approach (see Section 9.2).

(a)

(b)

(c)

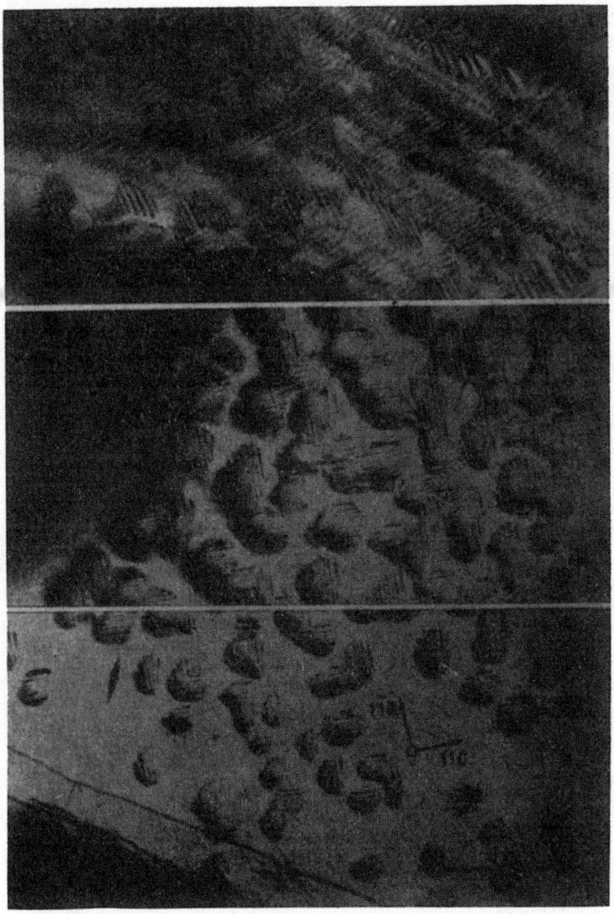

Figure 98. Electron micrographs of consecutive stages of the coarsening process in a Nb-O alloys (236), from (*a*) to (*c*). (Courtesy of J. van Landuyt.)

foil CoPt (239). In further growth during aging the particles cluster and coalesce (239, 240). Impingement of new phase monodomain particles produces assemblies of (110) twins later transformed into heterogeneous lamellae. Each lamella is a "sandwich" of monodomain platelets with the (110) habit and alternating directions of the tetragonal axis. The corresponding multidomain structure looks as if it were formed in (110) twinning [see Fig. 100 taken from (241)]. In principle the morphology of these thin-plate multidomain crystals coincides with the morphology of martensitic crystals. This morphology has, in particular, been observed in a cubic-to-orthorhombic ordering reaction in CuAu. The remarkable agreement between the crystallographic theory and the experimental results [242] is noteworthy.

Figure 99. Electron micrograph of internally twinned lenticular shaped crystals in Nb-O alloy (236). (Courtesy of J. van Landuyt.)

It should be stressed here again that the formation of all microstructures discussed does not involve actual (110) twinning. The twinlike structure is merely produced by the growth and coarsening mechanism described in Sections 11.2 to 11.5. Pennison et al. (239) have reported rather strong evidence that confirms this conclusion. A detailed discussion of these problems may be found in the review by Tanner and Leamy (243). One may also find there strong arguments for the key role of the strain energy in the succession of microstructures observed.

Stable martensitelike structures were observed in Zr-H interstitial solid solutions where fcc ordering of the fcc-based interstitial solution (δ hydride) and formation of tetragonal metastable-ordered γ and ε phases occurs (244–247). The martensitelike structure was also observed in ZrH_2 where ordering also results in transition from the cubic to the tetragonal phase. The hypothesis that these phase transformations involve martensitic shear is not necessary to account for all the crystallographic and morphological features observed. The macroscopic shear is, as a matter of fact, a secondary effect produced by crystal

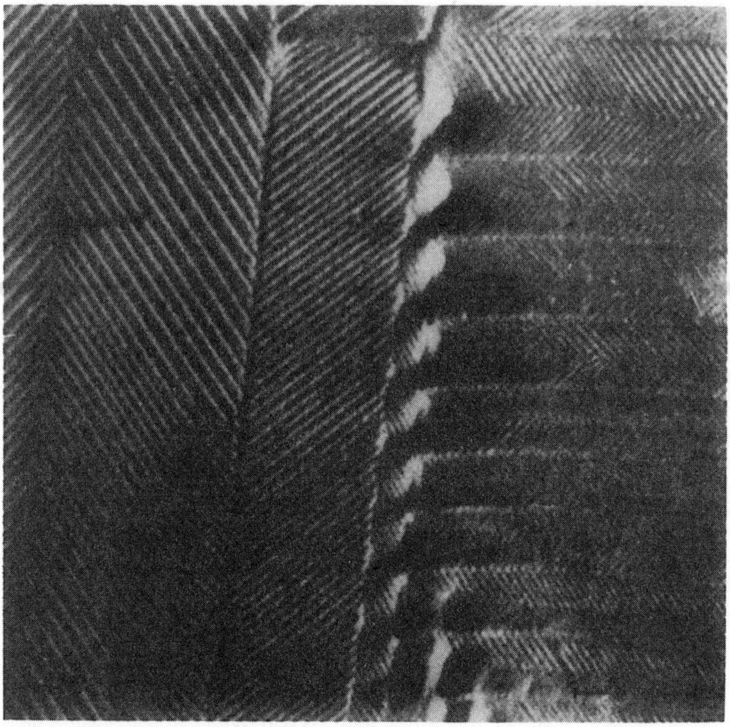

Figure 100. Electron micrograph of the multidomain structure of the tetragonal phase in a Cu-Au alloy (241). (Courtesy of V. I. Syutkina and E. S. Jakovleva.)

lattice expansion generated by interstitial atom ordering. This effect, however, is decisive for determining the stable microstructure of ordered phases. The typical martensitelike structures observed in ZrH alloys and interpreted in terms of the crystallographic theory of martensitic transformation (249, 250) may also be described as a result of strain relaxation involved in usual diffusional transformations.

The decomposition of a bcc Ta-O alloy provides one more example of a microstructural reconstruction. According to Hörz (251), short-time aging of Ta-O alloys at 523° K results in the formation of a fine tweedlike microstructure. Aging of a Ta-3 at.%O alloy for 6×10^4 min leads to a fully developed TaO_y platelet pattern.* The morphology of TaO_y plates was studied by Pawel, Cathcart, and Campbell (252) and van Landuyt and Wayman (228). Oxidation of Ta samples at higher temperatures (of about 700° K) yields directly thin-plate multidomain morphology with the $(320)_b$ habit. The mean twinning fraction

*It should be remembered that in this case coarsening is complicated by accompanying phase transformations in suboxides.

observed, w_0 is within the range 0.84 to 0.88. Using the crystal lattice parameters of the orthorhombic TaO_y phase,

$$a = 3.271\ \text{Å},\quad b = 3.201\ \text{Å},\quad c = 3.610\ \text{Å} \tag{11.6.1}$$

reported in (254) and the measured parent bcc-phase lattice parameter,

$$a_0 = 3.322\ \text{Å} \tag{11.6.2}$$

Wayman and van Landuyt calculated the habit plane orientation (229). They applied the crystallographic theory of the martensitic transformation (see Sections 6.3 to 6.6) to calculate the habit plane normal:

$$\mathbf{n}_0 = (0.818,\ 0.016,\ 0.575) \tag{11.6.3}$$

This departs from the normal to the (302) habit plane observed by 1.68°.

Since the stress-free transformation strain corresponding to the bcc→(TaO_y) transition is far below unity, the habit can also be determined from the linear theory described in Section 11.3. In fact the transformation strain components calculated from (11.6.1) and (11.6.2) are

$$\varepsilon_{11}^0 = \frac{a}{a_0} - 1 = \frac{3.271}{3.322} - 1 = -0.0154$$

$$\varepsilon_{22}^0 = \frac{b}{a_0} - 1 = \frac{3.201}{3.322} - 1 = -0.0364$$

$$\varepsilon_{33}^0 = \frac{c}{a_0} - 1 = \frac{3.610}{3.322} - 1 = 0.0867 \tag{11.6.4}$$

The analogue of Eq. (11.3.9) for an orthorhombic transformation strain is

$$\hat{\boldsymbol{\varepsilon}}^0 = w\hat{\boldsymbol{\varepsilon}}^0(1) + (1-w)\hat{\boldsymbol{\varepsilon}}^0(2)$$

$$= w\begin{pmatrix} \varepsilon_{33}^0 & 0 & 0 \\ 0 & \varepsilon_{11}^0 & 0 \\ 0 & 0 & \varepsilon_{22}^0 \end{pmatrix} + (1-w)\begin{pmatrix} \varepsilon_{11}^0 & 0 & 0 \\ 0 & \varepsilon_{33}^0 & 0 \\ 0 & 0 & \varepsilon_{22}^0 \end{pmatrix} \tag{11.6.5}$$

At

$$w = w_0 = -\frac{\varepsilon_{33}^0}{\varepsilon_{11}^0 - \varepsilon_{33}^0} = -\frac{0.0867}{-0.0154 - 0.0867} = 0.85 \tag{11.6.6}$$

Eq. (11.6.5) may be rewritten as

$$\hat{\boldsymbol{\varepsilon}}^0 = \begin{pmatrix} \varepsilon_{11}^0 + \varepsilon_{33}^0 & 0 & 0 \\ 0 & 0 & 0 \\ 0 & 0 & \varepsilon_{22}^0 \end{pmatrix} \tag{11.6.7}$$

The value $w_0 = 0.85$ compares well with the mean fraction of one of the types of domains composing the multidomain plates determined experimentally, 0.84 to 0.88 (228).

Comparison of the matrix (11.6.7) with (11.3.13), and application of the same procedure as in Section 11.3, yields for the orthorhombic case

$$\mathbf{n}_0 = \left(\sqrt{\frac{|\varepsilon_{11}^0 + \varepsilon_{33}^0|}{|\varepsilon_{11}^0 + \varepsilon_{33}^0| + |\varepsilon_{22}^0|}},\ 0,\ \sqrt{\frac{|\varepsilon_{22}^0|}{|\varepsilon_{11}^0 + \varepsilon_{33}^0| + |\varepsilon_{22}^0|}} \right) \tag{11.6.8}$$

Substitution of the numerical values (11.6.4) into (11.6.8) yields the habit plane normal

$$\mathbf{n}_0 = (0.8136, 0, 0.5813) \tag{11.6.9}$$

The habit plane normal to the vector (11.6.9) makes an angle of 1.8° with the (302) habit plane observed. The linear theory result (11.6.9) only differs from the result of the more accurate calculation (11.6.3) by 1.2°.

In conclusion, two points should be mentioned. As shown in Chapters 9 to 11, the analysis of strain-induced interactions provides a considerable progress in predicting microstructures of actual two-phase alloys. The strain-energy relaxation approach makes it possible to describe principal characteristics of two-phase states if the elastic properties and crystal lattice parameters of the phases are known. The general result of the strain theory is that the formation of single monodomain inclusions requires strain energy expenditures that are too high and therefore for kinetic reasons occurs only in the initial stage of the decomposition. The gathering of monodomain new phase particles to "grains," consisting of (110) monodomain platelets with alternating (110) twin-related directions of the tetragonal axis, allows a significant strain energy decrease. These "grains" look like small twinned new phase crystals. Further strain energy relaxation takes place in an aggregation of these "grains" into thin-plate crystals in the pattern of twinned new phase lamellae.

The other point to be stressed is that Sections 11.1 to 11.5 dealt only with stable configurations occurring in the final stages of strain-induced coarsening, because only more or less plausible assumptions can be made concerning the actual kinetics of macrostructure transformations. An attempt to build up the theoretical basis for prediction of the microstructure kinetics will be undertaken in the next chapter where computer simulation of the martensitic transformation and decomposition reactions will be described. The formulation of the next chapter is entirely based on studies undertaken by Prof. Morris and his group at the University of California, Berkeley, in 1977 to 1980.

Finally, the reader has certainly recognized the strong analogy between modulated "basket" structures formed in coarsening of cubic precipitates in cubic matrices and "tweed" structures occurring in coarsening of tetragonal precipitates in cubic matrices. Both phenomena have been treated similarly, in terms of the same formalism of the strain energy minimization. Even though the structures have different geometries, the physics of the phenomena proves to be the same.

12

COMPUTER SIMULATION OF PHASE TRANSFORMATION IN CRYSTALLINE SOLIDS

We have shown that the elastic strain relaxation is the dominant factor that determines the development and successive reconstruction of a two-phase microstructure. The pure thermodynamic approach based on the strain energy concept, and applied in Chapters 8 to 11, however, fails to describe all important features of phase transformations. This is particularly the case with the phase transformation kinetics. The kinetics aspect poses a difficult problem because of the voluminous calculations, which can hardly be done manually, needed to construct a more or less realistic model. Progress in this field can only be made by the use of high-speed computers, by developing a computer simulation technique that investigates the morphology transformation sequence in two-phase alloys. This technique was employed to simulate the decomposition and martensitic transformation processes in alloy by Wen, Morris, and Khachaturyan (233, 234, 84, 85).

12.1. MARTENSITE TRANSFORMATION (84, 85)

As described in Sections 6.1 to 6.7, the martensitic transformation is a diffusionless transformation that accomplishes a change in crystal structure. Since, at least in the initial stages of the transformation, the particles of the new phase are coherently connected to the parent matrix, the problem of elastic accomodation necessarily plays the major role in the thermodynamics of the transformation. It is now commonly thought that the martensite transformation nucleates heterogeneously and that its initiation is catalized by the presence of crystal lattice defects whose strain fields may partly compensate for the strain associated with the formation of the martensite particle. Second, the nature of the martensite transformation is such that a single nucleation process is generally insufficient to permit the transformation to proceed to completion. The repeated nucleation of individual martensite plates is necessary. The strain energy associated with the martensite transformation may also be important in promoting this secondary nucleation, by catalyzing nucleation in those places and orientations that serve to relieve the internal elastic field.

The influence of elastic strain on the growth and morphology of martensite plates is conditioned by the fact that the martensite particle may form in any one of several crystallographic variants, each of which exhibits the preferred crystallographic relation with the parent matrix.

The existence of several distinct orientational variants of the product phase provides a configurational freedom that may be used to reduce the strain energy. The martensite plate is free to grow as a composite particle of two or more orientational variants which are so configured as to allow a mutual relaxation of elastic strain.

Given the prominence of elastic effects in the nucleation, growth, and morphology of martensite, it seems reasonable to suppose that many of the other interesting and unique features of the martensite transformation will have their source in the need to accommodate elastic strain as the transformation proceeds. Tractable models of the developing process of the martensite transformation that include the predominant elastic effects should lead to new theoretical insight. One such model can be drawn directly from the elastic theory developed in Chapter 7. Since this theory permits us to calculate the strain energy of an arbitrary distribution of coherent inclusions, it may be used to compute the strain energy of a progressing martensite transformation in any hypothetical intermediate stage and to identify the incremental transformation steps that are most favorable with respect to the strain energy. While such a model is not likely to be tractable by manual computation, the equations are such that they can quite easily be phrased for solution by a computer.

A simple elastic model of the developing martensitic transformation is formulated in the balance of this section. It is obtained by a straightforward adaptation of the strain-induced interaction theory presented in Section 7.2, with minor modifications to show how the possibly finite size of the elementary martensite particles and their surface and twin boundary energies may be taken into account.

12.1.1. Description of Martensitic Transformation in Terms of Elementary Particles

Since a computer operates with discrete numbers, any computer simulation of the martensitic transformation should be also a discrete process. It requires the introduction of a minimum element of martensitic phase which we shall call the elementary martensite particle. Any distribution of martensite within a crystal will then be represented by a suitable combination of elementary martensite particles. In the linear elasticity approach that we shall use the elastic strain within the crystal is merely the sum of the strain associated with each of the elementary martensite particles; the elastic strain energy is the sum of their self-energies plus the energy associated with their mutual interaction.

The smaller the elementary particle, the more accurately an arbitrary distribution of the martensite phase can be modeled. It should be recognized, however, that there is a physical lower boundary on the size of elementary particles that can reasonably be assumed to exist in isolation. This boundary is set by the need for the elementary particle to have the crystallographic identity of the martensite phase; if the particle is too small, the interphase surface tension will dominate and destroy the right internal atomic arrangement. The linear size of the elementary particle must at least exceed the correlation length required for a coherent transition from the parent phase to that of the martensite. This length in martensitic transformations is on the order of interatomic distances.

As shown in Section 6.2, the martensitic transformation usually follows the nucleation-and-growth mechanism, and the size of the critical nucleus of the martensitic phase is commensurate with the correlation length. The growth of a martensitic crystal, as well as the classical crystal growth from the gas phase, involves a succession of elementary events. In both cases an elementary event is the thermofluctuation formation of a growth embryo whose size is also on the order of the correlation length magnitude. It has been emphasized in Section 6.2 that the growth mechanism of a martensitic crystal resembles closely the mechanism of the dislocation motion under applied stress. In dislocation the role of the growth embryo is played by a double kink (see Fig. 53) whose size is also commensurate with the correlation length related to the intrinsic characteristics of the dislocation.

Since the critical size nucleus and growth embryo have approximately the same volume, they may be described by elementary martensite particles whose size is commensurate with the correlation length. Therefore actual elementary particles that describe the development of the martensite transformation have the minimum possible size allowed by the requirement of the crystallographic identity of the martensite phase. In fact elementary particles represent the minimum size "bricks" of the martensite phase. If there are ν orientational variants (ν types of structure domains) of the martensite, we must include in our consideration ν kinds of elementary particles and therefore ν kinds of "bricks." Using these "bricks" as "construction" material, we can always determine any desirable spatial distribution of arbitrary inclusions of the new phase. To do this, we must also introduce a sort of grid whose unit supercell coincides with the elementary particle shape and size. The grid thus becomes a framework whose unit supercells can be filled by elementary particles in any desirable order (Fig. 101).

Each of the ν kinds of elementary particles may be distinguished by its orientation and by its stress-free transformation strain, $\varepsilon_{ij}^0(p)\,(p=1, 2, \ldots, \nu)$.

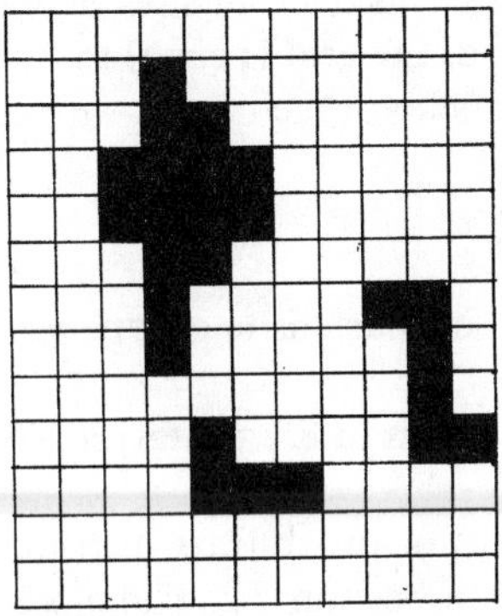

Figure 101. Schematic drawing of the grid lattice with unit supercells having the size and shape of an elementary martensite particle. Shaded supercells are filled by martensite particles.

Bearing in mind the definitions given, we may describe mathematically the spatial distribution of the martensitic phase by the function

$$\tilde{\zeta}_p(\mathbf{R}) = \begin{cases} 1 & \text{for an elementary particle of type } p \text{ at } \mathbf{R} \\ 0 & \text{otherwise} \end{cases} \tag{12.1.1}$$

where **R** is the translation vector of the grid. Since only one martensite particle may occupy a given cell of the grid, the $\tilde{\zeta}_p(\mathbf{R})$ satisfies the identity

$$\tilde{\zeta}(\mathbf{R}) = \sum_{p=1}^{\nu} \tilde{\zeta}_p(\mathbf{R}) = \begin{cases} 1 & \text{for } \mathbf{R} \text{ occupied by martensite} \\ 0 & \text{otherwise} \end{cases} \tag{12.1.2}$$

To establish the relation between the grid distribution function, $\tilde{\zeta}_p(\mathbf{R})$, and the shape function, $\tilde{\theta}_p(\mathbf{r})$, of the martensite phase orientational variant, p, let the shape of the grid supercell at **R** be described by the function

$$\tilde{\eta}_0(\mathbf{r}-\mathbf{R}) = \begin{cases} 1 & \text{if } \mathbf{r} \text{ is within the supercell at } \mathbf{R} \\ 0 & \text{otherwise} \end{cases} \tag{12.1.3}$$

where **r** is the vector labeling an arbitrary site of the parent phase lattice.

Using the definitions (12.1.1) and (12.1.3), we may represent the shape function, $\tilde{\theta}_p(\mathbf{r})$, of the martensite phase as

$$\tilde{\theta}_p(\mathbf{r}) = \sum_{\mathbf{R}} \tilde{\eta}_0(\mathbf{r}-\mathbf{R})\tilde{\zeta}_p(\mathbf{R}) \tag{12.1.4}$$

Taking the Fourier transform of both sides, we have

$$\theta_p(\mathbf{k}) = \int \tilde{\theta}_p(\mathbf{r}) e^{-i\mathbf{kr}} d^3r = \eta_0(\mathbf{k})\zeta_p(\mathbf{k}) \tag{12.1.5}$$

where

$$\eta_0(\mathbf{k}) = \int \tilde{\eta}_0(\mathbf{r}) e^{-i\mathbf{kr}}\, d^3r \tag{12.1.6}$$

$$\zeta_p(\mathbf{k}) = \sum_{\mathbf{R}} \tilde{\zeta}_p(\mathbf{R}) e^{-i\mathbf{kr}} \tag{12.1.7}$$

The summation in Eq. (12.1.7) is over all sites **R** of the grid superlattice. If the total number of supercells of the grid is equal to N, and if N_p is the number of supercells occupied by particles of the type p, we have

$$w_p = \frac{N_p}{N} = \bar{\zeta}_p \tag{12.1.8}$$

for the volume fraction of the martensite of type p.

12.1.2. Strain Energy in Terms of Elementary Particles

The theory of strain energy of an arbitrary distribution of new phase coherent inclusions was formulated in Section 7.2. The close form of equations for the strain energy is given by (7.2.46). To reformulate Eq. (7.2.46) in terms of elementary martensite particles, we must substitute Eq. (12.1.5) and (12.1.8) into (7.2.46). This gives

$$E_{\text{elast}} = \frac{1}{2} N \sum_{p=1}^{v} v\lambda_{ijkl}\varepsilon_{ij}^0(p)\varepsilon_{kl}^0(p)\bar{\zeta}_p - \frac{1}{2} N \sum_{p,q} v\lambda_{ijkl}\varepsilon_{ij}^0(p)\varepsilon_{kl}^0(q)\bar{\zeta}_p\bar{\zeta}_q$$
$$-\frac{1}{2}\sum_{p,q}\int (\mathbf{n}\hat{\sigma}^0(p)\hat{\mathbf{\Omega}}(\mathbf{n})\hat{\sigma}^0(q)\mathbf{n})|\eta_0(\mathbf{k})|^2\Delta\zeta_p(\mathbf{k})\Delta\zeta_q^*(\mathbf{k})\frac{d^3k}{(2\pi)^3} \tag{12.1.9}$$

where v is the grid supercell volume, $Nv = V$ is the total volume of the system,

$$\Delta\zeta_p(\mathbf{k}) = \sum_{\mathbf{R}} (\tilde{\zeta}_p(\mathbf{R}) - \bar{\zeta}_p)e^{-i\mathbf{kR}} \tag{12.1.10}$$

Ideally, Eq. (12.1.9) should be written in the real space but with space integrals renormalized to sums over grid superlattice sites $\{\mathbf{R}\}$. This would make it possible to calculate the free energy of a distribution of elementary martensite particles directly, without reference to real space integrals. To accomplish the renormalization to the grid superlattice, we rewrite the wave vectors, **k**, in the form

$$\mathbf{k} = \mathbf{K} + \boldsymbol{\kappa} \tag{12.1.11}$$

where **K** is the reciprocal lattice vector referred to the grid. The reduced wave vector κ is defined within the first Brillouin zone of the grid. The reciprocal lattice vector **K** meets the relation

$$\mathbf{KR} = 2\pi m \quad (m \text{ is an integer})$$

Substituting the definition (12.1.11) into (12.1.9), we may rewrite this equation in the identical form:

$$E_{\text{elast}} = \frac{1}{2} N \sum_{p=1}^{v} v\lambda_{ijkl}\varepsilon_{ij}^0(p)\varepsilon_{kl}^0(p)\bar{\zeta}_p - \frac{1}{2} N \sum_{p,q} v\lambda_{ijkl}\varepsilon_{ij}^0(p)\varepsilon_{kl}^0(q)\bar{\zeta}_p\bar{\zeta}_q$$
$$-\frac{1}{2}\sum_{p,q}\sum_{\mathbf{K}}\int_{(\text{B.z.})} \Gamma_{pq}(\mathbf{K}+\boldsymbol{\kappa})\Delta\zeta_p(\mathbf{K}+\boldsymbol{\kappa})\Delta\zeta_q^*(\mathbf{K}+\boldsymbol{\kappa})\frac{d^3\kappa}{(2\pi)^3} \tag{12.1.12}$$

where

$$\Gamma_{pq}(\mathbf{k})=\frac{\mathbf{k}}{k}\,\hat{\sigma}^0(p)\hat{\mathbf{\Omega}}\left(\frac{\mathbf{k}}{k}\right)\hat{\sigma}^0(q)\frac{\mathbf{k}}{k}\,|\eta_0(\mathbf{k})|^2 \tag{12.1.13}$$

The summation in the third term of Eq. (12.1.12) is over all reciprocal lattice vectors $\mathbf{K}$; the integration over κ is carried out within the first Brillouin zone of the grid.

Using the relation (12.1.11) in the definition (12.1.10), we obtain

$$\Delta\zeta_p(\mathbf{K}+\kappa)=\sum_{\mathbf{R}}(\tilde{\zeta}_p(\mathbf{R})-\bar{\zeta}_p)e^{-i(\mathbf{K}+\kappa)\mathbf{R}}=\sum_{\mathbf{R}}(\tilde{\zeta}_p(\mathbf{R})-\bar{\zeta}_p)e^{-i\kappa\mathbf{R}}=\Delta\zeta_p(\kappa) \tag{12.1.14}$$

Eqs. (12.1.14) and (12.1.12) give

$$E_{\text{elast}}=\frac{1}{2}N\sum_{p=1}^{v}v\lambda_{ijkl}\varepsilon_{ij}^0(p)\varepsilon_{kl}^0(p)\bar{\zeta}_p-\frac{1}{2}N\sum_{p,q}v\lambda_{ijkl}\varepsilon_{ij}^0(p)\varepsilon_{kl}^0(q)\bar{\zeta}_p\bar{\zeta}_q$$
$$-\frac{1}{2}\sum_{p,q}v\int_{(\text{B.z.})}\bar{\Gamma}_{pq}(\kappa)\Delta\zeta_p(\kappa)\Delta\zeta_q^*(\kappa)\frac{d^3\kappa}{(2\pi)^3} \tag{12.1.15}$$

where

$$\bar{\Gamma}_{pq}(\kappa)=\sum_{\mathbf{K}}\Gamma_{pq}(\kappa+\mathbf{K})$$
$$=\sum_{\mathbf{K}}\frac{\mathbf{K}+\kappa}{|\mathbf{K}+\kappa|}\,\hat{\sigma}^0(p)\hat{\mathbf{\Omega}}\left(\frac{\mathbf{K}+\kappa}{|\mathbf{K}+\kappa|}\right)\hat{\sigma}^0(q)\frac{\mathbf{K}+\kappa}{|\mathbf{K}+\kappa|}\frac{|\eta_0(\mathbf{K}+\kappa)|^2}{v} \tag{12.1.16}$$

We will now separate the part of the sum over the reciprocal space that is independent of the particular configuration of elementary martensite particles. Let us introduce the constant Q_{pq}

$$Q_{pq}=\langle\bar{\Gamma}_{pq}(\kappa)\rangle=v\int_{(\text{B.z.})}\bar{\Gamma}_{pq}(\kappa)\frac{d^3\kappa}{(2\pi)^3} \tag{12.1.17}$$

The third term in Eq. (12.1.15) may be rewritten in the identical form

$$-\frac{1}{2}\sum_{p,q}v\int_{(\text{B.z.})}\bar{\Gamma}_{pq}(\kappa)\Delta\zeta_p(\kappa)\Delta\zeta_q^*(\kappa)\frac{d^3\kappa}{(2\pi)^3}=\frac{1}{2}\sum_{p,q}v\int_{(\text{B.z.})}V_{pq}(\kappa)\Delta\zeta_p(\kappa)\Delta\zeta_q^*(\kappa)\frac{d^3\kappa}{(2\pi)^3}$$
$$+\frac{1}{2}\sum_{p,q}Q_{pq}\frac{1}{v}\int_{(\text{B.z.})}|\eta_0(\kappa)|^2\Delta\zeta_p(\kappa)\Delta\zeta_q^*(\kappa)\frac{d^3\kappa}{(2\pi)^3} \tag{12.1.18}$$

where

$$V_{pq}(\kappa)=-\bar{\Gamma}_{pq}(\kappa)+Q_{pq}\frac{|\eta_0(\kappa)|^2}{v^2} \tag{12.1.19}$$

$$|\eta_0(\kappa)|^2=\sum_{\mathbf{K}}|\eta_0(\mathbf{K}+\kappa)|^2 \tag{12.1.20}$$

The second term in Eq. (12.1.18) may be simplified by using the identity

$$\int_{(\mathrm{B.z.})} |\eta_0(\boldsymbol{\kappa})|^2 \Delta\zeta_p(\boldsymbol{\kappa})\Delta\zeta_q^*(\boldsymbol{\kappa}) \frac{d^3\kappa}{(2\pi)^3} = vN(\bar{\zeta}_p\delta_{pq} - \bar{\zeta}_p\bar{\zeta}_q) \tag{12.1.21}$$

To prove this identity, let us rewrite its left-hand side in the grid-site representation

$$\int_{(\mathrm{B.z.})} |\eta_0(\boldsymbol{\kappa})|^2 \Delta\zeta_p(\boldsymbol{\kappa})\Delta\zeta_q^*(\boldsymbol{\kappa}) \frac{d^3\kappa}{(2\pi)^3} = \sum_{\mathbf{R},\mathbf{R}'} M(\mathbf{R}-\mathbf{R}')\Delta\tilde{\zeta}_p(\mathbf{R})\Delta\tilde{\zeta}_q(\mathbf{R}') \tag{12.1.22}$$

where

$$M(\mathbf{R}-\mathbf{R}') = \int_{(\mathrm{B.z.})} |\eta_0(\boldsymbol{\kappa})|^2 e^{-i\boldsymbol{\kappa}(\mathbf{R}-\mathbf{R}')} \frac{d^3\kappa}{(2\pi)^3} \tag{12.1.23}$$

Substituting the definition (12.1.20) into (12.1.23) and taking into account (12.1.11), we obtain

$$M(\mathbf{R}-\mathbf{R}') = \sum_{\mathbf{K}} \int_{(\mathrm{B.z.})} |\eta_0(\mathbf{K}+\boldsymbol{\kappa})|^2 e^{-i(\mathbf{K}+\boldsymbol{\kappa})(\mathbf{R}-\mathbf{R}')} \frac{d^3\kappa}{(2\pi)^3}$$

$$= \iiint_{-\infty}^{\infty} |\eta_0(\mathbf{k})|^2 e^{-i\mathbf{k}(\mathbf{R}-\mathbf{R}')} \frac{d^3k}{(2\pi)^3} \tag{12.1.24}$$

Using the convolution theorem in the right-hand side of (12.1.24), we can write

$$M(\mathbf{R}-\mathbf{R}') = \int \tilde{\eta}_0(\mathbf{r}-\mathbf{R})\tilde{\eta}_0(\mathbf{r}-\mathbf{R}')d^3r \tag{12.1.25}$$

Since the shape function of the grid cell, $\tilde{\eta}_0(\mathbf{r}-\mathbf{R})$, is equal to unity within the grid cells at $\mathbf{R}$ and equal to zero outside them, the integral (12.1.25) is only nonzero if $\mathbf{R}=\mathbf{R}'$. Bearing in mind the definition of the shape function, $\tilde{\eta}_0(\mathbf{r}-\mathbf{R})$, we have in the latter case

$$M(0) = \int (\tilde{\eta}_0(\mathbf{r}-\mathbf{R}))^2 d^3r = \int \tilde{\eta}_0(\mathbf{r}-\mathbf{R})d^3r = v$$

where v is the volume of a grid cell.

Therefore

$$M(\mathbf{R}-\mathbf{R}') = v\delta_{\mathbf{R}\mathbf{R}'} \tag{12.1.26}$$

where $\delta_{\mathbf{R}\mathbf{R}'}$ is the Kronecker symbol. Substitution of (12.1.26) into (12.1.22) yields

$$\int_{(\mathrm{B.z.})} |\eta_0(\boldsymbol{\kappa})|^2 \Delta\zeta_p(\boldsymbol{\kappa})\Delta\zeta_q^*(\boldsymbol{\kappa}) \frac{d^3\kappa}{(2\pi)^3} = v\sum_{\mathbf{R}} \Delta\tilde{\zeta}_p(\mathbf{R})\Delta\tilde{\zeta}_q(\mathbf{R}) \tag{12.1.27}$$

Since $\tilde{\zeta}_p(\mathbf{R})$ is either zero or unity, and two elementary particles cannot occupy the same site **R**, we have

$$\sum_{\mathbf{R}} \tilde{\zeta}_p(\mathbf{R})\tilde{\zeta}_q(\mathbf{R}) = N_p\delta_{pq} = \bar{\zeta}_p N\delta_{pq}$$

$$\sum_{\mathbf{R}} \tilde{\zeta}_p(\mathbf{R}) = N_p = N\bar{\zeta}_p \tag{12.1.28}$$

where N_p is the total number of elementary particles of the type p, N is the total number of grid sites.

With the relations (12.1.28), Eq. (12.1.27) may be rewritten in the form

$$\begin{aligned}\int_{(\mathrm{B.z.})} |\eta_0(\boldsymbol{\kappa})|^2 \Delta\zeta_p(\boldsymbol{\kappa})\Delta\zeta_q^*(\boldsymbol{\kappa})\frac{d^3\kappa}{(2\pi)^3} &= v\sum_{\mathbf{R}}(\tilde{\zeta}_p(\mathbf{R})-\bar{\zeta}_p)(\tilde{\zeta}_q(\mathbf{R})-\bar{\zeta}_q)\\ &= v\sum_{\mathbf{R}}\tilde{\zeta}_p(\mathbf{R})\tilde{\zeta}_q(\mathbf{R}) - v\bar{\zeta}_q\sum_{\mathbf{R}}\tilde{\zeta}_p(\mathbf{R}) - v\bar{\zeta}_p\sum_{\mathbf{R}}\tilde{\zeta}_q(\mathbf{R})\\ &\quad + vN\bar{\zeta}_p\bar{\zeta}_q = vN(\delta_{pq}\bar{\zeta}_p - \bar{\zeta}_p\bar{\zeta}_q)\end{aligned} \tag{12.1.29}$$

The expression (12.1.29) proves the identity (12.1.21).

Substitution of the identity (12.1.21) into (12.1.18) yields

$$\begin{aligned}-\frac{1}{2}\sum_{p,q} v\int_{(\mathrm{B.z.})}\bar{\Gamma}_{pq}(\boldsymbol{\kappa})\Delta\zeta_p(\boldsymbol{\kappa})\Delta\zeta_q^*(\boldsymbol{\kappa})\frac{d^3k}{(2\pi)^3} &= \frac{1}{2}\sum_{p,q} v\int_{(\mathrm{B.z.})} V_{pq}(\boldsymbol{\kappa})\Delta\zeta_p(\boldsymbol{\kappa})\Delta\zeta_q^*(\boldsymbol{\kappa})\frac{d^3\kappa}{(2\pi)^3}\\ &\quad + \frac{N}{2}\sum_p Q_{pp}\bar{\zeta}_p - \frac{N}{2}\sum_{p,q} Q_{pq}\bar{\zeta}_p\bar{\zeta}_q\end{aligned} \tag{12.1.30}$$

Substitution of (12.1.30) into (12.1.15) gives

$$\begin{aligned}E_{\mathrm{elast}} &= \frac{1}{2}N\sum_{p=1}^{\nu}(v\lambda_{ijkl}\varepsilon_{ij}^0(p)\varepsilon_{kl}^0(p) - Q_{pp})\bar{\zeta}_p\\ &\quad - \frac{1}{2}N\sum_{p,q}(v\lambda_{ijkl}\varepsilon_{ij}^0(p)\varepsilon_{kl}^0(q) - Q_{pq})\bar{\zeta}_p\bar{\zeta}_q\\ &\quad + \frac{1}{2}\sum_{p,q} v\int_{(\mathrm{B.z.})} V_{pq}(\boldsymbol{\kappa})\Delta\zeta_p(\boldsymbol{\kappa})\Delta\zeta_q^*(\boldsymbol{\kappa})\frac{d^3\kappa}{(2\pi)^3}\end{aligned} \tag{12.1.31}$$

Using the cyclic boundary conditions, we may write the integral over the first Brillouin zone as a sum over all quasi-continuum points $\boldsymbol{\kappa}$ permitted by the cyclic boundary conditions:

$$v\int_{(\mathrm{B.z.})}\frac{d^3\kappa}{(2\pi)^3}(\cdots) = \frac{1}{N}\sum_{\boldsymbol{\kappa}}(\cdots) \tag{12.1.32}$$

Using Eq. (12.1.32) in (12.1.31), we obtain

$$\begin{aligned}E_{\mathrm{elast}} &= N\sum_{p=1}^{\nu} E_{\mathrm{self}}^{\mathrm{elast}}\bar{\zeta}_p + \frac{1}{2}N\sum_{p,q} V_{pq}(0)\bar{\zeta}_p\bar{\zeta}_q\\ &\quad + \frac{1}{2}\frac{1}{N}\sum_{\boldsymbol{\kappa}} V_{pq}(\boldsymbol{\kappa})\Delta\zeta_p(\boldsymbol{\kappa})\Delta\zeta_q^*(\boldsymbol{\kappa})\end{aligned} \tag{12.1.33}$$

where according to (12.1.19)

$$V_{pq}(\kappa)=\begin{cases}-\bar{\Gamma}_{pq}(\kappa)+Q_{pq}\dfrac{1}{v^2}|\eta_0(\kappa)|^2 & \text{if } \kappa\neq 0\\[2ex] -v\lambda_{ijkl}\varepsilon_{ij}^0(p)\varepsilon_{kl}^0(q)+Q_{pq} & \text{if } \kappa=0\end{cases} \tag{12.1.34}$$

and

$$E_{\text{self}}^{\text{elast}}=\frac{1}{2}(v\lambda_{ijkl}\varepsilon_{ij}^0(p)\varepsilon_{kl}^0(p)-Q_{pp}) \tag{12.1.35}$$

is the elastic self-energy required to introduce a single elementary particle into the parent phase.

Since

$$\sum_{\mathbf{R}} e^{i\kappa\mathbf{R}}=N\delta_{\kappa 0}$$

we have

$$\begin{aligned}\Delta\zeta_p(\kappa)&=\sum_{\mathbf{R}}(\tilde{\zeta}_p(\mathbf{R})-\bar{\zeta}_p)e^{-i\kappa\mathbf{R}}=\zeta_p(\kappa)-\bar{\zeta}_pN\delta_{\kappa 0}\\ &=\begin{cases}\zeta_p(\kappa) & \text{if } \kappa\neq 0\\ 0 & \text{if } \kappa=0\end{cases}\end{aligned} \tag{12.1.36}$$

Substitution of (12.1.36) into (12.1.33) yields

$$\begin{aligned}E_{\text{elast}}&=N\sum_{p=1}^{\nu}E_{\text{self}}^{\text{elast}}\bar{\zeta}_p+\frac{1}{2}N\sum_{p,q}V_{pq}(0)\bar{\zeta}_p\bar{\zeta}_q\\ &\quad+\frac{1}{2}\frac{1}{N}\sum_{\kappa}{}' V_{pq}(\kappa)\zeta_p(\kappa)\zeta_q^*(\kappa)\end{aligned} \tag{12.1.37}$$

where priming means that the term in the sum over κ corresponding to $\kappa=0$ is omitted.

Using the identity

$$\bar{\zeta}_p=\frac{1}{N}\sum_{\mathbf{R}}\tilde{\zeta}_p(\mathbf{R})=\frac{1}{N}\zeta_p(0)$$

we obtain

$$E_{\text{elast}}=N\sum_{p=1}^{\nu}E_{\text{self}}^{\text{elast}}\bar{\zeta}_p+\frac{1}{2N}\sum_{\kappa}V_{pq}(\kappa)\zeta_p(\kappa)\zeta_q^*(\kappa) \tag{12.1.38}$$

where priming is removed because the term $[\sum_{p,q}V_{pq}(0)\zeta_p(0)\zeta_q(0)]/2N$ related to the term with $\kappa=0$ is included.

The back Fourier transform of Eq. (12.1.38) yields eventually

$$E_{\text{elast}}=N\sum_{p=1}^{\nu}E_{\text{self}}^{\text{elast}}\bar{\zeta}_p+\frac{1}{2}\sum_{\mathbf{R},p}\sum_{\mathbf{R}',q}W_{pq}^{el}(\mathbf{R}-\mathbf{R}')\tilde{\zeta}_p(\mathbf{R})\tilde{\zeta}_q(\mathbf{R}') \tag{12.1.39}$$

where

$$W_{pq}^{el}(\mathbf{R}-\mathbf{R}')=\frac{1}{N}\sum_{\kappa} V_{pq}(\kappa)e^{i\kappa(\mathbf{R}-\mathbf{R}')}=\frac{1}{N}V_{pq}(0)+\frac{1}{N}\sum_{\kappa}{}' V_{pq}(\kappa)e^{i\kappa(\mathbf{R}-\mathbf{R}')} \tag{12.1.40}$$

gives the pairwise strain-induced interaction energies between elementary particles of the types p and q at the sites $\mathbf{R}$ and $\mathbf{R}'$.

According to the definitions (12.1.34)*

$$\begin{aligned} W_{pq}(0)&=\frac{1}{N}\sum_{\kappa} V_{pq}(\kappa)\simeq v\int_{(\text{B.z.})} V_{pq}(\kappa)\frac{d^3\kappa}{(2\pi)^3}\\ &=-v\int_{(\text{B.z.})}\bar{\Gamma}_{pq}(\kappa)\frac{d^3\kappa}{(2\pi)^3}+Q_{pq}\frac{1}{v}\int_{(\text{B.z.})}|\eta_0(\kappa)|^2\frac{d^3\kappa}{(2\pi)^3} \end{aligned} \tag{12.1.41}$$

Since

$$\frac{1}{v}\int_{(\text{B.z.})}|\eta_0(\kappa)|^2\frac{d^3\kappa}{(2\pi)^3}=1$$

[see Eqs. (12.1.24) and (12.1.26)], we have

$$W_{pq}(0)=0 \tag{12.1.42}$$

[see the definition (12.1.17)].

The relation (12.1.42) shows that the situations with two elementary particles occupying the same site of the grid which do not make any physical sense are automatically excluded from consideration. This is the reason why the constant Q_{pq} is introduced in Eq. (12.1.31). It follows from Eq. (12.1.37) that the first term describes the self-energy required to introduce noninteracting particles into the parent phase, the second term is related to the configurationally independent contribution from the image-force-induced interaction energy and the third term describes the configurationally dependent pairwise strain-induced interactions.

12.1.3. Chemical Free Energy

When an elementary particle has a finite size, it is reasonable to represent the chemical free energy as a conventional sum of bulk free energy and surface energy terms. The bulk free energy term is the same for all martensite variants. If we ignore the temperature dependence of the enthalpy and entropy of the transformation, which is reasonable at least at T near T_0, the transformation temperature, then the bulk free energy change, $\Delta\mu_{\text{bulk}}$, caused by an elementary

*The equality (12.1.41) is valid with an accuracy to terms of the order of $1/N$ that tend to zero as $N \to \infty$.

particle formation may be written in the familiar form

$$\Delta\mu_{\text{bulk}}=\frac{Q_0}{T_0}(T-T_0) \tag{12.1.43}$$

where Q_0 is the transformation heat release produced per elementary particle.

The representation of the surface energy is more complex. Two distinct surface energies are relevant to the martensite transformation: the free energy, $\Delta\mu_s$, of an element of coherent interphase with the parent phase and the free energy, $\Delta\mu_{pq}^s$, of an element of interphase between martensite elementary particles of the types p and q. Generally, both interfacial energies depend on the orientation. The total interfacial energy should be computed by integration over the surface of each distinct monodomain martensite inclusion. At least in certain cases, however, this integration may be replaced by a summation over the elementary particles.

Assume a cubic elementary particle (and thus a cubic grid superlattice), and let its facets have the energy $\Delta\mu_0^s$, if in contact with the matrix, $\Delta\mu_1^s$ if in contact with an elementary particle of a different type, and 0 if in contact with an elementary particle of the same type. It follows from the definition

$$W_{pq}^s(\mathbf{R}-\mathbf{R}')=\begin{cases}(\Delta\mu_1^s-2\Delta\mu_0^s)-\Delta\mu_1^s\delta_{pq} & \text{if } (\mathbf{R}-\mathbf{R}') \text{ is the nearest-neighbor distance}\\ 0 & \text{if otherwise}\end{cases} \tag{12.1.44}$$

that the surface energy may be written

$$F_{\text{surf}}=\sum_{p,\mathbf{R}} z\Delta\mu_0^s\tilde{\zeta}_p(\mathbf{R})+\tfrac{1}{2}\sum_{p,q}\sum_{\mathbf{R},\mathbf{R}'} W_{pq}^s(\mathbf{R}-\mathbf{R}')\tilde{\zeta}_p(\mathbf{R})\tilde{\zeta}_q(\mathbf{R}') \tag{12.1.45}$$

where z is the number of nearest neighbors in the grid ($z=6$ if the superlattice is cubic).

The total chemical energy, including the bulk chemical free energy (12.1.43) and the surface free energy (12.1.45), then becomes

$$\begin{aligned}F_{\text{chem}}=&\sum_{\mathbf{R}}\left[\frac{Q}{T_0}(T-T_0)+z\Delta\mu_0^s\right]\sum_{p=1}^{v}\tilde{\zeta}_p(\mathbf{R})\\&+\tfrac{1}{2}\sum_{p,q}\sum_{\mathbf{R},\mathbf{R}'}W_{pq}^s(\mathbf{R}-\mathbf{R}')\tilde{\zeta}_p(\mathbf{R})\tilde{\zeta}_q(\mathbf{R}')\end{aligned} \tag{12.1.46}$$

12.1.4. Total Energy and Thermoelastic Potential

Summing Eqs. (12.1.46) and (12.1.39) gives an expression for the total free energy of a configuration of elementary martensite particles specified by the set of v distribution functions $\tilde{\zeta}_p(\mathbf{R})(p=1,2,\ldots,v)$:

$$\begin{aligned}F=&\sum_{\mathbf{R}}\left(\frac{Q}{T_0}(T-T_0)+z\Delta\mu_0^s+E_{\text{self}}^{\text{elast}}\right)\sum_{p=1}^{v}\tilde{\zeta}_p(\mathbf{R})\\&+\tfrac{1}{2}\sum_{p,\mathbf{R}}\sum_{q,\mathbf{R}'}(W_{pq}^{el}(\mathbf{R}-\mathbf{R}')+W_{pq}^s(\mathbf{R}-\mathbf{R}'))\tilde{\zeta}_p(\mathbf{R})\tilde{\zeta}_q(\mathbf{R}')\end{aligned} \tag{12.1.47}$$

where we have used the fact that the self-energy of all martensite variants is the same.

Let us introduce the concept of thermoelastic potential. The thermoelastic potential, $\phi_p(\mathbf{R})$, is the free energy change on the introduction of an elementary martensite particle of the type p at the position $\mathbf{R}$ in the presence of the distribution $\{\tilde{\zeta}_q(\mathbf{R}')\}$ over sites other than $\mathbf{R}$. Therefore

$$\phi_p(\mathbf{R}, \{\tilde{\zeta}_q(\mathbf{R}')\}) = F(\{\tilde{\zeta}_q(\mathbf{R}') + \delta_{pq}\delta_{\mathbf{R}\mathbf{R}'}\}) - F(\{\tilde{\zeta}_q(\mathbf{R}')\})$$

$$= \frac{Q\Delta T}{T_0} + z\Delta\mu_0^s + E_{\text{self}}^{\text{elast}} + \sum_{q,\mathbf{R}'} (W_{pq}^{el}(\mathbf{R}-\mathbf{R}') + W_{pq}^{s}(\mathbf{R}-\mathbf{R}'))\tilde{\zeta}_q(\mathbf{R}') \tag{12.1.48}$$

If the elementary particle, $(p, \mathbf{R})$, is present in the initial distribution, the energy change associated with the removal of the particle is equal to

$$\Delta F = -\phi_p(\mathbf{R}, \{\tilde{\zeta}_q(\mathbf{R}')\})$$

where $\{\tilde{\zeta}_q(\mathbf{R}')\}$ again represents the distribution of elementary particles over sites other than $\mathbf{R}$.

It follows from the definition (12.1.48) that the thermoelastic potential is a function of the elementary particle distribution, in other words, depends on the specific spatial distribution of the martensitic phase, namely, on the morphology of the two-phase alloy.

12.2. PATH AND KINETICS OF MARTENSITIC TRANSFORMATION

A phase transformation may be described by specifying its path, or the sequence of intermediate microstructures that a body assumes between the time the transformation is initiated and the time it is completed. In the present case a martensitic transformation will be represented by the stepwise addition of elementary martensite particles to the lattice. The transformation path is given by the sequence of distinguishable configurations, $\{\tilde{\zeta}_q(\mathbf{R})\}$, adopted by the body. If the configurations are numbered in the order of their appearance, and if the transformation is assumed to occur in unit steps that involve the creation and annihilation of a single elementary particle, each configuration is related to its predecessor by the simple addition:

$$\{\tilde{\zeta}_q^{(\alpha+1)}(\mathbf{R})\} = \{\tilde{\zeta}_q^{(\alpha)}(\mathbf{R}) \pm \delta_{\mathbf{R}\mathbf{R}'}\delta_{qp}\} \tag{12.2.1}$$

where α is the number of the specific configuration of elementary particles.

If the sign in Eq. (12.2.1) is positive, the transformation step is the addition of an elementary particle of type p at $\mathbf{R}'$ to $\{\tilde{\zeta}_q^{(\alpha)}(\mathbf{R})\}$; a negative sign has the meaning that this elementary particle is eliminated from $\{\tilde{\zeta}_q^{(\alpha)}(\mathbf{R})\}$ by reverse transformation.

Given a configuration $\{\tilde{\zeta}_q^{(\alpha)}(\mathbf{R})\}$, the set of possible events connecting it with the succeeding configuration $\{\tilde{\zeta}_q^{(\alpha+1)}(\mathbf{R})\}$ contains either creation of an elementary particle of type p ($p=1, 2, \ldots, \nu$) at any grid cell, which is free of martensite, or the annihilation of any of the existing elementary particles. All the elementary events permitted with the configuration $\{\tilde{\zeta}_q^{(\alpha)}(\mathbf{R})\}$ may be numbered. Each of them, the creation and annihilation, results in a free energy change which may be expressed in terms of appropriate thermoelastic potentials:

$$\Delta G_q^{(\alpha)}(j)=\begin{cases} \phi_q^{(\alpha)}(\mathbf{R}_j) & \text{if } \tilde{\zeta}_q^{(\alpha)}(\mathbf{R}_j)=0, \quad q=1,2,\ldots,\nu \\ -\phi_q^{(\alpha)}(\mathbf{R}_j) & \text{if } \tilde{\zeta}_q^{(\alpha)}(\mathbf{R}_j)=1 \end{cases} \tag{12.2.2}$$

The index j in Eq. (12.2.2) designates the coordinates $\mathbf{R}_j$ of the grid sites where an elementary event may occur.

The spectrum of the values $\Delta G_q^{(\alpha)}(j)$ related to configuration α is the most important dynamic characteristics of the system because it predetermines the martensitic transformation kinetics. The thermodynamic stability of the configuration α is determined by the lowest value of $\Delta G_q^{(\alpha)}$. Let

$$\Delta G^{(\alpha)}=\min\,[\Delta G_q^{(\alpha)}(j)] \tag{12.2.3}$$

If $\Delta G^{(\alpha)}<0$, then there is at least one elementary change of $\{\tilde{\zeta}_q^{(\alpha)}(\mathbf{R})\}$ that will lead to a decrease of the free energy; hence $\{\tilde{\zeta}_q^{(\alpha)}(\mathbf{R})\}$ is thermodynamically unstable. If $\Delta G^{(\alpha)}>0$, then every elementary change in $\{\tilde{\zeta}_q^{(\alpha)}(\mathbf{R})\}$ leads to an increase of the free energy, and the configuration described by $\{\tilde{\zeta}_q^{(\alpha)}(\mathbf{R})\}$ is either stable or metastable.

The formation of an elementary particle through a thermally activated process will generally be opposed by an activation barrier of height Δg^*. Making the usual assumption that the total activation barrier opposing the event (j, q) is

$$\Delta G_q^{*(\alpha)}(j)=\tfrac{1}{2}\Delta G_q^{(\alpha)}(j)+g^* \tag{12.2.4}$$

we have the spectrum of activation barriers for all possible elementary events at the configuration α.

If the minimal kinetic barrier, $\Delta G^{*(\alpha)}$, is positive, then every elementary change in $\{\tilde{\zeta}_q^{(\alpha)}(\mathbf{R})\}$ is opposed by a finite activation barrier and requires a thermal fluctuation in order to happen. Such a configuration will be termed kinetically stable in the sense that it can be maintained by forbidding positive fluctuations in the energy. On the other hand, if $\Delta G^{*(\alpha)}<0$, then there is at least one elementary event that represents a thermodynamic instability of $\{\tilde{\zeta}_q^{(\alpha)}(\mathbf{R})\}$ and is not opposed by any activation barrier. This even will occur spontaneously even in the absence of thermal fluctuations. The configuration $\{\tilde{\zeta}_q^{(\alpha)}(\mathbf{R})\}$ is hence kinetically unstable. The analysis of the kinetics of transformation through kinetically unstable configurations poses problems that have their source in the finite speed of sound in real crystals.* The theory leading to the definition and evaluation of the free energy change, $\Delta G^{(\alpha)}$, is based on equations that assume a static elastic equilibrium. But, if an elementary transformation occurs at a point $\mathbf{R}$ within a crystal, the associated elastic disturbance propagates only at the speed of sound, c_s. It will not be sensed at a point $\mathbf{R}'$ until

*The same takes place for dislocation motion not controlled by thermal-activated processes.

after a time interval $\Delta t \sim |\mathbf{R}-\mathbf{R}'|/c_s$, and the modified static equilibrium state cannot be assumed until several of these time intervals have passed. The kinetics of transformations that evolve at speeds near that of sound constitutes the scope of strict applicability of the static equilibrium model. An error in the precise kinetics of unstable events is of no great concern, however, so long as the sequence of these events is reasonably well represented. In the usual time frame of interest in the kinetics of phase transformations, the distinction between a process that happens at the speed of sound and one that happens instantaneously is immaterial. The sensible kinetics of the phase transformation is controlled by those configurations along the transformation path that are kinetically stable and that consequently require finite time or additional supercooling before they transform. The nature of the kinetically stable configurations along the transformation path may depend on the sequence in which unstable configurations are sampled, but not on the kinetics of their evolution.

Given this analysis, we shall represent the martensitic transformation by a model that ignores the retarding associated with the finite speed of sound and assumes that elastic equilibrium is instantaneously reestablished after each elementary event.

The probability of each elementary event, (j, q), will then be proportional to the Boltzmann factor $\exp(-\Delta G_q^{*(\alpha)}(j)/\kappa T)$. The normalized probability $P_q(j)$ of the realization of the specific event (j, q) is then

$$P_q(j)=\frac{\exp(-\Delta G_q^{*(\alpha)}(j)/\kappa T)}{\sum_{p,l} \exp(-\Delta G_p^{*(\alpha)}(l)/\kappa T)} \qquad (12.2.5)$$

where the summation is over all the possible elementary events, $\{l, p\}$.

Referring to definition (12.2.4), we have

$$P_q(j)=\frac{\exp(-\Delta G_q^{(\alpha)}(j)/2\kappa T)}{\sum_{p,l} \exp(-\Delta G_p^{(\alpha)}(l)/2\kappa T)} \qquad (12.2.6)$$

The model described in this section and in Section 12.1 provides the basis for computer simulation of the martensitic transformation kinetics. The computer simulation should include the following steps:

1. Calculate all thermoelastic potentials of a configuration of elementary martensite particles α from Eq. (12.1.48).
2. Calculate the spectrum of the free energy changes, $\Delta G_q^{(\alpha)}(j)$, corresponding to all permitted elementary events in the configuration α [Eq. (12.2.2)].
3. Calculate the probabilities (12.2.6) of the permitted elementary events at the configuration α.
4. Find the specific elementary event (the creation or annihilation of an elementary particle) providing the transition from the αth to $(\alpha+1)$th configuration. In principle, all permitted elementary events may occur. Their realization, however, depends on the probability distribution (12.2.6) and can be simulated by the Monte Carlo method.

The preceding analysis reduces the problem of determining the sequence of assumed configurations as the transformation proceeds. Even with the help of computer simulation, however, this is a formidable problem in an analysis of a transformation that occurs at a finite temperature. The transformation path will be determined by what amounts to a random walk over the set of possible configurations. Particularly, if there are metastable intermediate states during the transformation, the transformation may not go at all monotonically toward completion but rather oscillate for a long time before making a net positive step.

To construct a tractable phase transformation, it is useful to have a representation of the transformation path that is reasonably accurate and relatively simple to analyze. The obvious choice is the minimum-energy path, which is taken at each step the transformation progresses in such a way as to minimize the total activation energy.

We naturally come to the minimum-energy path concept in the case where the typical difference between the closest elementary free energy changes, $\Delta G_q^{(\alpha)}(j)$, corresponding to the creation or annihilation of an elementary particle is higher than the thermal energy, κT. The probability distribution (12.2.6) then describes an almost deterministic process: the probability of the elementary event producing the minimum energy change, $\Delta G^{(\alpha)}$, tends to unity whereas the probabilities of all other events vanish. This tendency will become increasingly strong as the temperature goes down.

A simple way to locate the minimum-energy path in a computer simulation, which will be used in the next sections, is to assume that the transformation cannot reverse itself and then choose each transformation step so that the incremental energy is as low as possible. If we rule out the possibility of reverse transformations, then the computation of the transformation kinetics along the minimum-energy path is straightforward and simple. In the more realistic case where reverse transformations are allowed, the kinetics may be approximated by using equations analogous to diffusion through random walk processes, with the kinetic consequences of the reverse transformation gathered into a "correlation factor."

It should be noted that, when the transformation progression is determined by a minimum-energy path, complete temperature reversibility and reproducibility are observed. The reverse transformation reaction in heating proceeds as the back succession of transformation events: the first portion of martensite formed in cooling disappears last. In other words, there is somewhat like the "memory effect" when the reverse transformation in heating proceeds as the reverse succession of transformation events in cooling (the first portions of the martensite produced in cooling disappear at the end step). As for the reproducibility, each new cooling-heating cycle follows exactly the same transformation path. The thermoactivation nature of the transformation kinetics involves certain energy dissipation which determines the area of the hysteresis loop in a cooling-heating thermocycle.

Let us consider one more important particular case where the free energy changes corresponding to various transformation events in the αth configuration are commensurate with the thermal motion energy, κT. Here the transformation will not develop along the minimum-energy path because the thermally activated transitions involved in the various paths will give a substantial contribution to the transformation kinetics. Then the "memory" of the transformation path is lost, and the path of the reverse transformation in heating does not necessarily coincide with that of the direct transformation in cooling. In this connection, the foregoing analysis of the martensitic transformation yields a simple explanation of the memory effect in martensitic alloys in terms of elementary transformation events.

With thermal activation made easier, we should begin to observe a tendency for a significant amount of isothermal transformations. That is, if a system is held at a fixed value of temperature for a reasonable period of time, thermal activation may intervene to cause an increasing transformation with time. The character of isothermal transformation should strongly resemble that observed experimentally: a martensite plate nucleates and grows rapidly to completion, after which one observes a finite time period before the second thermal activation event occurs, after which the new plate grows rapidly until it reaches completion.

A final consequence of thermal activation will be to cause the martensitic start temperature, and the progress of the transformation, to become sensitive to the rate at which the system is cooled. A common experimental observation is that martensite transformation occurring at relatively high temperatures exhibit martensite start temperatures that depend on the quenching rate. As the rate of cooling is decreased, the martensite start temperature increases. This phenomenon would appear to be a straightforward result of a thermally activated process. In a thermally activated process the probability that an event will occur increases with time and temperature. Therefore, if the system is continuously cooled, then the temperature at which the event is observed to happen will increase as the cooling rate goes down.

12.3. COMPUTER SIMULATION OF PSEUDO-TWO-DIMENSIONAL MARTENSITIC TRANSFORMATION

In this section we shall give an account of the first attempt to simulate martensitic transformation, as described by Wen, Morris, and Khachaturyan (84, 85). In the specific model used, the crystal is assumed to be a pseudo-two-dimensional body. The pseudo-two-dimensional crystal is represented by a square grid of points, forty on a side, in a square array. Each unit cell of the grid plane is a cross section of a rod extending to infinity in both directions. An elementary martensite particle is then also represented as an infinite rod normal to the grid plane whose cross section coincides with the unit cell of the grid. The array is assumed to be periodic across each of its boundaries.

Let the solid have isotropic elastic constants, and let the martensitic transformation be described by a Bain-like transformation strain which is a simple shear involving an expansion along one of the axes of the two-dimensional grid and compensating contracting along the other, the third axis normal to the grid being unaffected. In this case there are two variants of elementary martensite particles that differ in the orientation of the expansion axis. The two variants of the stress-free transformation strain are then described by the two tensors:

$$\varepsilon_{ij}^{0}(1)=\varepsilon_0\begin{pmatrix}1 & 0 & 0\\ 0 & \bar{1} & 0\\ 0 & 0 & 0\end{pmatrix},\quad \varepsilon_{ij}^{0}(2)=\varepsilon_0\begin{pmatrix}\bar{1} & 0 & 0\\ 0 & 1 & 0\\ 0 & 0 & 0\end{pmatrix} \tag{12.3.1}$$

where ε_0 is the magnitude of the shear. Elementary martensite particles interact with one another through elastic strain fields, as discussed in the previous sections.

The martensitic transformation is assumed to occur through the discrete formation of elementary particles of the martensitic phase. The net energy change on the introduction of an elementary martensite particle is characterized by the thermoelastic potential. The thermoelastic potential is the sum of three terms:

$$\phi_p(\mathbf{R})=\frac{Q_0(T-T_0)}{T_0}+E_{\text{self}}^{\text{elast}}+\phi_p^{el}(\mathbf{R}) \tag{12.3.2}$$

where $Q_0(T-T_0)/T_0$ is the chemical bulk free energy (the driving force of the transformation) which is assumed to depend linearly on supercooling, $\Delta T=T-T_0$; $E_{\text{self}}^{\text{elast}}$ is the elastic self-energy which is equal to the elastic energy required to introduce an elementary particle into the single phase matrix:

$$\phi_p^{el}(\mathbf{R})=\sum_{q,\,\mathbf{R}'} W_{pq}^{el}(\mathbf{R}-\mathbf{R}')\tilde{\zeta}_q(\mathbf{R}') \tag{12.3.3}$$

is the elastic potential derived in Section 12.1 which includes the elastic two-body interactions with all other elementary particles present. Since the surface energy that includes the martensite-parent phase interphase energy and surface energy between two orientational variants possible is far less than the elastic strain energy, we neglect the surface energy terms in Eq. (12.3.2). The thermoelastic potential, $\phi_p(\mathbf{R})$, is calculated by the computer in accordance with Eq. (12.3.2) for the two-dimensional case, for all sites $\mathbf{R}$ of the grid and for $p=1$ or 2, corresponding to two possible variants of elementary particles. The code employs the stress-free strains (12.3.1) and carries out the integration over $\boldsymbol{\kappa}$ in Eq. (12.1.40) with $V_{pq}(\boldsymbol{\kappa})$ given by Eq. (12.1.34). Isotropic continuous elasticity is assumed.

To simplify the calculations, the function $|\eta_0(\mathbf{k})|^2$ in Eqs. (12.1.16) and (12.1.18) is approximated by a step function equal to v within the first Brillouin zone of the grid lattice and to zero outside it. Eq. (12.1.40) then becomes

$$W_{pq}^{el}(\mathbf{R}-\mathbf{R}')=-\frac{1}{N}\left[vTr(\hat{\sigma}^0(p)\hat{\varepsilon}^0(q))-Q_{pq}\right]$$
$$-\frac{1}{N}\sum_{\boldsymbol{\kappa}}{}'\left[v(\mathbf{n}\hat{\sigma}^0(p)\hat{\boldsymbol{\Omega}}(\mathbf{n})\hat{\sigma}^0(q)\mathbf{n})-Q_{pq}\right]e^{i\boldsymbol{\kappa}(\mathbf{R}-\mathbf{R}')} \tag{12.3.4}$$

where the summation is over all points κ within the first Brillouin zone (except for $\kappa=0$) permitted by the cyclic boundary conditions, Tr is the trace of the matrix, $\mathbf{n}=\boldsymbol{\kappa}/\kappa$,

$$Q_{pq}=\frac{v}{N}\sum_{\kappa}\mathbf{n}\hat{\sigma}^0(p)\hat{\boldsymbol{\Omega}}(\mathbf{n})\hat{\sigma}^0(q)\mathbf{n} \tag{12.3.5}$$

In the isotropic case

$$\hat{\boldsymbol{\Omega}}(\mathbf{n})=\frac{1}{\mu}\hat{\mathbf{I}}-\frac{1}{2\mu(1-\sigma_1)}\mathbf{n}*\mathbf{n} \tag{12.3.6}$$

where σ_1 is the Poisson ratio, μ the shear modulus, and

$$\hat{\sigma}^0(1)=2\mu\hat{\varepsilon}^0(1),\quad \hat{\sigma}^0(2)=2\mu\hat{\varepsilon}^0(2) \tag{12.3.7}$$

The calculations with Eq. (12.3.7) yield

$$Tr(\hat{\sigma}^0(p)\hat{\varepsilon}^0(q))=v2\mu Tr(\hat{\varepsilon}^0(p)\hat{\varepsilon}^0(q)) \tag{12.3.8}$$

The calculation Q_{pq} in (12.3.5) with Eq. (12.3.6) gives

$$Q_{11}=Q_{22}=-Q_{12}=4\mu\varepsilon_0^2 v\left(1-\frac{1}{4(1-\sigma_1)}\right) \tag{12.3.9}$$

Figure 102 shows a three-dimensional plot of strain-induced interaction energy, $W^{el}_{pp}(\mathbf{R})$, between two single elementary particles of the type p in the

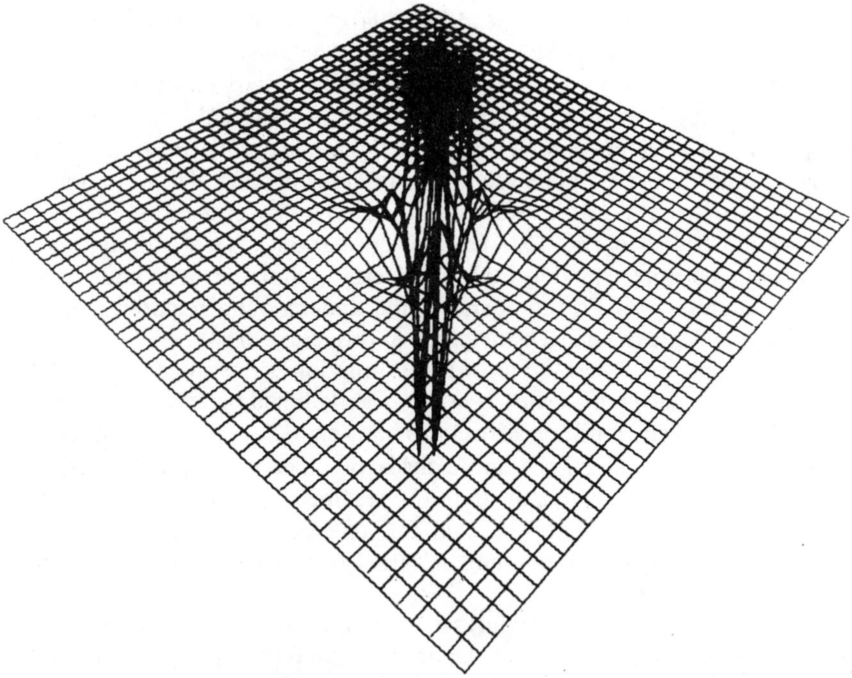

Figure 102. Three-dimensional plot of the thermoelastic potential $\phi_p(\mathbf{R})=W^{el}_{pp}(\mathbf{R})$ at every site $\mathbf{R}$ caused by a type p elementary particle at the center of the grid lattice, in the case of elastic isotropy.

case of elastic isotropy. The potential field varies dramatically with direction. The coordinate-dependent part of $W_{pq}^{el}(\mathbf{R}-\mathbf{R}')$ has a long-distance asymptote that decreases as $1/|\mathbf{R}-\mathbf{R}'|^2$.

For each configuration of elementary martensite particles, the computer calculates all thermoelastic potentials, $\phi_p(\mathbf{R})$, for every point, $\mathbf{R}$, on the grid and for both values p using Eq. (12.3.2), finds the minimum thermoelastic potential $\phi_0=\min \phi_p(\mathbf{R})=\phi_{p_0}(\mathbf{R}_0)$, and if ϕ_0 is negative, creates a new elementary particle of the kind p_0 at the point $\mathbf{R}_0$ [the event $(p_0, \mathbf{R}_0)$ corresponds to the minimum thermoelastic potential ϕ_0]. If ϕ_0 proves to be a positive value, the transformation is halted. Therefore ϕ_0 is the "quantum" of the free energy, the elementary martensite formation energy of one elementary particle of the kind of p_0 at $\mathbf{R}_0$. In all cases discussed below the energies are in the unit of $\mu\varepsilon_0^2 v$ where μ is the shear modulus, ε_0 is the amount of the transformation strain, v is the volume of an elementary particle. The procedure just described corresponds to a deterministic minimum energy path with a low temperature limit. This transformation path is defined by the dual requirement that transformation can only occur if the energy associated with the formation of an additional elementary particle is negative, and that, if several such particles with negative energies are possible, the one having the lowest energy will be formed. The transformation hence occurs so as to minimize the energy decrease per elementary step, and only occurs if the incremental energy change is negative.

As should be clear from Sections 12.1 and 12.2, most of the more restrictive assumptions cited above can be easily relaxed in the model. Since the martensitic transformation is known to be heterogeneously nucleated, the transformation of a defective lattice is assumed. For simplicity, the preexisting defects are taken to be a random distribution of elementary martensite particles. In the case illustrated here, ten such particles are randomly distributed over 40×40 grid. Under the assumptions listed, the martensitic transformation is athermal and occurs progressively on continuous cooling. The variation in a fraction of martensite, with undercooling (driving force) measured in energy units, is illustrated in Fig. 103. The transformation is seen to initiate at an undercooling of approximately 0.09 and to reach completion at an undercooling of approximately 0.6. For comparison, the undercooling required to homogeneously nucleate the martensitic phase in this case is 0.6447.

The development of the martensite transformation is illustrated in Fig. 104a to d. The transformation nucleates as a two-layer twinned plate and grows along the (110) habit plane until it nearly closes on itself. Before it does, however, the growing plate encounters the strain fields of other preexisting defects that cause the transformation to stop and to be resumed after further undercooling. This phenomenon illustrates the dual role of preexisting defects. The strain fields of these defects promote the nucleation of martensite but interfere with its growth.

The transformation develops further on decreasing temperature. Additional plates nucleate and grow and may be oriented parallel or perpendicular to the original martensite plate. These autocatalytically nucleated plates sometimes

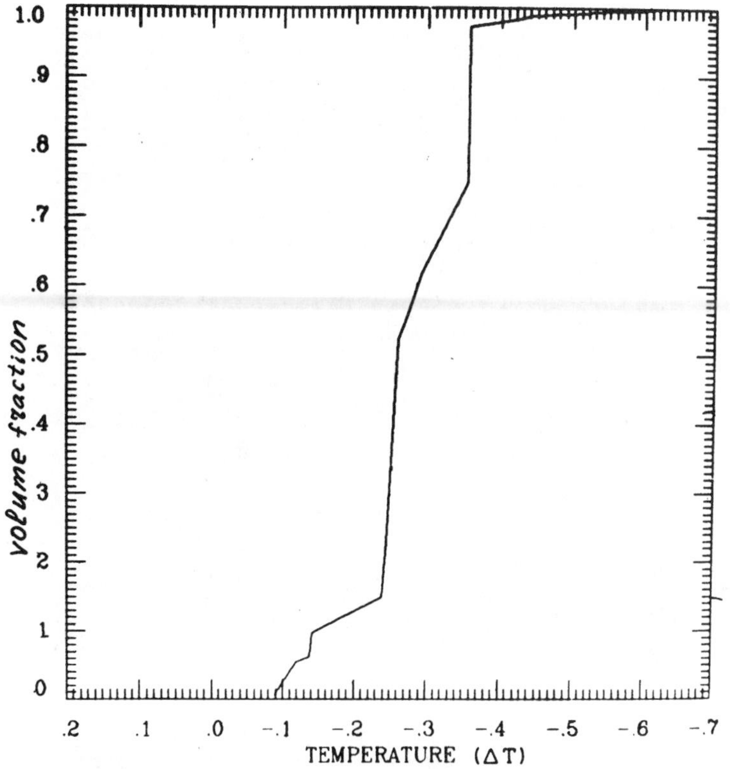

Figure 103. The prototype simulation of the martensitic transformation in the case of ten randomly distributed preexisting defects. The variation in the martensite volume fraction is calculated with undercooling. Undercooling measured in energy units, $\mu\varepsilon_0^2 v T_0/Q_0$, acts as a driving force.

initiate from preexisting defects and are sometimes homogeneously nucleated in defect-free regions of the lattice. Interestingly, the parallel martensite plates often form in an aggregate twin orientation to one another with a layer of the retained parent phase in the intervening space. An intermediate stage in the transformation illustrating some of these features is shown in Fig. 104c. As the transformation nears completion, only a small residue of isolated parent phase particles is retained. This residue of the parent phase is extremely stable, and a large undercooling is required to eliminate it and bring the transformation to completion.

The energetics of the transformation is illustrated in Fig. 105 which shows the magnitude of the chemical driving force as a horizontal line and the magnitude of the elastic energy change per step as an oscillating function for the first 400 transformation steps. Elastic energy is a noisy function that oscillates about the value zero. The transformation is stopped by large occasional excursions from zero that exceed the chemical driving force. These excursions identify the nucleation steps along the transformation path. They are responsible

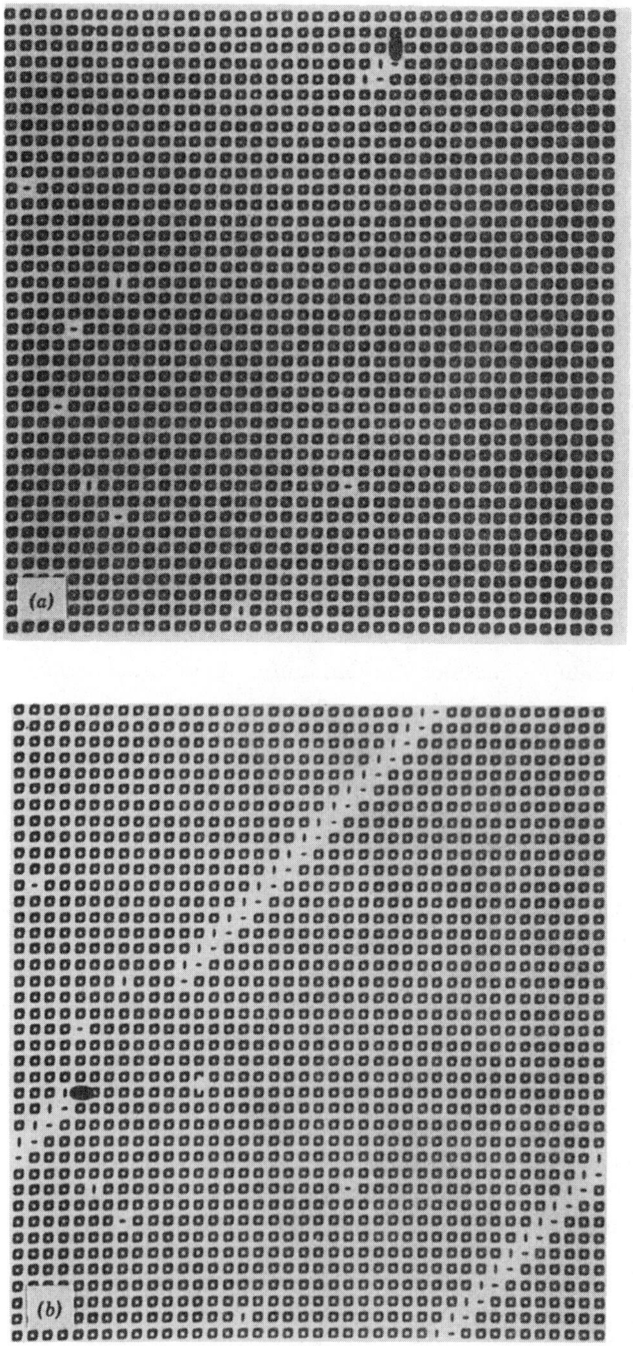

Figure 104. The successive stages of the development of the structure associated with the same martensitic transformation as described in Fig. 103: □ a vacant site of the grid related to the parent

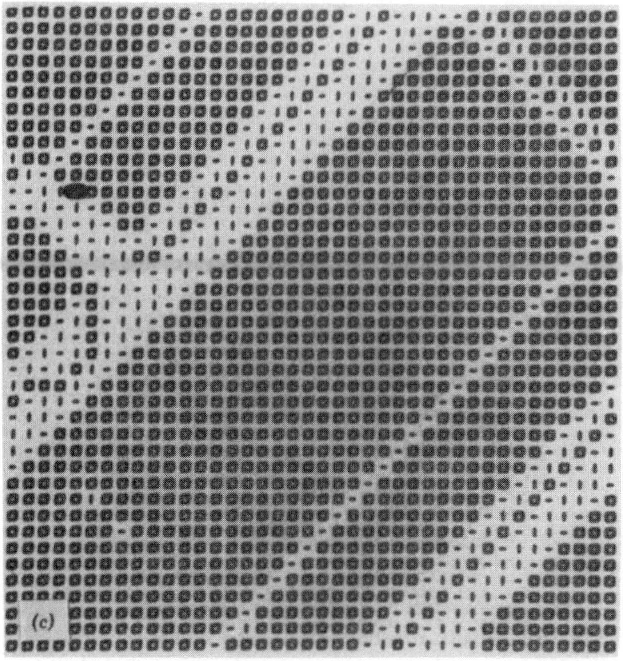

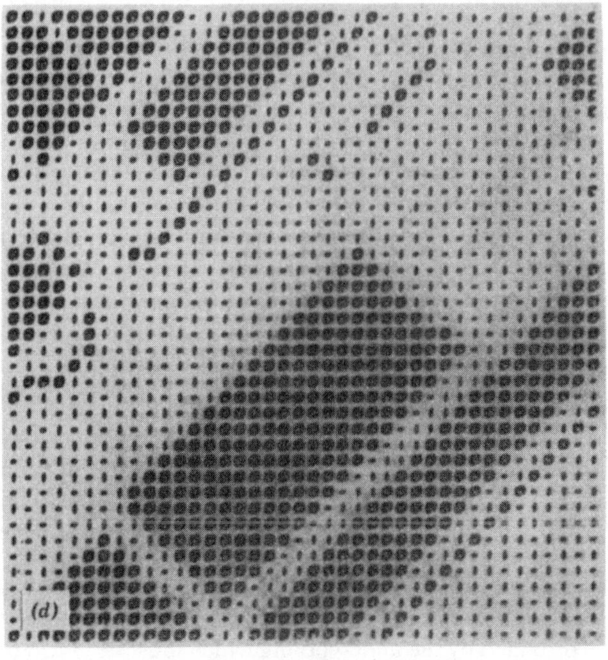

phase; horizontal and vertical dashes designate elementary martensite particles of the first and second orientational variants, respectively.

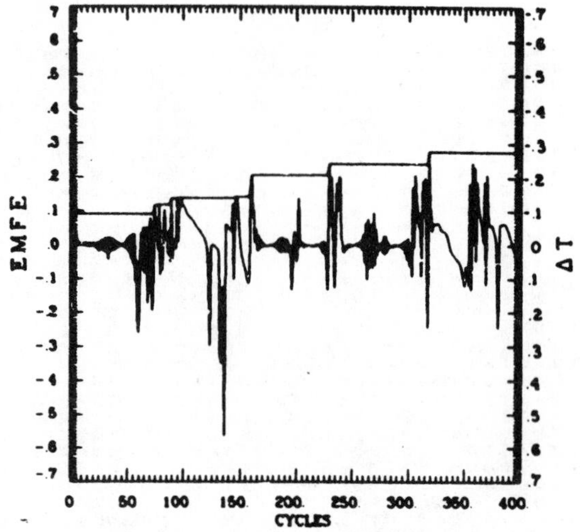

Figure 105. The strain energy change $\phi_0 - Q_0\Delta T/T_0$ associated with each successive elementary transformation event. Horizontal lines show the magnitude of the chemical driving force; the number of calculation cycles is equal to the number of elementary particles.

for the progression of the martensite transformation through a sequence of transformation bursts as the temperature is lowered and the chemical driving force raised.

Interestingly, the stress-free transformation strains (12.3.1) are invariant plane strains with the (110) and $(1\bar{1}0)$ invariant planes. Therefore the minimum strain energy should be martensite platelets with the (110) and $(1\bar{1}0)$ habits. The computer simulation results presented in Figs. 104 confirm this prediction, for all martensite plates generated do have the (110) and $(1\bar{1}0)$ habits.

The microstructures illustrated in Figs. 104 bear a very strong resemblance to those actually observed in a number of systems. Discrete internally twinned plates are present; they branch along the two variants of the (110) habit plane; and they are separated by regions of retained parent phase. At about half way through the transformation, all the preexisting defects have been consumed by the transformation. The undercooling required to accomplish the transformation is less than half that which would be required to initiate the transformation by a homogeneous nucleation process.

In summary, it should be noted that the model transformation described here reproduces most of the characteristic features of observed martensitic transformations in a simple and intriguing correspondence. The transformation proceeds through the growth of discrete martensitic plates that are twinned. In the earlier stages of the transformation, these plates are heterogeneously nucleated at preexisting defects in the lattice. The driving force for the transformation is provided by the undercooling which must be lowered continuously to keep the transformation going over a range of temperatures between the martensitic start and martensitic finish points. Some parent phase is retained in the microstructure until the undercooling becomes extremely large. As observed experimentally, this parent phase tends to be retained in thin lamella-separated adjacent martensite plates. Finally, the martensite transformation occurs spontaneously on cooling in situations where the self-energy of the martensite is large and positive.

12.4. COMPUTER SIMULATION OF STRAIN-INDUCED COARSENING OF TETRAGONAL PRECIPITATES IN CUBIC MATRIX: "TWEED" STRUCTURE FORMATION (233, 234)

It has been shown in Chapter 11 that the so-called "tweed" structure is often found in the early stage of precipitation when the parent phase is cubic and precipitates are tetragonal. The "tweed" structure has been observed in a variety of systems such as Cu-Be (255), dental Cu-Au-based alloy (256), Nb-O (236), Ni-V (257), and Ta-O (258). It involves an alignment of tetragonal precipitates along ⟨110⟩ directions of the cubic parent phase. An example is shown in Fig. 106.

As shown in Section 11.5, a tweedlike structure arises because of the natural tendency of the precipitate distribution to minimize the elastic energy. This approach is, however, of only limited applicability because, even if a tweedlike structure is found with an elastic energy minimum, it is not certain that it can be reached through the natural evolution of the system.

Let us examine a strain-induced coarsening model that assumes "tweed" forms through a reconfiguration of small precipitates which are initially in an almost random distribution. The approach described below involves modeling by computer simulation the progressive reconfiguration of a distribution of small tetragonal particles embedded in a cubic matrix. For simplicity in computation, a pseudo-two-dimensional model is used. The driving force of the reconfiguration is taken to be associated with a decrease of elastic energy. The elastic energy is computed in the long-wave approximation on the assumption that the difference in the elastic constants between the precipitate and matrix phases may be neglected (the model and all elastic energy calculations are the same as with the computer simulation of the martensitic transformation described in Sections 12.1 to 12.3). Special treatment of the reconfiguration kinetics is obviated by choosing the reconfiguration steps so as to maximize the decrease in the elastic strain energy. It will be seen that under these assumptions the model naturally leads to a tweedlike precipitate configuration that yields both diffraction patterns and "bright-field" representations in encouraging agreement with experiment.

Using the same approach as with the martensitic transformation (see Section 12.1.1), we shall define the minimum or "elementary" particle of the precipitate phase and introduce a reference grid superlattice for the matrix whose cell coincides with the elementary particle in size and shape. We may then represent

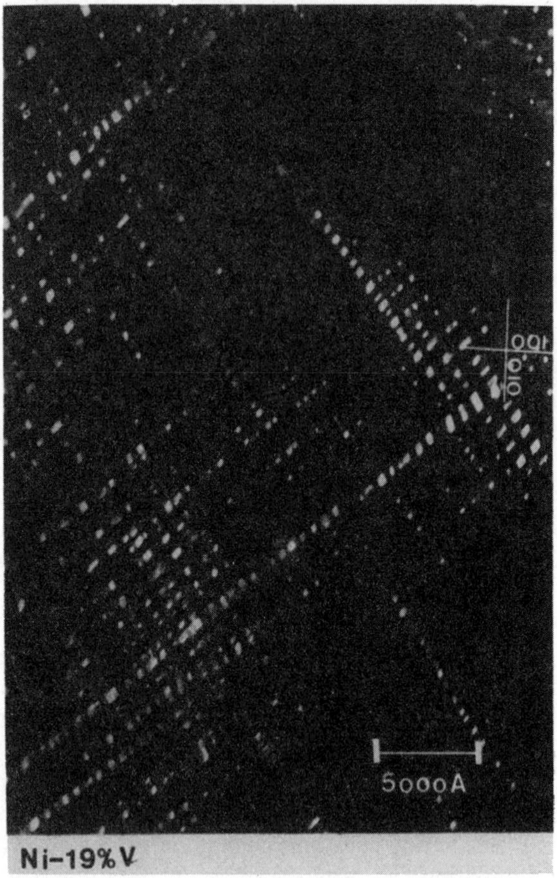

Figure 106. Electron micrograph of the tweed structure in Ni-19 at%V alloy (257). (Courtesy R. J. Rioja and D. E. Laughlin.)

an arbitrary distribution of the precipitate phase by filling appropriate cells of the grid superlattice with appropriate variants of precipitate phase particles. The distribution of new phase precipitates over the superlattice represented in terms of elementary particles was already considered in Section 12.1 to model martensitic transformation.

12.4.1. Elastic Potential

Diffusional reconfiguration of precipitates may be modeled as a sequence of elementary steps of three types: (1) translation of one elementary particle to a vacant nearest-neighbor site of the grid without a change in the particle's type, (2) a similar translation involving a change in type, (3) a change in type with no translation. Each of these events may be interpreted as an annihilation of an elementary particle and the creation of a new particle differing in type or position from the parent particle. Since the total number of elementary particles is conserved in this process, and since the chemical and elastic self-energies are independent of the precipitate particles' type, p (this is true if the particles are coherent tetragonal phase precipitates that differ only in the orientation of the tetragonal axis), the total change in energy is given by the second term of the elastic strain energy equation (12.1.39).

Hence we define the elastic potential by

$$\phi_p(\mathbf{R}) = \sum_{q,\,\mathbf{R}'} W_{pq}^{el}(\mathbf{R}-\mathbf{R}')\tilde{\zeta}_q(\mathbf{R}') \tag{12.4.1}$$

which gives the change in the configurational interaction energy caused by the creation of a particle of the type p at $\mathbf{R}$ in a given precipitate distribution specified by the function $\tilde{\zeta}_q(\mathbf{R}')$. An evolution step involving the substitution of the elementary particle $(p, \mathbf{R})$ for the particle $(q, \mathbf{R}')$ causes the total energy change

$$\Delta\phi = \phi_p(\mathbf{R}) - \phi_q(\mathbf{R}') \tag{12.4.2}$$

It is thermodynamically favored if $\Delta\phi$ is negative.

In a computer simulation of the reconfiguration process, the elastic potentials, $\phi_p(\mathbf{R})$, are easy to calculate once the interaction functions, $W_{pq}^{el}(\mathbf{R}-\mathbf{R}')$, are known and may be updated for changes of the precipitate configuration by mere addition. Given the list of the $\phi_p(\mathbf{R})$ potentials for the initial configuration of precipitates, the energy changes associated with all possible elementary events may be computed, and the favored event may be selected such that it provides the maximum energy decrease or is characterized by the highest statistical probability of thermal activation. Once the selected event is allowed to occur, the list of potentials, $\phi_p(\mathbf{R})$, may be updated by simple addition, a new event selected, and the evolution of the distribution of precipitates continued. This procedure provides a rapid, efficient computer simulation for a lattice of reasonable size.

12.4.2. Diffraction Pattern and Bright-Field Electron Images of Pseudo-Two-Dimensional Structure

To compute the diffraction pattern from a distribution of elementary precipitate particles, we need a solution for the elastic displacement field, $\mathbf{u}(\mathbf{R})$. The displacement field is the sum

$$u_i^{\text{total}}(\mathbf{R}) = \bar{\varepsilon}_{ij}R_j + u_i(\mathbf{R}) \tag{12.4.3}$$

where according to Eq. (7.2.26)

$$\bar{\varepsilon}_{ij} = \sum_p \varepsilon_{ij}^0(p)w_p \tag{12.4.4}$$

is the homogeneous strain which can be reduced by a mere change in the crystal lattice parameters, $\mathbf{u}(\mathbf{R})$ is the local displacement related to the heterogeneous strain.

It follows from Eqs. (7.2.35), (7.2.36a), and (12.1.4) that the Fourier transform of the local strain is given by

$$\mathbf{v}(\mathbf{k}) = \int \mathbf{u}(\mathbf{r}) e^{-i\mathbf{k}\mathbf{r}} d^3 r$$

$$= -i \sum_p \hat{\boldsymbol{\Omega}}(\mathbf{n}) \hat{\sigma}^0(p) \frac{\mathbf{k}}{k^2} \eta_0(\mathbf{k}) \zeta_p(\mathbf{k}) \tag{12.4.5}$$

Since all terms in Eq. (12.4.5) are known for a given distribution of precipitates, the internal displacement field can be found through the back Fourier transform

$$\mathbf{u}(\mathbf{R}) = \int \mathbf{v}(\mathbf{k}) e^{i\mathbf{k}\mathbf{R}} \frac{d^3 k}{(2\pi)^3} \tag{12.4.6}$$

Neglecting the difference between the atomic scattering factors of the atoms present in the crystal, the kinematic approximation for the diffraction intensity at the diffraction vector **k** is proportional to

$$I(\mathbf{k}) = \sum_{\mathbf{r},\mathbf{r}'} e^{i\mathbf{k}[\mathbf{r}+\mathbf{u}(\mathbf{r})-\mathbf{r}'-\mathbf{u}(\mathbf{r}')]} \tag{12.4.7}$$

where the summation is over all crystal lattice sites. If $\mathbf{u}(\mathbf{r})$ is assumed to be equal to $\mathbf{u}(\mathbf{R})$ within the volume where $\tilde{\eta}_0(\mathbf{r}-\mathbf{R})$ is nonzero,

$$I(\mathbf{k}) = \sum_{\mathbf{r},\mathbf{r}'} e^{i\mathbf{k}(\mathbf{r}-\mathbf{r}')} \sum_{\mathbf{R}} \sum_{\mathbf{R}'} \tilde{\eta}_0(\mathbf{r}-\mathbf{R}) \tilde{\eta}_0(\mathbf{r}'-\mathbf{R}') e^{i\mathbf{k}[\mathbf{u}(\mathbf{R})-\mathbf{u}(\mathbf{R}')]}$$

$$= \sum_{\mathbf{R}} \sum_{\mathbf{R}'} e^{i\mathbf{k}(\mathbf{R}-\mathbf{R}')} e^{i\mathbf{k}[\mathbf{u}(\mathbf{R})-\mathbf{u}(\mathbf{R}')]} |\eta_0(\mathbf{k})|^2 \tag{12.4.8}$$

Since the "tweed" microstructure derives its name from that characteristic appearing in bright-field electron micrographs, it may be interesting to simulate this effect with the pseudo-two-dimensional model. To accomplish this, the conventional column approximation for bright-field images should be modified somewhat because elementary particles have an infinite extension in the direction normal to the grid plane.

Eq. (12.4.8) may be rewritten

$$I(\mathbf{k}) = \sum_{\mathbf{R}} I(\mathbf{k}, \mathbf{R}) \tag{12.4.9}$$

where

$$I(\mathbf{k}, \mathbf{R}) = \sum_{\boldsymbol{\rho}} |\eta_0(\mathbf{k})|^2 \cos\{\mathbf{k}[\boldsymbol{\rho} + \mathbf{u}(\mathbf{R}) - \mathbf{u}(\mathbf{R}-\boldsymbol{\rho})]\} \tag{12.4.10}$$

may be interpreted as the diffraction intensity at the diffraction vector **k** from the supercell **R**, $\boldsymbol{\rho} = \mathbf{R} - \mathbf{R}'$. The fraction of this intensity captured by an aperture of angle θ_0 about the normal to the grid lattice is

$$I(\mathbf{R}) = \iint_{|\mathbf{k}|<k_0} I(\mathbf{k}, \mathbf{R}) \lambda_e^2 \frac{d^2 k}{4\pi^2} \tag{12.4.11}$$

where $\lambda_e^2 d^2 k / 4\pi^2$ is an element of the solid angle, $k_0 = 4\pi \sin\theta_0/\lambda e \approx 4\pi/\lambda_e \times \theta_0$,

λ_e is the wavelength of the incident electron radiation. Setting $\eta_0(\mathbf{k}) \approx 1$ for small k's, and using the relation

$$\iint_{|\mathbf{k}|<k_0} \cos \mathbf{kr}\, d^2k = 2\pi k_0^2 \frac{J_1(k_0 r)}{k_0 r} \tag{12.4.12}$$

where $J_1(k_0 r)$ is the Bessel function of order 1, we obtain

$$I(\mathbf{R}) = \frac{\lambda_e^2}{2\pi} k_0^2 \sum_{\mathbf{R}'} \frac{J_1(k_0|(\mathbf{R}-\mathbf{R}') + \mathbf{u}(\mathbf{R}) - \mathbf{u}(\mathbf{R}')|)}{k_0|(\mathbf{R}-\mathbf{R}') + \mathbf{u}(\mathbf{R}) - \mathbf{u}(\mathbf{R}')|} \tag{12.4.13}$$

for the bright-field intensity associated with the grid cell at $\mathbf{R}$ of the pseudo-two-dimensional lattice.

12.4.3. Computer Simulation Results

The example we are going to discuss (233) is chosen to illustrate a crude simulation of the Cu-Be case (255) which is the best studied "tweed" microstructure. The elastic constants were taken to be equal to those of pure copper:

$$\lambda_{iiii} = c_{11} = 17.6 \times 10^{11} \frac{\text{dyne}}{\text{cm}^2} \quad (i=1, 2, 3)$$

$$\lambda_{iijj} = c_{12} = 12.3 \times 10^{11} \frac{\text{dyne}}{\text{cm}^2} \quad (i \neq j=1, 2, 3)$$

$$\lambda_{ijij} = c_{44} = 8.17 \times 10^{11} \frac{\text{dyne}}{\text{cm}^2} \quad (i \neq j=1, 2, 3) \tag{12.4.14}$$

The two variants of the transformation strain are assumed to be

$$\hat{\varepsilon}^0(1) = \begin{pmatrix} 0.132 & 0 & 0 \\ 0 & -0.200 & 0 \\ 0 & 0 & 0 \end{pmatrix}, \quad \hat{\varepsilon}^0(2) = \begin{pmatrix} -0.200 & 0 & 0 \\ 0 & 0.132 & 0 \\ 0 & 0 & 0 \end{pmatrix} \tag{12.4.15}$$

In the initial state, fifty elementary particles of each variant were distributed over a 200×200 square grid lattice. The initial structure is shown in a halftone of a computer-made drawing (Fig. 107a); one variant is shown dark, and the other one light against the grey background of the cubic matrix. The unit translations of the grid lattice coincide with the [100] and [010] directions of the matrix lattice under consideration. To accomplish the simulation, we allow the system to evolve through discrete steps involving a change in the orientational variant, or a diffusional step to an unoccupied nearest-neighbor site, or a combination of the two. The energetic consequence of each event is computed using Eqs. (12.4.1) and (12.4.2).

The initial stages of the evolution involve possible steps leading to decrease in the total energy of the system. The most favorable of these—that which causes the maximum energy decrease—is selected and accomplished; then

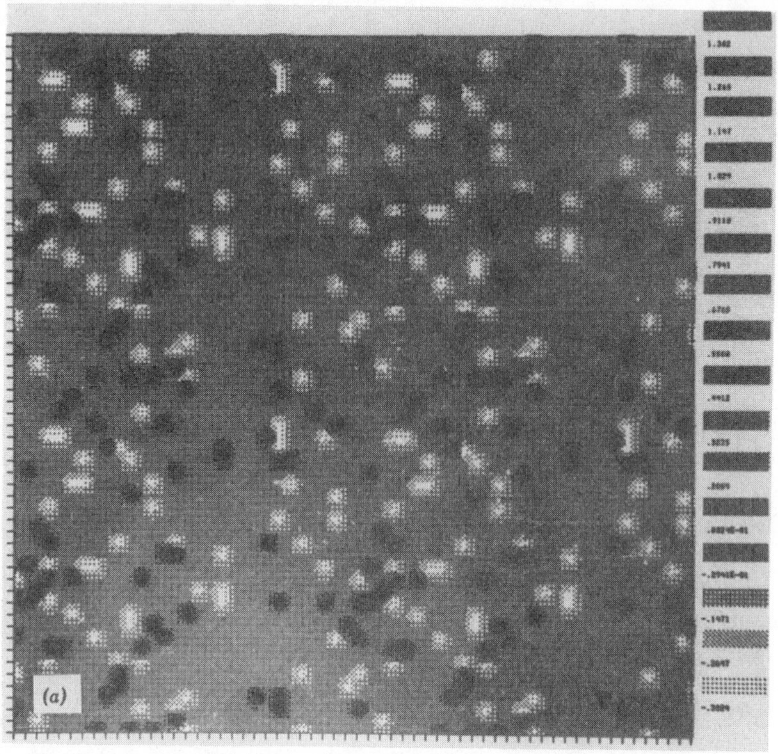
(a)

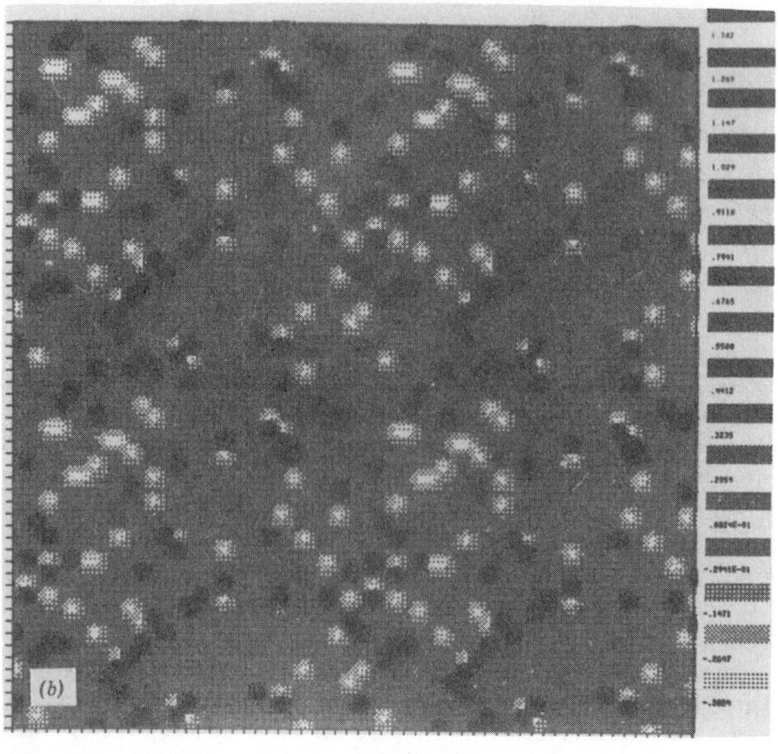
(b)

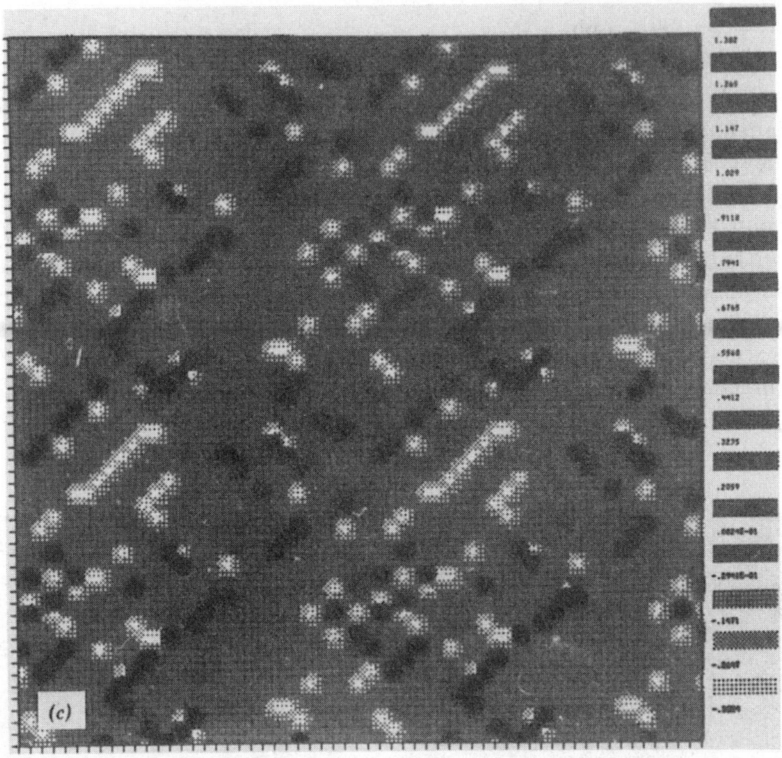

Figure 107. Sequence of computer-generated pictures showing the distribution of elementary particles in the tetragonal phase. (*a*) In the initial state; (*b*) after 100 relaxation steps; (*c*) after 500 relaxation steps. The two precipitate variants appear in black and white against the grey background matrix. The figure includes a 2 × 2 repetition of the 200 × 200 grid to show continuity across the periodic boundary.

the relevant elastic potentials are recomputed, and the selection process is repeated to continue the evolution.

In practice, however, the system has been found to reach rapidly a metastable state in which all possible elementary steps require an increase in the strain energy. A physical system evolving at a low temperature would of course remain trapped in the first metastable state; at a higher temperature thermal fluctuations would eventually provide an escape. Unless the temperature is very high, an escape from the metastable state will usually be preceded by a rather prolonged oscillation period. To simulate the fluctuation escape process, we adopted the strategy of selecting the set of the most favorable events, those whose energies differ by only a small percentage from the minimum. A random choice was then made among these events under the constraint that the system should not retrace the path of its evolution. Escape from the metastable state leads to a further spontaneous evolution. The system is observed to pass through a sequence of metastable configurations of decreasing energies and increasing order of the "tweedlike" type. The evolution of the system is illustrated in Figs. 107b and c which show the configuration of the particles after 100 and 500 evolution steps, respectively. The particles order rapidly to form single-variant lines along the (11) lines of the two-dimensional lattice which correspond to the (110) planes in the three-dimensional network. The ordering is apparent after 100 steps and grows to become striking after 500 steps by which time the computed configuration strongly resembles the tweed structure of Ni-V, as shown in the Rioja-Laughlin micrograph given in Fig. 106. The linear extension of the (022) diffraction spot and the crossed pattern of the (002) spot are common observations in the diffraction patterns of tweed microstructures (255, 256). These effects are apparent in the electron diffraction patterns shown in Fig. 108a.

Simulated bright-field electron micrographs from the array of precipitates are shown in Fig. 108b for the initial configuration and configurations formed

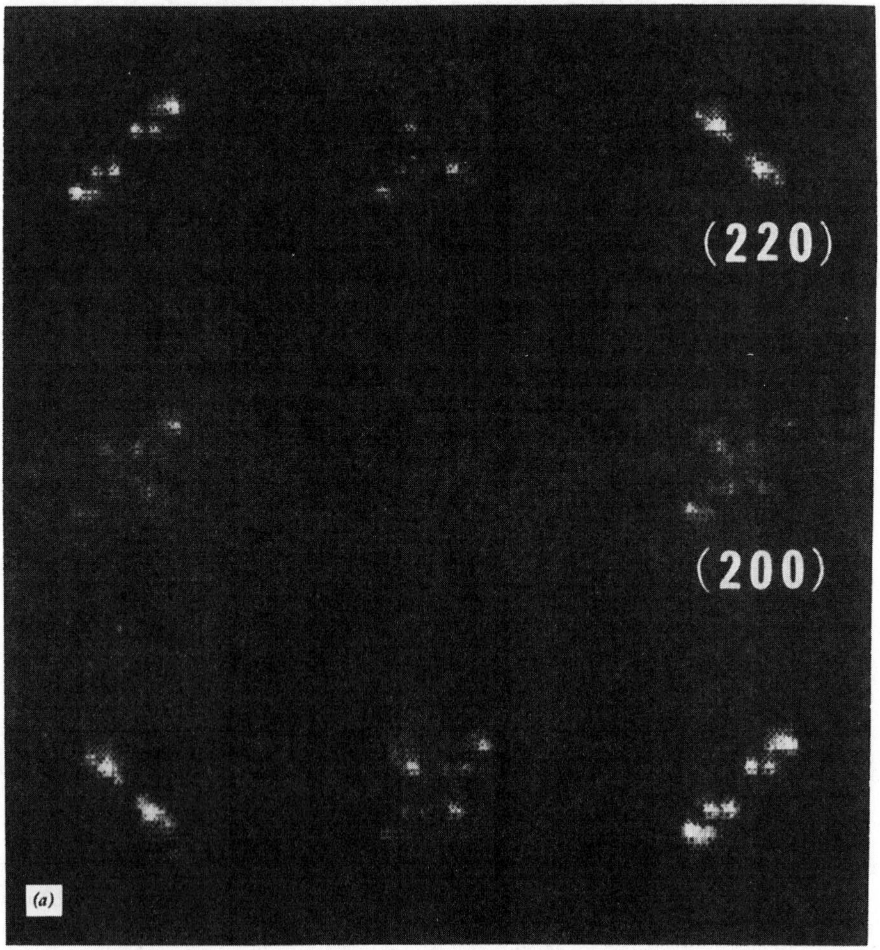

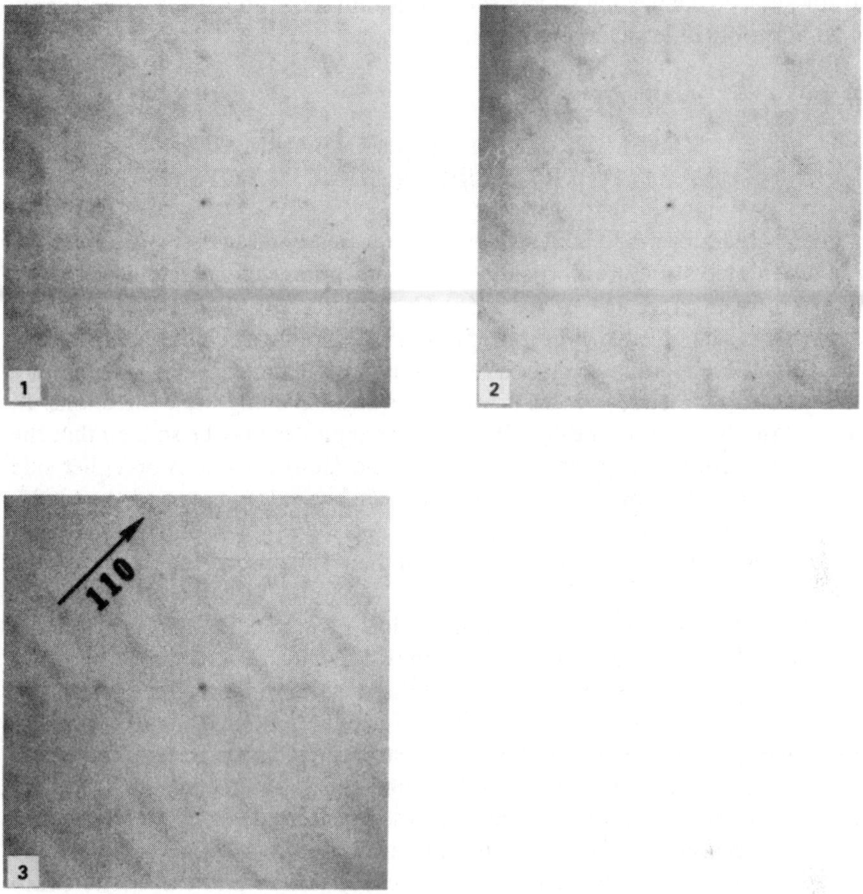

Figure 108. (*a*). Simulated diffraction pattern from the structure shown in Fig. 107c. (*b*) Simulated bright-field electron microscopic contrast: (1) in the initial state, (2) after 100 relaxation steps, (3) after 500 relaxation steps.

after 100 and 500 steps. Both of these micrographs show the modulated "tweed" structure from which the structures under discussion derive their name. The similarity of the experimental electron micrographs to the computer-generated ones is obvious.

The model employed in the computer simulation described in this section is obviously oversimplified in both its basic assumptions and treatment of the reconfiguration process. Nevertheless, the results show a striking and interesting correspondence with the prominent experimental features of the "tweed" reaction and closely reproduce the precipitate configuration, the diffraction pattern, and the "tweedy" appearance of bright-field electron images. The work referred to lends a strong support to the premise that "tweed" is generated in strain-induced coarsening which leads to the minimization of the strain energy of an assembly of tetragonal precipitates.

12.5. COMPUTER SIMULATION OF FORMATION OF MODULATED STRUCTURE IN CUBIC ALLOYS (234)

The problem of modulated structures was already discussed in Chapter 10. The modulated structures formed by cubic precipitates in cubic matrices were observed, for instance, in Cu-Ni-Fe (65), Cu-Mn-Al (223), Ni-Al (177), Fe-Cr-Co (259), nonstoichiometric β-brass (35), and other alloys. Unlike the "tweed" structure, these modulated structures display an alignment along $\langle 100 \rangle$ directions and show pseudoperiodic spatial distributions along these directions. In such a case the order in the distribution of precipitates may be so high that the diffraction patterns show extra spots along the $\langle 100 \rangle$ directions on either side of the matrix reflections. The distance between the extra spots and the fundamental reflections corresponds to the pseudoperiod of the precipitate distribution. Examples of modulated structures and their diffraction patterns are shown in Fig. 86.

The development of a regular distribution of precipitates during aging is presumed to start from a random distribution in the as-quenched state. The coarsening process that occurs during aging results in an aggregation of the initial precipitates, involving "uphill" diffusional motion to yield spatially correlated distributions, that is, modulated structures. This process is controlled by relaxation of the sum of the interphase and elastic strain energies. The latter term arises from the mismatch between the crystal lattices of the precipitate and parent phases. Unlike the chemical free energy which only depends on the total volume of precipitates, the interphase free energy and the elastic strain energy are explicit functions of the distribution of precipitates. Below we shall consider the computer simulation of the strain-induced coarsening process after Wen, Morris, and Khachaturyan (234). The computer simulation approach offers a quantitative treatment of the formation of the regular structures occurring in a simple coarsening process.

The quantitative treatment of an arbitrary distribution of precipitates of arbitrary shapes is conducted the same as for martensitic and "tweed" transformations, by introducing the concept of an elementary particle of the precipitate phase, the minimum-size new phase brick, and of the grid. The unit cells of the grid frame can be filled by elementary particles in any desirable order. Since the total volume of the precipitate phase does not vary during the coarsening, the total number of elementary particles is constant and the coarsening only affects the spatial redistribution of elementary particles over grid sites.

In cubic-to-cubic phase transformation each elementary particle is a carrier of the stress-free strain:

$$\varepsilon_{ij}^{0}=\varepsilon_0\delta_{ij} \tag{12.5.1}$$

We have the single type of new phase inclusion, and any arbitrary configuration (arbitrary substructure) of elementary particles can be described by the single distribution function

$$\tilde{\zeta}(\mathbf{R})=\begin{cases}1 & \text{if an elementary particle occupies the grid site } \mathbf{R}\\ 0 & \text{otherwise}\end{cases} \qquad (12.5.2)$$

Redistribution of elementary particles during the coarsening is a diffusional process. We can describe the diffusion motion if we assume that each elementary event in the reconfiguration of elementary particles is a translation of one of the elementary particles to a nearest-neighbor vacant grid site. It will easily be seen that each of these events may be interpreted as an annihilation of an elementary particle at the site $\mathbf{R}$ and the subsequent creation of an elementary particle at a nearest-neighbor site.

Any elementary event involving the annihilation of an elementary particle at $\mathbf{R}'$ and the creation of a new elementary particle at $\mathbf{R}$ (the $\mathbf{R}' \to \mathbf{R}$ translation of an elementary particle) results in the energy change

$$\Delta\phi(\mathbf{R}' \to \mathbf{R})=\phi(\mathbf{R})-\phi(\mathbf{R}') \qquad (12.5.3)$$

where $\mathbf{R}$ differs from $\mathbf{R}'$ by the nearest-neighbor distance, $\phi(\mathbf{R})$ is the thermoelastic potential, namely, the change of the total free energy (given by the sum of the bulk free energy, interphase energy, and elastic free energy) caused by the creation of an elementary particle at the site $\mathbf{R}$. An elementary event is favorable thermodynamically if the thermoelastic potential change, $\Delta\phi(\mathbf{R}' \to \mathbf{R})$, is negative, and it is unfavorable if $\Delta\phi(\mathbf{R}' \to \mathbf{R})$ is positive. We assume after (234) that the step-by-step evolution of an alloy in coarsening proceeds through a sequence of elementary events such that provide the maximum possible decrease of the total free energy, with each event giving the minimum possible value of the thermoelastic potential.

To save computation time, the first attempt to simulate the coarsening has been made with a pseudo-two-dimensional model. The term pseudo-two-dimensional means that all elementary particles (and therefore all precipitates) are infinite rods normal to the quadratic plane grid whose cross section coincides with the grid unit cell. The strain ε_{ij}^0 in this case can be represented by the two-dimensional unit tensor:

$$\varepsilon_{ij}^0=\varepsilon_0\begin{pmatrix}1 & 0\\ 0 & 1\end{pmatrix} \qquad (12.5.4)$$

The thermoelastic potential is calculated according to the equation

$$\phi(\mathbf{R})=\phi_0+\sum_{\mathbf{R}'} W^{el}(\mathbf{R}-\mathbf{R}')\tilde{\zeta}(\mathbf{R}') \qquad (12.5.5)$$

where ϕ_0 is the bulk energy contribution (it has a constant value), $W^{el}(\mathbf{R}-\mathbf{R}')$ are the strain-induced interaction energies calculated from Eq. (12.1.40) with ε_{ij}^0 given by Eq. (12.5.4) and c_{11}, c_{12}, c_{44} given by (12.4.14).

The inclusion of the interphase energy proves to have only a small qualitative

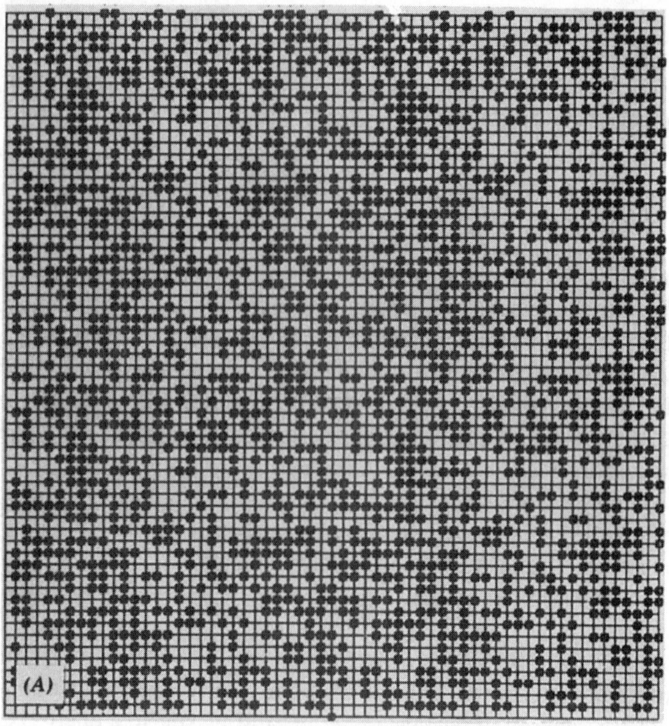
(A)

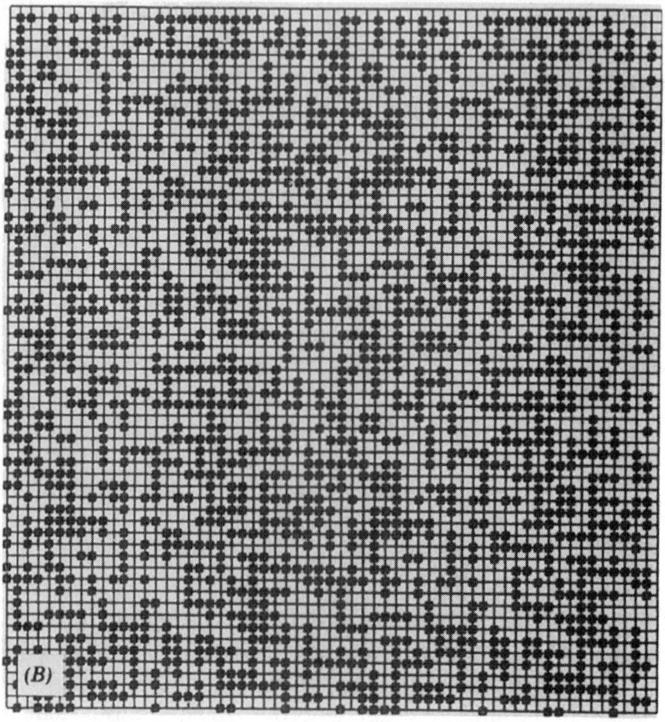
(B)

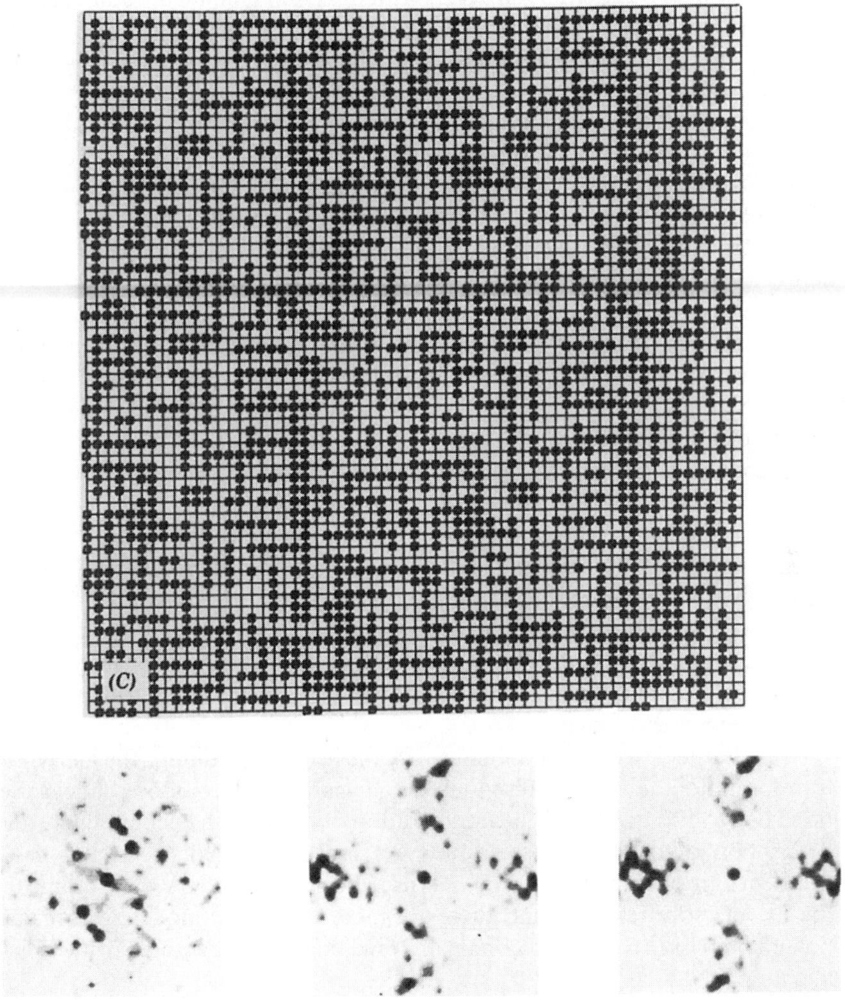

A′ B′ C′

Figure 109. (a) Sequence of computer-generated pictures showing the distribution of elementary particles of the cubic phase in the cubic matrix: the random distribution A is transformed into the basketlike distributions B and C, The figure includes 2×2 repetition of the grid. (b) Simulated diffraction pattern A′, B′, C′ are diffraction patterns corresponding to the structures A, B, C.

effect. Below we shall present the computer simulation results obtained without an interphase energy term.

The calculation generates a random distribution of elementary particles taken to reproduce the initial "as-quenched" state (Fig. 109a). Figures 109b and c demonstrate the development of the modulated structure during coarsening and the formation of a basketlike structure with the ⟨100⟩ alignment. Comparison of Figs. 109c and 86 shows the computer simulation results to agree with the electron microscopic observations. The computed diffraction patterns (Figs. 109a′, b′, c′) corresponding to the structures presented in Figs. 109a to c demonstrate the most typical feature of modulated structures: the development of side-bands on the diffraction pattern (extra spots along the directions ⟨100⟩ on both sides of the fundamental reflections). We can conclude from this that the observed morphology of two-phase alloys arising during aging can be modeled reasonably well as the result of a relaxation of the elastic strain caused by the mismatch between the crystal lattices of the precipitate and parent phases.

It is particularly important that the spinodal-like structure obtained in the cubic-cubic case (Fig. 109) is formed from the initial random distribution of precipitates through direct coarsening. It does not require continuous development of periodic concentration waves as is accepted in conventional spinodal decomposition theory.

To conclude this section, it is worthwhile emphasizing the following. The conventional analytical treatment of actual systems can in fact only be successfully performed in the simplest cases. It fails when a more or less realistic model of multiphase alloys is considered. As a matter of fact, computer simulation is a new approach to the problem and is applicable to real alloys. The computer makes it possible to solve problems that have been considered unsolvable. The use of high-speed computers for modeling the processes occurring in alloys cannot be obviated, and I believe that computer-simulated research will form a new field in the material sciences.

Even the first attempts to simulate, with the help of computers, the martensitic transformation and the strain-induced coarsening in decomposed alloys (Sections 12.3 to 12.5) using but idealized crude models gave an encouragingly good description of those processes, and the results obtained are sometimes in excellent agreement with electron microscopic observations. We may hope that the use of more realistic (and therefore more elaborate) models will make it possible to come close to theoretically predicting the structures of two-phase alloys applied in industry.

There is another problem that should be mentioned. The physical processes occurring in alloys are affected by varied factors that sometimes conceal the important characteristics of the phenomenon. Eliminating them may be a difficult problem which can hardly be solved in all cases. From this standpoint computer simulation serves as an "experiment" carried out under ideal conditions in the absence of any interfering factors.

13

MICROSCOPIC ELASTICITY THEORY OF MACROSCOPICALLY HOMOGENEOUS SOLID SOLUTIONS

13.1. INTRODUCTION

As shown in Chapters 7 to 12, strain-induced interactions in a two-phase alloy are the major factor determining its equilibrium morphology and the kinetics of reconfiguration of its microstructure during coarsening. Coherent new phase inclusions considered in the preceding chapters are macroscopic particles whose typical size considerably exceeds interatomic distances. Thanks to this, all the calculations can be made in terms of continuum elasticity. The case of elastic displacement field generated by solute atoms or other types of point defects closely resembles that of a two-phase alloy. The only difference in principle is that, unlike inclusion-generated elastic strain, both the diameters of and typical separations between solute atoms are of the same order of magnitude as interatomic distances. Obviously, any adequate mathematical treatment of the problem should be made in terms of microscopic elasticity (crystal lattice statics), which takes into account the discrete nature of the crystal lattice.

The calculation of the elastic displacement field generated by point defects in an infinite crystal lattice was proposed by Matsubara in 1952 (260), Kanzaki in 1957 (261), Krivoglaz and Tikhonova in 1958 (262).

A calculation of the elastic energy produced by a group of point defects (solute atoms) within an infinite crystal body was made by Krivoglaz and Tikhonova in 1958 (262) and Khachaturyan in 1962 (263). Later similar results were obtained by Hardy and Bullough in 1967 and 1968 (264, 265) for stain-induced interactions of pairs of vacancies in Al and Cu.

The complexity of the microscopic elasticity problem increases when solute atoms form a solid solution within a finite body of a solvent crystal, that is, when the solute to solvent atomic ratio becomes nonzero. The first successful attempt to solve this problem belongs to Eshelby (266), in 1956. He suggested a simple model that treats substitutional solute atoms as continuum spherical inclusions within a finite elastically isotropic matrix. The diameter of these spherical "inclusions" was assumed to differ from the diameter of the appropriate holes in the matrix. In effect Eshelby's approximation coincides with the well-known Debye model. According to Debye crystal lattice vibration frequencies are approximated by the vibration frequencies of an elastically isotropic continuous body, the maximum wave number being cut off by the sphere whose volume in the **k**-space is equal to that of the first Brillouin zone of the crystal. The significance of Eshelby's remarkable work (266) derives from that it has provided an understanding of such delicate phenomena pertaining to the problem of the elastic energy of a finite solid solution crystal as the image-force effect. Also Eshelby's results may serve as a criterion for the correctness of more realistic models because any consistent microscopic theory should yield Eshelby's expression for elastic energy following the limit transition to continuum elasticity and elastic isotropy.

The microscopic theory of a solid solution was proposed by Khachaturyan (267), in 1967. The theory is formulated in purely microscopic terms and yields Eshelby's results after the limit transition to isotropic continuum (the Debye approximation). Two years later the microscopic theory (267) was reformulated by Cook and de Fontaine (268). The final expression for the elastic energy derived by Cook and de Fontaine does not, however, give the limit transition to Eshelby's theory because the elastic relaxation term associated with uniform strain relaxation was lost in the derivation.*

Returning to the elastic energy of a solid solution, it should be noted that elastic interactions of solute atoms through the displacement field (strain-induced interactions) produce atomic redistributions and sometimes lead to long-range ordering which ensures relaxation of the elastic energy. It seems reasonable to expect that interstitial solutions should exhibit particularly strong strain-induced interactions.

Zener was the first to assume that strain-induced interactions are responsible for ordering in the Fe-C martensite which is an interstitial solution based on a bcc host lattice of α-Fe (270). Now thirty years later the idea that strain-induced interaction is a dominant contribution to the ordering energy in most of the interstitial solutions is generally accepted.

The order of magnitude of strain-induced interaction energies is

$$W \approx \lambda a_0^3 u_0^2$$

where λ and u_0 are typical quantities of the elastic modulus and the distortion

*The details will be discussed later, in Section 13.5.5.

tensor (a tensor of concentration coefficients of the host lattice expansion), a_0 is the lattice parameter of the host lattice.

Since for interstitial alloys $\lambda \approx 10^{12}$ erg/cm^3, $a_0^3 \approx 10^{-23}$ cm^3, $u_0^2 \approx 0.1$, one obtains $W \approx 1$ eV. The energy of the order of 1 eV is about two orders of magnitude higher than the usual values of the chemical interchange energies, and therefore the thermodynamic behavior of an interstitial solution is determined by the strain-induced interaction. In this situation there are optimal configurations that provide the minimum host lattice displacements due to the mutual compensation of displacements induced by each interstitial and consequently the minimum elastic energy. In particular, such atomic configurations can form a periodic sublattice whose periods are a few corresponding periods of the host lattice. The process of transition to such an optimal periodic configuration is termed strain-induced ordering.

The theory of strain-induced interaction (267) enables one to calculate the Fourier transforms of the long-range interaction energies, employing independently found parameters: vibration frequencies of the host lattice and the components of tensor of concentration coefficients of the host lattice expansion due to solving atoms. If Fourier transforms of interaction energies are known, one can always employ the method of static concentration waves formulated in Chapter 3 to predict the atomic arrangement of an ordered phase and the order-disorder temperature. In doing so, it should be remembered that the pairwise strain-induced interaction potential has a long range and decays asymptotically as r^{-3} (r being the distance between a pair of interstitials). This means that the concentration wave technique is a unique way to treat the order-disorder problem since the traditional treatment of this problem is possible only in the framework of the model of nearest- and next nearest-neighbor interaction. The theory of strain-induced interaction may also be efficiently applied to the spinodal decomposition of substitutional and interstitial solutions, to the formation of complexes of defects and impurity atoms, and so on.

13.2. ELASTIC ENERGY OF SOLID SOLUTIONS

We shall now describe the microscopic theory of elastic energy of a solid solution proposed by Khachaturyan (267). In this we shall follow the line of reasoning suggested by Wen in the theory of multicomponent substitutional solutions (85). This will make it possible to express more clearly the physical results (267) and to elucidate the separation of chemical free energy, elastic energy, and physically distinct contributions to the elastic energy.

For definiteness, let us consider a binary interstitial solution whose host lattice can be generated by translating a single atom (the results that will be obtained directly apply to a binary substitutional solution and only require minor modifications to describe a multicomponent substitutional solutions). The lattice thus constructed is one of the fourteen Bravais lattices. We also assume that all interstices that can be occupied by interstitial atoms are crystallographically equivalent; in other words, they may be obtained from one of them by translations, reflections, and rotations making up the space group of the host crystal.

Let ν interstitial positions (interstices) lie within one primitive unit cell of the host lattice and let them be indexed with subscripts $p=1, 2, \ldots, \nu$. In a Bravais host lattice the position of any interstice $(p, \mathbf{r})$ may be characterized by specifying the coordinate of the host atom $\mathbf{r}$ which determines the position of the primitive unit cell and the index p of the interstice within this cell.

As shown in Section 7.2, a heterogeneous two-phase state may be generated by the application of a cycle of six operations diagrammed in Fig. 60. A similar procedure may be employed to obtain an interstitial (or substitutional) solid solution from pure one-component solvent. Let us perform the following operations shown in Fig. 110:

Step 1. Taking the pure solvent crystal, isolate ν clusters of which the pth contains N_p solvent atoms, and cut each cluster out of the solvent lattice. If the solvent and each of the clusters are sufficiently large for surface effects to be ignored, the energy of the assembly is not changed in the process.

Step 2. Inject N_p interstitial atoms into N_p interstices of type p of the pth cluster, and let the cluster relax to its stress-free shape. This procedure transforms ν pure solvent lattice clusters into ν clusters (ν orientational variants) of a completely ordered stoichiometric interstitial phase with an ideal lattice. All primitive unit cells of each of the transformed clusters contains one interstitial atom apiece situated in the interstices of type 1 in the first cluster, in the interstices of type 2 in the second cluster, . . . , in the interstices of type p in the pth cluster, . . . , and in the interstices of type ν in the νth cluster.

This transformation will involve the free energy change

$$\Delta F_2^{\text{chem}} = \sum_{p=1}^{\nu} \Delta F^{\text{chem}}(p) \tag{13.2.1}$$

where $\Delta F^{\text{chem}}(p)$ is the chemical free energy difference between the transformed cluster of type p and the initial pure host lattice cluster of type p.

The free expansion of the pth solvent lattice cluster under its transformation into the pth cluster of the completely ordered stoichiometric interstitial compound involves no elastic energy change because the cluster is stress-free by definition. Actually, the internal stress is zero because the transformed clusters are microscopically homogeneous ideal crystals.* They consist of strictly identical primitive unit cells with one interstitial atom per unit cell. The external stress is also zero because of free expansion of the cluster lattice under alloying.

This stress-free expansion is described by the strain tensor, $u_{ij}^0(p)$ ($p=1, 2, \ldots, \nu$) defined in Section 7.1.

*This is particularly obvious in the case of a multicomponent substitutional solid solution for which Step 2 would mean a complete replacement of all solvent atoms of the pth cluster by solute atoms of the p kind.

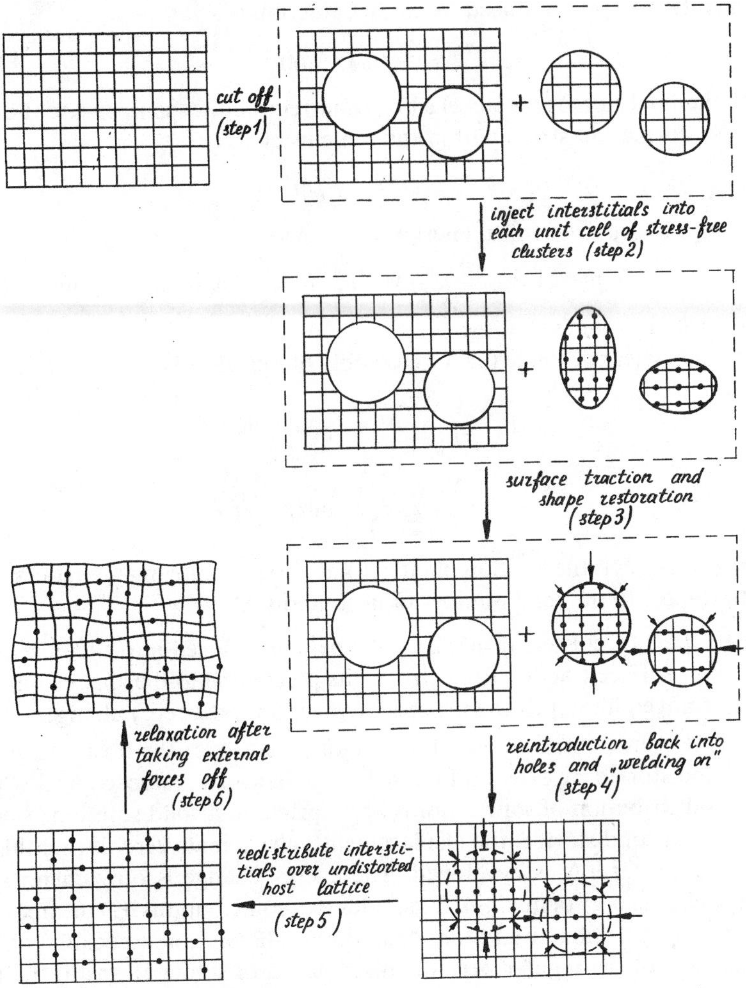

Figure 110. The six-step cycle to create an interstitial solid solution, proceeding from a pure solvent crystal and gas of interstitial atoms as a reference state: ● an interstitial atom.

Eq. (13.2.1) may be rewritten

$$\Delta F_2^{\text{chem.}} = \sum_{p=1}^{\nu} \mu_{\text{int}} N_p \tag{13.2.2}$$

where μ_{int} is the chemical potential of interstitial atoms in the cluster produced in Step 2 from the pure solvent as a reference state, N_p is the number of interstitial atoms in the pth cluster.

Step 3. Let surface traction be applied to each transformed cluster to restore it to the shape it had before the transformation. This requires the strain $\varepsilon_{ij} = -u_{ij}^0(p)$. Since each cluster is macroscopic, the deformation may be described by continuum elasticity.

The internal stress induced by surface traction is

$$\sigma_{ij} = \lambda_{ijkl}(-u^0_{kl}(p)) \tag{13.2.3}$$

Since the final homogeneous elastic strain required to compensate the stress-free transformation strain within the pth cluster is

$$\varepsilon_{ij} = -u^0_{ij}(p) \tag{13.2.4}$$

the change in the mechanical energy is

$$\Delta F_3(p) = \tfrac{1}{2} V_p \lambda_{ijkl}(-u^0_{ij}(p))(-u^0_{kl}(p)) = \tfrac{1}{2} N_p v_0 \lambda_{ijkl} u^0_{ij}(p) u^0_{kl}(p) \tag{13.2.5}$$

where v_0 is the atomic volume.

The total energy change for the assembly of v clusters is

$$\begin{aligned} \Delta F_3 &= \frac{1}{2} \sum_{p=1}^{v} N_p v_0 \lambda_{ijkl} u^0_{ij}(p) u^0_{kl}(p) \\ &= \frac{N}{2} \sum_{p=1}^{v} v_0 \lambda_{ijkl} u^0_{ij}(p) u^0_{kl}(p) \bar{c}_p \end{aligned} \tag{13.2.6}$$

where $\bar{c}_p = N_p/N$ is the fraction of interstitial atoms occupying interstices of type p with respect to the total number of host atoms N.

Step 4. Let the clusters ($p = 1, 2, \ldots, v$) be reintroduced into the host crystal. Since after surface traction each cluster fits precisely into the space from which it was removed in Step 1, the process involves no free energy change.

Step 5. Let interstitial solute atoms be dispersed over the host lattice maintained undistorted by external forces. This generates a macroscopically homogeneous distribution of solute atoms appropriate to a solid solution. Since the host lattice is undistorted, the displacements, $\mathbf{u}(\mathbf{r})$, remain zero during this step, and no strain energy is developed. The energy change is only chemical (the chemical free energy of mixing). It includes two contributions: (1) the free energy change, $\Delta\mu^s_p$, per interstitial atom caused by a dissolution associated with the replacement of the regular periodic distribution of solute atoms in the transformed clusters with the distribution over the whole crystal corresponding to the solid solution; (2) a configurational free energy change caused by chemical interactions of solute atoms. For the pairwise interaction model, the resulting free energy change is

$$\Delta F_5^{\text{chem}} = \sum_p N_p \Delta\mu^s_p + \tfrac{1}{2} \sum_{p,\mathbf{r}} \sum_{q,\mathbf{r}'} W^{\text{chem}}_{pq}(\mathbf{r} - \mathbf{r}') c(p, \mathbf{r}) c(q, \mathbf{r}') \tag{13.2.7}$$

where the summation is over all host crystal lattice sites $\mathbf{r}$ and $\mathbf{r}'$ and all types of interstices p and q,

$$c(p, \mathbf{r}) = \begin{cases} 1 & \text{when the interstice of type } p \text{ within a unit cell } \mathbf{r} \text{ is occupied with an interstitial atom} \\ 0 & \text{otherwise} \end{cases} \tag{13.2.8}$$

$W_{pq}^{\text{chem}}(\mathbf{r}-\mathbf{r}')$ is the pairwise interaction energy for a pair of interstitials, one of which occupies the interstice $(p, \mathbf{r})$ and the other one is at $(q, \mathbf{r}')$.

Step 6. Lastly, let the distributed atoms relax to generate the equilibrium elastic strain. During the relaxation each solute atom acts as a centre of force distorting the host lattice around it. The displacements are opposed by the elastic resistance of the lattice. The associated relaxation energy may therefore be written

$$\Delta F_6^{\text{relax}} = -\sum_{p=1}^{\nu} \sum_{\mathbf{r},\mathbf{r}'} c(p, \mathbf{r}')\mathbf{f}(p, \mathbf{r}-\mathbf{r}')\mathbf{u}(\mathbf{r}) + \tfrac{1}{2}\sum_{\mathbf{r},\mathbf{r}'} A_{ij}(\mathbf{r}-\mathbf{r}')u_i(\mathbf{r})u_j(\mathbf{r}') \tag{13.2.9}$$

which is a counterpart of Eq. (7.2.8). Subscripts $i, j = 1, 2, 3$ are the Cartesian coordinate indexes, $A_{ij}(\mathbf{r}-\mathbf{r}')$ is the Born-von Karman tensor of the host lattice (86) which characterizes the rigidity of the host lattice against displacements. As is known, the set of Born-von Karman constants, $A_{ij}(\mathbf{r}-\mathbf{r}')$, is the fundamental characteristic of the dynamic properties of a crystal because it determines the vibration frequency spectrum. The constants $\mathbf{f}(p, \mathbf{r}-\mathbf{r}')$ are the so-called Kanzaki forces with which the solute atom $(p, \mathbf{r})$ act on a host atom at site $\mathbf{r}'$ of the undistorted host lattice.* The forces $\mathbf{f}(p, \mathbf{r}-\mathbf{r}')$ are the material constants that describe the ability of a solute atom to distort the environmental host lattice.

It is easy to see that Eq. (13.2.9) is but a second-order Taylor expansion of the relaxation elastic energy in powers of the displacement $\mathbf{u}(\mathbf{r})$. This "harmonic" approximation corresponds to linear microscopic elasticity. The inclusion of higher-order terms in $\mathbf{u}(\mathbf{r})$ and other than linear terms in $c(p, \mathbf{r})$ would mean going beyond the superposition approximation. According to the superposition approximation, the total displacement field $\mathbf{u}(\mathbf{r})$ generated by a group of several solute atoms is a mere sum of the fields generated by the atoms making up the group. The higher-order terms in $\mathbf{u}(\mathbf{r})$ and $c(p, \mathbf{r})$ should describe multiparticle elastic interactions between solute atoms, but this would complicate the theory considerably.

Since the elastic relaxation in Step 5 is spontaneous, it may be written.

$$\Delta F_6^{\text{relax}} \leqslant 0 \tag{13.2.10}$$

The physical meaning of the elastic relaxation is to remove a part of the energy of elastic distortion introduced when the transformed stress-free clusters were deformed in Step 3 to fit properly into the host lattice.

Summing the free energy contributions, ΔF_1 to $\Delta F_6^{\text{relax}}$, from all six steps of the cycle gives the total energy expenditure in the formation of a macroscopically

*For a multicomponent substitutional solution the constant $\mathbf{f}(p, \mathbf{r}-\mathbf{r}')$ is the force with which a solute atom of type p at site $\mathbf{r}$ acts on an atom at site $\mathbf{r}'$ of the undistorted lattice.

homogeneous solid solution from a pure solvent:

$$\Delta F = \Delta F_2^{\text{chem}} + \Delta F_3 + \Delta F_5^{\text{chem}} + \Delta F_6^{\text{relax}}$$

$$= \sum_p N_p[\mu_{\text{int}} + \Delta\mu_p^s + \tfrac{1}{2}v_0\lambda_{ijkl}u_{ij}^0(p)u_{kl}^0(p)]$$

$$+ \tfrac{1}{2}\sum_{p,\mathbf{r}}\sum_{q,\mathbf{r}'} W_{pq}^{\text{chem}}(\mathbf{r}-\mathbf{r}')c(p,\mathbf{r})c(q,\mathbf{r}')$$

$$- \sum_{p,\mathbf{r},\mathbf{r}'} \mathbf{f}(p,\mathbf{r}-\mathbf{r}')c(p,\mathbf{r})\mathbf{u}(\mathbf{r}') + \tfrac{1}{2}\sum_{\mathbf{r},\mathbf{r}'} A_{ij}(\mathbf{r}-\mathbf{r}')u_i(\mathbf{r})u_j(\mathbf{r}') \tag{13.2.11}$$

The term linear in the total number of solute atoms, N_p, occupying interstices of type p represents the self-energy (the sum of the chemical and elastic energies) of solute atoms, the term quadratic in $c(p, \mathbf{r})$ includes the chemical energy of mixing, the the last term includes the elastic relaxation energy and determines response of the solution to elastic displacements. In a simpler notation

$$\Delta F = F_{\text{chem}} + E_{\text{elast}} \tag{13.2.12}$$

is the total free energy change required to produce a macroscopically homogeneous solution starting from the pure solvent crystal and dilute gas of solute atoms (a system of noninteracting solute atoms). In Eq. (13.2.12)

$$F_{\text{chem}} = \sum_{p=1}^{\nu} N_p(\mu_{\text{int}} + \Delta\mu_p^s) + \tfrac{1}{2}\sum_{p,\mathbf{r}}\sum_{q,\mathbf{r}'} W_{pq}^{\text{chem}}(\mathbf{r}-\mathbf{r}')c(p,\mathbf{r})c(q,\mathbf{r}') \tag{13.2.13}$$

and

$$E_{\text{elast}} = \tfrac{1}{2}N\sum_p \bar{c}_p v_0\lambda_{ijkl}u_{ij}^0(p)u_{kl}^0(p) - \sum_{p,\mathbf{r},\mathbf{r}'} c(p,\mathbf{r}')\mathbf{f}(p,\mathbf{r}-\mathbf{r}')\mathbf{u}(\mathbf{r})$$

$$+ \tfrac{1}{2}\sum_{\mathbf{r},\mathbf{r}'} A_{ij}(\mathbf{r}-\mathbf{r}')u_i(\mathbf{r})u_j(\mathbf{r}') \tag{13.2.14}$$

is the elastic energy change.

The derivation just given, as a matter of fact, defines the chemical pairwise interaction potentials $W_{pq}^{\text{chem}}(\mathbf{r}-\mathbf{r}')$ as interaction energies between two interstitial atoms at the interstices $(p, \mathbf{r})$ and $(q, \mathbf{r}')$ within an undistorted host lattice whose crystal lattice site positions coincide with those of the host lattice free of solute atoms. It also gives a mathematically consistent quantitative definition for the intuitive idea of elastic energy of a solid solution.

It should be mentioned here that Eqs. (13.2.12) to (13.2.14) are also valid for a multicomponent substituational solid solution. The subscript p already designates a kind of solute atom. A consistent microscopic treatment of a multicomponent substitutional solution was given by Wen (85).

To obtain a multicomponent substitutional solution from pure one-component solvent, the thermodynamic cycle of the six operations for interstitial solutions should be modified slightly: the transformation of the host lattice cluster in Step 2 should be initiated by substituting an atom of the pth solute for each host atom of the pth host lattice cluster. All the other operations are the same as considered above.

The coefficients in Eq. (13.2.14) of course depend on the restrictions imposed by the solvent crystal symmetry and by the natural requirement that E_{elast} be invariant under arbitrary rigid body translations **T**. A rigid body translation is given by the change of variables

$$\mathbf{u}(\mathbf{r}) \rightarrow \mathbf{u}(\mathbf{r}) + \mathbf{T}$$

It is easy to see that the change of variables does not affect Eq. (13.2.14) if

$$\sum_{\mathbf{r},\mathbf{r}'} A_{ij}(\mathbf{r}-\mathbf{r}') \equiv 0 \qquad (13.2.15a)$$

and

$$\sum_{\mathbf{r}'} \mathbf{f}(p, \mathbf{r}-\mathbf{r}') \equiv 0 \qquad (13.2.15b)$$

Certain additional relations can be derived from the condition of elastic energy invariance under rigid body rotation. The invariance of elastic energy under infinitesimal rigid body rotation requires Eq. (13.2.14) to be unaffected by the change of variables

$$u_i(\mathbf{r}) \rightarrow u_i(\mathbf{r}) + \omega_{ij} r_j$$

where ω_{ij} is the asymmetric infinitesimal tensor describing the displacement, $\omega_{ij} r_j$, in rigid body rotation. Using the asymmetry condition, $\omega_{ij} = -\omega_{ji}$, we may write additional constraints

$$\sum_{\mathbf{r}'} \mathbf{f}(p, \mathbf{r}-\mathbf{r}') \times \mathbf{r}' \equiv 0 \qquad (13.2.16a)$$

and

$$\sum_{\mathbf{r},\mathbf{r}'} A_{ij}(\mathbf{r}-\mathbf{r}') r_l r_k \equiv \sum_{\mathbf{r},\mathbf{r}'} A_{lj}(\mathbf{r}-\mathbf{r}') r_i r'_k \equiv \sum_{\mathbf{r},\mathbf{r}'} A_{ik}(\mathbf{r}-\mathbf{r}') r_l r'_j \qquad (13.2.16b)$$

13.2.1. Determination of Equilibrium Displacement and Strain Energy

Expression (13.2.14) yields "the elastic Hamiltonian" in which the displacements, $\mathbf{u}(\mathbf{r})$, are dynamic variables. In other words, Eq. (13.2.14) enables us to calculate the static elastic energy corresponding to any arbitrary set of atomic displacements, $\{\mathbf{u}(\mathbf{r})\}$. In crystal lattice statics we are, however, interested in equilibrium displacements rather than in arbitrary ones. Computation of the equilibrium lattice displacement field within a solid solution is not only important by itself but also, as will be shown, permits us to determine the expression for the equilibrium elastic energy that pertains to the crystal lattice statics problem.

Equilibrium displacement should be determined from a minimization of elastic energy (13.2.14) with respect to the displacement field, $\mathbf{u}(\mathbf{r})$. To do this, it is convenient to represent the static displacement field, $\mathbf{u}(\mathbf{r})$, as the sum of two terms:

$$u_i(\mathbf{r}) = \varepsilon_{ij} r_j + v_i(\mathbf{r}) \tag{13.2.17}$$

where ε_{ij} is the constant describing uniform macroscopic strain. The term $\varepsilon_{ij} r_j$ is singled out because this term, unlike the local displacement field, $\mathbf{v}(\mathbf{r})$, affects the shape of the crystal body. The field $\mathbf{v}(\mathbf{r})$, which does not produce macroscopic effects, must be defined in such a way that it vanishes over the surface of the body.

The concentration field may also be decomposed into its mean and variation:

$$c(p, \mathbf{r}) = \bar{c}_p + \Delta c(p, \mathbf{r}) \tag{13.2.18}$$

With these definitions, the elastic energy (13.2.14) may be rewritten

$$E_{\text{elast}} = E^{00}_{\text{elast}} - N\varepsilon_{ij} \sum_p \Lambda_{ij}(p)\bar{c}_p + \tfrac{1}{2} N \Lambda_{ijkl} \varepsilon_{ij} \varepsilon_{kl}$$

$$- \sum_p \sum_{\mathbf{r},\mathbf{r}'} \mathbf{f}(p, \mathbf{r}-\mathbf{r}')\mathbf{v}(\mathbf{r}')\Delta c(p, \mathbf{r}') + \tfrac{1}{2} \sum_{\mathbf{r},\mathbf{r}'} A_{ij}(\mathbf{r}-\mathbf{r}') v_i(\mathbf{r}) v_j(\mathbf{r}') \tag{13.2.19}$$

where

$$E^{00}_{\text{elast}} = \tfrac{1}{2} N \sum_p \bar{c}_p v_0 \lambda_{ijkl} u^0_{ij}(p) u^0_{kl}(p) \tag{13.2.20}$$

is the term independent of the elastic displacement. The tensors $\Lambda_{ij}(p)$ and Λ_{ijkl} are given by the summations:

$$\Lambda_{ij}(p) = \sum_{\mathbf{r}} f_i(p, \mathbf{r}) r_j = \tfrac{1}{2} \sum_{\mathbf{r}} (f_i(p, \mathbf{r}) r_j + f_j(p, \mathbf{r}) r_i) \tag{13.2.21}$$

$$\Lambda_{ijkl} = \sum_{\mathbf{r}} A_{ij}(\mathbf{r}) r_l r_m = \tfrac{1}{4} \sum_{\mathbf{r}} (A_{ij}(\mathbf{r}) r_l r_m$$

$$+ A_{lm}(\mathbf{r}) r_i r_j + A_{il}(\mathbf{r}) r_j r_m + A_{jm}(\mathbf{r}) r_i r_l) \tag{13.2.22}$$

In the definitions (13.2.21) and (13.2.22) the invariance relations (13.2.16) are used.

The cross terms depending on $\mathbf{v}(\mathbf{r})\bar{c}_p$ and $\varepsilon_{ij}\Delta c(p, \mathbf{r})$ vanish if the definitions (13.2.17) and (13.2.18) are taken into account.

13.2.2. Uniform Strain Relaxation

According to (13.2.17) uniform strain ε_{ij} describes the shape deformation of a solid solution body because, by definition, the local displacement field, $\mathbf{v}(\mathbf{r})$, vanishes at the external boundary. In general when a solid solution body is constrained, the minimization of elastic energy (13.2.19) should be carried out under an additional condition of conservation of the strain ε_{ij}. This condition may be introduced by the Lagrange multiplier method with the help of the generating functional (the thermodynamic potential):

$$\Omega_{\text{elast}} = E_{\text{elast}} - \sigma_{ij} \int_{(V)} \varepsilon_{ij} d^3r = E_{\text{elast}} - \sigma_{ij}\varepsilon_{ij} V \tag{13.2.23}$$

where σ_{ij} is the Lagrange multiplier, V is the volume of the solid solution body, E_{elast} is given by Eq. (13.2.19). The minimization of the function (13.2.23) with respect to the variable ε_{ij} yields

$$\frac{\delta\Omega_{\text{elast}}}{\delta\varepsilon_{ij}} = 0$$

or, in terms of (13.2.23),

$$\frac{\delta E_{\text{elast}}}{\delta\varepsilon_{ij}} - \sigma_{ij} V = 0 \tag{13.2.24}$$

Substituting Eq. (13.2.19) into (13.2.24), we obtain

$$\sigma_{ij} = -\sum_p \frac{1}{v_0} \Lambda_{ij}(p)\bar{c}_p + \frac{1}{v_0} \Lambda_{ijkl}\varepsilon_{kl} \tag{13.2.25}$$

where $v_0 = V/N$ is the atomic volume. Defining the tensor, $\bar{\varepsilon}_{ij}$, by the relation

$$\Lambda_{ijkl}\bar{\varepsilon}_{kl} = \sum_p \Lambda_{ij}(p)\bar{c}_p \tag{13.2.26}$$

we have

$$\sigma_{ij} = \frac{1}{v_0} \Lambda_{ijkl}(\varepsilon_{kl} - \bar{\varepsilon}_{kl}) \tag{13.2.27}$$

Comparison with the Hooke's law

$$\sigma_{ij} = \lambda_{ijkl}\varepsilon_{kl}^{\text{elast}} \tag{13.2.28}$$

where σ_{ij} is the stress, $\varepsilon_{kl}^{\text{elast}}$ is the part of the elastic strain caused by externally imposed stress, σ_{ij}, gives the equalities

$$\lambda_{ijkl} = \frac{1}{v_0} \Lambda_{ijkl} \tag{13.2.29}$$

$$\varepsilon_{kl}^{\text{elast}} = \varepsilon_{kl} - \bar{\varepsilon}_{kl} \tag{13.2.30}$$

and shows that the Lagrange multiplier σ_{ij} is just the homogeneous stress. In the stress-free state $\sigma_{ij}=0$ and $\varepsilon_{kl}^{\text{elast}}=0$; hence at $\sigma_{ij}=0$

$$\varepsilon_{kl} = \bar{\varepsilon}_{kl} \tag{13.2.31}$$

is the uniform relaxation strain introduced by alloying the crystal in the stress-free state. Eq. (13.2.31) solves the problem of the determination of the equilibrium uniform strain, ε_{ij}, generated by alloying the host crystal.

The crystal deformation, $\bar{\varepsilon}_{ij}$, caused by alloying may be expressed in terms of the solute content. Using the relation (13.2.29) in Eq. (13.2.26), we obtain

$$\lambda_{ijkl}\bar{\varepsilon}_{kl}=\sum_p \frac{1}{v_0}\Lambda_{ij}(p)\bar{c}_p \tag{13.2.32}$$

This equation may be solved for $\bar{\varepsilon}_{ij}$:

$$\bar{\varepsilon}_{ij}=\sum_p \frac{1}{v_0} s_{ijkl}\Lambda_{kl}(p)\bar{c}_p \tag{13.2.33}$$

where s_{ijkl} is the elastic compliance tensor which is inverse to λ_{ijkl}.

Eq. (13.2.33) shows that the strain energy approximation (13.2.14) results in Vegard's law, in the linear dependence of the crystal lattice expansion with respect to each of the solute species. An introduction of unharmonic terms and a higher order in $c(p, \mathbf{r})$ terms in Eq. (13.2.14) would yield the deviation from Vegard's law.

The coefficients $\Lambda_{kl}(p)$ in Eq. (13.2.33) may be estimated by extrapolating to the limit $\bar{c}_p=1$. In this case

$$\bar{\varepsilon}_{ij}=u^0_{ij}(p)$$

by definition, and Eq. (13.2.33) may be rewritten

$$\bar{\varepsilon}_{ij}=u^0_{ij}(p)=\frac{1}{v_0} s_{ijkl}\Lambda_{kl}(p) \tag{13.2.34}$$

Convolution of Eq. (13.2.34) with the elastic modulus tensor, λ_{ijkl}, yields

$$\lambda_{ijkl}\left[u^0_{kl}(p)-\frac{1}{v} s_{klnm}\Lambda_{nm}(p)\right]=0$$

or, combining the definition of the compliance tensor and Eq. (7.2.16),

$$\sigma^0_{ij}(p)=\lambda_{ijkl}u^0_{kl}(p)=\frac{1}{v}\Lambda_{ij}(p) \tag{13.2.35}$$

The stress $\sigma^0_{ij}(p)$ is a material constant, the "transformation stress," and is merely the negative of the elastic stress required to reverse the strain caused by the stress-free transformation of the pth host lattice cluster into the stoichiometric interstitial-ordered phase cluster (see Step 2).

With substitution of the equilibrium uniform strain $\bar{\varepsilon}_{ij}$ for ε_{ij} in Eq. (13.2.19), using Eq. (13.2.33) and (13.2.34), we have in the absence of an external stress

$$E_{\text{elast}}=E^{00}_{\text{elast}}-\tfrac{1}{2}N\sum_{p,q} v_0\lambda_{ijkl}u^0_{ij}(p)u^0_{kl}(q)\bar{c}_p\bar{c}_q$$
$$-\sum_p\sum_{\mathbf{r},\mathbf{r}'}(\mathbf{f}(p,\mathbf{r}-\mathbf{r}')\mathbf{v}(\mathbf{r}))\Delta c(p,\mathbf{r}')+\tfrac{1}{2}\sum_{\mathbf{r},\mathbf{r}'}A_{ij}(\mathbf{r}-\mathbf{r}')v_i(\mathbf{r})v_j(\mathbf{r}') \tag{13.2.36}$$

13.2.3. Local Displacement Relaxation

Eq. (13.2.36) describes the elastic energy minimized with respect to the variables ε_{ij}. To find the equilibrium elastic energy, we must also minimize Eq. (13.2.36) with respect to the local displacement, $\mathbf{v}(\mathbf{r})$. The equilibrium local displacement field $\mathbf{v}(\mathbf{r})$, is found from the minimum condition

$$\frac{\delta E_{\text{elast}}}{\delta \mathbf{v}(\mathbf{r})} = 0 \tag{13.2.37}$$

Substituting Eq. (13.2.36) into (13.2.37) yields the equilibrium equation

$$\sum_{\mathbf{r}'} A_{ij}(\mathbf{r}-\mathbf{r}')v_j(\mathbf{r}') = \sum_{p,\mathbf{r}'} f_i(p, \mathbf{r}-\mathbf{r}')\Delta c(p, \mathbf{r}') \tag{13.2.38}$$

Because the local displacement, $\mathbf{v}(\mathbf{r})$, vanishes at the boundary of the crystal body, by definition, we may apply the cyclic boundary conditions.* It should be emphasized that the local displacement, $\mathbf{v}(\mathbf{r})$, vanishes at the boundary only in the case of a macroscopically homogeneous crystal—if the crystal dimensions far exceed typical lengths characterizing the spatial distribution of solute atoms.

Taking Fourier transforms of both sides of Eq. (13.2.38), we obtain

$$\tilde{A}_{ij}(\mathbf{k})\tilde{v}_j(\mathbf{k}) = \sum_{p=1}^{\nu} F_i(p, \mathbf{k})\Delta\tilde{c}(p, \mathbf{k}) \tag{13.2.39}$$

where

$$\tilde{A}_{ij}(\mathbf{k}) = \sum_{\mathbf{r}} A_{ij}(\mathbf{r})e^{-i\mathbf{k}\mathbf{r}} \tag{13.2.40}$$

is the dynamic matrix,

$$\mathbf{F}(p, \mathbf{k}) = \sum_{\mathbf{r}} \mathbf{f}(p, \mathbf{r})e^{-i\mathbf{k}\mathbf{r}} \tag{13.2.41}$$

is the Fourier transform of the Kanzaki forces, $\mathbf{f}(p, \mathbf{r})$, and

$$\Delta\tilde{c}(p, \mathbf{k}) = \sum_{\mathbf{r}} \Delta c(p, \mathbf{r})e^{-i\mathbf{k}\mathbf{r}} \tag{13.2.42}$$

The solution to Eq. (13.2.39) for the Fourier components $\tilde{\mathbf{v}}(\mathbf{k})$ is

$$\tilde{v}_i(\mathbf{k}) = G_{ij}(\mathbf{k}) \sum_{p=1}^{\nu} F_j(p, \mathbf{k})\Delta\tilde{c}(p, \mathbf{k}) \tag{13.2.43}$$

where $G_{ij}(\mathbf{k})$ is Green's tensor which is the inverse of the dynamic matrix

*The decomposition (13.2.17) enables us to introduce the displacement $\mathbf{v}(\mathbf{r})$ vanishing at the boundary and therefore to use the cyclic boundary conditions for $\mathbf{v}(\mathbf{r})$. The cyclic boundary conditions cannot be applied to the total displacement field $\mathbf{u}(\mathbf{r})$ because $\mathbf{u}(\mathbf{r})$ does not vanish at the boundary owing to uniform strain.

$\tilde{A}_{ij}(\mathbf{k})$:

$$G_{il}(\mathbf{k})\tilde{A}_{lj}(\mathbf{k})=\delta_{ij} \tag{13.2.44}$$

The Hermittian tensor $G_{ij}(\mathbf{k})$ can be written in terms of eigenvectors, $\mathbf{e}_s(\mathbf{k})$, and eigenvalues, $m\omega_s^2(\mathbf{k})$, of the dynamic matrix, $\tilde{A}_{ij}(\mathbf{k})$. These are the solutions to the equation:

$$\tilde{A}_{ij}(\mathbf{k})e_s^j(\mathbf{k})=m\omega_s^2(\mathbf{k})e_s^i(\mathbf{k}) \tag{13.2.45}$$

where m is the mass of the host atoms, ω_s is the vibration frequency related to the branch s ($s=1, 2, 3$), $\mathbf{k}$ is the wave vector. Green's tensor, $G_{ij}(\mathbf{k})$, may then be written

$$G_{ij}(\mathbf{k})=\sum_{s=1}^{3}\frac{e_s^i(\mathbf{k})e_s^{*j}(\mathbf{k})}{m\omega_s^2(\mathbf{k})} \tag{13.2.46}$$

Substituting the back Fourier transforms,

$$\mathbf{v}(\mathbf{r})=\frac{1}{N}\sum_{\mathbf{k}}\tilde{\mathbf{v}}(\mathbf{k})e^{i\mathbf{k}\mathbf{r}} \tag{13.2.47}$$

$$\Delta c(p, \mathbf{r})=\frac{1}{N}\sum_{\mathbf{k}}\Delta\tilde{c}(p, \mathbf{k})e^{i\mathbf{k}\mathbf{r}} \tag{13.2.48}$$

where the summations are over all N points in the first Brillouin zone of the host lattice permitted by the cyclic boundary conditions, into Eq. (13.2.36), we obtain

$$\begin{aligned} E_{\text{elast}} &= E_{\text{elast}}^{00} -\tfrac{1}{2}N\sum_{p,q} v_0\lambda_{ijkl}u_{ij}^0(p)u_{kl}^0(q)\bar{c}_p\bar{c}_q \\ &-\frac{1}{N}\sum_p\sum_{\mathbf{k}}(\mathbf{F}(p, \mathbf{k})\tilde{\mathbf{v}}(\mathbf{k}))\Delta\tilde{c}^*(p, \mathbf{k})+\frac{1}{2N}\sum_{\mathbf{k}}\tilde{A}_{ij}(\mathbf{k})\tilde{v}_i(\mathbf{k})\tilde{v}_j^*(\mathbf{k}) \end{aligned} \tag{13.2.49}$$

Eq. (13.2.49) gives the **k**-space representation of the "elastic Hamiltonian" in terms of amplitudes of the displacement and concentration waves, $\Delta\tilde{c}(p, \mathbf{k})$ and $\tilde{\mathbf{v}}(\mathbf{k})$.

The minimum value of the elastic energy (13.2.49) corresponding to the mechanical equilibrium may be calculated if the crystal lattice displacement field obtained by solving the static equilibrium equation is used. Using the solution 13.2.43) for the internal displacement field in Eq. (13.2.49) and (13.2.20), we obtain for the elastic energy

$$\begin{aligned} E_{\text{elast}} &=\frac{1}{2}N\sum_p v_0\lambda_{ijkl}u_{ij}^0(p)u_{kl}^0(p)\bar{c}_p \\ &-\frac{1}{2}N\sum_{p,q} v_0\lambda_{ijkl}u_{ij}^0(p)u_{kl}^0(q)\bar{c}_p\bar{c}_q \\ &-\frac{1}{2}\frac{1}{N}\sum_{p,q}\sum_{\mathbf{k}}(F_i(p, \mathbf{k})G_{ij}(\mathbf{k})F_j^*(q, \mathbf{k}))\Delta\tilde{c}(p, \mathbf{k})\Delta\tilde{c}^*(q, \mathbf{k}) \end{aligned} \tag{13.2.50}$$

Eq. (13.2.50) is the total elastic energy produced in alloying an unconstrained solvent crystal; the relaxation of the elastic energy caused by the total volume expansion during the transformation process is taken into account.

13.2.4. Separation of Physically Distinct Contributions to Elastic Energy

Equation (13.2.50) gives the total strain energy expended to form a stress-free macroscopically homogeneous alloy from a pure solvent crystal and dilute (ideal) gas of solute elements. As has been mentioned, Eq. (13.2.50) is valid for both an interstitial solution with ν interstices per host atom and a ν-component substitutional solution. Physically, the elastic contribution to the free energy of the crystal can be divided into three parts:

1. Elastic self-energy which is a mere sum of one-particle energies. Each one-particle energy is the mechanical work required to introduce a solute atom into the pure host lattice. According to this definition elastic self-energy does not include strain-induced interaction between solute atoms and is therefore proportional to the total number N_p of solute atoms of type p.
2. Configuration-independent pairwise interaction energy whose magnitude depends on only the total number of interstitials in the various types of interstices rather than on their mutual arrangement; this interaction is indirect, and its source is the elastic image forces that arise from relaxation of the unconstrained crystal boundary considered above.
3. Configuration-dependent pairwise interaction energy that results from direct elastic interactions of solute atoms. The correct separation of these physically distinct contributions to elastic energy may prove to be of importance. For example, in treating the elastic energy effect on the decomposition of a binary alloy, we have to include only the direct pairwise interactions because the self-energy and image-force contributions are not affected by the reconfiguration of a fixed number of solute atoms. It will be shown in Section 13.5.2 that the direct pairwise interaction vanishes in an elastically isotropic binary solution, so only the self-energy and image-force contributions to elastic energy assume non zero values.

As already mentioned, Eq. (13.2.50) is valid for both an interstitial binary solid solution with ν interstices per host atom and a ν-component substitutional solid solution. In the first case the index p enumerates the types of interstices, and in the second case, the kinds of solute atoms. Eq. (13.2.50) does not in fact separate the elastic energy into physically distinct terms; part of both the self-energy and image-force contributions are included into the third term on the right-hand side of the equation.

The separation of the terms depends on the microscopic nature of the solid solution. For example, the ν-component substitutional solutions differ from binary interstitial solutions with ν interstices per host atom despite the fact that both types of solutions are described by the same equation (13.2.50).

We shall consider first interstitial solutions. The third term of Eq. (13.2.50) may be then rewritten in an identical form:

$$\frac{1}{2N}\sum_{p,q}\sum_{\mathbf{k}}(F_i(p,\mathbf{k})G_{ij}(\mathbf{k})F_j^*(q,\mathbf{k}))\Delta\tilde{c}(p,\mathbf{k})\Delta\tilde{c}^*(q,\mathbf{k})$$
$$=\frac{1}{2N}\sum_{p,q}\sum_{\mathbf{k}}[F_i(p,\mathbf{k})G_{ij}(\mathbf{k})F_j^*(q,\mathbf{k})-Q\delta_{pq}]\Delta\tilde{c}(p,\mathbf{k})\Delta\tilde{c}^*(q,\mathbf{k})$$
$$+\frac{1}{2}\sum_{p}Q\frac{1}{N}\sum_{\mathbf{k}}|\Delta\tilde{c}(p,\mathbf{k})|^2 \tag{13.2.51}$$

where

$$Q=\frac{1}{N}\sum_{\mathbf{k}}F_i(p,\mathbf{k})G_{ij}(\mathbf{k})F_j^*(p,\mathbf{k})=\langle F_i(p,\mathbf{k})G_{ij}(\mathbf{k})F_j^*(p,\mathbf{k})\rangle \tag{13.2.52}$$

is the constant obtained by averaging over the first Brillouin zone of the host lattice. The constant Q does not depend on the index p labeling interstices because all the interstices are crystallographically equivalent—may be brought in coincidence with each other by the symmetry operations of the host lattice.

Using Parseval's formula

$$\frac{1}{N}\sum_{\mathbf{k}}|\Delta\tilde{c}(p,\mathbf{k})|^2\equiv\sum_{\mathbf{r}}(\Delta c(p,\mathbf{r}))^2 \tag{13.2.53}$$

we obtain

$$\sum_{\mathbf{r}}(\Delta c(p,\mathbf{r}))^2=\sum_{\mathbf{r}}(c(p,\mathbf{r})-\bar{c}_p)^2=\sum_{\mathbf{r}}(c(p,\mathbf{r}))^2-2\bar{c}_p\sum_{\mathbf{r}}c(p,\mathbf{r})+\bar{c}_p^2N \tag{13.2.54}$$

Since, by the definition $c(p, \mathbf{r})$,

$$(c(p,\mathbf{r}))^2\equiv c(p,\mathbf{r}) \tag{13.2.55}$$

we have

$$\sum_{\mathbf{r}}(c(p,\mathbf{r}))^2=N_p \tag{13.2.56}$$

Substituting (13.2.56) into (13.2.54) yields

$$\sum_{\mathbf{r}}(\Delta c(p,\mathbf{r}))^2=N_p-2\bar{c}_pN_p+\bar{c}_p^2N=N\bar{c}_p(1-\bar{c}_p) \tag{13.2.57}$$

Using the relation (13.2.57), we may rewrite Eq. (13.2.53):

$$\frac{1}{N}\sum_{\mathbf{k}}|\Delta\tilde{c}(p,\mathbf{k})|^2\equiv N\bar{c}_p(1-\bar{c}_p) \tag{13.2.58}$$

Inserting this identity in Eq. (13.2.51), we obtain

$$\frac{1}{N}\sum_{p,q}\sum_{\mathbf{k}}(F_i(p,\mathbf{k})G_{ij}(\mathbf{k})F_j^*(q,\mathbf{k}))\Delta\tilde{c}(p,\mathbf{k})\Delta\tilde{c}^*(q,\mathbf{k})$$
$$=\frac{1}{2}NQ\sum_{p}\bar{c}_p(1-\bar{c}_p)+\frac{1}{2N}\sum_{p,q}\sum_{\mathbf{k}}[F_i(p,\mathbf{k})G_{ij}(\mathbf{k})F_j^*(q,\mathbf{k})-Q\delta_{pq}]\Delta\tilde{c}(p,\mathbf{k})\Delta\tilde{c}^*(q,\mathbf{k}) \tag{13.2.59}$$

Substituting Eq. (13.2.59) into Eq. (13.2.50) gives the equation

$$E_{\text{elast}} = \frac{1}{2} N \sum_p [v_0 \lambda_{ijkl} u^0_{ij}(p) u^0_{kl}(p) - Q] \bar{c}_p - \frac{N}{2} \sum_{p,q} [v_0 \lambda_{ijkl} u^0_{ij}(p) u^0_{kl}(q) - Q\delta_{pq}] \bar{c}_p \bar{c}_q$$

$$- \frac{1}{2N} \sum_{p,q} \sum_{\mathbf{k}} [F_i(p, \mathbf{k}) G_{ij}(\mathbf{k}) F^*_j(q, \mathbf{k}) - Q\delta_{pq}] \Delta\tilde{c}(p, \mathbf{k}) \Delta c^*(q, \mathbf{k}) \qquad (13.2.60)$$

Let us derive the following relations:

$$\lim_{\mathbf{k}\to 0} \Delta\tilde{c}(p, \mathbf{k}) = \lim_{\mathbf{k}\to 0} \sum_{\mathbf{r}} (c(p, \mathbf{r}) - \bar{c}_p) e^{-i\mathbf{kr}} = \sum_{\mathbf{r}} (c(p, \mathbf{r}) - \bar{c}_p) = N_p - \bar{c}_p N = 0$$

$$(13.2.61)$$

Since

$$\sum_{\mathbf{r}} e^{-i\mathbf{kr}} = N\delta_{\mathbf{k}0}$$

where $\delta_{\mathbf{k}0}$ is 1 if $\mathbf{k}=0$ and zero otherwise, we have

$$\Delta\tilde{c}(p, \mathbf{k}) = \begin{cases} \tilde{c}(p, \mathbf{k}) & \text{if } \mathbf{k} \neq 0 \\ 0 & \text{if } \mathbf{k} = 0 \end{cases} \qquad (13.2.62)$$

where

$$\tilde{c}(p, \mathbf{k}) = \sum_{\mathbf{r}} c(p, \mathbf{r}) e^{-i\mathbf{kr}} \qquad (13.2.63)$$

Taking into account Eq. (13.2.62), we may rewrite Eq. (13.2.60) in the form

$$E_{\text{elast}} = \frac{1}{2} N \sum_p [v_0 \lambda_{ijkl} u^0_{ij}(p) u^0_{kl}(p) - Q] \bar{c}_p - \frac{N}{2} \sum_{p,q} [v_0 \lambda_{ijkl} u^0_{ij}(p) u^0_{kl}(q) - Q\delta_{pq}] \bar{c}_p \bar{c}_q$$

$$- \frac{1}{2N} \sum_{p,q} \sum_{\mathbf{k}}{}' [F_i(p, \mathbf{k}) G_{ij}(\mathbf{k}) F^*_j(q, \mathbf{k}) - Q\delta_{pq}] \tilde{c}(p, \mathbf{k}) \tilde{c}^*(q, \mathbf{k}) \qquad (13.2.64)$$

where priming means that the term corresponding to $\mathbf{k}=0$ is ommitted from the summation. Eq. (13.2.64) may be simplified by substitution:

$$V_{pq}(\mathbf{k}) = \begin{cases} -v_0 \lambda_{ijkl} u^0_{ij}(p) u^0_{kl}(q) + Q\delta_{pq} & \text{if } \mathbf{k} = 0 \\ -F_i(p, \mathbf{k}) G_{ij}(\mathbf{k}) F^*_j(q, \mathbf{k}) + Q\delta_{pq} & \text{if } \mathbf{k} \neq 0 \end{cases} \qquad (13.2.65\text{a})$$

As will be shown below, the term

$$V^{\text{config}}_{pq}(\mathbf{k}) = -F_i(p, \mathbf{k}) G_{ij}(\mathbf{k}) F^*_j(q, \mathbf{k}) + Q\delta_{pq} \qquad (13.2.65\text{b})$$

is the Fourier transform of the configuration-dependent pairwise strain-induced potential.

Bearing in mind the relation

$$\tilde{c}(p, 0) = \sum_{\mathbf{r}} c(p, \mathbf{r}) = N\bar{c}_p = N_p$$

we may represent Eq. (13.2.64) in the dense form:

$$E_{\text{elast}} = N_{\text{int}} E^0_{\text{self}} + \frac{1}{2N} \sum_{\mathbf{k}} V_{pq}(\mathbf{k}) \tilde{c}(p, \mathbf{k}) \tilde{c}^*(q, \mathbf{k}) \tag{13.2.66}$$

where $N_{\text{int}} = N \sum_p \bar{c}_p$ is the total number of interstitials,

$$E^0_{\text{self}} = \tfrac{1}{2}[v_0 \lambda_{ijkl} u^0_{ij}(p) u^0_{kl}(p) - Q] \tag{13.2.67}$$

is the elastic self-energy required to introduce a single solute atom into the pure host lattice [E^0_{self} in Eq. (13.2.67) does not depend on p because all interstices p are crystallographically equivalent]; $V_{pq}(\mathbf{k})$ is given by Eq. (13.2.65a).

Eq. (13.2.66) may also be rewritten in the crystal lattice site representation

$$E_{\text{elast}} = N_{\text{int}} E^0_{\text{self}} + \tfrac{1}{2} \sum_{\mathbf{r}, \mathbf{r}'} W^{el}_{pq}(\mathbf{r} - \mathbf{r}') c(p, \mathbf{r}) c(q, \mathbf{r}') \tag{13.2.68}$$

where

$$W^{el}_{pq}(\mathbf{r} - \mathbf{r}') = \frac{1}{N} \sum_{\mathbf{k}} V_{pq}(\mathbf{k}) e^{i\mathbf{k}(\mathbf{r} - \mathbf{r}')}$$

or

$$W^{el}_{pq}(\mathbf{r} - \mathbf{r}') = -\frac{1}{N}(v_0 \lambda_{ijkl} u^0_{ij}(p) u^0_{kl}(p) - Q\delta_{pq})$$
$$-\frac{1}{N} \sum_{\mathbf{k}}{}' (F_i(p, \mathbf{k}) G_{ij}(\mathbf{k}) F^*_j(q, \mathbf{k}) - Q\delta_{pq}) e^{i\mathbf{k}(\mathbf{r} - \mathbf{r}')} \tag{13.2.69}$$

The second term in Eq. (13.2.68) describes the pairwise interaction energy only if the term with equal summation indexes ($p = q$ and $\mathbf{r} = \mathbf{r}'$ simultaneously) vanishes; that is, if the summation in Eq. (13.2.68) is over all the pairs of solute atoms separated by a nonzero distance. This is the case if the coefficients, $W^{el}_{pq}(\mathbf{r} - \mathbf{r}')$, vanish at $\mathbf{r} - \mathbf{r}'$ and $p = q$; that is, when

$$W^{el}_{pp}(0) = 0 \tag{13.2.70}$$

As will be shown below, the constant Q given by (13.2.52) is chosen so that the requirement (13.2.70) be fulfilled. Let us verify Eq. (13.2.70).

It follows from Eq. (13.2.69) that

$$W^{el}_{pp}(0) = -\frac{1}{N}(v_0 \lambda_{ijkl} u^0_{ij}(p) u^0_{kl}(p) - Q) - \frac{1}{N} \sum_{\mathbf{k}}{}' (F_i(p, \mathbf{k}) G_{ij}(\mathbf{k}) F^*_j(q, \mathbf{k}) - Q) \tag{13.2.71}$$

Recalling definition (13.2.52), and the fact that the sum over $\mathbf{k}$ contains N terms, we obtain

$$W^{el}_{pq}(0) = -\frac{1}{N}[v_0 \lambda_{ijkl} u^0_{ij}(p) u^0_{kl}(p) - Q] \to 0 \tag{13.2.72}$$

because the number of crystal lattice sites N is a macroscopically large value.

Let us now consider the important particular case of a binary interstitial solution with a single interstice per host atom. The index p in Eq. (13.2.60) may be ommitted in this case. Eq. (13.2.60) then becomes

$$E_{\text{elast}}=\frac{1}{2}N(v_0\lambda_{ijkl}u_{ij}^0u_{kl}^0-Q)\bar{c}(1-c)-\frac{1}{2N}\sum_{\mathbf{k}}[(F_i(\mathbf{k})G_{ij}(\mathbf{k})F_j^*(\mathbf{k}))-Q]|\Delta\tilde{c}(\mathbf{k})|^2 \tag{13.2.73}$$

where u_{ij}^0 is by definition the concentration coefficient of the host lattice expansion.

It is important that Eq. (13.2.73) is equally well applicable to both the binary interstitial solution with a single interstice per host atom considered above and the binary substitutional solution.

Eq. (13.2.73) will be applied in the next section to obtain limit transitions to all known theories of elastic energy of binary substitutional solid solutions.

13.2.5. Elastic Energy of Multicomponent Substitutional Solid Solutions

The separation of physically distinct terms in the elastic energy of a multicomponent substitutional solid solution was suggested by Wen (85). She also started from the general Eq. (13.2.50) for the total elastic energy. According to Wen the separation of physically distinct terms in Eq. (13.2.50) may be achieved by rewriting the third term of that equation in the form

$$\begin{aligned}\frac{1}{2N}&\sum_{p,q}\sum_{\mathbf{k}}(F_i(p,\mathbf{k})G_{ij}(\mathbf{k})F_j^*(q,\mathbf{k}))\Delta\tilde{c}(p,\mathbf{k})\Delta\tilde{c}^*(q,\mathbf{k})\\ &=\frac{1}{2N}\sum_{p,q}\sum_{\mathbf{k}}[(F_i(p,\mathbf{k})G_{ij}(\mathbf{k})F_j^*(q,\mathbf{k}))-Q_{pq}]\Delta\tilde{c}(p,\mathbf{k})\Delta\tilde{c}^*(q,\mathbf{k})\\ &\quad+\frac{1}{2}\sum_{p,q}Q_{pq}\frac{1}{N}\sum_{\mathbf{k}}\Delta\tilde{c}(p,\mathbf{k})\Delta\tilde{c}^*(q,\mathbf{k})\end{aligned} \tag{13.2.74}$$

where the index p denotes the kind of a solute atom (unlike the case of an interstitial solution),

$$Q_{pq}=\frac{1}{N}\sum_{\mathbf{k}}F_i(p,\mathbf{k})G_{ij}(\mathbf{k})F_j^*(q,\mathbf{k}) \tag{13.2.75}$$

[compare with Eqs. (13.2.51) and (13.2.52)].

Using the relation

$$c(p,\mathbf{r})c(q,\mathbf{r}')=\delta_{pq}c(p,\mathbf{r}) \tag{13.2.76}$$

which follows from the obvious point that two solute atoms of different kinds cannot occupy the same crystal lattice site, $\mathbf{r}$, we may write the the identity

$$\frac{1}{N}\sum_{\mathbf{k}}\Delta\tilde{c}(p,\mathbf{k})\Delta\tilde{c}^*(q,\mathbf{k})\equiv N\delta_{pq}\bar{c}_p-N\bar{c}_p-N\bar{c}_p\bar{c}_q \tag{13.2.77}$$

[the identity (13.2.77) is not valid for interstitial solutions]. Using this identity in the last term of Eq. (13.2.74), substituting the result into (13.2.50), and grouping together like terms, we obtain

$$E_{\text{elast}}=\frac{1}{2}N\sum_{p}(v_0\lambda_{ijkl}u^0_{ij}(p)u^0_{kl}(p)-Q_{pp})\bar{c}_p$$
$$+\frac{1}{2}N\sum_{p,q}(-v_0\lambda_{ijkl}u^0_{ij}(p)u^0_{kl}(q)+Q_{pq})\bar{c}_p\bar{c}_q$$
$$+\frac{1}{2N}\sum_{p,q}\sum_{\mathbf{k}}[-(F_i(p,\mathbf{k})G_{ij}(\mathbf{k})F^*_j(q,\mathbf{k})+Q_{pq}]\Delta\tilde{c}(p,\mathbf{k})\Delta\tilde{c}^*(q,\mathbf{k}) \quad (13.2.78)$$

The first term is the sum of the elastic self-energies

$$E^0_{\text{self}}(p)=\tfrac{1}{2}(v_0\lambda_{ijkl}u^0_{ij}(p)u^0_{kl}(p)-Q_{pp}) \quad (13.2.79)$$

required to introduce a solute atom of type p into the pure solvent crystal.

The second term in Eq. (13.2.78) is the configurationally independent indirect binary interaction (the image-force term) which depends on the cross products $\bar{c}_p\bar{c}_q$, and the third term is the configuration-dependent pairwise direct interaction.

13.3. CALCULATION OF STRAIN-INDUCED INTERACTION IN BCC AND FCC SUBSTITUTIONAL AND INTERSTITIAL SOLID SOLUTIONS

Eqs. (13.2.60) and (13.2.78) enable us to calculate physically distinct contributions to the elastic energy of both binary interstitial and multicomponent substitutional solutions. This energy is expressed in terms of material constants, crystal lattice parameters, concentration expansion coefficients, Kanzaki forces, frequencies of host lattice vibrations, and elastic moduli of pure solvent crystals. All these characteristics may be determined by independent experiments. The determination of the Kanzaki forces, $\mathbf{f}(p, \mathbf{r})$, is the most difficult problem. It follows from Eqs. (13.2.21) and (13.2.35) that the forces, $\mathbf{f}(p, \mathbf{r})$, obey the following summation rule:

$$\sigma^0_{ij}(p)=\lambda_{ijkl}u^0_{kl}(p)=\frac{1}{v_0}\sum_{\mathbf{r}}f_i(p,\mathbf{r})r_j=\frac{1}{2v_0}\sum_{\mathbf{r}}(f_i(p,\mathbf{r})r_j+f_j(p,\mathbf{r})r_i) \quad (13.3.1)$$

Expression (13.3.1) may be treated as the set of simultaneous linear equations in the components $f_i(p, \mathbf{r})$. The number of these equations is equal to the number of different components of the tensor, $\sigma^0_{ij}(p)$. Therefore Eq. (13.3.1) may be solved if the number of unknowns, $f_i(p, \mathbf{r})$, is equal to the number of equations [to the number of different components of the tensor, $\sigma^0_{ij}(p)$]. One may reduce the number of the unknowns, $f_i(p, \mathbf{r})$, using the nearest or nearest and next-nearest interaction approximation, by assuming the forces to vanish beyond the first or second coordination shells.

Let us consider a few examples of the determination of the forces, $\mathbf{f}(p, \mathbf{r})$.

13.3.1. Substitutional FCC Solid Solution

A solute atom in a fcc substitutional solid solution is a dilation point defect because a crystal lattice site in the fcc lattice has the cubic point symmetry (any second-rank tensor having the cubic symmetry is proportional to the Kronecker symbol, δ_{ij}). Crystal lattice expansion caused by alloying atoms of type p may then be written in the form

$$\bar{\varepsilon}_{ij}=u^0_{ij}(p)\bar{c}_p$$

where

$$u^0_{ij}(p)=\frac{da}{ad\bar{c}_p}\,\delta_{ij} \tag{13.3.2}$$

$da/(ad\bar{c}_p)$ is the linear expansion coefficient related to alloying with atoms of type p. The values $\sigma^0_{ij}(p)$ may be obtained from Eq. (13.3.2) as follows:

$$\sigma^0_{ij}(p)=\lambda_{ijkl}u^0_{kl}(p)=\lambda_{ijkl}\,\frac{da}{ad\bar{c}_p}\,\delta_{kl}=\lambda_{ijkk}\,\frac{da}{ad\bar{c}_p}=(c_{11}+2c_{12})\,\frac{da}{ad\bar{c}_p}\,\delta_{ij} \tag{13.3.3}$$

Here and below, the Cartesian coordinate system related to the [100], [010], and [001] cubic lattice axes is used.

Since all the nonzero components of the tensor, $\sigma^0_{ij}(p)$, are equal to each other, Eq. (13.3.1) enables us to determine the single component of the forces, $\mathbf{f}(p, \mathbf{r})$. Let us assume that the force, $\mathbf{f}(p, \mathbf{r})$, vanishes beyond the first coordination shell. The nearest-neighbor atoms in the fcc lattice are separated by twelve translations $\langle\frac{1}{2}\frac{1}{2}0\rangle$. Since the segments $\langle\frac{1}{2}\frac{1}{2}0\rangle$ linking the nearest neighbors in the fcc lattice lie along the symmetry directions $\langle 110\rangle$, a force acting from a solute atom at the site O on an atoms at site $\langle\frac{1}{2}\frac{1}{2}0\rangle$ should be also directed along the $\langle 110\rangle$ axis. In other words, the forces, $\mathbf{f}(p, \mathbf{r})$, may be represented in the form

$$\mathbf{f}(p,\mathbf{r})=\begin{cases}(\pm f_1(p), \pm f_1(p), 0) & \text{if } \mathbf{r}=\left(\pm\frac{a_0}{2}, \pm\frac{a_0}{2}, 0\right)\\ (\pm f_1(p), 0, \pm f_1(p)) & \text{if } \mathbf{r}=\left(\pm\frac{a_0}{2}, 0, \pm\frac{a_0}{2}\right)\\ (0, \pm f_1(p), \pm f_1(p)) & \text{if } \mathbf{r}=\left(0, \pm\frac{a_0}{2}, \pm\frac{a_0}{2}\right)\\ 0 & \text{otherwise}\end{cases} \tag{13.3.4}$$

where $f_1(p)$ is the projection of the force, $\mathbf{f}(p, \mathbf{r})$ at $\mathbf{r}=(a_0/2, a_0/2, 0)$ on the x-axis, a_0 is the crystal lattice parameter of the solvent.

The Fourier transform (13.2.41) of the forces in (13.3.4) is therefore

$$\mathbf{F}(p,\mathbf{k})=-4if_1(p)\left[\sin\frac{k_xa_0}{2}\left(\cos\frac{k_ya_0}{2}+\cos\frac{k_za_0}{2}\right),\right.$$

$$\left.\frac{\sin k_ya_0}{2}\left(\cos\frac{k_xa_0}{2}+\cos\frac{k_za_0}{2}\right), \sin\frac{k_za_0}{2}\left(\cos\frac{k_xa}{2}+\cos\frac{k_ya}{2}\right)\right] \tag{13.3.5}$$

where $\mathbf{k}=(k_x, k_y, k_z)$, $i=\sqrt{-1}$ is the imaginary unit. To determine the value of $f_1(p)$, we substitute Eq. (13.3.4) into (13.3.1). Referring to the relation (13.3.3), we obtain

$$(c_{11}+2c_{12})\frac{da}{ad\bar{c}_p}\delta_{ij}=-\frac{i}{v_0}4f_1(p)a_0\delta_{ij} \tag{13.3.6}$$

Since $v_0=a_0^3/4$ in the fcc lattice, it follows from Eq. (13.3.6) that

$$f_1(p)=\frac{c_{11}+2c_{12}}{16}a_0^2\frac{da}{ad\bar{c}_p} \tag{13.3.7}$$

Using Eq. (13.3.7) in Eq. (13.3.5), we obtain for a substitional fcc solution (262)

$$\begin{aligned}
F_x(p,\mathbf{k})&=-i\frac{c_{11}+2c_{12}}{4}a_0^2\frac{da}{ad\bar{c}_p}\sin\frac{k_xa_0}{2}\left(\cos\frac{k_ya_0}{2}+\cos\frac{k_za_0}{2}\right)\\
F_y(p,\mathbf{k})&=-i\frac{c_{11}+2c_{12}}{4}a_0^2\frac{da}{ad\bar{c}_p}\sin\frac{k_ya_0}{2}\left(\cos\frac{k_xa_0}{2}+\cos\frac{k_za_0}{2}\right)\\
F_z(p,\mathbf{k})&=-i\frac{c_{11}+2c_{12}}{4}a_0^2\frac{da}{ad\bar{c}_p}\sin\frac{k_za_0}{2}\left(\cos\frac{k_xa_0}{2}+\cos\frac{k_ya_0}{2}\right)
\end{aligned} \tag{13.3.8}$$

13.3.2. Substitutional BCC Solid Solution

A solute atom of type p in a bcc substitutional solid solution is, like in a fcc substitutional solution, a dilation center because crystal lattice sites of a bcc solution also have a cubic point symmetry. The forces, $\mathbf{f}(p, \mathbf{r})$, can in this case also be found using the nearest-neighbor interaction model. The only difference with the fcc solution is that in an bcc solution eight nearest-neighbor sites lie on the $\langle 111\rangle$ directions and have the coordinates $(\pm a_0/2, \pm a_0/2, \pm a_0/2)$. The directions $\langle 111\rangle$ are the threefold symmetry axes, and the corresponding forces, $\mathbf{f}(p, \mathbf{r})$, are also directed along these axes. Calculations similar to those made for the fcc case yield

$$\begin{aligned}
F_x(p,\mathbf{k})&=-i(c_{11}+2c_{12})a_0^2\frac{da}{ad\bar{c}_p}\sin\frac{k_xa_0}{2}\cos\frac{k_ya_0}{2}\cos\frac{k_za_0}{2}\\
F_y(p,\mathbf{k})&=-i(c_{11}+2c_{12})a_0^2\frac{da}{ad\bar{c}_p}\sin\frac{k_ya_0}{2}\cos\frac{k_xa_0}{2}\cos\frac{k_za_0}{2}\\
F_z(p,\mathbf{k})&=-i(c_{11}+2c_{12})a_0^2\frac{da}{ad\bar{c}_p}\sin\frac{k_za_0}{2}\cos\frac{k_xa_0}{2}\cos\frac{k_ya_0}{2}
\end{aligned} \tag{13.3.9}$$

13.3.3. Interstitial FCC-Based Solution with Octahedral Positions of Interstitial Atoms

The fcc lattice contains one octahedral interstice per primitive unit cell (per host atom), so the index p may be omitted. Octahedral interstices in the fcc lattice have cubic point symmetry, and thus the injection of solute atoms into these interstices gives rise to an isotropic crystal lattice expansion described by the strain

$$\bar{\varepsilon}_{ij} = u_{ij}^0 \bar{c}$$

where

$$u_{ij}^0 = \frac{da}{ad\bar{c}}\,\delta_{ij} \tag{13.3.10}$$

An interstitial atom in an octahedral interstice has six nearest-neighbor host atoms along the [100], [010], and [001] directions which are the fourfold symmetry axes. The coordinates of the nearest atoms are $(\pm a_0/2, 0, 0)$, $(0, \pm a_0/2, 0)$, $(0, 0, \pm a_0/2)$. The force, $\mathbf{f}(\mathbf{r})$, between the interstitial atom and a nearest host atom is therefore directed by the symmetry condition along the segment linking these atoms. Calculations similar to those performed for the fcc-based substitutional alloy yield

$$\mathbf{F}(\mathbf{k}) = -i\frac{a_0^2}{2}(c_{11}+2c_{12})\frac{da}{ad\bar{c}}\left(\sin\frac{k_x a_0}{2}, \sin\frac{k_y a_0}{2}, \sin\frac{k_z a_0}{2}\right) \tag{13.3.11}$$

13.3.4. Interstitial BCC-Based Solution with Occupancy of Octahedral Interstices

As shown in Section 6.8, the bcc lattice contains octahedral interstices of three types, O_x, O_y, and O_z, designated by the index p ($p = 1, 2, 3$). These interstices lie between pairs of nearest host atoms along the [100], [010], and [001] directions, respectively. Since the octahedral interstices have the tetragonal point symmetry, the insertion of interstitial atoms produces tetragonal uniform strain.

The concentration expansion coefficients

$$u_{ij}^0(1) = \begin{pmatrix} u_{33}^0 & 0 & 0 \\ 0 & u_{11}^0 & 0 \\ 0 & 0 & u_{11}^0 \end{pmatrix}, \qquad u_{ij}^0(2) = \begin{pmatrix} u_{11}^0 & 0 & 0 \\ 0 & u_{33}^0 & 0 \\ 0 & 0 & u_{11}^0 \end{pmatrix},$$

$$u_{ij}^0(3) = \begin{pmatrix} u_{11}^0 & 0 & 0 \\ 0 & u_{11}^0 & 0 \\ 0 & 0 & u_{33}^0 \end{pmatrix} \tag{13.3.12}$$

describe expansion produced by the introduction of interstitial atoms into the O_x, O_y, and O_z interstices, respectively.

We shall now determine the Fourier transform $\mathbf{F}(3, \mathbf{k})$ following (262). The value of $\sigma_{ij}^0(3)$ is given by

$$\sigma_{ij}^0(3) = \lambda_{ijkl} u_{kl}^0(3) = \begin{pmatrix} \sigma_{11}^0 & 0 & 0 \\ 0 & \sigma_{11}^0 & 0 \\ 0 & 0 & \sigma_{33}^0 \end{pmatrix} \tag{13.3.13}$$

where*

$$\sigma_{11}^0 = (c_{11} + c_{12})u_{11}^0 + c_{12}u_{33}^0 \qquad (13.3.14)$$
$$\sigma_{33}^0 = 2c_{12}u_{11}^0 + c_{11}u_{33}^0$$

$$u_{11}^0 = \frac{da}{adn_3}, \quad u_{33}^0 = \frac{dc}{adn_3} \qquad (13.3.15)$$

Since the components of the tensor, $\sigma_{ij}^0(3)$, in (13.3.13) assume only two different nonzero values, Eq. (13.3.1) enables one to determine two components of the forces, $\mathbf{f}(3, \mathbf{r})$.

Any O_z interstice has two nearest-neighbor host atoms at a distance of $(0, 0, \pm a_0/2)$ and four next-nearest host atoms removed by $(\pm a_0/2, \pm a_0/2, 0)$. These vectors lie along the symmetry directions, and therefore the corresponding forces, $\mathbf{f}(3, \mathbf{r})$, are directed along them. Each force has one independent component, which allows us to determine the forces corresponding to two coordination shells. Calculations with Eqs. (13.3.1) and (13.3.14) give

$$F_x^{oct}(3, \mathbf{k}) = -ia_0^2\sigma_{11}^0 \exp\left(-\tfrac{1}{2}ik_za_0\right) \sin\frac{k_xa_0}{2} \cos\frac{k_ya_0}{2}$$

$$F_y^{oct}(3, \mathbf{k}) = -ia_0^2\sigma_{11}^0 \exp\left(-\tfrac{1}{2}ik_za_0\right) \sin\frac{k_ya_0}{2} \cos\frac{k_xa_0}{2} \qquad (13.3.16)$$

$$F_z^{oct}(3, \mathbf{k}) = -ia_0^2\sigma_{33}^0 \exp\left(-\tfrac{1}{2}ik_za_0\right) \sin\frac{k_za_0}{2}$$

The vectors $\mathbf{F}^{oct}(2, \mathbf{k})$ and $\mathbf{F}^{oct}(1, \mathbf{k})$ may be obtained from (13.3.16) by the cyclic permutation of the coordinate indexes x, y, z and the index p. For instance, $\mathbf{F}^{oct}(1, \mathbf{k})$ is given by the permutations $(p=3) \to (p=1)$, $x \to y$, $y \to z$, $z \to x$:

$$\mathbf{F}^{oct}(1, \mathbf{k}) = -ia_0^2 \exp\left(-i\frac{k_xa_0}{2}\right)\left(\sigma_{33}^0 \sin\frac{k_xa_0}{2},\right.$$
$$\left.\sigma_{11}^0 \sin\frac{k_ya_0}{2} \cos\frac{k_za_0}{2},\ \sigma_{11}^0 \sin\frac{k_za_0}{2} \cos\frac{k_ya_0}{2}\right) \qquad (13.3.17)$$

13.3.5. Interstitial BCC Solution with Occupancy of Tetrahedral Interstices

The bcc lattice has six tetrahedral interstices (109) per host lattice site which may be labeled by the indexes $p = 1, \bar{1}, 2, \bar{2}, 3, \bar{3}$. These interstices are displaced

*The term n_3 is the occupation probability of finding an interstitial atom in O_z site.

from the nearest host atom by the vectors $\mathbf{h}_p$:

$$\mathbf{h}_1 = a_0(\tfrac{1}{4}, \tfrac{1}{2}, 0) \quad \mathbf{h}_2 = a_0(0, \tfrac{1}{4}, \tfrac{1}{2}) \quad \mathbf{h}_3 = a_0(\tfrac{1}{2}, 0, \tfrac{1}{4})$$
$$\mathbf{h}_{\bar{1}} = a_0(\tfrac{\bar{1}}{4}, \tfrac{1}{2}, 0) \quad \mathbf{h}_{\bar{2}} = a_0(0, \tfrac{\bar{1}}{4}, \tfrac{\bar{1}}{2}) \quad \mathbf{h}_{\bar{3}} = a_0(\tfrac{1}{2}, 0, \tfrac{\bar{1}}{4}) \qquad (13.3.18)$$

Let the types of tetrahedral interstices be designated T_1, T_2, T_3, $T_{\bar{1}}$, $T_{\bar{2}}$, $T_{\bar{3}}$. Tetrahedral interstices have a tetragonal point symmetry. An interstitial atom in a tetrahedral interstice therefore produces a tetragonal distortion of the host lattice. The tensors of the concentration coefficients of the bcc lattice expansion corresponding to the cubic axes of the bcc lattice have the form (13.3.12). The occupation of the tetrahedral sites of the types T_1 and $T_{\bar{1}}$ is in agreement with the concentration expansion matrix, $u_{ij}^0(1)$; the occupation of the sites T_2 and $T_{\bar{2}}$ or T_3 and $T_{\bar{3}}$ gives the matrices $u_{ij}^0(2)$ and $u_{ij}^0(3)$, respectively. The tensor $\sigma_{ij}^0(p)$ related to the tensor $u_{ij}^0(p)$ by the equation

$$\sigma_{ij}^0(p) = \lambda_{ijkl} u_{kl}^0(p)$$

has two different nonzero components, σ_{11}^0 and σ_{33}^0 [see Eq. (13.3.14)]. Eq. (13.3.1) therefore allows one to find two components of the vector, $\mathbf{f}(p, \mathbf{r})$, only if the nearest neighbor model is adopted. The calculation similar to those performed above yields

$$\mathbf{F}^{\text{tetr}}(3, \mathbf{k}) = -ia_0^2 \exp\left(-i\frac{k_x a_0}{2}\right)$$
$$\times\left\{\sigma_{11}^0 \sin\frac{k_x a_0}{2}, \sigma_{11}^0 \sin\frac{k_y a_0}{2} e^{-i(k_z a_0/2)}, i\sigma_{33}^0\left(\cos\frac{k_y a_0}{2} e^{-i(k_z a_0/2)} - \cos\frac{k_x a_0}{2}\right)\right\}$$
$$(13.3.19a)$$

$$\mathbf{F}^{\text{tetr}}(\bar{3}, \mathbf{k}) = -ia_0^2 \exp\left(i\frac{k_x a_0}{2}\right)$$
$$\times\left\{\sigma_{11}^0 \sin\frac{k_x a_0}{2}, \sigma_{11}^0 \sin\frac{k_y a_0}{2} e^{i(k_z a_0/2)}, -i\sigma_{33}^0\left(\cos\frac{k_y a_0}{2} e^{i(k_z a_0/2)} - \cos\frac{k_x a_0}{2}\right)\right\}$$
$$(13.3.19b)$$

The vectors $\mathbf{F}^{\text{tetr}}(p, \mathbf{k})$ for $p = 1$, $\bar{1}$ and 2, $\bar{2}$ can be obtained from (13.3.19) by a cyclic permutation of both vector components and the indexes, p.

The Fourier transforms (13.3.8), (13.3.9), (13.3.11), (13.3.16), and (13.3.19) are expressed in terms of the concentration expansion coefficients that can be found from the concentration dependence of the crystal lattice parameters of the alloy. This provides the possibility to carry out any elastic energy calculation based on Eqs. (13.2.60) or (13.2.78) if the Born-von Karman matrices, $A_{ij}(\mathbf{r})$, are known from an independent neutron inelastic scattering experiment. In numerical computations it is often convenient to use the representation

$$v_i^{*0}(q, \mathbf{k}) = G_{ij}(\mathbf{k}) F_j^*(q, \mathbf{k}) \qquad (13.3.20)$$

where by definition (13.3.20), $\mathbf{v}^0(p, \mathbf{k})$ satisfies the set of linear equations

$$\tilde{A}_{ij}(\mathbf{k})v_j^{*0}(q, \mathbf{k}) = F_i^*(q, \mathbf{k}) \tag{13.3.21}$$

The quadratic form $F_i(p, \mathbf{k})G_{ij}(\mathbf{k})F_j^*(q, \mathbf{k})$ in Eqs. (13.2.60) and (13.2.78) may then be replaced by the scalar product $[\mathbf{F}(p, \mathbf{k})\mathbf{v}^{*0}(q, \mathbf{k})]$:

$$F_i(p, \mathbf{k})G_{ij}(\mathbf{k})F_j^*(q, \mathbf{k}) = (\mathbf{F}(p, \mathbf{k})\mathbf{v}^{*0}(q, \mathbf{k})) \tag{13.3.22}$$

Substitution (13.3.22) is made because the solution of the set of three linear equations (13.3.21) in three components of the vector $\mathbf{v}^{*0}(q, \mathbf{k})$ is usually simplier technically than the determination of the Green function, $G_{ij}(\mathbf{k})$.

The components of the dynamic matrix, $\tilde{A}_{ij}(\mathbf{k})$, are easy to express in terms of the Born-von Karman constants. For instance, assuming the constants, $A_{ij}(\mathbf{r})$, to be zero beyond eight coordination shells, we obtain the following representation for the bcc lattice:

$$\begin{aligned}
\tilde{A}_{11}(\mathbf{k}) = {} & 8\alpha_1[1 - \cos \pi h \cos \pi k \cos \pi l] + [4\alpha_2 \sin^2 \pi h + 4\beta_2(\sin^2 \pi k + \sin^2 \pi l)] \\
& + [4\alpha_3(2 - \cos 2\pi h \cos 2\pi k - \cos 2\pi h \cos 2\pi l) + 4\beta_3(1 - \cos 2\pi k \cos 2\pi l)] \\
& + [8\alpha_4(1 - \cos 3\pi h \cos \pi k \cos \pi l) \\
& + 8\beta_4(2 - \cos 3\pi k \cos \pi h \cos \pi l - \cos 3\pi l \cos \pi h \cos \pi k)] \\
& + 8\alpha_5(1 - \cos 2\pi h \cos 2\pi k \cos 2\pi l) \\
& + [4\alpha_6 \sin^2 2\pi h + 4\beta_6(\sin^2 2\pi k + \sin^2 2\pi l)] \\
& + [8\beta_7(1 - \cos \pi h \cos 3\pi k \cos 3\pi l) \\
& + 8\alpha_7(2 - \cos \pi k \cos 3\pi h \cos 3\pi l - \cos \pi l \cos 3\pi h \cos 3\pi k)] \\
& + [4\alpha_8(2 - \cos 4\pi h \cos 2\pi k - \cos 4\pi h \cos 2\pi l) \\
& + 4\beta_8(2 - \cos 2\pi h \cos 4\pi k - \cos 2\pi h \cos 4\pi l) \\
& + 4\gamma_8(2 - \cos 4\pi k \cos 2\pi l - \cos 2\pi k \cos 4\pi l)]
\end{aligned} \tag{13.3.23a}$$

$$\begin{aligned}
\tilde{A}_{12}(\mathbf{k}) = {} & 8\beta_1 \sin \pi h \sin \pi k \cos \pi l + 4\gamma_3 \sin 2\pi h \sin 2\pi k \\
& + [8\delta_4(\sin 3\pi h \sin \pi k + \sin 3\pi k \sin \pi h)\cos \pi l \\
& + 8\gamma_4 \sin \pi h \sin \pi k \cos 3\pi l] + 8\beta_5 \sin 2\pi h \sin 2\pi k \cos 2\pi l \\
& + [8\delta_7(\sin \pi h \sin 3\pi k + \sin \pi k \sin 3\pi h)\cos 3\pi l \\
& + 8\gamma_7 \sin 3\pi h \sin 3\pi k \cos \pi l] \\
& + 4\delta_8(\sin 4\pi h \sin 2\pi k + \sin 2\pi h \sin 4\pi k)
\end{aligned} \tag{13.3.23b}$$

where

$$\pi h = \frac{k_x a_0}{2}, \quad \pi k = \frac{k_y a_0}{2}, \quad \pi l = \frac{k_z a_0}{2}$$

and the designations of the Born-von Karman constants are the same as in (271). The other components of the tensor, $\tilde{A}_{ij}(\mathbf{k})$ may be determined from (13.3.23) by a cyclic permutation of all the Cartesian indexes.

The numerical calculations are convenient to carry out as follows:

1. Solve Eq. (13.3.21) in $\mathbf{v}^{*0}(q, \mathbf{k})$ at each vector $\mathbf{k}$ within the first Brillouin zone of the solvent lattice using the representations for $\tilde{A}_{ij}(\mathbf{k})$ and $\mathbf{F}(p, \mathbf{k})$ derived in this section as coefficients of linear equations.
2. Calculate the value of the parameter Q using the equation

$$Q = \frac{1}{N}\sum_{\mathbf{k}} \mathbf{F}(p, \mathbf{k})\mathbf{v}^{*0}(p, \mathbf{k}) \tag{13.3.24a}$$

(in the case of an interstitial solution) or the value of the parameter Q_{pq} using the equation

$$Q_{pq} = \frac{1}{N}\sum_{\mathbf{k}} \mathbf{F}(p, \mathbf{k})\mathbf{v}^{*0}(q, \mathbf{k}) \tag{13.3.24b}$$

(in the case of a multicomponent substitutional solution). The summations in Eqs. (13.3.24) are over all N points of the first Brillouin zone allowed by the cyclic boundary conditions.
3. Calculate the Fourier transforms of the strain-induced interaction energies

$$V_{pq}(\mathbf{k}) = -(\mathbf{F}(p, \mathbf{k})\mathbf{v}^{*0}(q, \mathbf{k})) + Q\delta_{pq} \tag{13.3.25a}$$

for an interstitial solution or

$$V_{pq}(\mathbf{k}) = -(\mathbf{F}(p, \mathbf{k})\mathbf{v}^{*0}(q, \mathbf{k})) + Q_{pq} \tag{13.3.25b}$$

for a multicomponent substitutional one.
4. Calculate the back Fourier transform to the pairwise strain-induced interaction potentials

$$W_{pq}(\mathbf{r}) = \frac{1}{N}\sum_{\mathbf{k}}{}' (-\mathbf{F}(p, \mathbf{k})\mathbf{v}^{*0}(q, \mathbf{k}) + Q\delta_{pq})e^{i\mathbf{k}\mathbf{r}} \tag{13.3.26a}$$

in the case of an interstitial solution and

$$W_{pq}(\mathbf{r}) = \frac{1}{N}\sum_{\mathbf{k}}{}' (-\mathbf{F}(p, \mathbf{k})\mathbf{v}^{*0}(q, \mathbf{k}) + Q_{pq})e^{i\mathbf{k}\mathbf{r}} \tag{13.3.26b}$$

in the case of a substitutional solution.

13.4. STRAIN-INDUCED INTERACTION OF PAIRS OF SOLUTE ATOMS IN BCC SOLUTIONS BASED ON αFe, Ta, Nb, AND V

In this Section we shall give an account of the computer calculations carried out by Blanter and Khachaturyan (272) which illustrate potentialities of the microscopic elasticity theory formulated in Sections 13.2 and 13.3. The microscopic theory will be applied in calculating the interaction energies of interstitials and vacancies and the displacements of host atoms around interstitials and vacancies in four metals having the bcc crystal lattice, α-Fe, V, Nb, and Ta. We shall consider the cases when interstitial atoms occupy both octahedral and tetrahedral interstices.

As mentioned previously, an interstitial atom in octahedral and tetrahedral interstices produces a uniform tetragonal crystal lattice distortion which is described by the concentration expansion coefficients (13.3.12). To be more precise, let us consider two examples: an interstitial occupying an octahedral interstice O_z and an interstitial occupying a tetrahedral interstice T_3. In both cases the concentration coefficients of crystal lattice expansion are described by the tensor

$$u_{ij}^0(3)=\begin{pmatrix} u_{11}^0 & 0 & 0 \\ 0 & u_{11}^0 & 0 \\ 0 & 0 & u_{33}^0 \end{pmatrix} \tag{13.4.1}$$

The Fourier transforms of the Kanzaki forces for octahedral and tetrahedral occupancies are given by Eqs. (13.3.16) and (13.3.19a). According to these equations the Fourier transforms $\mathbf{F}^{\text{oct}}(3, \mathbf{k})$ and $\mathbf{F}^{\text{tetr}}(3, \mathbf{k})$ are linear functions of the components σ_{11}^0 and σ_{33}^0 defined by Eq. (13.3.14). These values may be rewritten

$$\sigma_{11}^0=(c_{11}+c_{12})u_{11}^0+c_{12}u_{33}^0=c_{44}u_{33}^0\left[\frac{c_{12}}{c_{44}}+\frac{(c_{11}+c_{12})}{c_{44}}t_1\right]$$

$$\sigma_{33}^0=c_{11}u_{33}^0+2c_{12}u_{11}^0=c_{44}u_{33}^0\left[\frac{c_{11}}{c_{44}}+\frac{2c_{12}}{c_{44}}t_1\right] \tag{13.4.2}$$

where

$$t_1=\frac{u_{11}^0}{u_{33}^0} \tag{13.4.2a}$$

is the tetragonality factor. Therefore the Fourier transforms $\mathbf{F}^{\text{oct}}(3, \mathbf{k})$ and $\mathbf{F}^{\text{tetr}}(3, \mathbf{k})$ are linear functions of the tetragonality factor t_1. It follows from Eq. (13.2.65) that at $N \to \infty$ the Fourier transform of the interaction potential is a quadratic form of $\mathbf{F}(p, \mathbf{k})$:

$$V_{pq}(\mathbf{k})=-F_i(p, \mathbf{k})G_{ij}(\mathbf{k})F_j^*(q, \mathbf{k})+Q\delta_{pq} \tag{13.4.3}$$

where

$$Q=\frac{1}{N}\sum_{\mathbf{k}} F_i(p, \mathbf{k})G_{ij}(\mathbf{k})F_j^*(p, \mathbf{k}) \tag{13.4.4}$$

Since $\mathbf{F}(p, \mathbf{k})$ is the linear function of the tetragonality factor t_1, one can see that Q and therefore $V_{pq}(\mathbf{k})$ are also quadratic functions of t_1. In other words, using the energy unit $c_{44}a_0^3(u_{33}^0)^2$, we may always represent Eq. (13.4.3) in the form

$$\frac{V_{pq}(\mathbf{k})}{c_{44}a_0^3(u_{33}^0)^2}=\tilde{n}_{pq}(\mathbf{k})+\tilde{b}_{pq}(\mathbf{k})t_1+\tilde{d}_{pq}(\mathbf{k})t_1^2 \tag{13.4.5}$$

where $\tilde{n}_{pq}(\mathbf{k})$, $\tilde{b}_{pq}(\mathbf{k})$, and $\tilde{d}_{pq}(\mathbf{k})$ are dimensionless constants. The back Fourier transform of Eq. (13.4.5) yields pairwise interaction energies in the form

$$\frac{W_{pq}^{el}(\mathbf{r})}{c_{44}a_0^3(u_{33}^0)^2}=n_{pq}(\mathbf{r})+b_{pq}(\mathbf{r})t_1+d_{pq}(\mathbf{r})t_1^2 \tag{13.4.6}$$

where $n_{pq}(\mathbf{r})$, $b_{pq}(\mathbf{r})$, and $d_{pq}(\mathbf{r})$ are material constants of pure solvent depending only on the Born-von Karman constants (on the vibrational spectrum of the host crystal). Eq. (13.4.6) is convenient because all information about the kind of interstitial atom is contained in the tetragonality factor t_1, and the coefficients $n_{pq}(\mathbf{r})$, $b_{pq}(\mathbf{r})$, and $d_{pq}(\mathbf{r})$ may be calculated and listed for any solvent element.

Without the loss of generality, we may confine our analysis to pairs of atoms, one of which occupies an O_z position, $(q, \mathbf{r}')=(3, 0)$ with the coordinates $(0, 0, a_0/2)$, and the other one any position $(p, \mathbf{r})$ with the coordinate $\mathbf{h}_p+\mathbf{r}$. The interstitials making up the pair are separated by

$$\mathbf{h}=\mathbf{h}_p+\mathbf{r}-\mathbf{h}_3=(x, y, z)a_0 \tag{13.4.7}$$

where $\mathbf{h}_p$ is a vector linking the pth interstice and the nearest host atom, (x, y, z) are the coordinates of the vector $\mathbf{h}$ separating the two interacting atoms. Using (13.4.7), we may simplify Eq. (13.4.6) and rewrite it as follows

$$\frac{W(x, y, z)}{c_{44}a_0^3(u_{33}^0)^2}=n(x, y, z)+b(x, y, z)t_1+d(x, y, z)t_1^2 \tag{13.4.8}$$

The values $n(x, y, z)$, $b(x, y, z)$, and $d(x, y, z)$ are the universal coefficients that can be used with any interstitial solid solution based on the relevant host lattice. The specific characteristics of interstitials are taken into account through the tetragonality factor t_1 only.

The constants $n(x, y, z)$, $b(x, y, z)$, and $d(x, y, z)$ were calculated for four bcc materials (272): α-Fe, V, Nb, and Ta, with the numerical values of the Born-von Karman constants in eight coordination shells taken from the neutron inelastic diffraction studies (273–275, 271) and with the dynamic matrix representation (13.3.23). The calculations were carried out for both the octahedral and tetrahedral occupancy cases. The calculation results are listed in Tables 13.1 and 13.2.

To exemplify the calculation procedure based on the Tables 13.1 and 13.2, let us consider the computation of the interaction energy of a pair of C-atoms separated by the vector $(\frac{1}{2}, 0, \frac{1}{2})$ in α-Fe (here and below all atomic coordinates are given in a_0 units). Carbon atoms occupy octahedral interstices in α-Fe. Since, by definition, the coordinates of the origin C-atom are $(0, 0, \frac{1}{2})$, the position of the second C-atom of the pair is given by

$$(0, 0, \tfrac{1}{2})+(\tfrac{1}{2}, 0, \tfrac{1}{2})=(\tfrac{1}{2}, 0, 1)$$

The interstitial site with the $(\frac{1}{2}, 0, 1)$ coordinates is an O_x site.

Table 13.1 gives for α-Fe and the separation distance of $(\frac{1}{2}, 0, \frac{1}{2})$ the constants:

$$n(\tfrac{1}{2}, 0, \tfrac{1}{2})=-0.038, \quad b(\tfrac{1}{2}, 0, \tfrac{1}{2})=-0.216, \quad d(\tfrac{1}{2}, 0, \tfrac{1}{2})=-0.277 \tag{13.4.9}$$

Table 13.1 Coefficients for Calculation of Pairwise Strain-Induced Interaction Energies of Interstitials in Octahedral Interstices

	V			Nb			Ta			Fe		
(*x*, *y*, *z*)	*n*	*b*	*d*	*n*	*b*	*d*	*n*	*b*	*d*	*n*	*b*	*d*
$(0, \frac{1}{2}, 0)$	−0.619	−2.892	−2.090	−1.020	−4.612	−3.407	−0.390	−1.689	−1.407	−0.162	−0.713	−0.546
$(\frac{1}{2}, 0, \frac{1}{2})$	−0.101	−0.671	−0.948	−0.160	−1.109	−1.531	−0.072	−0.424	−0.566	−0.038	−0.216	−0.277
$(\frac{1}{2}, \frac{1}{2}, \frac{1}{2})$	−0.103	−0.199	−0.266	−0.192	−0.391	−0.419	−0.047	−0.110	−0.133	−0.032	−0.080	−0.089
(0, 0, 1)	+0.664	+1.121	+0.547	+0.868	+1.611	+0.880	+0.338	+0.656	+0.335	+0.122	+0.168	+0.022
(1, 0, 0)	+0.016	+0.032	−0.073	+0.017	−0.003	−0.252	−0.010	−0.028	−0.054	−0.018	−0.039	−0.045
$(1, \frac{1}{2}, 0)$	+0.110	+0.561	+0.363	+0.519	+0.777	+0.165	+0.067	+0.318	+0.248	+0.013	+0.077	+0.042
$(\frac{1}{2}, 1, \frac{1}{2})$	−0.011	−0.004	+0.207	−0.037	−0.058	+0.251	+0.005	+0.054	+0.155	−0.005	−0.006	+0.048
(1, 1, 0)	+0.075	+0.352	+0.614	+0.139	+0.598	+0.981	+0.045	+0.227	+0.360	−0.002	+0.043	+0.131
(1, 0, 1)	−0.067	−0.126	−0.078	−0.092	−0.165	−0.077	−0.021	−0.052	−0.058	−0.003	−0.015	−0.021
$(1, \frac{1}{2}, 1)$	−0.008	−0.138	−0.130	−0.007	−0.173	−0.192	+0.005	−0.031	−0.043	+0.009	+0.005	−0.011
$(0, \frac{3}{2}, 0)$	−0.030	−0.088	−0.063	−0.068	−0.242	−0.200	−0.023	−0.087	−0.071	−0.007	−0.040	−0.022
$(\frac{1}{2}, 0, \frac{3}{2})$	−0.018	−0.043	−0.015	−0.039	−0.094	−0.046	−0.014	−0.036	−0.035	−0.006	−0.020	−0.026
$(\frac{3}{2}, \frac{1}{2}, \frac{1}{2})$	−0.012	−0.045	−0.068	−0.001	−0.070	−0.145	−0.004	−0.033	−0.061	−0.006	−0.017	−0.015
$(\frac{1}{2}, \frac{1}{2}, \frac{3}{2})$	+0.033	+0.015	−0.022	+0.036	+0.054	+0.050	+0.015	+0.005	+0.009	+0.012	+0.005	−0.011
(1, 1, 1)	−0.025	−0.025	−0.085	−0.017	−0.008	−0.100	−0.003	+0.020	+0.005	+0.005	+0.033	+0.020
$(1, \frac{3}{2}, 0)$	+0.005	−0.008	−0.002	+0.004	−0.005	−0.022	0.000	−0.007	−0.010	−0.003	−0.012	−0.006
(2, 0, 0)	−0.020	−0.074	+0.006	−0.029	−0.070	+0.086	−0.004	−0.023	+0.005	−0.003	−0.008	−0.005
(0, 0, 2)	+0.049	−0.024	−0.057	+0.246	+0.263	+0.012	+0.014	−0.004	+0.002	+0.010	0.000	−0.010
$(2, \frac{1}{2}, 0)$	−0.011	+0.007	−0.023	+0.001	+0.101	+0.044	−0.004	0.000	−0.005	−0.002	−0.002	−0.007
$(1, \frac{3}{2}, 1)$	−0.015	−0.040	+0.022	−0.017	−0.032	+0.038	−0.007	−0.015	+0.010	+0.001	+0.007	+0.024
$(\frac{1}{2}, 2, \frac{1}{2})$	−0.004	−0.015	+0.021	−0.001	+0.028	+0.093	−0.001	−0.005	+0.003	−0.002	−0.009	−0.001
$(\frac{3}{2}, 0, \frac{3}{2})$	+0.010	+0.025	+0.025	+0.007	0.000	+0.023	+0.004	−0.002	−0.001	+0.002	0.000	−0.003
$(\frac{3}{2}, \frac{3}{2}, \frac{1}{2})$	+0.009	+0.040	+0.084	+0.026	+0.098	+0.123	+0.002	+0.010	+0.033	0.000	+0.010	+0.029
$(\frac{3}{2}, \frac{1}{2}, \frac{3}{2})$	−0.010	−0.012	−0.004	−0.029	−0.047	−0.020	−0.009	−0.017	−0.022	0.000	0.000	−0.004
(2, 1, 0)	−0.004	−0.008	+0.009	−0.011	−0.009	+0.058	0.000	−0.008	−0.020	−0.003	−0.007	+0.002
(0, 1, 2)	+0.023	+0.026	+0.001	−0.005	−0.038	−0.029	+0.011	+0.017	+0.016	+0.002	−0.003	−0.007
(2, 0, 1)	+0.006	+0.052	+0.056	+0.021	+0.079	+0.062	0.000	+0.011	+0.010	−0.002	−0.003	−0.003

Table 13.2 Coefficients for Calculations of Strain-Induced Pairwise Interaction Energies of Interstitials in Tetrahedral Interstices [The Initial Interstitial Atom is in a $(\frac{1}{2}, 0, \frac{1}{4})$ Position]

	V			Nb			Ta		
(x, y, z)	n	b	d	n	b	d	n	b	d
$\frac{1}{4}, 0\ \frac{1}{4}$	−1,053	−4,920	−5,304	−1,757	−8,875	−7,989	−0,672	−2,986	−3,197
$0, 0, \frac{1}{2}$	1,128	1,062	−1,399	2,032	4,730	−4,020	0,453	0,414	−0,753
$\frac{\bar{1}}{4}, \frac{1}{2}, \frac{\bar{1}}{4}$	−0,642	−2,827	−2,031	−1,059	−5,293	−2,562	−0,405	−1,668	−1,315
$\frac{\bar{1}}{2}, \frac{1}{2}, 0$	−1,252	−2,407	−1,338	−1,859	−5,308	−0,454	−0,661	−1,500	−0,921
$\frac{\bar{3}}{4}, 0, \frac{1}{4}$	0,051	0,025	0,201	0,102	0,226	0,465	0,014	−0,003	0,069
$\frac{1}{2}, \frac{1}{2}, \frac{1}{2}$	0,485	0,938	0,294	0,688	2,010	−0,256	0,245	0,564	0,243
$\frac{\bar{1}}{4}, \frac{1}{2}, \frac{3}{4}$	0,242	1,175	0,805	0,413	2,230	1,093	0,157	0,678	0,521
$1, 0, 0$	−0,415	−0,125	0,951	−0,674	−1,247	1,994	−0,215	−0,141	0,411
$0, 0, 1$	−0,075	−0,266	0,106	−0,356	−1,204	0,404	0,028	−0,022	0,042
$\bar{1}, 0, \frac{\bar{1}}{2}$	0,733	1,895	1,899	1,165	3,754	1,963	0,394	1,161	1,140
$\frac{3}{4}, \frac{1}{2}, \frac{3}{4}$	0,023	−0,355	−0,411	0,010	−0,589	−0,942	0,013	−0,113	−0,162
$\frac{\bar{1}}{2}, \frac{1}{2}, 1$	0,040	−0,092	−0,187	0,110	0,086	−0,424	0,049	0,037	−0,043

Table 13.3 Experimental Data used in Calculations

	a_0 (in Å)	c_{11} (in 10^{12} dyne/cm^2)	c_{12}	c_{44}	Alloy	u_{11}^0	u_{33}^0	t_1
Fe	2.865	2.42	1.465	1.12	Fe-C	−0.09	0.86	−0.10
					Fe-N	−0.07	0.83	−0.08
V	3.026	2.28	1.19	0.426	V-O	−0.10	0.66	−0.15
					V-N	−0.14	0.69	−0.2
					V-H	0.025	0.074	0.3
Nb	3.300	2.46	1.34	0.287	Nb-O	−0.06	0.50	−0.12
					Nb-N	−0.05	0.60	−0.08
					Nb-H	0.019	0.032	0.6
Ta	3.303	2.67	1.61	0.825	Ta-O	−0.04	0.47	−0.08
					Ta-N	−0.05	0.56	−0.09
					Ta-H	0.034	0.051	0.7

It follows from Table 13.3 that for α-Fe-C alloy (110) we have

$$u_{11}^0 = -0.09, \quad u_{33}^0 = 0.86, \quad t_1 = -0.104 \tag{13.4.10}$$

Using the numerical values for α-Fe—$c_{44} = 1.12 \times 10^{12}$ dyne/cm^2, $a_0 = 2.865 \times 10^{-8}$ cm, and $u_{33}^0 = 0.86$—we may calculate the energy unit

$$c_{44} a_0^3 (u_{33}^0)^2 = 1.12 \times 10^{12} (2.865 \times 10^{-8})^3 (0.86)^2 = 19.48 \times 10^{-12} \text{ erg} = 12.15 \text{ eV} \tag{13.4.11}$$

Using the values (13.4.9), (13.4.11), and (13.4.10) in Eq. (13.4.8), we obtain

$$W(\tfrac{1}{2}, 0, \tfrac{1}{2}) = 12.15[-0.038 - 0.216(-0.104) - 0.277(-0.104)^2] = -0.225 \text{ eV} \tag{13.4.12}$$

It is easy to see from Eqs. (13.4.3) and (13.4.4) that in the interaction of different kinds of interstitial atoms (α and β kinds of interstitials) their interaction energy

$$\frac{W(\alpha, \beta; x, y, z)}{c_{44} a_0^3 u_{33}^0(\alpha) u_{33}^0(\beta)} = n(x, y, z) + b(x, y, z) \frac{t_1(\alpha) + t_1(\beta)}{2} + d(x, y, z) t_1(\alpha) t_1(\beta) \tag{13.4.13}$$

where $t_1(\alpha) = u_{11}^0(\alpha)/u_{33}^0(\alpha)$, $t_1(\beta) = u_{11}^0(\beta)/u_{33}^0(\beta)$, and the coefficients $n(x, y, z)$, $b(x, y, z)$, and $d(x, y, z)$ are the same as in Eq. (13.4.8).

The interaction energies calculated from Eq. (13.4.8) for particular interstitials occupying octahedral interstices are listed in Table 13.4.

The universal equation which is valid for any interstitial solution based on the relevant solvent element can also be derived to describe atomic displacements generated by interstitial atoms.

Table 13.4 The Strain-Induced Pairwise Interaction Energies $W(x, y, z)$, in eV of Specific Interstitials Located in Octahedral Interstices

(x, y, z)	V-O	V-N	Nb-O	Nb-N	Ta-O	Ta-N	Fe-C	Fe-N	V-O-N	Nb-O-N	Ta-O-N	Fe-C-N
$(0, \frac{1}{2}, 0)$	−0.73	−0.41	−0.82	−1.52	−1.05	−1.44	−1.18	−1.27	−0.57	−1.12	−1.23	−1.23
$(\frac{1}{2}, 0, \frac{1}{2})$	−0.07	−0.01	−0.08	−0.18	−0.16	−0.22	−0.23	−0.26	−0.04	−0.12	−0.19	−0.25
$(\frac{1}{2}, \frac{1}{2}, \frac{1}{2})$	−0.26	−0.26	−0.24	−0.37	−0.17	−0.22	−0.31	−0.31	−0.26	−0.30	−0.18	−0.31
$(0, 0, 1)$	+1.60	+1.59	+1.11	+1.71	+1.19	+1.62	+1.33	+1.38	+1.60	+1.37	+1.39	+1.33
$(1, 0, 0)$	+0.03	+0.02	+0.02	+0.04	−0.04	−0.05	−0.18	−0.18	+0.03	+0.03	−0.04	−0.18
$(1, \frac{1}{2}, 0)$	+0.11	+0.04	+0.12	+0.24	+0.17	+0.23	+0.07	+0.08	+0.07	+0.17	+0.20	+0.08
$(\frac{1}{2}, 1, \frac{1}{2})$	−0.02	−0.01	−0.04	−0.07	0	+0.01	−0.05	−0.05	−0.01	−0.05	0	−0.05
$(1, 1, 0)$	+0.11	+0.10	+0.13	+0.22	+0.11	+0.16	−0.06	−0.06	+0.10	+0.17	+0.13	−0.06
$(1, 0, 1)$	−0.16	−0.15	−0.12	−0.18	−0.07	−0.10	−0.02	−0.02	−0.16	−0.15	−0.08	−0.02
$(1, \frac{1}{2}, 1)$	+0.03	+0.05	+0.02	+0.01	+0.03	+0.04	+0.11	+0.10	+0.04	+0.02	+0.04	+0.10
$(0, \frac{3}{2}, 0)$	−0.06	−0.05	−0.07	−0.11	−0.07	−0.09	−0.04	−0.05	−0.05	−0.09	−0.08	−0.04
$(\frac{1}{2}, 0, \frac{3}{2})$	−0.04	−0.03	−0.04	−0.07	−0.05	−0.06	−0.05	−0.05	−0.04	−0.06	−0.05	−0.05
$(\frac{3}{2}, \frac{1}{2}, \frac{1}{2})$	−0.02	−0.02	+0.01	+0.01	−0.01	−0.01	−0.06	−0.06	−0.02	+0.01	−0.01	−0.06
$(\frac{1}{2}, \frac{1}{2}, \frac{3}{2})$	+0.10	+0.10	+0.05	+0.07	+0.06	+0.08	+0.14	+0.14	+0.10	+0.06	+0.07	+0.14
$(1, 1, 1)$	−0.07	−0.08	−0.03	−0.04	−0.02	−0.03	+0.02	+0.03	−0.07	−0.03	−0.02	+0.03
$(1, \frac{3}{2}, 0)$	+0.02	+0.02	+0.01	+0.01	0	0	−0.02	−0.02	+0.02	+0.01	+0.01	−0.02
$(\frac{3}{2}, 1, \frac{1}{2})$	+0.01	0	+0.03	+0.04	+0.01	+0.01	+0.02	+0.02	0	+0.03	+0.01	+0.02
$(2, 0, 0)$	−0.03	−0.02	−0.03	−0.05	−0.01	−0.01	−0.03	−0.03	−0.02	−0.04	−0.01	−0.03
$(0, 0, 2)$	+0.16	+0.18	+0.35	+0.52	+0.06	+0.08	+0.13	+0.12	+0.17	+0.42	+0.07	+0.12
$(2, \frac{1}{2}, 0)$	−0.04	−0.05	−0.02	−0.02	−0.02	−0.03	−0.02	−0.02	−0.04	−0.02	−0.02	−0.02
$(1, \frac{1}{2}, 1)$	−0.03	−0.02	−0.02	−0.03	−0.02	−0.03	+0.01	+0.01	+0.02	−0.03	−0.03	−0.01
$(\frac{1}{2}, 2, \frac{1}{2})$	0	0	0	−0.01	0	−0.01	−0.01	−0.02	0	−0.01	−0.01	−0.01
$(\frac{3}{2}, 0, \frac{3}{2})$	0	+0.02	+0.01	+0.02	+0.02	+0.03	+0.03	+0.02	+0.02	+0.01	+0.02	+0.02
$(\frac{3}{2}, \frac{3}{2}, \frac{1}{2})$	+0.02	+0.01	+0.03	+0.04	0	0	−0.01	−0.01	+0.01	+0.03	0	−0.01
$(\frac{3}{2}, \frac{1}{2}, \frac{3}{2})$	−0.02	−0.03	−0.04	−0.06	−0.03	−0.06	0	0	−0.02	−0.05	−0.04	0
$(2, 1, 0)$	−0.01	−0.01	−0.01	−0.02	0	0	−0.03	−0.03	−0.01	−0.02	−0	−0.03
$(0, 1, 2)$	+0.01	+0.01	0	0	+0.04	+0.05	+0.03	+0.03	+0.06	−0.01	+0.05	+0.03
$(2, 0, 1)$	0	0	+0.02	+0.04	0	−0.01	−0.02	−0.02	0	+0.03	0	−0.02

In the case of only one interstitial at interstice $(p, \mathbf{r})=(p, 0)$, we have

$$c(q, \mathbf{r})=\delta_{qp}\delta_{\mathbf{r}0}$$

and therefore at $N \to \infty$

$$\Delta c(q, \mathbf{k})=\delta_{qp} \tag{13.4.14}$$

Substituting Eq. (13.4.14) into (13.2.43), we obtain

$$\tilde{v}_i^0(p, \mathbf{k})=G_{ij}(\mathbf{k})\sum_q F_j(q, \mathbf{k})\Delta\tilde{c}(q, \mathbf{k})=G_{ij}(\mathbf{k})\sum_q F_j(q, \mathbf{k})\delta_{qp}=G_{ij}(\mathbf{k})F_j(p, \mathbf{k}) \tag{13.4.15}$$

The displacement field is given by the back Fourier transform of Eq. (13.4.15):

$$v_i^0(p, \mathbf{r})=\frac{1}{N}\sum_{\mathbf{k}} G_{ij}(\mathbf{k})F_j(p, \mathbf{k})e^{i\mathbf{k}\mathbf{r}} \tag{13.4.16}$$

Since $\mathbf{F}(p, \mathbf{k})$ is a linear function of the tetragonality factor t_1, it follows from Eq. (13.4.16) that the displacements, $\mathbf{v}^0(p, \mathbf{r})$, can also be represented as a linear function of t_1:

$$\frac{\mathbf{v}^0(p, \mathbf{r})}{a_0 u_{33}^0}=\mathbf{h}(p, \mathbf{r})+\mathbf{g}(p, \mathbf{r})t_1 \tag{13.4.17}$$

Without the loss of generality, the interstitial atom may be placed at $(p, \mathbf{r})=(3, 0)$, its coordinates being $(0, 0, \frac{1}{2})$. Eq. (13.4.17) may then be rewritten

$$\frac{\mathbf{v}^0(x, y, z)}{a_0 u_{33}^0}=\mathbf{h}(x, y, z)+\mathbf{g}(x, y, z)t_1 \tag{13.4.18}$$

where (x, y, z) are the coordinates of host atoms. The vectors $\mathbf{h}(x, y, z)$ and $\mathbf{g}(x, y, z)$ are listed in Tables 13.5 and 13.6.

To calculate the vacancy-vacancy and vacancy-interstitial interactions, one should know the Kanzaki forces, $\mathbf{f}^{\text{vac}}(\mathbf{r})$, produced by a vacancy in the host lattice. These forces, for instance, may be calculated from the model pairwise interaction energies between host atoms proposed by Machlin (276):

$$E=-\frac{A}{r^4}+\frac{B}{r^8} \tag{13.4.19}$$

where

$$A=0.0496\times a_0^4\varepsilon, \quad B=0.0305\times\varepsilon a_0^8$$

and ε is the cohesion energy.

The forces $\mathbf{f}^{\text{vac}}(\mathbf{r})$ may then be calculated as follows

$$\mathbf{f}^{\text{vac}}(\mathbf{r})=-\frac{\partial}{\partial r}(-E) \tag{13.4.20}$$

Their Fourier transform is

$$\mathbf{F}^{\text{vac}}(\mathbf{k})=\sum_{\mathbf{r}}\frac{\partial E(\mathbf{r})}{\partial \mathbf{r}}e^{-i\mathbf{k}\mathbf{r}} \tag{13.4.21}$$

Table 13.5 Coefficients for Calculation of Host Atom Displacement caused by Interstitial in Octahedral Interstice $(0, 0, \frac{1}{2})$

		V		Nb		Ta		Fe	
R		**h**	**g**	**h**	**g**	**h**	**g**	**h**	**g**
(0, 0, 0)	$v_x^0/(a_0u_{33}^0)$	0	0	0	0	0	0	0	0
	$v_y^0/(a_0u_{33}^0)$	0	0	0	0	0	0	0	0
	$v_z^0/(a_0u_{33}^0)$	−0.307	−0.184	−0.305	−0.216	−0.225	−0.265	−0.187	−0.134
$(\frac{1}{2}, \frac{1}{2}, \frac{1}{2})$	$v_x^0/(a_0u_{33}^0)$	0.038	0.218	0.049	0.333	0.042	0.188	0.023	0.128
	$v_y^0/(a_0u_{33}^0)$	0.038	0.218	0.049	0.333	0.042	0.188	0.023	0.128
	$v_z^0/(a_0u_{33}^0)$	0	0	0	0	0	0	0	0
(1, 0, 0)	$v_x^0/(a_0u_{33}^0)$	−0.008	−0.010	−0.016	−0.013	−0.009	−0.001	−0.004	+0.012
	$v_y^0/(a_0u_{33}^0)$	0	0	0	0	0	0	0	0
	$v_z^0/(a_0u_{33}^0)$	0.003	0.012	0.004	0.025	−0.003	0.004	−0.008	0.000
(1, 1, 1)	$v_x^0/(a_0u_{33}^0)$	0.006	0.043	0.014	0.083	0.006	0.035	+0.006	0.035
	$v_y^0/(a_0u_{33}^0)$	0.006	0.043	0.014	0.083	0.006	0.035	+0.006	0.035
	$v_z^0/(a_0u_{33}^0)$	−0.004	0.010	−0.008	0.000	−0.001	0.014	0.006	0.021

Table 13.6 Coefficients for Calculation of Host Atom Displacement Caused by Interstitial in Tetrahedral Interstice ($\frac{1}{2}$, 0, $\frac{1}{4}$)

		V		Nb		Ta	
$\mathbf{R}=(x, y, z)$	$\mathbf{v}^0/(a_0u_{33}^0)$	h	g	h	g	h	g
0, 0, 0	$v_x^0/(a_0u_{33}^0)$	−0,109	−0,422	−0,116	−0,426	−0,113	−0.380
	$v_y^0/(a_0u_{33}^0)$	0	0	0	0	0	0
	$v_z^0/(a_0u_{33}^0)$	−0,289	−0,247	−0,299	−0,274	−0,238	−0,238
$\frac{1}{2}, \frac{1}{2}, \frac{1}{2}$	$v_x^0/(a_0u_{33}^0)$	0	0	0	0	0	
	$v_y^0/(a_0u_{33}^0)$	+0,109	+0,422	+0,116	+0,426	+0,113	+0,380
	$v_z^0/(a_0u_{33}^0)$	+0,289	+0,247	+0,299	+0,274	+0,238	+0,238
0, 1, 0	$v_x^0/(a_0u_{33}^0)$	+0,017	+0,001	+0,019	+0,008	+0,011	−0,009
	$v_y^0/(a_0u_{33}^0)$	−0,018	+0,011	−0,021	+0,004	−0,016	+0,007
	$v_z^0/(a_0u_{33}^0)$	+0,045	+0,027	+0,071	+0,060	+0,024	+0,010
0, 0, 1	$v_x^0/(a_0u_{33}^0)$	−0,013	+0,044	−0,006	+0,051	−0,008	+0,034
	$v_y^0/(a_0u_{33}^0)$	0	0	0	0	0	0
	$v_z^0/(a_0u_{33}^0)$	+0,014	−0,050	−0,004	−0,072	0,025	−0,033
0, 1, 1	$v_x^0/(a_0u_{33}^0)$	−0,018	−0,022	−0,021	−0,020	−0,018	−0,022
	$v_y^0/(a_0u_{33}^0)$	+0,032	+0,066	+0,034	+0,080	+0,025	+0,060
	$v_z^0/(a_0u_{33}^0)$	+0,041	+0,054	+0,062	+0,082	+0,032	+0,046

The pairwise interaction potentials for vacancies and the elastic displacement field generated by a vacancy are described by the equation

$$W_{\text{vac-vac}}(\mathbf{r})=\frac{1}{N}\sum_{\mathbf{k}}[-F_i^{\text{vac}}(\mathbf{k})G_{ij}(\mathbf{k})F_j^{*\text{vac}}(\mathbf{k})+Q]e^{i\mathbf{kr}} \tag{13.4.22}$$

where

$$Q=\frac{1}{N}\sum_{\mathbf{k}}F_i^{\text{vac}}(\mathbf{k})G_{ij}(\mathbf{k})F_j^{*\text{vac}}(\mathbf{k})$$

and by the equation

$$v_i^0(\mathbf{r})=\frac{1}{N}\sum_{\mathbf{k}}G_{ij}(\mathbf{k})F_j^{\text{vac}}(\mathbf{k})e^{i\mathbf{kr}} \tag{13.4.23}$$

The calculation results are listed in Table 13.7.

The interaction of a vacancy with an interstitial atom in an octahedral site is described by the equation

$$W_{\text{vac-inter}}(\mathbf{r})=\frac{1}{N}\sum_{\mathbf{k}}[-F_i^{\text{vac}}(\mathbf{k})G_{ij}(\mathbf{k})F_j^{*\text{oct}}(p,\mathbf{k})]e^{i\mathbf{kr}} \tag{13.4.24}$$

Since $\mathbf{F}^{\text{oct}}(p, \mathbf{k})$ is a linear function of t_1, the interaction potentials (13.4.24) are also linear functions of t_1:

$$\frac{W_{\text{vac-inter}}(x, y, z)}{u_{33}^0}=l(x, y, z)+m(x, y, z)t_1 \tag{13.4.25}$$

Table 13.7 Calculations of Vacancy and Bivacancy Characteristics

Metal	Nearest Atom Displacement v^0/a_0	Ratio of Vacancy Volume to Atom Volume	Interaction Energy of Two Vacancies (in eV): First Coordination Shell	Interaction Energy of Two Vacancies (in eV): Second Coordination Shell
V	0.038	0.87	−0.029	−0.109
Nb	0.037	0.88	−0.030	−0.169
Ta	0.033	0.89	−0.036	−0.155
Fe	0.024	0.92	−0.028	−0.068

Table 13.8 Coefficients for Calculation of Strain-Induced Interaction Energies, in eV, of a Vacancy Located in Coordinate Origin and Interstitial Located in Octahedral Site with Coordinates (*x*, *y*, *z*)

	V		Nb		Ta		Fe	
(x, y, z)	l	m	l	m	l	m	l	m
$(0, 0, \frac{1}{2})$	−0.881	−2.776	−1.290	−3.892	−1.329	−3.624	−0.537	−1.463
$(\frac{1}{2}, \frac{1}{2}, \frac{1}{2})$	−0.861	−0.908	−1.155	−1.168	−1.191	−1.448	−0.497	−0.590
$(1, 0, \frac{1}{2})$	−0.049	−0.021	−0.068	−0.013	−0.070	−0.082	−0.058	−0.084
$(\frac{1}{2}, \frac{1}{2}, 1)$	0.322	0.174	0.404	0.294	0.489	0.442	0.159	0.075
$(1, 1, \frac{1}{2})$	0.102	0.454	0.132	0.575	0.021	0.706	0.046	0.243
$(1, 1, \frac{3}{2})$	0.020	−0.070	0.031	−0.034	0.036	−0.036	0.044	0.017
$(\frac{1}{2}, \frac{3}{2}, 2)$	0.013	0.034	0	0.070	0.016	0.006	0.010	0.001
$(2, 1, \frac{1}{2})$	−0.001	−0.009	0.015	0.075	−0.005	−0.030	−0.007	−0.004

Table 13.9 Interaction Energy, in eV, of a Vacancy Located in Coordinate Origin and Specific Interstitial Located in Octahedral Interstice with Coordinates (*x*, *y*, *z*)

(x, y, z)	V-O	V-N	Nb-O	Nb-N	Ta-O	Ta-N	Fe-C	Fe-N
$(0, 0, \frac{1}{2})$	−0.299	−0.221	−0.398	−0.560	−0.479	−0.563	−0.330	−0.344
$(\frac{1}{2}, \frac{1}{2}, 0)$	−0.479	−0.469	−0.518	−0.645	−0.501	−0.589	−0.374	−0.377
$(1, 0, \frac{1}{2})$	−0.034	−0.033	−0.037	−0.048	−0.030	−0.035	−0.042	−0.042
$(\frac{1}{2}, \frac{1}{2}, 1)$	0.194	0.196	0.187	0.230	0.212	0.249	0.130	0.127
$(1, 1, \frac{1}{2})$	0.020	0.006	0.035	0.055	0.068	0.080	0.018	0.021
$(1, 1, \frac{3}{2})$	0.020	0.023	0.017	0.019	0.018	0.021	0.036	0.035
$(\frac{1}{2}, \frac{1}{2}, 2)$	0.005	0.005	−0.006	−0.007	0.007	0.008	0.009	0.008
$(2, 1, \frac{1}{2})$	0	0.001	0.005	0.006	−0.001	−0.002	−0.006	−0.006

where (x, y, z) are the coordinates of the separation vector between the vacancy and the interstitial atom. The calculated coefficients $l(x, y, z)$ and $m(x, y, z)$ are listed in Table 13.8.

Substitution of the values u_{33}^0 and t_1 for various interstitial atoms into Eq. (13.4.25) gives the values of interaction energies listed in Table 13.9.

Table 13.10 Coefficients for Calculation of Strain-Induced Interaction Energies of Interstitial Atom in Octahedral Interstice and Substitutional Atom in Host BCC Lattice

Coordination Shell Number	Separation Distance (x, y, z)	V		Nb		Ta		α-Fe	
		l_1	m_1	l_1	m_1	l_1	m_1	l_1	m_1
1	$(\frac{1}{2}\ 0\ 0)$	−0.203	−0.660	−0.212	−0.674	−0.218	−0.628	−0.166	−0.195
2	$(0\ \frac{1}{2}\ \frac{1}{2})$	−0.223	−0.206	−0.218	−0.103	−0.200	−0.188	−0.162	−0.188
3	$(\frac{1}{2}\ 1\ 0)$	−0.030	−0.037	−0.031	−0.041	−0.022	−0.024	−0.029	−0.034
4	$(1\ \frac{1}{2}\ \frac{1}{2})$	0.072	0.016	0.063	−0.024	0.071	0.024	0.042	0.048
5	$(\frac{3}{2}\ 0\ 0)$	−0.026	0.001	−0.037	0.005	−0.029	−0.020	−0.024	−0.028
6	$(\frac{1}{2}\ 1\ 1)$	0.016	0.098	0.016	0.103	0.027	0.113	0.007	0.009
7	$(\frac{3}{2}\ 1\ 0)$	0.008	−0.004	0.002	−0.003	0.005	−0.007	0.009	0.010
8	$(\frac{3}{2}\ 1\ 1)$	0.006	−0.018	0.005	−0.018	0.006	−0.014	0.013	0.015

The microscopic theory of elastic interactions also provides the possibility to calculate pairwise interaction energies between substitutional and interstitial solute atoms occupying octahedral sites in the V, Nb, Ta, and α-Fe host lattices.

The equation for pairwise elastic interaction energies will be identical to Eq. (13.4.24). It has the form

$$W^{\text{subst-int}}(p, \mathbf{r}) = -\frac{1}{N}\sum_{\mathbf{k}} F_i^{\text{subst}}(\mathbf{k})G_{ij}(\mathbf{k})F_j^{*\text{oct}}(\mathbf{k})e^{i\mathbf{k}\mathbf{r}} \tag{13.4.26}$$

Without the loss of generality, interstitial atoms may be assumed to occupy O_z interstices. Taking into account Eqs. (13.3.9) and (13.3.16) and the fact that the vector $\mathbf{F}^{\text{oct}}(3, \mathbf{k})$ is a linear function of t_1, we may rewrite Eq. (13.4.26) as

$$\frac{W^{\text{subst-int}}(x, y, z)}{(c_{11}+2c_{12})a_0^3u_{33}^0u^0} = l_1(x, y, z) + m_1(x, y, z)t_1 \tag{13.4.27}$$

where (x, y, z) are the coordinates of the separation vector between the substitutional and interstitial atoms and $u^0 = da/ad\bar{c}$ is the concentration coefficient of crystal lattice expansion produced by substitutional atoms.

The coefficients $l_1(x, y, z)$ and $m_1(x, y, z)$ for various substitutional solutes are listed in Table 13.10. The numerical data given in that table make it possible to calculate interaction energies between arbitrary substitutional and interstitial atoms, provided the concentration expansion coefficients for these atoms are known.

13.5. LIMIT TRANSITION TO CONTINUUM THEORY: ESHELBY'S THEORY OF SOLID SOLUTION; ELASTIC ENERGY AND SPINODAL DECOMPOSITION; DISCUSSION OF COOK-DE FONTAINE'S VERSION; "ELASTIC ENERGY PARADOX"; LIMIT TRANSITION TO COHERENT INCLUSIONS

In this section we shall consider the limit transition from the microscopic theory of a binary substitutional solution to its continuum counterpart, for instance, to the theory proposed by Eshelby (266).

The elastic energy of a binary substitutional solution (267) is given by Eq. (13.2.73)

$$E_{\text{elast}} = \frac{1}{2}N(v_0\lambda_{ijkl}u_{ij}^0u_{kl}^0 - Q)\bar{c}(1-\bar{c}) + \frac{1}{2}\frac{1}{N}\sum_{\mathbf{k}}[-F_i(\mathbf{k})G_{ij}(\mathbf{k})F_j^*(\mathbf{k}) + Q]|\Delta\tilde{c}(\mathbf{k})|^2 \tag{13.5.1}$$

which is equally good for both binary substitutional solution and binary interstitial solution with one interstice per host atom.

Here

$$Q = \frac{1}{N}\sum_{\mathbf{k}} F_i(\mathbf{k})G_{ij}(\mathbf{k})F_j^*(\mathbf{k}) \tag{13.5.2}$$

As has been demonstrated in Section 13.2, the first term in Eq. (13.5.1),

$$E^0_{\text{elast}} = \tfrac{1}{2}N(v_0\lambda_{ijkl}u^0_{ij}u^0_{kl} - Q)\bar{c}(1-\bar{c}) \tag{13.5.3}$$

describes the configuration-independent contribution to the elastic energy which cannot be affected by spatial redistribution of solute atoms. The second term of (13.5.1),

$$\Delta E^{\text{heter}}_{\text{elast}} = \frac{1}{2N}\sum_{\mathbf{k}} [-F_i(\mathbf{k})G_{ij}(\mathbf{k})F^*_j(\mathbf{k}) + Q]|\Delta\tilde{c}(\mathbf{k})|^2 \tag{13.5.4}$$

on the other hand; describes the configuration-dependent contribution to the elastic energy. It gives the elastic energy change associated with a spatial redistribution of solute atoms and can be interpreted as pairwise elastic energy of solute atom interaction (elastic interaction Hamiltonian). The pairwise elastic interaction energy (13.5.4) plays the key role in the theory of phase transformations because this is the only energy that determines the effect of the crystal lattice distortion on the coherent stage of the decomposition and ordering.

Eq. (13.5.1) may, however, be represented in a different form in which both physically distinct terms are combined to give one term. Actually, making use of the identity (13.2.58), which in the case of a binary substitutional solution reads

$$\frac{1}{N}\sum_{\mathbf{k}} |\Delta\tilde{c}(\mathbf{k})|^2 \equiv N\bar{c}(1-\bar{c}) \tag{13.5.5}$$

and employing (13.5.5) in the first term of Eq. (13.5.1), we come to the following compact form for the total elastic energy

$$E_{\text{elast}} = \frac{1}{2}\frac{1}{N}\sum_{\mathbf{k}} [v_0\lambda_{ijkl}u^0_{ij}u^0_{kl} - F_i(\mathbf{k})G_{ij}(\mathbf{k})F^*_j(\mathbf{k})]|\Delta\tilde{c}(\mathbf{k})|^2 \tag{13.5.6}$$

13.5.1. Limit Transition to the Continuum Theory

As mentioned in Section 13.1, the limit transition to the continuum elasticity occurs if the functions $\mathbf{F}(\mathbf{k})$ and $G_{ij}(\mathbf{k})$ entering the summand in Eq. (13.5.1) are replaced by the first nonvanishing term of the Taylor series expansion in $\mathbf{k}$. It also requires that the summation over N points of the first Brillouin zone should be replaced by the integration over the sphere in the $\mathbf{k}$-space whose volume, $(2\pi)^3/v_0$, is equal to the volume of the first Brillouin zone.

The power series expansion of $\mathbf{F}(\mathbf{k})$ in $\mathbf{k}$ may be represented in the form

$$F_i(\mathbf{k}) = \sum_{\mathbf{r}} f_i(\mathbf{r})e^{-i\mathbf{k}\mathbf{r}} = \sum_{\mathbf{r}} f_i(\mathbf{r})[1 - ik_jr_j + \cdots] \approx \sum_{\mathbf{r}} f_i(\mathbf{r}) - ik_j\sum_{\mathbf{r}} f_i(\mathbf{r})r_j + \cdots \tag{13.5.7}$$

It follows from Eqs. (13.2.15b) and (13.2.21) that the first nonvanishing term of the power series expansion in **k**, (13.5.7), may be rewritten

$$F_i(\mathbf{k}) \cong -i\Lambda_{ij}k_j \tag{13.5.8}$$

Using the relation (13.2.35), we may represent Eq. (13.5.8) in its final form

$$F_i(\mathbf{k}) \cong -iv_0\sigma^0_{ij}k_j \tag{13.5.9}$$

where

$$\sigma^0_{ij} = \lambda_{ijkl}u^0_{kl} \tag{13.5.10}$$

As for the Green function, $G_{ij}(\mathbf{k})$, the first nonvanishing term at $\mathbf{k}\to 0$ is

$$G_{ij}(\mathbf{k}) \cong \frac{1}{v_0k^2}\Omega_{ij}(\mathbf{n}) \tag{13.5.11}$$

where $\mathbf{n}=\mathbf{k}/k$ and $\Omega_{ij}(\mathbf{n})$ is determined as the inverse to the tensor

$$\Omega^{-1}_{ij}(\mathbf{n}) = \lambda_{iklj}n_kn_l$$

[see Eq. (8.1.3)]. The approximation (13.5.11) in fact means neglecting the spacial dispersion of the host lattice vibration frequencies. To obtain the long-wavelength limit of the value Q, we should:

1. Replace the summation over the first Brillouin zone by the integration over the sphere whose volume, $(2\pi)^3/v_0$, is equal to the volume of the first Brillouin zone. The transition to the integral is in accordance with the relation
$$\frac{1}{N}\sum_{\mathbf{k}}(\cdots) = v_0\int\frac{d^3k}{(2\pi)^3}(\cdots)$$
2. Substitute the long-wave transitions (13.5.9) and (13.5.11) into (13.5.2).

The second procedure yields

$$\begin{aligned} Q &= \frac{1}{N}\sum_{\mathbf{k}} F_i(\mathbf{k})G_{ij}(\mathbf{k})F^*_j(\mathbf{k}) = v_0\int_{\text{(sphere)}}\frac{d^3k}{(2\pi)^3}\, v_0 n_i\sigma^0_{ij}\Omega_{jk}(\mathbf{n})\sigma^0_{kl}n_l \\ &= v_0^2\int_0^{k_D}\frac{k^2dk\,dO_{\mathbf{n}}}{(2\pi)^3}\, n_i\sigma^0_{ij}\Omega_{jk}(\mathbf{n})\sigma^0_{kl}n_l \\ &= v_0^2\int_0^{k_D}\frac{4\pi k^2dk}{(2\pi)^3}\oint n_i\sigma^0_{ij}\Omega_{jk}(\mathbf{n})\sigma^0_{kl}n_l\,\frac{dO_{\mathbf{n}}}{4\pi} \end{aligned} \tag{13.5.13}$$

where k_D is the Debye cutoff radius defined by the relation

$$\frac{4\pi}{3}(k_D)^3 = \frac{(2\pi)^3}{v_0}$$

$dO_{\mathbf{n}}$ is the solid angle element normal to the unit vector **n**.

The integration over k yields

$$Q = v_0 \oint n_i \sigma_{ij}^0 \Omega_{jk}(\mathbf{n}) \sigma_{kl}^0 n_l \frac{dO_\mathbf{n}}{4\pi} \tag{13.5.14}$$

where the integration $\oint dO_\mathbf{n}(\cdots)$ is taken over all directions of $\mathbf{n}$, that is, over the surface of the unit radius sphere.

Introducing the definitions

$$L(\mathbf{n}) = n_i \sigma_{ij}^0 \Omega_{jk}(\mathbf{n}) \sigma_{kl}^0 n_l \tag{13.5.15}$$

and the mean value

$$\langle L(\mathbf{n}) \rangle_\mathbf{n} = \frac{1}{4\pi} \oint dO_\mathbf{n} L(\mathbf{n}) \tag{13.5.16}$$

and using them in Eq. (13.5.14), we have

$$Q = v_0 \langle L(\mathbf{n}) \rangle_\mathbf{n} \tag{13.5.17}$$

Substitution of the representations (13.5.9), (13.5.11), and (13.5.17) into (13.5.1) ensures the long-wave limit transition in the elastic energy equation (13.5.1):

$$E_{\text{elast}} = \frac{1}{2} V[\lambda_{ijkl} u_{ij}^0 u_{kl}^0 - \langle L(\mathbf{n}) \rangle_\mathbf{n}] \bar{c}(1-\bar{c}) - \frac{1}{2} \frac{v_0}{N} \sum_\mathbf{k} [L(\mathbf{n}) - \langle L(\mathbf{n}) \rangle_\mathbf{n}] |\Delta\tilde{c}(\mathbf{k})|^2 \tag{13.5.18}$$

where $V = Nv_0$ is the total volume of the crystal, or

$$E_{\text{elast}} = E_{\text{elast}}^0 + \Delta E_{\text{elast}}^{\text{heter}}$$

where

$$E_{\text{elast}}^0 = \frac{1}{2} V[\lambda_{ijkl} u_{ij}^0 u_{kl}^0 - \langle L(\mathbf{n}) \rangle_\mathbf{n}] \bar{c}(1-\bar{c}) \tag{13.5.19}$$

and

$$\Delta E_{\text{elast}}^{\text{heter}} = -\frac{v_0}{2N} \sum_\mathbf{k} [L(\mathbf{n}) - \langle L(\mathbf{n}) \rangle_\mathbf{n}] |\Delta\tilde{c}(\mathbf{k})|^2 \tag{13.5.20}$$

Let us consider a cubic substitutional solution. In this case solute atoms are dilation centers and thus result in an isotropic concentration expansion

$$u_{ij}^0 = \frac{da}{ad\bar{c}} \delta_{ij} \tag{13.5.21}$$

where $da/(ad\bar{c})$ is the concentration expansion coefficient. Using (8.6.1) and (13.5.21) in the definition of σ_{ij}^0, we have

$$\sigma_{ij}^0 = \lambda_{ijkl} u_{kl}^0 = \frac{da}{ad\bar{c}} (c_{11} + 2c_{12}) \delta_{ij} \tag{13.5.22}$$

Substituting (13.5.22) and (8.6.3) into (13.5.15) yields

$$L(\mathbf{n}) = \left(\frac{da}{ad\bar{c}} \right)^2 \frac{(c_{11} + 2c_{12})^2}{c_{11}} \phi(\mathbf{n}) \tag{13.5.23a}$$

where

$$\phi(\mathbf{n})=\frac{1+2\xi(n_1^2n_2^2+n_1^2n_3^2+n_2^2n_3^2)+3\xi^2n_1^2n_2^2n_3^2}{c_{11}+\xi(c_{11}+c_{12})(n_1^2n_2^2+n_1^2n_3^2+n_2^2n_3^2)+\xi^2(c_{11}+2c_{12}+c_{44})n_1^2n_2^2n_3^2}c_{11} \tag{13.5.23b}$$

Using Eqs. (13.5.23), (13.5.21) in (13.5.19) and (13.5.20), we have

$$E^0_{\text{elast}}=\frac{1}{2}V(c_{11}+2c_{12})\left(\frac{da}{d\bar{c}}\right)^2\left[3-\frac{c_{11}+2c_{12}}{c_{11}}\langle\phi(\mathbf{n})\rangle_{\mathbf{n}}\right]\bar{c}(1-\bar{c}) \tag{13.5.24}$$

and

$$\Delta E^{\text{heter}}_{\text{elast}}=-\frac{v_0}{2N}\frac{(c_{11}+2c_{12})^2}{c_{11}}\left(\frac{da}{d\bar{c}}\right)^2\sum_{\mathbf{k}}[\phi(\mathbf{n})-\langle\phi(\mathbf{n})\rangle_{\mathbf{n}}]|\Delta\tilde{c}(\mathbf{k})|^2 \tag{13.5.25}$$

where

$$\langle\phi(\mathbf{n})\rangle_{\mathbf{n}}=\frac{1}{4\pi}\oint\phi(\mathbf{n})dO_{\mathbf{n}} \tag{13.5.26}$$

To estimate the mean value (13.5.26), we may employ the extrapolation similar to that used in Eq. (9.4.25):

$$\phi(\mathbf{n})\cong 1-4\frac{\Delta}{c_{11}+c_{12}+2c_{44}}(n_1^2n_2^2+n_1^2n_3^2+n_2^2n_3^2)$$
$$-54\frac{\Delta^2n_1^2n_2^2n_3^2}{(c_{11}+c_{12}+2c_{44})(c_{11}+2c_{12}+4c_{44})} \tag{13.5.27}$$

where $\Delta=c_{11}-c_{12}-2c_{44}$. Then

$$\langle\phi(\mathbf{n})\rangle_{\mathbf{n}}\cong 1-\frac{4}{5}\frac{c_{11}-c_{12}-2c_{44}}{c_{11}+c_{12}+2c_{44}}-\frac{(c_{11}-c_{12}-2c_{44})^2\times 18}{35(c_{11}+2c_{12}+4c_{44})(c_{11}+c_{12}+2c_{44})} \tag{13.5.28}$$

since $\langle n_1^2n_2^2\rangle=\langle n_1^2n_3^2\rangle=\langle n_2^2n_3^2\rangle=1/15$, $\langle n_1^2n_2^2n_3^2\rangle=1/105$. Eq. (13.5.27) gives the exact values $\phi(\mathbf{n})$ at the symmetry directions, $\langle 100\rangle$, $\langle 110\rangle$, $\langle 111\rangle$, and a very good extrapolation for intermediate directions (within 1 to 3 percent).

The difference of (13.5.27) and (13.5.28) is thus equal to

$$\phi(\mathbf{n})-\langle\phi(\mathbf{n})\rangle_{\mathbf{n}}=\frac{c_{11}-c_{12}-2c_{44}}{c_{11}+c_{12}+2c_{44}}\left[-4\left(n_1^2n_2^2+n_1^2n_3^2+n_2^2n_3^2-\frac{1}{5}\right)\right.$$
$$\left.-54\frac{c_{11}-c_{12}-2c_{44}}{c_{11}+2c_{12}+4c_{44}}\left(n_1^2n_2^2n_3^2-\frac{1}{105}\right)\right] \tag{13.5.29}$$

13.5.2. The Limit Transition to the Eshelby Theory

As mentioned in Section 13.1, Eshelby's theory is based on the sphere-in-hole elastically isotropic continuum model. Therefore the results of the Eshelby theory should follow from Eqs. (13.5.19) and (13.5.20) if the limit transition to the elastically isotropic continuum is made. As is known [see Eq. (8.3.1)], this limit transition occurs when

$$\Delta = c_{11} - c_{12} - 2c_{44} = 0, \quad c_{11} = 2\mu \frac{1-\sigma_1}{1-2\sigma_1}$$

$$c_{12} = 2\mu \frac{\sigma_1}{1-2\sigma_1}, \quad c_{44} = \mu \tag{13.5.30}$$

where σ_1 is the Poisson ratio, μ is the shear modulus.

Substitution of relations (13.5.30) into (13.5.23b) results in

$$\phi(\mathbf{n}) \equiv 1 \quad \text{and thus} \quad \langle \phi(\mathbf{n}) \rangle_{\mathbf{n}} = 1 \tag{13.5.31}$$

Making use of (13.5.31) and (13.5.30) in Eqs. (13.5.24) and (13.5.25), we have

$$E^0_{\text{elast}} = V\mu \frac{1+\sigma_1}{1-\sigma_1} \left(\frac{da}{ad\bar{c}} \right)^2 \bar{c}(1-\bar{c}) \tag{13.5.32a}$$

$$\Delta E^{\text{heter}}_{\text{elast}} = 0 \tag{13.5.32b}$$

In other words, the total elastic energy is

$$E_{\text{elast}} = E^0_{\text{elast}} = V\mu \frac{1+\sigma_1}{1-\sigma_1} \left(\frac{da}{ad\bar{c}} \right)^2 \bar{c}(1-\bar{c}) \tag{13.5.33}$$

The limit transition (13.5.33) is in complete agreement with the Eshelby result for an isotropic solid solution. The result (13.5.32b) is also in agreement with the so-called Crum theorem, according to which dilation centers in an isotropic matrix do not interact with each other. The interaction through image forces does not depend on the mutual location of the dilation centers and enters the configuration-independent term E^0_{elast}, whose value is

$$-V\mu \frac{1+\sigma_1}{1-\sigma_1} \left(\frac{da}{ad\bar{c}} \right)^2 \bar{c}^2$$

Therefore the total elastic energy (13.5.33) of an isotropic solid solution in the long-wave approximation cannot be changed by atomic redistribution; it depends on the total number of solute atoms rather than on the concentration distribution.* For this case the elastic strain energy cannot have any effect on the decomposition thermodynamics. We should, however, bear in mind that the latter conclusion only holds when the decomposition does not break the coherency between the crystal lattices of the phase transformation product and matrix (and concentration redistribution does not break the coherency by

*All these conclusions are valid if atomic distributions do not break the macroscopic homogeneity, that is, if the largest typical distance characterizing the distribution of solute element atoms is far smaller than the size of the crystal.

itself). The stress relaxation during the breaking of coherency promotes the decomposition reaction, even in the case of an elastically isotropic solution.

The conclusion that the elastic energy has no effect on the decomposition that does not break coherency seems to contradict the conventional theory of spinodal decomposition (42). According to this theory the elastic energy contribution in an elastically isotropic alloy affects both spinodal kinetics and thermodynamics. The basis for this difference between the continuum theory of spinodal decomposition proposed by Cahn (42, 43) and the above-formulated theory that proceeds from the microscopic elastic Hamiltonian deserves a more detailed discussion.*

13.5.3. Elastic Energy in Spinodal Decomposition

A change of elastic energy due to the formation of a concentration heterogeneity in an elastically isotropic solid solution can be obtained by averaging of the configuration-dependent part of the elastic interaction Hamiltonian (13.5.25):

$$\langle \Delta E^{\text{heter}}_{\text{elast}} \rangle_0 = -\frac{v_0}{2N}\left(\frac{da}{ad\bar{c}}\right)^2 \frac{(c_{11}+2c_{12})^2}{c_{11}} \sum_{\mathbf{k}} [\phi(\mathbf{n}) - \langle \phi(\mathbf{n}) \rangle_{\mathbf{n}}] \langle |\Delta \tilde{c}(\mathbf{k})|^2 \rangle_0 \qquad (13.5.34)$$

where $\langle \cdots \rangle_0$ is averaging over an essembly. Neglecting concentration fluctuations (short-range order effect), we have

$$\langle |\Delta \tilde{c}(\mathbf{k})|^2 \rangle_0 \cong |\langle \Delta \tilde{c}(\mathbf{k}) \rangle_0|^2 = |\Delta \tilde{n}(\mathbf{k})|^2 \qquad (13.5.35)$$

where

$$\Delta n(\mathbf{k}) = \sum_{\mathbf{r}} \langle c(\mathbf{r}) - \bar{c} \rangle_0 e^{-i\mathbf{k}\mathbf{r}} = \sum_{\mathbf{r}} (n(\mathbf{r}) - \bar{c}) e^{-i\mathbf{k}\mathbf{r}}$$

$$n(\mathbf{r}) = \langle c(\mathbf{r}) \rangle_0 \qquad (13.5.36)$$

As shown in Chapter 3, Eq. (13.5.35) corresponds to the so-called mean-field approximation, or self-consistent field approximation. Substitution of (13.5.35) into (13.5.34) yields

$$\langle \Delta E^{\text{heter}}_{\text{elast}} \rangle_0 = -\frac{v_0}{2N}\left(\frac{da}{ad\bar{c}}\right)^2 \frac{(c_{11}+2c_{12})^2}{c_{11}} \sum_{\mathbf{k}} [\phi(\mathbf{n}) - \langle \phi(\mathbf{n}) \rangle_{\mathbf{n}}] |\Delta \tilde{n}(\mathbf{k})|^2 \qquad (13.5.37)$$

The sum of the chemical free energy

$$F^{\text{chem}} = F_0^{\text{chem}}(\bar{c}) + \frac{1}{2N} \sum_{\mathbf{k}} \left(mk^2 + \frac{d^2 f(\bar{c})}{d\bar{c}^2} \right) |\Delta \tilde{n}(\mathbf{k})|^2 \qquad (13.5.38)$$

[see Eqs. (5.3.10) and (5.3.11)], and the elastic energy terms (13.5.24) and

*This difference is in particular closely related to the so-called "elastic energy paradox" which was originally pointed out by Cahn (277) and will be considered below.

(13.5.37) results in the total free energy of coherent heterogeneous solid solution:

$$F=F_0(\bar{c})+\frac{1}{2N}\sum_{\mathbf{k}}\left[\left(mk^2+\frac{d^2f(\bar{c})}{d\bar{c}^2}\right)\right.$$
$$\left.-v_0\frac{(c_{11}+2c_{12})^2}{c_{11}}\left(\frac{da}{ad\bar{c}}\right)^2(\phi(\mathbf{n})-\langle\phi(\mathbf{n})\rangle_{\mathbf{n}})\right]|\Delta\tilde{n}(\mathbf{k})|^2 \qquad (13.5.39)$$

where

$$F_0(\bar{c})=F_0^{\text{chem}}(\bar{c})+\frac{1}{2}V(c_{11}+2c_{12})\left(\frac{da}{ad\bar{c}}\right)^2\left[3-\frac{c_{11}+2c_{12}}{c_{11}}\langle\phi(\mathbf{n})\rangle_{\mathbf{n}}\right]\bar{c}(1-\bar{c}) \qquad (13.5.40)$$

represents the free energy of the homogeneous solution which is not affected by concentration heterogeneities. The second term of Eq. (13.5.39)

$$\Delta F=F-F_0(\bar{c})=\frac{1}{2N}\sum_{\mathbf{k}}\left[mk^2+\frac{d^2f(\bar{c})}{d\bar{c}^2}\right.$$
$$\left.-v_0\frac{(c_{11}+2c_{12})^2}{c_{11}}\left(\frac{da}{ad\bar{c}}\right)^2(\phi(\mathbf{n})-\langle\phi(\mathbf{n})\rangle_{\mathbf{n}})\right]|\Delta\tilde{n}(\mathbf{k})|^2 \qquad (13.5.41)$$

is the total free energy change including both chemical and elastic energy changes.

Equation (13.5.41) correctly describes the effect of concentration heterogeneities on the total free energy and therefore should be applied to the spinodal decomposition theory. It is worthwhile noting that Eq. (13.5.41) differs from the corresponding equation proposed by Cahn (43). Let us dwell on this difference and show why Cahn's equation should be revised and replaced by Eq. (13.5.41).

Making use of the designations utilized in this book, one may rewrite Cahn's equation of free energy change (43) as follows

$$\Delta F_{\text{Cahn}}=\frac{1}{2N}\sum_{\mathbf{k}}\left[mk^2+\frac{d^2f(\bar{c})}{d\bar{c}^2}+v_0\left(\frac{da}{ad\bar{c}}\right)^2Y(\mathbf{n})\right]|\Delta\tilde{n}(\mathbf{k})|^2 \qquad (13.5.42)$$

where

$$Y(\mathbf{n})=(c_{11}+2c_{12})\left[3-\frac{c_{11}+2c_{12}}{c_{11}}\phi(\mathbf{n})\right] \qquad (13.5.43)$$

It is clear that Eq. (13.5.42) does not coincide with (13.5.41) and, what is more, has different physical results. Actually, since $Y(\mathbf{n})$ in Eq. (13.5.43) is always a positive value, any concentration heterogeneity should, according to Eq. (13.5.42), increase the elastic energy. Therefore in Cahn's theory, elastic energy stabilizes a homogeneous solid solution. On the other hand, Eq. (13.5.41) derived in this section results in just the opposite conclusion: the formation of a concentration heterogeneity along a "soft" direction where $\{\phi(\mathbf{n})-\langle\phi(\mathbf{n})\rangle_{\mathbf{n}}\}>0$ decreases the elastic energy, that is, destabilizes the homogeneous state with respect to "soft-mode" heterogeneities and thus promotes the decomposition.

To elucidate the origin of this discrepancy between the elastic energy change

$$\langle \Delta E^{\text{heter}}_{\text{Cahn}} \rangle_0 = \frac{v_0}{2N} \sum_{\mathbf{k}} \left(\frac{da}{ad\bar{c}} \right)^2 Y(\mathbf{n}) |\Delta \tilde{n}(\mathbf{k})|^2 \qquad (13.5.44)$$

suggested by Cahn and the elastic energy change (13.5.37) derived above, we shall demonstrate that Cahn's equation (13.5.44) directly follows from the microscopic theory if the total elastic Hamiltonian (13.5.6) is averaged. It should, however, be emphasized that the total elastic Hamiltonian (13.5.6) includes both the configuration-dependent energy (13.5.4) and configuration-independent energy (13.5.3).

Actually, using the long-wave approximations (13.5.9) and (13.5.11) in the total elastic Hamiltonian (13.5.6), and taking into account the definitions (13.5.21) and (13.5.22), we have from (13.5.6)

$$E_{\text{elast}} \cong \frac{v_0}{2N} \sum_{\mathbf{k}} \left(\frac{da}{ad\bar{c}} \right)^2 Y(\mathbf{n}) |\Delta \tilde{c}(\mathbf{k})|^2 \qquad (13.5.45)$$

One may readily see that averaging of the total elastic Hamiltonian (13.5.45) and using the mean-field approximation (13.5.35) for $\langle |\Delta \tilde{c}(\mathbf{k})|^2 \rangle_0$ yields Cahn's equation (13.5.44).

Now the physical reason for the discrepancy between Eqs. (13.5.44) and (13.5.37) becomes obvious. Cahn's equation (13.5.44) is really a mean value of the *total* elastic Hamiltonian which contains both the configuration-independent energy, E^0_{elast}, and configuration-dependent energy, $\Delta E^{\text{heter}}_{\text{elast}}$, whereas Eq. (13.5.37) is a mean value of only the configuration-dependent energy, $\Delta E^{\text{heter}}_{\text{elast}}$. In reality, the configuration-dependent energy, $\Delta E^{\text{heter}}_{\text{elast}}$, is just the one that should be used to obtain the elastic energy change produced by a macroscopic concentration heterogeneity because heterogeneities do not affect the configuration independent energy, E^0_{elast}.

It should, however, be emphasized that the corrections that should be introduced into Cahn's equation (13.5.44) to obtain Eq. (13.5.37) are quite insignificant. The function $Y(\mathbf{n})$ in Eq. (13.5.44) differs from the correct function

$$-\frac{(c_{11}+2c_{12})^2}{c_{11}} (\phi(\mathbf{n}) - \langle \phi(\mathbf{n}) \rangle_{\mathbf{n}})$$

in Eq. (13.5.37) only by a constant. This additional constant merely results in some shift in the spinodal temperature and does not affect the most important conclusion of the Cahn theory (43), the appearance of the "soft" directions $\langle 100 \rangle$ in the spinodal decomposition at $c_{11} - c_{12} - 2c_{44} < 0$. It should also be pointed out that the remarkable series of papers by Cahn (42, 43, 277) has really established the elastic energy concept as a basis for the present-day understanding of phase transformation mechanisms in crystalline solids.

13.5.4. Discussion of Cook-de Fontaine's Version of the Microscopic Elastic Theory

We now shall dwell on the Cook and de Fontaine version of the theory (268) because it is usually referred to when the microscopic elasticity of a solid solution is discussed. It should, however, be mentioned that the major result of the Cook and de Fontaine paper, the microscopic elastic energy equation, differs from that derived two years earlier by the author (267). This discrepancy is worth discussing in more detail because both theories, (267) and (268), are formulated in terms of the same harmonic model, and thus the same final calculation results should be expected [the theory (267) is equally applicable to binary substitutional and interstitial solutions]. The Cook-de Fontaine equation also differs from that derived by Hoffman (278).*

Obviously, the reason for this discrepancy consists in the fact that the term corresponding to homogeneous strain relaxation (the calculation of this term may be found in Section 13.2.2) taken into account in (267) and (278) was lost in the Cook-de Fontaine derivation (268). It happened because of the erroneous assumption made in (268). According to this assumption the work done by overall change in volume of the lattice with the addition of solute to pure solvent is zero when no stress is applied at the surface. In reality, this work is just equal to the homogeneous strain relaxation discussed in Section 13.2.2.

Using the designations employed above, the elastic Hamiltonian derived by Cook and de Fontaine may be rewritten in the following form:

$$E_{\text{elast}}^{\text{C.F.}} = \frac{1}{2N} \sum_{\mathbf{k}} [\Omega(\mathbf{k}) - F_i(\mathbf{k}) G_{ij}(\mathbf{k}) F_j^*(\mathbf{k})] |\Delta\tilde{c}(\mathbf{k})|^2 \tag{13.5.46}$$

where $\Omega(\mathbf{k})$ is, by definition, some scalar function of $\mathbf{k}$. The scalar function $\Omega(\mathbf{k})$ was not consistently derived in (268) but was formally introduced to match the well-known long-wave limit (43). According to Cook-de Fontaine's theory [see Eq. (41) in (268)]

$$\Omega(\mathbf{k}) = \sum_{\mathbf{r}} \psi(\mathbf{r}) e^{-i\mathbf{k}\mathbf{r}} \tag{13.5.47}$$

where

$$\psi(\mathbf{r}) = f_i(\mathbf{r}) u_{ij}^0 r_j + \tfrac{1}{4} s_{ijkl} A_{ij}(\mathbf{r}) u_{ik}^0 u_{jl}^0 r_k r_l \tag{13.5.48}$$

and

$$s_{ijkl} = \frac{2\lambda_{ijkl}}{\lambda_{ikjl} + \lambda_{iljk}} \tag{13.5.49}$$

are the weight coefficients. The Einstein summation convention does not apply to Eqs. (13.5.48) and (13.5.49). It should, however, be pointed out that the

*Despite the differences in the point of derivation and manipulation, Hoffman's elastic energy equation (278) is the same as that derived by Khachaturyan (267).

introduction of the weight factors (13.5.49) and violation of the Einstein summation convention in Eqs. (13.5.48) and (13.5.49) lead to a physically meaningless result: the coupling parameters, $\psi(\mathbf{r})$, defined by Eq. (13.5.48) are not scalar invariants. The values, $\psi(\mathbf{r})$, prove to be dependent on the choice of the Cartesian coordinate system. The correct result

$$\Omega(\mathbf{k}) \equiv v_0 \lambda_{ijkl} u_{ij}^0 u_{kl}^0 = \text{constant} \tag{13.5.50}$$

follows from comparison of Eq. (13.5.46) with Eq. (13.5.6). Eq. (13.5.50) was also obtained by Hoffman [Eq. (28) in (278)].

It may be noted that the function $\Omega(\mathbf{k})$ in Eq. (13.5.46) does not even provide the long-wave limit transition at an arbitrary direction $\mathbf{k}$. Indeed, at $\mathbf{k} \to 0$ Eq. (13.5.47) yields

$$\Omega(0) = \sum_{\mathbf{r}} \psi(\mathbf{r}) = u_{ij}^0 \sum_{\mathbf{r}} f_i(\mathbf{r}) r_j + \tfrac{1}{4} \sum_{\mathbf{r}} s_{ijkl} A_{ij}(\mathbf{r}) r_k r_l u_{ik}^0 u_{jl}^0 \tag{13.5.51}$$

Using the definition (13.3.1) in Eq. (13.5.51), we have

$$\Omega(0) = v_0 \sigma_{ij}^0 u_{ij}^0 + \tfrac{1}{4} \sum_{\mathbf{r}} s_{ijkl} A_{ij}(\mathbf{r}) r_k r_l u_{ik}^0 u_{jl}^0 \tag{13.5.52}$$

or

$$\Omega(0) = v_0 \lambda_{ijkl} u_{ij}^0 u_{kl}^0 + \tfrac{1}{4} \sum_{\mathbf{r}} s_{ijkl} A_{ij}(\mathbf{r}) r_k r_l u_{ik}^0 u_{jl}^0 \tag{13.5.53}$$

As has been shown, however, the correct limit transition to the continuum elasticity results is ensured if $\Omega(0)$ is equal to $v_0 \lambda_{ijkl} u_{ij}^0 u_{kl}^0$ rather than to the value (13.5.53). In other words, the limit transition to the continuum at any direction of $\mathbf{k}$ in Cook-de Fontaine's theory might only be provided if the second term in Eq. (13.5.53) is identically equal to zero. This is not the case, however.

With regard to the second term, the summand in Eq. (13.5.46), it correctly describes the local displacement relaxation energy. This term was first obtained in (263).

The recent review by de Fontaine (269) does not elucidate the situation since the incorrect elastic term $\Omega(\mathbf{k})$ in Eq. (13.5.46) is just included into the Fourier transform of the direct "chemical" pairwise interaction energy without these two physically distinct contributions to this Fourier transform being specified.

13.5.5. "Elastic Energy Paradox"

According to the Eshelby treatment of the sphere-in-hole model of a solid solution, its elastic free energy is described by the equation

$$E_{\text{elast}} = V\mu \frac{1+\sigma_1}{1-\sigma_1} \left(\frac{da}{a d\bar{c}} \right)^2 \bar{c}(1-\bar{c}) \tag{13.5.54}$$

which was rederived in this section [see Eq. (13.5.33)]. Since the second derivative of the elastic energy (13.5.54),

$$\frac{d^2E_{\text{elast}}}{d\bar{c}^2}=-2V\mu\frac{1+\sigma_1}{1-\sigma_1}\left(\frac{da}{ad\bar{c}}\right)^2$$

is a negative value, the elastic energy contribution to the free energy promotes the decomposition.

Yet, according to Cahn's treatment of the spinodal decomposition (43), the elastic energy contribution leads to just the opposite result. It stabilizes the homogeneous solid solution, since the function $Y(\mathbf{n})$ entering Cahn's equation (13.5.42) is positive, and thus the formation of any macroscopic heterogeneity (the formation of the spectrum of nonnegative values $|\Delta\tilde{n}(\mathbf{k})|^2$) yields an elastic energy increase.

This contradiction has been pointed out by Cahn (277) and was later named "the elastic energy paradox" (269). The microscopic elastic theory formulated in this chapter enables us to readily understand this contradiction between the theories based on the same model. One should only take into account three important circumstances.

1. As shown above, Cahn's equation (13.5.42) should be revised and replaced with Eq. (13.5.39).

2. As shown in Eq. (13.5.32), the total elastic energy of an elastically isotropic solution is configuration-independent ($\Delta E_{\text{elast}}^{\text{heter}}=0$) and can thus be affected only by a change in the total number of solute atoms in the crystal. The latter cannot, however, be realized since the decomposition reaction occurs within the same crystal body. Nevertheless, we may always imagine it in terms of a situation where the crystal is cut into two pieces. This procedure ensures vanishing of elastic energy that otherwise would have been generated by the crystal lattice mismatch between two coherently adjacent pieces of the crystal with different solute atom contents. Therefore the cutting reduces the elastic energy (13.5.54) and thus provides an additional driving force for the decomposition. This separation of the crystal into two pieces is, as a matter of fact, equivalent to the breaking of coherency between crystal lattices which results in stress relaxation. Therefore Eshelby's configuration-independent elastic energy term really promotes the decomposition. But this only occurs if the decomposition proceeds according to the incoherent (stable) diagram involving the formation of stress-free incoherent phases.

3. The formation of a smooth macroscopic concentration heterogeniety which is considered in Cahn's theory of the spinodal decomposition, on the other hand, does not break the crystal lattice coherency because it is reduced to an atomic redistribution over the crystal lattice sites and thus does not violate the crystal lattice site systematics (topology). In other words, the formation of the concentration heterogeneity is energetically an absolutely different process from that considered above. This process is related to the coherent decomposition that occurs at the stressed state without the violation of the coherency between the matrix and the phase transformation product. As shown in Section 13.5.3 [see eqn. (13.5.34)], the formation of a smooth concentration heterogeneity along the elastically "soft" direction results in a decrease rather than an increase in elastic energy, as assumed by Cahn.*

Summing up this discussion, one may see that the two cases considered above are related to two physically different situations and thus cannot be directly compared. But in both cases the elastic energy contribution to free energy promotes the decomposition reaction.

13.5.6. Limit Transition to the Theory of Coherent Inclusions

Clustering of solute atoms can form coherent inclusions in the parent phase matrix which have been considered in Chapters 7 and 8. We shall demonstrate that the agglomeration of solute atoms to produce coherent inclusions ensures the limit transition of the microscopic elastic theory to the theory of coherent inclusions (134, 136) considered in Section 8.1.

An atomic distribution of solute atoms $c(\mathbf{r})$ that describes a coherent inclusion may be represented as follows:

$$c(\mathbf{r})=\tilde{\theta}(\mathbf{r})=\begin{cases}1 & \text{if } \mathbf{r} \text{ is within the inclusion}\\ 0 & \text{otherwise}\end{cases} \tag{13.5.55}$$

Then, by definition,

$$\bar{c}=\frac{1}{N}\sum_{\mathbf{r}} c(\mathbf{r})=\frac{1}{N}\sum_{\mathbf{r}} \tilde{\theta}(\mathbf{r})=\langle\tilde{\theta}(\mathbf{r})\rangle \tag{13.5.56}$$

and

$$\Delta c(\mathbf{r})=c(\mathbf{r})-\bar{c}=\tilde{\theta}(\mathbf{r})-\langle\tilde{\theta}(\mathbf{r})\rangle \tag{23.5.57}$$

Since

$$\sum_{\mathbf{r}} \langle\tilde{\theta}(\mathbf{r})\rangle e^{-i\mathbf{k}\mathbf{r}}=\langle\tilde{\theta}(\mathbf{r})\rangle \sum_{\mathbf{r}} e^{-i\mathbf{k}\mathbf{r}}=\langle\tilde{\theta}(\mathbf{r})\rangle N\delta_{\mathbf{k}0}=\delta_{\mathbf{k}0}\sum_{\mathbf{r}} \tilde{\theta}(\mathbf{r})$$

The Fourier transform of Eq. (13.5.57) yields

$$\Delta\tilde{c}(\mathbf{k})=\sum_{\mathbf{r}} \Delta c(\mathbf{r})e^{-i\mathbf{k}\mathbf{r}}=\sum_{\mathbf{r}} (\tilde{\theta}(\mathbf{r})-\langle\tilde{\theta}(\mathbf{r})\rangle)e^{-i\mathbf{k}\mathbf{r}}=\begin{cases}(1/v_0)\theta(\mathbf{k}) & \text{if } \mathbf{k}\neq 0\\ 0 & \text{if } \mathbf{k}=0\end{cases} \tag{13.5.58}$$

where

$$\theta(\mathbf{k})=v_0\sum_{\mathbf{r}} \tilde{\theta}(\mathbf{r})e^{-i\mathbf{k}\mathbf{r}}\approx\int \tilde{\theta}(\mathbf{r})e^{-i\mathbf{k}\mathbf{r}}d^3r \tag{13.5.59}$$

*In the case of the elastically isotropic solid solution, the formation of a smooth coherent macroscopic concentration heterogeneity does not in the least affect the elastic energy [see Eq. (13.5.32b)].

Substituting Eq. (13.5.58) into (13.5.6) results in

$$E_{\text{elast}} = \frac{1}{2N} \sum_{\mathbf{k}}' \left[v_0 \lambda_{ijkl} u_{ij}^0 u_{kl}^0 - F_i(\mathbf{k}) G_{ij}(\mathbf{k}) F_j^*(\mathbf{k}) \right] \frac{|\theta(\mathbf{k})|^2}{v_0^2} \tag{13.5.60}$$

where the prime in the summation over $\mathbf{k}$ implies deletion of the $\mathbf{k}=0$ term.

Replacing the summation over N quasi-continuum points $\mathbf{k}$ of the first Brillouin zone with integration according to the usual rule

$$\frac{1}{N} \sum_{\mathbf{k}} (\cdots) = v_0 \int \frac{d^3k}{(2\pi)^3} (\cdots)$$

we have

$$E_{\text{elast}} = \frac{v_0}{2} ⨍ \frac{d^3k}{(2\pi)^3} \left[v_0 \lambda_{ijkl} u_{ij}^0 u_{kl}^0 - F_i(\mathbf{k}) G_{ij}(\mathbf{k}) F_j^*(\mathbf{k}) \right] \frac{|\theta(\mathbf{k})|^2}{v_0^2} \tag{13.5.61}$$

The integral (13.5.61) is taken over the first Brillouin zone of the crystal; the sign $⨍$ has the meaning that the volume, $(2\pi)^3/V$, about $\mathbf{k}=0$ is to be excluded from the integration. When the total volume of the crystal, V, tends to infinity, this exclusion defines the "principal value" of the integral. To complete the identification of Eq. (13.5.61) with Eq. (8.1.1) for a coherent precipitate, we shall note that the shape function, $|\theta(\mathbf{k})|^2$, of an inclusion of volume, V_{incl} ($V_{\text{incl}} \ll V$), will have a significant magnitude within a $\mathbf{k}$-space volume of approximately $\Delta^3\mathbf{k} \approx (2\pi)^3/V_{\text{incl}}$ about $\mathbf{k}=0$, that is, within the range

$$\Delta k \approx 2\pi / L_{\text{incl}} \tag{13.5.62}$$

where L_{incl} is the typical size of the inclusion. When the size L_{incl} is much larger than the crystal lattice parameter, a (this is the case when the inclusion can be considered macroscopic), we may rewrite the relation (13.5.62) in the following form:

$$\frac{\Delta k a}{2\pi} \approx \frac{2\pi}{L_{\text{incl}}} \frac{a}{2\pi} = \frac{a}{L_{\text{incl}}} \ll 1 \tag{13.5.63}$$

The inequality (13.5.63) demonstrates that the main contribution to the integral (13.5.61) is provided by the integration over small $\mathbf{k}$ whose value ensures the transition to the continuum elasticity. The inequality (13.5.63) justifies the replacement of the functions $\mathbf{F}(\mathbf{k})$ and $G_{ij}(\mathbf{k})$ with their long-wavelength limits (13.5.9) and (13.5.11). Using the long-wavelength limits (13.5.9) and (13.5.11) in (13.5.61), we have

$$E_{\text{elast}} = \frac{1}{2} ⨍ \frac{d^3k}{(2\pi)^3} \left[\lambda_{ijkl} u_{ij}^0 u_{kl}^0 - n_i \sigma_{ij}^0 \Omega_{jk}(\mathbf{n}) \sigma_{kl}^0 n_l \right] |\theta(\mathbf{k})|^2 \tag{13.5.64}$$

This equation is identical to the solution (8.1.1) for the elastic energy of a coherent inclusion in an elastic continuum.

13.6. ROLE OF ELASTIC ENERGY AND VACANCIES IN THERMODYNAMICS OF STABLE SEGREGATIONS: K-STATE

The elastic energy contribution to the thermodynamics of a solid solution enables us to explain one more interesting phenomenon, the formation of the so-called K-state in some single-phase alloys (279). The K-state is signified by a considerable drop in electroconductivity during low-temperature tempering of alloys which were previously quenched from high temperatures and a plastically deformed state. Recently, x-ray and electron microscopic studies have demonstrated that the K-state is associated with the formation of extremely fine particles whose composition differs from that of the matrix (280–285). The major peculiarity that makes these particles different from new phase precipitates arising in the two-phase field of the equilibrium diagram is that these particles do not grow during tempering. The K-state has been detected in the single-phase field of the Cu-Al (280), Cu-Zn (281), Fe-Al (282, 283, 286), and Ni-Mo (284, 285) equilibrium diagrams. The recent x-ray single-crystal study of the short-range order kinetics in Fe-Al alloys seems to confirm the electron microscopic observations (287). A nonmonotonic dependence of the diffuse intensity associated with short-range order over time has been reported. At the first stage of annealing the usual increase in diffuse intensity was observed, caused by the short-range order establishing an equilibrium. At the second stage, however, the diffuse intensity decreased. The latter effect was related to the formation of concentration segregations and the corresponding increase in intensity around the fundamental reflections at the expense of a decrease in the short-range order intensity (287). The nonmonotonic dependence of short-range order diffuse intensity on time was also observed in the earlier x-ray studies of polycrystalline Ni-Pt (288, 289), Cu-Al (290, 291), Ni-Si (292), Fe-Al (293) alloys.

The electron microscopic observation of the K-state in the single-phase field of the Fe-Al diagram, which elucidates the mechanism behind the formation of stable segregations, should be singled out as being of particular importance (294, 286). It was reported that the formation of stable segregations (K-state) in Fe-Al alloys depends on the thermal "history" of the sample. Segregations arise only in the cases where upon quenching the alloy contains a large number of excess vacancies resulting from a high-temperature state. Otherwise, if an alloy is cooled slowly to reduce the amount of vacancies to an equilibrium value, when excess vacancies are absent, the K-effect is not observed. The simple estimation of the diffusion kinetics shows that the role of excess vacancies cannot be reduced to the conventional acceleration of diffusion. Indeed, even when reduced to an equilibrium number, the vacancies would maintain a sufficiently high diffusion rate to provide segregation formation during $\sim 10^2$ sec [diffusivity $D_{Al} \sim \exp\left(-\dfrac{44{,}000\text{ cal}}{2T^0}\right)$ in cm^2/sec units, the typical segregation size $l \sim 30$ Å, $T = 623°$ K].

Another argument against the idea of vacancies as accelerators of diffusion can be found if we consider the Fe-Al alloy with $c_{Al} > c_{Al}^0 = 19.6$ at.% Al (at 300°C). These alloys undergo fast decomposition and, in contrast to alloys with lower Al content, demonstrate insensitivity to the "thermal history" of the sample which, in fact, determines the amount of excess vacancies.

Summing up the foregoing, we arrive at the important conclusion that vacancies play the key role in the thermodynamics of the K-state formation rather than kinetics. Following the theoretical consideration presented in (295), the origin of stable nongrowing segregations in the single-phase field of the equilibrium diagram can be explained if the elastic stress generated by the mismatch between the crystal lattices of the segregation and the matrix and the role of vacancies in stress relaxation are taken into account simultaneously. As has been shown in Section 13.5.2 [Eq. (13.5.33)], Eshelby's sphere-in-hole model of a cubic elastically isotropic solid solution yields the elastic energy in the form

$$E_{\text{elast}} = V\mu \frac{1+\sigma_1}{1-\sigma_1}\left(\frac{da}{ad\bar{c}}\right)^2 \bar{c}(1-\bar{c}) \tag{13.6.1}$$

The elastic energy (13.6.1) depends on the total number of solute atoms rather than on their mutual arrangement. Therefore the elastic energy of an isotropic solid solution cannot be affected by atomic redistributions over crystal lattice sites, including those that result in the decomposition. The decomposition reaction that occurs due to atomic redistribution over a crystal lattice site maintains, by definition, the coherency between the crystal lattices of new phase precipitates and the matrix, and thus is described by a coherent equilibrium diagram. Since the formation of a two-phase coherent mixture is accompanied by internal stress caused by the crystal lattice mismatch, the coherent diagram is always metastable.

In other words, the elastic energy contribution cannot affect the decomposition thermodynamics if the decomposition occurs according to the coherent metastable equilibrium diagram. It is determined by the chemical free energy only. If, however, the decomposition results in the formation of incoherent new phase precipitates, internal stress is removed and the decomposition occurs according to a stable incoherent equilibrium diagram.

To calculate the incoherent diagram, we should consider the combination of the chemical and elastic-free energies:

$$F_{\text{inch}} = \int_{(V)} \left[f_{\text{chem}}(\bar{c}(\mathbf{r})) + \mu \frac{1+\sigma_1}{1-\sigma_1}\left(\frac{da}{ad\bar{c}}\right)^2 \bar{c}(\mathbf{r})(1-\bar{c}(\mathbf{r})) \right] d^3r \tag{13.6.2}$$

where the concentration $\bar{c}(\mathbf{r})$ of solute atoms assumes two values

$$\bar{c}(\mathbf{r}) = \begin{cases} c_{\alpha'}^{\text{inch}} & \text{if } \mathbf{r} \text{ is within the parent phase } \alpha' \\ c_{\alpha''} & \text{if } \mathbf{r} \text{ is within the precipitate phase } \alpha'' \end{cases} \tag{13.6.3}$$

It follows from Eq. (13.6.2) that the elastic-free energy promotes incoherent decomposition. Both types of diagrams, metastable coherent and stable incoherent, are depicted in Fig. 111. The two-phase field of the incoherent diagram is wider than that of the coherent one and contains the coherent diagram two-phase field within itself.

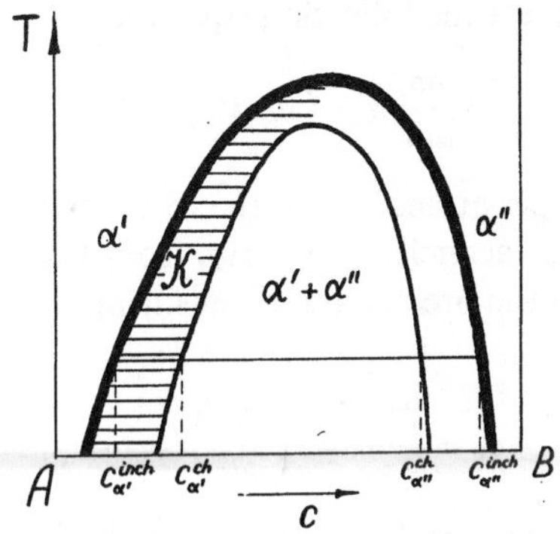

Figure 111. Metastable coherent and stable incoherent diagrams with a miscibility gap. The shaded field is related to the K-state. The solid line is a solvus for the incoherent stable diagram. The thin line is a solvus for the metastable coherent diagram. The atomic radius of the B-atom is larger than that of A-atoms.

An interesting situation arises when a representative point of the alloy is within the shaded field in Fig. 111—when it is between two nearest solvus curves related to the coherent and incoherent diagrams. Then the alloy is within the single-phase field of the metastable coherent diagram and within the two-phase field of the stable incoherent diagram. In this situation the decomposition can occur only if a mechanism exists that breaks the crystal lattice coherency between the phases, thus allowing the transition from the coherent to the stable incoherent diagram. Stress relaxation caused by breaking the coherency can be provided by a vacancy "pump" at the interphase boundary. Let us consider the case where the atomic diameter of a solute atom B is larger than that of a solvent atom A. The formation of a segregation of B atoms then results in a situation where the volume of the segregation (new phase precipitate) proves to be larger than the volume of the appropriate hole in the matrix. The elastic accommodation of this difference in volumes at the coherent fitting of the crystal lattices generates elastic stress. If, however, we bring the vacancies up to the surface layer of the precipitate and thus annihilate its excess volume, the interphase boundary becomes completely incoherent, but the system reaches a stress-free state. In this situation the alloy will be described by an incoherent stable diagram.

The excess volume of the precipitate formed due to the solute atom segregation is

$$\Delta V = 3\,\frac{da}{a d\bar{c}}\,(c_{\alpha''} - c_{\alpha'})V_{\text{prec}} \tag{13.6.4}$$

where $c_{\alpha''}$ and $c_{\alpha'}$ are the atomic fractions of the solute element in the precipitate and matrix phases, respectively, V_{prec} is the precipitate volume.

To remove the excess volume ΔV by means of vacancies, we should bring up the N_v vacancies to the precipitate surface layer. In doing so, the number of vacancies, N_v, should meet the following equation

$$3\frac{da}{ad\bar{c}}(c_{\alpha''}-c_{\alpha'})V_{\text{prec}}=N_v v_0 \tag{13.6.5}$$

where v_0 is the volume per atom, $N_v v_0$ is the removed volume of the precipitate after the delivery of N_v vacancies. It follows from Eq. (13.6.5) that the atomic fraction of vacancies in the precipitate, c_v^0, ensuring the complete stress relaxation is

$$c_v^0=\frac{N_v}{N_{\text{prec}}}=3\left(\frac{da}{ad\bar{c}}\right)(c_{\alpha''}-c_{\alpha'}) \tag{13.6.6}$$

where $N_{\text{prec}}=V_{\text{prec}}/v_0$ is the total number of atoms in the precipitate.

The alternative mechanism of stress relaxation is the formation of vacancy dislocation loops enveloping the precipitate. This mechanism also results in the equation

$$c_v^0=\xi\left(\frac{da}{ad\bar{c}}\right)(c_{\alpha''}-c_{\alpha'}) \tag{13.6.7}$$

similar to Eq. (13.6.6) where ξ is the dimensionless constant of the order of unity.

The typical value $da/(ad\bar{c})$ in a substitutional solution is of about 0.01, $(c_{\alpha''}-c_{\alpha'})$ ~0.1. Using these values in Eq. (13.6.7), we can estimate

$$c_v^0\sim 10^{-3}$$

If the representative point of the solution is within the shaded area in Fig. 111, the decomposition can occur only in accordance with the incoherent diagram since the point lies within the single-phase field of the coherent diagram. Therefore the decomposition can only be realized if all precipitates formed are in a stress-free state, that is, if their excess volume is removed with vacancies. In this situation the total volume of the precipitate phase will be limited by the total number of excess vacancies rather than by the usual lever rule: the transformation will cease when the excess vacancy supply is exhausted.

The vacancy balance equation reads

$$c_v^0 N_{\text{prec}}=N\bar{c}_v \tag{13.6.8}$$

where N_{prec} and N are the total number of atoms in all precipitates and the total number of atoms in the alloy, respectively, $\bar{c}_v$ is the atomic fraction of excess vacancies in the initial homogeneous solution. The left-hand side of Eq. (13.6.8) is the total number of vacancies consumed by all the precipitates formed. The right-hand side of Eq. (13.6.8) is the total number of excess vacancies in the initial solution. Eq. (13.6.8) yields

$$w=\frac{N_{\text{prec}}}{N}=\frac{\bar{c}_v}{c_v^0} \tag{13.6.9}$$

where w is the volume fraction of the precipitate α'' phase.

Therefore, if within the field the representative point of the alloy is between the solubility curves of the coherent and incoherent equilibrium diagrams, the volume fraction of the precipitate phase depends on the concentration of excess vacancies (on the "thermal history" of the sample) and the crystal lattice mismatch parameter $da/(ad\bar{c})$ [see Eq. (13.6.6)] rather than on the location of the representative point in the incoherent diagram. An increase in the volume fraction, w, up to its equilibrium value in the incoherent diagram, would as determined by the lever rule, take an extremely long time since it requires the delivery of vacancies from the surface and other vacancy sources to each point of the crystal. This stage of the decomposition is not detected by structural methods. It should be noted that the concentration gap corresponding to the shaded area in Fig. 111 (the K-state field) depends only on the elastic energy contribution. The gap is wider, the larger the crystal lattice mismatch $da/(ad\bar{c}) \times (c_{\alpha''} - c_{\alpha'})$, that is, the larger the difference between the atomic diameters of solute and solvent atoms. Therefore the K-state should be expected in alloys whose solute and solvent atoms have a large difference in atomic diameters.

Let us estimate the width of the gap between the solubility limits of the coherent and incoherent equilibrium diagrams corresponding to the K-state range. Eq. (13.6.2) in this case may be written in the form

$$F_{\text{inch}} = \int_{(V)} \left\{ -\frac{1}{2}\left[V^{\text{chem}}(0) - 2\mu v_0 \frac{1+\sigma_1}{1-\sigma_1}\left(\frac{da}{ad\bar{c}}\right)^2 \right] \bar{c}(\mathbf{r})(1-\bar{c}(\mathbf{r})) + \kappa T[\bar{c}(\mathbf{r}) \ln \bar{c}(\mathbf{r}) + (1-\bar{c}(\mathbf{r})) \ln (1-\bar{c}(\mathbf{r}))] \right\} \frac{dV}{v_0} \tag{13.6.10}$$

where Eq. (4.1.13) is taken into account.

It follows from Eq. (13.6.10) that we can introduce a new energy coefficient $V(0)$,

$$V(0) = V^{\text{chem}}(0) - 2\mu v_0 \frac{1+\sigma_1}{1-\sigma_1}\left(\frac{da}{ad\bar{c}}\right)^2 \tag{13.6.11}$$

which enters Eq. (4.1.18) describing the solubility limit. The elastic energy term in Eq. (13.6.11) ensures the shift $\Delta c(\mathbf{r})$ of the solubility limit of the coherent diagram, c_{ch}, to the new value, c_{inch}, corresponding to the incoherent diagram. The shift will also be described by Eq. (4.1.18) where the value c is substituted by $c_{\text{inch}} = c_{\text{ch}} + \Delta c$:

$$\frac{-V^{\text{chem}}(0) + 2\mu v_0 \dfrac{1+\sigma_1}{1-\sigma_1}\left(\dfrac{da}{ad\bar{c}}\right)^2}{\kappa T} = \frac{1}{c_{\text{ch}} + \Delta c - \frac{1}{2}} \ln \frac{c_{\text{ch}} + \Delta c}{1 - c_{\text{ch}} - \Delta c} \tag{13.6.12}$$

The power series expansion of (13.6.12) in Δc and retention of the first-order term yields

$$-\frac{V^{\text{chem}}(0)}{\kappa T} + \frac{2\mu v_0}{\kappa T}\frac{1+\sigma_1}{1-\sigma_1}\left(\frac{da}{ad\bar{c}}\right)^2 \cong \frac{\ln(c_{\text{ch}}/(1-c_{\text{ch}}))}{c_{\text{ch}} - \frac{1}{2}} + \frac{1}{(c_{\text{ch}} - \frac{1}{2})}\left[\frac{1}{c_{\text{ch}}(1-c_{\text{ch}})} - \frac{1}{c_{\text{ch}} - \frac{1}{2}} \ln \frac{c_{\text{ch}}}{1-c_{\text{ch}}}\right]\Delta c \tag{13.6.13}$$

Since, by definition,

$$-\frac{V^{\text{chem}}(0)}{\kappa T}=\frac{1}{c_{\text{ch}}-\frac{1}{2}}\ln\frac{c_{\text{ch}}}{1-c_{\text{ch}}}$$

we may rewrite Eq. (13.6.13):

$$\frac{2\mu v_0}{\kappa T}\frac{1+\sigma_1}{1-\sigma_1}\left(\frac{da}{ad\bar{c}}\right)^2=-\left[\frac{1}{c_{\text{ch}}-\frac{1}{2}}\ln\frac{c_{\text{ch}}}{1-c_{\text{ch}}}-\frac{1}{c_{\text{ch}}(1-c_{\text{ch}})}\right]\frac{1}{c_{\text{ch}}-\frac{1}{2}}\Delta c$$

or

$$c_{\text{inch}}=c_{\text{ch}}-\frac{2\mu v_0}{\kappa T}\frac{1+\sigma_1}{1-\sigma_1}\left(\frac{da}{ad\bar{c}}\right)^2(c_{\text{ch}}-\tfrac{1}{2})\left[\frac{1}{c_{\text{ch}}-\frac{1}{2}}\ln\frac{c_{\text{ch}}}{1-c_{\text{ch}}}-\frac{1}{c_{\text{ch}}(1-c_{\text{ch}})}\right]^{-1} \tag{13.6.14}$$

Eq. (13.6.14) enables us to estimate the K-state range in the case of Fe-Al alloys. This estimation is rather rough since, by definition, Eq. (13.6.14) is only valid for systems with a miscibility gap, which is not the case with the Fe-Al diagram. Using the values $v_0=a_0^3/2=11.81\times10^{-24}$, $\mu\sim10^{12}$ dyne/cm^2, $da/(ad\bar{c})\simeq0.02$, $\sigma_1\sim0.3$, $T=573°\text{K}$, $c_{\text{ch}}\approx0.20$, $\kappa=1.38\times10^{16}$ erg/grad in Eq. (13.6.14), we have $c^{\text{inch}}\approx0.09=9$ at.% Al.

The approximate location of the K-state field in the Fe-Al diagram determined by the electron microscopic method was found in (283). According to these data $c_{\text{inch}}\sim0.10$ which, considering the rough nature of the theoretical estimation, is in an agreement with the value $c_{\text{inch}}\sim0.09$.

The foregoing discussion refers to the case where a solute atom diameter is larger than that of a solvent atom. The K-state field in the equilibrium diagram thus should adjoin the coherent diagram solvus which lies at the side of the diagram near the component with the smaller atomic diameter (only in this case vacancies remove the excess volume of precipitates). Otherwise, the volume of precipitates is smaller than the appropriate volume of their holes in the matrix, and thus the stress relaxation can occur only by means of a far less efficient mechanism, by the delivery of interstitial atoms to the surface layer of precipitates. The fact that the K-state was found in Fe-Al (based on Fe), Cu-Al (based on Cu), and Ni-Mo (based on Ni) alloys confirms these conclusions since the atomic diameters of Fe, Cu, and Ni are smaller than the diameters of Al and Mo.

Concluding this section, we shall make one more estimation based on the concept of the K-state. Usually, high-temperature quenching preceding the K-state annealing occurs from 800°C. The energy of the vacancy formation in Fe-Al is ~18 kcal/mole (296). Therefore the equilibrium fraction of vacancies inherited from the 800°C state is

$$\bar{c}_v=\exp\left(-\frac{18{,}000}{2T}\right)=\exp\left(-\frac{18{,}000}{2\times1073}\right)\approx2\times10^{-4}$$

The vacancy concentration in precipitates ensuring the complete stress relaxation is given by Eq. (13.6.6). At $c_{\alpha''} \approx 0.3$, $c_{\alpha'} \sim 0.1$ and $da/(ad\bar{c}) \simeq 0.02$, it gives

$$c_v^0 = 3 \times 0.02(0.3 - 0.1) = 1.2 \times 10^{-2}$$

Making use of Eq. (13.6.9), we can estimate the volume fraction of the precipitate phase:

$$w = \frac{\bar{c}_v}{c_v^0} = \frac{2 \times 10^{-4}}{1.2 \times 10^{-2}} \approx 1.6 \times 10^{-2} \qquad (13.6.15)$$

Yet direct electron microscopic observations have demonstrated that the diameter of precipitates is about 50 Å, the precipitate density is $\rho \approx 10^{17}$ to 10^{18} cm^{-3} (283). These data enable us to determine the volume fraction of the precipitate phase. It is given by the equation

$$w = \rho V_{\mathrm{prec}} = 10^{18} \frac{4\pi}{3} \left(\frac{50 \times 10^{-8}}{2} \right)^3 \simeq 6.5 \times 10^{-2} \qquad (13.6.16)$$

The latter estimation is in agreement with the estimation (13.6.15) based on the above-formulated concept of the K-state.

14

APPLICATION OF MICROSCOPIC ELASTIC THEORY TO THERMODYNAMICS OF PHASE TRANSFORMATIONS

14.1. SCREENING OF PAIRWISE INTERACTION IN SOLID SOLUTIONS

As is known, the Coulomb potential of fixed electrical charges immersed in a gas of charge carriers proves to be screened. The screening is caused by the correlation effect—by the formation of a field of opposite sign charge carriers around the external charge. The resulting screened Coulomb potential,

$$\Phi(\mathbf{r})=\frac{e_1}{r}\exp\left(-\frac{r}{r_D}\right) \tag{14.1.1}$$

differs from the unscreened conventional electrostatic potential

$$\Phi_0(r)=\frac{e_1}{r}$$

where e_1 is the charge introduced, r_D is the Debye screening radius, r the distance. This phenomenon is known as the Debye screening. Physically, it is identical to the short-range order effect.

We shall demonstrate that the screening of pairwise interaction is not the prerogative of the Coulomb charge carrier system. The Debye screening is a particular case of a widespread phenomenon realized in any nonideal gas and, specifically, in a nonideal solid solution.

To calculate the screening of the externally applied potential, $W_0(\mathbf{r})$, by solute atoms, we should determine the perturbation effect caused by this potential in the solute atom distribution around the source of the potential and combine the additional potential generated by this perturbation with the original non-screened potential, $W_0(\mathbf{r})$. The perturbation effect of the potential, $W_0(\mathbf{r})$, is reduced to the formation of a field of solute atoms around the source of the potential. According to the Le Chatelier-Braun principle the response of the system to the applied external potential is a counteraction to compensate this potential. In other words, this reaction is the above-discussed screening of the external potential. These conclusions can be generalized further: the introduction of an external potential into any polarizable medium should result in the screening of the potential.

Now we shall return to the calculation of the perturbation produced by an external source. We shall proceed from Eqn. (3.4.4). In the mean-field approximation the resultant potential at point $\mathbf{r}$ is the sum of the "bare" potential $W_0(\mathbf{r})$ generated by the external source and the potential (3.4.6) produced by all atoms of the solid solution:

$$\Phi(\mathbf{r})=W_0(\mathbf{r})+\sum_{\mathbf{r}'} W(\mathbf{r}-\mathbf{r}')n(\mathbf{r}') \tag{14.1.2}$$

Substitution of Eq. (14.1.2) into Eq. (3.4.4) yields

$$n(\mathbf{r})=\left[\exp\left(\frac{-\mu+W_0(\mathbf{r})+\sum_{\mathbf{r}'} W(\mathbf{r}-\mathbf{r}')n(\mathbf{r}')}{\kappa T}\right)+1\right]^{-1} \tag{14.1.3}$$

Let us assume that the initial solid solution is stable in the disordered state. Then

$$n_0(\mathbf{r})=c \tag{14.1.4}$$

where $n_0(\mathbf{r})$ is the atomic distribution of the initial solid solution, c is the atomic fraction of the solute element. The mean-field equation (3.6.2) is at $n_0(\mathbf{r})=c$

$$c=\left[\exp\left(\frac{-\mu+V(0)c}{\kappa T}\right)+1\right]^{-1} \tag{14.1.5}$$

where $V(0)=\sum_{\mathbf{r}} W(\mathbf{r})$.

If

$$n(\mathbf{r})=c+\delta n(\mathbf{r}) \tag{14.1.6}$$

where $\delta n(\mathbf{r})$ is the perturbation of the distribution (14.1.4) produced by the external potential $W_0(\mathbf{r})$, Eq. (14.1.3) may be rewritten

$$c+\delta n(\mathbf{r})=\left[\exp\left(\frac{-\mu+W_0(\mathbf{r})+V(0)c+\sum_{\mathbf{r}'} W(\mathbf{r}-\mathbf{r}')\delta n(\mathbf{r}')}{\kappa T}\right)+1\right]^{-1} \tag{14.1.7}$$

The expansion of the right-hand side of Eq. (14.1.7) in powers of

$[W_0(\mathbf{r})+\sum_{\mathbf{r}'} W(\mathbf{r}-\mathbf{r}')\delta n(\mathbf{r}')]/\kappa T$ and retention of the first-order term yields

$$c+\delta n(\mathbf{r})\cong c-\frac{c(1-c)}{\kappa T}\left[W_0(\mathbf{r})+\sum_{\mathbf{r}'} W(\mathbf{r}-\mathbf{r}')\delta n(\mathbf{r}')\right] \tag{14.1.8}$$

where Eq. (14.1.5) has been used. Eq. (14.1.8) corresponds to the approximation in which the externally applied potential $W_0(\mathbf{r})$ is considerably lower than the total potential generated by all atoms. This condition can be expressed by the following relation

$$\frac{\max|W_0(\mathbf{r})|}{\max|V(\mathbf{k})|}\ll 1 \tag{14.1.9}$$

Eq. (14.1.8) may be rewritten

$$\delta n(\mathbf{r})=-\frac{c(1-c)}{\kappa T}\left[W_0(\mathbf{r})+\sum_{\mathbf{r}'} W(\mathbf{r}-\mathbf{r}')\delta n(\mathbf{r}')\right] \tag{14.1.10}$$

The Fourier transform of Eq. (14.1.10) yields

$$\delta\tilde{n}(\mathbf{k})=-\frac{c(1-c)}{\kappa T}[V_0(\mathbf{k})+V(\mathbf{k})\delta n(\mathbf{k})] \tag{14.1.11}$$

where

$$V_0(\mathbf{k})=\sum_{\mathbf{r}} W_0(\mathbf{r})e^{-i\mathbf{k}\mathbf{r}} \tag{14.1.12a}$$

$$V(\mathbf{k})=\sum_{\mathbf{r}} W(\mathbf{r})e^{-i\mathbf{k}\mathbf{r}} \tag{14.1.12b}$$

The solution of Eq. (14.1.11) results in

$$\delta n(\mathbf{k})=-\frac{[c(1-c)/\kappa T]V_0(\mathbf{k})}{1+[c(1-c)/\kappa T]V(\mathbf{k})} \tag{14.1.13}$$

The resultant screened potential at point $\mathbf{r}$ produced by the externally applied potential, $W_0(\mathbf{r})$, and the combined potential generated by the perturbation of the solute atom distribution, $\delta n(\mathbf{r})$, is

$$W_{\text{scr}}(\mathbf{r})=W_0(\mathbf{r})+\sum_{\mathbf{r}'} W(\mathbf{r}-\mathbf{r}')\delta n(\mathbf{r}') \tag{14.1.14}$$

Its Fourier transform is

$$V_{\text{scr}}(\mathbf{k})=V_0(\mathbf{k})+V(\mathbf{k})\delta\tilde{n}(\mathbf{k}) \tag{14.1.15}$$

where

$$V_{\text{scr}}(\mathbf{k})=\sum_{\mathbf{r}} W_{\text{scr}}(\mathbf{r})e^{-i\mathbf{k}\mathbf{r}} \tag{14.1.16}$$

Substitution of Eq. (14.1.13) into (14.1.15) yields the final equation for the

Fourier transform of the resultant (screened) potential:

$$V_{scr}(\mathbf{k}) = V_0(\mathbf{k}) - \frac{[c(1-c)/\kappa T]V_0(\mathbf{k})V(\mathbf{k})}{1+[c(1-c)/\kappa T]V(\mathbf{k})}$$

or

$$V_{scr}(\mathbf{k}) = \frac{V_0(\mathbf{k})}{1+[c(1-c)/\kappa T]V(\mathbf{k})} \tag{14.1.17}$$

Eq. (14.1.17) is quite general. It yields the classic Debye screening potential as a particular case when $V_0(\mathbf{k})$ and $V(\mathbf{k})$ are the Fourier transforms of the Coulomb potentials, $W_0(\mathbf{r}) = \Phi_0(r) = e_1/r$ and $W(\mathbf{r}) = e/r$, respectively. Then*

$$V_0(\mathbf{k}) = \frac{4\pi}{v_0}\frac{e_1 e}{k^2}, \quad V(\mathbf{k}) = \frac{4\pi}{v_0}\frac{e^2}{k^2} \tag{14.1.18}$$

Substitution of Eqs. (14.1.18) into (14.1.17) yields

$$V_{scr}(\mathbf{k}) = \frac{4\pi}{v_0}\frac{e_1 e}{k^2 + [4\pi c(1-c)/(v_0 \kappa T)]e^2} \tag{14.1.19}$$

The back Fourier transform of Eq. (14.1.19) is

$$W_{scr}(\mathbf{r}) = \frac{1}{N}\sum_{\mathbf{k}} V_{scr}(\mathbf{k})e^{i\mathbf{kr}} \approx v_0/(2\pi)^3 \int V_{scr}(\mathbf{k})e^{i\mathbf{kr}}d^3k = \frac{e_1 e}{r}\exp\left(-\frac{r}{r_D}\right) \tag{14.1.20}$$

where

$$r_D = \sqrt{\frac{v_0 \kappa T}{4\pi c(1-c)e^2}} \tag{14.1.21}$$

is the Debye screening radius. Eq. (14.1.21) is the same as the one for the classical Debye screening radius if $c(1-c)/v_0 \approx c/v_0 = \rho_0$ (the low-density limit) where $\rho_0 = c/v_0$ is the charge carrier volume density.

If the configuration-dependent part (13.5.4) of the elastic energy is taken into consideration, we have

$$V(\mathbf{k}) = V^{\text{chem}}(\mathbf{k}) - (F_i(\mathbf{k})G_{ij}(\kappa)F_j^*(\mathbf{k}) - Q) \tag{14.1.22}$$

where the first term of Eq. (14.1.22) corresponds to the chemical pairwise interaction and the second term yields the Fourier transform of the configuration-dependent elastic pairwise interaction energies.

Substituting Eq. (14.1.22) into (14.1.17), we have

$$V_{scr}(\mathbf{k}) = \frac{V_0(\mathbf{k})}{1+[c(1-c)/\kappa T][V^{\text{chem}}(\mathbf{k}) - (F_i(\mathbf{k})G_{ij}(\mathbf{k})F_j^*(\mathbf{k}) - Q)]} \tag{14.1.23}$$

The interesting characteristics of the screening caused by the elastic interaction is the transformation of an externally applied short-range potential, $W_0(\mathbf{r})$,

*The representation (14.1.18) corresponds to the neglect of the discrete crystal lattice structure in Eqs. (14.1.12) and is valid in the long-wave limit ($k \rightarrow 0$).

to the long-range one. The Fourier transform of a short-range potential, $W_0(\mathbf{r})$, is, by definition, an analytical function of $\mathbf{k}$ at $\mathbf{k}=0$ and can always be represented as power series expansion

$$V_0(\mathbf{k}) = V_0(0) + \beta_{ij}k_ik_j + \cdots \qquad (14.1.24)$$

The long-range asymptotic of the screened potential, $W_{scr}(\mathbf{r})$ [it is the back Fourier transform of Eq. (14.1.23) at $r\to\infty$, is

$$W_{scr}(\mathbf{r}) = \frac{1}{N}\sum_{\mathbf{k}} V_{scr}(\mathbf{k})e^{i\mathbf{k}\mathbf{r}} \to \frac{1}{N}\sum_{\mathbf{k}} e^{i\mathbf{k}\mathbf{r}} \lim_{|\mathbf{k}|\to 0} V_{scr}(\mathbf{k}) \qquad (14.1.25)$$

Substituting Eq. (14.1.23) into (14.1.25), and using the asymptotics,

$$V_0(\mathbf{k})\to V_0(0) \quad \text{and} \quad V^{\text{chem}}(\mathbf{k})\to V^{\text{chem}}(0), \quad \text{at } \mathbf{k}\to 0$$

which follow from the analytical behavior of the functions $V_0(\mathbf{k})$ [see Eq. (14.1.24)] and $V^{\text{chem}}(\mathbf{k})$ at $\mathbf{k}=0$ as well as the long-wavelength asymptotics

$$F_i(\mathbf{k})\to -iv_0\sigma^0_{ij}k_j, \quad G_{ij}(\mathbf{k})\to \frac{1}{v_0k^2}\Omega_{ij}(\mathbf{n})$$

[see Eqs. (13.5.9) and (13.5.11)], we have

$$W_{scr}(\mathbf{r})\to v_0 \int \frac{d^3k}{(2\pi)^3} \frac{V_0(0)e^{i\mathbf{k}\mathbf{r}}}{1+[c(1-c)/\kappa T][V^{\text{chem}}(0) - v_0 n_i\sigma^0_{ij}\Omega_{jk}(\mathbf{n})\sigma^0_{kl}n_l + Q]} \qquad (14.1.26)$$

at $r\to\infty$. The mathematical procedure developed by Khachaturyan and Shatalov (136) to calculate integrals having the form (14.1.26) leads to the conclusion that

$$W_{scr}(\mathbf{r})\to \frac{P(\mathbf{m})}{r^3} \qquad (14.1.27)$$

where $P(\mathbf{m})$ is the function of the direction $\mathbf{m}=\mathbf{r}/r$. It is obvious that the short-range external potential $W_0(\mathbf{r})$ [its Fourier transform $V_0(\mathbf{k})$ is an analytical function of $\mathbf{k}$ at $\mathbf{k}=0$ only if $W_0(\mathbf{r})$ decays faster than e^{-r/r_0} does] screened by elastically interacting solute atoms is transformed into the long-range potential, $W_{scr}(\mathbf{r})$, whose decay is proportional to $1/r^3$.

14.2. ELASTIC ENERGIES AND ATOMIC STRUCTURE OF ORDERED BCC INTERSTITIAL SOLUTIONS

It was shown in Section 13.1 that the typical elastic interaction energy of interstitial atoms is of about 1 eV, about two orders of magnitude larger than the respective chemical energy. As such, the elastic energy contribution is the dominant factor that determines the phase transformations in the solid solution. The Fourier transform of the pairwise elastic interaction energies was calculated in Section 13.2.4. According to Eq. (13.2.65a) the Fourier transform of the interaction energies between interstitial atoms at the sites $(p, 0)$ and $(q, \mathbf{r})$ is

$$V_{pq}(\mathbf{k})=\begin{cases}-v_0\lambda_{ijkl}u^0_{ij}(p)u^0_{kl}(q)+Q\delta_{pq} & \text{at } \mathbf{k}=0\\ -F_i(p,\mathbf{k})G_{ij}(\mathbf{k})F^*_j(q,\mathbf{k})+Q\delta_{pq} & \text{at } \mathbf{k}\neq 0\end{cases} \tag{14.2.1}$$

where the equations for $\mathbf{F}(p, \mathbf{k})$ are presented in Section 13.3, the Green function, $G_{ij}(\mathbf{k})$, is inverse to the dynamical matrix, $\tilde{A}_{ij}(\mathbf{k})$, which is the Fourier transform of the Born-von Karman constants,

$$Q=\frac{1}{N}\sum_{\mathbf{k}} F_i(p,\mathbf{k})G_{ij}(\mathbf{k})F^*_j(p,\mathbf{k}) \tag{14.2.2}$$

The summation in (14.2.2) is taken over the first Brillouin zone of the bcc host lattice.

The concentration wave method formulated in Section 3.11 enables us to determine the structure of the most stable high-temperature ordered phase formed from a disordered solution if the Fourier transform, $V_{pq}(\mathbf{k})$, of the interaction energies is known. It has been shown [see Eq. (3.11.22)] that the most stable high-temperature phase is formed at the order-disorder transition temperature

$$T_0=-\frac{c(1-c)}{\kappa}\lambda_{\sigma_0}(\mathbf{k}_0) \tag{14.2.3}$$

where

$$\lambda_{\sigma_0}(\mathbf{k}_0)=\min \lambda_\sigma(\mathbf{k}) \tag{14.2.4}$$

is the minimum value of the eigenvalue spectrum, $\lambda_\sigma(\mathbf{k})$, of the matrix, $V_{pq}(\mathbf{k})$. The structure of the high-temperature phase is generated by the star of the vector, $\mathbf{k}_0$, and the "polarization vector" (eigenvector), $v_{\sigma_0}(p, \mathbf{k}_0)$, which corresponds to the absolute minimum of $\lambda_\sigma(\mathbf{k})$.

It has been stressed in Section 3.11 that the structure of the stable phase is described by a superposition of concentration modes, including the minimum number of long-range order parameters. As a rule this is a layer structure whose concentration mode representation is generated by the dominant mode, $v_{\sigma_0}(p, \mathbf{k}_0)e^{i\mathbf{k}_0\mathbf{r}}$. The concentration mode representation (3.11.27) in this case has a form

$$n(p,\mathbf{r})=c+\tfrac{1}{2}\sum_{s=1}^{s_{\max}}\left[e^{is\mathbf{k}_0\mathbf{r}}\sum_\sigma \eta_{s,\sigma}\gamma_{s,\sigma}v_\sigma(p,s\mathbf{k}_0)+e^{-is\mathbf{k}_0\mathbf{r}}\sum_\sigma \eta_{s,\sigma}\gamma^*_{s,\sigma}v^*_\sigma(p,s\mathbf{k}_0)\right] \tag{14.2.5}$$

where $\eta_{s,\sigma}$ are the long-range order parameters, $\gamma_{s,\sigma}$ are the normalization coefficients chosen to ensure all $\eta_{s,\sigma}$ to be equal to unity in the completely ordered state; $s_{\max}$ is the minimum positive integer which multiplied by $\mathbf{k}_0$ gives a fundamental reciprocal lattice vector $2\pi\mathbf{H}$ of the host lattice.

The simplest way to organize the representation (14.2.5), to find the minimum number of concentration modes generated by the dominant mode, $v_{\sigma_0}(p, \mathbf{k}_0)e^{i\mathbf{k}_0\mathbf{r}}$, is to raise the sum (3.11.26) to a power of s_{max}:

$$[c+\eta_{s0,\sigma_0}|\gamma|(e^{i\phi}v_{\sigma_0}(p, \mathbf{k}_0)e^{i\mathbf{k}_0\mathbf{r}}+e^{-i\phi}v^*_{\sigma_0}(p, \mathbf{k}_0)e^{-i\mathbf{k}_0\mathbf{r}})]^{s_{max}} \tag{14.2.6}$$

where $\gamma_{s0,\sigma_0}=|\gamma|e^{i\phi}$. All the concentration modes and their normalization coefficients, which enter the function (14.2.6) after removing of brackets, should also enter Eq. (14.2.5).

Following Khachaturyan and Shatalov (109) and Blanter and Khachaturyan (171), we shall consider the application of the elastic energy approach to the analysis of the structure transformations occurring during ordering of interstitial atoms in V, Nb, Ta, and α-Fe.

14.2.1. Ordering within Octahedral Interstices of V, Nb, Ta, and α-Fe

As mentioned in Section 6.8, there are three kinds of octahedral interstices in the bcc lattice, O_x, O_y, and O_z. They are displaced from the nearest host atom by the vectors $\mathbf{h}_1=a_0(\frac{1}{2}, 0, 0)$, $\mathbf{h}_2=a_0(0, \frac{1}{2}, 0)$, and $\mathbf{h}_3=a_0(0, 0, \frac{1}{2})$, respectively.

Making use of the same method as applied in Section 13.4, we shall introduce the tetragonality factor t_1 [see Eq. (13.4.2a)]. In this case the Fourier transforms (13.3.16) and (13.3.17) of coupling forces, $\mathbf{f}(p, \mathbf{r})$, will be linear functions of the tetragonality factor, t_1, and the elements of the elastic interaction matrix, $V_{pq}(\mathbf{k})$, will be quadratic forms of t_1. It follows from Eqs. (13.3.16) and (13.3.17) that all eigenvalues can be represented in the dimensionless form:

$$\frac{\lambda_\sigma(\mathbf{k})}{c_{44}a_0^3(u_{33}^0)^2}=\lambda_\sigma^0(\mathbf{k}, t_1) \tag{14.2.7}$$

The convenience of the representation (14.2.7) consists in the fact that the normalized eigenvalue, $\lambda_\sigma^0(\mathbf{k}, t_1)$, depends on only one parameter t_1 in which all the information concerning the kind of interstitial atoms is buried. In this situation the minimum eigenvalue, and hence the structure of ordered phases for each solvent metal, is a function of a single parameter, t_1 rather than the values of the concentration expansion coefficients, u_{11}^0 and u_{33}^0.

The minimum of the eigenvalue spectrum, $\lambda_\sigma^0(\mathbf{k}, t_1)$, was sought in a wide range, $-0.5\leq t_1\leq 0$ in the case of octahedral site occupation. The superlattice wave vector, $\mathbf{k}_0$, and eigenvector, $v_{\sigma_0}(p, \mathbf{k}_0)$, corresponding to the minimum eigenvalue, $\lambda_{\sigma_0}(\mathbf{k}_0)$, were found by sorting all eigenvalues, $\lambda_\sigma^0(\mathbf{k}, t_1)$, over 1332 points within the irreducible part of the first Brillouin zone of the host metal. The Born-von Karman constants used in the numerical calculations were found by inelastic neutron scattering measurements (271, 273, 274) and the x-ray diffuse scattering method (275). Four eigenvalues (14.2.7) represented in terms of the parameter, t_1, are listed in Table 14.1. In the case of V, Nb, Ta, and α-Fe-based alloys employment of the tetragonality factor, t_1, within a reasonable range, $-0.5\leq t_1\leq 0$, results in four different kinds of eigenvectors, $v_{\sigma_0}(p, \mathbf{k}_0)$, and corresponding wave vectors, $\mathbf{k}_0$, providing the absolute minimum of the eigenvalue spectrum, $\lambda_\sigma(\mathbf{k})$, of the matrix, $V_{pq}(\mathbf{k})$. Since each vector, $\mathbf{k}_0$, and the corresponding eigenvector, $v_{\sigma_0}(p, \mathbf{k}_0)$, determine the structure of the most stable

Table 14.1 Calculation of Eigenvalues, $\lambda_\sigma^0(\mathbf{k}, t_1) = n_0 + b_0 t_1 + d_0 t_1^2$, $Q/[c_{44} a_0^3 (u_{33}^0)^2] = n_1 + b_1 t_1 + d_1 t_1^2$ and $\mathbf{k}_0 = \pi(\mathbf{a}_1^* + \mathbf{a}_2^*)$

Host Metal	$\lambda_1^0(0, t_1)$			$\lambda_2^0(0, t_1)$			$\lambda_1^0(\mathbf{k}_0, t_1)$			$\lambda_2^0(\mathbf{k}_0, t_1)$			$\frac{Q}{c_{44} a_0^3 (u_{33}^0)^2}$		
	n_0	b_0	d_0	n_0	b_0	d_0	n_0	b_0	d_0	n_0	b_0	d_0	n_1	b_1	d_1
α-Fe	−1.963	−8.385	−8.392	0.001	2.020	0.734	−0.592	−2.391	−1.946	0.125	1.737	0.897	0.425	1.168	1.161
Ta	−2.549	−11.318	−11.328	0.377	4.235	2.307	−2.191	−9.160	−12.2	0.486	4.543	6.483	1.02	2.96	2.95
Nb	−2.429	−11.967	−14.128	1.760	10.46	6.47	−5.55	−22.87	−32.99	1.54	11.03	5.65	3.04	7.91	7.75
V	−7.064	−31.158	−31.208	−0.060	8.562	2.659	−4.196	−15.85	−12.66	0.619	7.091	3.093	1.89	4.66	4.61

high-temperature phase arising directly from the disordered solution, we conclude that the microscopic elastic energy approach provides the formation for four high-temperature phases (one of these proves to be of the same structure as the disordered solution, and its formation thus corresponds to the isostructural decomposition). Below we consider these structures.

1. The minimum eigenvalue, $\lambda_\sigma(\mathbf{k})=\lambda_1(0)$, corresponds to $\mathbf{k}=\mathbf{k}_0=0$ and is associated with the eigenvector*

$$v_{\sigma_0}(p, \mathbf{k}_0)=v_1(p, 0)=(v_1(1, 0), v_1(2, 0), v_1(3, 0))=\left(\frac{1}{\sqrt{3}}, \frac{1}{\sqrt{3}}, \frac{1}{\sqrt{3}}\right) \tag{14.2.8}$$

The eigenvalue, $\lambda_1(0)$, may be represented in terms of the interaction matrix, $V_{pq}(0)$, as follows

$$\lambda_1(0)=V_{11}(0)+2V_{12}(0) \tag{14.2.9}$$

since, by symmetry, the matrix, $V_{pq}(0)$, always has the form

$$V_{pq}(0)=\begin{pmatrix} V_{11}(0) & V_{12}(0) & V_{12}(0) \\ V_{12}(0) & V_{11}(0) & V_{12}(0) \\ V_{12}(0) & V_{12}(0) & V_{11}(0) \end{pmatrix} \tag{14.2.10}$$

Substituting the elements $V_{11}(0)$ and $V_{12}(0)$ from (14.2.1) to (14.2.9), we have for a bcc solution:

$$\lambda_1(0)=-\frac{a_0^3}{2}(c_{11}+2c_{12})(u_{33}^0+2u_{11}^0)^2+Q \tag{14.2.11}$$

The eigenvector (14.2.8) generates the distribution

$$n(p, \mathbf{r})=c+\eta\gamma v_{\sigma_0}(p, 0)=\begin{cases} c+\dfrac{\eta\gamma}{\sqrt{3}} & \text{at } p=1\ (O_x \text{ positions}) \\[2ex] c+\dfrac{\eta\gamma}{\sqrt{3}} & \text{at } p=2\ (O_y \text{ positions}) \\[2ex] c+\dfrac{\eta\gamma}{\sqrt{3}} & \text{at } p=3\ (O_z \text{ positions}) \end{cases} \tag{14.2.12}$$

The function (14.2.11) describes the disordered phase whose octahedral interstices are occupied by interstitial atoms with the same probability, $c+\eta\gamma/\sqrt{3}$. Since η may assume both positive and negative values, the phase transition associated with the dominant eigenvector (14.2.8) results in the isostructural decomposition into two disordered phases with different compositions. Such a decomposition occurs for Nb within the range $-0.07<t_1<0$, for Ta within the range $-0.14<t_1<0$, and for α-Fe within the range $-0.23<t_1<0$, but it is not registered for V at any t_1 value (see Fig. 112).

*We give the eigenvectors normalized according to the relation

$$\sum_p |v_{\sigma_0}(p, \mathbf{k}_0)|^2=1$$

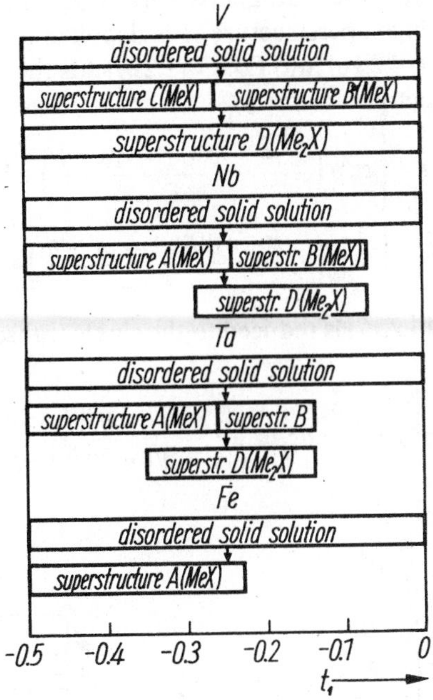

Figure 112. The stability ranges for interstitial superstructures in V, Nb, Ta, and α-Fe with respect to the tetragonality factor $t_1 = u_{11}^0/u_{33}^0$ (octahedral interstices). The atomic structures A, B, C and D are presented in Fig. 113.

The minimum eigenvalue, $\lambda_{\sigma_0}(\mathbf{k}_0) = \lambda_2(0)$, corresponds to $\mathbf{k}_0 = 0$, $\sigma_0 = 2$, and is associated with the eigenvector,

$$v_{\sigma_0}(p, \mathbf{k}_0) = v_2(p, 0) = (v_2(1, 0), v_2(2, 0), v_2(3, 0)) = \left(-\frac{1}{\sqrt{6}}, -\frac{1}{\sqrt{6}}, \frac{2}{\sqrt{6}}\right) \tag{14.2.13}$$

The eigenvalue, $\lambda_2(0)$, may be represented in terms of the elements of the interaction matrix (14.2.10) as follows:

$$\lambda_2(0) = V_{11}(0) - V_{12}(0) \tag{14.2.14}$$

Using Eq. (14.2.1), we may rewrite Eq. (14.2.14)

$$\lambda_2(0) = -\frac{a_0^3}{2}(c_{11} - c_{12})(u_{11}^0 - u_{33}^0)^2 + Q \tag{14.2.15}$$

The eigenvector (14.2.13) generates the ordered distribution

$$n(p, \mathbf{r}) = c + \eta\gamma v_2(p, 0) = \begin{cases} c - c\eta & \text{at } p = 1\ (O_x) \\ c - c\eta & \text{at } p = 2\ (O_y) \\ c + 2c\eta & \text{at } p = 3\ (O_z) \end{cases} \tag{14.2.16}$$

where $\gamma = \sqrt{6}c$. It follows from Eq. (14.2.16) that the dominant concentration mode (14.2.13) provides the redistribution of interstitials between three types of octahedral interstices: solute atoms pass from O_x and O_y interstices into O_z interstices. The resulting superstructure A is depicted in Fig. 113a.

Substitution of Eq. (14.2.16) into Eq. (3.11.2) yields the equilibrium equations for the long-range order parameter η:

$$c - c\eta = \left[\exp\left(\frac{-\mu + \lambda_1(0)c - \lambda_2(0)c\eta}{\kappa T}\right) + 1\right]^{-1}$$

$$c + 2c\eta = \left[\exp\left(\frac{-\mu + \lambda_1(0)c + 2\lambda_2(0)c\eta}{\kappa T}\right) + 1\right]^{-1} \qquad (14.2.17)$$

The set of equations (14.2.17) is reduced to the single transcendental equation in the unknown η:

$$T = -\frac{3c\lambda_2(0)}{\kappa}\eta\left\{\ln\frac{[(1-c)/c+\eta][1+2\eta]}{[(1-c)/c-2\eta][1-\eta]}\right\}^{-1} \qquad (14.2.18)$$

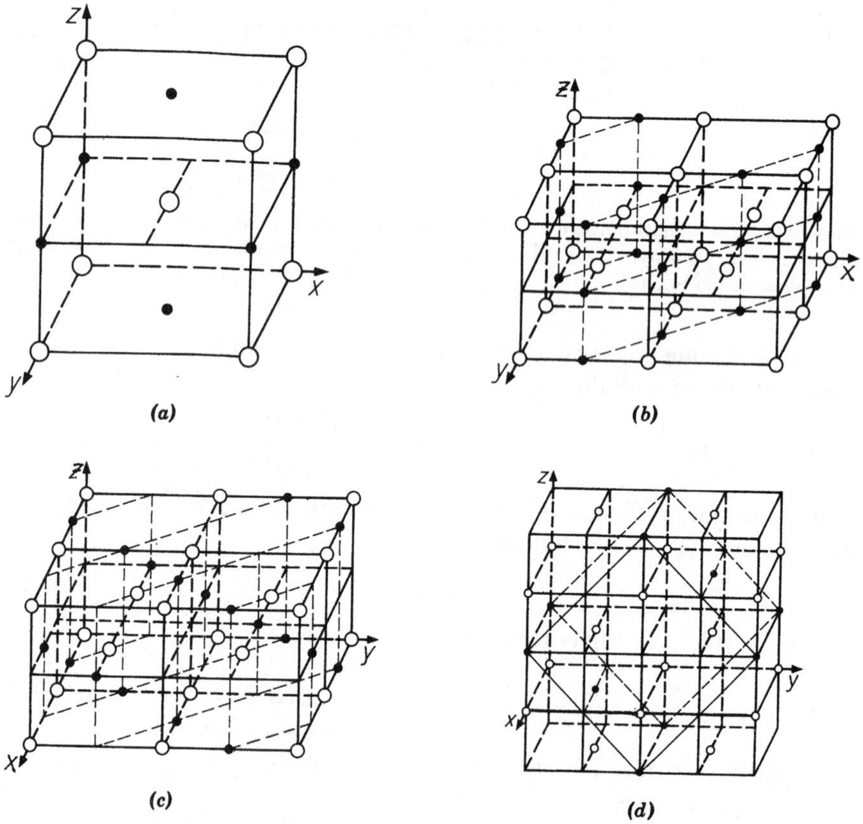

Figure 113. The stable superstructures formed by ordered distributions of interstitial atoms in octahedral sites of a bcc host lattices. (*a*), (*b*), and (*c*) are schematic drawings of the high-temperature superstructures A, B, and C, respectively, with the stoichiometric formula MeX; (*d*) is the secondary ordered superstructure D with the stoichiometric formula Me_2X: O-host atom. ●-interstitial atom.

which determines the temperature dependence of the long-range order parameter.

At $c = c_{st} = \frac{1}{3}$ and $\eta = 1$, Eq. (14.2.16) describes the superstructure A in the completely ordered state. Its stoichiometric formula is MeX. The superstructure A arises in Nb within the range $-0.5 < t_1 < -0.24$, in Ta within the range $-0.5 < t_1 < -0.26$, and in α-Fe within $-0.5 < t_1 < -0.23$. It does not arise in V at any value of t_1 (Fig. 112).

It should be noted that the elastic energy approach similar to that formulated above was applied by Alfeld (297, 298) to describe high-temperature decomposition and ordering in hydrides. The approach (297, 298) is applicable to the phase transformation associated with the dominant wave vector, $\mathbf{k}_0 = 0$. It is, however, noteworthy that the energetic constants which in the theory (297, 298) play the same role as the eigenvalues, $\lambda_1(0)$ and $\lambda_2(0)$, should be revised, the term Q missed in Alfeld's consideration should be introduced.

2. The minimum eigenvalue $\lambda_{\sigma 0}(\mathbf{k}_0) = \lambda_1(p, \mathbf{k}_0)$ corresponds to

$$\mathbf{k}_0 = \pi(\mathbf{a}_1^* + \mathbf{a}_2^*) \tag{14.2.19}$$

and the eigenvector

$$v_{\sigma 0}(p, \mathbf{k}_0) = v_1(p, \mathbf{k}_0) = (v_1(1, \mathbf{k}_0), v_1(2, \mathbf{k}_0), v_1(3, \mathbf{k}_0)) = \left(\frac{1}{\sqrt{2}}, \frac{1}{\sqrt{2}}, 0\right) \tag{14.2.20}$$

where $\mathbf{a}_1^*$, $\mathbf{a}_2^*$, and $\mathbf{a}_3^*$ are the reciprocal lattice vectors of the bcc host lattice in the [100], [010], and [001] directions, respectively.

The eigenvalue, $\lambda_1(\mathbf{k}_0)$, may be represented in terms of the interaction matrix, $V_{pq}(\mathbf{k}_0)$. It is

$$\lambda_1(\mathbf{k}_0) = V_{11}(\mathbf{k}_0) + V_{12}(\mathbf{k}_0) \tag{14.2.21}$$

since the symmetry considerations applied to the matrix, $V_{pq}(\mathbf{k}_0)$, always result in the form

$$V_{pq}(\mathbf{k}_0) = \begin{pmatrix} V_{11}(\mathbf{k}_0) & V_{12}(\mathbf{k}_0) & 0 \\ V_{12}(\mathbf{k}_0) & V_{11}(\mathbf{k}_0) & 0 \\ 0 & 0 & V_{33}(\mathbf{k}_0) \end{pmatrix} \tag{14.2.22}$$

where $\mathbf{k}_0$ is given by Eq. (14.2.19).

Substituting the elements $V_{11}(\mathbf{k}_0)$ and $V_{12}(\mathbf{k}_0)$ from (14.2.1), and using Eqs. (13.3.16) and (13.3.17) for $\mathbf{F}^{\text{oct}}(p, \mathbf{k}_0)$ as well as the bilinear representation (13.2.46) of the Green function, $G_{ij}(\mathbf{k}_0)$, we have

$$\lambda_1(\mathbf{k}_0) = -\frac{a_0^4[u_{11}^0(c_{11} + 2c_{12}) + u_{33}^0(c_{11} + c_{12})]^2}{m\omega_L^2(\mathbf{k}_0)} + Q \tag{14.2.23}$$

where $\omega_L(\mathbf{k}_0)$ is the frequency of the longitudinal vibration mode of the host lattice with the wave vector $\mathbf{k}_0$.

To determine all concentration modes entering Eq. (14.2.5) and generated by the dominant mode $v_1(p, \mathbf{k}_0)e^{i\mathbf{k}_0\mathbf{r}}$, we shall use the procedure described above.

One may readily see that in the case under consideration $s_{max}=2$ since the wave vector, $\mathbf{k}_0=\pi(\mathbf{a}_1^*+\mathbf{a}_2^*)$, being multiplied by the factor 2 yields the fundamental reciprocal lattice vector, $2\mathbf{k}_0=2\pi(\mathbf{a}_1^*+\mathbf{a}_2^*)$, corresponding to the (110) fundamental reciprocal point of the bcc host lattice.

In this case Eq. (14.2.6) becomes

$$(c+\eta\gamma v_1(p,\mathbf{k}_0)e^{i\mathbf{k}_0\mathbf{r}})^2=c^2+2\eta\gamma v_1(p,\mathbf{k}_0)e^{i\mathbf{k}_0\mathbf{r}}c+\eta^2\gamma^2(v_1(p,\mathbf{k}_0)e^{i\mathbf{k}_0\mathbf{r}})^2 \tag{14.2.24}$$

Since $2\mathbf{k}_0$ is the fundamental reciprocal lattice vector, we have by definition

$$e^{i2\mathbf{k}_0\mathbf{r}}\equiv 1$$

Taking the latter into account, we have

$$\begin{aligned}(v_1(p,\mathbf{k}_0)e^{i\mathbf{k}_0\mathbf{r}})^2&=(v_1(p,\mathbf{k}_0))^2e^{i2\mathbf{k}_0\mathbf{r}}=v_1^2(p,\mathbf{k}_0)=\left(\frac{1}{\sqrt{2}},\frac{1}{\sqrt{2}},0\right)^2=\left(\frac{1}{2},\frac{1}{2},0\right)\\&=\frac{1}{\sqrt{3}}\left(\frac{1}{\sqrt{3}},\frac{1}{\sqrt{3}},\frac{1}{\sqrt{3}}\right)-\frac{1}{\sqrt{6}}\left(-\frac{1}{\sqrt{6}},-\frac{1}{\sqrt{6}},\frac{2}{\sqrt{6}}\right)\\&=\frac{1}{\sqrt{3}}v_1(p,0)-\frac{1}{\sqrt{6}}v_2(p,0)\end{aligned} \tag{14.2.25}$$

where the eigenvectors $v_1(p,0)$ and $v_2(p,0)$ are given by Eqs. (14.2.8) and (14.2.13).

Substitution of Eq. (14.2.25) to (14.2.24) thus yields

$$(c+\eta\gamma v_1(p,\mathbf{k}_0)e^{i\mathbf{k}_0\mathbf{r}})^2=\left(\sqrt{3}c^2+\frac{\eta^2\gamma^2}{\sqrt{3}}\right)v_1(p,0)-\frac{\eta^2\gamma^2}{\sqrt{6}}v_2(p,0)+2c\eta\gamma v_1(p,\mathbf{k}_0)e^{i\mathbf{k}_0\mathbf{r}} \tag{14.2.26}$$

It follows from Eq. (14.2.26) that the distribution (14.2.5) describing the stable high-temperature superstructure generated by the dominant concentration mode, $v_1(p,\mathbf{k}_0)e^{i\mathbf{k}_0\mathbf{r}}$, should also include the modes, $v_2(p,0)$, that enter Eq. (14.2.26). Therefore Eq. (14.2.5) may be rewritten as follows:

$$n(p,\mathbf{r})=c-\eta_{0,2}\gamma_{0,2}v_2(p,0)+\eta_{1,1}\gamma_{1,1}v_1(p,\mathbf{k}_0)e^{i\mathbf{k}_0\mathbf{r}}$$

$$=\begin{cases}c+\dfrac{\eta_{0,2}\gamma_{0,2}}{\sqrt{6}}+\dfrac{\eta_{1,1}\gamma_{1,1}}{\sqrt{2}}e^{i\mathbf{k}_0\mathbf{r}} & \text{if } p=1\\[2ex] c+\dfrac{\eta_{0,2}\gamma_{0,2}}{\sqrt{6}}+\dfrac{\eta_{1,1}\gamma_{1,1}}{\sqrt{2}}e^{i\mathbf{k}_0\mathbf{r}} & \text{if } p=2\\[2ex] c-\dfrac{2\eta_{0,2}\gamma_{0,2}}{\sqrt{6}} & \text{if } p=3\end{cases}$$

or at $\gamma_{0,2}=\frac{1}{4}\sqrt{6}$ and $\gamma_{1,1}=1/\sqrt{2}$

$$n(p,\mathbf{r})=\begin{cases}c+\frac{1}{4}\eta_{0,2}+\frac{1}{2}\eta_{1,1}e^{i\mathbf{k}_0\mathbf{r}} & \text{if } p=1\\ c+\frac{1}{4}\eta_{0,2}+\frac{1}{2}\eta_{1,1}e^{i\mathbf{k}_0\mathbf{r}} & \text{if } p=2\\ c-\frac{1}{2}\eta_{0,2} & \text{if } p=3\end{cases} \tag{14.2.27}$$

Substitution of Eq. (14.2.27) into (3.11.2) yields the set of equations for the long-range order parameters $\eta_{0,2}$ and $\eta_{1,1}$:

$$c+\frac{1}{4}\eta_{0,2}+\frac{1}{2}\eta_{1,1}e^{i\mathbf{k}_0\mathbf{r}}=\left[\exp\left(\frac{-\mu+\lambda_1(0)c+\lambda_2(0)\frac{1}{4}\eta_{0,2}+\lambda_1(\mathbf{k}_0)\frac{1}{2}\eta_{1,1}e^{i\mathbf{k}_0\mathbf{r}}}{\kappa T}\right)+1\right]^{-1}$$

$$c-\frac{1}{2}\eta_{0,2}=\left[\exp\left(\frac{-\mu+\lambda_1(0)c-\lambda_2(0)\frac{1}{2}\eta_{0,2}}{\kappa T}\right)+1\right]^{-1} \quad (14.2.28a)$$

Taking into account the fact that the function $e^{i\mathbf{k}_0\mathbf{r}}$ at $\mathbf{k}_0=\pi(\mathbf{a}_1^*+\mathbf{a}_2^*)$ assumes the value of either 1 or -1, one may transform the latter set of the transcendental equations to the set

$$\ln\left\{\frac{[(c+\frac{1}{4}\eta_{0,2})^2-(\frac{1}{2}\eta_{1,1})^2]}{[(1-c-\frac{1}{4}\eta_{0,2})^2-(\frac{1}{2}\eta_{1,1})^2]}\,\frac{(1-c+\frac{1}{2}\eta_{0,2})^2}{(c-\frac{1}{2}\eta_{0,2})^2}\right\}=-\frac{3}{2}\frac{\lambda_2(0)\eta_{0,2}}{\kappa T} \quad (14.2.28b)$$

where

$$\left(\frac{1}{2}\eta_{1,1}\right)^2$$
$$=\frac{(1-c-\frac{1}{4}\eta_{0,2})^2(c-\frac{1}{2}\eta_{0,2})^2\exp[-\frac{3}{2}\lambda_2(0)(\eta_{0,2}/\kappa T)]-(c+\frac{1}{4}\eta_{0,2})^2(1-c+\frac{1}{2}\eta_{0,2})^2}{(c-\frac{1}{2}\eta_{0,2})^2\exp[-\frac{3}{2}\lambda_2(0)(\eta_{0,2}/\kappa T)]-(1-c+\frac{1}{2}\eta_{0,2})^2}$$
$$(14.2.28c)$$

Eqs. (14.2.28) determine the temperature dependence of the long-range order parameters $\eta_{0,2}$ and $\eta_{1,1}$.

At the stoichiometric composition and in the completely ordered state (when $c=c_{st}=\frac{1}{2}$, $\eta_{0,2}=\eta_{1,1}=1$), Eq. (14.2.27) describes the superstructure B with the stoichiometric formula MeX shown in Fig. 113b.

The numerical computations show that the elastic interaction results in the superstructure B in V if $-0.26<t_1<0$, in Nb if $-0.24<t_1<-0.07$, in Ta if $-0.26<t_1<-0.14$, and it is not realized in α-Fe (Fig. 112).

3. The minimum eigenvalue $\lambda_{\sigma_0}(\mathbf{k}_0)=\lambda_2(\mathbf{k}_0)$ corresponds to

$$\mathbf{k}_0=\pi(\mathbf{a}_1^*+\mathbf{a}_2^*)$$

and the eigenvector

$$v_2(p,\mathbf{k}_0)=\left(\frac{1}{\sqrt{2}},-\frac{1}{\sqrt{2}},0\right) \quad (14.2.29)$$

It follows from the matrix, $V_{pq}(\mathbf{k}_0)$ [see Eq. (14.2.22)], that

$$\lambda_2(\mathbf{k}_0)=V_{11}(\mathbf{k}_0)-V_{12}(\mathbf{k}_0) \quad (14.2.30)$$

Using Eqs. (14.2.1), (13.2.46), and (13.3.16) in Eq. (14.2.30) at $\mathbf{k}=\mathbf{k}_0$, we have

$$\lambda_2(\mathbf{k}_0)=-\frac{a_0^4(c_{11}-c_{12})^2(u_{33}^0-u_{11}^0)^2}{m\omega_T^2(\mathbf{k}_0)}+Q \quad (14.2.31)$$

where $\omega_T(\mathbf{k}_0)$ is the transversal vibration mode of the host lattice with the wave vector, $\mathbf{k}_0$. The corresponding polarization vector of the vibration mode is

$$\mathbf{e}_T(\mathbf{k}_0)=\left(\frac{1}{\sqrt{2}}, -\frac{1}{\sqrt{2}}, 0\right)$$

The treatment similar to that considered above results in the distribution (14.2.5) generated by the dominant concentration mode, $v_2(p, \mathbf{k}_0)e^{i\mathbf{k}_0\mathbf{r}}$. It has the form

$$n(p, \mathbf{r})=c-\eta_{0,2}\frac{\sqrt{6}}{4}v_2(p, 0)+\eta_{1,2}\frac{1}{\sqrt{2}}v_2(p, \mathbf{k}_0)e^{i\mathbf{k}_0\mathbf{r}}$$

$$=\begin{cases} c+\frac{1}{4}\eta_{0,2}+\frac{1}{2}\eta_{1,2}e^{i\mathbf{k}_0\mathbf{r}} & \text{at } p=1 \\ c+\frac{1}{4}\eta_{0,2}-\frac{1}{2}\eta_{1,2}e^{i\mathbf{k}_0\mathbf{r}} & \text{at } p=2 \\ c-\frac{1}{2}\eta_{0,2} & \text{at } p=3 \end{cases} \qquad (14.2.32)$$

at $\gamma_{0,2}=\sqrt{6}/4$ and $\gamma_{1,2}=1/\sqrt{2}$.

At the stoichiometric composition $c=c_{st}=\frac{1}{2}$ and at $\eta_{0,2}=\eta_{1,2}=1$, Eq. (14.2.32) describes the superstructure C with the stoichiometric formula MeX depicted in Fig. 113(c).

Substitution of Eqs. (14.2.32) into (3.11.2) results in the same set of the transcendental equations (14.2.28a) as obtained through the substitution of Eq. (14.2.27) into Eq. (3.11.2). The final equilibrium equations (14.2.28b) and (14.2.28c) are also valid therefore for the superstructure C.

In the case of vanadium elastic interaction results in the superstructure C only when $-0.5<t_1<-0.26$ (Fig. 112). The superstructures B and C are layer phases whose atomic arrangement is formed by alternating (110) planes. Each second (110) plane in the O_x and O_y octahedral bcc sublattices is occupied by interstitials. The mutual location of the filled planes, however, is different for the B and C superstructures. The tetragonal axial ratio of these superstructures is $c/a<1$, and their superlattice vector is orthogonal to the tetragonal axis.

14.2.2. Low-Temperature Secondary Superstructures Based on the Octahedral Interstice Occupation

All three high-temperature superstructures obtained here have the stoichiometric composition MeX. As shown in Section 3.10, if the composition of the superstructures is not stoichiometric, a decrease in temperature results in either the secondary ordering (ordering of interstitial atom vacancies within the high-temperature superstructures) or decomposition. In both cases the alloy attains a new stoichiometric composition.

To find the structure of a secondary phase, one has to analyze the first variation of the self-consistent field equation (3.11.2) with respect to the variations of the interstitial distribution, $\delta n(p, \mathbf{r})$, within the initial high-temperature superstructure. To do this, one has to substitute

$$n(p, \mathbf{r})=n_0(p, \mathbf{r})+\delta n(p, \mathbf{r})$$

into Eq. (3.11.2) where $n_0(p, \mathbf{r})$ is the equilibrium distribution of interstitials for the high-temperature superstructure. Expanding the resultant equation in a power series of $\delta n(p, \mathbf{r})$ up to the first-order terms, one has

$$\delta n(p, \mathbf{r}) = -\frac{n_0(p, \mathbf{r})(1-n_0(p, \mathbf{r}))}{\kappa T} \sum_{q, \mathbf{r}} W_{pq}(\mathbf{r}-\mathbf{r}')\delta n(q, \mathbf{r}') \qquad (14.2.33)$$

[compare with Eq. (3.10.2)). Eq. (14.2.33) is a linear set of homogeneous equations of $\delta n(p, \mathbf{r})$ which has the trivial solution $\delta n(p, \mathbf{r}) \equiv 0$. It means that the initial high-temperature structure is stable with respect to arbitrary infinitesimal variations of the distribution. The appearance of a nontrivial (nonzero) solution to Eq. (14.2.33) means that the homogeneous state of the high-temperature superstructure has lost its stability with respect to infinitesimal concentration fluctuations. The latter occurs when the determinant of the set (14.2.33) vanishes. The temperature condition that determines the vanishing of the determinant thus gives the temperature of the secondary ordering or decomposition.

After (109) let us consider the secondary phase transition in the nonstoichiometric high-temperature superstructure A.

According to Eq. (14.2.16) in this case

$$n_0(p, \mathbf{r}) = n_0(p) \qquad (14.2.34)$$

where $n_0(p)$ are constants independent of the crystal lattice site coordinates $\mathbf{r}$. Using the definitions (14.2.34) in Eq. (14.2.33), we have

$$\delta n(p, \mathbf{r}) = -\frac{n_0(p)(1-n_0(p))}{\kappa T} \sum_{q, \mathbf{r}'} W_{pq}(\mathbf{r}-\mathbf{r}')\delta n(q, \mathbf{r}') \qquad (14.2.35)$$

The Fourier transform of Eq. (14.2.35) yields

$$\delta\tilde{n}(p, \mathbf{k}) = -\frac{n_0(p)(1-n_0(p))}{\kappa T} \sum_{q} V_{pq}(\mathbf{k})\delta\tilde{n}(q, \mathbf{k})$$

or

$$\sum_{q} [n_0(p)(1-n_0(p))V_{pq}(\mathbf{k}) + \kappa T\delta_{pq}]\delta\tilde{n}(q, \mathbf{k}) = 0 \qquad (14.2.36)$$

where

$$\delta\tilde{n}(p, \mathbf{k}) = \sum_{\mathbf{r}} \delta n(p, \mathbf{r})e^{-i\mathbf{k}\mathbf{r}} \qquad (14.2.37)$$

Introducing the Hermitian matrix

$$\tilde{V}_{pq}(\mathbf{k}) = \sqrt{n_0(p)(1-n_0(p))}\, V_{pq}(\mathbf{k})\sqrt{n_0(q)(1-n_0(q))} \qquad (14.2.38)$$

we may rewrite Eq. (14.2.36)

$$\sum_{q} [\tilde{V}_{pq}(\mathbf{k}) + \kappa T\delta_{pq}] \frac{\delta\tilde{n}(q, \mathbf{k})}{\sqrt{n_0(q)(1-n_0(q))}} = 0 \qquad (14.2.39)$$

where $\delta n(q, \mathbf{k})/\sqrt{n_0(q)(1-n_0(q))}$ plays the part of new unknowns.

The nontrivial solution to Eq. (14.2.39) that determines the loss of stability of the phase A arises when the determinant of Eq. (14.2.39) vanishes; when

$$\mathrm{Det}\|\tilde{V}_{pq}(\mathbf{k})+\kappa T\delta_{pq}\| = \begin{vmatrix} \tilde{V}_{11}(\mathbf{k})+\kappa T & \tilde{V}_{12}(\mathbf{k}) & \tilde{V}_{12}(\mathbf{k}) \\ \tilde{V}^*_{12}(\mathbf{k}) & \tilde{V}_{22}(\mathbf{k})+\kappa T & \tilde{V}_{23}(\mathbf{k}) \\ \tilde{V}^*_{13}(\mathbf{k}) & \tilde{V}^*_{23}(\mathbf{k}) & \tilde{V}_{33}(\mathbf{k})+\kappa T \end{vmatrix} = 0 \qquad (14.2.40)$$

Eq. (14.2.40) is a cubic equation in κT. It follows from Eq. (14.2.40) that $-\kappa T$ is an eigenvalue of the matrix, $\tilde{V}_{pq}(\mathbf{k})$. Therefore the secondary ordering temperature is given by the minimum eigenvalue of the matrix, $\tilde{V}_{pq}(\mathbf{k})$;

$$T_1 = -\frac{1}{\kappa}\min \tilde{\lambda}_\sigma(\mathbf{k}) \qquad (14.2.41)$$

where $\tilde{\lambda}_\sigma(\mathbf{k})$ are the eigenvalues of the matrix, $\tilde{V}_{pq}(\mathbf{k})$. The occupation numbers, $n_0(p)$, of octahedral sites O_x, O_y, and O_z entering $\tilde{V}_{pq}(\mathbf{k})$ are determined by Eq. (14.2.16), the value of the long-range order parameter, $\eta_{0,2}$, being the solution of the equilibrium equation (14'2.18) at the temperature T.

To find the structure of a low-temperature secondary ordered phase, one should find the wave vector $\mathbf{k}_1$, the branch, $\sigma=\sigma_1$, and the eigenvector, $\tilde{v}_{\sigma_1}(p,\mathbf{k}_1)e^{i\mathbf{k}_1\mathbf{r}}$, which ensure the minimum of the eigenvalue, $\tilde{\lambda}_\sigma[\mathbf{k}, t_1, \eta_{0,2}(T)]$, of the matrix, $\tilde{V}_{pq}(\mathbf{k}, t_1, \eta_{0,2}(T))$. The dominant concentration mode, $\tilde{v}_{\sigma_1}(p,\mathbf{k}_1)e^{i\mathbf{k}_1\mathbf{r}}$, of the secondary phase, together with the dominant mode of the initial phase, $v_{\sigma_0}(p, 0)$, generate the structure of the secondary phase.

The calculation of the secondary ordering in the high-temperature superstructures B and C is more complicated (171). To obtain the temperature of the secondary ordering and the structure of the secondary ordering phase, one should substitute $n_0(p,\mathbf{r})$ from Eq. (14.2.27) or (14.2.32) into (14.2.35) and carry out the Fourier transformation of (14.2.35). This procedure results in a new 6×6 matrix corresponding to the set of linear equations obtained above:

$$\mathbf{D}_{pq}(\mathbf{k}) = \begin{pmatrix} a(p)V_{pq}(\mathbf{k})a(q)+b(p)V_{pq}(\mathbf{k}-\mathbf{k}_0)b(q) & \pm a(p)V_{pq}(\mathbf{k})b(q)\pm b(p)V_{pq}(\mathbf{k}-\mathbf{k}_0)a(q) \\ \pm a(p)V_{pq}(\mathbf{k}-\mathbf{k}_0)b(q)\pm b(p)V_{pq}(\mathbf{k})a(q) & a(p)V_{pq}(\mathbf{k}-\mathbf{k}_0)a(q)+b(p)V_{pq}(\mathbf{k})b(q) \end{pmatrix} \qquad (14.2.42)$$

where

$$a(p) = \begin{cases} \frac{1}{2}\sqrt{(c+\frac{1}{4}\eta_{0,2}+\frac{1}{2}\eta_{1,1})(1-c-\frac{1}{4}\eta_{0,2}-\frac{1}{2}\eta_{1,1})} & \\ \quad +\frac{1}{2}\sqrt{(c+\frac{1}{4}\eta_{0,2}-\frac{1}{2}\eta_{1,1})(1-c-\frac{1}{4}\eta_{0,2}+\frac{1}{2}\eta_{1,1})} & \text{if } p=1,2 \\ \sqrt{(c-\frac{1}{2}\eta_{0,2})(1-c+\frac{1}{2}\eta_{0,2})} & \text{if } p=3 \end{cases}$$

$$b(p) = \begin{cases} \frac{1}{2}\sqrt{(c+\frac{1}{4}\eta_{0,2}+\frac{1}{2}\eta_{1,1})(1-c-\frac{1}{4}\eta_{0,2}-\frac{1}{2}\eta_{1,1})} & \\ \quad -\frac{1}{2}\sqrt{(c+\frac{1}{4}\eta_{0,2}-\frac{1}{2}\eta_{1,1})(1-c-\frac{1}{4}\eta_{0,2}+\frac{1}{2}\eta_{1,1})} & \text{if } p=1,2 \\ 0 & \text{if } p=3 \end{cases} \qquad (14.2.43)$$

The plus sign corresponds to the high-temperature phase *B*, and the minus sign to the high-temperature phase *C*.

The secondary ordering temperature T_1 is determined from the condition

$$T_1 = -\frac{1}{\kappa} \min \tilde{\lambda}_\sigma(\mathbf{k})$$

where $\tilde{\lambda}_\sigma(\mathbf{k})$ is the eigenvalue spectrum of the matrix (14.2.42). The long-range order parameters, $\eta_{0,2}$ and $\eta_{1,1}$, entering (14.2.43) are determined by the equilibrium equations (14.2.28b) and (14.2.28c).

The minimum eigenvalue of the matrix, $\mathbf{D}_{pq}(\mathbf{k})$, determines the superlattice vector, $\mathbf{k}_1$, and the eigenvector, $\tilde{v}_\sigma(p, \mathbf{k})$, of the secondary phase. Naturally, numerical computations of secondary phase transitions are only possible by means of a high-speed computer.

1. Secondary ordering in the superstructure A.

The calculations for Nb and Ta demonstrate that the minimum eigenvalue of the matrix (14.2.38) is realized at the wave vector:

$$\mathbf{k}_1 = \pi(\mathbf{a}_2^* + \mathbf{a}_3^*)$$

within the range $-0.26 < t_1 < -0.24$ and $-0.35 < t_1 < -0.26$, respectively. Since in the basic superstructure *A* only the O_z octahedral interstices are occupied, the concentration wave $e^{i\mathbf{k}_1\mathbf{r}}$ should modulate the occupation probabilities of only the O_z sublattice. In a completely ordered state the occupation probabilities will thus be described by the equation

$$n(p, \mathbf{r}) = \begin{cases} 0 & \text{if } p=1\ (O_x \text{ interstices}) \\ 0 & \text{if } p=2\ (O_y \text{ interstices}) \\ \frac{1}{2}+\frac{1}{2}e^{i\mathbf{k}_1\mathbf{r}} & \text{if } p=3\ (O_z \text{ interstices}) \end{cases} \tag{14.2.44}$$

The distribution (14.2.44) describes the secondary superstructure *D* displayed in Fig. 113d. Its axial ratio is larger than unity ($c/a > 1$); the superlattice vector, $\mathbf{k}_1$, makes an angle of 45° with the tetragonal axis [001] in the bcc lattice. The stoichiometric formula of phase *D* is Me_2X.

2. Secondary ordering in the superstructures *D* and *C* (171).

The minimum eigenvalue of the matrix (14.2.42) occurs at $\mathbf{k}_1 = 0$ and $\tilde{v}_{\sigma_1}(p, \mathbf{k}_1) = (\tilde{v}_{\sigma_1}(1, 0), \tilde{v}_{\sigma_1}(2, 0), 0)$ for all values of the parameter t_1 corresponding to the phases *B* and *C* in V, Nb, and Ta. Since $\mathbf{k}_1 = 0$, the secondary dominant concentration mode does not change the periodicity of the initial structure. This merely results in the transition of interstitial atoms from the first octahedral sublattice (O_x) to the second one (O_y), or vice versa. In both cases the atomic redistribution in the secondary ordering reaction yields the superstructure *D* with the distribution

$$n(p, \mathbf{r}) = \begin{cases} \frac{1}{2}+\frac{1}{2}e^{i\mathbf{k}_1\mathbf{r}} & \text{if } p=1 \\ 0 & \text{if } p=2 \\ 0 & \text{if } p=3 \end{cases} \tag{14.2.45}$$

The distributions (14.2.44) and (14.2.45) are just the different orientation variants of the same structure D.

Therefore we face a very specific situation when the same secondary ordered phase D arises from different high-temperature ordered phases A, B, and C. The structure D is the only ordered secondary phase with the stoichiometric composition Me_2X in all three metals, V, Nb, and Ta.

Tertiary and higher-order ordering reactions may occur in phase D when the composition is less than that corresponding to the stoichiometric formula, Me_2X. The calculation of such structures can hardly be made by means of the static concentration wave formalism and requires the application of another technique [computer simulation of ordering kinetics (299)].

14.2.3. The Effect of the Nearest-Neighbor Interstitial Interaction

All the calculations described above are based on the premise that pairwise interactions between interstitial atoms are determined only by elastic interaction. It is of interest to analyze the chemical interaction effect on the results from the elastic interaction model.

In this connection we consider the short-range contact "chemical" interaction. Strong contact repulsion arises between the nearest and next-nearest interstitial atoms because of the "overlap" of their electron shells. The "overlap" effect is especially significant in such cases as Fe-C alloy where the nearest-neighbor distance between C atoms proves to be less than the distance between centers of two "touching" carbon atoms (than the atomic diameter of C atoms).

A simple calculation shows that the positive (repulsive) energy, W_1, of the nearest-neighbor interstitials (they are separated by the distances $\langle a_0/2, 0, 0\rangle$ and occupy different kinds of interstices) results in the following effect on the eigenvalues:

$$\begin{aligned}
\lambda_1(0) &= V_{11}(0)+2V_{12}(0)\rightarrow\lambda_1(0)+4W_1\\
\lambda_2(0) &= V_{11}(0)-V_{12}(0)\rightarrow\lambda_2(0)-2W_1\\
\lambda_1(\mathbf{k}_0) &= V_{11}(\mathbf{k}_0)+V_{12}(\mathbf{k}_0)\rightarrow\lambda_1(\mathbf{k}_0)+2W_1\\
\lambda_2(\mathbf{k}_0) &= V_{11}(\mathbf{k}_0)-V_{12}(\mathbf{k}_0)\rightarrow\lambda_2(\mathbf{k}_0)-2W_1
\end{aligned} \tag{14.2.46}$$

The relations (14.2.46) show that the "switch on" of the additional repulsion between the nearest-neighbor interstitials equally stabilizes the superstructures A and C associated with the eigenvalues $\lambda_2(0)$ and $\lambda_2(\mathbf{k}_0)$ with respect to the superstructure B associated with the eigenvalue $\lambda_1(\mathbf{k}_0)$. This repulsion also depresses the decomposition reaction in favor of ordering.

This effect of nearest-neighbor repulsions may readily be understood proceeding from a purely geometrical consideration. As follows from Figs. 113a, b, c, the superstructure B is depressed by nearest-neighbor repulsion since, unlike the superstructures A and C, it has interstitial atoms separated by the distance of $\langle a_0/2, 0, 0\rangle$.

Summing up the foregoing, we may conclude that the contact nearest-neighbor solute atom repulsion should renormalize the diagram in Fig. 112.

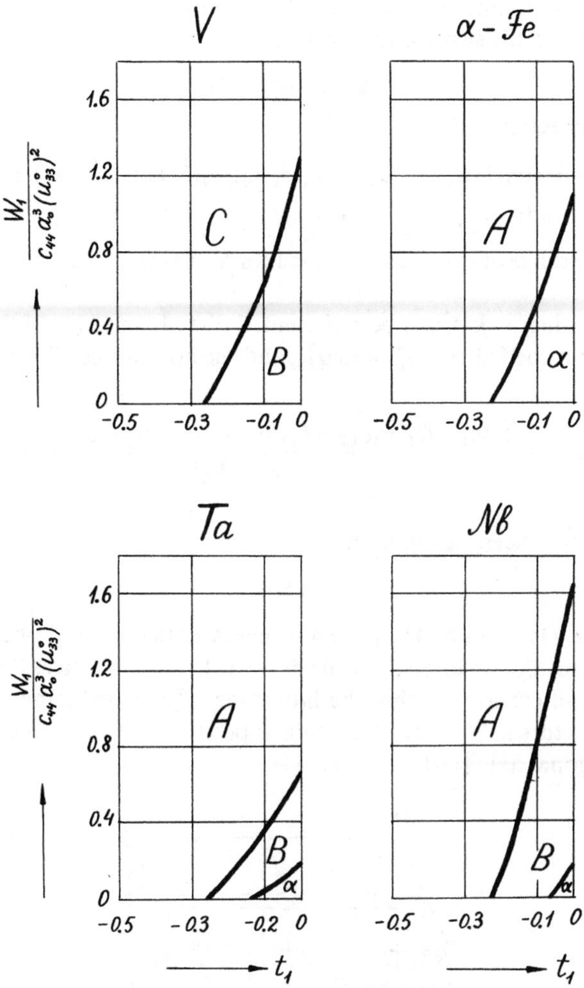

Figure 114. The stability fields of the A, B, and C phases on the W_1-t_1 diagram for V, Nb, Ta, and Fe. α is the field related to the isostructural decomposition.

This should expand the A and C phase stability ranges in t_1 at the expense of the B phase and the disordered phase. This effect is displayed in Fig. 114.

14.2.4. Ordering of Interstitials in Tetrahedral Interstices of the BCC Lattice of V, Nb, Ta (171)

The problem of finding of high-temperature superstructures in the case of tetrahedral interstice occupation is reduced to the determination of the eigenvalues of the 6×6 matrix $V_{pq}(\mathbf{k})$ given by Eq. (14.2.1), with $\mathbf{F}(p, \mathbf{k}) = \mathbf{F}^{\text{tetr}}(p, \mathbf{k})$ defined by (13.3.19).

Numerical computations show that the minimum eigenvalue, $\lambda_{\sigma_0}(\mathbf{k}_0)=\lambda_1(\mathbf{k}_0)$, corresponding to the superlattice vector,

$$\mathbf{k}_0=\pi(\mathbf{a}_1^*+\mathbf{a}_2^*) \tag{14.2.47}$$

and the eigenvector,

$$\begin{aligned} v_{\sigma_0}(p, \mathbf{k}_0) &= (v_1(1, \mathbf{k}_0), v_1(2, \mathbf{k}_0), v_1(3, \mathbf{k}_0), v_1(\bar{1}, \mathbf{k}_0), v_1(\bar{2}, \mathbf{k}_0), v_1(\bar{3}, \mathbf{k}_0)) \\ &= (v_1, v_1, v_3, v_1, v_1, v_3) \end{aligned} \tag{14.2.48}$$

within a wide range of t_1 values: $0<t_1<1$ for V, $0.3<t_1<1$ for Nb, $-0.3<t_1<1$ for Ta.

The eigenvalue, $\lambda_1(\mathbf{k}_0)$, can be represented directly in terms of the concentration expansion coefficients, u_{11}^0 and u_{33}^0, and the host lattice vibration frequency (109):

$$\lambda_1(\mathbf{k}_0) = -2a_0^4 \frac{[(c_{11}+3c_{12})u_{11}^0+(c_{12}+c_{11})u_{33}^0]^2+2[(c_{11}+c_{12})u_{11}^0+c_{12}u_{33}^0]^2}{m\omega_L^2(\mathbf{k}_0)}+Q \tag{14.2.49}$$

The dominant concentration mode

$$v_1(p, \mathbf{k}_0)e^{i\mathbf{k}_0\mathbf{r}}$$

where $v_1(p, \mathbf{k}_0)$ is given by Eq. (14.2.48) generates the ordered phase that arises due to the transition of interstitials from $p=1, \bar{1}, 2, \bar{2}$ sublattices into the $p=3, \bar{3}$ sublattices and ordering within the latter ones. The stoichiometric formula of the superstructure is MeX. Its dominant superstructure vector $\mathbf{k}_0$ is orthogonal to the tetragonal axis [001] (Fig. 115).

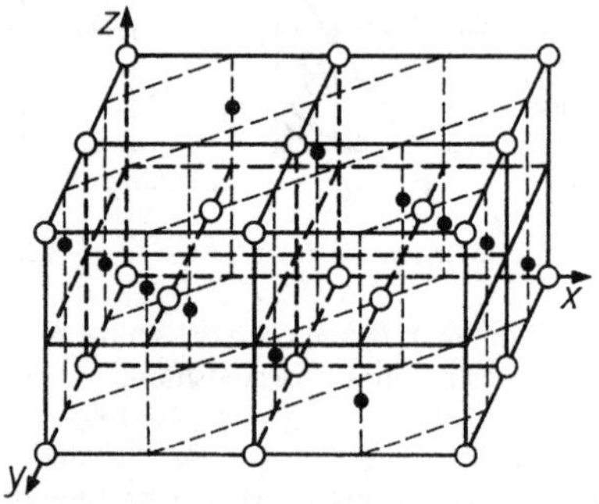

Figure 115. The stable high-temperature superstructures during the tetrahedral interstice occupation.

14.3. COMPARISON WITH EXPERIMENTAL OBSERVATIONS

All the foregoing consideration is valid on the assumption that the host lattice is mechanically stable with respect to the displacement field produced by interstitial atoms. Taking into account the fact that the crystal lattice distortion produced by an interstitial atom is much larger than that associated with a substitutional atom, we observe that the displacement effect on the mechanical stability of an interstitial solution is too large to be ignored in all cases.

The loss of mechanical stability of the host lattice in ordering limits the possibility of comparing theoretical predictions with experimental results. A direct comparison, however, is possible in cases where there is a small perturbation effect of solute atoms on the host lattice, in which case there is a small atomic diameter and a low atomic concentration of interstitials. The latter condition cannot be realized for stoichiometric high-temperature superstructures since their stoichiometric formula, MeX, corresponds to the high concentration of interstitials. Hydrides and deuterides are the only exception since the strain effect produced by hydrogen atoms, even at high hydrogen contents, proves to be too weak to destabilize the host atom lattice. Secondary ordering has been observed in systems other than hydrides and deuterides.

Fortunately, in most cases the crystal lattice rearrangement of the host lattice produced by mechanical instability can be predicted from the displacive modes in the initial host lattice which are generated by concentration waves of the ordered phases.* Moreover after the loss of stability of the host lattice, the complete structure of the ordered phase can be predicted if the assumption about the diffusionless (martensitic) character of the crystal lattice rearrangement is made.

As follows from Fig. 113a, the formation of the high-temperature phase A results in the transition of all interstitial atoms into O_z octahedral sites. This in turn leads to an increase in the axial ratio c/a which becomes larger than unity [see Eq. (6.8.22)]. The concentration dependence of the axial ratio of the A phase can be derived from Eq. (6.8.22):

$$\frac{c}{a}=\frac{a_0(1+u_{33}^0\bar{c})}{a_0(1+u_{11}^0\bar{c})}\approx 1+(u_{33}^0-u_{11}^0)\bar{c} \tag{14.3.1}$$

If the increase in the tetragonal axial ratio amounts to $\sqrt{2}$, the transition of the body-centered tetragonal (bct) host lattice to the fcc lattice occurs. The value of $\sqrt{2}$ is, however, the upper limit. Mechanical instability of the bct host lattice with respect to the fcc lattice may really take place at considerably smaller values of the axial ratio, c/a. This critical value, $(c/a)_{cr}$, determines the critical content, $\bar{c}_0$, of interstitials in the A phase. An increase in the interstitial atom content above the critical value $\bar{c}_0$ results in mechanical instability of the bct host lattice and its diffusionless rearrangement to form the fcc lattice. One may see from Fig. 113a that the bct→fcc diffusionless crystal lattice rearrangement of the A phase results in the formation of the NaCl-type structure.

To find the structure formed from the structure C as a result of the host lattice mechanical instability, we should determine displacive modes that provide the host lattice rearrangement associated with the instability. Displacements

*The term "host lattice instability" is applied to cases where the space group of the transformed host lattice is not a subgroup of the initial host lattice [this case is considered in detail in the review by Somenkov and Shil'stein (167)].

generated by concentration waves composing the structure C may be attributed to the host lattice rearrangement if their direction, value, and spatial distribution ensures the formation of one of the metastable modifications of the host crystal (for instance, hcp or fcc modifications). These displacements can readily be found since they are generated by concentration waves that produce an ordered distribution of interstitials. The distribution of interstitial atoms in the stoichiometric completely ordered phase C may be obtained from Eq. (14.2.32) if the values $\bar{c}=\bar{c}_{st}=\frac{1}{2}$ and $\eta_{0,2}=\eta_{1,2}=1$ are taken. The result is

$$n(p, \mathbf{r})=\begin{cases} \frac{1}{2}+\frac{1}{2}e^{i\mathbf{k}_0\mathbf{r}} & \text{at } p=1 \\ \frac{1}{2}-\frac{1}{2}e^{i\mathbf{k}_0\mathbf{r}} & \text{at } p=2 \\ 0 & \text{at } p=3 \end{cases} \tag{14.3.2}$$

One may clearly see that the distribution (14.3.2) is generated by the concentration mode

$$\Delta\tilde{n}(p, \mathbf{k}_0)=[\Delta\tilde{n}(1, \mathbf{k}_0), \Delta\tilde{n}(2, \mathbf{k}_0), \Delta\tilde{n}(3, \mathbf{k}_0)]=(\tfrac{1}{2}, -\tfrac{1}{2}, 0) \tag{14.3.3}$$

Substituting amplitudes (14.3.3) of the concentration waves into (13.2.43) yields

$$\begin{aligned}\tilde{v}_i(\mathbf{k}_0)&=G_{ij}(\mathbf{k}_0)(F_j^{\text{oct}}(1, \mathbf{k}_0)\Delta\tilde{n}(1, \mathbf{k}_0)+F_j^{\text{oct}}(2, \mathbf{k}_0)\Delta\tilde{n}(2, \mathbf{k}_0)+F_j^{\text{oct}}(3, \mathbf{k}_0)\Delta\tilde{n}(3, \mathbf{k}_0))\\ &=G_{ij}(\mathbf{k}_0)[F_j^{\text{oct}}(1, \mathbf{k}_0)\tfrac{1}{2}-F_j^{\text{oct}}(2, \mathbf{k}_0)\tfrac{1}{2}]\end{aligned} \tag{14.3.4}$$

where $\tilde{v}_i(\mathbf{k}_0)$ is the Fourier transform of the local displacement field produced by ordering in the structure C, $\mathbf{F}^{\text{oct}}(p, \mathbf{k})$ is given by Eqs. (13.3.16) and (13.3.17). Using the representations (13.3.16) and (13.3.17) at

$$\mathbf{k}=\mathbf{k}_0=\pi(\mathbf{a}_1^*+\mathbf{a}_2^*)$$

we have

$$\begin{aligned}&\mathbf{F}^{\text{oct}}(1, \mathbf{k}_0)\\ &=-ia_0^2e^{-i(k_x^0a_0/2)}\left(\sigma_{33}^0 \sin\frac{k_x^0a_0}{2}, \sigma_{11}^0 \sin\frac{k_y^0a_0}{2}\cos\frac{k_z^0a_0}{2}, \sigma_{11}^0 \sin\frac{k_z^0a_0}{2}\cos\frac{k_y^0a_0}{2}\right)\\ &=a_0^2(\sigma_{33}^0, \sigma_{11}^0, 0)\end{aligned} \tag{14.3.5a}$$

$$\begin{aligned}\mathbf{F}^{\text{oct}}(2, \mathbf{k}_0)=&-ia_0^2 \exp\left(-i\frac{k_y^0a_0}{2}\right)\\ &\times\left(\sigma_{11}^0 \sin\frac{k_x^0a_0}{2}\cos\frac{k_z^0a_0}{2}, \sigma_{33}^0 \sin\frac{k_y^0a_0}{2}, \sigma_{11}^0 \sin\frac{k_z^0a_0}{2}\cos\frac{k_x^0a_0}{2}\right)\\ =&\,a_0^2(\sigma_{11}^0, \sigma_{33}^0, 0)\end{aligned} \tag{14.3.5b}$$

It follows from (14.3.5) that

$$\begin{aligned}\mathbf{F}^{\text{oct}}(1, \mathbf{k}_0)-\mathbf{F}^{\text{oct}}(2, \mathbf{k}_0)&=a_0^2(\sigma_{33}^0-\sigma_{11}^0)(1, \bar{1}, 0)\\ &=a_0^2(c_{11}-c_{12})(u_{33}^0-u_{11}^0)(1, \bar{1}, 0)\end{aligned} \tag{14.3.6}$$

Eq. (14.3.4) can be simplified by using the bilinear representation (13.2.46) of the Green function at $\mathbf{k}=\mathbf{k}_0$:

$$G_{ij}(\mathbf{k}_0)=\frac{e_L^i(\mathbf{k}_0)e_L^j(\mathbf{k}_0)}{m\omega_L^2(\mathbf{k}_0)}+\frac{e_{T_1}^i(\mathbf{k}_0)e_{T_1}^j(\mathbf{k}_0)}{m\omega_{T_1}^2(\mathbf{k}_0)}+\frac{e_{T_2}^i(\mathbf{k}_0)e_{T_2}^j(\mathbf{k}_0)}{m\omega_{T_2}^2(\mathbf{k}_0)} \tag{14.3.7}$$

where $\mathbf{e}_L(\mathbf{k}_0)=(1/\sqrt{2},\ 1/\sqrt{2},\ 0)$, $\mathbf{e}_{T_1}(\mathbf{k}_0)=(1/\sqrt{2},\ \bar{1}/\sqrt{2},\ 0)$, $\mathbf{e}_{T_2}(\mathbf{k}_0)=(0,\ 0,\ 1)$ are the polarization vectors corresponding to the longitudinal and two transversal waves at $\mathbf{k}=\mathbf{k}_0=\pi(\mathbf{a}_1^*+\mathbf{a}_2^*)$, respectively, $\omega_L(\mathbf{k}_0)$, $\omega_{T_1}(\mathbf{k}_0)$, and $\omega_{T_2}(\mathbf{k}_0)$ are the corresponding vibration frequencies.

Combining Eqs. (14.3.7) and (14.3.6) with (14.3.4), we have

$$\tilde{\mathbf{v}}(\mathbf{k}_0)=\frac{a_0^2}{\sqrt{2}}\frac{(c_{11}-c_{12})(u_{33}^0-u_{11}^0)}{m\omega_{T_1}^2(\mathbf{k}_0)}\mathbf{e}_{T_1}(\mathbf{k}_0)=\frac{a^2}{2}\cdot\frac{(c_{11}-c_{12})(u_{33}^0-u_{11}^0)}{m\omega_{T_1}^2(\mathbf{k}_0)}(1,\ \bar{1},\ 0) \tag{14.3.8}$$

The displacements in the crystal lattice site representation are given by the back Fourier transform which in the case of (14.3.8) is reduced to

$$\mathbf{v}(\mathbf{r})=\begin{cases} v_x(\mathbf{r})=\dfrac{a_0^2}{2}\dfrac{c_{11}-c_{12}}{m\omega_{T_1}^2(\mathbf{k}_0)}(u_{33}^0-u_{11}^0)e^{i\mathbf{k}_0\mathbf{r}} \\ v_y(\mathbf{r})=-\dfrac{a_0^2}{2}\dfrac{c_{11}-c_{12}}{m\omega_{T_1}^2(\mathbf{k}_0)}(u_{33}^0-u_{11}^0)e^{i\mathbf{k}_0\mathbf{r}} \\ 0 \end{cases} \tag{14.3.9}$$

One may readily see that Eq. (14.3.9) describes a "shuffling" of (110) host lattice planes of the bcc lattice along the $[1\bar{1}0]$ direction. The shuffling results in the displacement of each second (110) plane by the value

$$\frac{a_0}{\sqrt{2}}\frac{c_{11}-c_{12}}{m\omega_{T_1}^2(\mathbf{k}_0)}(u_{33}^0-u_{11}^0)$$

in the direction $[1\bar{1}0]$. Such shuffling with the homogeneous tetragonal distortion produced by the interstitial distribution (14.3.2) and resulting in the axial ratio $c/a<1$ leads to the bcc→hcp crystal lattice rearrangement [the Burgers distortion (301)].

The diffusionless bcc→hcp crystal lattice rearrangement of the C structure results in the formation of the NiAs hcp structure in which interstitials occupy octahedral sites between the nearest $(001)_{\text{hcp}}$ host crystal planes of the hcp lattice. In this situation the ordered distribution of interstitial atoms is inherited from the C structure.* They form a primitive hexagonal lattice whose period c is twice as low as the period c of the host hcp lattice and whose period a coincides with the period of the hcp lattice (the WC-type structure).

All calculations in Section 14.2 proceeded from the premise that the pairwise interaction between interstitial atoms is determined only by elastic interaction.

*It should be remembered that the interstitial atom distribution resulting from the diffusionless crystal lattice rearrangement may be changed because of ordering within the newly formed host crystal lattice.

The chemical interaction was completely ignored. It was, however, shown in Section 14.2 that the nearest-neighbor "chemical" repulsion between interstitials depresses the *B* structure and the disordered structure and stabilizes the *A* and *C* structures. In this regard we can expect the formation of a high-temperature structure *A* in Nb-, Ta-, and Fe-based solutions (see Fig. 112) and the structure *C* in V-based alloys.

The superstructure *A* is, as shown above, transformed into the NaCl-type cubic phase. This phase is really observed in the Nb-O, Nb-N, Nb-C, Ta-O, Ta-C, and Fe-O systems. The effect of the bcc lattice instability results in the situation where we may expect the appearance of ordered phases only at low contents of interstitial atoms. For instance, the metastable superstructure *A* is really observed in the iron-carbon and iron-nitrogen martensites (it is followed by spinodal decomposition and secondary ordering, respectively).

Bcc-based superstructures can be thermodynamically stable at low temperatures if their stoichiometric compositions correspond to comparatively low interstitial atom contents. The latter takes place only for secondary ordered phases. The second condition of host lattice stability is stipulated for the interstitial atom diameter. To ensure the mechanical stability of the bcc host lattice of the secondary ordered phase, the atomic diameter of interstitials should be small. The latter condition is satisfied in metals for O-atoms and especially for H-atoms.

Experimental observations confirm the theoretical predictions for octahedral-based hydrides, deuterides, and oxides. The secondary ordered superstructures Ta_2O (23), Nb_2O (162), V_2H (170), and V_2D (168) which have the predicted structure displayed in Fig. 113d were really observed.

Interstitial alloys with C and N in Ta, Nb, and V have no predicted superstructures since the atomic diameters of carbon and nitrogen are too large. Ordering in these alloys would involve atomic distortion which results in the bcc host lattice instability. It is, however, of interest to note that the observed L3′ hexagonal superstructures in Nb-C, and V-C alloys and Ta_2C in the Ta-C system are closely related to the predicted bcc-based superstructure. The L3′ superstructures can be obtained from the bcc-based high-temperature *C* phase by the bcc→hcp diffusionless crystal lattice rearrangement. The Ta_2C superstructure can also be derived from the secondary ordered *D* phase if the bcc→hcp diffusionless transformation is applied.

The tetrahedral site occupation case was observed for hydrides and deuterides of V, Nb, and Ta only. The tetragonality factor for hydrides of these metals are $t_1=0.3$ for V-H, $t_1=0.6$ for Nb-H, and $t_1=0.7$ for Ta-H (163). These values fall within the stability range of the tetrahedral-based MeX superstructure displayed in Fig. 115 (see the earlier discussion of ordering in tetrahedral interstices). This predicted structure has in fact been observed in hydrides and deuterides of all three metals, V, Nb, and Ta (167).

The calculations made here demonstrate that elastic energy makes the dominant contribution to the mixing energy of interstitial alloys and thus controls the structure of interstitial superlattices. The concentration wave formalism enables us to predict the basic interstitial superlattices. The theory is not so successful in predicting the order-disorder temperatures and equilibrium diagrams. Attempts to calculate the critical temperature of ordering proceeding from the eigenvalues (14.2.1) give substantially larger values than those observed. The latter seems to be associated with short-range "chemical" repulsion of interstitial pairs ignored in the calculations. The repulsion is, in particular, caused by a screened Coulomb interaction among interstitial ions, and contact repulsion is caused by an "overlap" among electron shells of atoms. An accurate theoretical computation may be expected with further advances in the pseudopotential theory of transition metals.

Regardless of the fact that the elastic theory provides the dominant contribution to the binding energy of interstitial superstructures and yields stronger effects than short-range "chemical" interaction energy, the chemical free energy may be commensurable with differences between the elastic energies corresponding to various superstructures. In this connection "chemical" repulsion can substantially affect the stability fields of these superstructures. In other words, the elastic microscopic theory enables us to find several most probable "candidates" for a stable structure. To determine which of these "candidates" provides the most stable structure, we should take into account the "chemical" interaction and estimate the effect of this interaction on the stability of the superstructures under consideration. For instance, as shown at the end of Section 14.2, the "chemical" repulsion of the nearest-neighbor interstitials results in the stabilization of phases A and C and destabilization of phase B. It may readily be demonstrated that the next-nearest repulsion would make phase A more stable than C.

Concluding this section, we shall note that a very promising theory of ordered phase reconstruction associated with host lattice instability has been proposed by Somenkov, Irodova, and Shil'shtein (300). The theory (300) is based on the idea that a reconstructive phase transition is the result of ordering in the initial host lattice followed by a displacive host lattice rearrangement. The latter is produced by displacements caused by the ordered distribution of interstitials. The first process (ordering) is characterized by a reduction in the host latice symmetry, whereas the second one by an increase in symmetry. The important point is that the reciprocal lattices of both host lattices (before and after the host lattice rearrangement) and the ordered phase prove to be closely related to each other. A part of the reciprocal lattice points of one phase may be brought into coincidence with the reciprocal lattice points of another phase by means of a slight homogeneous crystal lattice distortion. In this connection the reciprocal lattice of any phase may be obtained from that of another one by either adding or removing the reciprocal lattice points, while the coinciding reciprocal lattice points determine the orientational relations between the crystal lattices. A symmetry increase in the reciprocal lattice representation can be ensured by the disappearance of certain reciprocal lattice points or by a change of the structure amplitudes belonging to various reciprocal lattice points which after transformation become crystallographically equivalent and thus equal to each other.

It has been reasonably assumed that the locations of disappearing reciprocal lattice sites satisfy Lifshitz's criterion for the reconstructed host lattice. Corresponding displacements within the initial host lattice ensure its reconstruction. The principles formulated above enable us to predict the crystal lattice structure or ordered phases after a reconstructive phase transition. Examples of ordered phase structure determinations based on this theory may be found in the review paper (167).

14.4. ORDERING IN IRON-CARBON MARTENSITE

X-ray studies of carbon steel martensite have led to the conclusion that martensite is a supersaturated interstitial solid solution of carbon in α-iron (105–107). Carbon atoms only occupy O_z interstices and thus form the nonstoichiometric superstructure A (see Fig. 113a). Such a distribution of carbon atoms results in a slight tetragonality of the host bcc lattice. But it remains undecided whether the tetragonality observed at room temperature is merely an effect of the diffusionless transformation (Bain deformation) or whether the ordered distribution is more advantageous than the disordered one from the thermodynamical point of view.

An increase in the tetragonality of carbon martensite of manganese steel (111, 112) (see also section 6.8) at room temperature convincingly solves this problem. It demonstrates without doubt that the ordered (tetragonal) distribution of C-atoms is more stable thermodynamically than the disordered one is.

Numerical calculations made in Section 14.2 showed that strain-induced interaction leads to the formation of an ordered superstructure A in α-Fe within the range $-0.5<t_1<-0.23$. Outside this range, at $t_1>-0.23$, strain-induced interaction results in isostructural decomposition into two disordered cubic phases. The range of stability of the A phase was calculated on the assumption that the interstitial-interstitial interaction is completely determined by pairwise elastic energies.

The idea that elastic (strain-induced) interaction in iron-carbon martensite may provide ordering was first proposed by Zener (270). The similar calculations were made in (303, 302). These results should, however, be revised since the analogues of Eq. (14.2.15) in (270, 302, 303) do not include the term Q (calculation of Q can only be made within the framework of microscopic elasticity). The accurate calculation of strain-induced ordering in iron-carbon martensite may be found in (304).

Now let us make estimations of the energetic parameters for iron-carbon martensite. The numerical calculation for α-Fe based on Eqs. (14.2.2) and (13.3.16) yields (see Table 14.1)

$$\frac{Q}{c_{44}a_0^3(u_{33}^0)^2}=0.425+1.168t_1+1.161t_1^2 \tag{14.4.1}$$

where the dependence (14.4.1) is valid for any iron-based interstitial alloy with octahedral site occupation. In the case of an iron-based interstitial alloy the following numerical values can be assumed:

$$c_{11}=2.42\times10^{12}\,\frac{\text{dyne}}{\text{cm}^2},\ c_{12}=1.465\times10^{12}\,\frac{\text{dyne}}{\text{nm}^2},\ c_{44}=1.12\times10^{12}\,\frac{\text{dyne}}{\text{cm}^2},$$

$$a_0=2.865\ \text{Å},\ u_{33}^0=0.86,\ u_{11}^0=-0.09,\ t_1=-0.1 \tag{14.4.2}$$

Substituting these data into Eq. (14.4.1), (14.2.11), and (14.2.15), we have

$$Q=3.88\ \text{eV}, \tag{14.4.3}$$

and the strain-induced eigenvalues (see Table 14.1)

$$\lambda_1(0)=-14.77\ \text{eV}$$
$$\lambda_2(0)=-2.47\ \text{eV} \tag{14.4.4}$$

The values (14.4.4) show that in the case of an Fe-C alloy the eigenvalue, $\lambda_1(0)$, corresponding to decomposition is lower than the value, $\lambda_2(0)$, corresponding to ordering. In this case an isostructural decomposition should be expected. Ordering that results in the formation of the A phase would occur if $\lambda_2(0)\leqslant\lambda_1(0)$. Therefore the elastic interaction by itself cannot provide stability of an ordered phase in a Fe-C alloy which is really observed. This circumstance, which is crucial to the strain-induced ordering model of the tetragonal iron-carbon martensite, was completely missed in the papers cited (270, 302–304).

To remove this discrepancy between the theory and the experimental observations, we should take into consideration the direct contact "chemical" repulsion of nearest-neighbor C-C pairs caused by an "overlapping" in the electron shells of the nearest carbon atoms. This effect definitely cannot be ignored since the distance between the nearest-neighbor octahedral sites in α-Fe, $\frac{1}{2}a_0=1.437$ Å, is smaller than the distance between two "touching" carbon atoms (the latter is equal to the carbon atom diameter, ~1.54 Å), and thus two nearest-neighbor carbon atoms should deform the electronic shells of each other.

We shall also assume that all chemical interactions other than the contact repulsion can be neglected.* Calculations of the atomic structure of organic compounds provide abundant material for C-C interaction potentials and, particularly, for the contact repulsion associated with electron shell "overlap" (305). The short-range contact repulsion of a C-C pair is well approximated by the potential (306)

$$W(r)=3603\times\exp\left(-\frac{r}{r_0}\right) \tag{14.4.5}$$

expressed in electron-volts where $r_0=0.2778$ Å. The repulsion (14.4.5), if combined with the van der Waals attraction of $-24.47/r^6$ eV, provides a good agreement between the calculated and observed structures of organic crystals.

*Other chemical interactions are of the same order of magnitude as the chemical interaction in a substitutional solution and should therefore be far weaker than both strain-induced interaction and direct contact repulsion in an interstitial solution.

It may, however, be reasonably assumed that the contact repulsion associated with the "overlap" of electron shells of the nearest atoms is not very sensitive to the type of chemical bond in a crystal. Of course the latter is not true for "chemical attraction" which is supposed to be highly affected by the bonding type, and for metals should be described in terms of the electronic theory of metals. Fortunately, in a metal interstitial solution all "chemical" interactions other than the contact repulsion can be ignored. Taking the foregoing discussion into account, we may assume that the contact repulsion (14.4.5) is reasonably good for the description of the short-range contact C-C interaction in a metal interstitial alloy. Using the radii of the first three coordination shells, $a_0/2 = 1.432$ Å, $a_0/\sqrt{2} = 2.025$ Å, $a_0\sqrt{3}/2 = 2.481$ Å, in Eq. (14.4.5), we have

$$W_1 = 20.78 \text{ eV}, W_2 = 2.46 \text{ eV}, W_3 = 0.47 \text{ eV} \tag{14.4.6}$$

Taking the potentials (14.4.6) into account, we may rewrite Eqs. (14.2.9) and (14.2.14) for the eigenvalues $\lambda_1(0)$ and $\lambda_2(0)$ as

$$\begin{aligned}\lambda_1(0) &= -\tfrac{1}{2}a_0^3(c_{11}+2c_{12})(2u_{11}^0+u_{33}^0)^2+Q+[4W_1+8W_2+8W_3]\\ \lambda_2(0) &= -\tfrac{1}{2}a_0^3(c_{11}-c_{12})(u_{33}^0-u_{11}^0)^2+Q+[-2W_1-4W_2+8W_3]\end{aligned} \tag{14.4.7}$$

Substitution of the numerical values (14.4.4) and (14.4.6) into (14.4.7) yields

$$\begin{aligned}\lambda_1(0) &= 91.78 \text{ eV}\\ \lambda_2(0) &= -50.11 \text{ eV}\end{aligned} \tag{14.4.8}$$

Since $\lambda_2(0)$ in (14.4.8) becomes smaller than $\lambda_1(0)$, one may conclude that the strain-induced interaction combined with contact short-range repulsion results in the ordered phase A.

The temperature dependence of the long-range order-parameter will be described by Eq. (14.2.18) which at $\bar{c} \ll 1$ may be reduced to

$$\frac{\kappa T}{\lambda_2(0)\bar{c}} = -\frac{3\eta}{\ln[(1+2\eta)/(1-\eta)]} \tag{14.4.9}$$

The temperature dependence of the long-range order parameter, η, with respect to the reduced temperature $\tau = \kappa T/(|\lambda_2(0)|3c)$ is depicted in Fig. 116.

The free energy of the ordered phase A in the mean-field approximation (3.11.3) may be represented in terms of the eigenvalues, $\lambda_1(0)$ and $\lambda_2(0)$:

$$\begin{aligned}F = \tfrac{1}{2}N\lambda_1(0)c^2 &+ 3N\lambda_2(0)c^2\eta^2 + \kappa TN\{2c(1-\eta)\ln[c(1-\eta)]\\ &+ c(1+2\eta)\ln[c(1+2\eta)] + 2[1-c(1-\eta)]\ln[1-c(1-\eta)]\\ &+ [1-c(1+2\eta)]\ln[1-c(1+2\eta)]\}\end{aligned} \tag{14.4.10}$$

where $\lambda_1(0)$ and $\lambda_2(0)$ are given by Eqs. (14.2.9) and (14.2.14).

At small c Eq. (14.4.10) may be simplified:

$$\begin{aligned}F(c,\eta) \simeq \frac{1}{2}N\lambda_1(0)c &+ 3N\lambda_2(0)c^2\eta^2 + \kappa TNc\ln\frac{c}{3}\\ &+ \kappa TNc[2(1-\eta)\ln(1-\eta) + (1+2\eta)\ln(1+2\eta)]\end{aligned} \tag{14.4.11}$$

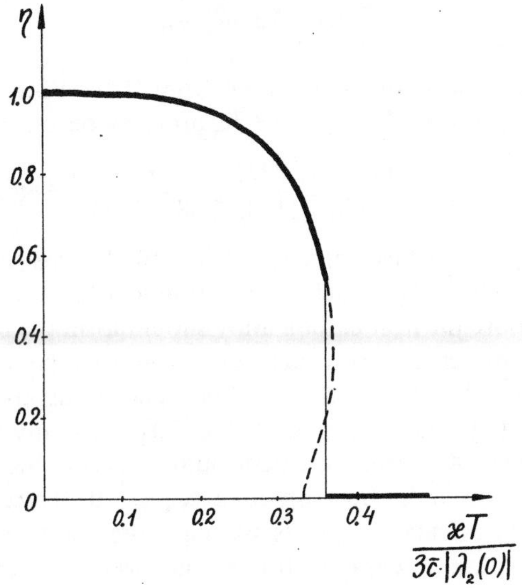

Figure 116. The temperature dependence of the long-range order parameter η with respect to T for the superstructure A. Dash line corresponds to the free energy maximum and should be ignored.

The equilibrium condition

$$F(c, \eta) = F(c, 0) \tag{14.4.12}$$

between the ordered and disordered phases taken at the equilibrium value of the long-range order parameter, η, determines the temperature of the order-disorder transition. The solution of the set of two transcendental equations (14.4.9) and (14.4.12) in η and T unknowns yields the value

$$\tau_0 = \frac{\kappa T_0}{\lambda_2(0)c} = -1.08 \tag{14.4.13}$$

for the reduced temperature of the order-disorder transition (T_0 is the order-disorder transition temperature) and shows that the long-range order parameter, η, has a discontinuity at $T = T_0$. It changes from zero at $T > T_0$ to ~ 0.5 at $T = T_0$. The latter shows ordering to be a first-order transition.

It follows from Eq. (14.4.13) that ordering temperature T_0 is determined by the equation

$$T_0 = -1.08 \frac{\lambda_2(0)c}{\kappa} \tag{14.4.14}$$

Ordering in martensite can be detected through the axial ratio measurements. According to Eqs. (6.8.5) and (14.2.16), we have

$$a = a_0[1 + u_{11}^0(n_2 + n_3) + u_{33}^0 n_1] = a_0[1 + (2u_{11}^0 + u_{33}^0)c - c\eta(u_{33}^0 - u_{11}^0)] \tag{14.4.15a}$$

$$c = a_0[1 + u_{11}^0(n_1 + n_2) + u_{33}^0 n_3] = a_0[1 + (2u_{11}^0 + u_{33}^0)c + 2c\eta(u_{33}^0 - u_{11}^0)] \tag{14.4.15b}$$

$$\frac{c}{a} \approx 1 + 3c\eta(u_{33}^{0} - u_{11}^{0}) \tag{14.4.15c}$$

Substituting the value $\lambda_2(0) = -50.11$ eV from (14.4.8) into (14.4.14), we find that at room temperature, $T = T_2 = 293°$K, ordering occurs at

$$c \geqslant c_0 = -\frac{\kappa T_2}{1.08 \times \lambda_2(0)} = \frac{293}{11606} \frac{1}{1.08 \times 50.11} = 4.66 \times 10^{-4}$$

Since $c = N_c/3N_{\text{Fe}}$, the critical value $c_0 = 4.66 \times 10^{-4}$ corresponds to the atomic fraction C/Fe $= 1.399 \times 10^{-3}$ which is 0.139 atomic percent C and 0.03 percent C by weight. Therefore the iron-carbon alloy should form the ordered tetragonal solution (the nonstoichiometric phase A) at room temperature if the carbon content is above 0.03 percent C by weight. The tetragonality of an α-Fe-C solution produced by ordering [see Eq. (14.4.15c)] can hardly be detected at such low carbon content. X-ray measurements allow one to observe the martensite phase tetragonality if the carbon content exceeds ~0.2 percent C by weight.

Summing up the calculation results, we may note that they are in agreement with x-ray observations of tetragonality in iron-carbon martensite. All attempts to find the order-disorder transition in martensite (the cubic-to-tetragonal phase transition) failed. Martensite observed at room temperature is always found to be tetragonal. At small carbon contents near 0.03 percent C by weight where the order-disorder transition may be expected, the ordered phase tetragonality should be too small to be detected.

At the end of this section a couple of remarks should be made. Regardless of the fact that contact repulsion is stronger than strain-induced interaction energies, it has no effect on the ordered phase thermodynamics when interstitial atoms within the ordered phase are separated by distances exceeding the third coordination shell radius for octahedral interstices, $a_0\sqrt{3}/2$. In such a case the atomic arrangement of the ordered phase is fully determined by strain-induced interaction. This is due to the radius of the short-range contact repulsion being on the order of $a_0\sqrt{3}/2$ whereas the radius of the elastic interaction is equal to infinity.

The second remark concerns the stability of the ordered phase A which is formed in an α-Fe-C alloy. Carbon content in iron-carbon martensite is substantially smaller than that corresponding to the stoichiometric composition of the A phase. As shown in Sections 3.11 and 14.2, in this situation an ordered solution cannot be stable. It will undergo a secondary phase transformation. It will be shown in the next section that the secondary phase transition results in a spinodal decomposition of the martensite phase.

14.5. SPINODAL DECOMPOSITION IN IRON-CARBON MARTENSITE

We have shown that the C-C interaction in an iron-carbon martensite results in the formation of the high-temperature ordered phase A. The atoms C are randomly distributed within the only "permitted" bcc sublattice of O_z octahedral sites in the phase A. The other bcc octahedral site sublattices, O_x and O_y, are vacant. Such an interstitial solution may be treated as a model bcc disordered binary "substitutional" solution whose crystal lattice sites are O_z octahedral interstices. In this model solution C atoms play the part of solute particles, whereas the rest of the sites which are carbon vacancies in the O_z sublattice may be interpreted as "solvent atoms." Thus consideration of the secondary phase transformation in the Fe-C martensite is substantially simplified since it is reduced to an analysis of the phase transformation in a disordered bcc binary substitutional solution with a known interaction potential.

According to the procedure described in Section 3.7, to decide whether the phase transformation is an ordering or a decomposition, we should find the minimum of the Fourier transform, $V(\mathbf{k})$, of pairwise interaction energies. If the minimum of $V(\mathbf{k})$ falls on $\mathbf{k}=0$, the phase transformation is a decomposition. Otherwise, if $V(\mathbf{k})$ assumes its minimal value at $\mathbf{k}=\mathbf{k}_0\neq 0$, ordering occurs. In the case under consideration, when C atoms occupy only O_z octahedral sites

$$V(\mathbf{k})=V_{33}(\mathbf{k})+V^{\text{chem}}(\mathbf{k}) \tag{14.5.1}$$

where according to Eq. (14.2.1)

$$V_{33}(\mathbf{k})=-F_i^{\text{oct}}(3,\mathbf{k})G_{ij}(\mathbf{k})F_j^{*\text{oct}}(3,\mathbf{k})+Q \tag{14.5.2}$$

and $V^{\text{chem}}(\mathbf{k})$ is the Fourier transform of the contact short-range repulsive energies (14.4.5):

$$V^{\text{chem}}(\mathbf{k})=\sum_{\mathbf{r}} W(\mathbf{r})e^{-i\mathbf{k}\mathbf{r}} \tag{14.5.3}$$

The summation in Eq. (14.5.3) is over only the bcc lattice sites. Since the potential (14.4.5) practically vanishes beyond the radius $a_0\sqrt{3}/2$ [see Eq. (14.4.6)], and $a_0\sqrt{3}/2$ is the first coordination shell radius of the bcc lattice, the nearest-neighbor interaction model for the contact repulsion should be applied.

As follows from (14.4.5), the nearest-neighbor interaction energies corresponding to the radius $a_0\sqrt{3}/2$ of the first coordination shell of the bcc lattice are equal to

$$W_3=3603\exp\left(-3.60\,\frac{a_0\sqrt{3}}{2}\right)(\text{eV}) \tag{14.5.4}$$

where a_0 is given in angstroms.

The sum (14.5.3) in the nearest-neighbor interaction model may be rewritten

$$V^{\text{chem}}(\mathbf{k})=8W_3\cos\left(\tfrac{1}{2}k_xa_0\right)\cos\left(\tfrac{1}{2}k_ya_0\right)\cos\left(\tfrac{1}{2}k_za_0\right) \tag{14.5.5}$$

where (k_x, k_y, k_z) are the components of the vector $\mathbf{k}$ in the Cartesian system related to axes [100], [010], and [001] of the bcc lattice.

Eq. (14.5.1) is thus

$$V(\mathbf{k})=-F_i^{\text{oct}}(3,\mathbf{k})G_{ij}(\mathbf{k})F_j^{*\text{oct}}(3,\mathbf{k})+Q+8W_3\cos\frac{k_xa_0}{2}\cos\frac{k_ya_0}{2}\cos\frac{k_za_0}{2} \tag{14.5.6}$$

where the Fourier transform $\mathbf{F}^{oct}(3, \mathbf{k})$ is given by Eq. (13.3.16) and in accordance with (14.4.3), $Q=3.88$ eV. The function $V(\mathbf{k})$ in (14.5.6) with a W_3 of 0.49 eV following from (14.5.4), assumes its minimum value at $\mathbf{k}=0$. The latter quantity indicates that the secondary phase transition results in the spinodal decomposition.

As shown in Section 2.2, a disordered solid solution is stable with respect to arbitrary infinitesimal concentration heterogeneities if the characteristic function $b(\mathbf{k}, T)$ related to the disordered state is positive at any wave vector, $\mathbf{k}$. In a mean-field approximation the characteristic function $b(\mathbf{k})$ is given by Eq. (5.1.26)

$$b(\mathbf{k}, T, \bar{n}) = V(\mathbf{k}) + \frac{\kappa T}{\bar{n}(1-\bar{n})} \tag{14.5.7}$$

where $\bar{n}$ is the fraction of the bcc lattice sites occupied by solute atoms. In the martensite case where C atoms occupy only O_z interstices, $\bar{n}=\text{C/Fe}=3c$ since, by definition, $c=(\text{C/Fe})/3$. Therefore, taking the definition (14.5.7) into account, we may represent the stability condition with respect to arbitrary infinitesimal heterogeneities in the form of the inequality

$$b(\mathbf{k}, T, \bar{n}) = V(\mathbf{k}) + \frac{\kappa T}{\bar{n}(1-\bar{n})} > 0 \tag{14.5.8}$$

which holds at any $\mathbf{k}$.

At the phase transformation temperature the stability condition (14.5.8) is violated at the point $\mathbf{k}_0$, ensuring the minimum value of $V(\mathbf{k})$, and we have

$$V(\mathbf{k}_0) + \frac{\kappa T_0}{\bar{n}(1-\bar{n})} = 0 \tag{14.5.9}$$

where $V(\mathbf{k}_0)=\min V(\mathbf{k})$. Eq. (14.5.9) as a matter of fact defines the phase transformation temperature, T_0. If the temperature is decreased below T_0, the characteristic function (14.5.7) becomes negative within some region about the wave vector, $\mathbf{k}_0$, where $V(\mathbf{k})$ assumes its minimum value:

$$b(\mathbf{k}, T, \bar{n}) = V(\mathbf{k}) + \frac{\kappa T}{\bar{n}(1-\bar{n})} \leqslant 0 \tag{14.5.10}$$

at a range of $\mathbf{k}$ around $\mathbf{k}_0$ below T_0. We shall call the region in $\mathbf{k}$-space where the inequality (14.5.10) holds the instability range since, as follows from Eq. (2.2.12), the appearance of an amplitude in any concentration wave whose wave vector, $\mathbf{k}$, is within the instability range results in a reduction in free energy. The boundary of the instability range will be determined by the surface in the $\mathbf{k}$-space where the sign of $b(\mathbf{k}, T, \bar{n})$ changes, that is, where

$$b(\mathbf{k}, T, \bar{n}) = V(\mathbf{k}) + \frac{\kappa T}{\bar{n}(1-\bar{n})} = 0 \tag{14.5.11}$$

In the case under consideration, when the minimum of $V(\mathbf{k})$ is assumed at $\mathbf{k}\to 0$, the region where $b(\mathbf{k}, T, \bar{n})$ becomes negative encircles the point $\mathbf{k}=\mathbf{k}_0=0$.

Therefore the instability range is located about $\mathbf{k}=0$, and we have the case of decomposition rather than ordering. According to Eq. (5.1.27), at the early stage of spinodal decomposition, amplitudes of all concentration waves whose wave vectors belong to the instability range will increase exponentially.

Since diffuse scattering intensity at a point $\mathbf{k}$ in the reciprocal space counted from the nearest fundamental reflection spot is proportional to the squared module of the amplitude of the static concentration wave with the wave vector $\mathbf{k}$, the shape of the diffuse scattering maxima in the vicinity of fundamental reflections is the same as the shape of the instability range where concentration wave amplitudes grow with time. Therefore the shape of diffuse maxima around fundamental reflections of martensite should be described by Eq. (14.5.11).

Substituting Eq. (14.5.6) to (14.5.11), we have

$$-F_i^{\text{oct}}(3,\mathbf{k})G_{ij}(\mathbf{k})F_j^{*\text{oct}}(3,\mathbf{k})+Q+8W_3\cos\frac{k_xa_0}{2}\cos\frac{k_ya_0}{2}\cos\frac{k_za_0}{2}+\frac{\kappa T}{\bar{n}(1-\bar{n})}=0 \tag{14.5.12}$$

Since the instability range in the spinodal decomposition case corresponds to small $\mathbf{k}$ value, we can rewrite Eq. (14.5.12):

$$-F_i^{\text{oct}}(3,\mathbf{k})G_{ij}(\mathbf{k})F_j^{*\text{oct}}(3,\mathbf{k})+Q+8W_3+\frac{\kappa T}{\bar{n}(1-\bar{n})}=0$$

or

$$F_i^{\text{oct}}(3,\mathbf{k})G_{ij}(\mathbf{k})F_j^{*\text{oct}}(3,\mathbf{k})=Q+8W_3+\frac{\kappa T}{\bar{n}(1-\bar{n})} \tag{14.5.13}$$

It is convenient to represent Eq. (14.5.13) in dimensionless form:

$$\frac{F_i^{\text{oct}}(3,\mathbf{k})G_{ij}(\mathbf{k})F_j^{*\text{oct}}(3,\mathbf{k})}{c_{44}a_0^3(u_{33}^0)^2}=\frac{Q+8W_3+[\kappa T/(\bar{n}(1-\bar{n}))]}{c_{44}a_0^3(u_{33}^0)^2} \tag{14.5.14}$$

The relation (14.5.14) is the equation of the instability range boundary surface in $\mathbf{k}$-space. Let us estimate the constant in the right-hand side of Eq. (14.5.14) for a Fe alloy containing 1 percent C by weight. The nearest-neighbor distance in the teragonally distorted bcc lattice of this Fe alloy becomes slightly larger than $a_0\sqrt{3}/2=2.481$ Å. It is equal to 2.560 Å. Using this value in Eq. (14.5.4), we have

$$W_3=0.358 \text{ eV} \tag{14.5.15}$$

Comparing this value with W_3 of 0.476 eV, which corresponds to the nearest-neighbor distance in the bcc lattice of pure iron, one may see that the strong dependence of the contact potential (14.5.4) on the interatomic distance results in a considerable correction due to the concentration-induced crystal lattice expansion that cannot be ignored. At $T=293^\circ K$ and $\bar{n}=0.0466$ (this corresponds to 1 percent C by weight) we have

$$Q+8W_3+\frac{\kappa T}{\bar{n}(1-\bar{n})}=3.880+8\times0.358+0.567=7.311 \text{ eV}$$

Using the dimensionless variables, we have

$$\frac{Q+8W_3+[\kappa T/(\bar{n}(1-\bar{n}))]}{c_{44}a_0^3(u_{33}^0)^2}=\frac{7.311}{12.17}=0.60 \tag{14.5.16}$$

The plot

$$\frac{F_i^{\text{oct}}(3,\mathbf{k})G_{ij}(\mathbf{k})F_j^{*\text{oct}}(3,\mathbf{k})}{c_{44}a_0^3(u_{33}^0)^2}=0.56 \tag{14.5.17}$$

where $\mathbf{F}^{\text{oct}}(3, \mathbf{k})$ is given by Eq. (13.3.16) is depicted in Fig. 117, as given in the paper (307). The corresponding shape of the surface described by Eq. (14.5.17) is depicted in Fig. 118. The region around each fundamental reciprocal lattice point of martensite limited by this surface is the instability range where the amplitudes of concentration waves grow with time. It thus gives the shape of diffuse scattering intensity maxima which are associated with concentration heterogeneities produced by spinodal decomposition.

The carbon atom redistribution process in a freshly formed martensite that does not affect the tetragonal axial ratio was observed by Izotov and Utevsky (308) in selected area electron diffraction patterns. They found diffuse intensity

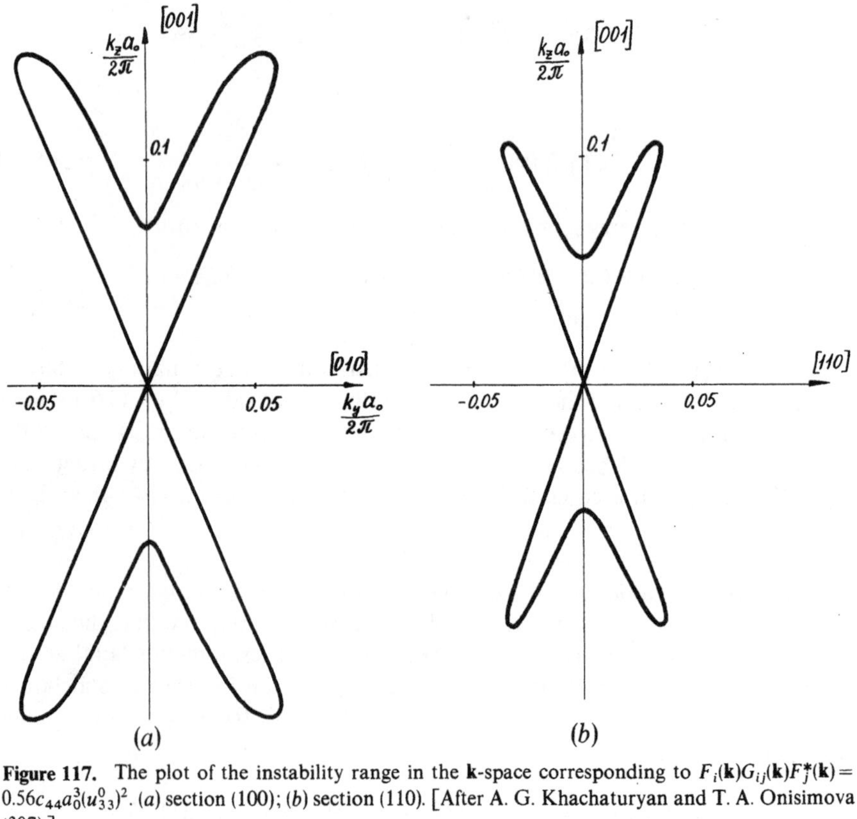

Figure 117. The plot of the instability range in the **k**-space corresponding to $F_i(\mathbf{k})G_{ij}(\mathbf{k})F_j^*(\mathbf{k})=0.56c_{44}a_0^3(u_{33}^0)^2$. (*a*) section (100); (*b*) section (110). [After A. G. Khachaturyan and T. A. Onisimova (307).]

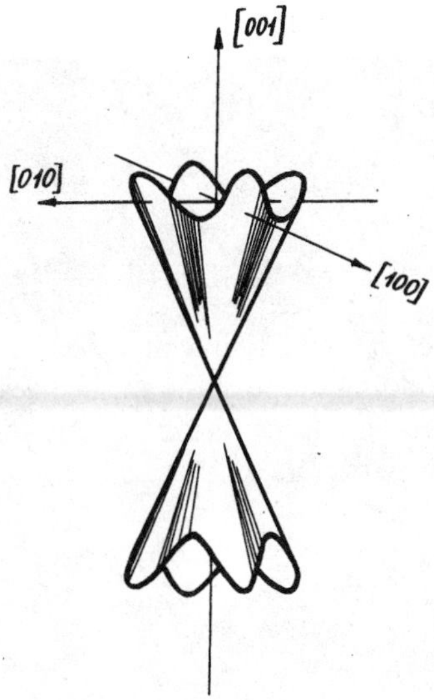

Figure 118. The spatial shape of diffuse scattering maxima near the reciprocal lattice site of martensite described by the surface $F_i(\mathbf{k})G_{ij}(\mathbf{k})F_j^*(\mathbf{k})=0.56c_{44}a_0^3(u_{33}^0)^2$. [After A. G. Khachaturyan and T. A. Onisimova (307).]

maxima near the martensite reciprocal lattice points at room temperature, Fig. 119a and b. Subsequent experiments (309) showed that the diffuse electron scattering from freshly formed martensite is absent at temperatures below $-60°$C. The diffuse scattering effects appeared after tempering at 10°C for 30 min and at 20°C for 5 min. These measurements suggest that the diffuse effects in the electron diffraction patterns are caused by carbon redistribution processes. Furthermore these effects appear to occur in the same O_z octahedral sublattice because the diffuse effect does not influence the tetragonal axial ratio.

Comparison of the calculated (Fig. 120a, b) and observed (Fig. 119a, b) diffraction patterns demonstrates the striking similarity between them. The idea that the observed effects in iron-carbon martensite may be associated with strain-induced interactions was proposed by Khachaturyan and Onisimova in the paper (307) where the shape of diffuse maxima corresponding to Eq. (14.5.17) was calculated for the first time.

It should be noted that an agreement between the calculated and observed shapes of diffuse maxima was achieved without fitting parameter. All the parameters used were taken from the results of independent measurements. This is especially substantial since the calculation results are very sensitive to the value of the energetic parameter in the right-hand side of Eq. (14.5.14). A deviation of this parameter that drives it out of a comparatively narrow range of 0.5 to 0.6 results in a change in the shape of the calculated diffraction maxima which then differs from the shapes observed in selected area electron diffraction patterns.

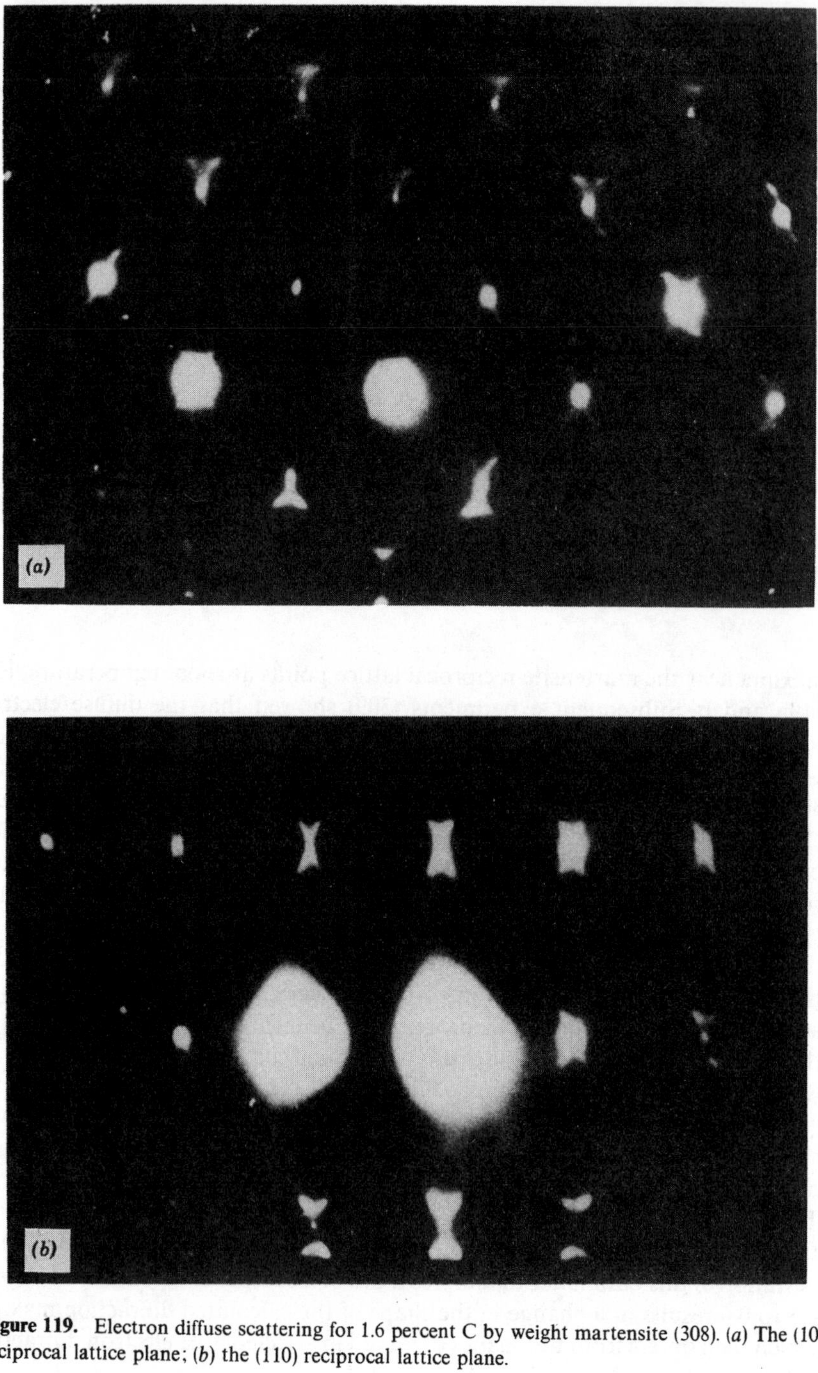

Figure 119. Electron diffuse scattering for 1.6 percent C by weight martensite (308). (*a*) The (100) reciprocal lattice plane; (*b*) the (110) reciprocal lattice plane.

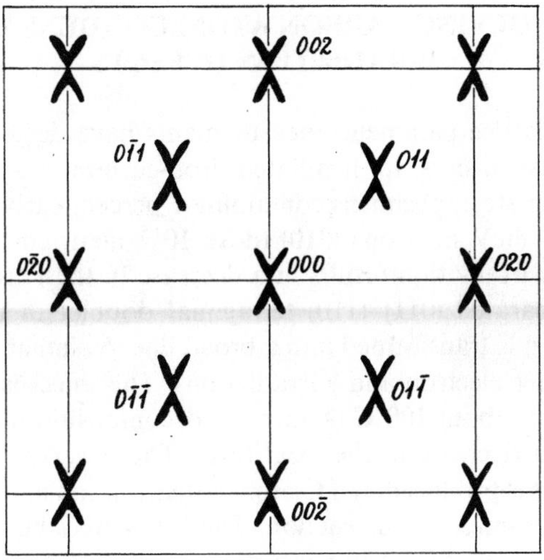

(a)

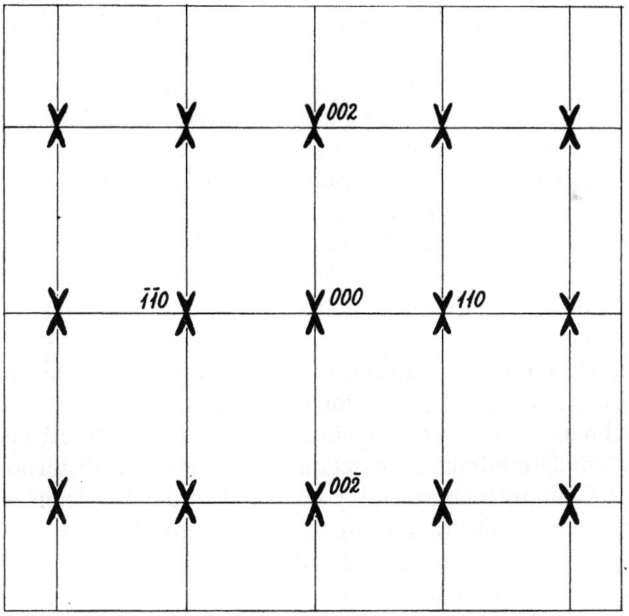

(b)

Figure 120. Calculated diffuse scattering pattern for 1 percent C by weight martensite (307). (*a*) The (100) reciprocal lattice plane; (*b*) the (110) reciprocal lattice plane.

14.6. PHASE TRANSFORMATION IN MARTENSITE OF CARBON STEEL INVOLVING CARBON ATOM CONDENSATION INTO IRRADIATION DEFECTS

X-ray crystal lattice parameter measurements have detected an interesting phenomenon that occurs in irradiated iron-carbon martensite (310–313). Quenched carbon-steel specimens containing 1 percent carbon by weight were irradiated by 15 meV neutrons (310) (dose 10^{15} neutr/cm^2) at temperatures below 60°C. The irradiation results in a decrease in the tetragonal axial ratio c/a: the well-separated (011)–(110) tetragonal doublet in the x-ray powder diffraction pattern is transformed into a broad line. A similar phenomenon was also observed after electron and γ irradiations. This effect is usually observed at temperatures of about 100°C (before the decomposition) and is apparently associated with a decrease in the axial ratio. The observed change in crystal lattice parameters is produced by a C-atom redistribution in the alloy and not by the martensite decomposition reaction. The latter may be deduced from the fact that tempering at room temperature leads to the opposite behavior than in the case of decomposition: first to an increase in the (011)–(110) line width in the powder diffraction pattern and then to a restoration of the (011)–(110) tetragonal doublet. This process goes faster than martensite decomposition. Heating of electron-irradiated martensite below 100°C shows that the rate of axial ratio restoration increases with temperature. The axial ratio c/a increases at 60 to 70°C in 30 min (312).

An effect still more unexpected was observed in martensite irradiated by 2 meV electrons and then cooled to temperatures below 0°C. During 1 min of specimen holding at temperatures near -50°C, the axial ratio of tetragonal martensite decreased from that observed at room temperature. Subsequent tempering at room temperature caused an increase in the c/a ratio. A repeated cooling to -50°C again reduced the axial ratio to its previous small value. Thus the axial ratio revealed thermodynamic reversibility. This effect was never observed in cooling nonirradiated martensite and irradiated martensite with a higher carbon content.

An interpretation of this phenomenon was proposed by Khachaturyan and Shatalov (314). According to (314) the tetragonality decrease in the irradiated carbon-steel martensite can be explained if one assumes that carbon atoms are trapped by irradiation-induced defects, and that the lattice distortion produced by captured C-atoms is considerably smaller than the distortion generated by "free" C-atoms in octahedral interstices. Accordingly, the reversible axial ratio decrease upon cooling is associated with a first-order transition involving C-atom "condensation" from normal O_z octahedral interstices into the irradiation defects ("traps"). The condensation process occurs at temperatures below the critical temperature. When the temperature increases to the critical point, a spontaneous "vaporization" of the "condensate" occurs, and C-atoms are transferred from the traps to their usual positions in O_z interstices. The temperature dependence of the fraction of C-atoms in octahedral sites not captured by the traps is represented in Fig. 121. This dependence will be calculated below after (314). According to (314) the first-order transition does not take place in all cases. Depending on the relation between the energetic parameters, two extreme cases are possible. The first one corresponds to the situation where the binding energy between C-atoms and traps is much larger than the typical interaction energies of C-atoms in octahedral interstices (Fig. 121a). Since the C-atoms will be captured at all reasonable temperatures, the first-order transition cannot occur. The second case arises when the interaction of C-atoms in octahedral sites is so strong that these sites are energetically favored over the "trap" positions (Fig. 121c). The first-order transition occurs in an intermediate case where binding between C-atoms and traps is commensurate with C-C interactions of atoms in octahedral sites (Fig. 121b).

The quantitative description of this phenomenon is based on the following model. Let an irradiated martensite crystal contain a certain number N_{tr} of

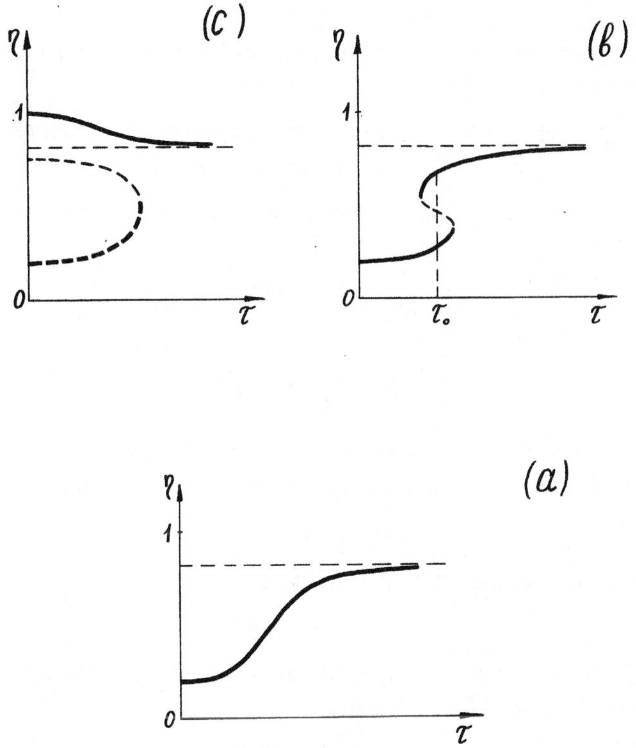

Figure 121. The typical temperature dependences of the fraction of C-atoms in O_z octahedral sites for irradiated martensite [see Eq. (14.6.6)]. (*a*) Strong binding energy between C-atoms and traps (case *A*); (*b*) typical interaction energy of C-atoms in octahedral sites is comparable with the binding energy of C-atom and trap (the phase transition case *B*); (*c*) small binding energy between C-atoms and traps (case *C*). [After A. G. Khachaturyan and G. A. Shatalov (314).]

octahedral sites that may be regarded as "traps" whose fraction related to the number of Fe-atoms, N_{Fe}, is

$$c_{tr} = \frac{N_{tr}}{N_{Fe}}$$

The binding energies between C-atoms and traps are suggested to be the same. They are equal to $-\Phi_{tr}(\Phi_{tr} > 0)$. "Free" carbon atoms are randomly distributed over O_z octahedral sites. Since C-atoms in traps produce no elastic strain, the strain-induced elastic interaction of C-atoms captured by traps with other C-atoms can be ignored. In this case the mean-field approximation to the internal energy is

$$U = U_0 + \tfrac{1}{2} N_{Fe} V_{33}(0) \bar{n}_{oct}^2 - \Phi_{tr} N_{tr} \bar{n}_{tr} \tag{14.6.1}$$

where $\bar{n}_{oct}$ and $\bar{n}_{tr}$ are the occupation probabilities of finding a carbon atom in an octahedral site O_z and in a trap, respectively, U_0 is the part of the internal energy that does not depend on a redistribution of interstitials between traps and octahedral sites, $V_{33}(0)$ is the Fourier transform of the elastic interaction energies at $\mathbf{k} = 0$. It is given by Eq. (14.2.1).

The local potential, $\Phi(\mathbf{r})$, created by traps and a disordered distribution of C-atoms over O interstices is

$$\Phi(\mathbf{r}) = \begin{cases} V_{33}(0)\bar{n}_{oct} & \text{if a C-atom occupies an } O_z \text{ interstice} \\ -\Phi_{tr} & \text{if a C-atom occupies a trap} \end{cases} \tag{14.6.2}$$

Using the local potential (14.6.2) in the mean-field equation (3.4.4), we obtain two equations:

$$\bar{n}_{oct} = \left[\exp\left(\frac{-\mu + V_{33}(0)\bar{n}_{oct}}{\kappa T} \right) + 1 \right]^{-1} \tag{14.6.3a}$$

$$\bar{n}_{tr} = \left[\exp\left(\frac{-\mu - \Phi_{tr}}{\kappa T} \right) + 1 \right]^{-1} \tag{14.6.3b}$$

where μ is the chemical potential. The chemical potential is determined by the condition of the conservation of the number of interstitial atoms, N_c:

$$N_{tr}\bar{n}_{tr} + N_{Fe}\bar{n}_{oct} = N_c$$

Dividing the latter equation by N_{Fe} we have

$$c_{tr}\bar{n}_{tr} + \bar{n}_{oct} = n_c^0 \tag{14.6.3c}$$

Excluding the values μ and $\bar{n}_{tr}$ from the simultaneous set of the transcendental equations (14.6.3a) to (14.6.3c), we obtain

$$\frac{\kappa T}{\Phi_{tr}} = \frac{1 + [V_{33}(0)/\Phi_{tr}]\bar{n}_{oct}}{\ln[(1 - \bar{n}_{oct})(n_c^0 - \bar{n}_{oct})/(\bar{n}_{oct}(c_{tr} - n_c^0 + \bar{n}_{oct}))]} \tag{14.6.4}$$

The relation (14.6.4) is the thermodynamic equation of state that describes the distribution of C-atoms between octahedral interstices and traps. Making use of the definitions

$$t_2 = \frac{c_{tr}}{n_c^0} = \frac{N_{tr}}{N_c} \quad \text{and} \quad \eta = \frac{N_{oct}}{N_c} = \frac{\bar{n}_{oct}}{n_c^0} \tag{14.6.5}$$

where N_{oct} is the number of C-atoms in octahedral sites, we may rewrite Eq. (14.6.4) in the dimensionless form:

$$\ln \frac{(1-n_c^0\eta)(1-\eta)}{n_c^0\eta(t_2+\eta-1)} = \frac{1}{\tau}(1-|\bar{\gamma}|\eta) \tag{14.6.6}$$

where

$$\tau = \frac{\kappa T}{\Phi_{tr}} \tag{14.6.7}$$

is the reduced temperature and

$$\bar{\gamma} = \frac{V_{33}(0)}{\Phi_{tr}} n_c^0 \tag{14.6.8}$$

is the only energetic parameter characterizing interactions in the system, η plays the part of the long-range order parameter [by definition (14.6.5), $0<\eta<1$]. The parameter η is the fraction of C-atoms that remains in octahedral interstices after the transition of other C-atoms to the traps. Eq. (14.6.6) reflects the fact that $\bar{\gamma}<0$. This is the most interesting range where phase transformation is possible. The opposite case, $\bar{\gamma}>0$, does not result in any interesting phenomenon since it corresponds to repulsion between C-atoms and irradiation-induced defects, which in this case cannot be regarded as traps.

Since the tetragonal axial ratio is proportional to the fraction of C-atoms remaining in octahedral interstices, η, and meets the equation

$$\frac{c}{a} = 1 + (u_{33}^0 - u_{11}^0)n_c^0\eta$$

the function $\eta=\eta(T)$ determined by Eq. (14.6.6) also yields the temperature dependence of the c/a ratio.

The analysis of Eq. (14.6.6) leads to three physically distinct ranges of the parameter $\bar{\gamma}$. Within range A where

$$|\bar{\gamma}| = \frac{|V_{33}(0)|}{\Phi_{tr}} n_c^0 < \xi(t_2) \tag{14.6.9a}$$

the temperature dependence of η on the reduced temperature τ is depicted in Fig. 121a. The lower limit value $\xi(t_2)$ is plotted in Fig. 122. The dependence of η on τ described by Fig. 121b corresponds to $|\bar{\gamma}|$, within the B range,

$$\xi(t_2) \leqslant |\bar{\gamma}| \leqslant 1 + n_c^0 t_2 \tag{14.6.9b}$$

The third range C is given by

$$|\bar{\gamma}| > 1 + n_c^0 t_2 \tag{14.6.9c}$$

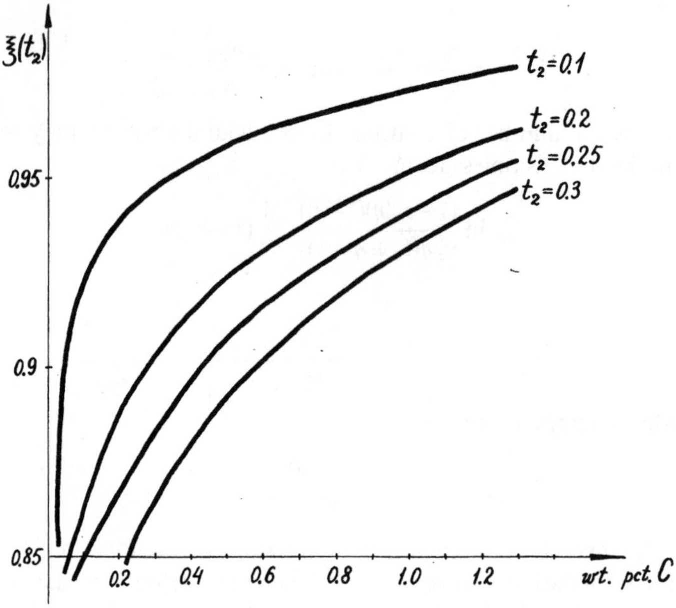

Figure 122. The plot $\xi(t_2)$ with respect to the percent C by weight for irradiated carbon martensite at different t_2.

If $\bar{\gamma}$ meets the inequality (14.6.9c), the temperature dependence of the atomic fraction η of "free" C-atoms in octahedral sites is described by the dependence displayed in Fig. 121c.

One may readily see from Fig. 121a that range A corresponds to the situation where carbon atoms captured by traps form a stable "condensate" that is in equilibrium with the "dilute vapor" of carbon atoms randomly distributed over octahedral sites. Such a state is manifested by a reduced axial ratio. The inequality (14.6.9a) holds in the case of strong binding between carbon atoms and traps [large Φ_{tr} and small $|V_{33}(0)|$]. According to Fig. 121c the range C corresponds to the situation where carbon atoms preferably occupy octahedral sites. This state is characterized by a normal axial ratio which is inherent to the non-irradiated martensite. The inequality (14.6.9c) is ensured by weak binding between carbon atoms and traps [small Φ_{tr} and large $|V_{33}(0)|$]. It should, however, be emphasized that the phase transition is absent in both cases A and C but takes place in the intermediate case B where the parameter $\bar{\gamma}$ is within the range (14.6.9b).

The concentration ranges corresponding to the regions A, B, and C can also be expressed in terms of the energy parameters $V_{33}(0)$ and Φ_{tr}. Let us designate the lower and upper concentration limits of the phase transformation region B by n_* and n_{**}, respectively:

$$n_* < n_c^0 < n_{**}$$

The lower limit, n_*, may be found from Eq. (14.6.9b). By definition, this gives

$$\frac{|V_{33}(0)|}{\Phi_{tr}} n_* = \xi(t_2)$$

or

$$n_* = \xi(t_2) \frac{\Phi_{tr}}{|V_{33}(0)|} \tag{14.6.10a}$$

The upper limit, n_{**}, may also be found from the inequality (14.6.9b). This yields

$$\frac{|V_{33}(0)|}{\Phi_{tr}} n_{**} = 1 + n_{**} t_2$$

The solution to the latter equation for n_{**} results in

$$n_{**} = \frac{\Phi_{tr}}{|V_{33}(0)| - t_2 \Phi_{tr}} \tag{14.6.10b}$$

Therefore the concentration range A (C-atoms are captured by traps) is

$$n_c^0 < n_* = \xi(t_2) \frac{\Phi_{tr}}{|V_{33}(0)|} \tag{14.6.11a}$$

the range B (the trap-to-octahedral site phase transition) is

$$\xi(t_2) \frac{\Phi_{tr}}{|V_{33}(0)|} \leqslant n_c^0 \leqslant \frac{\Phi_{tr}}{|V_{33}(0)| - t_2 \Phi_{tr}} \tag{14.6.11b}$$

and the range C (the usual martensite with C-atoms in octahedral sites) is

$$n_c^0 > \frac{\Phi_{tr}}{|V_{33}(0)| - t_2 \Phi_{tr}} = n_{**} \tag{14.6.11c}$$

The inequalities (14.6.9) may also be rewritten as

$$|\bar{\gamma}| = \frac{|V_{33}(0)|}{\Phi_{tr}} n_c^0 < \xi(t_2) \qquad \text{(within the range } A\text{)} \tag{14.6.12a}$$

$$\xi(t_2) \leqslant |\bar{\gamma}| \leqslant \frac{|V_{33}(0)|}{|V_{33}(0)| - t_2 \Phi_{tr}} \qquad \text{(within the range } B\text{)} \tag{14.6.12b}$$

$$|\bar{\gamma}| = \frac{|V_{33}(0)|}{\Phi_{tr}} n_c^0 > \frac{|V_{33}(0)|}{|V_{33}(0)| - t_2 \Phi_{tr}} \qquad \text{(within the range } C\text{)} \tag{14.6.12c}$$

The plot η versus τ depicted in Fig. 121b and corresponding to the range B shows the discontinuity in the temperature dependence of η. The latter has a jump at the critical value of τ_0 which is the first-order transition reduced temperature. Thus the relation

$$\tau_0 = \frac{\kappa T_0}{\Phi_{tr}} \tag{14.6.13}$$

gives the real temperature of the phase transition. At $\tau = \tau_0$ an avalanchelike condensation of carbon atoms into traps occurs. This effect, which is associated with a sharp decrease in the number of carbon atoms in octahedral interstices, should also result in a dramatic decrease in the tetragonal axial ratio of martensite. Such a phenomenon seems to be observed in a cooling irradiated martensite (310–313).

Since the parameter $\bar{\gamma}=n_c^0[V_{33}(0)/\Phi_{tr}]$ is proportional to the carbon content n_c^0, the axial ratio behavior is sensitive to the carbon concentration, the first-order transition being actually realized within a rather small range of carbon content. Let us estimate the parameters entering the equilibrium equation (14.6.6). The number of traps may be estimated from the magnitude of the maximum decrease in the axial ratio upon cooling. The estimation yields (314)

$$t_2=\frac{c_{tr}}{n_c^0}\approx 0.25 \tag{14.6.14}$$

It is based on the assumption that all traps are occupied by carbon atoms in the state of a minimum axial ratio and that captured carbon atoms do not distort the martensite lattice. It follows from the plot $\xi(t_2)$ in Fig. 122 that at about 1 percent C by weight

$$\xi(0.25)=0.932 \tag{14.6.15}$$

Let us estimate the parameter $|V_{33}(0)|\Phi_{tr}$. The Fourier transform $V_{33}(0)$ taken at $\mathbf{k}=0$ is determined by Eq. (14.2.1):

$$V_{33}(0)=-\frac{a_0^3}{2}\lambda_{ijkl}u_{ij}^0(3)u_{kl}^0(3)+Q$$

$$=-\frac{a_0^3}{2}[2(c_{11}+c_{12})(u_{11}^0)^2+c_{11}(u_{33}^0)^2+4c_{12}u_{11}^0u_{33}^0]+Q \tag{14.6.16}$$

Substituting the numerical values (14.4.2) and (14.4.3) for iron-carbon martensite into (14.6.16), we obtain

$$V_{33}(0)=-6.50\text{ eV} \tag{14.6.17}$$

To estimate the binding energy, Φ_{tr}, the nature of traps should be specified. Let us assume that the traps in martensite are created by irradiation-induced iron vacancies. As shown in Section 13.4 (see Table 13.9), iron vacancies strongly attract carbon atoms in the nearest and next-nearest octahedral sites. Carbon atoms in the six sites of the nearest and twelve sites of the next-nearest coordination shells have binding energies of -0.330 eV and -0.374 eV, respectively. If the mean value of the binding energy over the sites of the two coordination shells,

$$\Phi_{tr}=\frac{0.330\times 6+0.374\times 12}{18}=0.36\text{ eV} \tag{14.6.18}$$

is adopted, one may assume that each iron vacancy produces 18 traps.

As follows from Eqs. (14.6.17) and (14.6.18),

$$\frac{|V_{33}(0)|}{\Phi_{tr}}=\frac{6.50}{0.36}=18.05 \tag{14.6.19}$$

and

$$|\bar{\gamma}|=18.05n_c^0 \tag{14.6.20}$$

Combining the values (14.6.14), (14.6.15) and (14.6.19), the first-order transition range (14.6.11b) is reduced to

$$0.0516<n_c^0<0.05617 \tag{14.6.21}$$

which is equivalent to the weight percentage range of 1.107 to 1.205 percent C by weight. The latter fact demonstrates that the theory is highly sensitive to the carbon content.

It is of interest that the content, 1 percent C by weight, at which the phase transformation is observed in irradiated martensite is indeed close to the predicted lower concentration limit, 1.10 percent C by weight, for the phase transition. The observation of irradiated martensite containing 1.6 percent C by weight is also in agreement with theoretical predictions since with even large doses (of 2.6×10^9 P) do not produce any visible decrease in the axial ratio c/a (315); on the other hand, irradiation of 1 percent C by weight steel with even a dose one order of magnitude lower results in a considerable reduction in the axial ratio. Irradiation with a 3×10^9 P dose yields a complete confluence of lines of the (011)–(110) tetragonal doublet in a diffraction powder pattern. These facts may readily be understood in terms of theory if one notes that the carbon content, 1.6 percent C by weight, exceeds the upper limit, 1.205 percent C by weight, of the phase transformation region. According to the theory formulated above, the content 1.6 percent C by weight corresponds to the range C where carbon atoms occupy octahedral sites O_z and cannot be captured by traps.

Irradiation-induced defects depress the decomposition of martensite in tempering. The reduction in the axial ratio caused by carbide precipitation occurs faster in nonirradiated martensite (315). Binding of carbon atoms with traps prove to be stronger than that in carbides of low-temperature tempering. Irradiation of martensite previously tempered at 100° to 125°C even seems to result in a reverse dissolution of carbides during succeeding tempering at room temperature.

The estimations made here concern the concentration limits of the phase transformation in irradiated martensite. The theory proposed also enables us to carry out a quantitative comparison of the predicted temperature behavior of martensite with the observed results. To do this, we shall calculate the dependence of η on the reduced temperature $\kappa T/(0.0466\Phi_{tr})$, at $V_{33}(0)/\Phi_{tr}=-18.05$ and $n_c^0=0.0508$ (1.09 percent C by weight), 0.0526 (1.13 percent C by weight), and 0.0559 (1.20 percent C by weight) from Eq. (14.6.6). The corresponding plots are depicted in Fig. 123. The arrows in Fig. 123 indicate the reduced temperature related to 20°C and −50°C (these temperatures were used in the experimental

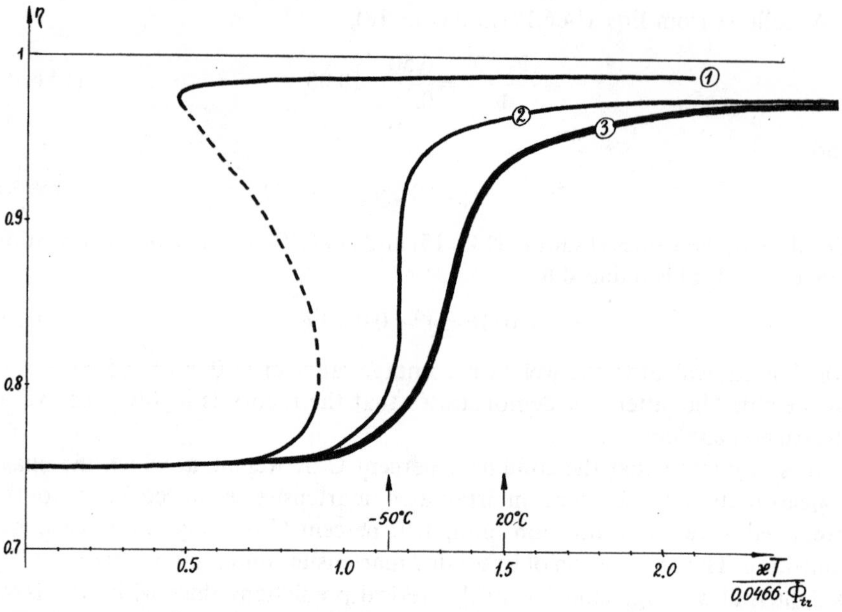

Figure 123. The predicted temperature dependence of the fraction of C-atoms in O_z interstices for the calculated values $\Phi_{tr} = 0.36$ eV and $V_{33}(0) = -6.50$ eV: (1) 1.2 percent C by weight, (2) 1.13 percent C by weight, (3) 1.09 percent C by weight. Arrows indicate the temperature range where the phase transformation was observed. The heavy line (case 3) may seemingly be related to the observed transformation.

observations). The calculation results displayed in Fig. 123 show that a fraction of C-atoms in octahedral sites, η, undergoes a dramatic drop upon cooling from 20° to -50°C in alloys containing 1.09 and 1.13 percent C by weight. This is, however, not the case with the alloy containing 1.2 percent C by weight. The phase transition temperature cannot be observed at 1.2 percent C by weight since with such carbon content the predicted phase transition temperature is lowered to -130°C.

At such low temperature the diffusion of C-atoms proves to be "frozen," so the condensation of C-atoms into traps cannot occur. Unlike irradiated martensites containing 1.09 and 1.13 percent C by weight, the martensite with 1.2 percent C by weight will therefore reveal no axial ratio abnormalities. The latter conclusion is in quantitative agreement with experimental data.

Summing up the foregoing and taking into account the fact that the theory includes no fitting parameters, we may conclude that the theory is in a very good agreement with the observation results.

APPENDIX 1

BASIC DEFINITIONS OF MATRIX ALGEBRA

We formulate in this appendix certain definitions of the algebra of operators (matrices) that have been used in the book.

1. An operator $\hat{\mathbf{A}}$ applied to a vector $\mathbf{r}$ transforms the latter into the vector $\mathbf{r}'$ according to the equation

$$\hat{\mathbf{A}}\mathbf{r}=\mathbf{r}' \tag{A.1.1}$$

The matrix representation of Eq. (A.1.1) (its suffix form) is

$$A_{ij}r_j=r_i' \tag{A.1.2}$$

where A_{ij} is the matrix representation of the operator $\hat{\mathbf{A}}$; the summation over twice-repeated indexes is implied. The transposed conjugate of $\hat{\mathbf{A}}$ is denoted by $\hat{\mathbf{A}}^+$. Its matrix elements are defined by the equation

$$(\hat{\mathbf{A}}^+)_{ij}=A_{ji}^* \tag{A1.3}$$

where the asterisk indicates complex conjugation. It often occurs (e.g., in quantum mechanics) that

$$\hat{\mathbf{A}}^+=\hat{\mathbf{A}} \tag{A1.4a}$$

or, taking into account the definition (A.1.3),

$$A_{ij}=A_{ji}^* \tag{A1.4b}$$

Such operators are called Hermitian operators.

Matrix elements of a product of two operators, $\hat{\mathbf{A}}$ and $\hat{\mathbf{B}}$, are determined by the sum

$$(\hat{\mathbf{A}}\hat{\mathbf{B}})_{ij}=A_{ik}B_{kj} \tag{A.1.5}$$

Using the matrix representation of the operator product, one may readily prove that

$$(\hat{\mathbf{A}}\hat{\mathbf{B}})^+=\hat{\mathbf{B}}^+\hat{\mathbf{A}}^+ \tag{A.1.6}$$

Indeed, according to the definitions (A.1.3) and (A.1.5) we have

$$(\hat{\mathbf{A}}\hat{\mathbf{B}})^{+}_{ij}=(A_{jk}B_{ki})^{*}=A^{*}_{jk}B^{*}_{ki}=(\hat{\mathbf{A}}^{+})_{kj}(\hat{B}^{+})_{ik}=(\hat{\mathbf{B}}^{+})_{ik}(\hat{\mathbf{A}}^{+})_{kj}=(\hat{\mathbf{B}}^{+}\hat{\mathbf{A}}^{+})_{ij} \qquad \text{(A.1.7)}$$

The equality (A.1.7),

$$(\hat{\mathbf{A}}\hat{\mathbf{B}})^{+}_{ij}=(\hat{\mathbf{B}}^{+}\hat{\mathbf{A}}^{+})_{ij}$$

proves the operator relation (A.1.6).

2. The cofactor, a_{ij}, of the matrix element, A_{ij}, is defined as $(-1)^{i+j}$ times the determinant of the submatrix obtained from the matrix, A_{ij}, by removing the ith row and the jth column of $\hat{\mathbf{A}}$. This definition enables us to write the following equation:

$$A_{ik}a_{jk}=\det\|\hat{\mathbf{A}}\|\delta_{ij} \qquad \text{(A1.8)}$$

which is given here without proof, where $\det\|\hat{\mathbf{A}}\|$ is the determinant of the matrix, A_{ij}. If the determinant of the matrix, A_{ij}, is not zero (the nonsingular matrix case), Eq. (A1.8) may be rewritten

$$A_{ik}\frac{a_{jk}}{\det\|\hat{\mathbf{A}}\|}=\delta_{ij} \qquad \text{(A.1.9)}$$

We are thus led to the inverse of the matrix, $\hat{\mathbf{A}}$, which will be denoted as, $\hat{\mathbf{A}}^{-1}$. Its matrix elements will be given by the equation

$$(\hat{\mathbf{A}}^{-1})_{kj}=\frac{a_{jk}}{\det\|\hat{\mathbf{A}}\|} \qquad \text{(A.1.10)}$$

Substituting Eq. (A.1.10) into (A.1.9), we have

$$(\hat{\mathbf{A}})_{ik}(\hat{\mathbf{A}}^{-1})_{kj}=\delta_{ij} \qquad \text{(A.1.11)}$$

or

$$\hat{\mathbf{A}}\hat{\mathbf{A}}^{-1}=\hat{\mathbf{I}} \qquad \text{(A.1.12)}$$

where $\hat{\mathbf{I}}$ is the identity operator. The operator equation (A.1.12) may also be regarded as the definition of the operator $\hat{\mathbf{A}}^{-1}$ inverse to $\hat{\mathbf{A}}$.

3. Let us introduce the definition of a unitary operator. An operator, $\hat{\mathbf{R}}$, is called a unitary operator if when applied to an arbitrary vector, $\mathbf{r}$, it does not change the length of the vector. In other words, a unitary operator, $\hat{\mathbf{R}}$, by its definition, meets the following equation

$$\mathbf{r}^{*\prime}\mathbf{r}'=\mathbf{r}^{*}\mathbf{r} \qquad \text{(A.1.13)}$$

where

$$\mathbf{r}'=\hat{\mathbf{R}}\mathbf{r} \qquad \text{(A.1.14)}$$

Eq. (A.1.13) may be rewritten using Eq. (A.1.14) as follows:

$$(\hat{\mathbf{R}}\mathbf{r})^{*}\hat{\mathbf{R}}\mathbf{r}=\mathbf{r}^{*}\mathbf{r} \qquad \text{(A.1.15a)}$$

or, using the suffix form,

$$R_{ik}^{*} r_{k}^{*} R_{ij} r_{j} = r_{i}^{*} r_{i} \tag{A.1.15b}$$

Combining the definitions (A.1.3) and (A.1.5) with Eq. (A.1.15b), we have

$$r_{k}^{*}(\hat{\mathbf{R}}^{+})_{ki}(\hat{\mathbf{R}})_{ij} r_{j} = r_{k}^{*}(\hat{\mathbf{R}}^{+}\hat{\mathbf{R}})_{kj} r_{j} = r_{i}^{*} r_{i}$$

that is,

$$r_{k}^{*}(\hat{\mathbf{R}}^{+}\hat{\mathbf{R}})_{kj} r_{j} = r_{i}^{*} r_{i} \tag{A.1.16}$$

The identity (A.1.16) holds at any vector **r** if

$$(\hat{\mathbf{R}}^{+}\hat{\mathbf{R}})_{kj} = \delta_{kj} \tag{A.1.17a}$$

or, using the operator form,

$$\hat{\mathbf{R}}^{+}\hat{\mathbf{R}} = \hat{\mathbf{I}} \tag{A.1.17b}$$

Comparing Eqs. (A.1.17b) and (A.1.12), one may see that

$$\hat{\mathbf{R}}^{+} = \hat{\mathbf{R}}^{-1} \tag{A.1.18}$$

Eq. (A.1.18) proves the important conclusion that the transpose conjugate of unitary matrix $\hat{\mathbf{R}}$ is inverse to $\hat{\mathbf{R}}$.

All symmetry operations of a crystal lattice, such as rotations, reflections, and translations, are unitary operators since they conserve the lengths of vectors.

4. A value λ and a vector **e** are an eigenvalue and an eigenvector of an operator, $\hat{\mathbf{A}}$, respectively, if they meet the following relation

$$\hat{\mathbf{A}}\mathbf{e} = \lambda \mathbf{e} \tag{A.1.19a}$$

or

$$(\hat{\mathbf{A}} - \lambda \hat{\mathbf{I}})\mathbf{e} = 0 \tag{A.1.19b}$$

Eq. (A.1.19b) is a set of linear homogeneous equations in the components of vector **e** whose number is equal to the dimension, n, of the space under consideration. As is known this set has a nontrivial solution if its determinant is equal to zero: if

$$\det \|\hat{\mathbf{A}} - \lambda \hat{\mathbf{I}}\| = 0 \tag{A.1.20}$$

Eq. (A.1.20) is the so-called secular equation. Since Eq. (A.1.20) is a power equation of the degree n (the order of the dimension of the space), it has n roots, $\lambda(1), \lambda(2), \ldots, \lambda(n)$. These roots are the eigenvalues of the operator, $\hat{\mathbf{A}}$. If the eigenvalues, $\lambda(1), \lambda(2), \ldots, \lambda(n)$, are found, the corresponding eigenvectors, $\mathbf{e}(1), \mathbf{e}(2), \ldots, \mathbf{e}(n)$, can readily be determined from Eq. (A.1.19).

5. Let us prove an important theorem that the eigenvalues of a Hermitian operator are real values and the corresponding eigenvectors mutually orthogonal.

Let $\hat{\mathbf{A}}^{+} = \hat{\mathbf{A}}$, and

$$\hat{\mathbf{A}}\mathbf{e}(1) = \lambda(1)\mathbf{e}(1) \quad \text{and} \quad \hat{\mathbf{A}}\mathbf{e}(2) = \lambda(2)\mathbf{e}(2) \tag{A.1.21}$$

where $\lambda(1) \neq \lambda(2)$. The scalar multiplication of Eqs. (A.1.21) by $\mathbf{e}^*(2)$ and $\mathbf{e}^*(1)$ yields

$$[\mathbf{e}^*(2)\hat{\mathbf{A}}\mathbf{e}(1)] = \lambda(1)[\mathbf{e}^*(2)\mathbf{e}(1)] \tag{A.1.22a}$$

$$[\mathbf{e}^*(1)\hat{\mathbf{A}}\mathbf{e}(2)] = \lambda(2)[\mathbf{e}^*(1)\mathbf{e}(2)] \tag{A.1.22b}$$

The suffix form of Eq. (A.1.22a) is

$$e_i^*(2)A_{ij}e_j(1) = \lambda(1)e_i^*(2)e_i(1) \tag{A.1.23}$$

The complex conjugation of Eq. (A.1.23) yields

$$e_i(2)A_{ij}^*e_j^*(1) = \lambda^*(1)e_i(2)e_i^*(1) \tag{A.1.24}$$

Renaming the indexes, $i \rightarrow j$ and $j \rightarrow i$, in the left-hand side of Eq. (A.1.24), we have

$$e_j(2)A_{ji}^*e_i^*(1) = \lambda^*(1)e_i^*(1)e_i(2) \tag{A.1.25}$$

Recalling definition (A.1.3), we may rewrite Eq. (A.1.25) as

$$e_i^*(1)(\hat{\mathbf{A}}^+)_{ij}e_j(2) = \lambda^*(1)e_i^*(1)e_i(2) \tag{A.1.26}$$

The invariant form of Eq. (A.1.26) is

$$\mathbf{e}^*(1)\hat{\mathbf{A}}^+\mathbf{e}(2) = \lambda^*(1)[\mathbf{e}^*(1)\mathbf{e}(2)] \tag{A.1.27}$$

Subtracting Eq. (A.1.27) from (A.1.22b) and taking into account the definition of the Hermitian operator, $\hat{\mathbf{A}}^+ = \hat{\mathbf{A}}$, we have

$$[\lambda(2) - \lambda^*(1)][\mathbf{e}^*(1)\mathbf{e}(2)] = 0 \tag{A.1.28}$$

Since $\lambda(2) - \lambda^*(1) \neq 0$, $[\mathbf{e}^*(1)\mathbf{e}(2)] = 0$, in other words, the vectors, $\mathbf{e}(1)$ and $\mathbf{e}(2)$, are orthogonal to each other. If $\mathbf{e}(2) = \mathbf{e}(1)$, the same consideration yields

$$[\lambda(1) - \lambda^*(1)][\mathbf{e}^*(1)\mathbf{e}(1)] = 0 \tag{A.1.29}$$

instead of Eq. (A.1.28). Since, by definition, $\mathbf{e}^*(1)\mathbf{e}(1)$ is a nonzero value, Eq. (A.1.29) gives

$$\lambda(1) = \lambda^*(1) \tag{A.1.30}$$

Eq. (A.1.30) shows that the eigenvalue $\lambda(1)$ is a real number. The same line of reasoning enables us to prove that all eigenvalues of a Hermitian operator $\hat{\mathbf{A}}$ are real numbers.

5. An arbitrary operator, $\hat{\mathbf{A}}$, can always be represented as a direct product of a unitary and a Hermitian operator:

$$\hat{\mathbf{A}} = \hat{\mathbf{F}}\hat{\mathbf{R}} \tag{A.1.31}$$

where $\hat{\mathbf{R}}$ is a unitary operator, $\hat{\mathbf{F}}$ is a Hermitian operator.

APPENDIX 2

BILINEAR REPRESENTATION OF A HERMITIAN OPERATOR

1. Let a Hermitian operator, $\hat{\mathbf{F}}$, have eigenvalues, $\lambda(1), \lambda(2), \ldots, \lambda(n)$, and the corresponding normalized eigenvectors, $\mathbf{e}(1), \mathbf{e}(2), \ldots, \mathbf{e}(n)$. Since the operator, $\hat{\mathbf{F}}$, is Hermitian, its eigenvalues are real numbers, and its eigenvectors form an orthonormalized basis comprising n vectors. The operator, $\hat{\mathbf{F}}$, and its matrix representation, F_{ij}, may be presented in terms of the eigenvalues, $\lambda(1), \ldots, \lambda(n)$, and eigenvectors, $\mathbf{e}(1), \ldots, \mathbf{e}(n)$, as a bilinear expansion:

$$F_{ij} = \lambda(1)e_i(1)e_j(1) + \lambda(2)e_i(2)e_j(2) + \cdots + \lambda(n)e_i(n)e_j(n)$$

$$= \sum_{s=1}^{n} \lambda(s)e_i(s)e_j(s) \tag{A.2.1}$$

The invariant form of the representation (A.2.1) is

$$\hat{\mathbf{F}} = \lambda(1)\mathbf{e}(1) * \mathbf{e}(1) + \lambda(2)\mathbf{e}(2) * \mathbf{e}(2) + \cdots + \lambda(n)\mathbf{e}(n) * \mathbf{e}(\mathbf{n})$$

$$= \sum_{s=1}^{n} \lambda(s)\mathbf{e}(s) * \mathbf{e}(s) \tag{A.2.2}$$

where the asterisk is the symbol of the dyadic multiplication of eigenvectors. Eq. (A.2.2) may be verified by applying the matrix $\hat{\mathbf{F}}$ to an eigenvector $\mathbf{e}(q)$:

$$\hat{\mathbf{F}}\mathbf{e}(q) = \sum_{s=1}^{n} \lambda(s)\mathbf{e}(s)[\mathbf{e}(s)\mathbf{e}(q)] \tag{A.2.3}$$

Since all the eigenvectors of a Hermitian operator are mutually orthogonal and are chosen to be normalized, we have

$$[\mathbf{e}(s)\mathbf{e}(q)] = \delta_{sq} \tag{A.2.4}$$

Making use of Eq. (A.2.4) in (A.2.3), we have the correct equation for the eigenvalue, $\lambda(q)$, and the eigenvector, $\mathbf{e}(\mathbf{q})$:

$$\hat{\mathbf{F}}\mathbf{e}(q) = \lambda(q)\mathbf{e}(q)$$

Let us consider a particular case when $\hat{\mathbf{F}} = \hat{\mathbf{I}}$ where $\hat{\mathbf{I}}$ is an identity operator.

Since, by definition, $(\hat{\mathbf{I}})_{ij} = \delta_{ij}$, we arrive at the conclusion that all the eigenvalues of the operator $\hat{\mathbf{I}}$ are equal to unity, that is, $\lambda(1) = \lambda(2) = \cdots = \lambda(n) = 1$. Choosing an arbitrary orthonormalized basis in n-dimensional space (one may readily see that any vector is an eigenvector of the identity operator $\hat{\mathbf{I}}$), we may write Eq. (A.2.2) in the form

$$\hat{\mathbf{I}} = \mathbf{e}(1) * \mathbf{e}(1) + \mathbf{e}(2) * \mathbf{e}(2) + \cdots + \mathbf{e}(n) * \mathbf{e}(n) \tag{A.2.5a}$$

or

$$e_i(1)e_j(1) + e_i(2)e_j(2) + \cdots + e_i(n)e_j(n) = \delta_{ij} \tag{A.2.5b}$$

2. Let us consider the operator $\mathbf{f}(\hat{\mathbf{F}})$ where $\hat{\mathbf{F}}$ is a Hermitian operator and $\mathbf{f}(\hat{\mathbf{F}})$ is an analytical function of $\hat{\mathbf{F}}$ that can be expanded in the Laurent series of $\hat{\mathbf{F}}$:

$$\mathbf{f}(\hat{\mathbf{F}}) = \sum_{m=-m_0}^{\infty} a_m \hat{\mathbf{F}}^m \tag{A.2.6}$$

where a_m are the expansion coefficients, m_0 is a positive integer. If $\lambda(1), \lambda(2), \ldots, \lambda(n)$ and $\mathbf{e}(1), \mathbf{e}(2), \ldots, \mathbf{e}(n)$ are the eigenvalues and eigenvectors, respectively, of the operator $\hat{\mathbf{F}}$, we have

$$\mathbf{f}(\hat{\mathbf{F}})\mathbf{e}(q) = \sum_m a_m \hat{\mathbf{F}}^m \mathbf{e}(q) = \sum_m a_m \lambda^m(q)\mathbf{e}(q) = \mathbf{f}(\lambda(q))\mathbf{e}(q)$$

We have thus proved that if $\lambda(q)$ is an eigenvalue of the operator $\hat{\mathbf{F}}$, the value $\mathbf{f}(\lambda(q))$ is an eigenvalue of the operator $\mathbf{f}(\hat{\mathbf{F}})$. Using this fact, we may represent the operator $\mathbf{f}(\hat{\mathbf{F}})$ as a bilinear expansion (A.2.2) of its eigenvectors:

$$\begin{aligned} \mathbf{f}(\hat{\mathbf{F}}) &= \mathbf{f}(\lambda(1))\mathbf{e}(1) * \mathbf{e}(1) + \mathbf{f}(\lambda(2))\mathbf{e}(2) * \mathbf{e}(2) + \cdots + \mathbf{f}(\lambda(\mathbf{n}))\mathbf{e}(n) * \mathbf{e}(n) \\ &= \sum_{s=1}^{n} \mathbf{f}(\lambda(s))\mathbf{e}(s) * \mathbf{e}(s) \end{aligned} \tag{A.2.7}$$

For example, if $\mathbf{f}(\hat{\mathbf{F}}) = \hat{\mathbf{F}}^{-1}$, Eq. (A.2.7) reads

$$\hat{\mathbf{F}}^{-1} = \sum_{q=1}^{n} \frac{1}{\lambda(q)} \mathbf{e}(q) * \mathbf{e}(q) \tag{A.2.8}$$

APPENDIX 3

CALCULATION OF THE ENERGY E_{edge}

According to Eq. (8.7.3)

$$E_{\text{edge}} = \frac{1}{2}\int_{-\infty}^{\infty} \frac{dk_z}{2\pi} \frac{4\sin^2 \frac{1}{2}k_z D}{k_z^2} \iint_{-\infty}^{\infty} \frac{d^2\tau}{(2\pi)^2} \Delta B\left(\frac{\mathbf{k}}{k}\right) \int_{(S)} d^2\rho \int_{(S)} d^2\rho' e^{-i\tau(\rho-\rho')} \tag{A.3.1}$$

where $\mathbf{k}=(\boldsymbol{\tau}, k_z)$, $\boldsymbol{\tau}=(k_x, k_y)$, k_z is the projection of the vector $\mathbf{k}$ on the direction $\mathbf{n}_0$ normal to the habit plane, $\boldsymbol{\rho}(x, y)$ is the projection of the vector $\mathbf{r}$ on the habit plane, $\mathbf{r}=(\boldsymbol{\rho}, z)$ where z is the projection of $\mathbf{r}$ on $\mathbf{n}_0$. The integration in Eq. (A.3.1) is over the area S of the habit plane limited by the curve $y=y(x)$ (see Fig. 63).

Since $D/L \ll 1$, the integration in (A.3.1) is actually carried out over a long narrow rod emerging from the origin, $\mathbf{k}=0$, and directed normal to the habit plane (along the vector $\mathbf{n}_0$). The characteristic length of this rod is of the order of $\Delta k_z \sim 2\pi/D$, the thickness $\Delta\tau \approx 2\pi/L$ (see Fig. 61). It follows that the requirement for the rod to be sufficiently narrow and long is reduced to the inequality

$$\frac{\Delta\tau}{\Delta k_z} \sim \frac{2\pi/L}{2\pi/D} = \frac{D}{L} \ll 1$$

When the integral is taken over $\mathbf{k}$, the main contribution to E_{edge} thus comes from the integration over the region where the vector $\mathbf{k}$, and consequently the vector $\mathbf{n}=\mathbf{k}/k$, departs slightly from the direction of the unit vector $\mathbf{n}_0$ (the rod axis) of the normal to the habit plane. The function $\Delta B(\mathbf{n})$ which enters the integrand (A.3.1) can thus be represented in the form of the first term of the expansion (8.7.9) in the variables $\delta\mathbf{n}$:

$$\Delta B(\mathbf{n}) \simeq \beta_{ij}(\mathbf{n}_0)\delta n_i \delta n_j + \cdots \tag{A.3.2}$$

where $\delta\mathbf{n}$ is the projection of the deviation $\Delta\mathbf{n}=\mathbf{n}-\mathbf{n}_0$ on the habit plane, $\beta_{ij}(\mathbf{n}_0)$ is the second-rank tensor which plays the part of second-order expansion coefficients.

Since by definition the vector $\delta\mathbf{n}$ is normal to the unit vector $\mathbf{n}_0$, it can be written in the form

$$\delta\mathbf{n}=\frac{\tau}{k}=\frac{\tau}{\sqrt{\tau^2+k_z^2}}=(k_x, k_y)\frac{1}{\sqrt{\tau^2+k_z^2}} \tag{A.3.3}$$

Substituting (A.3.3) to (A.3.2), we have

$$\Delta B(\mathbf{n})=\beta_{xx}(\mathbf{n}_0)\frac{k_x^2}{k_z^2+\tau^2}+2\beta_{xy}(\mathbf{n}_0)\frac{k_xk_y}{k_z^2+\tau^2}+\beta_{yy}(\mathbf{n}_0)\frac{k_y^2}{k_z^2+\tau^2} \tag{A.3.4}$$

Selecting the x- and y-axes in the habit plane so that the new coordinate axes coincide with the principal axes of the tensor $\beta_{ij}(\mathbf{n}_0)$, we may simplify Eq. (A.3.4):

$$\Delta B(\mathbf{n})=\beta_1\frac{k_x^2}{k_z^2+\tau^2}+\beta_2\frac{k_y^2}{k_z^2+\tau^2} \tag{A.3.5}$$

where β_1 and β_2 are the eigenvalues of the tensor $\beta_{ij}(\mathbf{n}_0)$.

It should be noted that within the region of small k_z $[|k_z|<(2\pi/L)\alpha_0$ where $\alpha_0>1$ is a factor$]$ the vector $\mathbf{n}$ departs substantially from $\mathbf{n}_0$, since the maximum angle of the deviation is of the order of $\Delta\tau/\Delta k_z\sim 2\pi/(Lk_z)$. We can compute this effect accurately by using the following procedure:

1. Exclude the region $(-2\pi/L)\alpha_0<k_z<(2\pi/L)\alpha_0$ from the integration (A.3.1).
2. Make this region tend to zero (make the length L tend to infinity while maintaining the thickness D constant).

The latter procedure corresponds to the limit transition, $D/L\to 0$. It enables us to obtain the asymptotic equation for the elastic energy E_{edge}:

$$E_{\text{edge}}\to\int_{(2\pi/L)\alpha_0}^{\infty}\frac{4\sin^2\frac{1}{2}k_zD}{k_z^2}\frac{dk_z}{2\pi}\iint_{-\infty}^{\infty}\Delta B(\mathbf{n})\frac{d^2\tau}{(2\pi)^2}\int_{(S)}d^2\rho\int_{(S)}e^{-i\tau(\rho-\rho')}d^2\rho' \tag{A.3.6}$$

Substituting (A.3.5) to (A.3.6), we have

$$E_{\text{edge}}\to\int_{(2\pi/L)\alpha_0}^{\infty}\frac{4\sin^2\frac{1}{2}k_zD}{k_z^2}\frac{dk_z}{2\pi}[I_1(k_z)+I_2(k_z)] \tag{A.3.7}$$

where

$$I_1(k_z)=\beta_2\iint_{-\infty}^{\infty}\frac{dk_xk_y}{(2\pi)^2}\frac{k_y^2}{k_x^2+k_y^2+k_z^2}\iint_{(S)}dxdy\iint_{(S)}dx'dy'e^{-ik_x(x-x')-ik_y(y-y')} \tag{A.3.8a}$$

$$I_2(k_z)=\beta_1\iint_{-\infty}^{\infty}\frac{dk_xdk_y}{(2\pi)^2}\frac{k_x^2}{k_x^2+k_y^2+k_z^2}\iint_{(S)}dxdy\iint_{(S)}dx'dy'e^{-ik_x(x-x')-ik_y(y-y')} \tag{A.3.8b}$$

$d^2\rho = dxdy$, $d^2\rho' = dx'dy'$. The integrals must be calculated under the condition $D/L \ll 1$. Consider the calculation of the integral $I_1(k_z)$ in detail.

The integration over k_x yields

$$I_1(k_z) = \frac{\beta_2}{2}\int_{-\infty}^{\infty}\frac{dk_y}{2\pi}\frac{k_y^2}{\sqrt{k_z^2+k_y^2}}\iint_{(S)} dxdy \iint_{(S)} dx'dy' \exp\left[-\sqrt{k_z^2+k_y^2}\,|x-x'| - ik_y(y-y')\right] \tag{A.3.9}$$

Integrating (A.3.9) over y and y' between the limits $y_-(x)$ and $y_+(x)$ (see Fig. 63), we obtain

$$I_1(k_z) = \frac{\beta_2}{2}\int_{-\infty}^{\infty}\frac{dk_y}{2\pi}\frac{1}{\sqrt{k_z^2+k_y^2}}\int dx \int dx' e^{-\sqrt{k_z^2+k_y^2}\,|x-x'|}$$
$$\times\left(e^{-ik_y y_+(x)} - e^{-ik_y y_-(x)}\right)\left(e^{ik_y y_+(x')} - e^{ik_y y_-(x')}\right) \tag{A.3.10}$$

The transition from the variables (x, x') to the new variables (x, ξ), where $\xi = x - x'$, enables us to rewrite Eq. (A.3.10) as follows:

$$I_1(k_z) = \frac{\beta_2}{2}\int_{-\infty}^{\infty}\frac{dk_y}{2\pi}\frac{1}{\sqrt{k_z^2+k_y^2}}\int dx \int d\xi e^{-\sqrt{k_z^2+k_y^2}\,|\xi|}$$
$$\times\left(e^{-ik_y y_+(x)} - e^{-ik_y y_-(x)}\right)\left(e^{ik_y y_+(x-\xi)} - e^{ik_y y_-(x-\xi)}\right) \tag{A.3.11}$$

Since $\sqrt{k_z^2+k_y^2} \approx \sqrt{D^{-2}+L^{-2}} \approx 1/D (D/L \ll 1)$, the presence of the multiplier

$$e^{-\sqrt{k_z^2+k_y^2}\,|\xi|} \approx e^{-|\xi|/D}$$

in the integrand of (A.3.11) enables us to extend the limits of the integral over ξ, which are on the order of L from $-\infty$ to ∞. With the effective value ξ being on the order of D, the differences $y_\pm(x-\xi) - y_\pm(x)$ can be replaced by the first non-vanishing term of the series:

$$y_\pm(x-\xi) - y_\pm(x) \approx \frac{dy_\pm(x)}{dx}\xi + \cdots \tag{A.3.12}$$

The accuracy of Eq. (A.3.12) is determined by the ratio of the first omitted term of the power series in ξ [this is the second-order term $\frac{1}{2}(d^2y_\pm/dx^2)\xi^2$] to the first-order term $(dy_\pm/dx)\xi$ under consideration: by the ratio

$$\frac{[d^2y_\pm(x)/dx^2]\xi^2}{(dy_\pm/dx)\xi} \tag{A.3.13}$$

As is known, the second derivative, d^2y/dx^2, can be expressed in terms of the radius of curvature, R, of the curve $y = y(x)$ as follows:

$$\frac{d^2y_\pm}{dx^2} = \frac{1}{R}\left[1 + \left(\frac{dy_\pm}{dx}\right)^2\right]^{3/2} \tag{A.3.14}$$

At $dy_\pm/dx \sim 1$ and $R \sim L$ [a smooth $y = y(x)$ contour] Eq. (A.3.14) yields

$$\frac{d^2y_\pm}{dx^2} \sim \frac{1}{L} \tag{A.3.15}$$

Combining the relations $[dy_\pm(x)/dx]\sim 1$, $[d^2y_\pm(x)/dx^2]\sim 1/L$, and $\xi\sim D$, we may rewrite the ratio (A.3.13)

$$\frac{[d^2y_\pm(x)/dx^2]\xi^2}{[dy_\pm(x)/dx]\xi}\approx\frac{(1/L)D^2}{D}=\frac{D}{L}\ll 1$$

The latter inequality shows that the representation (A.3.12) holds to an accuracy of $D/L\ll 1$. Substituting (A.3.12) into (A.3.11), we obtain

$$I_1(k_z)=\frac{\beta_2}{2}\int dx\int\frac{dk_y}{2\pi}\frac{1}{\sqrt{k_z^2+k_y^2}}\int_{-\infty}^{\infty}d\xi e^{-\sqrt{k_z^2+k_y^2}|\xi|}$$
$$\times\left[e^{ik_y[dy_+(x)/dx]\xi}+e^{ik_y[dy_-(x)/dx]\xi}-2\cos k_y(y_+(x)-y_-(x))\right]\qquad\text{(A.3.16)}$$

In Eq. (A.3.16) the limits of integrating over ξ have been extended to $-\infty$ and $+\infty$. We also neglected the terms of the form $(dy_\pm/dx)\xi\sim D$ in comparison with the term $y_+(x)-y_-(x)\sim L$ in the argument of the cosine in (A.3.16). The accuracy of this procedure is also on the order of $D/L\ll 1$. The integration of (A.3.16) over ξ yields

$$I_1(k_z)=\frac{1}{2|k_z|}\beta_2\int dx\int_{-\infty}^{+\infty}\frac{dk_y}{2\pi}\left\{\frac{1}{k_z^2+k_y^2[1+(dy_+/dx)^2]}\right.$$
$$\left.+\frac{1}{k_z^2+k_y^2[1+(dy_-/dx)^2]}-4\frac{\cos k_y[y_+(x)-y_-(x)]}{k_z^2+k_y^2}\right\}\qquad\text{(A.3.17)}$$

Integrating Eq. (A.3.17) over k_y, we obtain

$$I_1(k_z)=\frac{1}{2|k_z|}\beta_2\int dx\left\{\frac{1}{\sqrt{1+(dy_+(x)/dx)^2}}+\frac{1}{\sqrt{1+(dy_-(x)/dx)^2}}\right\}$$
$$=\frac{1}{2|k_z|}\beta_2\oint dx\frac{1}{\sqrt{1+(dy/dx)^2}}=\frac{1}{2}\frac{1}{|k_z|}\beta_2\oint\frac{dl}{1+(dy/dx)^2}$$
$$\text{(A.3.18)}$$

where $dl=\sqrt{dx^2+dy^2}=\sqrt{1+(dy/dx)^2}\,dx$ is the linear element of the curve $y=y(x)$. The curvilinear integral in (A.3.18) is taken along the closed contour $y=y(x)$ describing the precipitate perimeter. The term

$$2\frac{\exp[-|k_z|(y_+(x)-y_-(x))]}{|k_z|}$$

which arises as a result of integrating the third term in (A.3.17) is negelected since at $k_z\sim 2\pi/D$, $y_+(x)-y_-(x)\sim L$ and $L/D\gg 1$ we have

$$2\frac{\exp[-|k_z|(y_+(x)-y_-(x)]}{|k_z|}\sim D\exp\left(-\frac{L}{D}\right)\to 0$$

as $L/D\to\infty$.

We may calculate the integral $I_2(k_z)$ in the same way. It is equal to

$$I_2(k_z)=\frac{1}{2|k_z|}\beta_1\oint\frac{dy}{\sqrt{1+(dy/dx)^2}}=\frac{\beta_1}{2|k_z|}\oint\frac{(dy/dx)^2dl}{1+(dy/dx)^2}\qquad\text{(A.3.19)}$$

Substituting (A.3.18) and (A.3.19) into (A.3.7), we have

$$E_{\text{edge}} = \oint \left\{ \beta_1 \left[\frac{-dy/dx}{\sqrt{1+(dy/dx)^2}} \right]^2 + \beta_2 \left[\frac{1}{\sqrt{1+(dy/dx)^2}} \right]^2 \right\} dl$$
$$\times \frac{1}{2} \int_{2\pi\alpha_0/L}^{\infty} \frac{4\sin^2 \frac{1}{2}k_z D}{k_z^3} \frac{dk_z}{2\pi} \qquad \text{(A.3.20)}$$

As $D/L \to 0$, the asymptotic value of the integral over k_z in (A.3.20) is equal to

$$\frac{1}{2} \int_{2\pi\alpha_0/L}^{\infty} \frac{4\sin^2 \frac{1}{2}k_z D}{k_z^3} \frac{dk_z}{2\pi} \to \frac{D^2}{4\pi} \ln \frac{L}{D} \qquad \text{(A.3.21)}$$

The dependence of the right-hand side of (A.3.21) on α_0 can be neglected if $\ln L/D \gg \ln \alpha_0$.

Bearing in mind the asymptotic form (A.3.21), we may rewrite Eq. (A.3.20) in the final form:

$$E_{\text{edge}} = \oint \delta(\mathbf{m}) dl_{\mathbf{m}} \qquad \text{(A.3.22)}$$

where

$$\delta(\mathbf{m}) = (\beta_1 m_x^2 + \beta_2 m_y^2) \frac{D^2}{4\pi} \ln \frac{L}{D}$$

$$m_x = -\frac{dy/dx}{\sqrt{1+(dy/dx)^2}}$$

$$m_y = \frac{1}{\sqrt{1+(dy/dx)^2}} \qquad \text{(A.3.22a)}$$

give the general form of the components of the unit vector $\mathbf{m} = (m_x, m_y)$ normal to the curve $y = y(x)$ at the point x, $dl_{\mathbf{m}}$ is the line element of the contour $y = y(x)$ normal to the vector $\mathbf{m}$.

A general (nondiagonal) form of $\delta(\mathbf{m})$ is thus

$$\delta(\mathbf{m}) = \frac{D^2}{4\pi} \ln \frac{L}{D} \beta_{ij}(\mathbf{n}_0) m_i m_j \qquad \text{(A.3.23)}$$

REFERENCES

1. M. S. Wechsler, D. S. Lieberman, and T. A. Read, *Trans. Met. Soc. AIME*, **197**, 1503 (1953).
2. J. S. Bowles and J. K. Mackenzie, *Acta Met.*, **2**, 129, 138, 224 (1954).
3. A. G. Khachaturyan, *Phys. Met. Metallog.* (Engl. transl.), **13**, 493 (1962).
4. A. G. Khachaturyan, *Sov. Phys. Solid State* (Engl. transl.), **5**, 16 (1963).
5. A. G. Khachaturyan, *Sov. Phys. Solid State* (Engl. transl.), **5**, 548 (1963).
6. D. A. Badalyan, A. G. Khachaturyan, and A. J. Kitaigordsky, *Sov. Phys. Crystallogr.* (English transl.), **14**, 333 (1970).
7. D. A. Badalyan and A. G. Khachaturyan, *Sov. Phys. Solid State* (Engl. transl.) **12**, 346 (1970).
8. D. A. Badalyan and A. G. Khachaturyan, *Sov. Phys. Solid State* (Engl. transl.), **14**, 2270 (1973).
9. A. G. Khachaturyan, *Ordering in Substitutional and Interstitial Solid Solutions*, Progress in Materials Science, Pergamon Press, ed. by B. Chalmers, J. W. Christian and T. B. Massalshi, v. 22, 1978, pp. 1–150.
10. V. Gorsky, *Z. Phys.*, **50**, 64 (1928).
11. W. L. Bragg and E. J. Williams, *Proc. Roy. Soc.*, **A145**, 699 (1934).
12. W. L. Bragg and E. J. Williams, *Proc. Roy. Soc.*, **A152**, 231 (1935).
13. R. H. Fouler and E. A. Guggenheim, *Statistical Thermodynamics*, Cambridge University Press, 1939.
14. H. A. Bethe, *Proc. Roy Soc.*, **A150**, 552 (1935).
15. R. Peierls, *Proc. Roy Soc.*, **A154**, 207 (1936).
16. J. G. Kirkwood, *J. Chem. Phys.*, **6**, 70 (1938).
17. R. Kikuchi, *Phys. Rev.*, **81**, 988 (1951).
18. L. D. Landau, *Sov. Phys.*, **11**, 26, 545 (1937).
19. E. M. Lifshitz, *Fiz. Zh.*, **7**, 61, 251 (1942).
20. L. D. Landau, Zh. *Eksp. Teor. Fiz.*, **7**, 627 (1937).
21. R. A. Suris, *Fiz. Tverd. Tela*, **4**, 1154 (1962).
22. V. G. Vaks, A. J. Larkin, and C. A. Pikin, *Zh. Eksp Teor. Fiz.*, **51**, 361 (1966).
23. M. P. Usikov and A. G. Khachaturyan, *Sov. Phys. Crystallogr.* (Engl. transl.), **13**, 910 (1969).
24. B. I. Pokrovski, E. V. Isaeva, *Acta Crystallogr.*, **B34**, 1051 (1978).
25. B. I. Pokrovski, V. F. Kozlovski, *Zh. Neorg. Khim.*, **22**, 2035 (1979) (in Russian).
26. A. G. Khachaturyan, *Sov. Phys. JETP* (Engl. trans.), **36**, 753 (1973).
27. N. Terao, *J. Phys. Soc. Japan*, **15**, 227 (1960).
28. D. de Fontaine, *Acta Met.*, **23**, 553 (1975).
29. A. Lasserre, F. Reynaud, and P. Coulomb, "Order-Disorder Transformations in Alloys," *Proc. International Symposium of Order-Disorder Transformations in Alloys, ed. by H. Warlimont, Springer, Tübingen*, 1974, p. 545.

30. P. R. Ocamoto and G. Thomas, *Acta Met.*, **19**, 825 (1971).
31. M. Jamamoto, S. Nenno, I. Suburi, and Y. Mizutani, *Trans. Japan. Inst. Met.*, **11**, 120 (1970).
32. P. S. Swann, W. R. Duff, and R. M. Fisher, *Metal. Trans.*, **3**, 403 (1972).
33. L. Guttman, H. C. Schyders, and J. Arai, *Phys. Rev. Letters*, **22**, 517 (1969).
34. V. A. Somenkov, *Ber. Bunsenges. Phys. Chem.*, **76**, 733 (1972).
35. H. Kubo and C. M. Wayman, *Metal. Trans.* **10A**, 633 (1979).
35a. H. Kubo, I. Cornelis and C. M. Wayman, *Acta Met.*, **28**, 405 (1980).
36. H. Kubo and C. M. Wayman, *Acta Met.*, **28**, 395 (1980).
37. K. Enami, J. Hasunuma, A. Nagasawa, and S. Nenno, *Scripta Met.*, **10**, 879 (1976).
38. W. Vandermeulen and A. Deruyttere, *Met. Trans.*, **4**, 1659 (1973).
39. C. L. Corey and K. M. Tatteff, *Sci. Met.*, **10**, 909 (1976).
40. H. Kubo, K. Shimizu, and C. M. Wayman, *Met. Trans.*, **8A**, 493 (1977).
41. S. M. Allen and J. W. Cahn, *Acta Met.*, **23**, 1017 (1975); **24**, 425 (1976).
42. J. W. Cahn, *Acta Met.*, **9**, 795 (1961).
43. J. W. Cahn, *Acta Met.*, **10**, 179 (1962).
44. L. S. Ornstein and F. Zernike, *Proc. Amst. Sci.*, **17**, 793 (1914); **18**, 1520 (1916).
45. L. S. Ornstein and F. Zernike, *Phys. Z.*, **19**, 134 (1918); **27**, 761 (1926).
46. A. G. Khachaturyan and R. A. Suris, *Sov. Phys. Crystallogr.* (Engl. transl.), **13**, 63 (1968).
47. L. D. Landau and E. M. Lifshitz, *Quantum Mechanics*, 3d Ed., Pergamon Press, Oxford, 1977.
48. M. Volmer, *Zh. Electrochem.*, **35**, 555 (1929).
49. R. Becker and W. Döring, *Ann. Phys.*, **24**, 719 (1935).
50. J. S. Langer, *Ann. Phys.* (N.Y.), **65**, 53 (1971).
51. J. W. Cahn and J. E. Hilliard, *J. Chem. Phys.*, **31**, 688 (1959).
52. A. G. Khachaturyan, *Fiz. Tverd. Tela*, **9**, 2595 (1967); *Sov. Phys. Solid State*, **9**, 2040 (1968).
53. H. E. Cook, D. de Fontaine, and J. E. Hilliard, *Acta Met.*, **17**, 765 (1969).
54. D. de Fontaine and H. E. Cook, *Critical Phenomena in Alloys, Magnets and Superconductors*, ed. by R. E. Mills, E. Ascher, and R. I. Jaffe, McGraw-Hill, New York, 1971, p. 257.
55. H. E. Cook and J. E. Hilliard, *J. Appl. Phys.*, **40**, 2191 (1969).
56. E. M. Philofsky and J. E. Hilliard, *J. Appl. Phys.*, **40**, 2198 (1969).
57. W. M. Paulson, Ph.D. diss., Northwestern University, Evanston, Ill., 1972.
58. T. Tsakalakos, Ph.D. diss., Northwestern University, Evanston, Ill., 1977.
59. V. K. S. Shante and S. Kirkpatric, *Advan. Phys.*, **20**, 325 (1971).
60. L. C. C. da Silva and R. F. Mehl, *Trans. Met. Soc. AIME*, **191**, 155 (1951).
61. R. W. Balluffi and L. L. Seigle, *J. Appl. Phys.*, **25**, 607 (1954).
62. M. A. Krivoglaz, *Zh. Eksp. Teor. Fiz.*, **40**, 1812 (1961).
63. S. Chandrasekhar, *Rev. Mod. Phys.*, **15**, 1 (1943).
64. J. W. Morris and A. G. Khachaturyan, Proc. 106th Annual Meeting of AIME, Denver, 1978.
65. E. F. Butler and G. Thomas, *Acta Met.*, **18**, 347 (1970).
66. M. A. Krivoglaz, *Zh. Eksp. Teor. Fiz.*, **34**, 355 (1958) (in Russian).
67. R. J. Rioja and D. E. Laughlin, *Met. Trans.*, **8A**, 1257 (1977).
68. H. E. Cook, *Acta Met.*, **18**, 297 (1970).
69. A. G. Khachaturyan, *Sov. Phys. Solid State* (Engl. transl.), **11**, 2959 (1970).
70. S. V. Semenovskaya, Ph.D. diss., Metal Physics Institute, Kiev, 1976.
70a. M. M. Naumova, S. V. Semenovskaya, and Ya. S. Umanski, *Sov. Phys. Solid State* (Engl. transl.), **12**, 764 (1971).

71. M. M. Naumova, S. V. Semenovskaya, *Sov. Phys. Solid State* (Engl. transl.), **12**, 2954 (1971).

72. G. V. Kurdjumov and G. Sacks, *Zh. Phys.*, **64**, 325 (1930).

73. G. V. Kurdjumov, *Phys. Z. der Sowietunion*, **4**, 488 (1933).

74. E. Z. Kaminsky and G. V. Kurdjumov, *Zh. Tekh. Fiz.*, **6**, 984 (1936) (in Russian).

75. G. V. Kurdjumov, V. I. Miretzskii, and T. I. Stelletskaya, *Zh. Tekh. Fiz.*, **2**, 1959 (1939) (in Russian).

76. G. V. Kurdjumov, *Zh. Tekh. Fiz.*, **18**, 933 (1948) (in Russian).

77. G. V. Kurdjumov, *Dokl. Akad. Nauk SSSR*, **60**, 1543 (1948) (in Russian).

78. G. V. Kurdjumov and L. G. Khandros, *Dokl. Akad. Nauk SSSR*, **66**, 211 (1949) (in Russian).

79. G. V. Kurdjumov and O. P. Maksimova, *Dokl. Akad. Nauk SSSR*, **61**, 83 (1948) (in Russian).

80. L. D. Landau and E. M. Lifshitz, *Statistical Physics*, Addison-Wesley, Reading, Ma., 1958.

81. G. B. Olson and M. Cohen, *Metal, Trans.*, **7A**, 1897 (1976).

82. C. L. Magee, *Phase Transformations*, ASM, 1970, p. 115.

83. J. W. Christian, *Mechanisms of Phase Transformation in Crystalline Solids*, Institute of Metals, Monograph, N 33, 1969, p. 129.

84. S. Wen, A. G. Khachaturyan, and J. W. Morris, Proc. International Conference on Martensitic Transformativis, Cambridge, Ma., 1979, p. 94.

85. S. Wen, Ph.D. diss., University of California, Berkeley, 1979.

86. M. Born and K. Huang *Dynamical Theory of Crystal Lattices*, Clarendon Press, Oxford, 1954.

87. E. Scheil, *Z. Anorg. Allg. Chem.*, **207**, 21 (1932).

88. D. B. Novotny and J. F. Smith, *Acta Met.*, **13**, 881 (1965).

89. B. W. Batterman and C. S. Barrett, *Phys. Rev. Letters*, **13**, 390 (1964).

90. K. R. Keller, and J. J. Hanak, *Phys. Letters*, **21**, 263 (1966).

91. A. L. Roitburd and N. S. Kosenko, *Sci. Met.*, **11**, 1039 (1977).

92. J. W. Christian, *J. Inst. Metals, Bull. Met. Rev.*, **84**, 386 (1956).

93. J. E. Breedis and C. M. Wayman, *Trans. Met. Soc. AIME*, **224**, 1128 (1962).

94. C. M. Wayman, *Introduction to the Crystallography of Martensitic Transformations*, Macmillan Series in Material Science, Macmillan, New York, 1964.

95. A. B. Greninger and A. R. Troiano, *Trans. Met. Soc. AIME*, **185**, 590 (1949).

96. M. Watanabe and C. M. Wayman, *Met. Trans.*, **2**, 2221, 2229 (1971).

97. J. M. Marder and A. R. Marder, *Trans. ASM*, **62**, 1 (1969).

98. A. R. Marder and G. Krauss, *Trans. ASM*, **62**, 957 (1969).

99. K. Wakasa and C. M. Wayman, *Proc. International Conference on Martensitic Transformations*, Cambridge, Ma., 1979, p. 34.

100. G. Thomas and B. Rao, Proc. International Conference on Martensite, ICOMAT-77, Kiev, 1978, p. 57.

101. D. P. Dunne and C. M. Wayman, *Met. Trans.*, **4**, 147 (1973).

102. T. Tadaki and K. Shimizu, *Trans. J. Inst. Metals*, **16**, 105 (1975).

103. J. F. Breedis, *Trans. Met. Soc. AIME*, **230**, 1583 (1964).

104. G. S. Ansell, M. J. Carr and J. R. Strife, *Proc. International Symposium on Martensitic Transformations, J. Inst. Metals*, Kobe, Japan, 1976, p. 53.

105. W. L. Fink and E. D. Campbell, *Trans. Am. Soc. Steel Treast.*, **9**, 717 (1926).

106. N. Seliakov, G. V. Kurdjumov, and N. Goudtsov, *Z. Phys.*, **45**, 384 (1927).

107. G. V. Kurdjumov and E. Z. Kaminsky, *Nature*, **122**, 475 (1928).

108. E. Bain, *Trans. Met. Soc. AIME*, **9**, 751 (1924).

109. A. G. Khachaturyan and G. A. Shatalov, *Acta Met.*, **23**, 1089 (1975).

110. C. S. Roberts, *Trans. Met. Soc. AIME*, **197**, 203 (1953).

111. L. I. Lyssak and Ja. N. Vovk, *Phys. Met. Metallog.* (Engl. transl.), **20**, 540 (1965).

112. L. I. Lyssak and B. J. Nikolin, *Phys. Met. Metallog.* (Engl. transl.), **22**, 730 (1965).

113. L. I. Lyssak, Ja. N. Vovk and Ju. M. Polishtchuk, *Phys. Met. Metallog.* (Engl. transl.), **23**, 898 (1967).

114. Ju. L. Alshevsky and G. V. Kurdjumov, *Phys. Met. Metallog.* (Engl. transl.), **22**, 730 (1966).

115. L. I. Lyssak and L. R. Andruschtuk, *Phys. Met. Metallog.* (Engl. transl.), **28**, 348 (1969).

116. L. I. Lyssak and V. E. Danilchenko, *Phys. Met. Metallog.* (Engl. transl.), **24**, 299 (1967).

117. L. I. Lyssak and Ja. N. Vovk, *Phys. Met. Metallog.* (Engl. transl.), **31**, 646 (1971).

118. F. E. Fujita, *J. Japan Inst. Metals*, **39**, 1082 (1975).

119. L. I. Lyssak, S. P. Kondratiev, and Ju. M. Polishtchuk, *Phys. Met. Metallog.*, **36**, 546 (1973).

120. A. G. Khachaturyan, A. F. Ruminina, and G. V. Kurdjumov, *Proc. International Conference on Martensitic Transformations*, Cambridge, Ma., 1979, p. 71.

121. A. L. Roitburd and A. G. Khachaturyan, *Phys. Met. Metallog.*, **30**, 68 (1970).

122. G. V. Kurdjumov and A. G. Khachaturyan, *Met. Trans.*, **3**, 1069 (1972).

123. G. V. Kurdjumov and A. G. Khachaturyan, *Acta Met.*, **23**, 1077 (1975).

124. I. R. Entin, V. A. Somenkov, and, S. Sh. Shil'shtein, *Dokl. Akad. Nauk SSSR*, **206**, 1096 (1972) (in Russian).

125. D. P. Dunne and J. S. Bowles, *Acta Met.*, **17**, 201 (1969).

126. R. Oshima and C. M. Wayman, *Scripta Met.*, **8**, 223 (1974).

127. R. Oshima, H. Azuma, and F. E. Fujita, "New Aspects of Martensitic Transformations," Proc. International Symposium, Kobe, Japan, 1976, p. 293.

128. V. I. Izotov and L. M. Utevsky, *Microscopic electronique*, Résume des communications Congress intern., **2**, Paris, 1970, p. 505.

129. M. Oka and C. Wayman, *Trans. ASM*, **62**, 370 (1969).

130. K. Shimuzu, M. Oka, and C. Wayman, *Acta Met.*, **19**, 1 (1971).

131. G. Metauer and J. M. Shissler, *Memoires Sci. Rev. Met.*, **71**, 295 (1974).

132. J. D. Eshelby. *Proc. Roy Soc.* (*A*), **241**, 376 (1957).

133. J. D. Eshelby, *Proc. Roy. Soc.* (*A*), **252**, 561 (1959).

134. A. G. Khachaturyan, *Fiz. Tverd. Tela*, **8**, 2710 (1966); *Sov. Phys. Solid State*, **8**, 2163 (1967).

135. A. L. Roitburd, *Kristallogr.*, **12**, 567 (1967) (in Russian).

136. A. G. Khachaturyan and G. A. Shatalov, *Sov. Phys. Solid State*, **11**, 118 (1969).

137. S. L. Sass, T. Mura, and J. B. Cohen, *Phil. Mag.*, **16**, 679 (1967).

138. G. Faivre, *Phys. Status Solidi*, **35**, 249 (1969).

139. R. Sankaran and C. Laird, *J. Mech. Phys. Sol.*, **24**, 251 (1976).

140. J. K. Lee and W. C. Johnson, *Phys. Status Solidi*, **46**, 267 (1978).

141. J. K. Lee, D. M. Barnett, and H. I. Aaronson, *Metal. Trans.* (*A*), **8**, 963 (1977).

142. L. D. Landau and E. M. Lifshitz, *The Theory of Elasticity*, 2nd ed., Pergamon Press, Oxford, 1970.

143. R. W. James, *Optical Principles of the Diffraction of X-rays*, London, 1950.

144. L. J. Valpole, *Proc. Roy. Soc.* (*A*), **300**, 270 (1967).

145. J. R. Willis, "Assymmetric Problems of Elasticity," Adams Prize Essay, University of Cambridge, 1970.

146. J. M. Silcock, *J. Inst. Metals*, **61**, 89 (1960).

147. A. Lutts, *Acta Met.*, **9**, 577 (1961).

148. A. G. Khachaturyan and G. A. Shatalov, *Sov. Phys. JETP* (Engl. transl.), **29**, 557 (1969).

149. A. G. Khachaturyan and A. F. Ruminina, *Phys. Status Solidi* (*A*), **45**, 393 (1978).

150. J. W. Morris, A. G. Khachaturyan and S. H. Wen, "The Elastic Contribution to The Thermodynamics of Phase Transformations in Solids" (to be published in 1983 in Proceedings of International Conference on Solid-Solid Phase Transformations held at Pittsburgh, Pa, August 1981).

151. I. S. Gradstein and I. W. Ryzhik, *Tables of Integrals, Series and Products*, Academic Press, New York, 1965.

152. S. Wen, E. Kostlan, M. Hong, A. Khachaturyan, and J. W. Morris, *Acta Met.*, **29**, 124 (1981).

153. A. L. Roitburd and N. S. Kosenko, *Phys. Status Solidi* (*A*) **35**, 735 (1976).

154. A. G. Khachaturyan and V. N. Hairapetyan, *Phys. Status Solidi* (*B*), **57**, 801 (1973).

155. R. J. Azaro, J. P. Hirth, and D. M. Barnett, *J. Lothe Phys. St. Sol.* (*B*), **60**, 261 (1973).

156. V. L. Indenbom and V. I. Alshitz, *Phys. Status Solidi* (*B*), **63**, *K* 125 (1974).

157. J. W. Morris, M. Hong, J. Wedge, and A. G. Khachaturyan (forthcoming).

158. T. Bell and W. S. Owen, *J. Iron Steel Inst.*, **205**, 428 (1967).

159. K. H. Jack, *Proc. Roy. Soc.*, (*A*) **208**, 216 (1951).

160. A. V. Suyazov, M. P. Usikov, and V. M. Mogutnov, *Phys. Met. Metallogr* (Engl. transl.), **42**, 755 (1976).

161. M. P. Usikov and A. G. Khachaturyan, *Phys. Met. Metallogr.*, **30**, 614 (1970).

162. J. van Landuyt, R. Gevers, and S. Amelinckx, *Phys. Status Solidi*, **13**, 467 (1966).

163. M. S. Blanter and A. G. Khachaturyan, *Met. Trans.* (*A*), **9**, 753 (1978).

164. R. K. Viswanadham and C. A. Wert, *J. Less-Common Met.*, **48**, 1351 (1976).

165. D. R. Dierks and C. A. Wert, *Met. Trans.*, **3**, 1699 (1972).

166. N. L. Ryan, W. A. Soffa, and R. S. Crawford, *Metallogr.*, **1**, 195 (1968).

167. V. A. Somenkov and S. S. Shil'shtein, *Phase Transformations of Hydrogen in Metals*, Progress in Materials Science, v. 24, Pergamon Press, ed. by J. W. Christian, P. Haasen and T. B. Massalski, 1979, pp. 267–335.

168. V. A. Somenkov, I. R. Entin, A. Ju. Chervyakov, S. Sh. Shil'shtein, and A. A. Chertkov, *Sov. Phys. Solid State*, **13**, 2172 (1972).

169. H. Asano and M. Hirabayashi, *Phys. Status Solidi* (*A*), **15**, 267 (1973).

170. H. Asano, Y. Abe, and M. Hirabayashi, *J. Phys. Soc. Jap.*, **41**, 974 (1976).

171. M. S. Blanter and A. G. Khachaturyan, *Phys. Status Solidi* (*A*), **51**, 291 (1979).

172. D. C. Westlake, *J. Less-Common Met.*, **23**, 89 (1971).

173. M. P. Cassidy, B. C. Muddle, T. E. Scott, C. M. Wayman, and J. S. Bowles, *Acta Met.*, **25**, 829 (1977).

174. J. S. Bowles, B. C. Muddle, and C. M. Wayman, *Acta Met.*, **25**, 513 (1977).

175. A. J. Maeland, *J. Phys. Chem.*, **68**, 2197 (1964).

176. A. Kelly and R. B. Nicholson, *Precipitation Hardening*, Progress in Materials Science, v. 10, ed. by B. Chalmers, Macmillan, New York, 1963.

177. A. Ardell and R. B. Nicholson, *Acta Met.*, **14**, 1295 (1966).

178. B. A. Parker and D. R. F. West, *J. Austr. Inst. Met.*, **14**, 102 (1969).

179. E. Hornbogen and M. Roth, *Z. Metallk*, **59**, 157 (1968).

180. A. K. Chakraborty and E. Hornbogen, *Z. Metallk*, **57**, 28 (1966).

181. M. E. Hargreaves, *Acta Crystallogr.*, **4**, 301 (1951).

182. A. Guinier, *Heterogeneities in Solid Solutions*, Solid State Physics, v. 9, ed. F. Seitz and D. Turnbull, Academic Press, New York, 1959.

183. V. Gerold, *Phys. Status Solidi*, **1**, 37 (1961).

184. V. Gerold and W. Schweizer, *Z. Metallk*, **52**, 76 (1961).

185. R. Bauer and V. Gerold, *Z. Metallk*, **52**, 671 (1961).

186. R. Bauer and V. Gerold, *Acta Met.*, **10**, 637 (1962).

187. M. J. Marcinkowski and L. Zwell, *Acta Met*, **11**, 373 (1963).

188. V. Gerold, *Acta Crystallogr.*, **11**, 236 (1958).

189. K. Doi, *Acta Crystallogr.*, **13**, 45 (1960).

190. F. Sebilleau, Publications ONERA, No. 87, Paris, 1957.

191. J. M. Silkok, T. J. Heal, and H. K. Hardly, *J. Inst. Met.*, **82**, 239 (1953).

192. L. F. Mondolfo, N. A. Gjostein, and D. W. Levinson, *J. Metals*, **8**, 1378 (1956).

193. T. Maki, K. Tsuzaki, and I. Tamura, *Proc. International Conference on Martensitic Transformations*, Cambridge, Ma., 1979, p. 22.

194. G. Hausch and H. Warlimont, *Acta Met.*, **21**, 401 (1973).

195. A. J. Bradley, *Proc. Phys. Soc.*, **52**, 80 (1940).

196. V. Daniel and H. Lipson, *Proc. Roy. Soc.*, **181**, 368 (1943); **182**, 378 (1944).

197. K. Biederman and E. Kneller, *Z. Metallk*, **47**, 290 (1956).

198. M. Hillert, M. Cohen, and B. L. Averbach, *Acta Met.*, **9**, 536 (1961).

199. Y. Fukano, *J. Phys. Soc. Jap.*, **16**, 1195 (1961).

200. T. J. Tiedema, J. Bouman, and W. G. Burgers, *Acta Met.*, **5**, 310 (1957).

201. L. M. Magat, *Phys. Met. Metallogr.* (Engl. transl.), **19**, 521 (1965); **20**, 478 (1965).

202. R. G. Davies and R. H. Richman, *Trans. Met. Soc. AIME*, **236**, 1551 (1966).

203. V. Heubner, *Arch. Eisenhüttenw.*, **34**, 547 (1963).

204. M. V. Heimendahl and V. Heubner, *Acta Met.*, **11**, 1115 (1963).

205. E. Hornbogen and M. Roth., *Z. Metallk*, **58**, 842 (1967).

206. C. Buckle and J. Manenk, *Rev. Met.*, **57**, 436 (1960).

207. E. G. Nesterenko and K. V. Chuistov, *Phys. Met. Metallogr.* (Engl. transl.), **9**, 140 (1960).

208. A. H. Geisler, *Trans. ASM*, **43**, 70 (1951).

209. K. J. de Vos, *J. Appl. Phys.*, **37**, 1100 (1966); "The Relationship between Microstructure and Magnetic Properties of ALNICO Alloys," thesis, 1966.

210. E. G. Knizhnik, V. S. Kraposhin, Ja. L. Linetskii, and B. G. Livshitz, *Phys. Met. Metallogr.*, **25**, 425 (1968).

211. Ja. L. Linetskii, E. G. Knizhnik, and B. G. Livshitz, *Phys. Met. Metallogr.*, **29**, 265 (1970).

212. B. J. Livak and G. Thomas, *Acta Met.*, **19**, 497 (1970).

213. A. G. Khachaturyan, *Phys. Status Solidi*, **35**, 119 (1969).

214. A. G. Khachaturyan, *IEEE Trans. Magnetics*, **6**, 233 (1970).

215. A. G. Khachaturyan, *Sov. Phys. JETP*, **31**, 98 (1970).

216. C. Kittel, *Introduction to Solid State Physics*, Wiley, New York, 1956.

217. H. Warlimont, "Transformations Involving Coherent Phases and Effects on Mechanical Properties of Alloys," Proc. 5th International Symposium, University of California, Berkeley, 1972.

218. E. G. Krizhnik, V. S. Kraposhin, Ja. L. Linetskii, and B. G. Livshitz, *Electron Microscopic Studies of Crystalline Samples and Biological Object*, Nauka, Moscow, 1968.

219. B. G. Livshitz, E. G. Knizhnik, V. L. Kraposhin, and Ja. L. Linetskii, *IEEE Trans. Magnetics*, **6**, 237 (1970).

220. V. I. Sumin, A. A. Fridman, P. P. Pashkov, Ju. M. Rabinovich, and V. A. Altman, *Phys. Met. Metallogr.*, **43**, 652 (1974).

221. A. G. Khachaturyan, *The Theory of Phase Transformations and the Structure of Solid Solutions*, Nauka, Moscow, 1974, p. 321 (in Russian).

222. E. G. Povolotskii, *Phys. Met. Metallogr.* (Engl. transl.), **34**, 834 (1972).

223. M. Bouchard and G. Thomas, *Acta Met.*, **23**, 1485 (1975).

224. B. G. Livshitz, Ja. L. Linetskii, and I. M. Milyaev, *Dokl. Akad. Nauk SSR*, **170**, 3 (1966) (in Russian).

225. Ja. L. Linetskii, Ph.D. diss., Moscow Institute of Steel and Alloys, Moscow, 1966.

226. A. H. Geisler and J. B. Newkirk, *Trans. AIME*, **180**, 101 (1949).

227. A. L. Roitburd, *Fiz. Tverd. Tela*, **10**, 3619 (1968).

228. J. van Landuyt and C. M. Wayman, *Acta Met.*, **16**, 803 (1968).

229. C. M. Wayman and J. van Landuyt, *Acta Met.*, **16**, 815 (1968).

230. T. Mura, T. Mori, and M. Kato, *J. Mech. Phys. Solids*, **24**, 305 (1976).

231. T. Maki and C. M. Wayman, "Transformation Twin Width Variation in Fe-Ni and Fe-Ni-C Martensites," in *Proc. International Symposium on New Aspects of Martensitic Transformations*, Institute of Metals, Kobe, Japan, 1976, p. 69.

232. L. Yu. Vinnikov, I. Yu. Georgieva, and L. G. Maisterenko, "Metallofizika," *Naukova Dumka* (Kiev), **55**, 24 (1974).

233. S. Wen, J. W. Morris, and A. G. Khachaturyan, "Computer Simulation of a 'Tweed' Transformation in an Idealized Elastic Crystal," Proc. 108th Annual Meeting AIME, New Orleans, 1979, p. 126; *Met. Trans.*, **12A**, 581 (1981).

234. S. Wen, J. W. Morris, and A. G. Khachaturyan, "The Computer Simulation of the Formation of the 'Tweed' Structure and Modulated Structure in the Decomposition Reactions," Proc. International Symposium on Modulated Structures, Kona, Hawaii, 1979, pp. 168 to 172.

235. J. W. Christian, "The Mechanisms of Phase Transformations in Crystalline Solids," v. 33, Institute of Metals Monograph, London, 1969.

236. J. van Landuyt, *Phys. Status Solidi*, **6**, 957 (1964).

237. M. Hirabayashi and S. Weissman, *Acta Met.*, **10**, 25 (1962).

238. V. S. Arunachalain and R. S. Cahn, *J. Met. Sci.*, **2**, 160 (1967).

239. J. M. Pennison, A. Bourret, and P. Euren, *Acta Met.*, **19**, 1195 (1971).

240. L. E. Tanner, *Phys. Status Solidi*, **30**, 685 (1968).

241. V. I. Syutkina and E. S. Yakovleva. *Phys. Status Solidi*, **21**, 465 (1967).

242. R. Smith and J. S. Bowles, *Acta Met.*, **8**, 405 (1960).

243. L. E. Tanner and H. J. Leamy, "The Microstructure of Order-Disorder Transitions," in *Proc. International Symposium of Order-Disorder Transition in Alloys*, ed. by H. Warlimont, Springer, Tübingen, 1974, p. 181.

244. R. Chang, *J. Nucl. Mater.*, **2**, 335 (1960).

245. R. L. Beck, *Trans. ASM*, **55**, 542 (1962).

246. K. G. Barraclough and C. J. Beevers, *J. Nucl. Mater.*, **34**, 125 (1970).

247. G. C. Carpenter, J. F. Watters, and R. W. Gilbert, *J. Nucl. Mater.*, **48**, 267 (1973).

248. R. K. Edwards and E. Veleckis, *J. Phys. Chem.* (Ithaka), **66**, 1657 (1962).

249. M. P. Cassidy and C. M. Wayman, *Met. Trans. (A)*, **11**, 57 (1980).

250. M. P. Cassidy and C. M. Wayman, *Met. Trans. (A)*, **11**, 47 (1980).

251. G. Hörz, *Acta Met.*, **27**, 1893 (1979).

252. R. E. Pawell, J. V. Cathcart, and J. J. Cambell, *Acta Met.*, **10**, 149 (1962).

253. A. L. Roitburd, *Sov. Phys. USP*, **17**, 326 (1974); *Solid State Physics*, v. 33, Academic Press, New York, 1978, pp. 317–390.

254. G. Brayer, H. Müller, and G. Kühner, *J. Less-Common Met.*, **4**, 533 (1962).

255. L. E! Tanner, *Phil. Mag.*, **14**, 111 (1966).

256. K. Yosuda and Y. Kanzawa, *Trans. Jap. Inst. Met.*, **18**, 46 (1972).

257. R. J. Rioja and D. Laughlin, private communication, 1979; R. J. Rioja, Ph.D. diss., Department of Metallurgy, Carnegie-Mellon University, 1979.

258. F. F. Millilo and D. I. Potter, *Met. Trans.* (*A*), **9**, 283 (1978).

259. M. Okada, Ph.D. diss., University of California, Berkeley, 1978.

260. T. J. Matsubara, *J. Phys. Soc. Jap.*, **7**, 270 (1952).

261. H. Kanzaki, *J. Phys. Chem. Solids*, **2**, 24 (1957).

262. M. A. Krivoglaz and E. A. Tikhonova, *Ukr. Fiz. Zh.*, **3**, 297 (1958).

263. A. G. Khachaturyan, *Sov. Phys. Solid State* (Engl. transl.), **4**, 2081 (1963).

264. J. R. Hardy and R. Bullough, *Phil. Mag.*, **15**, 237 (1967).

265. R. Bullough and J. R. Hardy, *Phil. Mag.*, **17**, 833 (1968).

266. J. D. Eshelby, "Continuum Theory of Defects," *Solid State Physics*, **3**, 79 (1956).

267. A. G. Khachaturyan, *Fiz. Tverd. Tela*, **9**, 2595 (1967); *Sov. Phys. Solid State* (Engl. transl.), **9**, 2040 (1968).

268. H. Cook and D. de Fontaine, *Acta Met.*, **17**, 915 (1969).

269. D. de Fontaine, "Configurational Thermodynamics of Solid Solutions", *Solid State Physics*, **34**, 73 (1979).

270. C. Zener, *Phys. Rev.*, **74**, 639 (1948).

271. A. D. B. Woods, *Phys. Rev.* (*A*), **136**, 781 (1964).

272. M. S. Blanter and A. G. Khachaturyan, *Met. Trans.* (*A*), **9**, 753 (1978).

273. C. van Dijk and J. Bergsma, *Neutron Inelastic Scattering*, **1**, 233 (1968).

274. R. Colella and B. W. Batterman, *Phys. Rev.*, **B1**, 3913 (1970).

275. Y. Nakagawa and A. D. B. Woods, *Phys. Rev. Letters*, **11**, 271 (1963).

276. E. S. Machlin, *Acta Met.*, **22**, 95 (1974).

277. J. W. Cahn, "The Mechanism of Phase Transformations in Crystalline Solids," p. 1, Institute of Metals, London, 1969.

278. D. W. Hoffman, *Acta Met.*, **18**, 819 (1970).

279. H. Thomas, *Z. Phys.*, **129**, 219 (1951).

280. W. Gaudig and H. Warlimont, *Z. Metallk*, **60**, 488 (1969).

281. W. Gaudig, G. Thomas, and H. Warlimont, Proc. 3rd Bolton Landing Conference, Baton Rouge, 1970, p. 347.

282. H. Warlimont, *Microscopic Electronique*, v. 2, Grenoble, 1970, p. 177.

283. H. Warlimont and C. Thomas, *Metal Science J.*, **4**, 47 (1970).

284. J. E. Spruel and E. E. Stansbury, *J. Phys. Chem. Solids*, **26**, 811 (1965).

285. E. Ruedle, P. Delavignette, and S. Amelinckx, *Phys. Status Solidi*, **28**, 305 (1968).

286. D. Watanabe, H. Morita, H. Saito, and S. Ogawa, *J. Phys. Soc. Jap.*, **39**, 722 (1970).

287. M. M. Naumova and S. V. Semenovskaya, *Fiz. Tverd. Tela*, **13**, 381 (1971) (in Russian).

288. A. A. Kutznelson, *Ukr. Fiz. Zh.*, **8**, 251 (1963) (in Russian).

289. V. I. Iveronova and A. A. Kutznelson, *Kristallogr.*, **11**, 576 (1966) (in Russian).

290. A. A. Kutznelson and G. P. Revkevich, *Kristillogr.*, **10, 572** (1965) (in Russian).

291. V. I. Iveronova, A. A. Kutznelson, and P. Sh. Dazhaev, *Fiz. Met. Metallogr.*, **23**, 171 (1967).

292. V. I. Iveronova, A. A. Kutznelson, and G. P. Revkevich, *Fiz. Met. Metallogr.*, **26**, 1054 (1968).

293. V. I. Iveronova and A. A. Kutznelson, *Fiz. Met. Metallogr.*, **19**, 686 (1965).

294. R. C. Davies, *J. Phys. Chem. Solids*, **24**, 985 (1963).

295. A. G. Khachaturyan, *Fiz. Tverd. Tela*, **13**, 2417 (1971) (in Russian).

296. J. Bransky and P. S. Rudman, *Trans. ASM*, **55**, 335 (1962).

297. G. Alefeld, *Phys. Status Solidi* (*A*), **32**, 67 (1969).

298. G. Alefeld, *Ber. Bunsenges Phys. Chem.*, **76**, 746 (1972).

299. M. S. Blanter and A. G. Khachaturyan, *Phys. Status Solidi* (*A*), **60**, 641 (1980).

300. V. A. Somenkov, A. V. Irodova, and S. Sh. Shil'shtein, *Dokl. Akad. Nauk SSSR*, **206**, 1096 (1972) (in Russian).

301. W. Burgers, *Physica*, **1**, 561 (1934).

302. H. Sato, *J. Jap. Inst. Met*, **17**, 601 (1953).

303. J. C. Fisher, *Acta Met.*, **6**, 13 (1958).

304. A. G. Khachaturyan and G. A. Shatalov, *Phys. Met. Metallogr.*, **32**, 5 (1971).

305. A. I. Kitaigorodsky, *Molecular Crystals and Molecules*, Physical Chemistry Series, ed. by E. M. Loebl, Academic Press, New York, 1973, p. 170.

306. D. E. Williams, *J. Chem. Phys.*, **47**, 4680 (1967).

307. A. G. Khachaturyan and T. A. Onisimova, *Phys. Met. Metallogr.*, **26**, 12 (1968).

308. V. I. Isotov and L. M. Utevskii, *Phys. Met. Metallogr.*, **25**, 86 (1968).

309. G. V. Kurdjumov, A. V. Suyazov, and M. P. Usikov, *Dokl. Akad. Nauk SSSR*, **195**, 595 (1970) (in Russian).

310. V. K. Kritzkaya and V. A. Il'ina, *Dokl. Akad. Nauk SSSR*, **185**, 1273 (1969) (in Russian).

311. V. K. Kritzkaya, V. A. Il'ina, and L. N. Bystrov, *Dokl. Akad. Nauk SSSR*, **186**, 89 (1970) (in Russian).

312. V. K. Kritzkaya and A. V. Narkhov, *Fiz. Metalov i Metalloved.*, **29**, 1293 (1970).

313. V. K. Kutzkaya, M. P. Kulikovskaya, I. M. Vjunik, and A. V. Narkhov, *Fiz. Metallov i Metalloved.*, **34**, 503 (1972).

314. A. G. Khachaturyan and G. A. Shatalov, *Fiz. Tverd. Tela*, **12**, 2969 (1970).

315. V. K. Kritzkaya, G. V. Kurdjumov, and A. V. Narkhov, *Dokl. Akad. Nauk SSSR*, **206**, 590 (1972).

INDEX

ERRATA

Page 68 In third line from the bottom should be $\dfrac{\partial V(\boldsymbol{k})}{\partial \boldsymbol{k}}$ instead of $\dfrac{\partial b(\boldsymbol{k})}{\partial \boldsymbol{k}}$

Page 71 After Eq. (3.8.11): instead of bcc-based superstructure CuAu I should be fcc-based superstructure CuAu I

Page 85 In Eq. (3.10.17) in the first line instead of $\delta n(\boldsymbol{k}\text{-}2\boldsymbol{k}_o) \equiv n(\boldsymbol{k})$ should be $\delta n(\boldsymbol{k}\text{-}2\boldsymbol{k}_o) \equiv \delta n(\boldsymbol{k})$

Page 86 In the first line of the paragraph that is before Eq.(3.10.26) remove word "very"

Page 87 In Eq.(3.10.27) should be removed $\dfrac{1}{2}$ coefficient in each term.

Page 101 In Eq. (4.1.9) instead of $\left(\dfrac{df}{dc}\right)_{c=c_\beta}$ must be $\left(\dfrac{df}{dc}\right)_{c=c_\alpha}$

Page 106 In Eq. (4.2.15) insert the coefficient $\dfrac{1}{2}$, i.e.

$$m_{ij} = -\frac{1}{2}\int \frac{d^3\rho}{v}\, \rho_i\, \rho_j\, W(\rho) \qquad (4.2.15)$$

Page 107 Eq. (4.2.18) must be rewritten as follows:

$$m_{ij} = -\frac{1}{2}\int \frac{d^3\rho}{v}\, \rho_i\, \rho_j\, W(\rho) = -\frac{1}{6}\delta_{ij}\int \frac{d^3\rho}{v}\, \rho^2\, W(\rho) \qquad (4.2.18)$$

In Eqs. (4.2.19) and (4.2.24) instead of multiplier $\dfrac{1}{3}$ must be $\dfrac{1}{6}$.

Page 108 In Eq. (4.2.26) the last line should be $-\dfrac{1}{3}\delta_{ij}\sum_{\rho} \rho^2\, W(\rho) = 2m\,\delta_{ij}$

Eq. (4.2.26a) should be $m = -\dfrac{1}{6}\sum_{\boldsymbol{r}} r^2\, W(\boldsymbol{r})$

Eq. (4.2.27) should be $V(\boldsymbol{k}) = V(0) + m\, k^2$

Page 174 In Eq. (6.5.9) the sign in last term must be changed, in accordance with Eq. (1.4.9)

$$\left(\hat{\boldsymbol{R}}\right)_{ij} = \delta_{ij}\cos\phi + p_i p_j\left(1-\cos\phi\right) + \delta_{ijk} p_k\, |\sin\phi| \qquad (6.5.9)$$

Page 179 Eq. (6.5.47) must have a form:

$$y_o \approx \frac{-\varepsilon^o_{11} - \varepsilon^o_{33} + |\,\varepsilon^o_{11} - \varepsilon^o_{33}\,|}{2} = \frac{-\varepsilon^o_{11} - \varepsilon^o_{33} + (\varepsilon^o_{33} - \varepsilon^o_{11})}{2} = -\varepsilon^o_{11}$$

Page 181 Before Eq. (6.6.11): instead of $\hat{A}_i^{-1}$, it should be $\hat{A}_1^{-1}$

Page 200 The first word in third line from the top is: "elastically"

Page 208 In Eq. (7.2.40): remove 2 round parentheses, i.e. make it as:

$$E^{heter}_{relax} = \int \frac{d^3k}{(2\pi)^3}\left[-i\sum_{p=1}^{v} \sigma^o_{ij}(p)\Delta\theta_p(\boldsymbol{k}) k_j v_i^*(\boldsymbol{k}) + \frac{1}{2}\lambda_{ijkl} k_j k_l v_i(\boldsymbol{k}) v_k^*(\boldsymbol{k})\right]$$

In Eq. (7.2.41), remove 2 round parentheses and insert coefficient $\dfrac{1}{2}$ in second term, i.e. make it as:

$$E^{heter}_{relax} = \int \frac{d^3k}{(2\pi)^3}\left[-i\sum_{p=1}^{v}\left(\boldsymbol{k}\hat{\sigma}^o(p)\boldsymbol{v}^*(\boldsymbol{k})\right)\Delta\theta_p(\boldsymbol{k}) + \frac{1}{2}\left(\boldsymbol{v}(\boldsymbol{k})\hat{\boldsymbol{G}}^{-1}(\boldsymbol{k})\boldsymbol{v}(\boldsymbol{k})^*\right)\right]$$

Page 209 Expression following Eq. (7.2.51) should have minus sign in the exponent. It should be:

$$\Delta\theta_p(\boldsymbol{k}) = \int dV\left[\tilde{\theta}_p(\boldsymbol{r}) - \frac{V_p}{V}\right]\exp(-i\boldsymbol{k}\cdot\boldsymbol{r}) = \begin{cases}\theta_p(\boldsymbol{k}) \text{ if } \boldsymbol{k}\neq 0\\ 0 \text{ otherwise}\end{cases}$$

Page 211 The last sentence on the bottom should be: "Using (7.3.7) in (7.3.6) gives "

Page 221 In 9th line from the top (1st paragraph) instead of (8.1.32) should be (8.1.27)

Page 242 In Eq. (8.5.10) misses the sign =. This Eq. has a form:

$$\boldsymbol{r}' = \hat{\boldsymbol{A}}\boldsymbol{r} = \left(\hat{\boldsymbol{I}} + \hat{\boldsymbol{S}}(\boldsymbol{n}_o)\otimes\boldsymbol{n}_o\right)\boldsymbol{r} = \boldsymbol{r} + \hat{\boldsymbol{S}}(\boldsymbol{n}_o)(\boldsymbol{n}_o\cdot\boldsymbol{r})$$

Page 244

Eq.(8.6.2) should be corrected as:

$$\Omega_{ii}^{-1}(\boldsymbol{n}) = c_{44} + (c_{11} - c_{44})n_i^2$$

$$\Omega_{ij}^{-1}(\boldsymbol{n}) = c_{44} + (c_{11} - c_{44})n_i n_j \qquad \text{if } i\neq j$$

Page 246 In Eq. (8.6.11g): $\varepsilon_{ij}^0\,\delta_{kl}^0$ should be replaced by $\varepsilon_{ij}^0\,\delta_{ij}^0$

Page 254 In the first term within braces in Eq. (8.7.17), letter p should be change for d so that this term should look like:

$$\int\left\{\frac{d}{dx}\frac{d}{dp_+}\left(\frac{\beta_1 p_+^2 + \beta_2}{\sqrt{1+p_+^2}}\right) - \mu\right\}\delta y_+ dx + \ldots$$

Page 262 After Eq. (8.7.52) the next line is: "Substituting Eq.(8.7.52) into (8.7.51) gives"

Page 263 Left part of Eq. (8.7.54) has a form: $K^{5/3}\ ln(eK) = \ldots\ldots$

Page 376 Eq.(11.2.30b) has a form:

$$\Delta\hat{\sigma}^o = \hat{\sigma}^o(1) - \hat{\sigma}^o(2)$$

Page 379 Corrected Eq. (11.3.12) is:

$$\frac{\varepsilon_{33}^o}{\varepsilon_{11}^o - \varepsilon_{33}^o} < 0$$

Corrected equations (11.3.15) and (11.3.16) as:

$$\varepsilon_{33}^o > 0,\quad \varepsilon_{11}^o < -\varepsilon_{33}^o \qquad (11.3.15)$$

$$\varepsilon_{33}^o < 0,\quad \varepsilon_{11}^o > -\varepsilon_{33}^o \qquad (11.3.16)$$

Page 394 The second line of Eq.(11.4.69) should read as:

$$\times\frac{|\varepsilon_{11}^o| + 2|\varepsilon_{11}^o + \varepsilon_{33}^o|(c_{11}+c_{12})/(c_{11}+c_{12}+2c_{44})}{\sqrt{1+\frac{1}{2}|\varepsilon_{11}^o+\varepsilon_{33}^o|/|\varepsilon_{33}^o|\ \left(|\varepsilon_{11}^o+\varepsilon_{33}^o|+\varepsilon_{11}^o\right)}}$$

Page 395 Eq. (11.4.71) should read as:

$$\varsigma_o = \frac{1}{2}\frac{\sqrt{2}}{(2\pi)^3}\alpha(w)\frac{|\varepsilon_{11}^o| + 2|\varepsilon_{11}^o + \varepsilon_{33}^o|(c_{11}+c_{12})/(c_{11}+c_{12}+2c_{44})}{\sqrt{1+\frac{1}{2}|\varepsilon_{11}^o+\varepsilon_{33}^o|/|\varepsilon_{33}^o|\ \left(|\varepsilon_{11}^o+\varepsilon_{33}^o|+\varepsilon_{11}^o\right)}} \qquad (11.4.71)$$

Page 560(Appendix3) In Eq.(1.3.19) change the current term

$$\frac{1}{2|k_z|}\beta_1\oint\frac{dy}{\sqrt{1+(dy/dx)^2}} \quad \text{for} \quad \frac{1}{2|k_z|}\beta_1\oint\frac{dy}{\sqrt{1+(dx/dy)^2}}$$